EDITORIAL TEAM

Editor, U.S. Environmental Protection Agency: ....Susan Herrod Julius
Editor, U.S. Environmental Protection Agency: ....Jordan M. West
Technical Advisor, Climate Change Science Program Office: .....David J. Dokken
Technical Editor, ICF International: ....Susan Asam
Technical Editor, ICF International: ....Anne Choate
Technical Reviewer, Manomet Center for Conservation Sciences: Hector Galbraith
Copy Editor, ICF International: ....Brad Hurley
Senior Graphic Designer, ICF International: ....Debby Bunting
Graphics Specialist, ICF International: ....Dana Allison
Reference Coordinator, ICF International: ....Sarah Shapiro
Logistical and Technical Support: ICF International: ....Joe Herr, Kathryn Maher, Sandy Seymour

FEDERAL EXECUTIVE TEAM

Director, Climate Change Science Program: ....William J. Brennan
Director, Climate Change Science Program Office: ....Peter A. Schultz

Lead Agency Principal Representative to CCSP, National Program Director for the Global Change Research Program, U.S. Environmental Protection Agency: ....Joel D. Scheraga

Product Lead, Global Ecosystem Research and Assessment Coordinator, Global Change Research Program, U.S. Environmental Protection Agency: ....Susan Herrod Julius

Chair, Synthesis and Assessment Product Advisory Group Associate Director, National Center for Environmental Assessment, U.S. Environmental Protection Agency: ....Michael W. Slimak

Synthesis and Assessment Product Coordinator, Climate Change Science Program Office: ....Fabien J.G. Laurier

Special Advisor, National Oceanic and Atmospheric Administration: ....Chad McNutt

Image on cover of bleached Acropora coral, Fagatele Bay National Marine Sanctuary, American Samoa, courtesy of Eric Mielbrecht, Emerald Coast Environmental Consulting

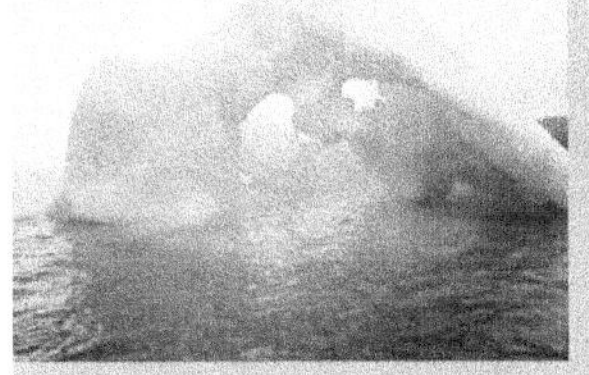

Preliminary Review of Adaptation Options for Climate-Sensitive Ecosystems and Resources

Final Report, Synthesis and Assessment Product 4.4
Report by the U.S. Climate Change Science Program
and the Subcommittee on Global Change Research

EDITORS:

Susan Herrod Julius, U.S. Environmental Protection Agency
Jordan M. West, U.S. Environmental Protection Agency

AUTHORS:

Jill S. Baron, U.S. Geological Survey and Colorado State University
Linda A. Joyce, U.S.D.A. Forest Service
Brad Griffith, U.S. Geological Survey
Peter Kareiva, The Nature Conservancy
Brian D. Keller, National Oceanic and Atmospheric Administration
Margaret Palmer, University of Maryland
Charles Peterson, University of North Carolina
J. Michael Scott, U.S. Geological Survey and University of Idaho

June 20, 2008

Members of Congress:

On behalf of the National Science and Technology Council, the U.S. Climate Change Science Program (CCSP) is pleased to transmit to the President and the Congress this Synthesis and Assessment Product (SAP), *Preliminary Review of Adaptation Options for Climate-Sensitive Ecosystems and Resources.* This is part of a series of 21 SAPs produced by the CCSP aimed at providing current assessments of climate change science to inform public debate, policy, and operational decisions. These SAPs are also intended to help the CCSP develop future program research priorities. This SAP is issued pursuant to Section 106 of the Global Change Research Act of 1990 (Public Law 101-606).

The CCSP's guiding vision is to provide the Nation and the global community with the science-based knowledge needed to manage the risks and capture the opportunities associated with climate and related environmental changes. The SAPs are important steps toward achieving that vision and help to translate the CCSP's extensive observational and research database into informational tools that directly address key questions being asked of the research community.

This SAP focuses on adaptation options for climate-sensitive ecosystems and resources on Federally owned and managed lands. It was developed with broad scientific input and in accordance with the Guidelines for Producing CCSP SAPs, the Federal Advisory Committee Act, the Information Quality Act, Section 515 of the Treasury and General Government Appropriations Act for fiscal year 2001 (Public Law 106-554), and the guidelines issued by the Environmental Protection Agency pursuant to Section 515.

We commend the report's authors for both the thorough nature of their work and their adherence to an inclusive review process.

Sincerely,

Carlos M. Gutierrez
Secretary of Commerce Chair,
Committee on Climate Change
Science and Technology
Integration

Samuel W. Bodman
Secretary of Energy
Vice Chair, Committee on
Climate Change Science and
Technology Integration

John H. Marburger III
Director, Office of Science
and Technology Policy
Executive Director,
Committee on Climate Change
Science and Technology
Integration

TABLE OF CONTENTS

AUTHOR TEAMS FOR THIS REPORT

Chapter 1 **Lead Authors:** Susan Herrod Julius, U.S. EPA; Jordan M. West, U.S. EPA
Contributing Authors: Geoffrey M. Blate, AAAS Fellow, U.S. EPA; Jill S. Baron, USGS and Colo. State Univ.; Linda A. Joyce, USDA Forest Service; Peter Kareiva, The Nature Conservancy; Brian D. Keller, NOAA; Margaret A. Palmer, Univ. Md.; Charles H. Peterson, Univ. N. Car.; J. Michael Scott, USGS and Univ. Id.

Chapter 2 **Lead Authors:** Susan Herrod Julius, U.S. EPA; Jordan M. West, U.S. EPA; Geoffrey M. Blate, AAAS Fellow, U.S. EPA

Chapter 3 **Lead Author:** Linda A. Joyce, USDA Forest Service
Contributing Authors: Geoffrey M. Blate, AAAS Fellow, U.S. EPA; Jeremy S. Littell, JISAO CSES Climate Impacts Group, Univ. Wa.; Steven G. McNulty, USDA Forest Service; Constance I. Millar, USDA Forest Service; Susanne C. Moser, NCAR; Ronald P. Neilson, USDA Forest Service; Kathy A. O'Halloran, USDA Forest Service; David L. Peterson, USDA Forest Service

Chapter 4 **Lead Author:** Jill S. Baron, USGS and Colo. State Univ.
Contributing Authors: Craig D. Allen, USGS; Erica Fleishman, NCEAS, Univ. Southern Calif.; Lance Gunderson, Emory Univ.; Don McKenzie, USDA Forest Service; Laura Meyerson, Univ. Rhode Is.; Jill Oropeza, Colo. State Univ.; Nate Stephenson, USGS

Chapter 5 **Lead Authors:** J. Michael Scott, USGS and Univ. Id.; Brad Griffith, USGS
Contributing Authors: Robert S. Adamcik, USFWS; Daniel M. Ashe, USFWS; Brian Czech, USFWS; Robert L. Fischman, Indiana Univ. School of Law; Patrick Gonzalez, The Nature Conservancy; Joshua J. Lawler, Univ. Wa.; A. David McGuire, USGS; Anna Pidgorna, Univ. Id.

Chapter 6 **Lead Author:** Margaret A. Palmer, Univ. Md.
Contributing Authors: Dennis Lettenmaier, Univ. Wa.; N. LeRoy Poff, Colo. State Univ.; Sandra Postel, Global Water Policy Project; Brian Richter, The Nature Conservancy; Richard Warner, Kinni Consulting

Chapter 7 **Lead Author:** Charles H. Peterson, Univ. N. Car.;
Contributing Authors: Richard T. Barber, Duke Univ.; Kathryn L. Cottingham, Dartmouth College; Heike K. Lotze, Dalhousie Univ.; Charles A. Simenstad, Univ. Wa.; Robert R. Christian, East Car. Univ.; Michael F. Piehler, Univ. N. Car.; John Wilson, U.S. EPA

Chapter 8 **Lead Author:** Brian D. Keller, NOAA
Contributing Authors: Satie Airamé, Univ. Ca. Santa Barbara; Billy Causey, NOAA; Alan Friedlander, NOAA; Daniel F. Gleason, Ga. Southern Univ.; Rikki Grober-Dunsmore, NOAA; Johanna Johnson, Great Barrier Reef MPA; Elizabeth McLeod, The Nature Conservancy; Steven L. Miller, Univ. N. Car. Wilmington; Robert S. Steneck, Univ. Maine; Christa Woodley, Univ. Ca. Davis

Chapter 9 **Lead Author:** Peter Kareiva, The Nature Conservancy
Contributing Authors: Carolyn Enquist, The Nature Conservancy; Ayana Johnson, Univ. Ca. San Diego; Susan Herrod Julius, U.S. EPA; Joshua Lawler, Oregon State Univ.; Brian Petersen, Univ. Ca. Santa Cruz; Louis Pitelka, Univ. Md.; Rebecca Shaw, The Nature Conservancy; Jordan M. West, U.S. EPA

Annex A

Editors: Susan Herrod Julius, U.S. EPA; Jordan M. West, U.S. EPA;
Lead Authors: Jill S. Baron, USGS and Colo. State Univ.; Brad Griffith, USGS; Linda A. Joyce, USDA Forest Service; Brian D. Keller, NOAA; Margaret A. Palmer, Univ. Md.; Charles H. Peterson, Univ. N. Car.; J. Michael Scott, USGS and Univ. Id.

National Forests Case Studies

Tahoe National Forest

Constance I. Millar, USDA Forest Service; Linda A. Joyce, USDA Forest Service; Geoffrey M. Blate, AAAS Fellow, U.S. EPA

Olympic National Forest

David L. Peterson, USDA Forest Service; Jeremy S. Littell, JISAO CSES Climate Impacts Group, Univ. Wa.; Kathy A. O'Halloran, USDA Forest Service

Uwharrie National Forest

Steven G. McNulty, USDA Forest Service

National Parks Case Study

Rocky Mountain National Park

Jill S. Baron, USGS and Colo. State Univ.; Jill Oropeza, Colo. State Univ.

National Wildlife Refuges Case Study

Alaska and the Central Flyway

Brad Griffith, USGS; A. David McGuire, USGS

Wild and Scenic Rivers Case Studies

Wekiva River

Rio Grande River

Upper Delaware River

Margaret A. Palmer, Univ. Md.; Dennis Lettenmaier, Univ. Wa.; N. LeRoy Poff, Colo. State Univ.; Sandra Postel, Global Water Policy Project; Brian Richter, The Nature Conservancy; Richard Warner, Kinni Consulting

National Estuaries Case Study

The Albemarle-Pamlico Estuarine System

Robert R. Christian, E. Car. Univ.; Charles H. Peterson, Univ. of N. Car.; Michael F. Piehler, Univ. of N. Car.; Richard T. Barber, Duke Univ.; Kathryn L. Cottingham, Dartmouth College; Heike K. Lotze, Dalhousie Univ.; Charles A. Simenstad, Univ. Wa.; John W. Wilson, U.S. EPA

Marine Protected Areas Case Studies

The Florida Keys National Marine Sanctuary

Billy Causey, NOAA; Steven L. Miller, Univ. N. Car. Wilmington; Brian D. Keller, NOAA

The Great Barrier Reef Marine Park

Johanna Johnson, Great Barrier Reef Marine Park Authority

Papahānaumokuākea (Northwestern Hawaiian Islands) Marine National Monument

Alan Friedlander, NOAA

The Channel Islands National Marine Sanctuary

Satie Airamé, Univ. Ca. Santa Barbara

Annex B

Editors: Susan Herrod Julius, U.S. EPA; Jordan M. West, U.S. EPA;
Lead Authors: Jill S. Baron, USGS and Colo. State Univ.; Brad Griffith, USGS; Linda A. Joyce, USDA Forest Service; Brian D. Keller, NOAA; Margaret A. Palmer, Univ. Md.; Charles H. Peterson, Univ. N. Car.; J. Michael Scott, USGS and Univ. Id.

ACKNOWLEDGMENTS

Federal Advisory Committee: Adaptation for Climate-Sensitive Ecosystems and Resources Advisory Committee (ACSERAC)

We would like to thank the following members of ACSERAC who provided excellent reviews of this report that resulted in an improved document, and to the Designated Federal Officials:

Chair
Paul G. Risser, University of Oklahoma

Vice-Chair
Reed F. Noss, University of Central Florida

Members
Joe Arvai, Michigan State University
Eric Gilman, IUCN Global Marine Programme
Carl Hershner, Virginia Institute of Marine Science
George Hornberger, University of Virginia
Elizabeth Malone, Joint Global Change Research Institute
David Patton, University of Arizona
Daniel Tufford, University of South Carolina
Robert Van Woesik, Florida Institute of Technology

Designated Federal Officials
Joanna Foellmer
Janet Gamble (back-up)

Thank you to Sharon Moxley (EPA) and to Versar, Inc for their support to the ACSERAC.

Additional External (Public) Reviewers

We would like to thank the many individuals who provided useful comments during the public review period. The draft manuscript, public review comments, and response to public comments are publicly available at: http://www.climatescience.gov/Library/sap/sap4-4/default.php

Stakeholder workshop participants and key reviewers are acknowledged in individual chapters.

US Environmental Protection Agency Internal Reviewers

We would like to thank the following internal reviewers who provided valuable comments on this report in preparation for external and public review:

Peter Beedlow, Office of Research and Development
Paul Bunje, American Association for the Advancement of Science Fellow
David Burden, Office of Research and Development
Barry Burgan, Office of Water
Ben DeAngelo, Office of Air and Radiation
Dominic Digiulio, Office of Research and Development
Anne Fairbrother, Office of Research and Development
Bill Fisher, Office of Research and Development
Eric Jorgensen, Office of Research and Development
Chris Laab, Office of Water
Michael Lewis, Office of Research and Development
Jeremy Martinich, Office of Air and Radiation
Larry Merrill, Region 3
Dave Olszyk, Office of Research and Development
Kathryn Parker, Office of Air and Radiation
Amina Pollard, Office of Research and Development
Jackie Poston, Region 10
Kathryn Saterson, Office of Research and Development
Karen Scott, Office of Air and Radiation
Jim Titus, Office of Air and Radiation
Jim Wigington, Office of Research and Development
Wendy Wiltse, Region 9
Steve Winnett, Region 1
Jeff Yang, Office of Research and Development
3 anonymous reviewers, Office of Research and Development

National Center for Environmental Assessment (NCEA), Global Change Research Program

We would like to thank our colleagues in the Global Change Research Program who contributed thoughtful insights, reviewed numerous drafts, and helped with the production of this report: Amanda Babson (AAAS Fellow), Britta Bierwagen, Geoffrey Blate (AAAS Fellow), Anne Grambsch, Thomas Johnson, Chris Pyke[1], Michael Slimak, Chris Weaver. We would also like to acknowledge the administrative support and oversight provided by the National Center for Environmental Assessment.

ICF International (support contractor to NCEA)

We would like to thank our colleagues at ICF International for their logistical and technical support for this report, with special thanks to: Susan Asam, Anne Choate, Randall Freed, Joseph Herr, Bradford Hurley, Kathryn Maher, Sandra Seymour, and Sarah Shapiro. Stakeholder workshops were organized by ICF and facilitated by Bill Dennison of the University of Maryland, Center for Environmental Science.

It was an honor and a pleasure to work with all of the people above as well as the many other colleagues we have encountered in the science and management communities who are working to address climate change impacts. We hope that this document will be a positive step forward in our collective effort to apply adaptation principles for climate-sensitive ecosystems and resources.

[1] Currently with CTG Energetics.

RECOMMENDED CITATIONS

For the Report as a Whole:

CCSP, 2008: *Preliminary review of adaptation options for climate-sensitive ecosystems and resources.* A Report by the U.S. Climate Change Science Program and the Subcommittee on Global Change Research. [Julius, S.H., J.M. West (eds.), J.S. Baron, B. Griffith, L.A. Joyce, P. Kareiva, B.D. Keller, M.A. Palmer, C.H. Peterson, and J.M. Scott (Authors)]. U.S. Environmental Protection Agency, Washington, DC, USA, 873 pp.

For Chapter 1:

Julius, S.H., J.M. West, G.M. Blate, J.S. Baron, B. Griffith, L.A. Joyce, P. Kareiva, B.D. Keller, M.A. Palmer, C.H. Peterson, and J.M. Scott, 2008: Executive Summary. In: *Preliminary review of adaptation options for climate-sensitive ecosystems and resources.* A Report by the U.S. Climate Change Science Program and the Subcommittee on Global Change Research [Julius, S.H., J.M. West (eds.), J.S. Baron, B. Griffith, L.A. Joyce, P. Kareiva, B.D. Keller, M.A. Palmer, C.H. Peterson, and J.M. Scott (Authors)]. U.S. Environmental Protection Agency, Washington, DC, USA, pp. 1-1 to 1-6.

For Chapter 2:

Julius, S.H., J.M. West, and G.M. Blate, 2008: Introduction. In: *Preliminary review of adaptation options for climate-sensitive ecosystems and resources.* A Report by the U.S. Climate Change Science Program and the Subcommittee on Global Change Research [Julius, S.H., J.M. West (eds.), J.S. Baron, B. Griffith, L.A. Joyce, P. Kareiva, B.D. Keller, M.A. Palmer, C.H. Peterson, and J.M. Scott (Authors)]. U.S. Environmental Protection Agency, Washington, DC, USA, pp. 2-1 to 2-24.

For Chapter 3:

Joyce, L.A., G.M. Blate, J.S. Littell, S.G. McNulty, C.I. Millar, S.C. Moser, R.P. Neilson, K. A. O'Halloran, and D.L. Peterson, 2008: National Forests. In: *Preliminary review of adaptation options for climate-sensitive ecosystems and resources.* A Report by the U.S. Climate Change Science Program and the Subcommittee on Global Change Research [Julius, S.H., J.M. West (eds.), J.S. Baron, B. Griffith, L.A. Joyce, P. Kareiva, B.D. Keller, M.A. Palmer, C.H. Peterson, and J.M. Scott (Authors)]. U.S. Environmental Protection Agency, Washington, DC, USA, pp. 3-1 to 3-127.

For Chapter 4:

Baron, J.S., C.D. Allen, E. Fleishman, L. Gunderson, D. McKenzie, L. Meyerson, J. Oropeza, and N. Stephenson, 2008: National Parks. In: *Preliminary review of adaptation options for climate-sensitive ecosystems and resources.* A Report by the U.S. Climate Change Science Program and the Subcommittee on Global Change Research [Julius, S.H., J.M. West (eds.), J.S. Baron, B. Griffith, L.A. Joyce, P. Kareiva, B.D. Keller, M.A. Palmer, C.H. Peterson, and J.M. Scott (Authors)]. U.S. Environmental Protection Agency, Washington, DC, USA, pp. 4-1 to 4-68.

For Chapter 5:

Scott, J.M., B. Griffith, R.S. Adamcik, D.M. Ashe, B. Czech, R.L. Fischman, P. Gonzalez, J.J. Lawler, A.D. McGuire, and A. Pidgorna, 2008: National Wildlife Refuges. In: *Preliminary review of adaptation options for climate-sensitive ecosystems and resources.* A Report by the U.S. Climate Change Science Program and the Subcommittee on Global Change Research [Julius, S.H., J.M. West (eds.), J.S. Baron, B. Griffith, L.A. Joyce, P. Kareiva, B.D. Keller, M.A. Palmer, C.H. Peterson, and J.M. Scott (Authors)]. U.S. Environmental Protection Agency, Washington, DC, USA, pp. 5-1 to 5-100.

For Chapter 6:

Palmer, M.A., D. Lettenmaier, N.L. Poff, S. Postel, B. Richter, and R. Warner, 2008: Wild and Scenic Rivers. In: *Preliminary review of adaptation options for climate sensitive ecosystems and resources.* A Report by the U.S. Climate Change Science Program and the Subcommittee on Global Change Research [Julius, S.H., J.M. West (eds.), J.S. Baron, B. Griffith, L.A. Joyce, P. Kareiva, B.D. Keller, M.A. Palmer, C.H. Peterson, and J.M. Scott (Authors)]. U.S. Environmental Protection Agency, Washington, DC, USA, pp. 6-1 to 6-73.

For Chapter 7:

Peterson, C.H., R.T. Barber, K.L. Cottingham, H.K. Lotze, C.A. Simenstad, R.R. Christian, M.F. Piehler, and J. Wilson, 2008: National Estuaries. In: *Preliminary review of adaptation options for climate-sensitive ecosystems and resources.* A Report by the U.S. Climate Change Science Program and the Subcommittee on Global Change Research [Julius, S.H., J.M. West (eds.), J.S. Baron, B. Griffith, L.A. Joyce, P. Kareiva, B.D. Keller, M.A. Palmer, C.H. Peterson, and J.M. Scott (Authors)]. U.S. Environmental Protection Agency, Washington, DC, USA, pp. 7-1 to 7-108.

For Chapter 8:

Keller, B.D., S. Airamé, B. Causey, A. Friedlander, D.F. Gleason, R. Grober-Dunsmore, J. Johnson, E. McLeod, S.L. Miller, R.S. Steneck, and C. Woodley, 2008: Marine Protected Areas. In: *Preliminary review of adaptation options for climate-sensitive ecosystems and resources.* A Report by the U.S. Climate Change Science Program and the Subcommittee on Global Change Research [Julius, S.H., J.M. West (eds.), J.S. Baron, B. Griffith, L.A. Joyce, P. Kareiva, B.D. Keller, M.A. Palmer, C.H. Peterson, and J.M. Scott (Authors)]. U.S. Environmental Protection Agency, Washington, DC, USA, pp. 8-1 to 8-95.

For Chapter 9:

Kareiva, P., C. Enquist, A. Johnson, S.H. Julius, J. Lawler, B. Petersen, L. Pitelka, R. Shaw, and J.M. West, 2008: Synthesis and Conclusions. In: *Preliminary review of adaptation options for climate-sensitive ecosystems and resources.* A Report by the U.S. Climate Change Science Program and the Subcommittee on Global Change Research [Julius, S.H., J.M. West (eds.), J.S. Baron, B. Griffith, L.A. Joyce, P. Kareiva, B.D. Keller, M.A. Palmer, C.H. Peterson, and J.M. Scott (Authors)]. U.S. Environmental Protection Agency, Washington, DC, USA, pp. 9-1 to 9-66.

For Annex A:

Julius, S.H., J.M. West, S. Airamé, R.T. Barber, J.S. Baron, G.M. Blate, B. Causey, R.R. Christian, K.L. Cottingham, A. Friedlander, B. Griffith, J. Johnson, L.A. Joyce, B.D. Keller, D. Lettenmaier, J.S. Littell, H.K. Lotze, A.D. McGuire, S.G. McNulty, C.I. Millar, S.L. Miller, K.A. O'Halloran, J. Oropeza, M.A. Palmer, D.L. Peterson, C.H. Peterson, M.F. Piehler, N.L. Poff, S. Postel, B. Richter, C.A. Simenstad, R. Warner, and J.W. Wilson, 2008: Annex A: Case Studies. In: *Preliminary review of adaptation options for climate-sensitive ecosystems and resources.* A Report by the U.S. Climate Change Science Program and the Subcommittee on Global Change Research [Julius, S.H., J.M. West (eds.), J.S. Baron, B. Griffith, L.A. Joyce, P. Kareiva, B.D. Keller, M.A. Palmer, C.H. Peterson, and J.M. Scott (Authors)]. U.S. Environmental Protection Agency, Washington, DC, USA, pp. A-1 to A-170.

For Annex B:

Julius, S.H., J.M. West, J S. Baron, B. Griffith, L.A. Joyce, B.D. Keller, M.A. Palmer, C.H. Peterson, and J.M. Scott, 2008: Annex B: Confidence Estimates for SAP 4.4 Adaptation Approaches. In: *Preliminary review of adaptation options for climate-sensitive ecosystems and resources.* A Report by the U.S. Climate Change Science Program and the Subcommittee on Global Change Research [Julius, S.H., J.M. West (eds.), J.S. Baron, B. Griffith, L.A. Joyce, P. Kareiva, B.D. Keller, M.A. Palmer, C.H. Peterson, and J.M. Scott (Authors)]. U.S. Environmental Protection Agency, Washington, DC, USA, pp. B-1 to B-36.

PREFACE

The U.S. Government's Climate Change Science Program (CCSP) is responsible for providing the best science-based knowledge possible to inform management of the risks and opportunities associated with changes in the climate and related environmental systems. To support its mission, the CCSP has commissioned 21 "synthesis and assessment products" (SAPs) to advance decision-making on climate change-related issues by providing current evaluations of climate change science and identifying priorities for research, observation, and decision support. This Report—SAP 4.4—focuses on federally managed lands and waters to provide a "Preliminary Review of Adaptation Options for Climate-Sensitive Ecosystems and Resources." It is one of seven reports that support Goal 4 of the CCSP Strategic Plan to understand the sensitivity and adaptability of different natural and managed ecosystems and human systems to climate and related global changes.

The purpose of SAP 4.4 is to provide useful information on the state of knowledge regarding adaptation options for key, representative ecosystems and resources that may be sensitive to climate variability and change. As its title suggests, this report is a preliminary review, defined as "the process of collecting and reviewing available information about known or potential adaptation options." The Intergovernmental Panel on Climate Change (IPCC) notes that there are few demonstrated examples of ecosystem-focused adaptation options (see IPCC Fourth Assessment Report, 17.4.2.1 and 4.6.2). Thus, the authors of this SAP found it necessary to examine adaptation options in the context of a desired ecosystem condition or natural resource management goal, as set forth by the resource management entity. Therefore, this report explores potential adaptation options that could be used by natural resource managers within the context of the legislative and administrative mandates of the six systems examined: National Forests, National Parks, National Wildlife Refuges, Wild and Scenic Rivers, National Estuaries, and Marine Protected Areas. Case studies throughout this report examine in greater detail some of the issues and challenges associated with implementation of adaptation options, but are not intended to be geographically comprehensive or representative of the full breadth of ecosystems that exist or adaptation options that are available.

The management systems selected for this report are meant to be representative of a variety of ecosystem types and management goals, in order to be useful to managers who work at different spatial and organizational scales. Time and resource constraints do not allow for a comprehensive coverage of all federally owned and managed lands and waters, which means that some important management systems (e.g., Bureau of Land Management lands, Department of Defense lands, tribal lands, research reserves) are not featured in this report. However, this preliminary review of existing adaptation knowledge does contain science-based adaptation strategies that are broadly applicable to not only other federal lands, but also state, local, territorial, tribal, and non-governmental holdings. Adaptive Management, a key tool recognized in this report, is an important concept within the Department of the Interior, and an Adaptive Management Technical Guide[1] was released in the spring of 2007. It provides a robust analytical framework that is based on the experience, in-depth consultation, and best practices of scientists and natural resource managers. The information in this SAP combined with Interior's Technical Guide is available for managers to consider and discuss. Additional work is needed to refine and add to this body of knowledge, including conducting detailed analyses of adaptation options on a case-by-case basis.

It must be noted that a discussion of the cost and benefits of implementing the adaptation options, either individually or collectively, was not a component of the SAP prospectus and is not included in this report. Relative to ecosystems, the IPCC noted that information is very limited on the economic and social costs and benefits of adaptation measures, especially the non-market costs and benefits of adaptation measures involving ecosystem protection, among others. Since this is a preliminary report, additional information on the costs and benefits is certainly warranted.

[1] **Williams**, B.K., R.C. Szaro, and C.D. Shapiro. 2007. Adaptive Management: The U.S. Department of the Interior Technical Guide. Adaptive Management Working Group, U.S. Department of the Interior, Washington, DC.

While SAPs 4.1, 4.2 and 4.3 analyze the impacts literature, this report focuses on the current science available on adaptation responses. This report synthesizes climate change research with the experience of on-the-ground ecosystem and resource managers to suggest adaptation options that consist of: 1) adjustments to current practices to ensure their effectiveness given climate change interactions with "traditional stressors," and 2) creation of new practices. The level of confidence in each of the adaptation approaches was evaluated by the authors based on their experience and assessment of the peer-reviewed literature on climate change impacts, current management techniques, and ecological responses. The adaptation approaches and measures suggested in this report are presented as options, not as prescriptive directives, standards, or rules.

Michael W. Slimak
Associate Director for Ecology
National Center for Environmental Assessment
Office of Research and Development
U.S. Environmental Protection Agency

CHAPTER 1

Executive Summary

Authors: Susan Herrod Julius, U.S. Environmental Protection Agency; Jordan M. West, U.S. Environmental Protection Agency; Geoff Blate, AAAS Fellow at U.S. Environmental Protection Agency; Jill S. Baron, U.S. Geological Survey and Colorado State University; Brad Griffith, U.S. Geological Survey; Linda A. Joyce, U.S.D.A. Forest Service; Peter Kareiva, The Nature Conservancy; Brian D. Keller, National Oceanic and Atmospheric Administration; Margaret Palmer, University of Maryland; Charles Peterson, University of North Carolina; J. Michael Scott, U.S. Geological Survey and University of Idaho

Climate variables are key determinants of geographic distributions and biophysical characteristics of ecosystems, communities, and species. Climate *change*[1] is therefore affecting many species attributes, ecological interactions, and ecosystem processes. Because changes in the climate system will continue into the future regardless of emissions mitigation, strategies for protecting climate-sensitive ecosystems through management will be increasingly important. While there will always be uncertainties associated with the future path of climate change, the response of ecosystems to climate impacts, and the effects of management, it is both possible and essential for adaptation to proceed using the best available science.

This report provides a preliminary review of adaptation options for climate-sensitive ecosystems and resources in the United States. The term "adaptation" in this document refers to adjustments in human social systems (*e.g.*, management) in response to climate stimuli and their effects. Since management always occurs in the context of desired ecosystem conditions or natural resource management goals, it is instructive to examine particular goals and processes used by different organizations to fulfill their objectives. Such an examination allows for discussion of specific adaptation options as well as potential barriers and opportunities for implementation. Using this approach, this report presents a series of chapters on the following selected management systems: National Forests, National Parks, National Wildlife Refuges, Wild and Scenic Rivers, National Estuaries, and Marine Protected Areas. For these chapters, the authors draw on the literature, their own expert opinion, and expert workshops composed of resource management scientists and representatives of managing agencies. The information drawn from across these chapters is then analyzed to develop the key synthetic messages presented below.

Many existing best management practices for "traditional" stressors of concern have the added benefit of reducing climate change exacerbations of those stressors. Changes in temperature, precipitation, sea level, and other climate-related factors can often exacerbate problems that are already of concern to managers. For example, increased intensity of precipitation events can further increase delivery of non-point source pollution and sediments to rivers, estuaries, and coasts. Fortunately, many management practices that exist to address such "traditional" stressors can also address climate change impacts. One such practice with multiple benefits is the construction of riparian buffer strips that (1) manage pollution loadings

[1] Climate change refers to any change in climate over time, whether due to natural variability or as a result of human activity. This usage differs from that in the United Nations Framework Convention on Climate Change, which defines "climate change" as: "a change of climate which is attributed directly or indirectly to human activity that alters the composition of the global atmosphere and which is in addition to natural climate variability observed over comparable time periods."

from agricultural lands into rivers today and (2) establish protective barriers against increases in both pollution and sediment loadings due to climate changes in the future. While multiple benefits may result from continuing with today's best practices, key adjustments in their application across space and time may be needed to ensure their continued effectiveness in light of climate change.

Seven "adaptation approaches" can be used for strategic adjustment of best management practices to maximize ecosystem resilience to climate change. As defined in this report, the goal of adaptation is to reduce the risk of adverse environmental outcomes through activities that increase the resilience of ecological systems to climate change. Here, resilience refers to the amount of change or disturbance that a system can absorb without undergoing a fundamental shift to a different set of processes and structures. Managers' past experiences with unpredictable and extreme events have already led to some existing approaches that can be adjusted for use in adapting to longer-term climate change. The specific "adaptation approaches" described below are derived from discussions of existing (and new) management practices to maintain or increase ecosystem resilience, drawn from across the chapters of this report.

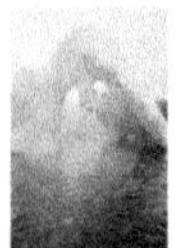

Protecting key ecosystem features involves focusing management protections on structural characteristics, organisms, or areas that represent important "underpinnings" or "keystones" of the overall system. **Reducing anthropogenic stresses** is the approach of minimizing localized human stressors (*e.g.*, pollution, fragmentation) that hinder the ability of species or ecosystems to withstand climatic events. **Representation** refers to protecting a portfolio of variant forms of a species or ecosystem so that, regardless of the climatic changes that occur, there will be areas that survive and provide a source for recovery. **Replication** centers on maintaining more than one example of each ecosystem or population such that if one area is affected by a disturbance, replicates in another area provide insurance against extinction and a source for recolonization of affected areas. **Restoration** is the practice of rehabilitating ecosystems that have been lost or compromised. **Refugia** are areas that are less affected by climate change than other areas and can be used as sources of "seed" for recovery or as destinations for climate-sensitive migrants. **Relocation** refers to human-facilitated transplantation of organisms from one location to another in order to bypass a barrier (*e.g.*, urban area).

Each of these adaptation approaches ultimately contributes to resilience, whether at the scale of individual protected area units, or at the scale of regional/national systems. The approaches above are not mutually exclusive and may be implemented jointly. The specific management activities that are selected under one or more approaches above should then be based on considerations such as: the ecosystem management goals, type and degree of climate effects, type and magnitude of ecosystem responses, spatial and temporal scales of ecological and management responses, and social and economic factors.

Levels of confidence in these adaptation approaches vary and are difficult to assess, yet are essential to consider in adaptation planning. Due to uncertainties associated with climate change projections as well as uncertainties in species and ecosystem responses, there is also uncertainty as to how effective the different adaptation approaches listed above will be at supporting resilience. It is therefore important to assess the confidence within the expert community that these approaches will support a degree of resilience that may allow ecosystems to persist without major losses of ecosystem processes or functions. Using one of the methodologies presented in the Intergovernmental Panel on Climate Change's guidelines[2] for estimating uncertainties, the authors of this report developed their confidence estimates by considering two separate but related elements of confidence. The first element is the amount of available evidence (high or low) to support the determination that the effectiveness of a given adaptation approach is well-studied and understood. Evidence might consist of any of the following sources: peer-reviewed and gray literature, data and observations, model results, and the authors'

[2] Guidance on uncertainty from *Climate Change 2007: Impacts, Adaptation and Vulnerability. Contribution of Working Group II to the Fourth Assessment Report of the Intergovernmental Panel on Climate Change*, M.L. Parry, O.F. Canziani, J.P. Palutikof, P.J. van der Linden and C.E. Hanson, Eds., Cambridge University Press, Cambridge, UK, 976pp.

own experience with each adaptation approach. The second element is the level of agreement or consensus throughout the scientific community about the different lines of evidence on the effectiveness of the adaptation approach.

The resulting confidence estimates vary, both across approaches and across management systems. Reducing anthropogenic stresses is one approach for which there is considerable scientific confidence in its ability to promote resilience for virtually any situation. Confidence in the other approaches—including protecting key ecosystem features, representation, replication, restoration, identifying refuges, and especially relocation—is much more variable. Despite this variability, many of the individual adaptation options under these approaches may still be effective. In these cases, a more detailed assessment of confidence for individual adaptation options is needed, based on a clearer understanding of how the ecosystem in question functions, the extent and type of climate change that will occur there, the resulting ecosystem impacts, and the projected ecosystem response to the adaptation option.

One method for integrating confidence estimates into resource management given uncertainty is adaptive management. Adaptive management is a process that promotes flexible decision-making so that adjustments are made in decisions as outcomes from management actions and other events are better understood. This method supports managers in taking action today using the best available information while also providing the possibility of ongoing future refinements through an iterative learning process.

The success of adaptation strategies may depend on recognition of potential barriers to implementation and creation of opportunities for partnerships and leveraging. In many cases, perceived barriers associated with legal or social constraints, restrictive management procedures, limitations on human and financial capital, and gaps in information may be converted into opportunities. For example, there may be a possibility to address difficulties associated with information or capacity shortages through leveraging of human capital. Existing staff could receive training on addressing climate change issues within the context of their current job descriptions and management frameworks, but a critical requirement for success of this activity would be to ensure that employees feel both valued as "climate adaptation specialists" and empowered by their institutions to develop and implement innovative adaptive management approaches that might be perceived as "risky." As a second example, partnerships among managers, scientists, and educators can go a long way toward efficiently closing information gaps. With good communication and coordination, scientists can target their research to better inform management challenges, resource managers can share data and better design monitoring to test scientific hypotheses, and outreach specialists can better engage the public in understanding and supporting adaptation activities. Two additional categories of opportunities that are especially promising are highlighted below.

The Nation's adaptive capacity can be increased through expanded collaborations among ecosystem managers. When managers seize opportunities to link with other managers to coordinate adaptation planning, they are able to broaden the spatial and ecological scope of potential adaptation options with a shared vision for increasing adaptive capacity. For example, many management units are nested within or adjacent to other systems. Collaboration across systems allows individual units to be, in effect, extended beyond their official boundaries to encompass entire ecosystems or regions; the result is a larger array of options for responding to future climate change impacts. Collaboration may also enhance research capacity and offer opportunities to share data, models, and experiences. In addition to overcoming limiting factors such as inadequate resources and mismatches of management unit size with ecosystem extent, collaborations may also be used to create flexible boundaries that follow unanticipated changes in ecosystems or species in response to climate change. Exercising opportunities for collaboration has the advantage of reducing uncertainties associated with attaining management goals under climate change because (1) the increase in the geographic range over which resources can be managed and the associated increase in available adaptation options makes success more likely, and (2) the increase in the resource base, in research capabilities, and in the size of data sets through data sharing and coordinated monitoring reduces statistical uncertainties and increases the probability of success.

The Nation's adaptive capacity can be increased through creative re-examination of program goals and authorities. Anticipated climate-induced changes in ecosystems and species and the uncertain nature of some of those changes will necessitate dynamic management systems that can accommodate and address such changes. Existing management authorities may be malleable enough to allow for changing conditions and dynamic responses, and with creative re-examination of those authorities their full capabilities could be applied. For example, federal land and water managers may be able to strategically apply traditional legislative authorities in non-traditional ways to coordinate management outside of jurisdictional boundaries. Similarly, while management policies can sometimes be limiting, the iterative nature of management planning may allow priorities and plans to be revisited on a cyclical basis to allow for periodic adjustments. Greater agility in program planning can increase the probability of meeting management goals by overcoming implementation barriers associated with narrowly defined and interpreted authorities.

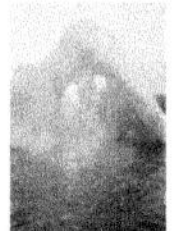

Establishing current baselines, identifying thresholds, and monitoring for changes will be essential elements of any adaptation approach. Climate changes may cause ecological thresholds to be exceeded, leading to abrupt shifts in the structure of ecosystems. Threshold changes in ecosystems have profound implications for management because such changes may be unexpected, large, and difficult to reverse. If these ecosystems cannot then be restored, actions to increase their resilience will no longer be viable. Understanding where thresholds have been exceeded in the past and where (and how likely) they may be exceeded in the future allows managers to plan accordingly and avoid tipping points where possible. Activities taken to prevent threshold changes include establishing current baseline conditions, modeling a range of possible climate changes and system responses, monitoring to identify relevant ecological changes, and responding by implementing adaptation actions at appropriate scales and times. Current baselines capture a benchmark set of conditions for the ecological attributes or processes that are critical for maintaining that system and the current set of ecosystem services that the public has come to expect from that system. Developing a range of quantitative or qualitative visions of the future (scenarios) and planning adaptation responses for that range provide an approach for addressing the large uncertainties associated with any single projection of the future. Sensitivity analyses for any given scenario explore key attributes of the system and their response to systematic changes in the climate drivers. Such analyses may allow managers to identify thresholds beyond which key management goals may become unattainable. Directed monitoring then supports managers' ability to detect changes in baseline conditions, informs their decisions about the timing of adaptation actions, and helps them evaluate the effectiveness of their actions. With such information, a program that has the authority to, for example, acquire land interests and water rights to restore a river to its historic flows would better be able to determine how, when, and where to use this authority.

Beyond "managing for resilience," the Nation's capability to adapt will ultimately depend on our ability to be flexible in setting priorities and "managing for change." Prioritizing actions and balancing competing management objectives at all scales of decision making is essential, especially in the midst of shifting budgets and rapidly changing ecosystems. Using a systematic framework for priority setting would help managers catalog information, design strategies, allocate resources, evaluate progress, and inform the public. This priority-setting could happen in an ongoing way to address changing ecological conditions and make use of new information. Over time, our ability to "manage for resilience" of current systems in the face of climate change will be limited as temperature thresholds are exceeded, climate impacts become severe and irreversible, and socioeconomic costs of maintaining existing ecosystem structures, functions, and services become excessive. At this point, it will be necessary to "manage for change," with a re-examination of priorities and a shift to adaptation options that incorporate information on projected ecosystem changes. Both "managing for resilience" and "managing for change" require more observation and experimentation to fill knowledge gaps on how to adapt to climate change. This report presents a preliminary review of existing adaptation knowledge to support managers in taking immediate actions to meet their management goals in the context of climate change. However, this is only a first step in better understanding this burgeoning area of research in adaptation science and management. It will be necessary to continuously refine and add to this body of knowledge in order to meet the challenge of preserving the Nation's lands and waters in a rapidly changing world.

Introduction

Authors: Susan Herrod Julius, U.S. Environmental Protection Agency; Jordan M. West, U.S. Environmental Protection Agency; Geoffrey M. Blate, AAAS Fellow at U.S. Environmental Protection Agency

Strategies for protecting climate-sensitive ecosystems will be increasingly important for management, because impacts resulting from a changing climate system are already evident and will persist into the future regardless of emissions mitigation. Climate is a dominant factor influencing the distributions, structures, functions, and services of ecosystems. Changes in climate can interact with other environmental changes to affect biodiversity and the future condition of ecosystems (*e.g.*, McCarty, 2001; IPCC, 2001; Parmesan and Yohe, 2003). The extent to which ecosystem condition may be affected will depend on the amount of climate change, the degree of sensitivity of the ecosystem to the climate change, and the availability of adaptation options for effective management responses. This Synthesis and Assessment Product (SAP), SAP 4.4, is charged with reviewing adaptation options for ecosystems that are likely to be sensitive to continuing changes in climate. SAP 4.4 is one of 21 SAPs commissioned by the U.S. government's Climate Change Science Program, seven of which examine the sensitivity and adaptability of different natural and managed ecosystems and human systems to climate and related global changes.

Adaptation is defined as an adjustment in natural or human systems to a new or changing environment. Adaptation to climate change refers to adjustment in natural or human systems in response to actual or expected climatic stimuli or their effects, which moderates harm or exploits beneficial opportunities (IPCC, 2001). In biological disciplines, adaptation refers to the process of genetic change within a population due to natural selection, whereby the average state of a character becomes better suited to some feature of the environment (Groom, Meffe, and Carroll, 2006). This type of adaptation, also referred to as autonomous adaptation (IPCC, 2001), is a reactive biological response to climate stimuli and does not involve intervention by society. Planned adaptation, on the other hand, refers to strategies adopted by society to manage systems based on an awareness that conditions are about to change or have changed, such that action is required to meet management goals (adapted from IPCC, 2001). This report focuses on the latter form of adaptation, with all subsequent uses of the term "adaptation" referring to strategies for management of ecosystems in the context of climate variability and change.

The purpose of adaptation strategies is to reduce the risk of adverse outcomes through activities that increase the resilience of ecological systems to climate change stressors (Scheffer *et al.*, 2001; Turner, II *et al.*, 2003; Tompkins and Adger, 2004). A stressor is defined as any physical, chemical, or biological entity that can induce an adverse response (U.S. Environmental Protection Agency, 2000). Resilience refers to the amount of change or disturbance that can be absorbed by a system before the system is redefined by a different set of processes and structures (Holling, 1973; Gunderson, 2000;

Bennett, Cumming, and Peterson, 2005). Potential adverse outcomes of climate change may vary for different ecosystems, depending on their sensitivity to climate stressors and their intrinsic resilience to climate change. The "effectiveness" of an adaptation option that is designed to boost ecosystem resilience will thus be case-dependent, and can be measured only against a desired ecosystem condition or natural resource management goal. This report evaluates the effectiveness of potential adaptation options for supporting natural resource management goals.

Adaptation options for enhancing ecosystem resilience include changes in management processes, practices, or structures to reduce anticipated damages or enhance beneficial responses associated with climate variability and change. In some cases, opportunities for adaptation offer stakeholders outcomes with multiple benefits, such as the addition of riparian buffer strips that (1) manage pollution loadings from agricultural land into rivers designated as "wild and scenic" today *and* (2) establish a protective barrier to increases in both pollution and sediment loadings associated with future climate change. Where there are multiple benefits to implementing specific adaptation options, this report seeks to identify those benefits.

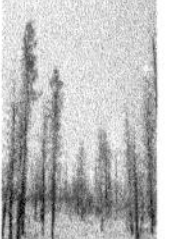

A range of adaptation options may be possible for many ecosystems, but a lack of information or resources may impede successful implementation. In some cases, managers may not have the knowledge or information available to address climate change impacts. In other instances, managers may understand the issues and have the relevant information but lack resources to implement adaptation options. Furthermore, even with improvement in the knowledge and communication of available and emerging adaptation strategies, the feasibility and effectiveness of adaptation will depend on the adaptive capacity of the ecological system or social entity. Adaptive capacity is defined as the potential or ability of a system, region, or community to counteract, adjust for, or take advantage of the effects of climate change (IPCC, 2001). Depending on the management goals, there may be biological, physical, economic, social, cultural, institutional, or technological conditions that enhance or hinder adaptation. To the extent possible, this report will address those factors that affect managers' ability to implement adaptation options.

2.1 GOAL AND AUDIENCE

The goal of SAP 4.4 is to provide useful information on the state of knowledge regarding adaptation options for key, representative ecosystems and resources that may be sensitive to climate variability and change. To provide such useful information, it is necessary to examine adaptation options in the context of a desired ecosystem condition or natural resource management goal. Therefore, this report explores potential adaptation options for supporting natural resource management goals in the context of management systems such as the National Park System or the National Wildlife Refuge System. Management systems such as these provide a framework of processes and procedures used to ensure that an organization's objectives are fulfilled.

Specifically, this report supports the stated goal by providing information on (1) the implications of the combined effects of climate changes and non-climate stressors on our ability to achieve specific resource management goals; (2) existing management options as well as new adaptation approaches that reduce the risk of negative outcomes; and (3) opportunities and barriers that affect successful implementation of management strategies to address climate change impacts. Through the provision of this information, the desired outcome of this report is an enhanced adaptive capacity to respond to future changes in climate.

The primary intended audience of this report is resource and ecosystem managers at federal, state, and local levels; tribes; nongovernmental organizations; and others involved in protected area management decisions. Additional audiences include scientists, engineers, and other technical specialists who will be able to use the information provided to set priorities for future research and to identify decision-support needs and opportunities. This information also may support tribes and government agencies at federal, state, and local levels in the development of policy decisions that promote adaptation and increase society's adaptive capacity for management of ecosystems and species within protected areas.

2.2 STAKEHOLDER INTERACTIONS

Stakeholder interactions play a key role in maximizing the relevance, usefulness, and credibility of assessments and encouraging ownership of the results (National Research Council, 2007). This may be especially true in the adaptation arena, where managers are challenged by both the technical aspects of adaptation and the constraints imposed by legal mandates and resource limitations. In these cases, participation by an appropriate array of stakeholders is important in order to ensure that proposed adaptation options are analyzed in light of both technical rigor and feasibility. Given this, the appropriate composition of stakeholders for SAP 4.4 includes: (1) those who wish to consider options for reducing the risk of negative ecological outcomes associated with climate variability and change; (2) researchers who study climate change impacts on ecosystems and topics relevant for adaptation to impacts of climate variability and change (*e.g.*, ecosystem restoration, sustainability); (3) science managers from the physical and social sciences who develop long-term research plans based on the information needs and decisions at hand; and (4) tribes and government agencies at federal, state, and local levels who develop and evaluate policies, guidelines, procedures, technologies, and other mechanisms to improve adaptive capacity.

The initial planning of SAP 4.4 involved engaging a narrowly defined targeted group of expert stakeholders to review the substance of the report. Small groups of no more than 20 people from the fields of adaptation science and resource management were asked to provide comments to the authors of the report on its content through participation in a series of six workshops (one for each "management system" chapter; see below). Chapter lead and contributing authors presented draft information on their chapters and case studies, and incorporated the expert input into their revisions.

Beyond the narrowly defined group of expert stakeholders mentioned above, a broader array of relevant stakeholders were invited to contribute to the shaping of this document through a public review process. Feedback was received from non-governmental organizations, industry, academia, state organizations, and private citizens, as well as federal government representatives. That feedback resulted in significant changes to this report. Final input was received from a Federal Advisory Committee composed primarily of academicians.

2.3 APPROACH FOR REVIEWING ADAPTATION OPTIONS FOR CLIMATE-SENSITIVE ECOSYSTEMS AND RESOURCES

This report examines federally protected and managed lands and waters as a context for reviewing adaptation options for climate-sensitive ecosystems and resources. The focus on federal holdings was chosen because (1) their protected status reflects the value placed on these ecosystems and resources by the American public; (2) the management goals for federal ecosystems are also representative of the range of goals and challenges faced by other ecosystem management organizations across the United States; and (3) adaptation options for federal ecosystems will require a variety of responses (equally applicable to non-federal lands) to ensure achievement of management goals over a range of time scales.

Approximately one-third of the nation's land base is managed by the federal government and administered by different agencies through a variety of "management systems." Since a comprehensive treatment of all federal holdings is beyond the scope of this report, the focus is on representative management systems that have clear management goals for which adaptation options can be discussed. Therefore, adaptation options are reviewed for six management systems: national forests, national parks, national wildlife refuges, wild and scenic rivers, national estuaries, and marine protected areas (especially national marine sanctuaries). By using a sample of management systems, the discussion of adaptation options can go beyond a general list to more specific options tailored to the management context and goals. This approach also allows exploration of any specific barriers and opportunities that may affect implementation. The array of adaptation options discussed should be useful to other resource managers, regardless of whether their management systems are represented in this report. Likewise, the types of barriers and

BOX 2.1. Case Study Selection Criteria.

The authors of this report, in consultation with agency representatives and stakeholders, used the following criteria for evaluation and selection of candidate case studies:

- Contains one or more ecosystem services or features that are protected by management goals;
- Management goals are sensitive to climate variability and change, and the potential impacts of climate variability and change are significant relative to the impacts of other changes;
- Adaptation options are available or possible for preserving a service or a physical or biological feature; and
- Adaptation options have potential for application in other geographic regions or for other ecosystem types.

In order to ensure that the entire collection of case studies would include broad representation across geographic areas, ecosystem types, and management goals and methods, the following characteristics were required of the group as a whole:

- Addresses a reasonable cross section of important, climate-sensitive ecosystems and/or ecosystem services and features;
- Addresses a range of adaptation responses (e.g., structural, policy, permitting);
- Distributed across the United States and valued by a national constituency; and
- Attributes allow for comparison of adaptation approaches and their effectiveness across the case studies (e.g., lessons learned about research gaps and about factors that enhance or impede implementation).

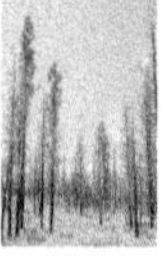

suggested methods for addressing those barriers should be sufficiently broad to be useful to a wider audience of resource managers. Other federally protected systems—such as wilderness preservation areas, biosphere reserves, research natural areas, natural estuarine research reserves, and public lands—could not be examined in this report because of limitations on time and resources. As a result, certain important and extensive management systems (*e.g.*, Bureau of Land Management) were not reviewed in this report. Thus, the material in this report represents only the beginning of what should be an ongoing effort to inform and support resource management decision making. Other management systems not represented in this report would also benefit from specific examination of important impacts and adaptation options.

For each of the six management systems selected, this report reviews (1) the historical origins of the management system and the formative factors that shaped its mission and goals, (2) key ecosystem components and processes upon which those goals depend, (3) stressors of concern for the key ecosystem characteristics, (4) management methods currently in use to address those stressors, (5) ways in which climate variability and change may affect attainment of management goals, and (6) options for adjusting current management strategies or developing new strategies in response to climate change. All of these elements vary considerably depending on the history and organizational structure of the management systems and the locations and types of ecosystems that they manage.

Specific management goals for the ecosystems in the different management systems vary based on the management principles or frameworks employed to reach targeted goals. Natural resource management goals are commonly expressed in terms of maintaining ecosystem integrity, achieving restoration, preserving ecosystem services, and protecting wildlife and other ecosystem characteristics. The achievement of management goals is thus dependant on our ability to protect, support, and restore the structure and functioning of ecosystems.

Changes in climate may affect ecosystems such that management goals are not achieved. Thus, the identified management goals from the literature review are analyzed for their sensitivity to climate variability and change, as well as to other stressors present in the system that may interact with climate change. Adaptive responses to climate variability and change are meant to reduce the risk of failing to achieve management goals. Therefore, each management system chapter discusses adaptation theories and frameworks, as well as options for modifying existing management actions and developing new approaches to address climate change impacts.

For each chapter, the above analysis of climate sensitivities and management responses includes one or more place-based case studies that explore the current state of knowledge regarding management options that could be used to adapt to the potential impacts of climate variability and change. The case studies—which were selected using a range of criteria (Box 2.1)—cover a variety of ecosystem types such as forests, rivers and streams, wetlands, estuaries, and coral reefs (Fig. 2.1). All case studies are presented in Annex A.

Taken together, the six management system chapters of this report offer an array of issues, viewpoints, and case studies to inform managers as they consider adaptation options. As such, they are not only useful individually but also serve as rich sources of "data" to inform the cross-cutting themes and synthetic approaches that comprise the "results" of the Synthesis and Conclusions chapter.

2.4 CLIMATE VARIABILITY AND CHANGE

Climate change is defined by the Intergovernmental Panel on Climate Change (IPCC) as any change in climate over time, whether due to natural variability or as a result of human activity (IPCC, 2007b). Climate variability refers to variations in the mean state and other statistics (such as standard deviations, the occurrence of extremes, etc.) of the climate on all temporal and spatial scales beyond that of individual weather events (IPCC, 2007b). The motivation for developing responses to projected changes in the climate system stems from observations of changes that have already occurred, as well as projected climate changes. The discussion below provides background information on observed climatic and ecological changes that have implications for management of ecosystems in the United States. For more detailed information, the reader is referred to recent publications of the IPCC (IPCC, 2007a; 2007b).

2.4.1 Increases in Surface Temperature

Evidence from observations of the climate system has led to the conclusion that human activities are contributing to a warming of the earth's atmosphere. This evidence includes an increase of 0.74 ± 0.18°C in global average surface temperature over the last century, and an even greater warming trend over the last 50 years than over the last 100 years. Eleven of the last 12 years (1995–2006) are among the 12

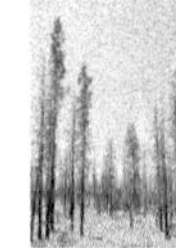

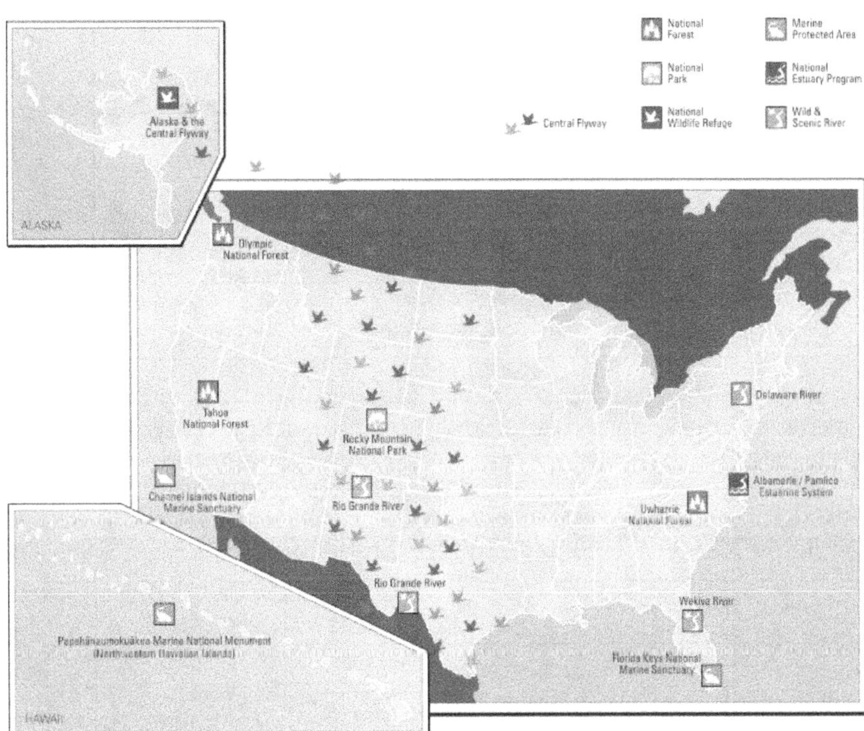

Figure 2.1. Map showing the geographic distribution in the United States of SAP 4.4 case studies.

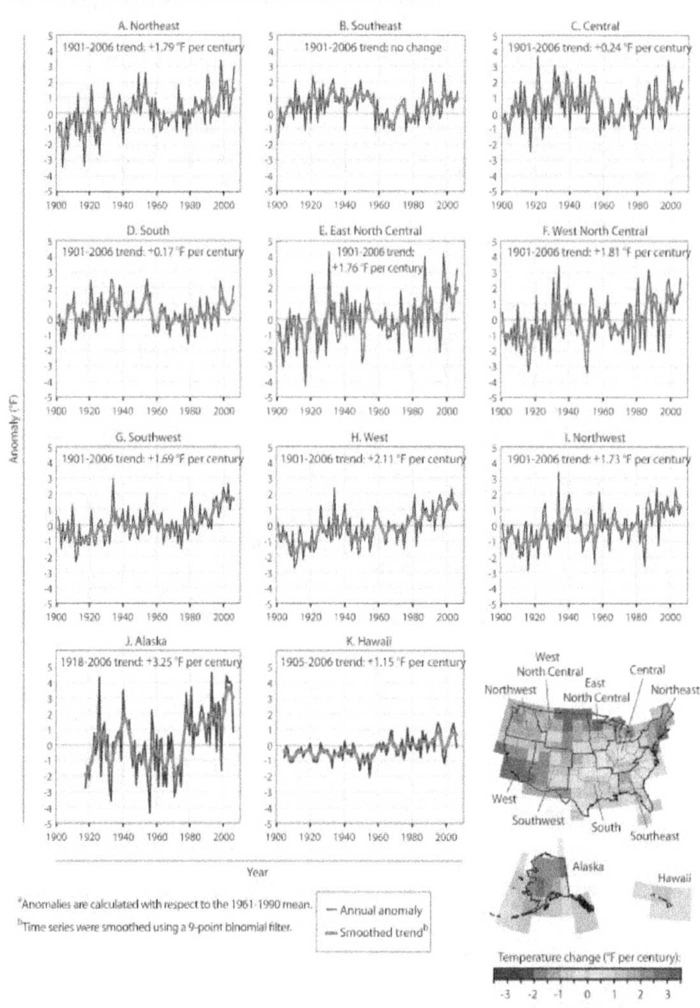

Figure 2.2. Annual mean temperature anomalies 1901–2006. Note: Red shades indicate warming over the period and blue shades indicate cooling over the period. Data courtesy of National Oceanic and Atmospheric Administration's National Climatic Data Center.

warmest years since the instrumental record of global surface temperature was started in 1850 (IPCC, 2007b).

In the continental United States, temperatures rose linearly at a rate of 0.06°C per decade during the first half of the 20th century. That rate increased to 0.33°C per decade from 1976 to the present. The degree of warming has varied by region (Fig. 2.2) across the United States, with the West and Alaska experiencing the greatest degree of warming (U.S. Environmental Protection Agency, 2007). These changes in temperature have led to an increase in the number of frost-free days. In the United States, the greatest increases have occurred in the West and Southwest (Tebaldi *et al.*, 2006).

2.4.2 Changes in Precipitation

Changes in climate have also been manifested in altered precipitation patterns. Over the last century, the amount of precipitation has increased significantly across eastern parts of North America and several other regions of the world (IPCC, 2007b). In the contiguous United States, this increase in total annual precipitation

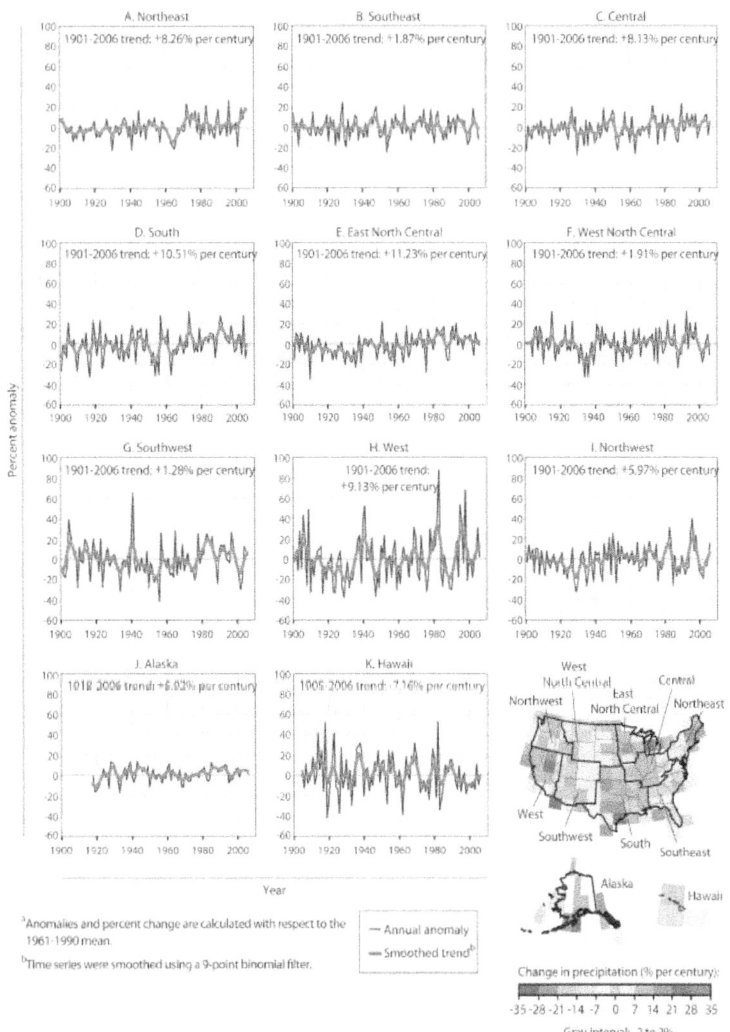

Figure 2.3. Annual precipitation anomalies 1901–2006. Note: Green shades indicate a trend towards wetter conditions over the period, and brown shades indicate a trend towards dryer conditions. Data courtesy of National Oceanic and Atmospheric Administration's National Climatic Data Center.

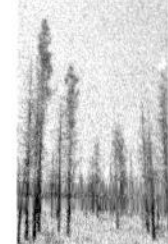

over the last century has been 6.1%. When looked at by region (Fig. 2.3), however, the direction and magnitude of precipitation changes vary, with increases of more than 10% observed in the East North Central and South, and a decrease of more than 7% in Hawaii (U.S. Environmental Protection Agency, 2007). The form of precipitation has also changed in some areas. For example, in the western United States, more precipitation has been falling as rain than snow over the last 50 years (Knowles, Dettinger, and Cayan, 2006).

2.4.3 Warming of the Oceans

Another manifestation of changes in the climate system is a warming in the world's oceans. The global ocean temperature rose by 0.10°C from the surface to 700 m depth from 1961–2003 (IPCC, 2007b). Observations of sea-surface temperatures, based on a reconstruction of the long-term variability and change in global mean sea-surface temperature for the period 1880–2005, show that they have reached their highest levels during the past three decades over

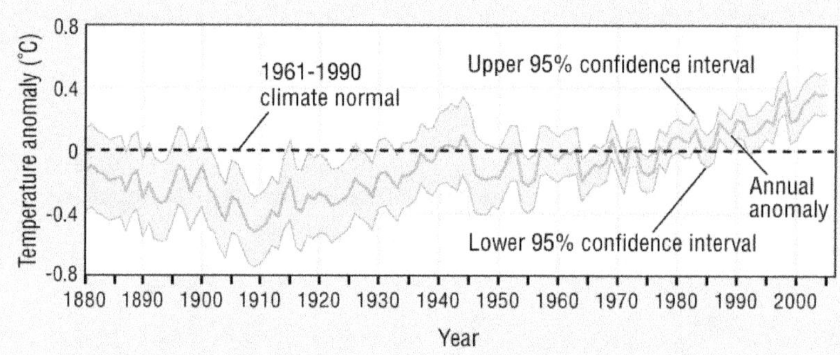

Figure 2.4. Annual global sea surface temperature anomaly, 1880–2005, compared with 1961–1990 climate normal (U.S. Environmental Protection Agency, 2007).

all latitudes (Fig. 2.4). Warming has occurred through most of the 20th century and appears to be independent of measured inter-decadal and short-term variability (Smith and Reynolds, 2005).

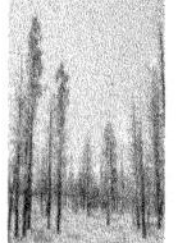

2.4.4 Sea Level Rise and Storm Intensity

Warming causes seawater to expand and thus contributes to sea level rise. This factor, referred to as thermal expansion, has contributed 1.6 ± 0.5 mm per year to global average sea level over the last decade (1993–2003). Other factors contributing to sea level rise over the last decade include a decline in mountain glaciers and ice caps (0.77 ± 0.22 mm per year), losses from the Greenland ice sheets (0.21 ± 0.07 mm per year), and losses from the Antarctic ice sheets (0.21 ± 0.35 mm per year) (IPCC, 2007c).

In the United States, relative sea levels have been rising along most of the coasts at rates of 1.5–3 mm per year (U.S. Environmental Protection Agency, 2007), which is consistent with the average rate globally for the 20th century (1.7±0.5 mm per year) (IPCC, 2007b). Relative sea level has risen 3–4 mm per year in the Mid-Atlantic states and 5–10 mm per year in the Gulf states, due to subsidence combined with accelerated global sea level rise (U.S. Environmental Protection Agency, 2007). On Florida's Gulf coast, relative sea level rise has led to a rate of conversion of about 2 meters of forest to salt marsh annually (Williams *et al.*, 1999).

The effects of sea level rise in coastal areas will be compounded if tropical cyclones become more intense. For the North Atlantic, there is observational evidence since about 1970 of an increase in intense tropical cyclone activity which is correlated with increases in tropical sea surface temperatures (IPCC, 2007b). Various high resolution global models and regional hurricane models also indicate that it is likely that some increase in tropical cyclone intensity will occur if the climate continues to warm (IPCC, 2007b). This topic remains an area of intense debate and investigation, with many competing opinions as to the accuracy of detection methods, the quality of historical data, and the strength of various modeling results (*e.g.*, see Donnelly and Woodruff, 2007; Landsea, 2007; Vecchi and Soden, 2007). Nevertheless, if the prospect of increasingly intense tropical cyclone activity is one plausible scenario for the future, then the possibility of intensified storm surges and associated exacerbation of sea level rise impacts may merit consideration and planning by managers.

2.4.5 Changes in Ocean pH

Between 1750 and 1994, the oceans absorbed about 42% of all emitted carbon dioxide (CO_2) (IPCC, 2007b). As a result, the total inorganic carbon content of the oceans increased by 118 ±19 gigatons of carbon over this period and is continuing to increase. This increase in oceanic carbon content caused calcium carbonate ($CaCO_3$) to dissolve at greater depths and led to a 0.1 unit decrease in surface ocean pH from 1750–1994 (IPCC, 2007b). The rate of decrease in pH over the past 20 years accelerated to 0.02 units per decade (IPCC, 2007b). A decline in pH, along with the concomitant decreased depth at which calcium carbonate dissolves, will likely impair the ability of marine organisms to use carbonate ions to build their shells or other hard parts (The Royal Society, 2005; Caldeira and Wickett, 2005; Doney, 2006; Kleypas *et al.*, 2006).

2.4.6 Warming in the Arctic

Other observations at smaller geographic scales lend evidence that the climate system is warming. For example, in the Arctic, average temperatures have increased and sea ice extent has shrunk. Over the last 100 years, the rate of increase in average Arctic temperatures has been almost twice that of the global average rate, and since 1978 the annual average sea ice extent has shrunk by 2.7 ± 0.6% per decade. The permafrost layer has also been affected in the Arctic, to the degree that the maximum area of ground frozen seasonally has decreased by about 7% in the Northern Hemisphere since 1900, with the spring realizing the largest decrease (up to 15%) (IPCC, 2007b).

2.4.7 Changes in Extreme Events

Whether they have become drier or wetter, many land areas have likely experienced an increase in the number and intensity of heavy precipitation (5 cm of rain or more) events (IPCC, 2007b). About half of the increase in total precipitation observed nationally has been attributed to the increase in intensity of storms (Karl and Knight, 1998). Heavy precipitation events are the principal cause of flooding in most of the United States (Groisman *et al.*, 2005).

The general warming trend observed in most of the United States was also accompanied by more frequent hot days, hot nights, and heat waves (IPCC, 2007b). Furthermore, higher temperatures along with decreased precipitation have been associated with observations of more intense and longer droughts over wider areas since the 1970s. Within the United States, the western region has experienced longer and more intense droughts, but these appear also to be related to diminishing snow pack and consequent reductions in soil moisture. In addition to the factors above, changes in sea-surface temperatures and wind patterns have been linked to droughts (IPCC, 2007b).

2.4.8 Changes in Hydrology

During the 20th century, the changes in temperature and precipitation described above caused important changes in hydrology over the continental United States. One change was a decline in spring snow cover. This trend was observed throughout the Northern Hemisphere starting in the 1920s and accelerated in the late 1970s (IPCC, 2007b). Declining snow cover is a concern in the United States, because many western states rely on snowmelt for their water use (Mote *et al.*, 2005). Less snow generally translates to lower reservoir levels. The earlier onset of spring snowmelt exacerbates this problem. Snowmelt started 2–3 weeks earlier in 2000 than it did in 1948 (Stewart, Cayan, and Dettinger, 2004).

Another important change, described in the preceding section, was the increase in heavy precipitation events documented in the United States during the past few decades. These changes have affected the timing and magnitude of streamflow. In the eastern United States, high streamflow measurements were associated with heavy precipitation events (Groisman, Knight, and Karl, 2001). Because of this association, there is a high probability that high streamflow conditions have increased during the 20th century (Groisman, Knight, and Karl, 2001). Increases in peak streamflow have not been observed in the West, most likely because of the reduction in snow cover (Groisman, Knight, and Karl, 2001).

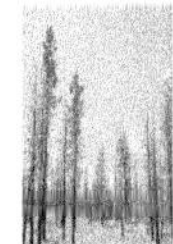

2.4.9 Observed Ecological Responses

An emerging but growing body of literature indicates that over the past three decades, the changes in the climate system described above—including the anthropogenic component of warming— have caused physical and biological changes in a variety of ecosystems (Root *et al.*, 2005; Parmesan, 2006; IPCC, 2007a) that are discernable at the global scale. These changes include shifts in genetics (Bradshaw and Holzapfel, 2006; Franks, Sim, and Weis, 2007), species' ranges, phenological patterns, and life cycles (reviewed in Parmesan, 2006). Most (85%) of these ecological responses have been in the expected direction (*e.g.*, poleward shifts in species distributions), and it is very unlikely that the observed responses are due to natural variability alone (IPCC, 2007a). The asynchronous responses of different species to climate change may alter species' interactions (*e.g.*, predator-prey relationships and competition) and have unforeseen consequences (Parmesan and Galbraith, 2004).

2.4.10 Future Anticipated Climate Change

Improvements in understanding of the anthropogenic influences on climate have led to greater confidence in most of the changes described in the previous section. This improved understanding, in combination with improvements in the models that simulate climate change processes, has also increased confidence in model projections of future climatic changes. The most recent models project future changes in the earth's climate system that are greater in magnitude and scope than those already observed. Based on annual average projections (from 21 global climate models), surface temperature increases by the end of the 21st century will range from 2°C near the coasts in the conterminous United States to at least 5°C in northern Alaska. Nationally, summertime temperatures are projected to increase by 3–5°C. Winter temperatures in Northern Alaska are projected to increase by 4.4–11°C. In addition, more extreme hot events and fewer extreme cold events are projected to occur (IPCC, 2007b).

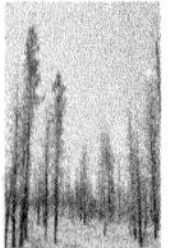

On average, annual precipitation will likely increase in the northeastern United States and will likely decrease in the Southwest over the next 100 years (IPCC, 2007b). In the western United States, precipitation increases are projected during the winter, whereas decreases are projected for the summer (IPCC, 2007b). As temperatures warm, precipitation will increasingly fall as rain rather than snow, and snow season length and snow depth are very likely to decrease in most of the country (IPCC, 2007b). More extreme precipitation events are also projected (Diffenbaugh *et al.*, 2005; Diffenbaugh, 2005), which, coupled with an anticipated increase in rain-on-snow events, would contribute to more severe flooding due to increases in extreme runoff (IPCC, 2007b).

The interaction of climate change with other stressors, as well as direct stressors from climate change itself, may cause more complicated responses than have so far been observed. In general, during the next 100 years, it is likely that many ecosystems will not be able to resist or recover from the combination of climate change, associated disturbances, and other global change drivers. Ecological responses to future climate change are expected with high confidence to negatively affect most ecosystem services. Major changes in ecosystem structure, composition, and function, as well as interspecific interactions, are very likely to occur where temperature increases exceed 1.5–2.5°C (IPCC, 2007a).

2.5 TREATMENT OF UNCERTAINTY: CONFIDENCE

In SAPs such as this report, evaluations of uncertainty are communicated for judgments, findings, and conclusions made in the text. Treatment of uncertainty involves characterization and communication of two distinct concepts: uncertainty in terms of *likelihood* or in terms of *confidence* in the science (IPCC, 2007b). Likelihood is relevant when assessing the chance of a specific future occurrence or outcome, and is often quantified as a probability. However, in this report, judgments and conclusions about adaptation will be associated with qualitative expressions of confidence rather than quantitative statements of likelihood.

Confidence is composed of two separate but related elements (IPCC, 2007b). The first element is the amount of evidence available to support the determination that the effectiveness of a given adaptation approach is well-studied and understood. The second element is the level of agreement or consensus within the scientific community about the different lines of evidence on the effectiveness of that adaptation approach. Thus, each of the synthetic adaptation approaches drawn from across the chapters of this report is assessed and given a ranking of "high" or "low" for each element (amount of evidence and amount of agreement). These assessments of confidence are presented and discussed in the Synthesis and Conclusions chapter.

2.6 THE ADAPTATION CHALLENGE: THE PURPOSE OF THIS REPORT

Understanding how to incorporate adaptation into strategic planning activities is an important challenge because: (1) the climate system is always changing and will continue to change; (2) those changes will affect attainment of management goals for ecosystems; and (3) there are varying levels of uncertainty associated with both the magnitude of climatic changes and the magnitude and direction of ecosystem responses. This report addresses where, when, and how adaptation strategies may be used to address climate change impacts on managed ecosystems, the barriers and opportunities that may be encountered while trying to implement those strategies, and potential long-term strategic shifts in management approaches that may be made to broaden the scope of adaptation strategies available to resource managers.

Different approaches are discussed to address adaptation in the planning process. These approaches generally fall into broad categories that may be distinguished by (1) timing of the management response: whether the response takes place prior to (proactively) or after (reactively) a climate event has occurred; and (2) intention of the managing agency: whether climate-induced changes are formally acknowledged and addressed in management plans (Box 2.2).

Given that management agencies' resources are likely to fluctuate over time, a key to the planning process will be to determine an approach that maximizes attainment of established short- and long-term goals, especially in light of the effect that climate change may have on those goals. This report provides a discussion of key questions, factors, and potential approaches to consider when setting priorities during the planning process, as well as examples of adaptation strategies that may be employed across different types of ecosystems and geographic regions of the country.

Addressing future changes is an imprecise exercise, fraught with uncertainties and unanticipated changes. Managers have to anticipate the interaction of multiple stressors, the interdependencies of organisms within an ecosystem, and the potential intertwined, cascading effects. Thus the ability to measure effectiveness of management options, *i.e.*, ecological outcomes of specific actions on the ground, is essential in order to continuously refine and improve adaptation. This report raises issues to consider when measuring management effectiveness for increasing the resilience of ecosystems to climate variability and change.

Another requirement for management effectiveness is successful implementation. Challenges to implementation may be associated with different organizational scales, operational tradeoffs, cost/benefit considerations, social/cultural factors, and planning requirements. The information in this report provides an improved understanding of barriers and opportunities associated with these challenges, including priority information gaps and technical needs.

Finally, some challenges to implementation of adaptation options and their ultimate success may require fundamental shifts in management approaches. This report will seek to identify and discuss possible short- and long-term shifts in management structures, approaches, and policies that increase the likelihood of effectiveness and success in implementation, and that may open the door to a greater array of adaptation options in the future.

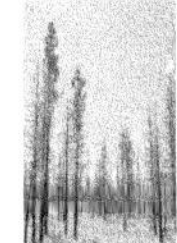

BOX 2.2. Approaches to Adaptation Planning.

1. No adaptation: future climate change impacts are not planned for by the managing agency and are not acknowledged as likely to occur.

2. Reactive adaptation: climate change impacts are not planned for by the managing agency, and adaptation takes place after the impacts of climate change have been observed.

3. Anticipatory adaptation:
   - Responsive: future climate change impacts are acknowledged as likely to occur by the managing agency, and responses to those changes are planned for when changes are observed.
   - Proactive: climate change impacts are acknowledged as likely to occur by the managing agency, and adaptation responses are planned for before the changes are observed.

National Forests

Lead Authors: Linda A. Joyce, U.S.D.A. Forest Service

Contributing Authors: Geoffrey M.Blate, AAAS Fellow at U.S. Environmental Protection Agency; Jeremy S. Littell, JISAO CSES Climate Impacts Group, University of Washington; Steven G. McNulty, U.S.D.A. Forest Service; Constance I. Millar, U.S.D.A. Forest Service; Susanne C. Moser, National Center for Atmospheric Research; Ronald P. Neilson, U.S.D.A. Forest Service; Kathy A. O'Halloran, U.S.D.A. Forest Service; David L. Peterson, U.S.D.A. Forest Service

3.1 SUMMARY

The National Forest System (NFS) is composed of 155 national forests (NFs) and 20 national grasslands (NGs), which encompass a wide range of ecosystems, harbor much of the nation's biodiversity, and provide myriad goods and services. The mission of the U.S. Forest Service (USFS), which manages the NFS, has broadened from water and timber to sustaining ecosystem health, diversity, and productivity to meet the needs of present and future generations. The evolution of this mission reflects changing societal values (e.g., increasing emphasis on recreation, aesthetics, and biodiversity conservation), a century of new laws, increasing involvement of the public and other agencies in NF management, and improved ecological understanding. Climate change will amplify the already difficult task of managing the NFS for multiple goals. This chapter offers potential adaptation approaches and management options that the USFS might adopt to help achieve its NF goals and objectives in the face of climate change.

KEY FINDINGS

Climate change will affect the NFS's ability to achieve its goals and objectives. Climate change will make the achievement of all seven strategic goals more challenging because they are all likely to be sensitive to the direct effects of climate change as well as the interactions of climate change with other major stressors.

Climate change will exacerbate the impact of other major stressors on NF and NG ecosystems. Wildfires, non-native and native invasive species, extreme weather events, and air pollution are the most critical stressors that climate change will amplify within NFS ecosystems. Reduced snowpack, earlier snowmelt, and altered hydrology associated with warmer temperatures and altered precipitation patterns are expected to complicate western water management and affect other ecosystem services that NFs provide (e.g., winter recreational opportunities). Drought will likely be a major management challenge across the United States. Ozone exposure and deposition of mercury, sulfur, and nitrogen already affect watershed condition, and their impacts will likely be exacerbated by climate change.

Both adaptation and mitigation strategies are needed to minimize potential negative impacts and to take advantage of possible positive impacts from climate change. Because mitigation options may have deleterious ecological consequences on local to regional scales and adaptation options may have associated carbon effects, it will be important to assess potential tradeoffs between the two approaches and to seek strategies that achieve synergistic benefits.

Developing an adaptation strategy will involve planning for and developing a suite of management practices to achieve multiple goals, along with evaluating different types of uncertainty (e.g., environmental conditions, models, data, resources, planning horizons, and public support), to support decisions about the most suitable adaptations to implement. Three different adaptation approaches are offered: no active adaptation, planned responses after a major disturbance event, and proactive steps taken in advance of a changing climate. The appropriateness of each strategy will likely vary across spatial and temporal scales of decision making; thus, selection of an approach will be influenced by specific management objectives and the adaptive capacity of the ecological, social, and economic environment. Although none of these approaches may successfully maintain extant ecosystems under a changing climate, the proactive approach is best suited to support natural adaptive processes (e.g., species migration) and maintain key ecosystem services. To succeed, proactive adaptation would require greater involvement and integration of managers at many levels to appropriately monitor ecosystem changes, adjust policies, and modify specific practices.

Reducing the impact of current stressors is a "no regrets" adaptation strategy that could be taken now to help enhance ecosystem resilience to climate change, at least in the near term. Increased effort and coordination across agencies and with private landowners to reduce these stressors (especially air pollution, drought, altered fire regimes, fragmentation, and invasive species) would benefit ecosystems now, begin to incorporate climate change incrementally into management and planning, and potentially reduce future interactions of these stressors with climate change. Approaches that quickly address problems that otherwise would become large and intractable (e.g., the Early Detection/Rapid Response program for invasive species) may also help managers reduce the impacts of climate-driven events such as floods, windstorms, and insect outbreaks. Consideration of post-disturbance management for short-term restoration and for long-term restoration under climate change prior to the disturbance (fire, invasives, flooding, hurricanes, ice storms) may identify opportunities and barriers. Large system-resetting disturbances offer the opportunity to influence the future structure and function of ecosystems through carefully designed management experiments in adapting to climatic change.

Incorporating climate change into the USFS planning process is an important step that could be taken now to help identify suitable management adaptations as well as ecological, social, and institutional opportunities and barriers to their implementation. Planning processes that include an evaluation of vulnerabilities (ecological, social, and economic) to climate change in the context of defining key goals and contexts (management, institutional, and environmental) might better identify suitable adaptive actions to be taken at present or in the short term, and better develop actions for the longer term. Coordination of assessments and planning efforts across the organizational levels in the USFS might better identify spatial and temporal scales for modeling and addressing uncertainty and risk linked to decision-making. Given the diversity of NFS ecosystems, a planning process that allows planners and managers to develop a toolbox of multiple adaptation options would be most suitable.

Better educating USFS employees about climate change and adaptation approaches is another step that could be implemented immediately. Developing adaptation options to climate change may require NF staff to have a more technical understanding of climate change as well as the adaptive capacity of social and economic environments. The challenge for NFs to keep up with the rapidly changing science also suggests the need to build on and strengthen current relationships between researchers (inside and outside of the USFS) and NF staff.

As climate change interacts with other stressors to alter NFS ecosystems, NFs may need to manage for change by increasing emphasis on managing for desired ecological processes by working with changes in structure and composition of NFs. The individual, disparate, and potentially surprising responses of species to climate change may preclude the preservation of current species assemblages over the long term. Under such a scenario, managing for change, despite uncertainty about its direction or magnitude, may be the most viable long-term option. Working toward the goal of desired future functions (e.g., processes, ecosystem services) would involve managing current and future conditions (e.g., structure, outputs), which may be dynamic through a changing climate, to sustain those future functions as climate changes.

Establishing priorities to address potential changes in population, species, and community abundances, structures, and ranges—including potential species extirpation and extinction— under climate change is an important adaptation that will require time and effort to develop. A careful examination of current prioritization methods would begin to identify opportunities and barriers to the analysis of tradeoffs and development of priorities under a changing climate. A tiered approach to priority-setting could include the "no regrets" actions mentioned above (reducing current stressors), "low regrets" actions that provide important benefits at little additional cost and risk, and "win-win" actions that reduce the impacts of climate change while also providing other benefits. Using triage to set priorities would acknowledge where limited resources might be more effective if focused on urgent but treatable problems.

As discussed in the three case studies (Tahoe NF, Olympic NF, and Uwharrie NF; see the Case Study Summaries and Annex A1), the USFS will need to overcome various barriers to take advantage of opportunities to implement adaptation options. The collaboration and cooperation with other agencies, national networks, and the public required to manage NF lands could be an opportunity or a barrier to adaptation. The ability of the USFS to adapt will be enhanced or hindered to the extent that these other groups recognize and address climate change. Adaptive management is also both an opportunity and a barrier. While it facilitates learning about ecosystem responses to management, it may not be useful when the ability to act adaptively is constrained by policies or public opinion, or when actions must be taken quickly.

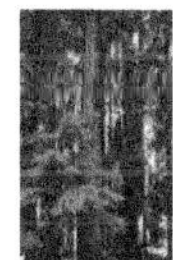

Applied research could help fill gaps in understanding and data while also providing enhanced tools for decision support. Research priorities include studies that assess the socioeconomic impacts of adaptation options, develop ways to reduce ecosystem vulnerability to disturbances that will be exacerbated by climate change (e.g., insects, fire, invasives), and show how climate change can be better incorporated into long-term forest planning (including improved communication). The USFS could also take advantage of current infrastructure and coordinate with other agencies to enhance monitoring and mapping efforts with climate change in mind.

There is a clear need for the USFS as a whole to respond to the potential impacts of climate change. While this report focuses on the NFS, climate change needs to be addressed across all functional lines and program areas (including state and private forestry, international programs, and research) of the USFS. Further enhancing the relationship between NFS managers, state and private forestry staffs, and scientists in the research branch should help the USFS address this challenge.

3.2 BACKGROUND AND HISTORY

3.2.1 Historical Context and Enabling Legislation

In the mid 1800s, the rapid western expansion of European-American settlement and the associated environmental impact of deforestation, human-caused wildfire, and soil erosion raised concerns about the sustainability of public lands (Rueth, Baron, and Joyce, 2002). At a meeting of the American Association for the Advancement of Science in 1873, Franklin Benjamin Hough described the environmental harm resulting from European forest practices and proposed that the United States take action to avoid such impacts. Congress directed the U.S. Department of Agriculture (USDA) to report on forest conditions, and in 1876 Hough—as the USDA special forestry agent—completed the first assessment of U.S. forests. In 1881, the Division of Forestry within USDA was created with the mission to provide information. Three years later, research was added to the mission.

With the passage of the Forest Reserve Act of 1891, President Harrison established the first timber land reserve (Yellowstone Park Timber Land Reserve, eventually to become the Shoshone National Forest) under the control of the General Land Office (Fig. 3.1). Over the next two years, Harrison designated more than 13 million acres (5.26 million ha) within 15 forest reserves in seven western states and Alaska (Rowley, 1985). The Forest Transfer Act of 1905 established the U.S. Forest Service, in USDA, and transferred the reserves from the General Land Office to USDA. With this legislation, the policy shifted from land privatization to federal forest protection, with integrated research and scientific information as an important element in the management for sustained timber yields and watershed protection (Rowley, 1985).[1] In 1907, the forest reserves were renamed to national forests (NFs). By 1909, the NFs had expanded to 172 million acres (70 million hectares) on 150 NFs.[2]

3.2.2 Evolution of National Forest Mission

In the 1891 act, the mission was to "improve and protect the forest within the boundaries, or for the purposes of securing favorable conditions of water flows, and to furnish a continuous supply of timber." In 1905, Secretary of Agriculture James Wilson wrote that questions of use must be decided "from the standpoint of the greatest good for the greatest number in the long run" (USDA Forest Service, 1993). The 1936 Report of the Chief recognized a greater variety of

[1] See also **MacCleery**, D., 2006: Reinventing the U.S. Forest Service: Evolution of the national forests from custodial management, to production forestry, to ecosystem management: A case study for the Asia-Pacific Forestry Commission. In: *Proceedings of the Reinventing Forestry Agencies Workshop*. Asia-Pacific Forestry Commission, FAO Regional Office for Asia and the Pacific, Thailand. 28 February, 2006. Manila, Philippines.

[2] **USDA Forest Service**. 2007. Table 21 National Forest Lands Annual Acreage (1891 to present). Report date October 10, 2007, http://www.fs.fed.us/land/staff/lar/2007/TABLE_21.htm, accessed on 11-28-2007.

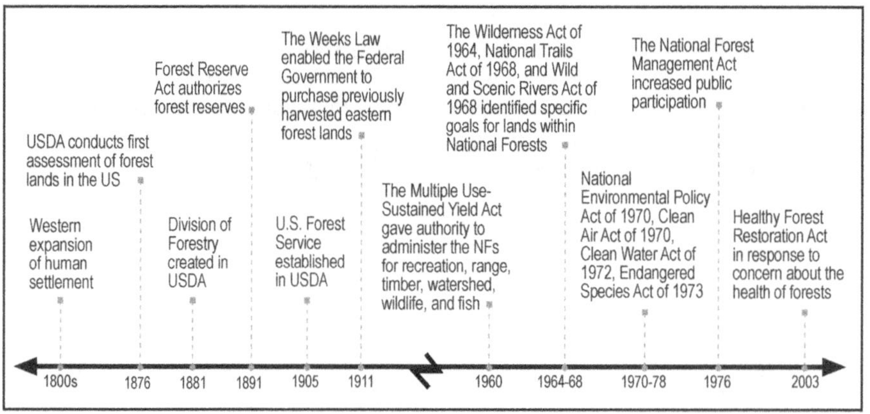

Figure 3.1. Timeline of National Forest System formation and the legislative influences on the mission of the national forests.

purposes for NFs including "timber production, watershed production, forage production, and livestock grazing, wildlife production, recreational use, and whatever combination of these uses will yield the largest net total public benefits."[1] In 1960, the Multiple Use-Sustained Yield Act officially broadened the mission to give the agency "permissive and discretionary authority to administer the national forest for outdoor recreation, range, timber, watershed, and wildlife and fish purposes."[3]

Specific management goals for land within national forest boundaries were identified by legislation in the 1960s: Wilderness Act of 1964, National Trails System Act of 1968, Wild and Scenic Rivers Act of 1968.[4] As these congressional designations encompassed land from many federal agencies, coordination with other federal and in some cases state agencies became a new component of the management of these designated NF lands. By 2006, 23 percent of the National Forest System's lands were statutorily set aside in congressional designations—the national wildernesses, national monuments, national recreation areas, national game refuges and wildlife preserves, wild and scenic rivers, scenic areas, and primitive areas.

Legislation of the 1970s established oversight by agencies other than the Forest Service for the environmental effect of land management within NFs. The Clean Air Act of 1970 and the Clean Water Act of 1972 gave the Environmental Protection Agency responsibility for setting air and water quality standards, and the states responsibility for enforcing these standards. Similarly, the U.S. Fish and Wildlife Service and the National Marine Fisheries Service were given a new responsibility through the required consultation process in the Endangered Species Act of 1973 to review proposed management on federal lands that could modify the habitat of listed species.

Additional legislation established greater public involvement in evaluating management impacts and in the forest planning process. The National Environmental Policy Act (NEPA) of 1970 required all federal agencies proposing actions that could have a significant environmental effect to evaluate the proposed action as well as a range of alternatives, and provide an opportunity for public comment. Increased public participation in the national forest planning process was provided for within the National Forest Management Act of 1976. Land management activities within the NFs were now, more than ever, in the local, regional, and national public limelight.

These laws and their associated regulations led to many changes within the organizational structure of the Forest Service, the composition of the skills within the local, regional, and national staffs, and the management philosophies used to guide natural resource management. Additionally, the public, environmental groups, internal agency sources, and the Forest Service's own research community were reporting that substantial changes were needed in natural resource management.[1] In 1992, Forest Service Chief Dale Robertson announced that "an ecological approach" would now govern the agency's management philosophy. In 1994, Chief Jack Ward Thomas issued the publication Forest Service Ethics and Course to the Future, which described the four components of ecosystem management: protecting ecosystems, restoring deteriorated ecosystems, providing multiple-use benefits for people within the capabilities of ecosystems, and ensuring organizational effectiveness. MacCleery[1] notes that this shift to ecosystem management occurred without explicit statutory authority, and as an administrative response to many factors such as public involvement in the planning processes, increased technical diversity within the Forest Service staffs, increased demand for recreational opportunities, and increased understanding in the natural resource sciences.

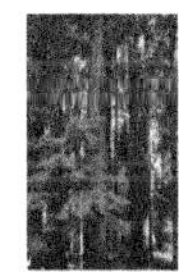

After the active wildfire season in 2000, federal agencies drafted the National Fire Plan to reduce the risk of wildfire to communities and natural resources. The Plan has focused prevention on the reduction of woody biomass (mechanical thinning, prescribed fire, wildland fire use, removal of surface fuels) and the restoration of ecosystems where past land use had altered fire regimes. The Healthy Forest Restoration Act of 2003 included provisions to expedite NEPA and other processes to increase the rate at which fuel treatments were implemented in the wildland-urban interfaces of at-risk communities, at-risk municipal watersheds, areas where fuel treatments could reduce the risk of fire in

[3] 16 U.S.C. § 528-531

[4] 16 U.S.C. § 1271-1287 P.L. 90-542

BOX 3.1. Strategic Plan Goals of the Forest Service, 2007–2012.

1. Restore, Sustain, and Enhance the Nation's Forests and Grasslands.
2. Provide and Sustain Benefits to the American People.
3. Conserve Open Space.
4. Sustain and Enhance Outdoor Recreation Opportunities.
5. Maintain Basic Management Capabilities of the Forest Service.
6. Engage Urban America with Forest Service Programs.
7. Provide Science-Based Applications and Tools for Sustainable Natural Resources Management.

habitat of threatened and endangered species, and where wind-throw or insect epidemics threaten ecosystem components or resource values.[5]

The 2007–2012 USDA Forest Service Strategic Plan describes the mission of the Forest Service, an agency with three branches: National Forest Systems, Research, and State and Private, as: "To sustain the health, diversity and productivity of the Nation's forest and grasslands to meet the needs of present and future generations" (USDA Forest Service, 2007b). The mission reflects public and private interests in the protection and preservation of natural resources, a century of laws passed to inform the management of NF lands, partnerships with states for stewardship of non-federal lands, and a century of research findings.

[5] H.R. 190

3.2.3 Interpretation of Goals

At the national level, the USDA Forest Service Strategic Plan identifies a set of strategic priorities that are implemented over a period of time through annual agency budgets. The strategic priorities or goals are based on national assessments of natural resources and in response to social and political trends (USDA Forest Service, 2007b) (Box 3.1). Within the NFS, these goals are interpreted in each level of the organization: national, regional, and individual administrative unit (forest, grassland, and prairie) (Fig. 3.2).

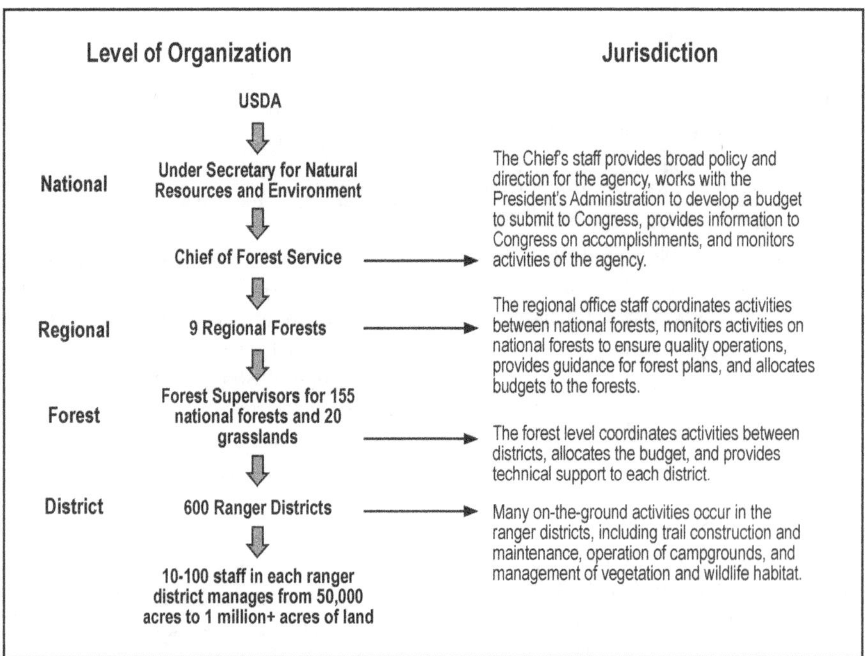

Figure 3.2. Jurisdiction and organizational levels within the National Forest System.

Individual unit planning (national forest, grassland or other units) provides an inventory of resources and their present conditions on a particular management unit. This inventory, coupled with the desired future condition for ecosystem services and natural resources within each national forest, is the basis for annual work planning and budgeting (USDA Forest Service, 2007b). Annual work planning identifies the projects that all units propose for funding within a fiscal year. This level of planning involves the final application of agency strategic direction into a unit's annual budget to move its resources toward its desired future condition. Project planning includes specific on-the-ground management for recreation, fisheries, restoration, vegetation management, and fuel treatments.

Individual administrative units have worked together to develop documents that guide management across several units. For example, the Pacific Northwest Forest Plan was initiated in 1993 to end an impasse over the management of federal lands within the range of the northern spotted owl. The area encompassed 24.5 million acres (~10 million ha); 17 NFs in Washington, Oregon, and California; and public lands in Oregon and Washington managed by the Bureau of Land Management.

3.3 CURRENT STATUS OF MANAGEMENT SYSTEMS

3.3.1 Key Ecosystem Characteristics Upon Which Goals Depend

The NFS (Fig. 3.3) includes a large variety of ecosystems with diverse characteristics. National Forests include ecosystem types ranging from evergreen broadleaf tropical forests within the Caribbean NF in Puerto Rico; alpine tundra on the Medicine Bow NF in Wyoming and the Arapaho NF in Colorado; oakbrush and piñon-juniper woodlands within the Manti-LaSal NF in Utah; northern hardwood forests on the White Mountains NF in New Hampshire; mixed hardwoods on the Wayne-Hoosier NF in Indiana; oak-hickory forests on the Pisgah NF in North Carolina; and ponderosa forests in the Black Hills NF of South Dakota, the Coconino and Sitgreaves NFs of Arizona, and the Lassen NF in California (Adams, Loughry, and Plaugher, 2004). The National Grasslands (NGs) include ecosystem types ranging from shortgrass prairie on the Pawnee NG in Colorado to tallgrass prairie on the Midewin NG in Illinois, and from tallgrass prairie on the Sheyenne NG to the stark badlands found in the Little Missouri NG, both in North Dakota.

Figure 3.3. One hundred fifty-five national forests and 20 national grasslands across the United States provide a multitude of goods and ecosystems services, including biodiversity.[6]

[6] **USDA Forest Service Geodata Clearinghouse**, 2007: FSGeodata Clearinghouse: other Forest Service data sets. USDA Forest Service Geodata Clearinghouse Website, Overlay created in ArcMap 8.1, boundary files are the alp_boundaries2 file set, http://fsgeodata.fs.fed.us/clearinghouse/other_fs/other_fs.html, accessed on

The NFs also includes aquatic systems (lakes, ponds, wetlands, and waterways). Considering its extent and diversity, the NFS is an important cultural and natural heritage and, as such, is valued by a wide variety of stakeholders.

National forests harbor much of the nation's terrestrial biodiversity. Specifically, NFs comprise three major attributes of biodiversity across multiple levels of organization (genes to landscapes) (see Noss, 1990): structural diversity (*e.g.*, genetic, population, and ecosystem structure), compositional diversity (*e.g.*, genes, species, communities, ecosystems, and landscape types), and functional diversity (*e.g.*, genetic, demographic, and ecosystem processes, life histories, and landscape-scale processes and disturbances). Biodiversity conservation has become an important goal of the USFS and is a consideration in planning.[7] National forests provide important habitat for many rare, threatened, and endangered plants and animals, ranging from charismatic species such as the grey wolf (*Canis lupus*) to lesser known species such as Ute ladies' tresses (*Spiranthes diluvialis*). Climate change will amplify the current biodiversity conservation challenge because it is already affecting and will continue to affect the relationships between climate and the various attributes and components (*i.e.*, genes, species, ecosystems, and landscapes) of biodiversity (Hansen *et al.*, 2001; Root *et al.*, 2003; Malcolm *et al.*, 2006; Parmesan, 2006).

National forests also provide myriad goods and services—collectively called ecosystem services (Millennium Ecosystem Assessment, 2005). Historically, timber, grazing, and fresh water have been the most important goods and services provided by NFs. Although timber harvest (Fig. 3.4) and domestic livestock grazing now occur at lower than historical levels (see also Mitchell, 2000; Haynes *et al.*, 2007), NFs harvested more than 2.2 billion board feet in 2006[8] and more than 7000 ranchers relied on NFs and national grasslands for grazing their livestock.[9] About 60 million Americans (20% of the nation's population in 3,400 towns and cities) depend on water that originates in national forest watersheds (USDA Forest

[7] For example see **USDA Forest Service**, 7-11-2007: Rocky Mountain region: species conservation program. USDA Forest Service Website, http://www.fs.fed.us/r2/projects/scp/, accessed on 7-30-2007.

[8] **USDA Forest Service**, 2006: FY1905-2006 annual national sold and harvest summary. Available from http://www.fs.fed.us/forestmanagement/reports/sold-harvest/documents/1905-2006_Natl_Sold_Harvest_Summary.pdf, USDA Forest Service Forest Management, Washington, DC.

[9] **USDA Forest Service**, 2007: *Grazing Statistical Summary 2005*. Washington, DC, pp.iii-108.

Figure 3.4. Historical harvest levels across the national forests.[8]

Service, 2007b). In addition, NFs contain about 3,000 public water supplies for visitors and employees (*e.g.*, campgrounds, visitor centers, and administrative facilities) (USDA Forest Service, 2007b). Thus, the condition of the watershed affects the quality, quantity, and timing of water flowing through it.[10] Climate change will almost certainly affect all three of these historical ecosystems services of NFs (see Section 3.3.4.2) and likely complicate the USFS's already formidable task of restoring, sustaining, and enhancing NFs and NGs while providing and sustaining benefits to the American people.

Over the past few decades, the USFS and the public have come to appreciate the full range of ecosystem services that NFs provide (see Box 3.2). The Millennium Ecosystem Assessment (2005) defines ecosystem services as the benefits people derive from ecosystems, and classifies these benefits into four general categories (Box 3.2): provisioning (*i.e.*, products from ecosystems), regulating (*i.e.*, regulation of ecosystem processes), cultural (*i.e.*, nonmaterial benefits), and supporting services (*i.e.*, services required for production of all other ecosystems services). Biodiversity can be treated as an ecosystem service in its own right, or can be seen as a necessary condition underpinning the long-term provision of other services (Millennium Ecosystem Assessment, 2005; Balvanera *et al.*, 2006; Díaz *et al.*, 2006). This report treats biodiversity as an ecosystem service. The growing importance of regulating services such as pest management, and watershed and erosion management (see Goal 1); provisioning services such as providing wood and energy (see Goal 2); and cultural services such as aesthetics and especially recreation (Goal 4) are reflected in the USFS national goals (see Box 3.1).

The achievement of strategic and tactical goals set forth by the USFS depends on conservation and enhancement of ecosystems services at various scales. Maintenance and enhancement of ecosystems services on NFs is considered within the context of all potential uses and values of individual NFs. Unlike federal lands afforded strict protection, NFs contain multiple resources to be used and managed for the

BOX 3.2. Ecosystem Services Described by the Millennium Ecosystem Assessment (2005).

Provisioning services—fiber, fuel, food, other non-wood products, fresh water, and genetic resources
Regulating services—air quality, climate regulation, water regulation, erosion regulation, water purification and waste treatment, disease regulation, pest regulation, pollination, and natural hazard regulation
Cultural services—cultural diversity, spiritual/religious values, knowledge systems, educational values, inspiration, aesthetic values, social relations, sense of place, cultural heritage values, recreation and ecotourism
Supporting services—primary production, soil formation, pollination, nutrient cycling, water cycling

benefit of current and future generations (see Multiple-Use Sustained-Yield Act of 1960). The USFS, as the steward of NFs and its resources, actively manages NFs to achieve the national goals outlined in Box 3.1 and the individual goals identified for each NF and NG.

3.3.2 Stressors of Concern on National Forests

3.3.2.1 Current Major Stressors

National forests are currently subject to many stressors that affect the ability of the USFS to achieve its goals. We define the term stressor as any physical, chemical, or biological entity that can induce an adverse response (U.S. Environmental Protection Agency, 2000). Stressors can arise from physical and biological alterations of natural disturbances within NFs, increased unmanaged demand for ecosystem services (such as recreation), alterations of the landscape mosaic surrounding NFs, chemical alterations in regional air quality, or from a legacy of past management actions (USDA Forest Service, 2007b).

Disturbances, both human-induced and natural, shape ecosystems by influencing their composition, structure, and function (Dale *et al.*, 2001). Over long time frames, ecosystems adapt and can come to depend on natural disturbances such as fire, hurricanes, windstorms, insects, and disease. For example, sites where fire has naturally occurred include plant species with seed cones that open only in response to heat from wildfire, and thick barked trees that

[10] **Brown**, T.C. and P. Froemke, 2006: *An Initial Ranking of the Condition of Watersheds Containing NFS Land: Approach and Methodology.* USDA Forest Service Rocky Mountain Research Station.

resist surface fire. When disturbances become functions of both natural and human conditions (*e.g.*, forest fire ignition and spread), the nature (*i.e.*, temporal and spatial characteristics) of the disturbance may change—such as when wildfire occurs outside of the recorded fire season. These altered disturbance regimes become stressors to ecosystems, and affect ecosystem services and natural resources within NF ecosystems (*e.g.*, fire, USDA Forest Service, 2007b).

Current Management Activities and the Legacy of Past Management

The legacy of past land-use can leave persistent effects on ecosystem composition, structure, and function (Dupouey *et al.*, 2002; Foster *et al.*, 2003). Depending on their scale and intensity, extractive activities such as timber harvesting, mining, and livestock grazing stress NF ecosystems, affecting their resilience and the services they provide. Current USFS management strategies emphasize mitigation of environmental impacts from these activities (see section 3.3.3). However, the legacy of extractive activities in the past (Rueth, Baron, and Joyce, 2002; Foster *et al.*, 2003) is a continuing source of stress in NFs. For example, past logging practices, in combination with fire suppression, fragmentation, and other factors, have homogenized forest species composition (including a shift from late- to early-successional species); created a unimodal age and size structure; and markedly reduced the number of large trees, snags, and coarse woody debris (Rueth, Baron, and Joyce, 2002; Foster *et al.*, 2003). The long-term ecological impacts of mining operations before the environmental regulations of the 1960s were promulgated have been similarly profound, including mortality of aquatic organisms from lethal concentrations of acid and toxic metals (*e.g.*, copper, lead, and cadmium) and alteration of aquatic and riparian food webs from bio-accumulation of these metals (Rueth, Baron, and Joyce, 2002). The uncontrolled grazing prevailing on federal lands (including areas that are now NFs) until the Taylor Grazing Act was enacted in the 1930s has left a similar environmental imprint. Overstocked rangelands contributed to widespread erosion, reduced soil productivity, and a shift in species composition, including the invasion of non-native species that have altered fire regimes (Rueth, Baron, and Joyce, 2002).

Land Use and Land Cover Change Surrounding National Forests

Changes in the land use and land cover surrounding NFs have been and continue to be associated with the loss of open space (subdivision of ranches or large timber holdings) (Birch, 1996; Sampson and DeCoster, 2000; Hawbaker *et al.*, 2006), the conversion of forestland to urban and built-up uses in the wildland-urban interface (WUI), and habitat fragmentation (related to increases in road densities and impervious surfaces). The amount of U.S. land in urban and built-up uses increased by 34% between 1982 and 1997, the result primarily of the conversion of croplands and forestland (Alig, Kline, and Lichtenstein, 2004). Subdivision of large timber holdings also results in a change in management, as private forest landowners no longer practice forest management (Sampson and DeCoster, 2000).

The WUI is defined as "the area where structures and other human developments meet or intermingle with undeveloped wildland" (Stewart, Radeloff, and Hammer, 2006). Between 1990 and 2000, 60% of all new housing units built in the United States were located in the WUI (Fig. 3.5), and currently 39% of all housing units are located in the WUI (Radeloff *et al.*, 2005). More than 80% of the

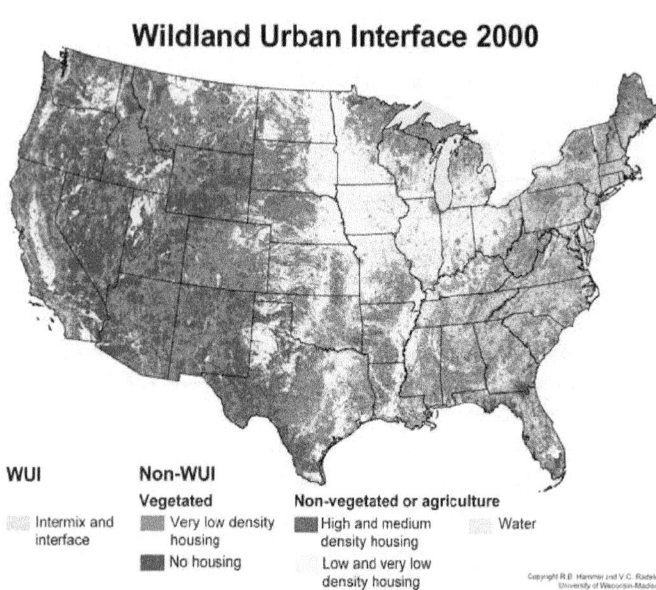

Figure 3.5. Wildland Urban Interface across the United States (Radeloff et al., 2005).

total land area in the United States is within about 1 km of a road (Riitters and Wickham, 2003). "Perforated" (*i.e.*, fragmented) forests with anthropogenic edges affect about 20% of the eastern United States. (Riitters and Coulston, 2005). These changes surrounding NFs can change the effective size of wildlife habitat, change the ecological flows (*e.g.*, fire, water, and plant and animal migrations) into and out of the NFs, increase opportunities for invasive species, increase human impact at the boundaries within the borders of NFs (Hansen and DeFries, 2007), and constrain management options (*e.g.*, fire use). In addition to these land use and land cover changes surrounding the large contiguous NFs, some NFs contain large areas of checkerboard ownership where sections of USFS lands and private ownership intermingle.

Invasive Species

A species is considered invasive if (1) it is non-native to the ecosystem under consideration, and (2) its introduction causes or is likely to cause economic or environmental harm, or harm to human health.[11] Invasive species have markedly altered the structure and composition of forest, woodland, shrubland, and grassland ecosystems. Non-native insects expanding their ranges nationally in 2004 include Asian longhorned beetle, hemlock woolly adelgid, the common European pine shoot beetle, and the emerald ash borer (USDA Forest Service Health Protection, 2005). Non-native diseases continuing to spread include beech bark disease, white pine blister rust, and sudden oak death.

[11] Executive Order 13112: Invasive Species

Within the Northeast, 350,000 acres (141,600 ha) of NFs are annually infested and affected by non-native species, including 165 non-native plant species of concern (USDA Forest Service, 2003). Plant species of greatest concern include purple loosestrife, garlic mustard, Japanese barberry, kudzu, knapweed, buckthorns, olives, leafy spurge, and reed and stilt grass (USDA Forest Service, 2003). Non-native earthworms have invaded and altered soils in previously earthworm-free forests throughout the northeastern United States (Fig. 3.6) (Hendrix and Bohlen, 2002; Hale et al., 2005; Frelich et al., 2006).

Non-native invasive plant species have altered fire regimes in the western United States, including Hawaii (Westbrooks, 1998; Mitchell, 2000), and consequently other important ecosystem processes (D'Antonio and Vitousek, 1992; Brooks et al., 2004). Cheatgrass (*Bromus tectorum*), now a common understory species in millions of hectares of sagebrush-dominated vegetation assemblages in the Intermountain West (Mack, 1981), alters the fuel complex, increases fire frequency, and reduces habitat provided by older stands of sagebrush (Williams and Baruch, 2000; Smith *et al.*, 2000; Ziska, Faulkner, and Lydon, 2004; Ziska, Reeves, and Blank, 2005).12 Similarly, buffelgrass (*Pennisetum ciliare*) and other African grasses

[12] See also **Tausch**, R.J., 1999: Transitions and thresholds: influences and implications for management in pinyon and juniper woodlands. In: *Proceedings: Ecology and Management of Pinyon-Juniper Communities Within the Interior West*, US Department of Agriculture, Forest Service, Rocky Mountain Research Station, pp. 361-365.

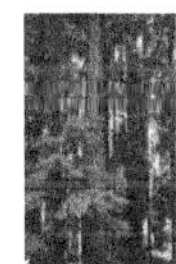

Figure 3.6. Influence of non-native earthworms on eastern forest floor dynamics (Frelich et al., 2006). Forest floor and plant community at base of trees before (a, left-hand photo) and after (b) European earthworm invasion in a sugar-maple-dominated forest on the Chippewa National Forest, Minnesota, USA. Photo credit: Dave Hansen, University of Minnesota Agricultural Experimental Station.

are now common in much of the Sonoran Desert, providing elevated fuel levels that could threaten cactus species with increased fire frequency and severity (Williams and Baruch, 2000). Fountain grass (Pennisetum setaceum), introduced to the island of Hawaii, greatly increases fire susceptibility in the dry forest ecosystems where fire was not historically frequent (D'Antonio, Tunison, and Loh, 2000). Cogongrass (Imperata cylindrica (L.) Beauv.) invasions have similarly altered fire regimes in pine savannas in the southeastern United States (Lippincott, 2000).

Air Pollution

Ozone, sulfur dioxide, nitrogen oxides (NOx), and mercury transported into NFs from urban and industrial areas across the United States affect resources such as vegetation, lakes, and wildlife. A combination of hot, stagnant summer air masses, expansive forest area, and high rates of NOx emissions combine to produce high levels of ozone, especially in the western, southern, and northeastern regions of the United States (Fiore *et al.*, 2002). Current levels of ozone exposure are estimated to reduce eastern and southern forest productivity by 5–10% (Joyce *et al.*, 2001; Felzer *et al.*, 2004). Elevated nitrogen deposition downwind of large, expanding metropolitan centers or large agricultural operations has been shown to affect forests when nitrogen deposited is in excess of biological demand (nitrogen saturation). Across the southern United States it is largely confined to high elevations of the Appalachian Mountains (Johnson and Lindberg, 1992), although recent increases in both hog and chicken production operations have caused localized nitrogen saturation in the Piedmont and Coastal Plain (McNulty *et al.*, forthcoming). In the western United States, increased nitrogen deposition has altered plant communities (particularly alpine communities in the Rocky Mountains) and reduced lichen and soil mychorriza (particularly in the Sierra Nevada mountains of Southern California) (Baron *et al.*, 2000; Fenn *et al.*, 2003). In Southern California, the interaction of ozone and nitrogen deposition has been shown to cause major physiological disruption in ponderosa pine trees (Fenn *et al.*, 2003). Mercury deposition negatively affects aquatic food webs as well as terrestrial wildlife, as a result of bioaccumulation, throughout the United States (Chen *et al.*, 2005; Driscoll *et al.*, 2007; Peterson *et al.*, 2007). In the Ottawa NF (Michigan), for example, 16 lakes and four streams have been contaminated by mercury that was deposited from pollution originating outside of NF borders (Ottawa National Forest, 2006).

Energy Activities

Of the estimated 99.2 million acres (40.1 million ha) of oil and gas resources on federal lands (USDA, USDI, and DOE, 2006), 24 million acres (9.7 million ha) are under USFS management. The Bureau of Land Management has the major role in issuing oil and gas leases and permits in NFs; however, the USFS determines the availability of land and the conditions of use, and regulates all surface-disturbing activities conducted under the lease (GAO, 2004). Principal causes of stress are transportation systems to access oil and gas wells, the oil and gas platforms themselves, pipelines, contamination resulting from spills or the extraction of oil and gas, and flue gas combustion and other activities in gas well and oil well productions. The extent to which these stressors affect forests depends on the history of land use and ownership rights to subsurface materials in the particular NF. For example, oil and gas development is an important concern in the Allegheny NF because 93% of the subsurface mineral rights are privately held, and because exploration and extraction have increased recently due to renewed interest in domestic oil supplies and higher crude oil prices (Allegheny National Forest, 2006).

Altered Fire Regimes

Fire is a major driver of forest dynamics in the West, South and Great Lakes region (Agee, 1998; Frelich, 2002), and fire regimes (return interval and severity) and other characteristics (season, extent, etc.) vary widely across the United States (Hardy *et al.*, 2001a; Schmidt et al., 2002). Fire and insect disturbances interact, often synergistically, compounding rates of change in forest ecosystems (Veblen *et al.*, 1994). Historical fire suppression has led to an increase in wildfire activity and altered fire regimes in some forests, resulting in increased density of trees and increased build-up of fuels (Covington *et al.*, 1994; Sampson *et al.*, 2000; Minnich, 2001; Moritz, 2003; Brown, Hall, and Westerling, 2004). Lack of fire or altered fire frequency and severity are considered sources of stress in those ecosystems dependent upon fire, such as forests dominated by ponderosa pine and lodgepole pine in the West, longleaf pine in the South, and oak and pine ecosystems in the East.

Unmanaged Recreation

National forests are enjoyed by millions of outdoor enthusiasts each year, but recreation—particularly unmanaged recreation—causes a variety of ecosystem impacts.[13] Recreational activities that can damage ecosystems include cutting trees for fire, starting fires in inappropriate places, damaging soil and vegetation through the creation of roads and trails, target practice and lead contamination, and pollution of waterways.[14] Impacts of these activities include vegetation and habitat loss from trampling, soil and surface litter erosion, soil compaction, air and water pollution, decreased water quality, introduction of non-native invasive species, and wildfires. The creation of unauthorized roads and trails by off-highway vehicle (OHVs) causes erosion, degrades water quality, and destroys habitat.[15]

Extreme Weather Events: Wind, Ice, Freeze-thaw events, Floods, and Drought

Severe wind is the principal cause of natural disturbance in many NFs (*e.g.*, Colorado, Veblen, Hadley, and Reid, 1991; Alaska, Nowacki and Kramer, 1998; northern temperate forests, Papaik and Canham, 2006). Wind is one of the three principal drivers (along with fire and herbivory) of forest dynamics in temperate forests of northeastern and north-central North America (for an example of a wind event, see Box 3.3) (Frelich, 2002). Turnover in northeastern forests depends on creation of gaps from individual trees falling down or being blown down by wind (Seymour, White, and deMaynadier, 2002). Winds from severe storms (*e.g.*, from tornadoes, hurricanes, derechos, and nor'easters) occurring at very infrequent intervals also replace stands at various spatial scales (0.2-3,785 ha; Seymour, White, and deMaynadier, 2002; see also McNulty, 2002). Worrall, Lee, and Harrington (2005) found that windthrow, windsnap, and chronic wind stress expand gaps initiated by insects, parasites, and disease in New Hampshire subalpine spruce-fir forests. Thus, wind, insects, and disease interact to cause chronic stress to forests, whereas extreme storms typically are stand-replacing events.

BOX 3.3. The "Boundary Waters-Canadian Derecho," a Straight-Line Wind Event in the Central United States and Canada.

During the pre-dawn hours on Sunday, July 4, 1999, thunderstorms were occurring over portions of the Dakotas. By 6 AM CDT, some of the storms formed into a bow echo and began moving into the Fargo, North Dakota area, with damaging winds. Thus would begin the "Boundary Waters-Canadian Derecho," which would last for more than 22 hours, travel more than 2,080 kilometers at an average speed almost 96 kph, and result in widespread devastation and many casualties in both Canada and the United States.

In the Boundary Waters Canoe Area (BWCA), winds estimated at 128-160 kph moved rapidly, causing serious damage to 1560 square kilometers of forest in the area. Tens of millions of trees were blown down. Sixty people in the BWCA were injured by falling trees, some seriously. Twenty of those injured were rescued by floatplanes flying to lakes within the forest.

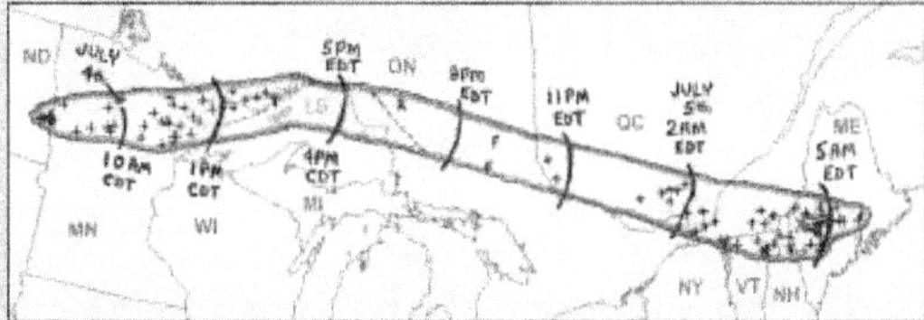

Area affected by the July 4–5, 1999 derecho event (outlined in blue). Curved purple lines represent the approximate locations of the "gust front" at three hourly intervals. "+" symbols indicate the locations of wind damage or estimated wind gusts above severe limits (93 kph or greater)[16].

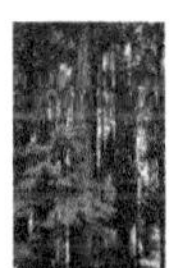

[13] Reviewed in **Leung**, Y.F. and J.L. Marion, 2000: Recreation impacts and management in wilderness: a state-of-knowledge review. In: Wilderness Ecosystems, Threats, and Management [Cole, D.N. (ed.)]. *Proceedings of the Wilderness science in a time of change conference*, 23, May 1999, U.S. Department of Agriculture, Forest Service, Rocky Mountain Research Station.

[14] **National Forest Foundation**, 2006: Recreation. National Forests Foundation Website, http://www.natlforests.org/consi_02_rec.html, accessed on 5-4-2007.

[15] **Foltz**, R.B., 2006: Erosion from all terrain vehicle (ATV) trails on National Forest lands. Proceedings of the 2006 ASABE Annual International Meeting, 9, July 2006, American Society of Agricultural and Biological Engineers, Portland Convention Center, Portland, OR. Available from http://asae.frymulti.com/request.asp?JID=5&AID=21056&CID=por2006&T=2.

[16] **NOAA's National Weather Service, Storm Prediction Center**, 2007: The Boundary Waters-Canadian Derechos. NOAA Website, http://www.spc.noaa.gov/misc/AbtDerechos/casepages/jul4-51999page.htm, accessed on 7-30-2007..

Ice storms are another important part of the natural disturbance regime (Irland, 2000; Lafon, 2006) that stress individual trees (Bruederle and Stearns, 1985), influence forest structure and composition (Rhoads *et al.*, 2002) and, when severe, can affect important ecosystem processes such as nitrogen cycling (Houlton *et al.*, 2003). The extent to which trees suffer from the stress and damage caused by ice appears to vary with species, slope, aspect, and whether severe winds accompany or follow the ice storm (Bruederle and Stearns, 1985; De Steven, Kline, and Matthiae, 1991; Rhoads *et al.*, 2002; Yorks and Adams, 2005). Growth form, canopy position, mechanical properties of the wood, and tree age and health influence the susceptibility of different species to ice damage (Bruederle and Stearns, 1985). Severe ice storms, such as the 1999 storm in New England, can shift the successional trajectory of the forest due to the interactions between the storm itself and effects of more chronic stressors, such as beech bark disease (Rhoads *et al.*, 2002).

Climate variability and extreme weather events also affect ecosystem response. Auclair, Lill, and Revenga (1996) identified the relationships between thaw-freeze and root-freeze events in winter and early spring and severe episodes of dieback in northeastern and Canadian forests. These extreme events helped trigger (and synchronize) large-scale forest dieback, because trees injured by freezing were more vulnerable to the heat and drought stress that eventually killed them. In northern hardwoods, freezing, as opposed to drought, was significantly correlated with increasing global mean annual temperatures and low values of the Pacific tropical Southern Oscillation Index (Auclair, Lill, and Revenga, 1996). Auclair, Eglinton, and Minnemeyer (1997) identified large areas in the Northeast and Canada where this climatic phenomenon affected several hardwood species. Lack of the insulating layer of snow was shown to increase soil freezing events in northern hardwood forests (Hardy *et al.*, 2001b).

Droughts (and even less-severe water stress) weaken otherwise healthy and resistant trees and leave them more susceptible to both native and non-native insect and disease outbreaks. Protracted droughts have already contributed to large-scale dieback of species such as ponderosa pine (see Box 3.4). Vegetation in NFs

BOX 3.4. Insects and Drought in Piñon-Juniper Woodlands in the Southwest United States.

Between 2002 and 2003, the southwestern United States experienced a sub-continental scale dieback of piñon pines (*Pinus edulis*), Ponderosa pines (*P. ponderosa*), and juniper (*Juniperus monosperma*), the dominant tree species in the region (Breshears *et al.*, 2005). Piñon pines were hit hardest, and suffered 40–80% mortality across an area spanning 12,000 km² of Colorado, Utah, Arizona, and New Mexico. Beetles (*Ips confusus* LeConte) were the proximate cause of death of the piñons, but the beetle infestation was triggered by a major "global-change type drought" that depleted soil water content for at least 15 months (Breshears *et al.*, 2005). Although a major drought occurred in the same region in the 1950s, mortality was apparently less extensive—most prominently Ponderosa and piñon pines older than 100 years and on the driest sites died (Allen and Breshears, 1998). In contrast, the more recent drought killed piñons across all size classes and elevations. It also killed 2–26% of the more drought-tolerant junipers, and reduced by about half the live basal cover of *Bouteloua gracilis*, a dominant grass in the piñon-juniper woodlands (Breshears *et al.*, 2005). The more recent drought also was characterized by warmer temperatures, which increased the water stress on the trees. This increased water stress was probably exacerbated by the increased densities of piñons that resulted from land use and anomalously high precipitation in the region from about 1978–1995 (Breshears *et al.*, 2005).

The scale of this dieback will greatly affect carbon stores and dynamics, runoff and erosion, and other ecosystem processes, and may also lead to an ecosystem type conversion (Breshears *et al.*, 2005). The possibility that vegetation diebacks at the scale observed in this example may become more common under climate change presents a major management challenge.

These photos—taken from similar vantages near Los Alamos, NM—show the large-scale dieback of piñon pines in 2002–2003 that resulted from a protracted drought and associated beetle infestation. In 2002, the pines had already turned brown from water stress, and by 2004, they had lost all their needles.

Photo credit: CD Allen, USGS

with sandy or shallow soils is more susceptible to drought stress than vegetation growing in deeper or heavier soils (Hanson and Weltzin, 2000), resulting in situations where soil type and drought interact to substantially increase fire risk. The extent and severity of fire impacts is closely associated with droughts; the most widespread and severe fires occur in the driest years (Taylor and Beaty, 2005; Westerling *et al.*, 2006). The temporal and spatial distribution of droughts also affects watershed condition by affecting surface water chemistry (Inamdar *et al.*, 2006).

Floods caused by extreme precipitation events—especially those that co-occur with or contribute to snowmelt—are another important stressor in NFs. In floodplain forests, periodic floods deposit alluvium, contribute to soil development, and drive successional processes (Bayley, 1995; Yarie *et al.*, 1998). Tree damage and mortality caused by inundation depends on several factors including season, duration, water levels, temperature and oxygen, mechanical damage, and concentration of contaminants. Floods in upland forests, however, are considered large, infrequent disturbances (Turner *et al.*, 1998; Michener and Haeuber, 1998) dominated by mechanical damage that affects geophysical and ecological processes (Swanson *et al.*, 1998). The physical damage to aquatic and riparian habitat from landslides, channel erosion, and snapped and uprooted trees can be extensive and severe, or quite heterogeneous (Swanson *et al.*, 1998). Flooding facilitates biotic invasions, both by creating sites for invasive species to become established and by dispersing these species to the sites (Barden, 1987; Miller, 2003; Decruyenaere and Holt, 2005; Truscott *et al.*, 2006; Watterson and Jones, 2006; Oswalt and Oswalt, 2007).

3.3.2.2 Stress Complexes in Western Ecosystems

A warmer climate is expected to affect ecosystems in the western United States by altering stress complexes (Manion, 1991)—combinations of biotic and abiotic stresses that compromise the vigor of ecosystems—leading to increased extent and severity of disturbances (McKenzie, Peterson, and Littell, forthcoming). Increased water deficit will accelerate the stress complexes experienced in forests, which typically involve some combination of multi-year drought, insects, and fire. Increases in fire disturbance superimposed on ecosystems with increased stress from drought and insects may have significant effects on growth, regeneration, long-term distribution and abundance of forest species, and carbon sequestration (Fig. 3.7).

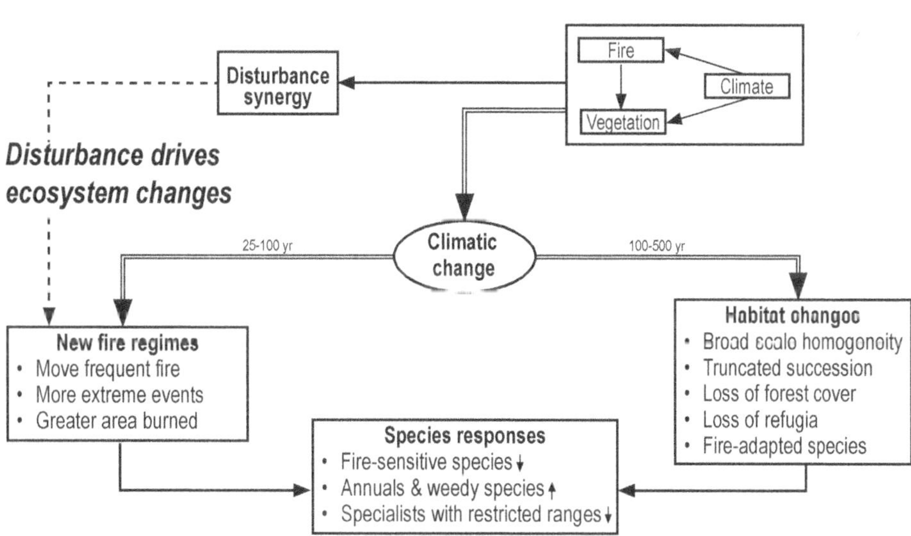

Figure 3.7. Conceptual model of the relative time scales for disturbance vs. climatic change alone to alter ecosystems. Times are approximate. Adapted from (McKenzie et al., 2004).

Forests of western North America can be classified into energy-limited vs. water-limited vegetation (Milne, Gupta, and Restrepo, 2002; Littell and Peterson, 2005). Energy-related limiting factors are chiefly light (*e.g.*, productive forests where competition reduces light to most individuals) and temperature (*e.g.*, high-latitude or high-elevation forests). Energy-limited ecosystems in general appear to be responding positively to warming temperatures over the past 100 years (McKenzie, Hessl, and Peterson, 2001). In contrast, productivity in water-limited systems may decrease with warming temperatures, as negative water balances constrain photosynthesis (Hicke *et al.*, 2002), although this may be partially offset if CO_2 fertilization significantly increases water-use efficiency in plants (Neilson *et al.*, 2005b). Littell (2006) found that most montane Douglas fir (*Pseudotsuga menziesii*) forests across the northwestern United States appear to be water limited; under current climate projections these limits would increase in both area affected and magnitude.

Temperature increases are a predisposing factor causing often lethal stresses on forest ecosystems of western North America, acting both directly through increasingly negative water balances (Stephenson, 1998; Milne, Gupta, and Restrepo, 2002; Littell, 2006) and indirectly through increased frequency, severity, and extent of disturbances—chiefly fire and insect outbreaks (Logan and Powell, 2001; McKenzie *et al.*, 2004; Logan and Powell, 2005; Skinner, Shabbar, and Flanningan, 2006). Four examples of forest ecosystems whose species composition and stability are currently affected by stress complexes precipitated by a warming climate are described below. Two cases involve the loss of a single dominant species, and the other two involve two or more dominant species.

Piñon-Juniper Woodlands of the American Southwest

Piñon pine (*Pinus edulis*) and various juniper species (*Juniperus* spp.) are among the most drought-tolerant trees in western North America, and piñon-juniper ecosystems characterize lower treelines across much of the West. Piñon-juniper woodlands are clearly water-limited systems, and piñon-juniper ecotones are sensitive to feedbacks from environmental fluctuations and existing canopy structure that may buffer trees against drought (Milne *et al.*, 1996) (Box 3.4). However, severe multi-year droughts periodically cause dieback of piñon pines, overwhelming any local buffering. Interdecadal climate variability strongly affects interior dry ecosystems, causing considerable growth during wet periods. This growth increases the evaporative demand, setting up the ecosystem for dieback during the ensuing dry period (Swetnam and Betancourt, 1998). The current dieback is historically unprecedented in its combination of low precipitation and high temperatures (Breshears *et al.*, 2005). Fig. 3.8

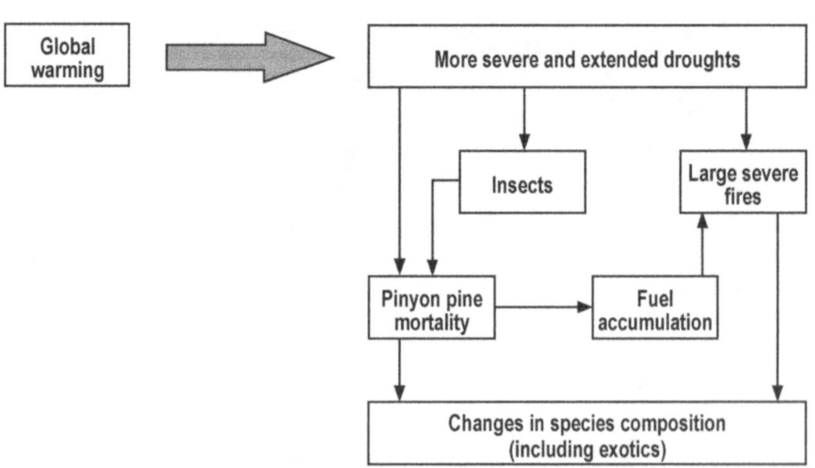

Figure 3.8. Stress complex in piñon-juniper woodlands of the American Southwest. From McKenzie *et al.* (2004).

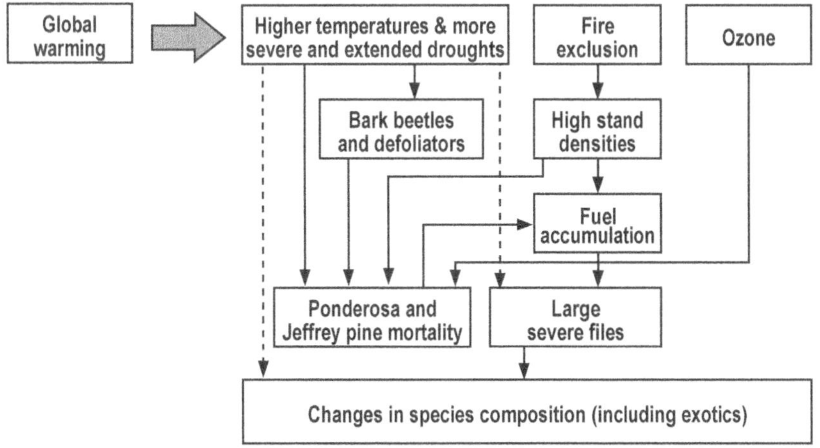

Figure 3.9. Stress complex in Sierra Nevada and southern Californian mixed-conifer forests. From McKenzie, Peterson, and Littell (forthcoming).

shows the stress complex associated with piñon-juniper ecosystems. Increased drought stress via warmer climate is the predisposing factor, and pinon pine mortality and fuel accumulations are inciting factors. Ecosystem change comes from large-scale severe fires that lead to colonization of invasive species (D'Antonio, 2000), which further compromises the ability of piñon pines to re-establish.

Mixed Conifer Forest of the Sierra Nevada and Southern California

These forests experience a Mediterranean climate with long, dry summers. Fire frequency and extent have not increased concomitantly with warmer temperatures, but instead have decreased to their lowest levels in the last 2,000 years. Stine (1996) attributed this decline to decreased fuel loads from sheep grazing, decreased ignition from the demise of Native American cultures, and fire exclusion. Continued fire exclusion has led to increased fuel loadings, and competitive stresses on individual trees as stand densities have increased (Van Mantgem *et al.*, 2004). Elevated levels of ambient ozone from combustion of fossil fuels affect plant vigor in the Sierra Nevada and the mountains of southern California (Peterson, Arbaugh, and Robinson, 1991; Miller, 1992). Sierra Nevada forests support endemic levels of a diverse group of insect defoliators and bark beetles, but bark beetles in particular have reached outbreak levels in recent years, facilitated by protracted droughts and biotic complexes that include bark beetles interacting with root diseases and mistletoes (Ferrell, 1996). Dense stands, fire suppression, and exotic pathogens such as white pine blister rust (*Cronartium ribicola*) can exacerbate biotic interactions (Van Mantgem *et al.*, 2004) and drought stress. Fig. 3.9 shows the stress complex associated with Sierra Nevada forest ecosystems, and is likely applicable to the mountain ranges east and north of the Los Angeles basin.

Interior Lodgepole Pine Forests

Lodgepole pine (*Pinus contorta var. latifolia*) is widely distributed across western North America, often forming nearly monospecific stands in some locations. It is the principal host of the mountain pine beetle (*Dendroctonus ponderosae*), and monospecific stands are particularly vulnerable to high mortality during beetle outbreaks. Recent beetle outbreaks have caused extensive mortality across millions of hectares (Logan and Powell, 2001; Logan and Powell, 2005), with large areas of mature cohorts of trees (age 70–80 yr) contributing to widespread vulnerability.[17] Warmer temperatures facilitate bark beetle outbreaks in two ways: (1) drought stress makes trees more

[17] Carroll, A., 2006: Changing the climate, changing the rules: global warming and insect disturbance in western North American forests. Proceedings of the 2006 MTNCLIM conference, Mt. Hood, Oregon. Accessed at http://www.fs.fed.us/psw/cirmount/meetings/mtnclim/2006/talks/pdf/carroll_talk_mtnclim2006.pdf.

vulnerable to attack, and (2) insect populations respond to increased temperatures by speeding up their reproductive cycles (e.g., to one-year life cycles). Warming temperatures would be expected to exacerbate these outbreaks and facilitate their spread northward and eastward across the continental divide (Logan and Powell, 2005; but see Moore *et al.*, 2006). Fig. 3.10 shows the stress complex for interior lodgepole pine forests. Warmer temperatures, in combination with beetle mortality, set up some ecosystems for shifts in species dominance that will be mediated by disturbances such as fire.

Alaskan Spruce Forests

The state of Alaska has experienced historically unprecedented fires in the last decade, including the five largest fires in the United States. More than 2.5 million hectares burned in the interior during 2004. During the 1990s, massive outbreaks of the spruce bark beetle (*Dendroctonus rufipennis*) occurred on and near the Kenai Peninsula (including the Chugach NF) in southern Alaska (Berg *et al.*, 2006). Although periodic outbreaks have occurred throughout the historical record, these most recent ones may be unprecedented in extent and percentage mortality (over 90% in many places; Ross *et al.*, 2001; Berg *et al.*, 2006). Both these phenomena are associated with warmer temperatures in recent decades (Duffy *et al.*, 2005; Berg *et al.*, 2006; Werner *et al.*, 2006).

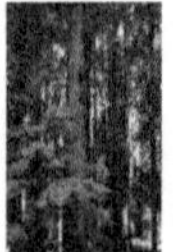

Although fire-season length in interior Alaska is associated with the timing of onset of late-summer rains, the principal driver of annual area burned is early summer temperature (Duffy *et al.*, 2005). In the interior of Alaska, white spruce (*Picea glauca*) and black spruce (*P. mariana*) are more flammable than their sympatric deciduous species (chiefly paper birch, *Betula papyrifera*). Similarly, conifers are the target of bark beetles, so in southern Alaska they will be disadvantaged compared with deciduous species. Fig. 3.11 shows the stress complex for Alaska forest ecosystems, suggesting a significant transition to deciduous life forms via more frequent and extensive disturbance associated with climate variability and change. This transition would be unlikely without changes in disturbance regimes, even under climate change, because both empirical and modeling studies suggest that warmer temperatures alone will not favor a life-form transition (Johnstone *et al.*, 2004; Bachelet *et al.*, 2005; Boucher and Mead, 2006).

3.3.3 Management Approaches and Methods Currently in Use to Manage Stressors

Management approaches addressing the current stressors are based on guidance from USFS manuals and handbooks, developed through planning processes that may occur after the

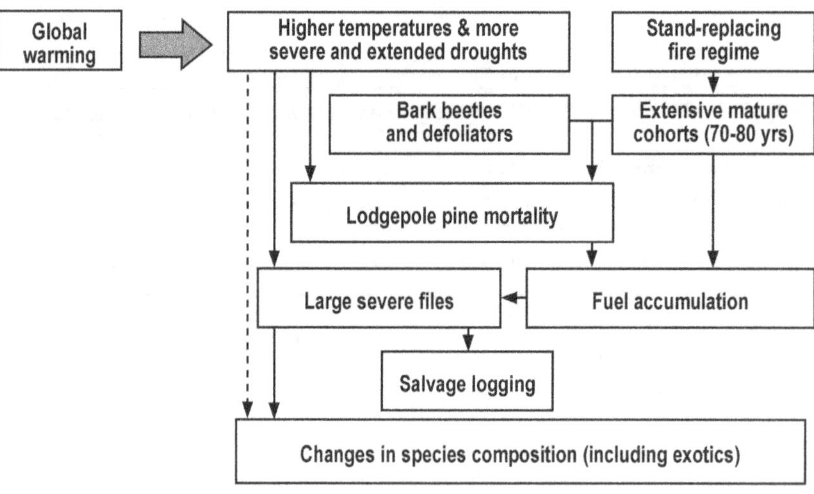

Figure 3.10. Stress complex in interior (British Columbia and United States) lodgepole pine forests. From McKenzie, Peterson, and Littell (forthcoming).

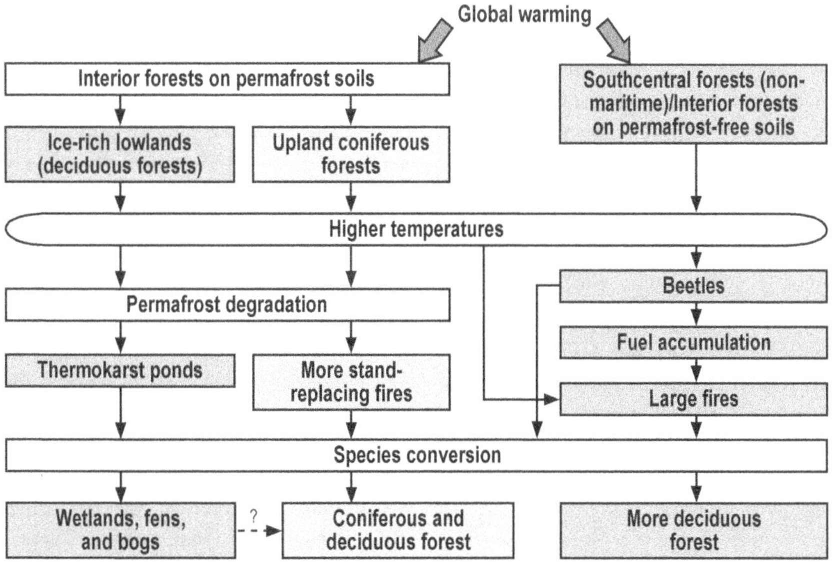

Figure 3.11. Stress complex in the interior and coastal forests of Alaska. From McKenzie, Peterson, and Littell (forthcoming).

disturbance (such as ice storms or wind events), and developed through regional scientific assessment and national planning efforts. For example, approaches for invasive species management are outlined in the National Strategy and Implementation Plan for Invasive Species Management; approaches for altered fire regimes are outlined in the National Fire Plan. Unmanaged recreation, particularly the use off-highway-vehicles, is being addressed through the new travel management plan. Management of native insects and pathogens that become problematic is the responsibility of the Forest Health Protection Program, working in cooperation with NFs. When extreme climate- or weather-related events occur, such as large wind blowdown events (see Box 3.3), management plans are developed in response to the stressor (such as after the blowdown event on the Superior National Forest).[18] Current USFS management strategies emphasize mitigation of environmental impacts from activities such as timber harvest and grazing through environmental analyses and the selection of the best management practices. Silvicultural practices are used to manipulate and modify forest stands for wildlife habitat, recreation, watershed management, and for fuels reductions, as well as for commercial tree harvests. Management approaches across the NFS are influenced by the local climate, physical environment (soils), plant species, ecosystem dynamics, and the landscape context (*e.g.*, WUI, proximity to large metropolitan areas for recreational use).

Adaptive management can be defined as a systematic and iterative approach for improving resource management by emphasizing learning from management outcomes (Bormann, Haynes, and Martin, 2007). An adaptive management approach was implemented through the Northwest Forest Plan to federal lands in the Pacific Northwest (Bormann, Haynes, and Martin, 2007). The Plan directed managers to experiment, monitor, and interpret as activities were applied both inside and outside adaptive management areas—and to do this as a basis for changing the Plan in the future. In that application, managers identified adaptive management areas; developed organizational strategies to apply the adaptive management process across the entire plan area (10 million acres); established a major regional monitoring program; and undertook a formal interpretive step that gathered what was learned and

[18] **USDA Forest Service**, 5-12-2006: Superior National Forests: lowdown on the blowdown. USDA Forest Service, http://www.fs.fed.us/r9/forests/superior/storm_recovery/, accessed on 5-7-2007.

translated new understanding for the use of decision makers (Haynes *et al.*, 2006). The Sierra Nevada Forest Plan Amendment (see Case Study Summary 3.1) contained a Sierra-wide adaptive management and monitoring strategy. This strategy is being implemented as a pilot project on two NFs in California. This seven-year pilot project, undertaken via a Memorandum of Understanding between the USFS, the U.S. Fish and Wildlife Service, and the University of California, applies scientifically rigorous design, treatment, and analysis approaches to fire and forest health, watershed health, and wildlife. Several watersheds of Tahoe NF are involved in each of the three issue areas of the adaptive management project.

Lessening the damages caused by native insects and pathogens is the goal of the USFS Forest Health Protection (FHP) program. This program includes efforts to control the native species of southern pine beetle and western bark beetles. FHP funds southern pine beetle suppression, prevention, and restoration projects on state lands, private lands, and NFs in the South. FHP's forest health monitoring program determines the status, changes, and trends in indicators of forest condition annually. The program uses data from ground plots and surveys, aerial surveys, and other biotic and abiotic data sources, and develops analytical approaches to address forest health issues.

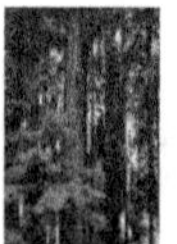

Reducing, minimizing, or eliminating the potential for introduction, establishment, spread and impact of invasive species across all landscapes and ownerships is the goal of the USFS National Strategy and Implementation Plan for Invasive Species Management (USDA Forest Service, 2004). The Plan encompasses four program elements: (1) prevention, (2) early detection and rapid response (EDRR), (3) control and management, and (4) rehabilitation and restoration. Activities in the Prevention element include regularly sanitizing maintenance equipment; requiring weed-free certified seed for restoration, and use of certified weed-free hay; training to identify invasive species; cooperating with other institutions and organizations to prevent the introduction of new forest pests from other countries; and providing technical assistance and funding for public education and prevention measures for invasive species on all lands, regardless of ownership. Activities in the EDRR program include the annual cooperative survey of federal, tribal, and private forestland for damage caused by forest insects and pathogens, and the establishment of the EDRR system for invasive insects in 10 ports and surrounding urban forests. Control and Management activities include treating invasive plants each year on federal, state, and private forested lands, and collaborating with biological control specialists to produce a guide to biological control of invasive plants in the eastern United States. Rehabilitation and Restoration activities highlight the importance of partnerships in such work as developing resistant planting stock for five-needle pine restoration efforts following white pine blister rust mortality, and coordinating at the national and regional levels to address the need for and supply of native plant materials (for example, seeds and seedlings) for restoration.

Reducing hazardous fuels and enhancing the restoration and post-fire recovery of fire-adapted ecosystems are two goals in the National Fire Plan. The two other goals focus on improving fire prevention and suppression, and promoting community assistance. The updated implementation plan (Western Governors' Association, 2006) emphasizes a landscape-level vision for restoration of fire-adapted ecosystems, the importance of fire as a management tool, and the need to continue to improve collaboration among governments and stakeholders at the local, state, regional, and national levels. Land managers reduce hazardous fuels through the use of prescribed fire, mechanical thinning, herbicides, grazing, or combinations of these and other methods. Treatments are increasingly being focused on the expanding WUI areas. Where fire is a major component of the ecosystem, wildland fire use—the management of naturally ignited fires—is used to achieve resource benefits. The appropriate removal and use of woody biomass, as described in the USFS Woody Biomass Strategy, has the potential to contribute to a number of the USFS's strategic goals while providing a market-based means to reduce costs.

In response to the expanded use of off-highway vehicles, the Forest Service's new travel management rule provides the framework for each national forest and grassland to designate a sustainable system of roads, trails, and areas

CASE STUDY SUMMARY 3.1

Tahoe National Forest, California
Pacific Southwest United States

Why this case study was chosen

The Tahoe National Forest:

- Is representative of the 18 national forests on the west slope of the Sierra Nevada range, which have great ecological value and a complex institutional context;
- Shares common geology, forest ecosystems, wildlife habitat, climate, snowpack characteristics, hydrological properties, elevation gradients, diversity of stakeholders, institutional contexts, recreational issues, and resource issues and conflicts with 18 other national forests on the west slope of the Sierra Nevada range;
- Can serve as a model for examining climate change impacts and adaptations for application across the entire Sierra Nevada.

Management context

The principal mission of the Tahoe National Forest (TNF) is to "serve as the public's steward of the land, and to manage the forest's resources for the benefit of all American people ...[and]...to provide for the needs of both current and future generations." The 1990 Tahoe National Forest Land and Resource Management Plan (TNF LRMP) details specific goals, objectives, desired future conditions, standards, and guidelines for a variety of resources including recreation, wilderness, wildlife, timber, water, air quality, minerals, and research.

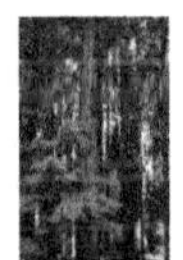

The Sierra Nevada Forest Plan Amendment (FPA; USFS, 2004) and the Herger-Feinstein Quincy Library Group Forest Recovery Act (US Congress, 1998) provide additional specific direction for the TNF. The FPA is a multi-forest plan that specifies goals and direction for (1) reducing buildup of woody fuels and minimizing fire risk, and (2) protecting old forests, wildlife habitats, watersheds, and communities on the national forests of the Sierra Nevada and Modoc Plateau. Forest practices, riparian management, and treatments to reduce the likelihood of severe fires specified in the FPA replace sections of the TNF LRMP. Adaptive management is a key component of the FPA, and the TNF plays a central role in the Sierra Nevada Adaptive Management Program.

The Herger-Feinstein Quincy Library Group Forest Recovery Act of 1998 also supersedes the TNF LRMP for specific resource and geographic areas in the Sierra Nevada, including the Sierraville Ranger District of the TNF. The Act was derived from an agreement by a broad coalition of local stakeholders to promote ecologic and economic health for selected federal lands and communities in the northern Sierra Nevada. The Act launched a pilot project to test a new adaptive management strategy for managing sensitive species as well as fire and woody fuels. In addition to implementing a riparian restoration program, the emphasis of the pilot project is to test, assess, and demonstrate the effectiveness of fuel-breaks, group selection, individual tree selection, and avoidance or protection of specified areas for managing sensitive species and wildfire.

Key climate change impacts

Projected increase of 2.3–5.8°C in annual temperatures by 2100;

- Projected decline in annual snowpack (97% at 1,000 m elevation and 89% for all elevations) by 2100;
- Observed increase in interannual and annual variability of precipitation;
- Observed increase in intensity of periodic multi-year droughts over the past century;
- Observed increase in large fire events in recent years;
- Projected increase in length of fire seasons and risk of uncharacteristically severe and widespread fire events;
- Expected increase in water temperatures in rivers and lakes and decrease in snow, water, and stream runoff in the warm season;
- Observed increase in severity of higher-elevation insect and disease outbreaks.

CASE STUDY SUMMARY 3.1 (CONTINUED)

Opportunities for adaptation

- Science-based rapid assessments of existing plans and policies would be a valuable first step toward understanding current levels of climate change preparedness and areas for potential improvements in operations.
- A revision of the comprehensive assessment of the Sierra Nevada Forest Plan Amendment could be pursued as an opportunity to integrate climate change considerations into management planning.
- The TNF could be a valuable addition to the U.S. Forest Service Ecosystem Services program as a pilot study.
- Increasing the sizes of management units for the forest would allow management of whole landscapes (watersheds, forest types) in a single resource plan, and may decrease administrative fragmentation.
- Actions to improve infiltration of water to groundwater reservoirs (such as decreasing road densities and modifying grazing practices to change surfaces from impervious to permeable) could be used to reduce losses from runoff and increase the quantity of stored groundwater for dry periods.
- Erosion and sediment loss following disturbances could be addressed by promptly reforesting affected areas and salvage-harvesting affected trees (where this activity will not cause further damage), so that a new forest canopy can be established before shrubs "capture" the site;
- A focus on reversing post-disturbance mortality and shrub invasion would increase the chances of successful forest regeneration, leading to restoration of key wildlife habitat and critical watershed protection functions.
- Fuel treatments could be implemented far beyond the season in which they have historically been employed, by further supporting and extending the seasonal tour of fire and fuels staff.
- TNF managers and staff have the expertise and are already prepared to seize adaptive opportunities that would be enabled by a regional biomass and biofuels industry, should a carbon market or regulatory environment develop to support these opportunities.
- Regular planning cycles afford a chance to build flexibility and responsiveness to climate change into management policies.
- "Climate-smart" capacity could be increased, when possible, through staff additions or staff training.
- Education and outreach activities can be used to increase awareness among policy makers, managers, the local public, and other stakeholders about the scientific bases for climate change, the implications for the northern Sierra Nevada and the TNF, and the need for active resource management

Conclusions

In many cases, best management practices (e.g., post-disturbance treatments) may be effective climate change adaptation strategies even though they may be intended to achieve other goals (e.g., maintain ecosystem health). This creates an opportunity for "win-win" strategies to be implemented, whereby benefits would accrue even if the climate did not change.

Barriers to adaptation include public opposition, insufficient funding, limited staff capacity, current large scope of on-the-ground needs, disjointed ownership patterns, and existing environmental legislation. Some barriers result from the interaction of individual barriers, such as when limited staff capacity and insufficient funding result in a continuous reactive approach to priority-setting, rather than a long-term planning process. Changing community demographics influence what landowners adjacent to the TNF accept in terms of ecosystem management, such as smoke from prescribed fires.

Opportunities exist for overcoming barriers to adaptation. Current or potential future opportunities include the possibility of year-round management for reducing woody fuels, active dialog with the public on adaptive management projects, the use of demonstration projects to respond to public concerns, and the potential of emerging carbon markets to promote the development of regional biomass and biofuels industries. Examples of promising areas for development include new management strategies that are operationally appropriate and practical to address climate change, scientifically supported practices for integrated management where resource management goals are integrated rather than partitioned into individual plans, prioritization tools for managing a range of species and diverse ecosystems, and dynamic landscape and project planning that incorporates probabilistic measures of habitat quality and availability in a temporal and spatial context.

open to motor vehicle use.[19] The rule aims to secure a wide range of recreational opportunities while ensuring the best possible care of the land. Designation includes class of vehicle and, if appropriate, time of year for motor vehicle use. Designation decisions are made locally, with public input and in coordination with state, local, and tribal governments.

The Federal Land Manager (broadly, the federal agency charged with protecting wilderness air quality; *e.g.*, the USFS or the National Park Service) has a responsibility to protect the Air Quality Related Values (AQRV) of Class I wilderness areas identified in and mandated by the Clean Air Act. Air resources managers develop monitoring plans for AQRV, such as pH and acid neutralizing capacity in high-elevation lakes. The Federal Land Manager must advise the air quality permitting agency if a new source of pollution, such as from an energy or industrial development, will cause an adverse impact to any AQRV.

3.3.4 Sensitivity of Management Goals to Climate Change

All USFS national goals (Box 3.1) are sensitive to climate change. In general, the direction and magnitude of the effect of climate change on each management goal depends on the temporal and spatial nature of the climate change features, their impact on the ecosystem, and the current status and degree of human alteration of the ecosystem (*i.e.*, whether the ecosystem has lost key components such as late-seral forests; free-flowing streams; or keystone species such as beaver, large predators, and native pollinators). The sensitivity of the management goals to climate change also will depend on how climate change interacts with the major stressors in each ecoregion and national forest. And finally, the sensitivity of the management goals to climate change will depend on the assumptions about climate that the management activities currently make. These assumptions range from the relationship between natural regeneration and climate to seasonal distributions of rainfall and stream flow and management tied to these distributions.

[19] 36 CFR Parts 212, 251, 261, and 295 Travel Management; Designated Routes and Areas for Motor Vehicle Use; Final Rule, November 9, 2005.

3.3.4.1 Goal 1: Restore, Sustain, and Enhance the Nation's Forests and Grasslands

The identified outcome for this goal is forests and grasslands with the capacity to maintain their health, productivity, diversity, and resistance to unnaturally severe disturbances (USDA Forest Service, 2007b). Ecosystem productivity and diversity are strongly influenced by climate. Changes in climatic variables, as well as the effects of interactions of climate change with other stressors (Noss, 2001; Thomas *et al.*, 2004; Millennium Ecosystem Assessment, 2005; Malcolm *et al.*, 2006), may affect all attributes and components of biodiversity (sensu Noss, 1990). Numerous effects of climate change on biodiversity components (*e.g.*, ecosystems, populations, and genes) and attributes (*i.e.*, structure, composition, and function of these components) have already been documented (reviewed in Parmesan, 2006). Natural disturbances such as fire regimes are tightly linked to key climate variables (*i.e.*, temperature, precipitation, and wind) (Agee, 1996; Pyne, Andrews, and Laven, 1996; McKenzie *et al.*, 2004). As a result, changes in weather and climate are quickly reflected in altered fire frequency and severity (Flannigan, Stocks, and Wotton, 2000; Dale *et al.*, 2001). Invasive species are currently contributing to a homogenization of the earth's biota (McKinney and Lockwood, 1999; Mooney and Hobbs, 2000; Rahel, 2000; Olden, 2006), increasing extinction risks for native species (Wilcove and Chen, 1998; Mooney and Cleland, 2001; Novacek and Cleland, 2001; Sax and Gaines, 2003), and harming the economy and human health (Pimentel *et al.*, 2000). Species that can shift ranges quickly and tolerate a wide range of environments, traits common to many invasive species, will benefit under a rapidly changing climate (Dukes and Mooney, 1999). Thus, this goal is sensitive to climate change.

Specific objectives related to this goal include reducing the risk to communities and natural resources from uncharacteristically severe wildfires; reducing adverse impacts from invasive non-native and native species, pests, and diseases; and restoring and maintaining healthy watersheds and diverse habitats.

Climate change and wildfire management

A continual reassessment of climate and land management assumptions may be necessary for effective wildfire management under future climate change. Future climate scenarios suggest a continued increase in fire danger across the United States (Flannigan, Stocks, and Wotton, 2000; Bachelet *et al.*, 2001; Brown, Hall, and Westerling, 2004; McKenzie *et al.*, 2004; Running, 2006) through increasing fire season length, potential size of fires, and areas vulnerable to fire, as well as by altering vegetation, which in turn will influence fuel loadings and consequently fire behavior. Future climate change may offer opportunities to conduct prescribed fire outside of traditional burn seasons, with increased accessibility in some areas in the winter (see Case Study Summary 3.1).

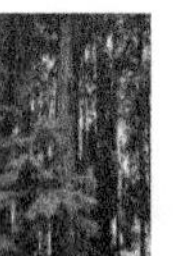

Since the mid-1980s, western forests have sustained more large wildfires, of longer duration, within a context of longer fire seasons, with 60% of the increase occurring at mid-elevations of the Northern Rocky Mountains (Westerling *et al.*, 2006). Land use influences do not appear to have altered fire regimes in high-elevation forests with long fire return intervals (Schoennagel, Veblen, and Romme, 2004). However, suppression of fires has led to the conversion of some lodgepole pine forests to fir and spruce. Some of these stand structures have changed significantly, which may increase their susceptibility to insect infestations (Keane *et al.*, 2002). Wildfire risk has increased in some ponderosa pine and mixed conifer forests (Schoennagel, Veblen, and Romme, 2004; Westerling *et al.*, 2006), where the exclusion of more frequent fires has led to denser stands and higher fuel loading. Future climate projections for western North America project June to August temperature increases of 2–5°C by 2040 to 2069, and precipitation decreases of up to 15% over that time period (Running, 2006). The potential for increased fire activity in high-elevation forests could be exacerbated by the increased fuel loads expected to result from enhanced winter survival of mountain pine beetles and similar pest species (Guarin and Taylor, 2005; Millar, Westfall, and Delany, forthcoming). Fires that occur in low- and mid-elevation forest types have potential for increasing fire severity (Keane *et al.*, 2002) as future burning conditions become more extreme.

Increases in the area burned or biomass burned under future climate scenarios are seen in a number of studies across the United States. Using historical data, warmer summer temperatures were shown to be significant in western state-level statistical models of area burned (McKenzie *et al.*, 2004). Using the IPCC B2 climate scenario and the Parallel Climate Model, wildfire activity was projected to increase from 1.5–4 times historical levels for all western states (except California and Nevada) by the 2070–2100 period. The highest increases were projected for Utah and New Mexico. The analysis of 19 climate models and their scenarios used in the Fourth IPCC Assessment Report (Seager *et al.*, 2007) show a consistency in the projections for increased drought in the Southwest, unlike any seen in the instrumental record. In Alaska, warmer and longer growing seasons and associated vegetation shifts under two future climate scenarios indicated an increase in the area of forests burned by a factor of two or three (Bachelet et al., 2005).

The combination of extended dry periods resulting from fewer, stronger rainfall events with warmer temperatures could render northeastern forests more susceptible to fire than they have been for the past 100 years of fire suppression (Scholze *et al.*, 2006). Similarly, drought may become an increasingly important stressor in eastern forests, which in turn may increase the risk of fire in areas that have experienced low frequency fire regimes during the past century or more (Lafon, Hoss, and Grissino-Mayer, 2005).

Some climate scenarios project less and others more precipitation for the southern United States (Bachelet *et al.*, 2001). Even under the wetter scenarios, however, the South is projected to experience an increase in temperature-induced drought and an increase in fires (Lenihan *et al.*, forthcoming). On average, biomass consumed by fire is expected to increase by a factor of two or three (Bachelet *et al.*, 2001; Bachelet *et al.*, forthcoming).

Climate Change and Invasive and Native Species Management

Invasive species are already a problem in many areas of the United States (Stein *et al.*, 1996; Pimentel *et al.*, 2000; Rahel, 2000; Von Holle and Simberloff, 2005). Climate change is expected to compound this problem, due to its direct influence on native species' distributions and the effects of its interactions with other stressors (Chornesky *et al.*, 2005). A continual reassessment of management strategies for invasive species may be necessary under a changing climate.

In general, the impacts of invasive species with an expanded range are difficult to predict, in part because the interactions among changing climate, elevated CO_2 concentrations, and altered nutrient dynamics are themselves still being elucidated (Simberloff, 2000). In some cases, however, the likely impacts are better understood. For example, future warming may accelerate the northern expansion of European earthworms, which have already substantially altered the structure, composition, and competitive relationships in North American temperate and boreal forests (Frelich *et al.*, 2006). In arid and semi-arid regions of the United States, increases in annual precipitation are expected to favor non-native invasive species at the expense of native vegetation on California serpentine soils (Hobbs and Mooney, 1991) and in Colorado steppe communities (Milchunas and Lauenroth, 1995). Understanding the potential to prevent and control invasives will require research on invasive species' population and community dynamics interacting with a changing ecosystem dynamic.

Increasing concentrations of CO_2 in the atmosphere may also create a competitive advantage to some invasive species (Dukes, 2000; Smith *et al.*, 2000; Ziska, 2003; Weltzin, Belote, and Sanders, 2003). These positive responses may require a re-evaluation of current management practices. Positive responses to elevated CO_2 have been reported for red brome, an introduced non-native annual grass in the Southwest (Smith *et al.*, 2000). Increasing presence of this exotic grass, along with its potential to produce fire fuel, suggest future vegetation shifts and increased fire frequency (Smith *et al.*, 2000) where vegetation has not evolved under frequent fire. The positive response to current (from pre-industrial) levels of atmospheric CO_2 by six invasive weeds—Canada thistle (*Cirsium arvense* (L.) Scop.), field bindweed (*Convolvulus arvensis* L.), leafy spurge (*Euphorbia esula* L.), perennial sowthistle (*Sonchus* L.), spotted knapweed (*Centaurea stoebe* L.), and yellow star-thistle (*Centaurea solstitialis* L.)—suggests that 20th century increases in atmospheric CO_2 may have been a factor in the expansion of these invasives (Ziska, 2003). Because increasing CO_2 concentrations allow invasive species to allocate additional carbon to root biomass, efforts to control invasive species with some currently used herbicides may be less effective under climate change (Ziska, Faulkner, and Lydon, 2004).

Further, the combination of elevated CO_2 concentrations and warmer temperatures is expected to exacerbate the current invasive species problem in the currently cooler parts of the United States (Sasek and Strain, 1990; Simberloff, 2000; Weltzin, Belote, and Sanders, 2003). The northward expansion of the range of invasive species currently restricted by minimum temperatures (*e.g.*, kudzu and Japanese honeysuckle) is a particular concern (Sasek and Strain, 1990; Simberloff, 2000; Weltzin, Belote, and Sanders, 2003). Invasive species with a C4 photosynthetic pathway (*e.g.*, itchgrass, *Rottboellia cochinchinensis*) are particularly likely to invade more northerly regions as frost hardiness zones shift northward (Dukes and Mooney, 1999). Although C3 species (*e.g.*, lamb's quarters, *Chenopodium album*) are likely to grow faster under elevated CO_2 concentrations (Bazzaz, 1990; Drake, Gonzalez-Meler, and Long, 1997; Nowak, Ellsworth, and Smith, 2004; Ainsworth and Long, 2005; Erickson *et al.*, 2007), C4 species seem to respond better to warmer temperatures (Alberto *et al.*, 1996; Weltzin, Belote, and Sanders, 2003), probably because the optimum temperature for photosynthesis is higher in C4 species (Dukes and Mooney, 1999).

Climate change will likely facilitate the movement of some native species into the habitats of others, and thus create novel species assemblages, potentially affecting current goods and services. Some of the dispersing native species will likely become problematic invaders that place many threatened and endangered species at greater risk of local extinction due to enhanced competition, herbivory, predation, and parasitism (Neilson *et al.*, 2005a; 2005b). For example, in the Pacific Northwest, barred owls (*Strix varia*), which are rapidly migrating generalists from eastern forests of the United States, have invaded the spotted owl's range in the Pacific Northwest and are now competing with the northern spotted owl (*Strix occidentalis caurina*) for nest sites (Kelly, Forsman, and Anthony, 2003; Noon and Blakesley, 2006; Gutierrez *et al.*, 2007). An increase of 3°C in minimum temperature could extend the southern pine beetle's northern distribution limit by 170 km, with insect outbreaks spreading into the mid-Atlantic states (Williams and Liebhold, 2002). Novel species assemblages may require a re-examination of management approaches for native species now acting as invasives; for threatened, endangered and rare species; and a re-evaluation of what ecosystem services can be managed within each NF.

Climate Change and Watershed Management

The hydrological regimes of NFs are closely linked to climate, as well as to the many other variables that climate change may affect. Changes in precipitation patterns, including declining snowpack, earlier snowmelt, more precipitation falling as rain vs. snow (Mote *et al.*, 2005), advances in streamflow timing (Stewart, Cayan, and Dettinger, 2004; Barnett, Adam, and Lettenmaier, 2005; Milly, Dunne, and Vecchia, 2005), and the increasing frequency and intensity of extreme precipitation events (Karl and Knight, 1998; Nearing, 2001; Groisman *et al.*, 2005) have affected the hydrology, and hence condition of watersheds and ecosystems throughout the United States (Dettinger *et al.*, 2004; Hayhoe *et al.*, 2004). Increases in flooding may occur as a result of the increased storm intensity projected by future climate models (IPCC, 2007). Changes in the distribution, form, and intensity of precipitation will make it more challenging to achieve the goal of improving watershed conditions.

Water shortages in some areas are projected, due to increasing temperatures and changing precipitation patterns, as well as to shifting demography and increased water demand (Arnell, 1999; Whiles and Garvey, 2004). National forest ecosystems in more arid parts of the country are expected to be particularly affected by projected climatic changes (Hayhoe *et al.*, 2004; Seager *et al.*, 2007). However, even in wetter regions (*e.g.*, the southeastern United States), hot temperatures and high evapotranspiration rates cause only 50% of annual precipitation to be available for streamflow (Sun *et al.*, 2005). Thus, future scenarios of climate and land-use change indicate that the water yield for this region will become increasingly variable.[20] In the Northeast, a temperature increase of 3°C was projected to decrease runoff by 11–13% annually, and to a greater extent during the summer months when flow is typically lowest (Huntington, 2003). Gains in water use efficiency from elevated CO_2 may be negated or overwhelmed by changes in the hydrological variables described above, leading to increased water stress for vegetation in NFs (Baron *et al.*, 2000; but see Huntington, 2003).

Climate Change and Biodiversity Management

Climate change affects biodiversity directly by altering the physical conditions to which many species are adapted. Although species with large geographic ranges have a wide range of physiological tolerance, species that are rare, threatened, endangered, narrowly distributed, and endemic, as well as those with limited dispersal ability, will be particularly at risk under climate change (Pounds *et al.*, 2006) because they may not be able to adapt in situ or migrate rapidly enough to keep pace with changes in temperature (Hansen *et al.*, 2001; Wilmking *et al.*, 2004; Neilson *et al.*, 2005b). Changes in precipitation patterns may disrupt animal movements and influence recruitment and mortality rates (Inouye *et al.*, 2000). The projected changes in fish habitat associated with increases in temperature and changes in hydrology (Preston, 2006) would cause shifts in the distributions of fish and other aquatic species (Kling *et al.*, 2003). Projected declines in suitable bird habitat of 62–89% would increase the extinction risk for Hawaiian honeycreepers (Benning *et al.*, 2002). Similar projected losses of suitable habitat in U.S. forests would decrease Neotropical migratory bird species richness by 30–57% (Price and Root, 2005). Interactions among species may also amplify or reverse the direct impacts of climate change on biodiversity (Suttle, Thompsen, and Power, 2007).

Tree species richness is projected to increase in the eastern United States as temperatures warm, but with dramatic changes in forest composition (Iverson and Prasad, 2001). Projections indicate

[20] Sun, G., S.G. McNulty, E. Cohen, J.M. Myers, and D. Wear, 2005: Modeling the impacts of climate change, landuse change, and human population dynamics on water availability and demands in the Southeastern US. Paper number 052219. Proceedings of the 2005 ASAE Annual Meeting, St. Joseph, MI.

CASE STUDY SUMMARY 3.2

Olympic National Forest, Washington
Pacific Northwest United States

Why this case study was chosen

The Olympic National Forest:

- Is located within a geographic mosaic of lands managed by federal and state agencies, tribal groups, and private land owners;
- Supports a diverse set of ecosystem services, including recreation, timber, water supply to municipal watersheds, pristine air quality, and abundant fish and wildlife—including several endemic species of plants and animals, as well as critical habitat for four threatened species of birds and anadromous fish;
- Is considered an urban forest because of its proximity to the cities of the greater Seattle area;
- Has numerous stakeholders and land management mandates associated with its natural and cultural resources.

Management context

The Olympic National Forest (ONF) is a "restoration forest" charged with managing large contiguous areas of second-growth forest. Natural resource objectives include managing for native biodiversity and promoting the development of late-successional forests; restoring and protecting aquatic ecosystems from the impacts of an aging road infrastructure; and managing for individual threatened and endangered species as defined by the Endangered Species Act or other policies related to the protection of rare species. Most management focuses on restoring old-growth forests, pristine waterways, and other important habitats; rehabilitating or restoring areas affected by unmaintained logging roads; invasive species control; and monitoring. Because the Northwest Forest Plan dictates that the ONF collaborate with other agencies, it will be important to reach consensus so that differing agency mandates, requirements, and strategies do not hinder adaptation to climate change.

Key climate change impacts

- Observed increase of 1.0°C in annual temperatures since 1920, with most warming in winters and since 1950;
- Observed decrease (30–60%) in spring snowpack, especially at lower elevations since 1950;
- Observed one-to-four-week advance in spring runoff in 2000 versus 1948;
- Projected increase in temperatures of 1.2–5.5°C by 2090, with greatest increases in summer;
- Projected decrease in snowpack, shifts in snowmelt and runoff timing, and increases in summer evapotranspiration;
- Expected negative consequences of higher temperatures and lower summer flows for resident fish species;
- Expected forest growth decrease at lower elevations and increase at higher elevations;
- Expected increase in floods and area burned by fire;
- Expected shift in species distribution and abundance.

Opportunities for adaptation

- The priorities for the ONF already emphasize management for landscape and biological diversity, and actions expected to be the most effective in this regard could be further promoted now as an important first step toward adaptation to climate change.
- The ONF's strategic plan leaves enough flexibility so that it can take immediate steps to incorporate climate change science into management actions and to enhance resilience to climate change, while at the same time fostering scientific research to support these actions.
- The early successional forests predominating in the ONF as a result of past timber management offer an opportunity to adapt to climate change with carefully considered management actions, because these early successional stages are most easily influenced.
- The ONF's experience collaborating with other agencies and organizations could be leveraged to develop innovative climate change adaptations that benefit multiple stakeholders; continued cooperation with existing and new partners in adapting to climate change will improve the likelihood of success by increasing the overall land base and resources.

CASE STUDY SUMMARY 3.2 (CONTINUED)

- By anticipating future impacts of climatic change on forest ecosystems, revised forest plans can become an evolving set of guidelines for forest managers.
- Coordinated revision of forest plans for the Olympic, Mt. Baker-Snoqualmie, and Gifford Pinchot National Forests offers an opportunity to develop regional-scale adaptations for similar ecosystems that are subject to similar stressors.

Conclusions

The management priorities for the ONF could facilitate managers' efforts to adapt to climate change and promote resilience to its impacts, but adaptive capacity is limited by the current allocation of scarce resources, policy environment, and lack of scientific information on the effects of climate change and the likely outcomes of adaptations. Increased support for adaptation, specific guidance on climate change impacts and adaptations for managers, and incorporating climate change explicitly into forest policies and planning at multiple scales are some of the ways these barriers can be overcome. In addition, the availability of regional climate and forest-climate research—and especially a proactive management-science partnership—set the stage for increases in adaptive capacity.

In the absence of more specific scientific guidance on how to adapt to climate change, and without new funding and additional staff, the ONF will likely manage for climate change by continuing to manage for biodiversity, which is a reasonable approach assuming that prioritizing landscape and biological diversity will confer adequate resilience to climate change over the long term. An adaptation strategy with more specific guidance could include a vision of what is needed; removal of as many barriers as possible; increased collaboration among agencies, managers, and scientists at multiple scales; and implementation of proven management actions (e.g., early detection/rapid response).

that spruce-fir forests in New England could be extirpated and maple-beech-birch forests greatly reduced in area, whereas oak-hickory and oak-pine forest types would increase in area (Bachelet *et al.*, 2001; Iverson and Prasad, 2001). Projected changes in temperature and precipitation suggest that southern ecosystems may shift dramatically. Depiction of the northern shift of the jet stream and the consequent drying of the Southeast (Fu *et al.*, 2006) varies among future climate scenarios, with some showing significant drying while others show increased precipitation (Bachelet *et al.*, 2001). However, even under many of the somewhat wetter future scenarios, closed-canopy forests the Southeast may revert, or in some areas, be converted to savanna, woodland, or grassland under temperature-induced drought stress and a significant increase in fire disturbance (Bachelet *et al.*, 2001; Scholze *et al.*, 2006).

Ecosystems at high latitudes and elevations (including many coniferous forests), as well as savannas, ecosystems with Mediterranean (*e.g.*, California) climates, and other water-limited ecosystems, are expected to be particularly vulnerable to climate change (Thomas *et al.*, 2004; Millennium Ecosystem Assessment, 2005; Malcolm *et al.*, 2006). Temperature-induced droughts in these ecosystems are expected to contribute to forest diebacks (Bugmann, Zierl, and Schumacher, 2005; Millar, Westfall, and Delany, forthcoming). Alpine ecosystems are also projected to decrease in area as temperatures increase (Bachelet *et al.*, 2001). Specifically, as treelines move upward in elevation, many species could be locally extirpated as they get "pushed" off the top of the mountains (Bachelet *et al.*, 2001). Also, given the strong species-area relationship that has been shown for the "island" habitats on the tops of western mountains, species diversity could be significantly reduced as these habitats become smaller or even disappear (McDonald and Brown, 1992).

Simulations of future vegetation distribution in the Interior West show a significant increase in woody vegetation as a result of enhanced water-use efficiency from elevated CO_2, moderate increases in precipitation, and a strengthening of the Arizona Monsoon (Neilson *et al.*, 2005a), with the greatest expansion of woody vegetation projected in the northern parts of the interior West (Lenihan *et al.*, forthcoming). The drier interior vegetation shows a large increase in savanna/woodland types, suggesting possibly juniper and yellow pine species range expansions. However, this region is also projected to be very susceptible to fire and drought-induced dieback, mediated by insect outbreaks (Neilson *et al.*, 2005a). Such outbreaks have already altered the species composition of much of this region (Breshears *et al.*, 2005).

A key predicted effect of climate change is the expansion of native species' ranges into biogeographic areas in which they previously could not survive (Simberloff, 2000; Dale *et al.*, 2001). This prediction is supported by the observed northward shift in the ranges of several species, both native and introduced, due to the reduction of cold temperature restrictions (Parmesan, 2006). In general, climate change would facilitate the movement of some species into the habitats of others, which would create novel species assemblages, especially during post-disturbance succession. An entire flora of frost-sensitive species from the Southwest may invade ecosystems from which they have been hitherto restricted, and in the process displace many extant native species over the course of decades to centuries (Neilson *et al.*, 2005b) as winter temperatures warm (Kim *et al.*, 2002; Coquard *et al.*, 2004) and hard frosts occur less frequently in the interior West (Meehl, Tebaldi, and Nychka, 2004; Tebaldi *et al.*, 2006). Similar migrations of frost-sensitive flora and fauna occurred during the middle-Holocene thermal maximum, which was comparable to the minimum projected temperature increases for the 21st century (Neilson and Wullstein, 1983).

Similarly increases in warm temperate/subtropical mixed forest are projected in the coastal mountains of both Oregon and Washington, with an increase in broadleaved species such as various oak species, tanoak, and madrone under many scenarios (Bachelet *et al.*, 2001; Lenihan *et al.*, forthcoming). However, slow migratory rates of southerly (California) species would likely limit their presence in Oregon through the 21st century (Neilson *et al.*, 2005b).

These potential shifts in species may or may not enhance the biodiversity of the areas into which they migrate. This shift will potentially confound management goals based on the uniqueness of species for which there are no longer habitats.

3.3.4.2 Goal 2: Provide and Sustain Benefits to the American People

The outcome for this goal is forests and grassland with sufficient long-term multiple socioeconomic benefits to meet the needs of society. Specific objectives are focused on providing a reliable supply of forest products and rangeland, with productivity that is consistent with achieving desired conditions on NFS lands and helps support local communities, meets energy resource needs, and promotes market-based conservation and stewardship of ecosystem services.

Co-benefits of joint carbon sequestration and biofuel production, along with other potential synergies, are certainly possible via forest management (Birdsey, Alig, and Adams, 2000; Richards, Sampson, and Brown, 2006), and would enable contribution to both the country's energy needs and its carbon sequestration and greenhouse gas mitigation goals. Forest management practices designed to achieve goals of removing and storing CO_2 are diverse, and the forestry sector has the potential for large contributions on the global to regional scales (Malhi, Meir, and Brown, 2002; Krankina and Harmon, 2006). Along with preventing deforestation, key activities include afforestation, reforestation, forest management, and post-harvest wood-product development (Harmon and Marks, 2002; Von Hagen and Burnett, 2006). Reducing deforestation (Walker and Kasting, 1992) and promoting afforestation provide important terrestrial sequestration opportunities (Nilsson and Schopfhauser, 1995),[21] as do many forest plantation and forest ecosystem management practices (*e.g.*, Briceno-Elizondo *et al.*, 2006). Many suggested approaches duplicate long-

[21] See also **Kadyszewski**, J., S. Brown, N. Martin, and A. Dushku, 2005: Opportunities for terrestrial carbon sequestration in the west. Winrock International. Presented at the Second Annual Climate Change Research Conference, From Climate to Economics and Back: Mitigation and Adaptation Options for California and the Western United States, 15, September 2005. Accessed at http://www.climatechange.ca.gov/events/2005_conference/presentations/2005-09-15/2005-09-15_KADYSZEWSKI.PDF.

recognized best forest management practices, where goals are to maintain healthy, vigorous growing stock, and keep sites as fully occupied as possible while still maintaining resistance to uncharacteristically severe fire, insects, and disease (Gottschalk, 1995). Projects planned to delay return of CO_2 to the atmosphere (*e.g.*, by lengthening rotations; Richards, Sampson, and Brown, 2006), both in situ (in the forest or plantation) and post-harvest, are most successful.

Climate change is expected to alter forest and rangeland productivity (Joyce and Nungesser, 2000; Aber *et al.*, 2001; Hanson *et al.*, 2005; Norby, Joyce, and Wullschleger, 2005; Scholze *et al.*, 2006). This alteration in forest productivity, in turn, will influence biomass available for wood products or for energy (Richards, Sampson, and Brown, 2006), whether as a direct energy source or for conversion to a biofuel. The interactions of climate change (*e.g.*, warming temperatures, droughts) and other stressors—including altered fire regimes, insects, invasive species, and severe storms—may affect the productivity of forests and rangelands. This alteration in forest productivity in turn would affect the volume of material that could be harvested for wood products or for energy, or the rate at which a forest would sequester carbon on site. The interactions of climate change with other stressors such as insects (Volney and Fleming, 2000; Logan, Regniere, and Powell, 2003), disease (Pounds *et al.*, 2006), and fire (Flannigan, Stocks, and Wotton, 2000; Whitlock, Shafer, and Marlon, 2003) will challenge the management of ecosystem services and biodiversity conservation in NF ecosystems. Indeed, Flannigan, Stocks, and Wotton (2000) noted that "the change in fire regime has the potential to overshadow the direct effects of climate change on species distribution and migration." Thus, this goal is sensitive to a changing climate.

Climate Change and Ecosystem Services

The distinctive structure and composition of individual NFs are key characteristics on which forest and rangeland products and ecosystem services depend, and that national forest managers seek to sustain using current management approaches. For example, efforts to achieve a particular desired forest structure, composition, and function have been based on an understanding of ecosystem dynamics as captured in historical references or baselines (*i.e.*, observed range of variation), and the now outdated theory that communities and ecosystems are at equilibrium with their environment (Millar and Woolfenden, 1999). Under a changing climate (increased temperatures; changes in rainfall intensity; and greater occurrence of extreme events, such as drought, flooding, etc.), such an approach may no longer be sensible. Ecosystem composition, structure, and function will change as species respond to these changes in climate. Thus, as climate change interacts with other stressors to alter NF ecosystems, it will be important to focus as much on maintaining and enhancing ecosystem processes as on achieving a particular composition. For these reasons, it will be increasingly important for the USFS to consider evaluating current management practices, their underlying climatic and ecological assumptions, and to consider managing ecosystems for change (discussed further in Sections 3.4–3.5).

Although forests are projected to be more productive under elevated CO_2 (Joyce and Birdsey, 2000; Hanson *et al.*, 2005; Norby, Joyce, and Wullschleger, 2005), productivity increases are expected to peak by 2030. Declines thereafter are likely to be associated with temperature increases, changes in precipitation, ozone effects, and other climate change stressors (Scholze *et al.*, 2006; Sitch *et al.*, 2007). Productivity increases may be offset especially where water and/or nutrients are limiting and increases in summer temperature further increase water stress (Angert *et al.*, 2005; Boisvenue and Running, 2006), and where ozone exposure reduces the capacity of forests to increase their productivity in response to elevated CO_2 (Karnosky, Zak, and Pregitzer, 2003; Hanson *et al.*, 2005; Karnosky *et al.*, 2005; King *et al.*, 2005). In cooler regions where water will not be a limiting resource, and where other stressors do not offset potential productivity increases, opportunities may increase for the production of biofuels and biomass energy. The feasibility of taking advantage of these opportunities may hinge on whether economic, political, and logistical barriers can be overcome (Richards, Sampson, and Brown, 2006). If, as projected, climate change enhances woody expansion and productivity for the near term in the intermountain West (Bachelet *et al.*, 2003), then forests and woodlands in that region could provide a source of fuel while mitigating the use of fossil fuels (Bachelet *et al.*, 2001).

Interactions of Climate Change with Other Stressors

Insect and disease outbreaks may become more frequent as the climate changes, because warmer temperatures may accelerate their life cycles (*e.g.*, Logan and Powell, 2001). As hardiness zones shift north[22] and frost-free days and other climatic extremes increase (Tebaldi *et al.*, 2006), the hard freezes that in the past slowed the spread of insect and disease outbreaks may become less effective, especially if the natural enemies (*e.g.*, parasitoids) of insects are less tolerant of the climate changes than are their hosts or prey (Hance *et al.*, 2007). In addition, previously confined southern insects and pathogens may move northward as temperatures warm (see Box 3.5) (Ungerer, Ayres, and Lombardero, 1999; Volney and Fleming, 2000; Logan, Regniere, and Powell, 2003; Parmesan, 2006), especially in the absence of predatory controls. While the expectation is for increased wildfire activity associated with increased fuel loads (*e.g.*, Fleming, Candau, and McAlpine, 2002), in some ecosystems (*e.g.*, subalpine forests in Colorado), insect outbreaks may decrease susceptibility to severe fires (*e.g.*, Kulakowski, Veblen, and Bebi, 2003).

Species, whether or not they are indigenous to the United States, may act invasively and increase the stress on ecosystems and on other native species. The rapid advance of the mountain pine beetle beyond its historic range (Logan and Powell, 2005) is a case in which a native species, indigenous to the American West, has begun to spread across large areas like an invasive species (as reflected by faster dispersal rates and greater range extension) because longer and warmer growing seasons allow it to more rapidly complete its lifecycle, and because warmer winters allow winter survival (Logan and Powell, 2001; Carroll *et al.*, 2004; Millar, Westfall, and Delany, forthcoming).

3.3.4.3 Goal 3: Conserve Open Space

The outcome for this goal is the maintenance of the environmental, social, and economic benefits of the Nation's forests and grasslands, protecting those forest and grasslands from conversion to other uses, and helping private landowners and communities maintain and manage their land as sustainable forests and grasslands. As described under Goals 1 and 2 above, the environmental benefits of forests and grasslands are influenced strongly by climate and changes in climate. Additionally, fragmentation and urbanization facilitate the spread of invasive species, and are key drivers contributing to biotic homogenization in the United States in general (Olden, 2006) Under a changing climate, landscape fragmentation may exacerbate or cause unexpected changes in species and ecosystems (Iverson and Prasad, 2001; Price and Root, 2005). Thus this goal will be sensitive to a changing climate.

Climate Change and Open Space

The loss of open space and land-use changes that are already problematic may be worsened under climate change, due to shifts in species' behaviors and changed habitat requirements. The loss of open space is of particular concern because it may impede species' migration and exacerbate edge effects (*e.g.*, windthrow, drought, and non-native invasive species) during extreme climatic events, and possibly result in increased population extirpation (Ewers and Didham, 2006). Fragmentation may result in the loss of larger management unit sizes, broad habitat corridors, and continuity of habitat. In this regard, enhancing coordination among the multiple agencies that manage adjacent lands to ensure habitat continuity will be essential (Malcolm *et al.*, 2006). Land-use change and invasive species are expected to exacerbate the effects of climate change, and hence make the goal of maintaining environmental benefits on forests and grasslands more challenging to achieve.

[22] **National Arbor Day Foundation**, 2006: Differences between 1990 USDA hardiness zones and 2006 arborday.org hardiness zones reflect warmer climate. Available at http://www.arborday.org/treeinfo/zonelookup.cfm.

BOX 3.5. Bark Beetles in Western North American Forests.

Bark beetles are native insects and important disturbance agents in western North American forests (Carroll *et al.*, 2004). Beetle outbreaks occur periodically when otherwise healthy trees are weakened from drought, injury, fire damage, and other stresses. Since 1996, bark beetles have infested and killed millions of pine, spruce, and fir trees over vast areas from Arizona to British Columbia. This outbreak, which is considered to be more extensive and damaging than any previously recorded in the West, is expected to continue without active management.[23]

The most "aggressive, persistent, and destructive bark beetle in the United States and western Canada" is the mountain pine beetle (*Dendroctonus ponderosae Hopkins*),[24] which will attack and kill most western pine species. The mountain pine beetle (MPB) infested 425,000 acres of Colorado's lodgepole pine (LP) forests in 2005 (Colorado Department of Natural Resources, 2005) and 660,000 acres (~40% of Colorado's LP forests) by 2006. The unprecedented scale of this outbreak in Colorado is attributable to a combination of factors, including large areas with even-age, monospecific stands (a result of fire suppression and other management practices), drought, and climate change (Colorado State Forest Service cited in Paulson, 2007).

Warmer winters have spurred extensive mountain pine beetle damage in the U.S. and Canadian Rockies. Left from Fox (2007); photo below is reprinted with permission from Colorado State University Extension, fact sheet no. 5.528, Mountain Pine Beetle, by D.A. Leatherman. and I. Aguayo.[25]

Despite the historic scale of the recent MPB outbreak in Colorado's lodgepole pine forests, periodic outbreaks, albeit on a smaller spatial scale, are considered normative (Logan and Powell, 2001). Lodgepole pine and MPB are co-evolved, and lodgepole pine is the MPB's most important host (Logan and Powell, 2001). Lodgepole pine has serotinous cones and is maintained by stand replacing fires that are facilitated by MPB-induced mortality. Dead needles from outbreaks are an important fuel, standing dead trees serve as fire ladders, and falling limbs and stems provide high fuel loads for high-intensity crown fires. Without such fires, more shade-tolerant species would eventually replace lodgepole pine in much of its range (Logan and Powell, 2001).

Other western pines, especially those growing at higher elevations such as whitebark pine, are not similarly co-evolved with MPB. Until recently, high elevation and high latitude habitats typically have been too harsh for MPB to complete its life cycle in one season. Because the ability to complete its life cycle in one season is central to the MPB's success (Amman, 1973),[26] MPB activity has historically been restricted to lower elevation pines, which are separated from high-elevation (3,000 m or 10,000 ft in Colorado) pines by non-host species.

Climate change will not only spur further MPB outbreaks, but will also likely facilitate the invasion of species currently restricted to more benign environments into whitebark pine and other high-elevation pine stands in the wake of MPB infestations (Logan and Powell, 2001). The fact that all aspects of the MPB's seasonality are controlled by seasonal temperature patterns (Logan and Bentz, 1999) supports this forecast. It is further supported by the finding that both the timing and synchrony of the beetle's life cycle are responsive to climate change (Logan and Powell, 2001). Specifically, Logan and Powell (2001)

[23] **Western Forestry Leadership Coalition**, 2007: Western bark beetle assessment: a framework for cooperative forest stewardship. Western Forestry Leadership Coalition Website, http://www.wflccenter.org/news_pdf/222_pdf.pdf, accessed on 7-31-2007.

[24] **The Bugwood Network**, 2007: Mountain Pine Beetle - Dendroctonus ponderosae (Hopkins). Bark and Boring Beetles of the World Website, http://www.barkbeetles.org/mountain/mpb.html, accessed on 7-30-2007.

[25] **Leatherman**, D.A. and I. Aguayo, 2007: Mountain Pine Beetle. Colorado State University Extension Website, http://www.ext.colostate.edu/pubs/insect/05528.html, accessed on 7-31-0007.

[26] See also **Safranyik**, L., 1978: Effects of climate and weather on mountain pine beetle populations. In: *Proceedings, Symposium: Theory and Practice of Mountain Pine Beetle Management in Lodgepole Pine Forests* [Berryman, A.A., G.D. Amman, and R.W. Stark (eds.)] University of Idaho Forest, Wildlife and Range Experiment Station, pp. 77-84.

showed that a 2°C increase in annual average temperature allows MPB populations to synchronously complete their life cycle in a single season. Such a shift from a two season, asynchronous life cycle confers the greatest chance for population success. Because the response of the MPB's life cycle to temperature is nonlinear, climate change-induced MPB outbreaks are likely to occur in high elevation pine ecosystems without warning.

In addition to creating ideal conditions for populations of MPB to reach epidemic levels, climate change has allowed the MPB to expand its range northward and eastward in recent decades (Carroll *et al.*, 2004). The current MPB range extends from northern Mexico through the American Rockies west and into British Columbia, Alberta, and Saskatchewan (Carroll *et al.*, 2004). The range of the MPB is constrained principally by climate rather than the availability of suitable hosts; lodgepole pine exists beyond the range of MPB (Logan and Powell, 2001; Carroll *et al.*, 2004). Evidence for the range expansion of MPB includes accelerating rates of infestation since 1970 into previously unsuitable habitats. Further range expansion is likely with additional warming (Carroll *et al.*, 2004). Logan and Powell (2001) predict a 7° northward shift in the range of MPB with a doubling of CO_2 and an associated temperature increase of 2.5°C. Such a shift would allow MPB to occupy previously unoccupied lodgepole pine habitat, and allow an invasion into jack pine ecosystems in both the United States and Canada, which have not been previously attacked by MPB (see map at right). The continuous habitat provided by lodgepole pine will facilitate this range shift. Although cold snaps and depletion of hosts caused previous large-scale MPB outbreaks to collapse, the current outbreak may not collapse because there is no shortage of host trees, and temperatures are expected to continue warming (Carroll *et al.*, 2004).

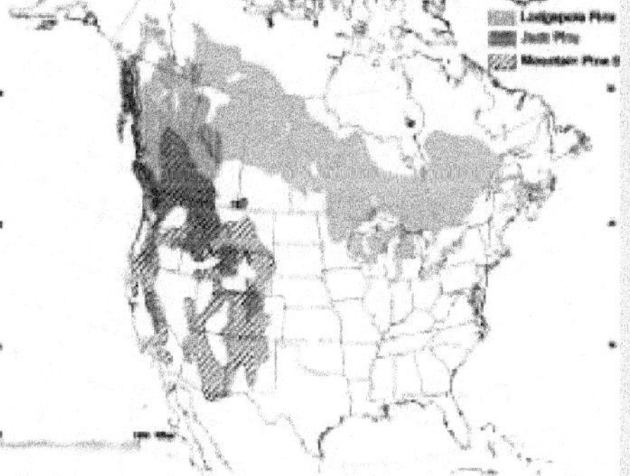

Geographic ranges of lodgepole pine (pink), mountain pine beetle (hatched), and jack pine (green). Source Logan and Powell (2001).

3.3.4.4 Goal 4: Sustain and Enhance Outdoor Recreation Opportunities

The outcome identified for this goal is high-quality outdoor recreational opportunities on the Nation's forests and grassland available to the public. Specific objectives include improving the quality and availability of outdoor recreation experiences, securing legal entry to NF lands and water, and improving the management of off-highway vehicle use. National forests across the United States are managed for a variety of outdoor recreational opportunities, capitalizing on the natural resources and ecosystem services available within each NF (Cordell *et al.*, 1999). The demands on NFs for recreation have diversified with population growth (local, regional, and national), preferences for different types of recreation, and technological influences on recreation (off-road motorized vehicles, mountain biking, snowboarding). Along with camping, hunting, and fishing, recreational activities now include skiing (downhill, cross-country), snowboarding, mountain biking, hiking, kayaking, rafting, and bird watching.

Climate Change and Recreation Management

Because individual recreational opportunities are often a function of climate (cold-water fisheries or winter snow), climate change may affect both the opportunity to recreate and the quality of recreation (Irland *et al.*, 2001), curtailing some recreational opportunities and expanding others.

Winter outdoor recreation—such as alpine and Nordic skiing, snowmobiling, skating, ice fishing, and other opportunities—may decrease and/or shift in location due to fewer cold days and reduced snowpack (National Assessment Synthesis Team, US Global Change Research Program, 2001). The costs of providing these opportunities (*e.g.*, increased snowmaking) are likely to rise (Irland *et al.*, 2001) or may result in potential conflicts with other uses (*e.g.*, water) (Aspen Global Change Institute, 2006). Other winter recreational activities (*e.g.*, ice skating, ice fishing, and ice climbing) may also become more restricted (both geographically and seasonally) as winter temperatures warm (National Assessment Synthesis Team, US Global Change Research Program, 2001), with limited opportunities for management to sustain these opportunities.

Altered streamflow patterns and warmer stream temperatures, observed trends that are projected to continue with future climate change (Regier and Meisner, 1990; Eaton and Scheller, 1996; Rahel, Keleher, and Anderson, 1996; Stewart, Cayan, and Dettinger, 2004; Barnett, Adam, and Lettenmaier, 2005; Milly, Dunne, and Vecchia, 2005), may change fishing opportunities from salmonids and other cold-water species to species that are less sensitive to warm temperatures (Keleher and Rahel, 1996; Melack *et al.*, 1997; Ebersole, Liss, and Frissell, 2001; Mohseni, Stefan, and Eaton, 2003) and altered streamflow (Marchetti and Moyle, 2001). One estimate indicates that cold-water fish habitat may decrease by 30% nationally and by 50% in the Rocky Mountains by 2100 (Preston, 2006). More precise estimates of the climate change impacts on fish populations will depend on the ability of modelers to consider other factors (*e.g.*, land use change, fire, invasive species, and disease) in addition to temperature and streamflow regimes (Clark *et al.*, 2001). The projected reductions in volume of free-flowing streams during summer months, due to advances in the timing of flow in these streams (Stewart, Cayan, and Dettinger, 2004; Barnett, Adam, and Lettenmaier, 2005; Milly, Dunne, and Vecchia, 2005), may also restrict canoeing, rafting, and kayaking opportunities (Irland *et al.*, 2001).

Climate change may also increase recreational opportunities, depending on the preferences of users, the specific climatic changes that occur, and the differential responses of individual species to those changes. Fewer cold days, for example, may encourage more hiking, biking, off-road vehicle use, photography, swimming, and other warm-weather activities. The different growth responses of closely related fish species to increases in temperature and streamflow (Guyette and Rabeni, 1995) may enhance opportunities for species favored by some anglers.

Interactions of Climate Change with Other Stressors

An increase in the frequency, extent, and severity of disturbances such as fire and severe storms also may affect the quality of recreation experienced by visitors to NFs during and after disturbances. Recreational opportunities may be curtailed if forest managers decide (for public safety or resource conservation reasons) to reduce access during and in the wake of major disturbances such as fire, droughts, insect outbreaks, blowdowns, and floods, all of which are projected to increase in frequency and severity during the coming decades (IPCC, 2007). Unlike smoke from prescribed fires, which is subject to NAAQS (National Ambient Air Quality Standards),[27] wildfire smoke is considered a temporary "natural" source by EPA and the departments of environmental quality in Montana, Idaho, and Wyoming, and is therefore not directly regulated. Within the Greater Yellowstone Ecosystem, prescribed fire smoke is managed to minimize smoke encroachment on sensitive areas (communities, Class 1 wilderness areas, high use recreation areas, scenic vistas) during sensitive periods.[22] After wildfire, the quality of the recreational experience has been shown to be affected by the need to travel through a historical fire area (Englin *et al.*, 1996) and by the past severity of fire (Vaux, Gardner, and Thomas, 1984). Groups experiencing different types of recreation (hiking versus mountain biking) react differently to wildfire, and reactions vary across geographic areas (Hesseln *et al.*, 2003). Changes in vegetation and other ecosystem components (*e.g.*, freshwater availability and quality) caused by droughts, insect and disease outbreaks (Rouault *et al.*, 2006), fires, and storms may alter the aesthetics, sense of place,

[27] **Story**, M., J. Shea, T. Svalberg, M. Hektner, G. Ingersoll, and D. Potter, 2005: *Greater Yellowstone Area Air Quality Assessment Update*. Greater Yellowstone Clean Air Partnership. Available at http://www.nps.gov/yell/planyourvisit/upload/GYA_AirQuality_Nov_2005.pdf.

and other cultural services that the public values.

The projected increases of pests and vector-borne diseases may also affect the quality of recreational experiences in NFs. Hard freezes in winter have been shown to kill more than 99% of pathogen populations annually (Burdon and Elmqvist, 1996; as cited in Harvell et al., 2002). The hard freezes necessary to slow the spread of insect and disease outbreaks may become less effective (Gutierrez *et al.*, 2007). In particular, warmer temperatures are expected to increase the development, survival, rates of disease transmission, and susceptibility of both human and non-human hosts (Harvell *et al.*, 2002; Stenseth *et al.*, 2006). Land-use change leading to conversion of forests adjacent to NFs may compound the effect of climate change on disease, because increases in disease vectors have been associated with loss of forests (Sutherst, 2004). Conversely, where climate change contributes to a decline in the impacts of pathogens—or in cases where species have demonstrated an ability to adapt to changes in disease prevalence (*e.g.*, Woodworth *et al.*, 2005)—the goal may become easier to achieve because visitors may have a positive experience.

3.3.4.5 Goal 5: Maintain Basic Management Capabilities of the Forest Service

The outcome identified for this goal is administrative facilities, information systems, and landownership management with the capacity to support a wide range of natural resources challenges. The means and strategies identified for accomplishing this goal include (and are not limited to) recruiting and training personnel to develop and maintain strong technical and leadership skills in Forest Service program areas to meet current and future challenges. Resource management is challenging in today's environment, and climate change will heighten that challenge. Maintaining technical skills associated with resource management will require the most current information on climate change and its potential impacts to ecosystems within the NFS, as well as its impacts on the ecological and socioeconomic systems surrounding the NFs. The depth of this technical understanding will influence policy development across all levels of the agency. Under a changing climate, ecosystem services will likely be altered within the NFs, resulting in the need to evaluate national policy as well as local land management objectives, relationships with current partnerships, and the need to develop new partnerships. Line officers and resource staff are faced with—and will continue to be faced with—the challenge of making decisions in an uncertain environment. This goal is sensitive to climate change.

Climate Change and Management Capabilities of the Forest Service

The capacity of the USFS to address climate change may require the staff within NFs to have a technical understanding of climate change impacts on ecological systems, to be able to share technical information and experiences (successes as well as failures) about managing under climate change efficiently and effectively, to be able to apply new knowledge to the development of management approaches, and to be able to develop and use planning tools with climate information. Current understanding about the relationships among climate and disturbances, ecosystem services, and forest and grassland products may no longer be appropriate under a changing climate. The climate sensitivity of best management practices, genetic diversity guidelines, restoration treatments, and regeneration guidelines may need to be revisited. Many forest managers are awaiting information from quantitative models about future climates and environments to guide climate-related planning, but adequate training and user-friendly interfaces will be needed before these can be implemented. Limited staff capacities within NFs, combined with the scope of current on-the-ground management needs, could slow the attainment of this goal.

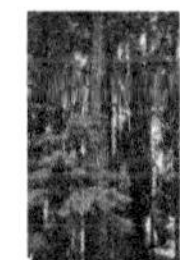

3.3.4.6 Goal 6: Engage Urban America with Forest Service Programs

The outcome identified for this goal is broader access by Americans to the long-term environmental, social, economic, and other types of benefits provided by the USFS. The climate change impacts associated with ecosystem services from NFs would suggest that this goal will be sensitive to climate change.

Climate Change and Urban America

Two objectives were identified for this goal: (1) promote conservation education and (2) improve the management of urban and community forests to provide a wide range of public benefits. The current goal of the conservation education program in the USFS is to "ensure that educational programs and materials developed or certified by the Forest Service incorporate the best scientific knowledge; are interdisciplinary and unbiased; support the Forest Service mission; and are correlated with appropriate national, State, and agency guidelines" (USDA Forest Service, 2007a). Incorporating the best scientific knowledge will require information on climate change and the potential impacts of climate change, necessitating a strong tie to and need for ongoing research on climate change and natural resource management.

Means and strategies identified for this goal include continuing urban forest inventory and analysis, to monitor the health and benefits of ecological and social services of urban forests and more effectively manage these complex landscapes; developing and disseminating strategies and options such as "green infrastructure," to effectively manage resources to maintain environmental quality and services in urban and urbanizing landscapes; helping communities increase professional urban forestry staffing, ordinances, management plans, and local advisory and advocacy groups for managing forest resources in cities, suburbs, and towns; developing and disseminating tools to ensure that urban trees and forests are strategically planned and managed to maximize ecosystem services and benefits; engaging partners and educators in conservation education and interpretive programs; developing methods to measure environmental literacy and techniques to engage urban residents in the management of urban forests; improving access by urban Americans to USFS resources and information; and developing partnerships with nontraditional partners to engage urban and underserved audiences.

The rapid and continuing growth of the WUI in both the eastern and western states is dramatically altering the strategic and tactical approaches to fire and forest management. Urban and urbanizing communities may need information on the changing dynamics of the surrounding wildland and urbanizing environment, as well as the need to manage the surrounding landscapes to reduce the risks from uncharacteristically severe wildfires, which are often related to drought and pest infestations. Urban and urbanizing communities' sense of place may have an important role in developing adaptation strategies for those environments.

3.3.4.7 Goal 7: Provide Science-based Applications and Tools for Sustainable Natural Resources Management

The outcome identified for this goal is that management decisions are informed by the best available science-based knowledge and tools. Means and strategies include developing and making available cost-effective methods for transferring scientific information, technologies, methods, and applications; providing information and science-based tools that are used by managers and policymakers; developing and implementing effective processes for engaging users in all phases of R&D study development; developing and deploying analysis and decision-support systems; developing tools for evaluating the efficiency and effectiveness of alternative management practices; and ensuring that current resource information is available to address the strategic, tactical, and operational business requirements of the USFS.

Under a changing climate, the need will arise for quantitative tools to address complex issues facing each forest and region, such as linkages between ecosystems; water resources; disturbances, including drought, fire, infestation and disease; regional migration patterns, including invasions of both native and exotic species; and local to regional carbon storage and carbon management, such as for biofuels. This goal will be sensitive to the impacts of a changing climate on ecosystems and the needs of resource managers.

Climate Change and Science-based Applications and Tools

As with any natural resource management issue, resource managers need access to current scientific information, qualitative/quantitative tools to use in decision support analyses at forest and project planning levels, and management strategies to guide on-the-ground

CASE STUDY SUMMARY 3.3

Uwharrie National Forest, North Carolina
Southeast United States

Why this case study was chosen

The Uwharrie National Forest:

- Consists of 61 separate parcels, intermingled within private land;
- Supports a wide variety of ecosystem services, including one of the greatest concentrations of archeological sites in the Southeast;
- Is currently seeing an increased demand for recreational opportunities associated with camping, hiking, fishing, boating, and hunting;
- Expects the regional changes in land use and population to amplify the challenges already faced by forest managers;
- Is in the process of incorporating climate change considerations into a revised forest plan.

Management context

The Uwharrie National Forest (UNF) consists of 61 separate fragments that provide key ecosystem services—recreation, fresh water, wildlife habitat, and wood products—to millions of people because of the UNF's close proximity to several major cities. This combination of fragmentation and high demand for goods and services already poses unique forest management challenges, which are expected to become more difficult as the regional population increases over the next 40 years. For example, climate change is expected to significantly affect regional water reserves, including Badin Lake, one of the largest water bodies in the region. Much of the area had been converted from drought and fire-resistant tree species to faster growing but less resistant tree species over the past 60 years. Conversion back to original vegetation is now under consideration in response to climate change.

Key climate change impacts

- Projected increase in wildfire risk and concerns about sustaining forest productivity;
- Projected increase in water shortages as biological and anthropogenic demand increases and supply decreases;
- Expected increase in soil erosion and stream sedimentation due to projected increase in frequency of intense storms;
- Projected increase in insect outbreaks due to longer growing season and drier forest conditions.

Opportunities for adaptation

- Re-establishment of more fire- and drought-tolerant longleaf pine through selective forest management and replanting could provide increased resistance to potential future drought and unusually severe wildlife events.
- Restoration of historical sites of longleaf pine savannas on the UNF through logging or controlled burning would result in reduced forest water use, water stress, wildfire fuel loads, and wildfire risk as the region continues to warm;
- Opportunities to relocate trails farther from streams, and thus increase the size of stream buffer zones, could minimize soil erosion and stream sedimentation under conditions of increasing storm intensity;
- Opportunities to engage in a dialogue with surrounding landowners on wildfire management might encourage clearing and removal of fuels around buildings and dwellings, and thus minimize risks to property and lives from the expected increase in wildfires within the landscape mosaic containing the UNF and these landowners.

Conclusions

Even without climate change, management of the UNF is a complex task. Continued increases in population and fragmentation of the landscape will only be compounded by climatic change and variability. While an extensive and well-maintained road network across the forest provides excellent access for wildfire suppression, and the patchy nature of the forest also helps to isolate fires, ecosystem services on the UNF are influenced by activities on the surrounding highly fragmented landscape. The forest's proximity to population centers increases the UNF's visibility and raises the public's awareness of the need for management action to mitigate negative impacts. The UNF could serve as a valuable example for other land managers on how forests can be managed to reduce climate change impacts through the modification of established forest management strategies and tools.

management. Scientific information is scattered across websites, scientific journals, regional assessments, government documents, and international reports, challenging attempts by resource managers to compile the best available information. At present, most established planning and operational tools within NFs, such as the Forest Vegetation Simulator, assume that climate will continue to reflect the historical climate. No climate information or dynamics are included in many of the currently available planning tools. Recognition that climate is an important element in natural resource management is beginning to occur in some of the natural resource management communities such as water resource planning. However, few analytical tools are available to incorporate uncertainty analyses into resource planning.

3.4 ADAPTING TO CLIMATE CHANGE

3.4.1 The Need for Anticipatory Adaptation

Climate is constantly changing at a variety of time scales, prompting natural and managed ecosystems to adjust to these changes. As a natural process, without human intervention, adaptation typically refers to the autonomous and reactive changes that species and ecosystems make in response to environmental change such as a climate forcing (Kareiva, Kingsolver, and Huey, 1993; Smit *et al.*, 2000; Davis and Shaw, 2001; Schneider and Root, 2002). Organisms respond to environmental change (including climate change) in one of three ways: adaptation, migration, or extinction. Adaptation typically refers to genetic changes, but also includes in situ acclimation (physiological adaptation to the changing environment while remaining in place) as well as phenological (*e.g.*, breeding, flowering, migration) and behavioral changes. This natural adaptation in the ecosystem is important to understand, so that the influence of management on these natural processes can be assessed. Space for evolutionary development under climate change may be important to incorporate into conservation and restoration programs under a changing climate (Rice and Emery, 2003).

We focus on adaptation as interventions and adjustments made by humans in ecological, social, or economic systems in response to climate stimuli and their effects, such as fire, wind damage, and so on. More specifically, in the social-science literature, the term adaptation refers to "a process, action, or outcome in a system (household, community, sector, region, country) in order for the system to better cope with, manage or adjust to some changing condition, stress, hazard, risk or opportunity" (Smit and Wandel, 2006).

Human adaptation to climate change impacts is increasingly viewed as a necessary complementary strategy to mitigation—reducing greenhouse gas emissions from energy use and land use changes in order to minimize the pace and extent of climate change (Klein *et al.*, 2007). Because adaptive strategies undertaken will have associated effects on carbon dynamics, it is important to consider carbon impacts of any proposed adaptive strategy. Forest management practices designed to achieve mitigation goals of reducing greenhouse gases (CO_2 in particular) are diverse, and have large potential mitigation contributions on the global to regional scales (Malhi, Meir, and Brown, 2002; Krankina and Harmon, 2006). Options for minimizing return of carbon to the atmosphere include storing carbon in wood products (Wilson, 2006), or using biomass as bioenergy, both electrical and alcohol-based. While many positive opportunities for carbon sequestration using forests appear to exist, evaluating specific choices is hampered by considerable difficulty in quantifying net carbon balance from forest projects (Cathcart and Delaney, 2006), in particular unintentional emissions such as wildfire and extensive forest mortality from insects and disease (Westerling *et al.*, 2003; Westerling and Bryant, 2005; Westerling *et al.*, 2006; Lenihan *et al.*, 2006). Adaptation and mitigation can have positive and negative influences on each other's effectiveness (Klein *et al.*, 2007). Management practices that lower vulnerabilities to uncharacteristically severe wildfire and non-fire mortality could meet multiple goals of mitigation and adaptation if such practices also reflected goals for other ecosystem services. Both strategies—adaptation and mitigation—are needed to minimize the potential negative impacts, and to take advantage of any possible positive impacts from climate variability and change

(Burton, 1996; Smit *et al.*, 2001; Moser *et al.*, forthcoming).

Several concepts related to adaptation are important to fully appreciate the need for successful anticipatory adaptation to climate-related stresses, as well as the opportunities and barriers to adaptation. The first of these is vulnerability. Vulnerability is typically viewed as the propensity of a system or community to experience harm from some stressor as a result of (a) being exposed to the stress, (b) its sensitivity to it, and (c) its potential or ability to cope with and/or recover from the impact (see review of the literature by Adger, 2006). Key vulnerabilities can be assessed by exploring the magnitude of the potential impacts, the timing (now or later) of impacts, the persistence and reversibility (or irreversibility) of impacts, the likelihood of impacts and confidence of those estimates, the potential for adaptation, the distributional aspect of impacts and vulnerabilities (disadvantaged sectors or communities), and the importance of the system at risk (Schneider *et al.*, 2007). Of particular importance here is a system's adaptive capacity: the ability of a system or region to adapt to the effects of climate variability and change. How feasible and/or effective this adaptation will be depends on a range of characteristics of the ecological system, such as topography and micro-refugia, soil characteristics, biodiversity; pre-existing stresses, such as the presence of invasive species or loss of foundation species or fragmentation of the landscape; the status of the local ecosystem, *e.g.*, early to late successional and its intrinsic "inertia" or responsiveness; and on characteristics of the social system interacting with, or dependent on, the ecosystem (Blaikie *et al.*, 1994; Wilbanks and Kates, 1999; Kasperson and Kasperson, 2001; Walker *et al.*, 2002; Adger, 2003).

As Smit and Wandel (2006) state in their recent review, "Local adaptive capacity is reflective of broader conditions (Yohe and Tol, 2002; Smit and Pilifosova, 2003). At the local level, the ability to undertake adaptations can be influenced by such factors as managerial ability; access to financial, technological, and information resources; infrastructure; the institutional environment within which adaptations occur; political influence, etc. (Blaikie, Brookfield, and Allen, 1987; Watts and Bohle, 1993; Adger, 1999; Handmer, Dovers, and Downing, 1999; Toth, 1999; Adger and Kelly, 2001; Smit *et al.*, 2001; Wisner *et al.*, 2004)." Adaptive capacity is determined mainly by local factors (*e.g.*, local forest managers' training in ecological processes, available staffing with appropriate skills, available financial resources, local stakeholder support) while other factors reflect more general socioeconomic and political systems (*e.g.*, federal laws, federal forest policies and regulations, state air quality standards, development pressures along the forest/urban interface, commodity market (timber, grazing) conditions, stakeholder support).

While the literature varies in the use of these and related concepts such as resilience and sustainability, adaptation in the context of NF management would be viewed as successful if stated management goals (see Section 3.3) were continued to be achieved under a changing climate regime while maintaining the ecological integrity of the nation's forests at various scales. For example, Section 3.3 identified the close relationship between ecosystem services and management goals, and their sensitivity to climate change. While these stated management goals are periodically updated or modified, this re-examination entails a risk of setting goals lower (*e.g.*, lower quality, quantity, or production) as environmental and climatic conditions deteriorate. For the purposes of this report it is assumed that the larger tenets of the cumulative laws directing NF management remain intact: "the greatest good of the greatest number in the long run…without impairment of the productivity of the land…[and] secure for the American people of present and future generations."

Below, we distinguish different adjustments of NF management approaches by reference to timing and intention. By "timing" we mean when the managing agency thinks about a management intervention: after a climate-driven, management-relevant event, or in advance of such an event. By "intention" we mean whether the managing agency acknowledges that a change is likely, anticipates possible impacts, and begins planning for a response prior to it occurring—for example, developing a monitoring or early warning system to detect changes as they occur (see Fig. 3.12). We distinguish three different adaptation scenarios: no active adaptation;

planned management responses to disturbances associated with changing climate regimes; and management responses in anticipation of future climate change, and in preparation for climate change now.

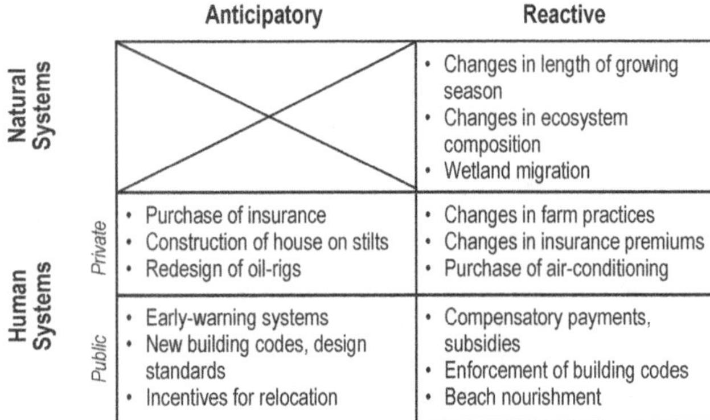

Figure 3.12. Anticipatory and reactive adaptation for natural and human systems (IPCC, 2001b).

3.4.1.1 No Active Adaptation

An approach of "no active adaptation" could describe two decision-making pathways. The event- or crisis-driven approach reacts to a climate or related environmental stimulus, without foresight and planning. No active adaptation could also result from the approach where consideration of the potential effects of climate change and management investment result in a conscious decision not to manage for climate change. The first approach would be without anticipatory planning, whereas the second, appearing as no active adaptation, would involve consideration of vulnerabilities and impacts. These reactions could be at any level of policy- or decision-making—national, regional, forest planning level, or project level.

The extent and severity of an extreme weather or climate event vis-à-vis the ecosystem's ability to naturally adjust to or recover from it, as well as the management agency's ability to quickly marshal the necessary response resources (money, staff, equipment, etc.) when the event occurs, will determine the ultimate impacts on the ecosystem and the cost to the managing agency. Depending on the extent of the impacts on the ecosystem and on the managing agency, future attainment of management goals may also be affected. While unforeseen opportunities may emerge, the cost of such unplanned reactive management is typically larger than if management tools can be put in place in a timely and efficient manner (a common experience with reactive vs. proactive resource or hazard management, *e.g.,* Tol, 2002; Multihazard Mitigation Council, 2006).

This reactive approach, which does not take into account changing climate conditions, is sometimes used when scientific uncertainty is considered too great to plan well for the future. There is a strong temptation to not plan ahead, because it avoids the costs and staff time needed to prepare for an event that is uncertain to occur. The risk to the agency of initiating expensive and politically challenging management strategies is large in the absence of a strong scientific consensus on vulnerabilities and climate change effects. However, not planning ahead also can mean incurring greater cost, and may bring with it great risk later on—risk that results from inefficiencies in the response when it is needed, wasted investments made in ignorance of future conditions, or potentially even greater damages because precautionary actions were not taken.

The reactive approach would also reflect a management philosophy that does not consider the likelihood of climate-driven changes and impacts. Most past forest planning documents typically described a multi-decadal future without climate variability or change. While the development of the National Fire Plan is an example of planning for increasingly challenging wildfires in a cost-efficient manner, the influence of climate change on wildfire is not considered. Addressing climate change in wildland fire management could include setting up pathways for information-sharing and coordination of climate change adaptation strategies of wildland fire agencies; considering climate change and variability when developing long-range wildland fire management plans and strategies; and incorporating the likelihood of more severe fire weather, lengthened wildfire seasons, and larger-sized fires when planning and allocating budgets.[28] Most management

[28] **National Association of State Foresters,** 2007: NASF Resolution No. 2007-1. Issue of Concern: The role that climate change plays in the severity and size of wildland fires is not explicitly recognized in the "National Fire Plan" and the Implementation Plan for its 120-year Strategy. htt://www.stateforesters.org/resolution/2007-01.pdf.

strategies or practices (*e.g.*, natural regeneration or cold-water fisheries restoration) assume a relatively constant climate or weather pattern. A careful study of the historical range of natural variability provides a wealth of information on ecological process—how diverse and variable past plant community dynamics have been (Harris *et al.*, 2006). However, pre-settlement patterns of vegetation dynamics (*e.g.*, a point in time such as the mid-1800s, the end of the so-called Little Ice Age) are associated with a climate that was much cooler, and may not adequately reflect the current climate or an increasingly warmer future climate and the associated vegetation dynamics. Many quantitative tools currently used do not include climate or weather in their dynamics. Growth and yield models, unmodified by growth and density control functions (Dixon, 2003), project forest growth without climate information. The past climate may not be an adequate guide to future climate (Williams, Jackson, and Kutzbach, 2007), and our understanding of the ecological assumptions underlying restoration management practices may also need to be revisited (Harris *et al.*, 2006).

An approach of no active adaptation could also result from consideration of the potential for climate change, and a conscious decision to not prepare for or adapt to it. Examples could include low-sensitivity ecosystems, short-term projects, or a decision to triage. For low-sensitivity ecosystems, vulnerability is low or the likely impacts of climate change are very low probability, or the effects of climate change are not undesired. Existing projects nearing completion, such as high-value short-rotation timber that is about to be harvested, could be considered not critical to prepare for climate change, assuming that the harvest will occur before any major threat of climate change or indirect effects of climate change emerge. The risk is deemed low enough to continue with current management. And finally, the decision to not manage for a particular species would reflect a strategy of no active adaptation. Most prioritizing methods rank all options with varying priorities. In contrast, proper and systematic triage planning includes the necessary option of not treating something that could/should be treated if more resources (time, money, staff, technology) were available. Issues needing treatment are relegated untreatable in triage planning when greater gain will ensue by allocating scarce resources elsewhere; i.e., in emergency situations where resources for treatment are limited, one cannot treat everything. Thus, conscious decisions are made for no action or no management.

Major institutional obstacles or alternative policy priorities can also lead to inattention to changing climatic and environmental conditions that affect land and resource management. Moreover, sometimes this approach is chosen unintentionally or inadvertently when climatic conditions change in ways that no one could have anticipated. Or, even if a "no action" plan is taken for the short run—say in anticipation of an impending harvest—the post-harvest plan may also inadvertently not take rapidly changing climate conditions into account for the "regeneration" of the next ecosystem.

3.4.1.2 Planned Management Responses to Changing Climate Regimes, Including Disturbances and Extreme Events

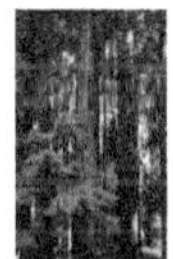

This approach to adaptation assumes that adjustments to historical management approaches are needed eventually, and are best made during or after a major climatic event. In this case, the managing agency would identify climate-change-cognizant management approaches that are to be implemented at the time of a disturbance, as it occurs, such as a historically unprecedented fire, insect infestation, or extreme windfall event, hurricanes, droughts and other extreme climatic events. A choice is made to not act now to prepare for climate change, but rather to react once the problem is evident. The rationale, again, could be that the climate change impacts are too uncertain to enact or even identify appropriate anticipatory management activities, or even that the best time for action from a scientific as well as organizational efficiency standpoint may be post-disturbance (*e.g.*, from the standpoint of managing successional processes within ecosystems and across the landscape).

For example, forest managers may see large disturbances (fire, flooding, insects, hurricanes) as opportunities to react to climate change. Those disturbances could be windows of opportunity for implementing adaptive practices, such as adjusting the size of management units to capture whole watersheds or landscapes, developing a prescribed fire plan for the post-fire treated landscape, addressing road and culvert needs to handle changes in erosion under climate change, revisiting objectives for even-age versus uneven-age management, reforesting with species tolerant to low soil moisture and high temperature, using a variety of genotypes in the nursery stock, and moving plant genotypes and species into the disturbed area from other seed zones. For example, where ecosystems move toward being more water-limited under climate change, populations from drier and warmer locations will be more resistant to such changing conditions. In practice, this typically means using trees from provenances that are farther south or at lower elevation than what is currently indicated for a particular geographic location (Ying and Yanchuk, 2006). Because local climate trends and variability will always be uncertain, managers can hedge their bets by managing for a variety of species and genotypes with a range of tolerances to low soil moisture and higher temperatures. In general, genetic diversity provides resilience to a variety of environmental stressors (Moritz, 2002; Reed and Frankham, 2003; Reusch *et al.*, 2005).

Furthermore, disturbed landscapes could be used as experiments in an adaptive management context that provide data for evaluating and improving approaches to adapt ecosystems to a warmer climate. An example may be to reforest an area after a fire or windfall event with a type of tree species that is better adjusted to the new or unfolding regional climate. This may be difficult to achieve, because the climate that exists during the early years of tree growth will be different from those that will persist during the later stages of tree growth.

Significant cost efficiencies, relative to the unplanned approach, may be achieved in this approach, as management responses are anticipated—at least generically—well in advance of an event, yet are implemented only when "windows of opportunity" open. Future constraints to implementing such changes will need to be anticipated and planned for, and, if possible removed in advance for timely adaptation to be able to occur when the opportunity arises. For example, managers could ensure that the genetic nursery stock is available for wider areas, or they could re-examine regulations restricting practices so that, immediately after a disturbance, management can act rapidly to re-vegetate and manage the site. Such an approach may be difficult to implement, however, as crises often engender political and social conditions that favor "returning to the status quo" that existed prior to the crisis rather than doing something new (*e.g.,* Moser, 2005).

3.4.1.3 Management Responses in Anticipation of Future Climate Change and in Preparation for Climate Change Now

The management approach that is most forward-looking is one that uses current information about future climate, future environmental conditions, and the future societal context of NF management to begin making changes to policy and on-the-ground management now and when future windows of opportunity open. Opportunities for such policy and management changes would include any planning or project analysis process in which a description of the changing ecosystem/disturbance regime as climate changes would be used to identify a proactive management strategy.

Relevant information for forest managers may include projections of regional or even local climates, including changes in average temperature, precipitation, changes in patterns of climatic extremes and disturbance patterns (*e.g.,* fire, drought, flooding), shifts in seasonally important dates (*e.g.,* growing degree-days, length of fire season), expected future distribution of key plant species, and changes in hydrological patterns. The ability of climate science to provide such information at higher spatial and temporal resolution has been improving steadily over recent years, and is likely to improve further in coming years (IPCC, 2007). Current model predictions have large uncertainties, which must be considered in making management adaptation decisions (see Sections 3.4.2.1 and 3.4.2.2 for other treatments of uncertainty). Other relevant information may be species-specific, such as the climatic conditions favored by certain plant or animal species over others, or the ways in which changed climatic conditions and the resultant habitats may become more or less favorable to particular species (*e.g.,* for threatened or endangered species). The overall goals of planned anticipatory management would be to facilitate adaptation in the face of the changing climate.

For example, based on the available information, large-scale thinnings might be implemented to reduce stand densities in order to minimize drought effects, avoid large wildfire events in areas where these are not typical, and manage the potential for increased insect and disease outbreaks under a changing climate. Widely spaced stands in dry forests are generally less stressed by low soil moisture during summer months (*e.g.*, Oliver and Larson, 1996). Disease and insect concerns are at least partially mitigated by widely spaced trees, because trees have less competition and higher vigor. Low canopy bulk densities in thinned stands, with concurrent treatments to abate surface fuels, can substantially mitigate wildfire risk (Peterson *et al.*, 2005). However, not all forest landscapes and stands are amenable to thinning, nor is it ecologically appropriate in some upper-elevation forest types. In these situations, shelterwood cutting that mitigates extreme temperatures at the soil surface can facilitate continued cover by forest tree species while mitigating risks of uncharacteristically severe fire, insects, and disease (Graham *et al.*, 1999). Again, it will be important to assess the tradeoffs between these silvicultural benefits and potential for genetic erosion resulting from the shelterwood treatment (Ledig and Kitzmiller, 1992). This approach is economically feasible in locations where wood removed through thinnings and shelterwood cuttings can be marketed as small-dimensional wood products or biomass (Kelkar *et al.*, 2006). To identify and provide the most relevant information to support such an anticipatory approach to adaptation, it is critical that scientists and managers work together to form a growing mutual understanding of information needs and research capabilities in the context of ongoing, trusted relationships (Slovic, 1993; Earle and Cvetkovich, 1995; Cash, 2001; Cash *et al.*, 2003; Cash and Borck, 2006; Vogel et al., forthcoming).[29] Further examples of such information needs are described in the next section and in the case studies (see Case Study Summaries and Annex A1).

[29] See also **Tribbia**, J. and S.C. Moser, in press: More than information: what California coastal managers need to prepare for climate change. *Environmental Science & Policy.*

Again, significant cost efficiencies and maybe even financial gains may be achieved in this approach, as management responses are anticipated well in advance and implemented at the appropriate time. If climatic changes unfold largely consistent with the scientific projections, this approach to adaptation may turn out to be the most cost-effective and ecologically effective (referred to as the "perfect foresight" situation by economists; see *e.g.*, Sohngen and Mendelsohn, 1998; Mastrandrea and Schneider, 2001; Yohe, Andronova, and Schlesinger, 2004). For example, analyses using forest sector economic models that assume "perfect foresight" have shown that when a diverse set of management options are available to managers under conditions of extensive mortality events from climate change, the economic impacts on the wood product sector, even with large-scale mortality events, are less costly than otherwise (Sohngen and Mendelsohn, 1998; Joyce, 2007).

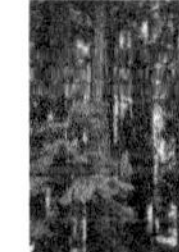

This approach may not be able to maintain ecosystems that currently exist (as those are better adapted to current climate regimes), but it may be best suited to support natural adaptive processes—such as planning corridor development to facilitate species migration to more appropriate climates, or managing for protection of viable habitats for threatened and endangered species to enhance or extend opportunities for adaptation (see Section 3.4.3.3). Under such a management approach, the specific management targets—such as outputs of particular rangeland and forest products, or maintenance of a particular species habitat—may themselves be adjusted over time, as the opportunities for those ecosystem services diminish under a changing climate and new opportunities for other services may have a greater chance of being met. The inability to maintain ecosystems that currently exist may suggest activities such as long-term seed bank storage with future options for re-establishing populations in new and more appropriate locations. Assessing the potential for this type of change will draw on ecological, economic, and social information. Importantly, such an approach would need to involve managers at various levels to monitor changes in the ecosystem (i.e., observed on the ground); coordinate and make appropriate changes in policies, regulations, plans, and programs at all relevant scales; and modify the on-the-ground practices needed to implement these higher-level policies. This degree of cross-scale integration is not typically achieved at present, and would need to occur in the future to effectively support such an approach to adaptation. Additionally, such considerations would need to involve the public, as well as stakeholders dependent upon the ecosystem services from NFs. On the local scale, the importance of establishing relationships with existing community organizations early on in a wildfire incident was identified in order to incorporate local knowledge into firefighting and rehabilitation efforts (Graham, 2003). This coordination was also important to establish a recovery base that continues once emergency personnel and resources have left the community. These partnerships should be developed as early as possible during the fire, and perhaps might best be developed before any fire in order to systematize actions, increase efficiency, and decrease potential contentions between locals and federal agencies by building trust (Graham, 2003). Lessons learned in integrating fire management across local to state to federal agencies may help in similar considerations of cross-scale integration of resource managers to address current and future resource management under a changing climate.

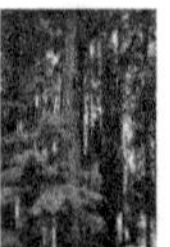

3.4.2 Approaches for Planning in the Context of Climate Change

3.4.2.1 Use of Models and Forecasting Information

Many forest managers are awaiting information from quantitative models about future climates and environments to guide climate-related planning. Increasingly sophisticated models are being developed at regional and finer spatial scales. In general, while model information will be important for planning, the best use of this information at local and regional scales currently is to help organize thinking, attain insight into the nature of potential processes, and understand qualitatively the range of magnitudes and likely direction and trends of possible future changes. Focusing on results that are similar across diverse models may indicate results of greater likelihood.

While science is progressing, uncertainty about climate projections are much greater at the local and regional scales important to land managers, because uncertainties amplify as data and model output are downscaled. Some climate parameters, such as changes in average annual temperature, may be more robust than others, such as changes in annual precipitation, which have higher uncertainties associated with them. Augmenting this uncertainty in physical conditions is the difficulty of modeling biological responses. Ecological response to climate-related changes is highly likely to be more difficult than climate to model accurately at local scales, because threshold and non-linear responses, lags and reversals, individualistic behaviors, and stochastic (involving probability) events are common (Webb, III, 1986; Davis, 1989). Models typically rely on directional shifts following equilibrium dynamics of entire plant communities (or, physiognomic community types), whereas especially in heterogeneous and mountainous regions, patchy environments increase the likelihood of complex, individualistic responses.

At the global scale, this uncertainty is dealt with through simultaneous analysis of multiple scenarios (IPCC, 2007), which yields a wide range of potential future climate conditions. Similarly, approaches at finer spatial scales could be developed to use scenario analysis (Peterson, Cumming, and Carpenter, 2003; Bennett *et al.*, 2003) (alternative future climate scenarios can be used to drive ecosystem and other natural resource models), thus examining the possible range of future conditions. Scenario analysis can help to identify potential management options that could be useful to minimize negative impacts and enhance the likelihood of positive impacts, within the range of uncertainty.

Uncertainty does not imply a complete lack of understanding of the future or a basis for a no action decision. Managing in the face of uncertainty will best involve a suite of approaches, including planning analyses that incorporate modeling with uncertainty, and short-term and long-term strategies that focus on enhancing ecosystem resistance and resilience, as well as actions taken that help ecosystems and resources move in synchrony with the ongoing changes that result as climates and environments vary.

3.4.2.2 Planning Analyses for Climate Change

RPA Assessment

The only legislatively required analysis with respect to climate change and USFS planning was identified in the 1990 Food Protection Act, which amended the 1974 Resources Planning Act (RPA). The 1990 Act required the USFS to assess the impact of climate change on renewable resources in forests and rangelands, and to identify the rural and urban forestry opportunities to mitigate the buildup of atmospheric CO_2. Since 1990, the RPA Assessments (*e.g.*, USDA Forest Service, 1993; USDA Forest Service, 2000; USDA Forest Service, forthcoming) have included an analysis of the vulnerability of U.S. forests to climate change, and the impact of climate change on ecosystem productivity, timber supply and demand, and carbon storage (Joyce, Fosberg, and Comanor, 1990; Joyce, 1995; Joyce and Birdsey, 2000; Haynes *et al.*, 2007). These analyses have identified several important aspects of the analysis of climate change impacts on the forest sector. Transient analyses, where annual dynamics are followed throughout the projection period, allow interactions between ecosystem responses to climate change and market responses to identify adaptation options to the changing climate. The forest sector trade at the global scale can influence the forest sector responses (price as well as products) within countries. National level analyses aggregate impacts across regions, and it remains important to identify the regional response, which may be greater, because that is where management decisions will be made (Joyce, 2007). Most critically, all of these analyses have stressed the importance of evaluating the ecological and the economic response in an integrated fashion

Adaptation strategies may vary based on the spatial and temporal scales of decision making. Planning at regional or national scales may involve acceptance of different levels of uncertainty and risk than appropriate at local (*e.g.*, NF or watershed) scales. National analyses associated with RPA offer the opportunity to develop potential approaches to link assessments at the national, regional, multi-forest, and NF scales. Such an approach could involve key questions, methods of assessment, approaches to uncertainty and risk, needed expertise and resources, responsibilities and timelines, and identification of spatial and temporal scales for modeling linked to decision making. The assessment would consider how vulnerabilities and sensitivities within these systems might be identified, given the available information, as well as identifying situations of high resilience to climate change or situations where the climate change effects might be locally buffered. Significant involvement by scientists, managers, policymakers, and stakeholders from local to national levels would be critical. Such a linked assessment could guide NFs and their partners in terms of a process to assess the impacts of climate change on natural resources and ecosystem services within their boundaries, across their boundaries, and at larger spatial scales such as regional and national.

Forest Planning and Project Analyses

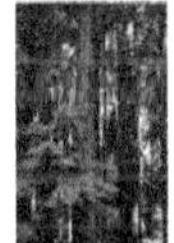

The following planning steps have been suggested as appropriate in a climate-change context when beginning a project (Spittlehouse and Stewart, 2003; see examples therein):

1. Define the issue (management situation, goals, and environmental and institutional contexts);
2. Evaluate vulnerabilities under changing conditions;
3. Identify suitable adaptive actions that can be taken at present or in the short term; and
4. Develop suitable adaptive actions that could be taken in the longer term.

In a survey of the forest plans available online in December 2006, 15 plans from a total of 121 individual forests had included references to climate change (terms "climate change," "climate variability," or "global warming") in the sections of the plan describing trends affecting management or performance risks, or, in earlier plans, as a concern in the environmental impact statement; both of these types of references are similar to Step 2 above (evaluating vulnerabilities).

Given the challenges of the uncertainty in climate scenarios at fine spatial scale (Section 3.4.2.1), a set of assumptions to be considered in planning has been proposed.[30] Specifically, the recommendations make use of an adaptive management approach to make adjustments in the use of historical conditions as a reference point. Flexibility to address the inherent uncertainty about local effects of climate change could be achieved through enhancing the resiliency of forests, and specific aspects of forest structure and function are mentioned (Box 3.6). These assumptions would allow the plan components to be designed in a way that allows for adaptability to climate change, even though the magnitude and direction of that change is uncertain. The assumptions to be examined (listed in Box 3.6) explore underlying premises about climate and climate change in the management processes.

One information-gathering option to help define the underlying assumptions and vulnerabilities to climate change might be to consider convening a science-based (*e.g.*, USFS research team) rapid assessment or "audit" of existing forest planning documents (e.g., the Forest Land Management Plan, or larger plans such as the Sierra Nevada Forest Plan amendment or the Northwest Forest Plan, and project plans). The purpose of the audit would be to determine the level of climate adaptedness, pitfalls, and areas for improvement in current forest plans and operations. Such an audit could focus on current management direction (written policy); current management practices (implementation); and priorities of species (*e.g.*, specific targeted species) and processes (fire, insects/disease). The audit would highlight concrete areas of the plans and projects that are poorly adapted to potential changes in climate, as well as those that are already climate-proactive. Audit recommendations would identify specific areas where changes are needed, and where improvements in forest planning or project-level planning and management could be made.

Information and tools needed to assist adaptation form the basis for a long-term, management-science partnership continually refining scientific information for resource management decisions. A wide suite of modeling approaches that project climate change impacts on ecosystems are available (for example, Melillo *et al.*, 1993; Joyce and Birdsey, 2000;

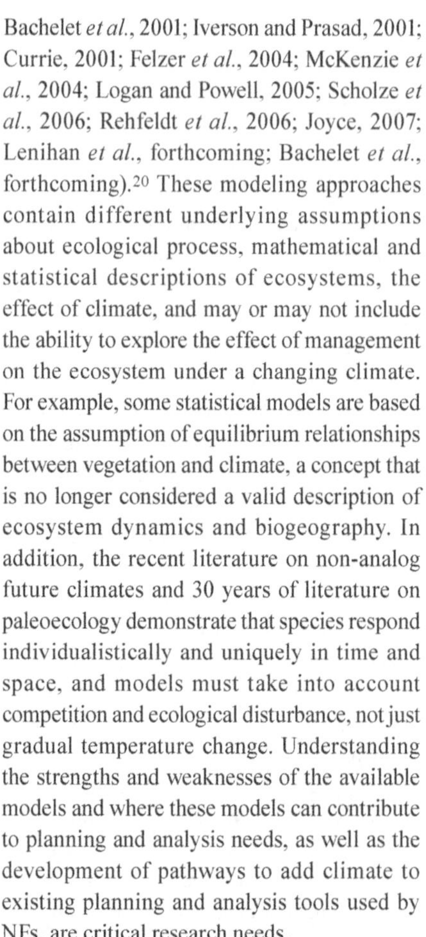

Bachelet *et al.*, 2001; Iverson and Prasad, 2001; Currie, 2001; Felzer *et al.*, 2004; McKenzie *et al.*, 2004; Logan and Powell, 2005; Scholze *et al.*, 2006; Rehfeldt *et al.*, 2006; Joyce, 2007; Lenihan *et al.*, forthcoming; Bachelet *et al.*, forthcoming).[20] These modeling approaches contain different underlying assumptions about ecological process, mathematical and statistical descriptions of ecosystems, the effect of climate, and may or may not include the ability to explore the effect of management on the ecosystem under a changing climate. For example, some statistical models are based on the assumption of equilibrium relationships between vegetation and climate, a concept that is no longer considered a valid description of ecosystem dynamics and biogeography. In addition, the recent literature on non-analog future climates and 30 years of literature on paleoecology demonstrate that species respond individualistically and uniquely in time and space, and models must take into account competition and ecological disturbance, not just gradual temperature change. Understanding the strengths and weaknesses of the available models and where these models can contribute to planning and analysis needs, as well as the development of pathways to add climate to existing planning and analysis tools used by NFs, are critical research needs.

In the short-term, natural resource managers could benefit from a manager's guide with current state-of-the art scientific concepts and techniques. Critical gaps in scientific understanding of the impacts of climate change, and of management on ecosystem services, hinder adaptation by limiting assessment of risks, efficacy, and sustainability of actions. Assistance and consultation on interpreting climate and ecosystem model output would provide the context and relevance of model predictions to be reconciled with managers' priorities for adaptation.

3.4.3 Approaches for Management in the Context of Climate Change

3.4.3.1 Toolbox of Management Approaches

A primary premise for adaptive approaches is that change, novelty, uncertainty, and uniqueness of individual situations are expected to define the planning backdrop of the future.

[30] **West**, 2005: *Letter and Attachments*. File Code 4070, letter dated July 26, 2005. Pacific Northwest Station.

BOX 3.6. Forest Planning Assumptions to Consider Regarding Climate Change. (Excerpted from West, 2005)[30]

Historic Conditions: We assume that historical conditions are a useful reference or point of comparison for current or future trends, in accord with the Healthy Forest Restoration Act, the 2005 planning rule, and LANDFIRE (and other national fire-related projects). However, we recognize that this assumption is likely to face substantial challenges as the effects of climate change on vegetation and disturbance regimes play out over the next several decades. Accordingly, an adaptive management approach can be used to test this assumption, make adjustments in the desired future condition, and plan goals and objectives as the local effects of climate change become apparent.

Flexibility and Considerations: Although climate and ecosystem forecast models have improved significantly, they cannot produce highly accurate local projections. Flexibility to address the inherent uncertainty about local effects of climate change could be achieved through enhancing the resiliency of forests by considering that:

- Diverse plantings will likely be more adaptable to changing conditions than will single species stands.
- Prescribed fire and thinning could be used to keep tree densities low to improve resistance to drought and pest infestations.
- Nitrogen-fixing species, intermixed in a stand, may facilitate regrowth after disturbance in a rapidly changing environment, although they may compete for water on droughty sites.
- Encouraging local industries that can adapt to or cope with variable kinds of forest products because of the uncertainty in which tree species will prosper under changed climate.
- Some vegetation types in vulnerable environments (e.g., ecotonal, narrow distribution, reliant on specific climate combinations, situations sensitive to insect/pathogens) will be highly sensitive to changes in climate and may undergo type conversions despite attempts at maintaining them (meadow to forest, treeline shifts, wetland loss). Some of these changes are likely to be inevitable.
- Reforestation after wildfire may require different species (i.e., diverse plantings, as mentioned above) than were present on the site pre-fire to better match site-type changes due to climate effects.
- Genetic diversity of planting stock may require different mixes than traditionally prescribed by seed zone guidelines.
- Massive forest diebacks may be clues to site transition issues.
- Behavior of invasive species is likely to be different as climates shift.
- Increasing interannual climate variability (e.g., dry periods followed by wet, as in alternating ENSO patterns) may set up increasingly severe fuels situations.
- Non-linear, non-equilibrium, abrupt changes in vegetation types and wildlife behavior may be more likely than linear, equilibrium, and gradual changes.
- Water supply and water quality issues might become critical, particularly if increased or prolonged drought or water quality changes are the local consequences of climate change.
- Carbon storage to reduce greenhouse gas and other effects might be important.

Adaptive Management: Effects due to climate change (e.g., wildfire severity/acreage trends, vegetation trends, insect and disease trends) may become more apparent as new information becomes available to NFs through regional or sub-regional inventories, data collection, and research. This information may be useful for adjusting desired conditions and guidelines as plans are implemented. Information of interest might include:

- The frequency, severity, and area trends of wildfire and insect/disease disturbances, stratified by environment
- The distribution of major forest types. For example, the lower and upper elevational limits of forests and woodlands might change as precipitation, temperature, and other factors change. These trends might be detected through a combination of permanent plots (e.g., Forest Inventory and Analysis plots) and remotely sensed vegetation data (e.g., gradient nearest neighbor analyses).
- Stream flow and other indicators of the forests' ability to produce water of particular quality and quantity.

Rapid changes that are expected in physical conditions and ecological responses suggest that management goals and approaches will be most successful when they emphasize ecological processes, rather than focusing primarily on structure and composition. Information needs (*e.g.*, projections of future climates, anticipated ecological responses) will vary in availability and accuracy at local spatial and temporal scales. Thus, strategic flexibility and willingness to work in a context of varying uncertainty will improve success at every level (Anderson *et al.*, 2003). Learning from experience and iteratively incorporating lessons into future plans—adaptive management in its broadest sense—is an appropriate lens through which natural-resource management is conducted (Holling, 2001; Noss, 2001; Spittlehouse and Stewart, 2003). Dynamism in natural conditions is appropriately matched by dynamic approaches to management and adaptive mindsets.

Given the nature of climate and environmental variability, the inevitability of novelty and surprise, and the range of management objectives and situations, a central dictum is that *no single approach will fit all situations* (Spittlehouse and Stewart, 2003; Hobbs *et al.*, 2006). From a toolbox of options such as those proposed below, appropriate elements (and modifications) should be selected and combined to fit the situation. Some applications will involve existing management approaches used in new locations, seasons, or contexts. Other options may involve experimenting with new practices.

A toolbox approach recognizes that strategies may vary based on the spatial and temporal scales of decision making. Planning at regional scales may involve acceptance of different levels of uncertainty and risk than appropriate at local (*e.g.*, NF or watershed) scales. The options summarized below fall under adaptation, mitigation, and conservation practices (Dale *et al.*, 2001; IPCC, 2001a). Based on the toolbox approach, an overall adaptive strategy will usually involve integrating practices that have different individual goals. An important consideration in building an integrative strategy is to first evaluate the various types of uncertainty: for example, uncertainty in present environmental and ecological conditions, including the sensitivity of resources; uncertainty in models and information sources about the future; uncertainty in support resources (staff, time, funds available); uncertainty in planning horizon (short- vs. long-term); and uncertainty in public and societal support. This evaluation would lead to a decision on whether it is best to develop reactive responses to changing disturbances and extreme events, or proactive responses anticipating climate change (see Section 3.4.1). The following options provide a framework for building management strategies in the face of climate change. Some examples of specific, on-the-ground, adaptation options are presented in Box 3.7 and are elaborated upon further in the sections that follow. Examples of institutional and planning adaptations, given in Box 3.8, are also elaborated upon further in the sections that follow.

3.4.3.2 Reducing Existing Stresses

The USFS implements a variety of management approaches to reduce the impact of existing stressors on NFs (see Section 3.3.3), and an increased emphasis on these efforts represents an important "no regrets" strategy. It is likely that the direct impacts of climate change on ecosystems and the effects of interactions of climate change with other major stressors may render NFs increasingly prone to more frequent, extensive, and severe disturbances, especially drought (Breshears *et al.*, 2005; Seager *et al.*, 2007), insect and disease outbreaks (Logan and Powell, 2001; Carroll *et al.*, 2004), invasive species, and wildfire (Logan and Powell, 2001; Brown, Hall, and Westerling, 2004; McKenzie *et al.*, 2004; Logan and Powell, 2005; Skinner, Shabbar, and Flanningan, 2006) (see also Section 3.3.2). The elevated water stress resulting from warmer temperatures in combination with greater variability in precipitation patterns and altered hydrology (*e.g.*, from less snowpack and earlier snowmelt, Mote *et al.*, 2005) would increase the frequency and severity of both droughts and floods (IPCC, 2001a). Air pollution can negatively affect the health and productivity of NFs, and the fragmented landscape in which many NFs are situated impedes important ecosystem processes, including migration. Efforts to address the existing stressors would address current management needs, and potentially reduce the future interactions of these stressors with climate change.

BOX 3.7. National Forest Adaptation Options.

- Facilitate natural (evolutionary) adaptation through management practices (e.g., prescribed fire and other silvicultural treatments) that shorten regeneration times and promote interspecific competition.
- Promote connected landscapes to facilitate species movements and gene flow, sustain key ecosystem processes (e.g., pollination and dispersal), and protect critical habitats for threatened and endangered species.
- Reduce the impact of current anthropogenic stressors such as fragmentation (e.g., by creating larger management units and migration corridors) and uncharacteristically severe wildfires and insect outbreaks (e.g., by reducing stand densities and abating fuels).
- Identify and take early proactive action against non-native invasive species (e.g., by using early detection and rapid response approaches).
- Modify genetic diversity guidelines to increase the range of species, maintain high effective population sizes, and favor genotypes known for broad tolerance ranges.
- Where ecosystems will very likely become more water limited, manage for drought- and heat-tolerant species and populations, and where climate trends are less certain, manage for a variety of species and genotypes with a range of tolerances to low soil moisture and higher temperatures.
- Spread risks by increasing ecosystem redundancy and buffers in both natural environments and plantations.
- Use the paleological record and historical ecological studies to revise and update restoration goals so that selected species will be tolerant of anticipated climate.
- Where appropriate after large-scale disturbances, reset succession and manage for asynchrony at the landscape scale by promoting diverse age classes and species mixes, a variety of successional stages, and spatially complex and heterogeneous vegetation structure.
- Use the paleological record and historical ecological studies to identify environments buffered against climate change, which would be good candidates for long-term conservation.
- Establish or strengthen long-term seed banks to create the option of re-establishing extirpated populations in new/more appropriate locations.

BOX 3.8. Examples of institutional and planning adaptations to improve the readiness of the USFS to cope with climate change.

- Rapidly assess existing USFS forest plans to determine the level of preparedness to climate change, examine underlying assumptions about climate, suggest improvements, and forge a long-term management-science partnership to continually refine information for resource management decisions.
- Anticipate and plan for more extreme events (e.g., incorporate likelihood of more severe fire weather and lengthened wildfire seasons in long-range fire management plans) that may lead to surprises and threshold responses and remove (if possible) future constraints to timely adaptive responses.
- Use climate and ecological models to organize thinking and understand potential changes in ecosystem processes, as well as the likely direction and magnitude of future climate trends and impacts, to explore adaptation options for climate change.
- Adjust management goals based on updated baseline conditions for species and ecosystems that have been significantly/cumulatively disturbed and are far outside of the historical range of variation.
- Use the federally mandated Resource Planning Assessment process to link assessments at the national, regional, and NF scales, and to provide guidance on assessing climate change impacts, uncertainty, vulnerability, and adaptation options.
- Coordinate with other agencies, as well as the private sector and other stakeholders, to reduce pollution and other landscape-scale anthropogenic stressors.

Drought has occurred across the United States in recent years, resulting most notably in large areas of forest mortality in the Southwest (see Section 3.3.2). Federal, state, and local governments, as well as private institutions, have drought management plans, but the National Drought Policy Commission Report (2000) stated that the current approach is patchy and uncoordinated. Climate change is likely to result in increased drought, with potential interactions with air quality and fire. Exposure to ozone may further exacerbate the effects of drought on both forest growth and stream health (McLaughlin *et al.*, 2007a; 2007b). Preparedness is an important element in reducing the potential impacts of drought on individuals, communities, and the environment. The development or refinement of drought plans that incorporate preparedness, mitigation, and response efforts would address the current stresses of drought, as well as begin to address potential adaptations to likely future droughts. Increased coordination among local, state, and federal government agencies on drought planning and drought-related policies (fire closures, recreation uses, and grazing management) would help in this regard. Coordination with the Bureau of Land Management, whose lands intermingle extensively with NF land, would be particularly beneficial. Enhancing the effectiveness of observation networks and current drought monitoring efforts would provide information on which to make management decisions, particularly in response to the impacts of drought on aquatic ecosystems, wildlife, threatened and endangered species, and forest health. Increased collaboration among scientists and managers would enhance the effectiveness of prediction, information delivery, and applied research, and would help develop public understanding of and preparedness for drought.

Invasive species are currently a problem throughout NFs, and disturbances such as fire, insects, hurricanes, ice storms, and floods create opportunities for invasive species to become established on areas ranging from multiple stands to landscapes. In turn, invasive plants alter the nature of fire regimes (Williams and Baruch, 2000; Lippincott, 2000; Pimentel *et al.*, 2000; Ziska, Reeves, and Blank, 2005)[12] as well as hydrological patterns (Pimentel et al., 2000), in some cases increasing runoff, erosion, and sediment loads (*e.g.*, Lacey, Marlow, and Lane, 1989). Potential increases in these disturbances under climate change will heighten the challenges of managing invasive species. Early detection/rapid response (EDRR, see Section 3.3.3) focuses on solving small problems before they become large, unsolvable problems, and recognizes that proactive management is more effective than long delays in implementation. The Olympic Land Management Plan, for example, recognizes that invasive species often become established in small, treatable patches, and are best addressed at early stages of invasion. Although designed for invasives, this EDRR approach may also be appropriate for other types of disturbances, because it could allow managers to respond quickly to the impacts of extreme events (disturbances, floods, windstorms, insect outbreaks), with an eye toward adaptation.

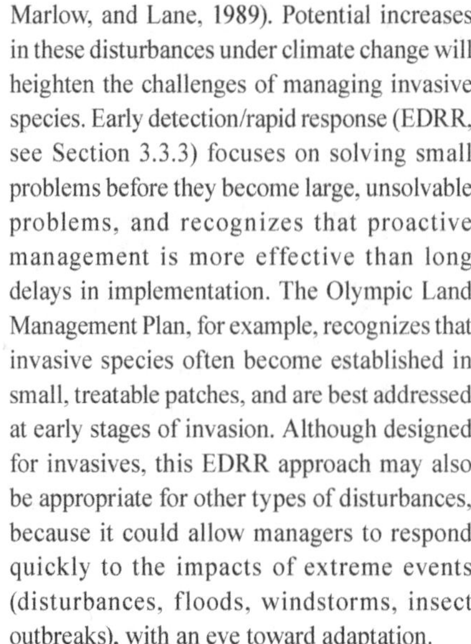

The USFS allocates considerable resources toward wildfire management (see Section 3.3.3). The projected increase in frequency, severity, and extent of fire under climate change is also likely to affect watershed condition, soil quality, erosional processes, and water quantity and quality in NFs (Wagle and Kitchen, Jr., 1972; Neary *et al.*, 1999; Spencer, Gabel, and Hauer, 2003; Certini, 2005; Guarin and Taylor, 2005; Neff, Harden, and Gleixner, 2005; Neary, Ryan, and DeBano, 2005; Murphy *et al.*, 2006; Deluca and Sala, 2006; Hauer, Stanford, and Lorang, 2007).

The National Fire Plan describes a wide variety of approaches to manage wildfire, the most prominent of which is hazardous fuels reduction. Fuel abatement approaches include prescribed fire, wildland fire use (see Section 3.3.3), and various mechanical methods such as crushing, tractor and hand piling, tree removal (to produce commercial or pre-commercial products), and pruning. Incorporation of additional climate information into fire management and planning may enhance current efforts to address wildfires.[31]

Air pollution from a variety of sources decreases forest productivity, diminishes watershed

[31] **National Association of State Foresters**, 2007: NASF Resolution No. 2007-1.Issue of Concern: The role that climate change plays in the severity and size of wildland fires is not explicitly recognized in the "National Fire Plan" and the Implementation Plan for its 120-year Strategy. http://www.stateforesters.org/resolution/2007-01.pdf.

condition, and deleteriously affects aquatic and terrestrial food webs in NFs (see Section 3.3.2). Although droughts and fires within NFs affect air quality, the USFS actively seeks to directly reduce these stressors and their impacts. In contrast, reducing the deposition of pollutants originating from outside NFs is beyond the agency's control, and thus the USFS mainly works to mitigate the impacts of these stressors. To directly reduce these stressors, the USFS would need to increase coordination with other agencies (federal, state, and local) and the private sector.

Efforts to reduce fragmentation and land use change near NFs by creating habitat corridors, increasing the size of management units, and identifying high-value conservation lands outside of NFs that could be managed in a coordinated way with the USFS will yield ecological benefits regardless of climate change. Large, connected landscapes will be even more critical as native species attempt to migrate or otherwise adapt to climate change. As is the case with air pollution, reducing these stressors with this approach will require increased coordination across federal, state, and local agencies as well as with private landowners.

One of the legacies of past management in NFs (see Section 3.3.2.1) is the presence of large landscapes consisting of even-aged stands, which are vulnerable to large-scale change by fire, insects, disease, and extreme weather events and their interactions. Management that emphasizes diverse, uneven age stands will benefit many NF ecosystems regardless of climate change. This approach would also likely enhance ecosystem resilience to climate change.

3.4.3.3 Adaptation Options

Forestalling Ecosystem Change

Create Resistance to Change

Notwithstanding the importance of dynamic approaches to change and uncertainty, one set of adaptive options is to manage ecosystems and resources so that they are better able to resist the influence of climate change (Parker *et al.*, 2000; Suffling and Scott, 2002). From rare species with limited available habitat to high-value forest plantation investments near rotation, maintaining the status quo for a limited period of time may be the only or best option in some cases. Creating resistance includes improving ecosystem defenses against climate effects per se, but also creating resistance against climate-exacerbated disturbance impacts. Conditions with low sensitivity to climate will be those most likely to accommodate resistance treatments, and high-sensitivity conditions will require the most intensive efforts to maintain current species and ecological functions.

For conditions with low sensitivity to climate, maintaining ecosystem health and biodiversity is an important adaptation approach, building on current understanding and management practices. Healthy forest stands recover more quickly from insect disturbances than do stressed stands, and conservation of biodiversity would aid in successful species migrations (Lemmen and Warren, 2004). Maintaining key processes, such as hydrological processes and natural disturbances, would be important. Management for resistance might require ensuring reasonable use of water from forests, and appropriate road closures to minimize invasive species transport (Christen and Matlack, 2006).

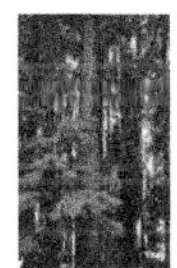

Fragmentation and land-use changes that are already problematic may be worsened under climate change due to shifts in species behaviors and changed habitat requirements. Anticipating these impacts for high-risk, high-value, and sensitive resources may require adopting landscape management practices that enable species movements. Creating larger management unit sizes, broad habitat corridors, and continuity of habitat would increase resistance of animal species to climate change by improving their ability to migrate. In this regard, enhancing coordination among the multiple agencies that manage adjacent lands to ensure habitat continuity will be essential (Malcolm *et al.*, 2006).

In the arid West, aggressive prophylactic actions may be needed to increase resistance of ecosystems from risks of climate-exacerbated disturbances such as drought, insect outbreak, and uncharacteristically severe wildfire. Resistance practices include thinning and fuels abatement treatments at the landscape scale to reduce crown fire potential and risk of insect epidemic, maintaining existing fuelbreaks, strategically placed area treatments that will reduce fuel continuity and drought susceptibility of forests, creating defensible fuel profile zones around high-value areas (such as WUI, critical habitat, or municipal watersheds), and similar treatments. Intensive and aggressive fuelbreaks may be necessary around highest-risk or highest-value areas, such as WUI or at-risk species, while mixed approaches may best protect habitat for biodiversity and general forest zones (Wheaton, 2001).

With respect to climate-related insect and disease outbreaks, traditional silvicultural methods may be applied creatively. These may involve intensive treatments, such as those used in high-value agricultural situations: resistance breeding, novel pheromone applications (such as sprayable micro-encapsulated methods), complex pesticide treatments, and aggressive fuelbreaks. Abrupt invasions, changes in behavior and population dynamics, and long-distance movements of native and non-native species may occur in response to changing climates. Monitoring non-native species, and taking aggressive early and proactive actions at key migration points to remove and block invasions, are important steps to increase resistance. However, monitoring species range distributions may indicate that native species, considered non-native to a particular area, may be migrating. Evaluating the original objectives and the changing local assemblages of species may be necessary before taking aggressive action. Conditions could be cumulatively adjusting to a changing climate, and maintenance of the status quo may not be feasible.

Efforts to increase resistance may be called for in other high-value situations. Building resistance to exacerbated effects of air pollution from climate change may require that aggressive thinning and age-control silvicultural methods are applied at broad landscape scales, that mixed species plantations be developed, that broader genetic parameters be used in plantations, or that plantations are switched to resistant species entirely (Papadopol, 2000).

Resisting climate change influences on natural forests and vegetation over time will almost always require increasingly aggressive treatments, accelerating efforts and investments over time, and a recognition that eventually these efforts may fail as conditions cumulatively change. Critical understanding of the changing environmental, social, and economic impacts of climate change will be needed to evaluate the success of management approaches to resist the influence of climate change. Creating resistance in most forest and rangeland situations to directional change is akin to "paddling upstream," and eventually conditions may change so much that resistance is no longer possible. For instance, climate change in some places will drive environments to change so much that site capacities shift from favoring one species to another, and a type conversion occurs.

Maintaining prior species may require significant extra and repeated efforts to supply needed nutrients and water, remove competing understory, fertilize young plantations, develop a cover species, thin, and prune. More seriously, forest conditions that have been treated to resist climate-related changes may cross thresholds and convert (*i.e.*, be lost) through extreme events such as wildfire, ice storm, tornado, insect epidemic, or drought, resulting in significant resource damage and loss. For this reason, in some situations, resistance options may best be applied in the short term and for projects with short planning horizons and high value, such as short-rotation biomass or biofuels plantings. Alternative approaches that work with processes of change, rather than against the direction of climate-related change, may enable inevitable changes to happen more gradually over time, and with less likelihood of cumulative, rapid, and catastrophic impact. For example, widely spaced thinning or shelterwood cuttings that create many niches for planted or naturally established seedlings may facilitate adaptation to change on some sites. In selecting these alternative approaches, a holistic analysis may be required to identify the break point beyond which intervention to natural selection and adaptation to climate changes may not be possible or cannot be managed at reasonable cost.

Promote Resilience to Climate Change

Resilient ecosystems are those that not only accommodate gradual changes related to climate, but resile (return to a prior condition of that ecosystem) after disturbance. Promoting resilience is the most commonly suggested adaptive option discussed in a climate-change context (*e.g.*, Dale et al., 2001; Spittlehouse and Stewart, 2003; Price and Neville, 2003), but

has its drawbacks as climate continues to change. Resilience can be increased through management practices similar to those described for resisting change, but applied more broadly, and specifically aimed at coping with disturbance (Dale *et al.*, 2001; Wheaton, 2001). As with any adaptation approach, land manager objectives will vary—*e.g.*, protection; management for endangered species, commodities, or low fire vulnerability—and these choices may or may not result in a decision to resile the system to a former state. An understanding of the ecological consequences of the changing climate is a critical component of identifying adaptation strategies.

An example of promoting resilience in forest ecosystems is a strategy that combines practices to reduce fire or insect and disease outbreaks (resistance) with deliberate and immediate plans to encourage return of the site, post-disturbance, to species reflective of its prior condition (resilience). Given that the plant establishment phases tend to be most sensitive to climate-induced changes in site potential, intensive management dedicated to the revegetation period through the early years of establishment may enable retention of the site by desired species, even if the site is no longer optimal for those species (Spittlehouse and Stewart, 2003). Practices could include widely spaced thinnings or shelterwood cuttings to promote resilience with living stands, and rapid treatment of forests killed by fire or insects. In forests killed by fire or other disturbance, resilience could be promoted by maintaining some degree of shade as appropriate for the forest type; intensive site preparation to remove competing vegetation; replanting with high-quality, genetically appropriate, and diverse stock; diligent stand-improvement practices; and minimizing invasion of non-native species (Dale *et al.*, 2001; Spittlehouse and Stewart, 2003). Many of these intensive forestry practices may have undesired effects on other elements of ecosystem health, and thus have often come under dispute. However, if the intent is to return a forest stand to its prior condition after disturbance under changing climate (*i.e.*, to promote resilience), then deliberate, aggressive, intensive, and immediate actions may be necessary.

Similar to the situation with regard to resistance options, the capacity to maintain and improve resilience will, for many contexts, become more difficult as changes in climate accumulate and accelerate over time. These options may best be exercised in projects that are short-term, have high value (*e.g.*, commercial plantations), or under ecosystem conditions that are relatively insensitive to the potential climate change effects (*e.g.*, warming temperatures). Climate change has the potential to significantly influence the practice and outcomes of ecological restoration (Harris *et al.*, 2006), where the focus is on tying assemblages to one place. A strategy that combines practices to restore vigor and redundancy (Markham, 1996; Noss, 2001) and ecological processes (Rice and Emery, 2003), so that after a disturbance these ecosystems have the necessary keystone species and functional processes to recover to a healthy state even if species composition changes, would be the goal of managing for ecosystem change.

Managing for Ecosystem Change

Enable Forests to Respond to Change

This suite of adaptation options intentionally plans for change rather than resisting it, with a goal of enabling forest ecosystems to naturally adapt as environmental changes accrue. Given that many ecological conditions will be moving naturally toward significant change in an attempt to adapt (*e.g.*, species migration, stand mortality and colonization events, changes in community

composition, insect and disease outbreaks, and fire events), these options seek to work with the natural adaptive processes. In so doing, options encourage gradual adaptation over time, thus hoping to avoid sudden thresholds, extreme loss, or conversion that may occur if natural change is cumulatively resisted.

Depending on the environmental context, management goals, and availability and adequacy of modeling information (climate and otherwise), different approaches may be taken. In this context, change is assumed to happen—either in known directions, with goals planned for a specific future, or in unknown directions, with goals planned directly for uncertainty. Examples of potential practices include the following:

1. Assist transitions, population adjustments, range shifts, and other natural adaptations. Use coupled and downscaled climate and vegetation models to anticipate future regional conditions, and project future ecosystems into new habitat and climate space. With such information, managers might plan for transitions to new conditions and habitats, and assist the transition—*e.g.*, as appropriate, move species uphill, plan for higher-elevation insect and disease outbreaks, reduce existing anthropogenic stresses such as air quality or land cover changes, anticipate species mortality events and altered fire regimes, or consider loss of species' populations on warm range margins and do not attempt restoration there (Ledig and Kitzmiller, 1992; Parker *et al.*, 2000; Spittlehouse and Stewart, 2003). Further examples might be to modify rotation lengths and harvest schedules, alter thinning prescriptions and other silvicultural treatments, consider replanting with different species, shift desired species to new plantation or forest locations, or take precautions to mitigate likely increases in stress on plantation and forest trees.

A nascent literature is developing on the advantages and disadvantages of "assisted migration," the intentional movement of propagules or juvenile and adult individuals into areas assumed to become their future habitats (Halpin, 1997; Collingham and Huntley, 2000; McLachlan, Hellmann, and Schwartz, 2007).

It is important to not generalize assumptions about habitat and climate change in specific areas. Local climate trajectories may be far different from state or regional trends, and local topography and microclimatology interact in ways that may yield very different climate conditions than those given by broad-scale models. In mountainous terrain especially, the climate landscape is patchy and highly variable, with local inversions, wind patterns, aspect differences, soil relations, storm tracks, and hydrology influencing the weather that a site experiences. Sometimes lower elevations may be refugial during warming conditions, as in inversion-prone basins, deep and narrow canyons, riparian zones, and north slopes. Such patterns, and occupation of them by plants during transitional climate periods, are corroborated in the paleoecological record (Millar and Woolfenden, 1999; Millar *et al.*, 2006). Additionally, land use change and agricultural practices can alter local and regional precipitation and climate patterns (Foley *et al.*, 2005; Pielke, Sr. *et al.*, 2006).

Despite the challenges in mountainous terrain, anticipating where climate and local species habitats will move will become increasingly important. On-the-ground monitoring of native species gives insight into what plants themselves are experiencing, and can suggest the directions of change and appropriate natural response at local scales. This can allow management strategies that mimic emerging natural adaptive responses. For instance, new species mixes (mimicking what is regenerating naturally), altered genotype selections, modified age structures, and novel silvicultural contexts (*e.g.*, selection harvest versus clearcut) may be considered.

2. Increase Redundancy and Buffers. This set of practices intentionally manages for an uncertain but changing future, rather than a specific climate future. Practices that involve spreading risks in diverse opportunities rather than concentrating them in a few are favored; using redundancy and creating diversity are key. Forest managers can facilitate natural selection and evolution by managing the natural regeneration process to enhance disturbances that initiate increased seedling development and genetic mixing, as has been suggested for white pines and white pine blister rust (Schoettle and Sniezko, forthcoming). Managers might also consider shortening generation times by increasing the frequency of regeneration, and increasing the effectiveness of natural selection by managing for high levels of intraspecific competition; in other words, by ensuring that lots of seedlings get established when stands are regenerated. This diversification of risk with respect to plantations can be achieved, for instance, by spreading plantations over a range of environments rather than within the historic distribution or within a modeled future location. Options that include using diverse environments and even species margins will provide additional flexibility. A benefit of redundant plantings across a range of environments is that they can provide monitoring information if survival and performance are measured and analyzed. Further, plantations originating as genetic provenance tests and established over the past several decades could be re-examined for current adaptations. This diversification of risk could also be achieved using natural regeneration and successional processes on NFs. A range of sites representing the diversity of conditions on a NF could be set aside after disturbance events to allow natural regeneration and successional processes to identify the most resistant species and populations. Other examples include planting with mixed species and age classes, as in agroforestry (Lindner, Lasch, and Erhard, 2000); increasing locations, sizes, and range of habitats for landscape-scale vegetation treatments; assuring that fuels are appropriately abated where vegetation is treated; and increasing the number of rare plant populations targeted for restoration, as well as increasing population levels within them (Millar and Woolfenden, 1999). In the same way, opportunistic monitoring, such as horticultural plantings of native species in landscaping, gardens, or parks, may provide insight into how species respond in different sites as climate changes, as well as engaging the public in such information gathering.

3. Expand Genetic Diversity Guidelines. Existing guidelines for genetic management of forest plantations and restoration projects dictate maintenance of and planting with local germplasm. In the past, small seed zones, used for collecting seed for reforestation or restoration, have been delineated to ensure that local gene pools are used and to avoid contamination of populations with genotypes not adapted to the local site. These guidelines were developed assuming that neither environments nor climate were changing—*i.e.*, a static background. Relaxing these guidelines may be appropriate under assumptions of changing climate (Ledig and Kitzmiller, 1992; Spittlehouse and Stewart, 2003; Millar and Brubaker, 2006; Ying and Yanchuk, 2006). In this case, options could be chosen based on the degree of certainty known about likely future climate changes and likely environmental changes (*e.g.*, air quality). If sufficient information is available, germplasm could be moved in the anticipated adaptive direction; for instance, rather than using local seed, seed from a warmer (often, downhill) current population would be used. By contrast, if an uncertain future is accepted, expanding seed zone sizes in all directions and requiring that seed collections be well distributed within these zones would be appropriate, as would relaxing seed transfer guidelines to accommodate multiple habitat moves, or introducing long-distance germplasm into seed mixes. Adaptive management of this nature is experimental by design, and will require careful documentation of treatments, seed sources, and outplanting locations in a corporate data structure to learn from both failures and successes of such mixes.

Traditional best genetic management practices will become even more important to implement under changing climates. Paying attention not only to the source but the balance of genetic diversity within seedlots and outplanting collections (*i.e.*, maintaining high effective population sizes) is prudent: approaches include maximizing the number of parents, optimizing equal representation by parents (*e.g.*, striving for equal numbers of seeds/seedlings per family), and thinning plantations such that existing genetic diversity is not greatly reduced. Genotypes known or selected for broad adaptations could also be favored. By contrast, although economic incentives may override, using a single or few genotypes (*e.g.*, a select clone or small clonal mix) is a riskier choice in a climate change context.

4. Manage for Asynchrony and Use Establishment Phase to Reset Succession to Current Conditions. Changing climates over paleoecologic timescales have repeatedly reset ecological community structure (species diversity) and composition (relative abundances) as plants and animals have adapted to natural changes in their environments. To the extent that climate acts as a region- and hemispheric-wide driver of change, the resulting shifts in biota often occur as synchronous changes across the landscape (Swetnam and Betancourt, 1998). At decadal and century scales, for instance, recurring droughts in the West and windstorms in the East have synchronized forest species, age composition, and stand structure across broad landscape. These then become further vulnerable to rapid shifts in climate, such as is occurring at present, which appear to be synchronizing forests through massive drought-insect-related diebacks. An opportunity exists to proactively manage the early successional stages that follow widespread mortality, by deliberately reducing synchrony.[32] Asynchrony can be achieved through a mix of activities that promotes diverse age classes, species mixes, stand diversities, genetic diversity, etc., at landscape scales. Early successional stages are likely the most successful (and practical) opportunities for resetting ecological trajectories that are adaptive to present rather than past climates, because this is the best chance for widespread replacement of plants. Such ecological resetting is evidenced in patterns of natural adaptation to historic climate shifts (Davis and Shaw, 2001).

5. Establish "Neo-Native" Plantations and Restoration Sites. Information from historic species ranges and responses to climate change can provide unique insight about species behaviors, ecological tolerances, and potential new habitats. For instance, areas that supported species in the past under similar conditions to those projected for the future might be considered sites for new plantations or "neo-native" stands of the species. These may be well outside the current species range, in locations where the species would otherwise be considered exotic. For instance, Monterey pine (*Pinus radiata*), endangered throughout its small native range, has naturalized along the north coast of California far disjunct from its present native distribution. Much of this area was paleohistoric range for the pine, extant during climate conditions that have been interpreted to be similar to expected futures in California (Millar, 1999). Using these locations specifically for "neo-native" conservation stands, rather than planning for the elimination of the trees as undesired exotics (which is the current management goal), is an example of how management thinking could accommodate a climate-change context (Millar, 1998). This option is relevant to both forest plantation and ecological restoration contexts.

6. Promote Connected Landscapes. Capacity to move (migrate) in response to changing climates is key to adaptation and long-term survival of plants and animals in natural ecosystems (Gates, 1993). Plants migrate, or "shift ranges" by dying in unfavorable sites and colonizing favorable edges, including internal species' margins. Capacity to do this is aided by managing for porous landscapes; that is, landscapes that contain continuous habitat with few physical or biotic restrictions, and through which species can move readily (recruit, establish, forage) (Halpin, 1997; Noss, 2001). Promoting large forested landscape units, with flexible management goals that can be modified as conditions change, will encourage species to respond naturally to changing climates (Holling, 2001). This enables managers to work with, rather than against, the flow of change. Evaluating and reducing fragmentation, and planning cumulative landscape treatments to encourage defined corridors as well as widespread habitat availability, is a proactive approach.

7. Realign Significantly Disrupted Conditions. Restoration treatments are often prescribed for forest species or ecosystems that have been significantly or cumulatively disturbed and are far outside natural ranges of current variation. Because historical targets, traditionally used as references for restoration, are often inappropriate in the face of changing climates, re-alignment with current process rather than restoration to historic pre-disturbance condition may be a preferred choice (Millar and Brubaker, 2006; Harris *et al.*, 2006; Willis and Birks, 2006). In this case, management goals seek to bring processes of the disturbed landscape into the range of current or anticipated future environments (Halpin, 1997). An example

[32] **Mulholland**, P., J. Betancourt, and D.D. Breshears, 2004: *Ecological Impacts of Climate Change: Report From a NEON Science Workshop.* American Institute of Biological Sciences, Tucson, AZ.

comes from the Mono Lake ecosystem in the western Great Basin of California (National Research Council, 1987; Millar and Woolfenden, 1999). A basin lake with no outlet, Mono Lake is highly saline, thus is naturally fishless but rich in invertebrate endemism and productivity, provides critical habitat for migratory waterfowl, and supports rich communities of dependent aquatic and adjacent terrestrial animal species. In 1941, the Los Angeles Department of Water and Power diverted freshwater from Mono Lake's tributaries; the streams rapidly dried and Mono Lake's level declined precipitously. Salinity increased, groundwater springs disappeared, and ecological thresholds were crossed as a series of unexpected consequences unfolded, threatening Mono Lake's aquatic and terrestrial ecosystems. An innovative solution involved a 1990 court-mediated re-alignment process. Rather than setting pre-1941 lake levels as a restoration goal, a water-balance model approach, considering current climates as well as future climatic uncertainties, was used to determine the most appropriate lake level for present and anticipated future conditions.[33]

Options Applicable to Both Forestalling Change and Managing for Change

Anticipate and Plan for Surprise and Threshold Effects

Evaluate potential for indirect and surprise effects that may result from cumulative climate changes or changes in extreme weather events. This may involve thinking outside the range of events that have occurred in recent history. For example, reductions in mountain snowpacks lead to more bare ground in spring, so that "average" rain events run off immediately rather than being buffered by snowpacks, and produce extreme unseasonal floods (e.g., Yosemite Valley, May 2005[34]). Similarly, without decreases in annual precipitation, and even with increasing precipitation, warming minimum temperatures are projected to translate to longer dry growing-season durations. In many parts of the West, especially Mediterranean climate regions, additional stresses of longer summers and extended evapotranspiration are highly likely to push plant populations over thresholds of mortality, as occurred in the recent multi-year droughts throughout much of the West (Breshears *et al.*, 2005). Evidence is accumulating to indicate that species interactions and competitive responses under changing climates are complex and unexpected (Suttle, Thompsen, and Power, 2007). Much has been learned from paleo-historic studies about likely surprises and rapid events as a result of climate change. Anticipating these events in the future means planning for more extreme ranges than in recent decades, and arming management systems accordingly (Millar and Woolfenden, 1999; Harris *et al.*, 2006; Willis and Birks, 2006).

Experiment with Refugia

Plant ecologists and paleoecologists recognize that some environments appear more buffered against climate and short-term disturbances, while others are sensitive. If such "buffered" environments can be identified locally, they could be considered sites for long-term retention of plants, or for new plantations (commercial or conservation). For instance, mountainous regions are highly heterogeneous environmentally; this patchiness comprises a wide range of micro-climates within the sites. Further, unusual and nutritionally extreme soil types (*e.g.*, acid podsols, limestones, etc.) have been noted for their long persistence of species and genetic diversity, resistance to invasive species, and long-lasting community physiognomy compared with adjacent fertile soils (Millar, 1989). During historic periods of rapid climate change and widespread population extirpation, refugial populations persisted on sites that avoided the regional climate impacts and the effects of large disturbance. For example, Camp (1995) reported that topographic and site characteristics of old-growth refugia in the Swauk Pass area of the Wenatchee National Forest were uniquely identifiable. These populations provided both adapted germplasm and local seed sources for advance colonization as climates naturally changed toward favoring the species. In similar fashion, a management goal might

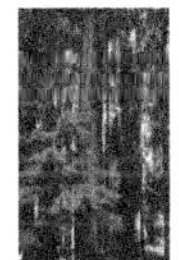

[33] **State of California**, 1994: *Decision and Order Amending Water Right Licenses to Establish Fisher Protection Flows in Streams Tributary to Mono Lake and to Protect Public Trust Resources at Mono Lake and in the Mono Lake Basin.* State Water Resources Board Decision 1631, pp.1-212.

[34] **Dettinger**, M., J. Lundquist, D. Cayan, and J. Meyer, 2006: The 16 May 2005 Flood in Yosemite National Park--A Glimpse into High-Country Flood Generation in the Sierra Nevada. Presentation at the American Geophysical Union annual meeting, San Francisco. http://www.fs.fed.us/psw/cirmount/meetings/agu/pdf2006/dettinger_etal_poster_AGU2006.pdf

focus specific attention to protect populations that currently exist in environmentally and climatically buffered, cooler, or unusually mesic environments.

3.4.4 Prioritizing Management Responses in Situations of Resource Scarcity

Species, plant communities, regional vegetation, and forest plantations will respond to changing climates individualistically. Some species and situations will be sensitive and vulnerable, while others will be naturally buffered and resilient to climate-influenced disturbances (Holling, 2001; Noss, 2001). Management goals for species and ecosystems across the spectrum of NFs also vary for many reasons. As a result, proactive climate planning will reflect a range of management intensities. Some species and ecosystems may require aggressive treatment to maintain viability or resilience, others may require reduction of current stressors, and others less intensive management, at least in the near future.

While evaluating priorities has always been important in resource management, the magnitude and scope of anticipated needs, combined with diminishing availability of human resources, dictate that priorities be evaluated swiftly, strictly, and definitively. A useful set of guidelines for certain high-demand situations comes from the medical practice of triage (Cameron *et al.*, 2000). Coming from the French triare, to sort, triage approaches were developed from the need to prioritize the care of injured soldiers in battlefield settings where time is short, needs are great, and capacity to respond is limited. Well-established emergency and disaster triage steps can be modified to fit resource needs when conditions cannot be handled with traditional planning or institutional capacity. Triage in a natural-resource context sorts management situations ("patients") into categories according to urgency, sensitivity, and capacity of available resources to achieve desired goals ("survival"). Cases are rapidly assessed and sorted into three to five major categories ("color tags") that determine further action:

1. Red: Significant ongoing emergency; immediate attention required. Cases in this category are extremely urgent, but may be successfully treated with immediate attention given available resources. Without attention, they will rapidly fail; in the medical sense, the patient will die soon if untreated. These cases receive the highest priority for treatment and use of available resources. Depending on available resources, some of these cases may be assigned black rather than red.

2. Yellow: Strong to medium potential for emergency. Cases in this category are sensitive to disruption, vulnerable due to history or disturbance (degree and extent of trauma), have the capacity with small additional disturbance to become rapidly worse, but are marginally stable at the time of assessment. These cases have medium priority.

3. Green: Low likelihood for emergency conditions. Cases in this category may have some problems but overall are relatively resistant to disturbance, have low stress or high capacity to deal with stress, a history of low vulnerability, and show signs of retaining stability at least in the short term with little need for intervention. These cases receive low priority, but conditions are monitored regularly for change.

4. Black: Conditions altered beyond hope of treatment. Cases in this category are so disrupted, altered, and weakened that chances of successfully treating them with available resources are nil. In medical context, patients are either dead or unable to be kept alive with existing capacity. These cases have the lowest priority in the short term, and alternative resolutions have to be developed.

While triage is valuable to practice under conditions of scarce resources or apparently overwhelming choice, it is not viable as a long-term or sole-use approach to priority-setting. Other approaches may be used for quick prioritizing of traditional management plans and practices. An example would be rapid assessments of current national forest land management plans, performed by teams of climate experts that visit NFs. Teams would rapidly review planning documents, interview staff, and visit representative field sites; they would conclude their visits with a set of recommendations on what aspects of the overall local forest management practices and plans are in (1) immediate need of significant revision, (2) need of revision in a longer time frame, and (3) no need of revision; already climate-savvy. Similar integrated threat assessment tools are being developed that help managers and decision-makers grasp categories of urgency.

In situations where available resources can be augmented, where time is not a critical factor, and where more information can be obtained, traditional evaluations and priority-setting will be most appropriate. Triage may be used, however, at any time and at any scale where urgency arises, and when demands become greater than normally managed. The common alternative under these conditions, reacting to crises chaotically and without rules of assessment, will achieve far less success in the long run than triage-based approaches.

3.4.5 Barriers to Adaptation Approaches

The USFS will need to overcome various barriers to take advantage of opportunities to implement adaptations to climate change. Insufficient resources, various uncertainties, checkerboard ownership patterns, lengthy planning processes, agency targets and reward systems, and air quality standards that restrict the use of prescribed fire are examples of such barriers. The need to coordinate with other agencies, the private sector, and the general public may either enhance or impede the ability of the USFS to implement management adaptations. How these other stakeholders perceive climate change and react to USFS management proposals will strongly influence how the USFS can ultimately adapt.

Developing innovative adaptations to climate change will require creative thinking, coupled with improved scientific understanding of proposed new approaches. The USFS may need to encourage planners and managers to relax perceptions about rules and other constraints that may, in reality, afford enough flexibility to try something new. Scientists would then need to be given the resources and support to test new approaches that are developed through this innovative process.

3.5 CONCLUSIONS AND RECOMMENDATIONS

3.5.1 Climate Change and National Forests

The mission of the NFs has broadened over time, from protecting water and producing timber to managing for multiple resources and now, to sustaining the health, diversity, and productivity of the nation's forests and grasslands to meet the needs of present and future generations. Increasingly ecosystem management, ecological integrity, resilience, and sustainability have become important concepts and goals of NF management.

The management of NF lands has broadened to include involvement by several other federal agencies, including EPA, the Fish and Wildlife Service, the National Marine Fisheries Service, and the Bureau of Land Management, as well as coordination on management of lands within NFs for national systems such as the Wilderness Preservation System, National Trails, National Monuments, and Wild and Scenic Rivers. The checkerboard ownership patterns of many of the western forests, the scattered private in-holdings of many NFs, and the scattered land parcels of the eastern forests result in the important need to coordinate with other federal and state agencies and with private land owners. Public involvement has increased. This broader level of participation—by the public and other federal and state agencies, as well as the assortment of different management units—is an asset, but also can be a challenge for coordinating and responding to novel situations such as climate change.

One of the challenges to the USFS will be the diversity of climatic changes experienced by NFs. Not only will each NF experience regional and site-specific changes in temperature and precipitation, but the forests are likely to experience changes in frequency, intensity, timing, and locations of extreme weather events such as the occurrence of ice storms; wind events such as derechos, tornados, and hurricanes; and flooding associated with high-intensity rainfall events or with shifts between rain and snow events. Local land management goals differ greatly by NF and grassland, and by management units within NFs (*e.g.*, wilderness, matrix working forests associated with the Northwest Forest Plan, ski areas, campgrounds, etc). Thus, no single approach to adaptation to climate change will fit all NFs. This diversity of climatic changes and impacts will interact with the diversity of stressors, the diversity of ecosystems, and the diversity of management goals across the NFs—in short, responses to climate change will need to reflect local and regional differences in climate, ecosystems, and the social and economic settings.

The NFs have, in many aspects, begun to address many of the challenges of climate variability and change—changes to historic disturbance regimes, historically unprecedented epidemics of native insects, large-scale forest mortality, extreme and unseasonal weather events, spread of non-native invasive species, drought, fuels accumulation, and ecosystem fragmentation. Current management approaches include landscape-scale planning and coordinated agency planning for fire suppression, regional water management, and coordinated agency efforts for invasive species, among others.

Adaptation options for climate-sensitive ecosystems encompass three approaches: no active planning for a changing environment, reaction to a changing disturbance regime, and anticipatory adaptation actions. The rationale for each adaptation approach involves consideration of the costs and benefits associated with the ecological, social, and economic components under the changing climate, the available information on future climatic conditions, and other technical and institutional concerns. In some cases, the choice of no active planning could reflect short-term goals on landscapes where the risk of climate change impacts may be minimal in the short term, for ecosystems with low sensitivity to climate change, where the uncertainty is great (climate variability large, potential impacts low), or where the resources to manage a particular ecosystem service jeopardized by climate change would be better used to manage other ecosystem resources. Responding to a climate-induced changing disturbance (*i.e.*, implementing adaptations after disturbances occur) might be justified in situations where managers determine that adjustments to historical management approaches are needed eventually, but are best made during or after a major climatic or disturbance event. In this instance, adaptive actions are incorporated after the disturbance occurs. The third option involves anticipating and specifically preparing for climate change opportunities and impacts. The choice involves using the best available information about future climate and environmental conditions, and the best available information about the societal context of forest management, to begin making changes to policy and on-the-ground management now, as well as when future windows of opportunity open. Each response may be appropriate in some circumstances and not in others.

3.5.2 Management Response Recommendations

3.5.2.1 Integrate Consideration of Climate Change across All Agency Planning Levels

Adaptation strategies may vary based on the spatial and temporal scales of decision making within the USFS. The integration of climate change and climate change impacts on ecosystem services into policy development and planning across all levels of the agency—USFS strategic goals, Resource Planning Act (RPA) Assessment, NF plans, multi-forest plans, project planning—could facilitate a cohesive identification of opportunities and barriers (institutional, ecological, social). Planning at regional or national scales may involve acceptance of different levels of uncertainty and risk than appropriate at local (*e.g.*, NF or watershed) scales. The current approach responds to the legislative requirement to address climate change analyses within the strategic national level through the RPA Assessment. National analyses associated with RPA offer the opportunity to develop potential approaches to link assessments at the scale of the national level, regional, multi-forest and NF. More quantitative approaches may be available at the national/regional scales, providing strategic guidance for broad consideration of climate change opportunities and impacts to management activities at finer scales.

3.5.2.2 Reframe the Role of Uncertainty in Land Management: Manage for Change

Current ecological conditions of NFs are projected to change under a changing climate, along with social and economic changes. The challenge for the USFS will be to determine which ecosystem services and which attributes and components of biodiversity can be sustained or achieved through management under a changing climate. There will be a need to anticipate and plan for surprise and threshold effects that are at once difficult to predict with certainty yet certain to result from the interaction of climate change and other stressors. Rather than targeting a single desired future condition, avoiding a range of undesirable future conditions may be more effective

There may also be a need to shift focus to managing for change, setting a goal of desired future function (processes, ecosystem services), and managing current and future conditions (structure, outputs), which may be quite dynamic because of a changing climate. Rapid changes that are expected in physical conditions and ecological responses suggest that management goals and approaches will be most successful when they emphasize ecological processes rather than focus on structure and composition. Under a changing climate, embracing uncertainty will necessitate a careful examination of various underlying assumptions about climate, climate change, ecological processes, and disturbances. Specifically, the USFS will need to re-evaluate (1) the dynamics of ecosystems under disturbances influenced by climate; (2) current management options as influenced by climate; and (3) important assumptions and premises about the nature of disturbances (*e.g.*, fire, insect outbreaks, diseases, extreme climate-related events, and the interactions among these disturbances) that influence management philosophy and approaches. Our assumptions about the climate sensitivity of best management practices, genetic diversity guidelines, restoration treatments, and regeneration guidelines may need to be revisited. Opportunities to test these assumptions through management activities and research experiments will be valuable. Current management approaches offer a good platform to reframe these strategies to address uncertain and varying climates and environments of the future.

3.5.2.3 Nurture and Cultivate Human Capital within the Agency

The USFS has a long tradition of attracting and retaining highly qualified employees. The capacity of the agency to address climate change may require the staff within NFs to have a more technical understanding of climate change, as well as building the adaptive capacity of the social and economic environments in which they work. Specifically, the USFS could provide opportunities to develop a better technical understanding of climate and its ecological and socioeconomic impacts, as well as options for adaptation and mitigation in NFs through the many training opportunities that currently exist within the USFS, including the silvicultural certification program, regional integrated resource training workshops, and regional training sessions for resource staff. New opportunities to share training of resource managers with other natural resource agencies could also enhance the ability of the USFS to address climate change in resource management. Additionally, increased awareness and knowledge of climate change could be transferred through the development of managers' guides, climate primers, management toolkits, a Web clearinghouse, and video presentations. Opportunities for managers to share information on the success or failure of different adaptation approaches will be critical.

The skill set necessary to address the challenge of managing natural resources under a changing climate may need to be examined. Staffing in areas such as silviculture, forest genetics and tree breeding, entomology (including

taxonomy), and insect control has declined. Access to this knowledge will be critical; the challenge will be how to staff internally, or to develop relationships with experts in other federal or state agencies, universities, or the private sector.

Resource management is challenging in today's environment, and climate change will increase that challenge. Line officers and resource staff are faced with—and will continue to be faced with—the challenge of making decisions in an uncertain environment. Facilitation of a learning environment, where novel approaches to addressing climate change impacts and ecosystem adaptation are supported by the agency, will support USFS employees as they attempt to achieve management goals in the face of climate uncertainty and change. Scientists and managers will sometimes be called upon to sift through apparently conflicting approaches to understanding climate impacts on ecosystems. What may appear as "mistakes" are, in fact, opportunities to learn the technical issues and conditions for assessing and using such approaches.

It may be that NF staff will not be able to keep up with the rapidly changing science. Thus, it is critical to build ongoing relationships between researchers (within and outside the USFS) and the NF staff. An example of such a partnership is the Regional Integrated Sciences and Assessments (RISA) program, which supports research that addresses complex climate-sensitive issues of concern to decision-makers and policy planners at a regional level. The RISA research team members are primarily based at universities, though some of the team members are based at government research facilities, non-profit organizations, or private sector entities. Traditionally the research has focused on the fisheries, water, wildfire, and agriculture sectors.

3.5.2.4 Develop Partnerships to Enhance Natural Resource Management under a Changing Climate

There is an urgent need for policy makers, managers, scientists, stakeholders, and the broader public to share the specific evidence of global climate change and its projected consequences on ecosystems, as well as their understanding of the choices, future opportunities, and risks. The dialogue on adaptation and mitigation might begin with the USFS and current partners. Changes in ecosystems service and biodiversity (*e.g.,* a loss of cold-water fisheries in some areas and the development of warm water fisheries) under a changing climate will likely reveal a need to develop new partnerships.

Education and outreach on the scale necessary will require new funding and educational initiatives. Effective efforts, informed by cutting-edge social science insights on effective communication, will involve diverse suites of educational media, including information delivery on multiple and evolving platforms. There will also be a need to educate landowners in the WUI about the potential for increased disturbances or changing patterns of disturbances in these areas, as well as the challenges of land ownership and protection of valued resources within this environment.

3.5.2.5 Increase Effective Collaboration Across Federally Managed Landscapes

Where federally managed land encompasses large landscapes, increasing collaboration will facilitate the accomplishment of common goals (*e.g.,* the conservation of threatened and endangered species), as well as adaptation and mitigation, that can only be attained on larger connected (or contiguous) landscapes. Common goals might include protection of threatened and endangered species habitats, integrated treatment of fuels or insect and disease conditions that place adjacent ownerships at risk, and developing effective strategies to minimize loss of life and property at the WUI.

While collaboration logically makes sense, and seems conceptually like the only way to manage complex ownerships, large landscapes, and across multiple jurisdictions, there are many challenges to such an approach. Attempting to collaborate multi-institutionally across large landscape scales can bring into focus unexpected institutional barriers and focus unanticipated societal responses. For example, large multi-forest landscapes have high investment stakes—with resulting political pressure from many different directions. Further, if collaboration is taken to mean equal participation and that each collaborator has an effective voice, then potential mismatches among laws, regulations, resources and staffing capacities can lead to situations in which collaboration by different groups is uneven and possibly unsuccessful. For example, the USFS, EPA, and the U.S. Fish and Wildlife Service each must obey its particular governing laws, and thus agency oversight can overrule attempts at equal participation and collaboration. Careful consideration of the challenges and expert facilitation may be necessary to successfully manage adaptation across large landscapes.

3.5.2.6 Establish Priorities for Addressing Potential Changes in Populations, Species, and Community Abundances, Structures, Compositions, and Ranges, Including Potential Species Extirpation and Extinction under Climate Change

A primary premise for adaptive approaches is that change, novelty, uncertainty, and uniqueness of individual situations are expected to define the planning backdrop of the future. Management goals for species and ecosystems across the spectrum of NFs also vary for many reasons. As a result, proactive climate planning will reflect a range of management intensities. Some species and ecosystems (already affected in the near-term) may require aggressive treatment to maintain viability or resilience; others may require reduction of current stressors, and others less intensive management, at least in the near future. While evaluating priorities has always been important in resource management, the magnitude and scope of anticipated needs, combined with diminishing availability of human resources, dictate that priorities may need to be evaluated swiftly, strictly, and definitively. Consideration of methods to establish these priorities before the crisis appears would facilitate decision-making. The medical metaphor of triage is appropriate here. Other approaches include developing strategies that establish options that are "win-win" or "no regrets," or those that gradually add options as resources and the need for change become apparent. These approaches are best developed jointly by neighboring land resource managers and private land owners, or regionally, to guide the management of currently rare or threatened and endangered species as well as of populations, species, communities, and ecosystems that expand and retreat across the larger landscape. These approaches could capitalize on the respective strengths of the various local, state, and federal land management agencies.

3.5.2.7 Reduce Current Stressors

The USFS implements a variety of management approaches to reduce the impact of existing stressors on NFs (see Section 3.3.3), and an increased emphasis on these efforts represents an important "no regrets" strategy. It is likely that the direct impacts of climate change on ecosystems, and the effects of interactions of climate change with other major stressors, may render NFs increasingly prone to more frequent, extensive, and severe disturbances, especially drought, insect and disease outbreaks, invasive species, and wildfire. Increased flooding is a likely possibility. Air pollution can negatively affect the health and productivity of NFs, and the fragmented landscape in which many NFs are situated impedes important ecosystem processes, including migration. Efforts to address the existing stressors would address current management needs, allow an incremental approach that begins to incorporate climate into management and planning, and potentially reduce the future interactions of these stressors with climate change.

3.5.2.8 Develop Early Detection and Rapid Response Systems for Post-Disturbance Management

Early detection and rapid response systems are a component in the current invasive species strategy of the USFS. Such an approach may have value for a broader suite of climate-induced stressors, for example using the current network of experimental forests and sites in an early detection and response system. Consideration of post-disturbance management for short-term restoration and for long-term restoration under climate change prior to the disturbance (fire, invasives, flooding, hurricanes, ice storms) may identify opportunities and barriers. Large system-resetting disturbances offer the opportunity to influence the future structure and function of ecosystems through carefully designed management experiments in adapting to climatic change. Current limitations (barriers) may need to be revisited so that restricted management practices are permitted.

3.5.3 Research Priorities

3.5.3.1 Conceptual (Research Gaps)

Global climate change will continually alter the dynamics of ecosystems, local climate, disturbances, and management, challenging not only the management options but also the current understanding of these dynamics within the scientific community. To address the long-term challenges, it will be valuable to establish strong management-research partnerships now to collaboratively explore the information and research needed to manage ecosystem services under a changing climate. These research-management partnerships could identify research studies on how forest planning can better adapt to climate change in the long-term, as well as in near-term project-level analyses. Further adaptation approaches could be tested, including improved communication of knowledge and research.

Climate change will interact with current stressors—air quality, native insects and diseases, non-native invasives, and fragmentation—in potentially surprising ways. Greater understanding of the potential interactions of multiple stressors and climate change is needed through field experiments, modeling exercises, and data mining and analysis of past forest history or even recent geological records. Such approaches could promote syntheses of disciplinary research related to climate and other stressors, and integrate the efforts of the research communities at universities, non-governmental organizations, state agencies, tribal organizations, and other federal agencies.

Climate change may also challenge current theories on ecosystem restoration. Current protocols about restoration may need further experimentation to determine the role and assumptions of climate in the current techniques, and how a changing climate might alter the application of these techniques.

Determining the baseline for monitoring, determining what to monitor, and evaluating whether current monitoring approaches will be adequate under a changing climate are critical research needs. These needs may be approached collaboratively with research institutions and other federal land management agencies.

Understanding ecosystem restoration practices—and what metrics to use for monitoring—will raise in importance the need for paleo-ecological research. Little of the current understanding of paleo-ecology is brought into current thinking about the dynamics of species, communities and landscapes. This knowledge, relevant to the present and future, provides a greater understanding of lessons about change, dynamism, thresholds, novelty, reversibility, individualistic responses, and non-analog conditions. Whether to manage for process or structure may be learned from studying past responses to historic climate change. A paleo approach places managers in the stream of change. Thus: what is a baseline? What are native species range distributions? What is natural?

The adaptive capacity of NFs and the surrounding social and economic systems is not well-understood. There is great need for social scientific research into the factors and processes that enhance NFs' adaptive capacity, as well as into the barriers and limits to potentially hinder effective and efficient adaptation. In addition, socioeconomic research and monitoring are needed on how social and economic variables and systems are changing, and are likely to change further, as climate change influences the opportunities and impacts within and surrounding NFs. The expansion of the urban and suburban environment into remote areas will likely be influenced by climate change—potentially shifting this expansion to higher elevations or to more northerly regions where winters may historically not have been as severe. Recreational choices are also likely to be influenced by climate changes, shifting outdoor activities across a spectrum of options from land-based to water-based, from lower/warmer regions to higher/cooler regions.

The need currently exists to develop tradeoff analyses for situations in which management actions taken now potentially could alter more serious impacts later, such as the tradeoffs of planned prescribed fire/air quality versus unplanned wildfire/smoke/air quality. Habitat restoration for threatened and endangered species under a changing climate might involve social, economic, and ecological impacts and opportunities on NF land, adjacent ownerships, or private land. Tradeoffs involve ecological benefits and consequences, as well as social and economic benefits and consequences. Similarly, the tradeoffs between mitigation and adaptation at present cannot be addressed in the available suite of decision-making and management tools.

These research priorities will be most useful to managers if they explicitly incorporate evaluations of uncertainty. Toward that end, new approaches for assessing (or evaluating) uncertainty with quantitative and qualitative management methods are needed.

3.5.3.2 Data Gaps (Monitoring/Mapping)

Information on the status of ecosystem services as climate changes will be important in ascertaining whether management goals are being attained under the changing climate. The Forest Inventory and Analysis data have informed historical analyses of productivity shifts as affected by recent climate variability and change at large spatial scales, and contributed to national accounting analyses of carbon in U.S. forests. Other potential analyses with these inventory data could include exploring the response of ecosystems to changing fire regimes and insect outbreaks. Opportunities exist to link the existing inventory networks within the USFS (Forest Inventory Analysis) with other existing and planned networks, such as the National Science Foundation's Long-term Ecological Research networks, the National Ecological Observation Network (NEON), and other monitoring programs within USGS and NASA. Increasingly, data are needed in a spatial format.

The Montreal Process Criteria and Indicators for Boreal and Temperate Forests have been used to describe sustainability of forests and rangelands by managers at several spatial scales. The use of Montreal Process Criteria and Indicators may also have value in assessing the opportunities and impacts on sustainability under a changing climate.

3.5.3.3 Tool Gaps (Models and Decision Support Tools)

There is a need to develop techniques, methods, and information to assess the consequences of climate change and variability on physical, biological, and socioeconomic systems at varying spatial scales, including regional, multi-forest, and NF scales. The analyses at the national scale in the RPA Assessment, particularly if extended beyond forest dynamics, could provide national-level information and set a larger context for the forest opportunities and impacts under climate change. Fine-scale analyses of the ecological and economic impacts of climate change will soon be available and could offer projections at the spatial scale of importance to managers.

There is a need to develop a toolbox for resource managers that can be used to quantify effects of climate change on natural resources, as a component of land management planning. This toolbox would have a suite of science-based products that deliver state-of-the-art information derived from data, qualitative models, and quantitative models in accessible formats, including a Web-based portal on climate-change science. Technology transfer through training packages on climate change that can be delivered through workshops and online tutorials would be valuable to internal staff and potentially to stakeholders.

Forest-scale decision support applications that incorporate the dynamics of climate,

climate variability, and climate change into natural resource management planning would enhance the information about climate used in management analyses. At present, most established planning and operational tools do not directly incorporate climate variability and change. These tools need to be informed by recent scientific data on climate trends and the relationship between climate and the resource of interest. Research can contribute immediately to the revision of popular tools such as the Forest Vegetation Simulator, thereby improving their accuracy for a variety of applications. A Web-based portal on climate change, customized for the needs of USFS users, will be an important component of the toolbox, providing one-stop shopping for scientific information, key publications, and climate-smart models. A training curriculum and tutorials will ensure that Forest Service managers receive current, consistent information on climate change issues.

It can not be overstated however, that effective decision support involves more than providing the right information and tools and the right time. Importantly, for climate change information to meet the needs of NF land managers at various scales of decision-making, and for that information to be used properly and effectively, it is highly advisable that ongoing relationships be built between those producing the relevant information (researchers) and those eventually using it (managers). Thus tools, Web-based tutorials, reports, and other written materials should always be viewed as decision-support products that must be embedded in an ongoing decision-support process.

3.5.3.4 Management Adjustments or Realignments

The development of management alternatives for adapting to and mitigating the effects of an uncertain and variable climate, and other stressors on natural resource outputs and ecosystem services, will require experimentation under the changing climate. Many proposed management alternatives may need to be established as small-scale pilot efforts, to determine the efficacy of such proactive approaches to adapting to climate change in various ecosystems and climates. Protocols for "assisted migration" of species need to be tested and established before approaches are implemented more broadly. Assumptions about the dynamics of ecosystems under climate change and alternative treatments may need to be revisited in field experiments. Regeneration and seedling establishment studies using a variety of vegetation management treatments under the changing climate may suggest that new approaches are needed to ensure ecosystem establishment and restoration.

New or innovative management options may need experiments or demonstration projects to explore their impact. For example, research is needed to increase our understanding of the impacts of active management on ecosystems—such as the effects of reintroducing species to disturbed ecosystems, or transferring species to areas outside of the current distribution but within areas of compatible climate. The potential for ex situ gene conservation techniques to remedy the impact of global change might be explored. These techniques (seed banks, common garden studies) conserve genetic diversity outside the environment where it exists at this time. Putting seed from diverse parents in diverse populations into long term storage will not prevent existing forest ecosystems from being disrupted, but it provides an opportunity to reestablish populations in new and more appropriate locations if needed. Establishing common garden studies with diverse materials at multiple locations can serve several purposes. Assuming the material planted in these plots survives, it can serve as a source of propagules for establishing new populations. The tests can also provide evidence of what sources of plant material are most adapted for the new conditions.

Research is needed to explore options to reduce both the short- and long-term vulnerability of ecosystems to disturbance altered by climate (insects, fire, disease, etc.). Many natural resource values can be enhanced by allowing fire to play its natural role where private property and social values can be protected. Research on new opportunities for ecosystem services within NFs is needed. Testing and developing a range of science-based management alternatives for adapting to and mitigating the effects of climate change on major resource values (water, vegetation, wildlife, recreation, etc.) may facilitate the attainment of these goals under a changing climate.

National Parks

Lead Author: Jill S. Baron, U.S. Geological Survey and Colorado State University

Contributing Authors: Craig D. Allen, U.S. Geological Survey; Erica Fleishman, National Center for Ecological Analysis and Synthesis; Lance Gunderson, Emory University; Don McKenzie, U.S. Department of Agriculture Forest Service; Laura Meyerson, University of Rhode Island; Jill Oropeza, Colorado State University; Nate Stephenson, U.S. Geological Survey

4.1 SUMMARY

Covering about 4% of the United States, the 338,000 km² of protected areas in the National Park System contain representative landscapes of all of the nation's biomes and ecosystems. The U.S. National Park Service Organic Act established the National Park System in 1916 "to conserve the scenery and the natural and historic objects and the wild life therein and to provide for the enjoyment of the same in such manner and by such means as will leave them unimpaired for the enjoyment of future generations."[1] Approximately 270 national park system areas contain significant natural resources. Current National Park Service policy for natural resource parks calls for management to preserve fundamental physical and biological processes, as well as individual species, features, and plant and animal communities. Parks with managed natural resources range from large intact (or nearly intact) ecosystems with a full complement of native species—including top predators—to those diminished by disturbances such as within-park or surrounding-area legacies of land use, invasive species, pollution, or regional manipulation of resources. The significance of national parks as representatives of naturally functioning ecosystems and as refugia for natural processes and biodiversity increases as surrounding landscapes become increasingly altered by human activities.

KEY FINDINGS

Addressing resilience to climate change in activities and planning will increase the ability of the National Park Service to meet the mission of the Organic Act. Climate has fundamentally defined national parks. Climate *change* is redefining these parks and will continue to do so. Rather than simply adding and ranking the importance of climate change against a host of pressing issues, managers are wise to begin to include climate change considerations into all activities and plans. There are a number of short-term approaches that may help to provide resilience over the next few decades. These include reducing habitat fragmentation and loss, invasive species, and pollution; protecting important ecosystem and physical features; restoring damaged systems and natural processes (recognizing that some restoration may not provide protection of dynamic systems); and reducing the risks of catastrophic loss through bet-hedging strategies such as establishing refugia, relocating valued species, replicating populations and habitats, and maintaining representative examples of populations and species. Short-term adaptation may

[1] 16 U.S.C. 1, 2, 3, and 4

involve prioritizing resources and determining which parks should receive immediate attention, while recognizing that the physical and biological changes that will accompany warming trends and increasing occurrences of extreme events will affect every one of the 270 natural national parks in the coming century.

Preparing for and adapting to climate change is as much a cultural and intellectual challenge as it is an ecological one. Successful adaptation begins by moving away from traditional ways of managing resources. Throughout its history, the National Park Service has changed its priorities and management strategies in response to increased scientific understanding. Today, confronted not only with climate change but with many other threats to natural resources from within and outside park boundaries, the Park Service again has the opportunity to revisit resource management practices and policies. Adaptation strategies include broadening the portfolio of management approaches to include scenario planning and adaptive management, increasing the capacity to learn from management successes and failures, and examining and responding to the multiple scales at which species and processes function.

Successful adaptation includes encouraging managers to take reasoned risks without concern for retribution. "Safe-to-fail" policies reward front-line managers for making decisions to protect resources under uncertainty. Although not desired, failures provide tremendous opportunities for learning. Learning from mistakes and successes is a critical part of adaptation to climate change. Learning is further enhanced by providing training opportunities, supporting continuous inquiry, promoting an atmosphere of respect, rewarding personal initiative, and as mentioned above, allowing for unintentional failure.

As climate change continues, thresholds of resilience will be overcome, increasing the importance of using methods that address uncertainty in planning and management. Technical or scientific uncertainty can be addressed through scenario-based planning and adaptive management approaches toward learning. First, scenario-based planning explores a wide set of possible or alternative futures. A finite number of scenarios (e.g., three to five) that depict a range of possible futures can be extremely useful for helping managers develop and implement plans, confront and evaluate the inevitable tradeoffs to be made when there are conflicting management goals, and minimize the anxiety or frustration that comes from having to deal with uncertainty. Scenarios that evaluate the feasibility of adaptation against ecological, social, or economic returns will be valuable in making difficult decisions, and in conveying results of decisions to the public. Public involvement in scenario building, from individual parks to national policy level, will prepare people for inevitable changes, and may build support for science-based management.

Second, adaptive environmental assessment and management employs a set of processes to integrate learning with management actions where uncertainty exists about the potential ecological responses. Adaptive management either establishes experiments to test the effectiveness of management approaches, or uses understanding gained from past management or science to plan and execute management actions. Both require iterative monitoring and interpretation to gauge the effectiveness of that action in achieving management goals.

Protecting natural resources and processes may continue to be achieved during the coming decades using science-based principles already familiar to Park Service managers. Protecting natural resources and processes in the near term begins with the need to first identify what is at risk. The next steps are to define the baselines (reference conditions) that constitute "unimpaired" in a changing world, decide the appropriate scales at which to manage the processes and resources, and set measurable targets of protection. Finally, monitoring of management results is important for understanding the degree to which management activities succeed or fail over time, and whether management activities need to be adjusted accordingly. In the long term, such science-based management principles will become more important when examples from the past may not serve as guides for future conditions. Some targets for adjusting to future conditions can be met by the National Park Service with internal strategies for managing park resources. For example, parks may manage visitor use practices or patterns differently to prevent people from inadvertently contributing to climate-change-enhanced damage, or remove infrastructure from floodplains or fire-prone areas to allow natural disturbances to proceed as naturally as possible.

Many management goals can only be achieved through regional interagency cooperation. The National Park Service can be a catalyst for regional collaboration with other land and resource management entities. For example, the National Park Service alone will not be able to protect and restore native species as distributions change in response to climate. The Natural Resource Challenge distinguishes between native and non-native plants, animals, and other organisms, and recommends non-natives are to be controlled where they jeopardize natural communities in parks. Regional partnerships with other land and resource management groups can anticipate, and even aid, the establishment of desirable climate-appropriate species that will take advantage of favorable conditions. By using species suited to anticipated future climates after disturbance or during restoration, protecting corridors or removing impediments to natural migration, and aggressively controlling unwanted species that threaten native species or impede current ecosystem function, managers may prevent establishment of less desirable species.

Climate change can best be met by engaging all levels of the National Park Service. While resource management is implemented at individual parks, planning and support can be provided at all management levels, with better integration between planners and resource management staff. A revision of the National Park Service Management Policies to incorporate climate change considerations would help to codify the importance of the issue. Park General Management Plans and resource management plans also could be amended to include the understanding, goals, and plans that address climate change issues. Climate change education and coordination efforts at the national level will be helpful for offering consistent guidance and access to information. Regional- and network-level workshops and planning exercises will be important for addressing issues at appropriate scales, as will interagency activities that address climate change impacts to physical and natural resources regardless of political boundaries.

4.2 BACKGROUND AND HISTORY

The U.S. national parks trace their distinctive origins to the early 19th century. The artist George Catlin is credited with initiating the uniquely American idea of protected national parks. While traveling through the Dakota territories in 1832, he expressed concern over the impact of westward expansion on wildlife, wilderness, and Indian civilization; he suggested they might be preserved "by some great protecting policy of government…in a magnificent park…A nation's park, containing man and beast, in all the wild and freshness of their nature's beauty" (Pitcaithley, 2001). In 1872, the U.S. Congress created the world's first national park, Yellowstone, in Wyoming and Montana territories "as a public park or pleasuring ground for the benefit and enjoyment of the people."[2] Other spectacular natural areas soon followed as Congress designated Sequoia, Yosemite, Mount Rainier, Crater Lake, and Glacier as national parks in an idealistic impulse to preserve nature (Baron, 2004).

The U.S. National Park System today includes a diverse set of ecological landscapes that form an ecological and cultural bridge between the past and the future. Covering about 4% of the United States, the 338,000 km² of protected areas in the park system contain representative landscapes of many of the world's biomes and

[2] H.R. 764

ecosystems. U.S. national parks are found across a temperature gradient from the tropics to the tundra, and across an elevational gradient from the sea to the mountains. These parklands are dynamic systems, containing features that reflect processes operating over time scales from seconds to millennia. For example, over millions of years, seasonal variation in flows and sediment in the Colorado River, which flows through Grand Canyon National Park, produced an unusual river ecosystem surrounded by rock walls that demonstrate countless annual cycles of snowmelt and erosion (Fig. 4.1). At the other end of the geologic spectrum are "new" park ecosystems such as the Everglades, which is less than 10,000 years old. Seasonal patterns of water coursing through the sloughs in the Everglades, as in the Grand Canyon, produced an ecosystem with plants and animals that requires the ebb and flow of water to persist (Fig. 4.2).

Figure 4.1. Looking up from the Colorado River at the Grand Canyon. Photo courtesy of Jeffrey Lovich, USGS.

As greenhouse gases continue to accumulate in the atmosphere, the effects of climate change on the environment will only increase. Ecological changes will range from the emergence of new ecosystems to the disappearance of others. Few natural ecosystems remain in the United States; the National Park Service (NPS) is steward of some of the most intact representatives of these systems. However, changes in climate that are now being driven by human activities are likely to profoundly alter national parks as we know them. Some iconic species are at high risk of extinction. For example, the Joshua tree is likely to disappear from both Joshua Tree National Monument and the southern two thirds of its range, where it is already restricted to isolated areas that meet its fairly narrow winter minimum temperature requirements (Fig. 4.3).[3] The distributions of many other species of plants and animals are likely to shift across the American landscape, independent of the borders of protected areas. National parks that have special places in the American psyche will remain parks, but their look and feel may change dramatically. For example, the glaciers in Glacier National Park are expected to melt by 2030 (Hall and Fagre, 2003). Therefore, the time is ripe for the NPS, the Department of the Interior, and the American public to revisit our collective vision of the purpose of parks.

Figure 4.2. Everglades National Park. Photo by Rodney Cammauf, courtesy of National Park Service.

[3] **Cole**, K.L., K. Larsen, P. Duffy, and S. Arundel, 2005: Transient dynamics of vegetation response to past and future major climatic changes in the Southwestern United States. *Proceedings of the Workshop on Climate Science in Support of Decision Making*, Online poster report, http://www.climatescience.gov/workshop2005/posters/P-EC4.2_Cole.pdf.

Figure 4.3. A Joshua Tree in Joshua Tree National Park. Photo courtesy of National Park Service.

Now is also the time to evaluate what can and should be done to minimize the effects of climate change on park resources, and to maximize opportunities for wildlife, vegetation, valued physical features, and the processes that support them to survive in the face of climate change. National parks increasingly are isolated by developed lands, and climate change is inseparable from the many other phenomena that degrade natural resources in national parks. Where national parks share boundaries with other federally or tribally managed lands, climate change can serve as a strong incentive to develop and implement regional efforts to manage ecosystems with a shared vision. Using climate change scenarios, we can realistically reevaluate current management efforts to reduce habitat fragmentation, remove or manage invasive species, maintain or restore natural disturbance regimes, and maximize air and water quality. Positive and negative feedbacks between contemporary changes in climate and resource management priorities must be carefully considered.

This chapter is directed specifically at the 270 national park areas with natural resource responsibilities, although many of the approaches we suggest are applicable to a diversity of resources and sites, including cultural and historical parks and other public and tribal lands. In this chapter, we suggest how national park managers might increase the probability that their resources and operations will adapt successfully to climate change. Successful adaptation begins by moving away from traditional ways of managing resources. We discuss strategies to stimulate proactive modes of thinking and acting in the face of climate change and other environmental changes. These strategies include broadening the portfolio of management approaches, increasing the capacity to learn from management successes and failures, and examining and responding to the multiple scales at which species and processes function. Strategies also include catalyzing ecoregional coordination among federal, state, and private entities, valuing human resources, and understanding what climate change means for interpreting the language of the NPS Organic Act. By modifying and expanding its current monitoring systems, NPS can expand its capacity to document and understand ecological responses to climate change and management interventions. By minimizing the negative effects from other current stressors, NPS may be able to increase the possibility that natural adjustments in habitats and processes can ease the transition to new climate regimes.

There are three critical messages this chapter is meant to convey:

1. We know climate has fundamentally defined our national parks. Their diversity and their stunning coastlines, caves, mountains and deserts are all the product of the interaction of temperature and precipitation, acting on the scale of days and seasons to eons. Climate *change* is redefining these parks, and will continue to do so. As such it cannot be considered merely as "one more stressor" to be considered and dealt with. Changing climate will undermine, or possibly enhance, efforts to reduce the damage done by other unnatural types of disturbances such as pollution, invasive species, or habitat fragmentation. *Starting now, the influence of changing climate must therefore be considered in conjunction with every resource management activity planned and executed in national parks.*

2. The adaptation approaches suggested in this chapter are meant to increase resilience,

which is defined as the amount of change or disturbance that a system can absorb before it undergoes a fundamental shift to a different set of processes or structures (Holling, 1973; Gunderson, 2000). *Because, however, the climate is changing and will continue to change, promoting resilience as a management strategy may only be effective until thresholds of resilience are overcome.* Our confidence in the effectiveness of the adaptation options proposed is based on near-term responses of perhaps the next several decades.

3. *Finally, and perhaps most importantly, the onset and continuance of climate change over the next century requires NPS managers to think differently about park ecosystems than they have in the past. Preparing for and adapting to climate change is as much a cultural and intellectual challenge as it is an ecological one.*

4.2.1 Legal History

The U.S. NPS Organic Act established the National Park System in 1916 "to conserve the scenery and the natural and historic objects and the wild life therein and to provide for the enjoyment of the same in such manner and by such means as will leave them unimpaired for the enjoyment of future generations."[5] This visionary legislation set aside lands in the public trust and created "a splendid system of parks for all Americans" (Albright and Schenck, 1999). The U.S. National Park System today includes more than 390 natural and cultural units, and has been emulated worldwide. The National Park System has the warm support of the American people, and parks are often the embodiment of widespread public sentiment for conservation and protection of the environment (Winks, 1997).

[5] 16 U.S.C. 1 2 3, and 4

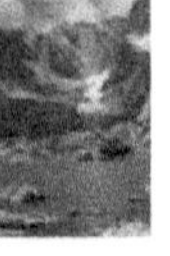

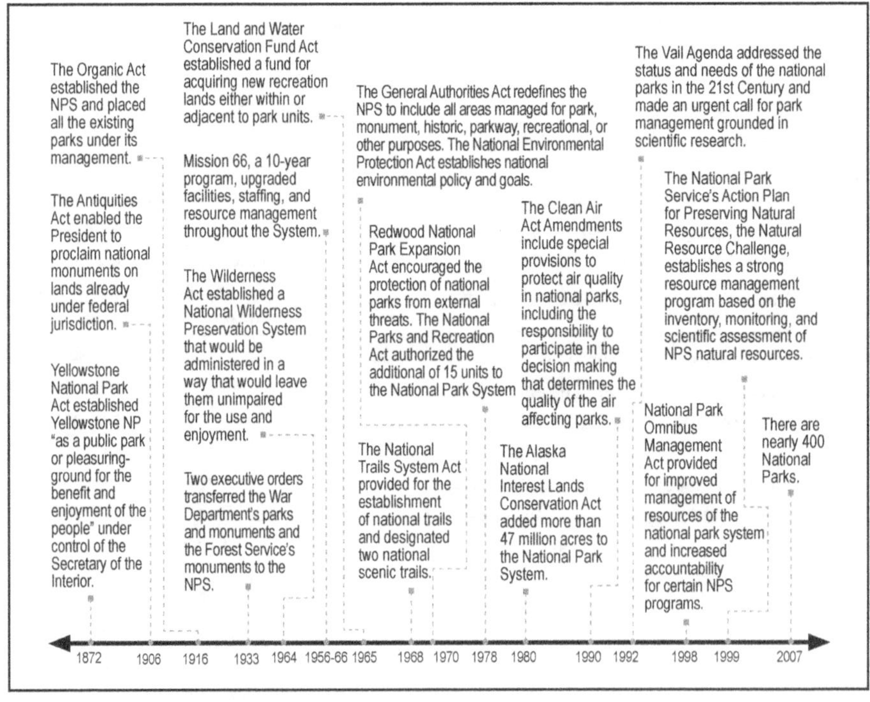

Figure 4.4. Historical timeline of the National Park Service.[4]

[4] Adapted from **National Park Service**, 2007: History. National Park Service, http://www.nps.gov/aboutus/history.htm, accessed on 4-10-2007.

The intent of Congress for management of national parks was initially set out in the Organic Act (see Fig. 4.4). The 1970 General Authorities Act and the 1978 "Redwood Amendment" to the Organic Act strengthened the Service's mission of conservation by clarifying that the "fundamental purpose" of the National Park System is the mandate to conserve park resources and values. This mandate is independent of the separate prohibition on impairment. Park managers have the authority to allow and manage human uses, provided that those uses will not cause impairment, which is an unacceptable impact. Enabling legislation and park strategic and general management plans are used to guide decisions about whether specific activities will cause impairment (National Park Service, 2006).

Other acts passed by Congress have extended the roles and responsibilities of national parks. National parks are included in the Wilderness Act of 1964 (for parks that include wilderness or proposed wilderness), the Wild and Scenic Rivers Act of 1968, the Clean Water Act of 1972, the Endangered Species Act of 1973, and the Clean Air Act of 1990. These acts, along with the Organic Act, are translated into management guidelines and policies in the 2006 Management Policies guide. Historian Robin Winks identified three additional acts that help to define the role of NPS in natural resource protection: the National Environmental Policy Act (NEPA) of 1972, the National Forest Management Act of 1976, and the Federal Land Policy and Management Act of 1976 (Winks, 1997).

Although its overarching mission has remained mostly unchanged, the NPS has undergone substantial evolution in management philosophy since 1916, and there are many examples that illustrate unconventional approaches to problems. For instance, national park status is not necessarily conferred in perpetuity. Twenty-four units of the National Park System were either deauthorized or transferred to other management custody for a number of reasons, demonstrating that designation of national park status is not necessarily permanent. While fifteen areas were transferred to other agencies because their national significance was marginal, others were deauthorized because their location was inaccessible to the public, and the management of five reservoirs was handed over to the Bureau of Reclamation.[6] Fossil Cycad National Monument in South Dakota, however, was deauthorized by Congress in 1957 due to near-complete loss of the fossil resource to collectors (National Park Service, 1998).

Prior to the 1960s, the NPS "practiced a curious combination of active management and passive acceptance of natural systems and processes, while becoming a superb visitor services agency" (National Park Service, 1999). The parks actively practiced fire suppression, aggressive wildlife management (which included culling some species and providing supplemental food to others), and spraying with pesticides to prevent irruptions of native insects. Development of ski slopes and golf courses within park boundaries was congruent with visitor enjoyment. During the 1960s, the Leopold Report on Wildlife Management in National Parks, the 1964 Wilderness Act, and the growth of the environmental movement ushered in a different management philosophy (Leopold, 1963). Managers began to consider natural controls on the size of wildlife populations. Some park managers decided skiing and golf were not congruent with their mission, and closed ski lifts and golf courses. The Wilderness Act of 1964 restricted mechanized and many other activities in designated or proposed wilderness areas within parks. Throughout its history, NPS has changed its priorities and management strategies in response to increased scientific understanding of ecological systems, public opinion, and new laws and administrative directives. Today, confronted not only with climate change but with many other threats to natural resources from within and outside park boundaries, the Park Service again has the opportunity to revisit resource management practices and policies.

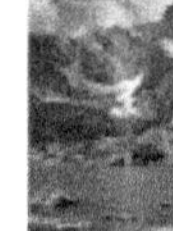

4.2.2 Interpretation of Goals

The aggregate federal laws described above strongly suggest that the intent of Congress is not only to "conserve unimpaired" but also to minimize human-caused disturbances, and to restore and maintain the ecological integrity of the national parks. The NPS mission

[6] **National Park Service**, 2003: National Park Service history: former National Park System units: an analysis. National Park Service, http://www.nps.gov/history/history/hisnps/NPSHistory/formerparks.htm, accessed on 7-13-2007.

BOX 4.1. The National Park Service Mission.

The National Park Service preserves unimpaired the natural and cultural resources and values of the National Park System for the enjoyment, education, and inspiration of this and future generations. The Park Service cooperates with partners to extend the benefits of natural and cultural resource conservation and outdoor recreation throughout this country and the world.

remains much as it was in 1916 (Box 4.1). In general, the Secretary of the Interior, and by extension, the Director of the NPS, have been given broad discretion in management and regulation provided that the fundamental purpose of conservation of park resources and values is met. Although individual park-enabling legislation may differ somewhat from park to park, all parks are bound by the NPS Organic Act, the Redwood National Park Expansion Act, and other legislation described above. The enabling language of the Organic Act creates a dilemma that complicates the Park Service's ability to define key ecosystem characteristics upon which the goals depend: for example, what is the definition of "unimpaired?" While "impair" is defined as "to cause to diminish, as in strength, value, or quality," it requires establishment of a baseline or reference condition in order to evaluate deviation from that condition.[7] Interpretations of how to manage parks to maintain unimpaired conditions have changed over time, from benign neglect early in the history of the national parks to restoring vignettes of primitive America and enhancing visitor enjoyment through much of the 20th century. The definition of "unimpaired" is central to how well NPS confronts and adapts its resources to climate change.

To accomplish its mission, NPS employs more than 14,000 permanent personnel and some 4,000 temporary seasonal employees (Fig. 4.5). Parks receive more than 270 million visitors each year. Operations and management occur at three levels of organization: national, regional, and individual park. Service-wide policy is

[7] "Impair" 2003: In: *The American Heritage® Dictionary of the English Language,* 4th ed. New York: Houghton Mifflin Company, 2000.

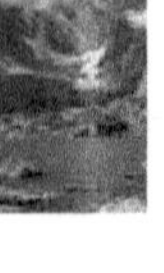

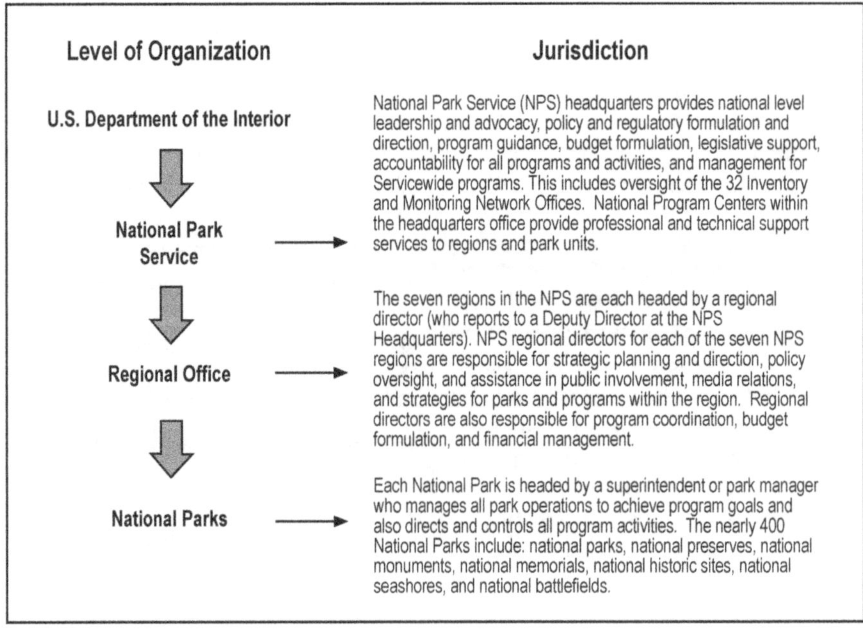

Figure 4.5. Organizational chart of National Park Service.[8]

[8] Adapted from **National Park Service**, 2007: Organization. National Park Service, http://www.nps.gov/aboutus/organization.htm, accessed on 4-10-2007.

issued by the Director of the NPS, and may also be issued by the President, Congress, the Secretary of the Interior, or the Assistant Secretary for Fish, Wildlife, and Parks. Many of the programs that make up or are supplemented by the Natural Resource Challenge, described below, are administered from the national headquarters, called the Washington Office. Seven regional offices divide the National Park System by geography (Northeast, National Capital, Southeast, Midwest, Intermountain, Pacific West, and Alaska Regions). Regional offices provide administrative services and oversight to parks, and serve as conduits for information between the Washington Office and parks. Two national-level offices, the Denver (Colorado) Service Center and the Interpretive Design Center at Harpers Ferry, West Virginia, provide professional architectural and engineering services, and media products (*e.g.*, publications, exhibits, interactive presentations, and audio-visual displays) to individual parks.

There are more than 14 different categories of park units within the National Park System, including national parks, national scenic rivers, lakeshores, seashores, historic sites, and recreation areas (Fig. 4.6). The parks in each category offer different experiences for visitors. In addition to the overarching NPS mission, certain activities can take place within individual park units depending on specific Congressional enabling legislation at the time of establishment. For example, public hunting is recognized as a legitimate recreational activity within the boundaries of many national lakeshores, seashores, recreation areas, and preserves because of the legislation that established those specific park units.

Approximately 270 National Park System areas contain significant natural resources. The Natural Resource Challenge, an action plan for preserving natural resources in national parks, was established in 2000 in the recognition that knowledge of the condition and trends of NPS

Figure 4.6. Map of the National Park System. Data courtesy of National Park Service, Harpers Ferry Center.[9]

[9] **National Park Service**, Harpers Ferry Center, 2007: Harpers Ferry Center: NPS maps. National Park Service, http://home.nps.gov/applications/hafe/hfc/carto-detail.cfm?Alpha=nps, accessed on 4-10-2007.

natural resources was insufficient to effectively manage them (National Park Service, 1999). The Natural Resource Challenge has already enabled a significant advancement in inventory, monitoring, and understanding of resources. There are four natural resource action plan goals (Box 4.2). These goals are aligned with the NPS Strategic Plan, which emphasizes the role of natural resource stewardship and has as its first goal the preservation of park resources. Central to the Natural Resource Challenge is the application of scientific knowledge to resource management.

The Natural Resource Challenge includes the Inventory and Monitoring Program (including NPS Resource Inventories and Vital Signs Monitoring Networks), the Biological Resources Management Program, and the Air Quality, Water Resources, and Geologic Resources Programs. Natural Resource Challenge programs mostly provide information, management guidance, and expertise to parks, as opposed to active management, although an exception is the Invasive Plant Management Teams. Individual parks set their own resource management agendas, which they carry out with permanent and seasonal staff and money from the park, the Natural Resource Preservation Program (a competitive research fund), and Park-Oriented Biological Support (a joint USGS/NPS program). Many parks also encourage or invite researchers to study specific issues facilitated by two NPS entities—the Cooperative Ecosystem Studies Units and the Research Learning Centers.

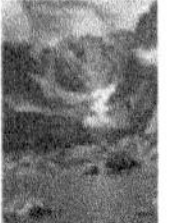

BOX 4.2. Natural Resource Action Plan Goals.

- National parks are preserved so that this generation and future generations can enjoy, benefit, and learn from them.
- Management of the national parks is improved through a greater reliance on scientific knowledge.
- Techniques are developed and employed that protect the inherent qualities of national parks and restore natural systems that have been degraded; collaboration with the public and private sectors minimizes degrading influences.
- Knowledge gained in national parks through scientific research is promulgated broadly by the National Park Service and others for the benefit of society.

Most parks operate under a General Management Plan, a broad planning document that creates a vision for the park for a 15- to 20-year period. The General Management Plan provides guidance for fulfilling the park's purpose and protecting the park's fundamental resources and values. As part of the General Management Plan, or sometimes developed as an addendum to the General Management Plan, Desired Conditions Plans articulate ideal future conditions that a park strives to attain. Individual parks may have up to 40 additional specific resource- or place-based management plans (an example is Rocky Mountain National Park's Elk and Vegetation Management Plan). These natural resource management plans are increasingly science driven. However, despite having guidance and policies for natural resource management planning, there are still many parks that have no planning documents identifying desired future conditions, and many of the General Management Plans are out of date.

Public input, review, and comment are encouraged, and increasingly required, in all park planning activities. Increasingly, park planning activities take place in regional contexts and in consultation with other federal, state, and private land and natural resource managers.

4.3 CURRENT STATUS OF MANAGEMENT SYSTEMS

4.3.1 Key Ecosystem Characteristics on Which Goals Depend

National parks are found in every major biome of the United States. Parks with managed natural resources range from large intact (or nearly intact) ecosystems with a full complement of native species—including top predators, (*e.g.*, some Alaskan parks, Yellowstone, Glacier; Stanford and Ellis, 2002)—to those diminished by disturbances such as within-park or surrounding-area legacies of land use, invasive species, pollution, or regional manipulation of resources (*e.g.*, hydrologic flow regimes).

Current NPS policy calls for management to preserve fundamental physical and biological processes, as well as individual species, features, and plant and animal communities (National Park Service, 2006). "The Service recognizes that natural processes and species are evolving, and NPS will allow this evolution to continue—minimally influenced by human actions" (National Park Service, 2006). Resources, processes, systems, and values are defined in NPS Management Policies (National Park Service, 2006) as:

- Physical resources such as water, air, soils, topographic features, geologic features, paleontological resources, and natural soundscapes and clear skies, both during the day and at night;
- Physical processes such as weather, erosion, cave formation, and wildland fire;
- Biological resources such as native plants, animals, and communities;
- Biological processes such as photosynthesis, succession, and evolution;
- Ecosystems; and
- Highly valued associated characteristics such as scenic views.

4.3.2 Stressors of Concern

Despite mandates to manage national parks to maintain their unimpaired condition, there are many contemporary human-caused disturbances (as opposed to natural disturbances) that create obstacles for restoring, maintaining, or approximating the natural conditions of ecosystems. The current condition of park resources can be a legacy of past human activities or can be caused by activities that take place outside park boundaries. We grouped the most widespread and influential of the disturbances that affect park condition into four broad classes: altered disturbance regimes, habitat fragmentation and loss, invasive species, and pollution.

These four classes of stressors interact. For example, alteration of the nitrogen cycle via atmospheric nitrogen deposition can facilitate invasion of non-native grasses. In terrestrial systems, invasion of non-native grasses can alter fire regimes, ultimately leading to vegetation-type conversions and effective loss or fragmentation of wildlife habitat (Brooks, 1999; Brooks *et al.*, 2004). Climate change is expected to interact with these pressures, exacerbating their effects. Climate change is already contributing to increasing frequency and intensity of wildfires in the western United States, potentially accelerating the rate of vegetation-type conversions that are being driven by invasive species (Mckenzie *et al.*, 2004; Westerling *et al.*, 2006). Two illustrations are presented in Boxes 4.3 and 4.4 of complex stressor interactions: fire and climate interactions in western parks, and myriad stressor interactions in the Everglades.

BOX 4.3. Interactions of Fire with Other Stressors and Resources.

Future increases in the size and severity of wildland fires are likely not just in the western park areas, but across the United States (Dale *et al.*, 2001). Such increases would have direct impacts on infrastructure and air quality. There would also be short- and long-term consequences for conservation of valued species and their habitats. McKenzie *et al.* (2004) presented a conceptual model of how interactions between naturally functioning ecosystems with some recurrence interval of fire can be perturbed under conditions of climate change (see below). Warmer and drier summers are likely to produce more frequent and more extensive fires. Trees and other vegetation are also likely to be stressed by drought and increasing insect attacks, since stressed vegetation is predisposed toward other stressors (Paine, Tegner, and Johnson, 1998). Insect-caused mortality can lead to large areas with accumulations of woody fuels, enhancing the probability of large fires. More frequent and more extensive fires will lead to greater area burned. Over time this can alter existing forest structure. Depending on the location, homogeneous forest stands can regenerate. Savannahs or grasslands may replace trees in some areas. Increased erosion on slopes may affect forest fertility and stream or lake water quality. Increased fire frequency—indeed, any kind of land disturbance—favors opportunistic and weedy species. Annual weeds, such as cheatgrass and buffelgrass in the western United States, regenerate rapidly after fire and produce abundant fuel for future fires. The number of native fire-sensitive species decreases. Vegetation types that are at risk from either fire or the combination of fire and invasive species put obligate bird, mammal, and insect species at risk of local or regional extinction (McKenzie *et al.*, 2004).

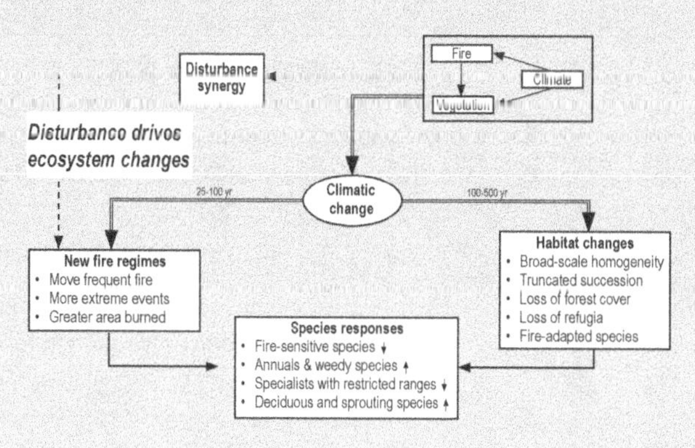

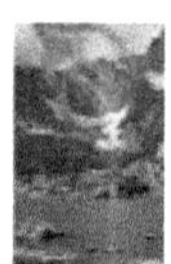

BOX 4.4. Altered Flow Regimes, Increased Nutrients, Loss of Keystone Species, and Climate Change.

From the freshwater marshes of the Everglades to the shallow waters of Florida Bay, human alterations have resulted in dramatic ecosystem changes—changes that are likely to become exaggerated by climate change. Nutrient enrichment of freshwater sawgrass marshes have led to marshes now dominated by cattails (Unger, 1999). The soil phosphorus content defines these alternate sawgrass or cattail states, and several types of disturbances (fires, drought, or freezes) can trigger a switch between states (Gunderson, 2001). Downstream, the Florida Bay system has flipped from a clear-water, seagrass-dominated state to one of murky water, algal blooms, and recurrently stirred-up sediments. Hurricane frequency, reduced freshwater flow entering the Bay, higher nutrient concentrations, removal of large grazers such as sea turtles and manatees, sea level rise, and construction activities that restrict circulation in the Bay have all contributed to the observed changes (Gunderson, 2001). A balance between freshwater inflows and sea levels maintains the salinity gradients necessary for mangrove ecosystems, which are important for mangrove fish populations, wood stork (*Mycteria americana*) and roseate spoonbill (*Platelea ajaja*) nesting colonies, and estuarine crocodiles.

Although there are intensive efforts to increase hydrologic flows to and through the Everglades, climate change is expected to increase the difficulty of meeting restoration goals. Interactions of fire, atmospheric CO_2, and hurricanes may favor certain tree species, possibly pushing open Everglades pine savannahs toward closed pine forests (Beckage, Gross, and Platt, 2006). Tree islands, which are hotspots of biodiversity, and peatlands that make up much of the Everglades landscape, may be additionally stressed by drought and peat fires. Animals that rely on these communities may see their habitat decrease (Smith *et al.*, 2003). Mangroves may be able to persist and move inland with climate change, but that will depend on the rates of sea level rise (Davis *et al.*, 2005).

4.3.2.1 Altered Disturbance Regimes

Natural disturbance processes such as fire, insect outbreaks, floods, avalanches, and forest blowdowns are essential drivers of ecosystem patterns (*e.g.*, species composition and age structure of forests) and processes (*e.g.*, nutrient cycling dynamics). Disturbance regimes are characterized by the spatial and temporal patterns of disturbance processes, such as the frequency, severity, and spatial extent of fire. Many natural disturbance regimes are strongly modulated by climate variability, particularly extreme climate events, as well as by human land uses. Thus, climate change is expected to alter disturbance regimes in ways that will profoundly change national park ecosystems. Three types of natural disturbances whose frequency and magnitude have been altered in the past century include fire, beach and soil erosion, and natural flow regimes.

Fire

Historic fire exclusion in or around many national parks has sometimes increased the potential for higher-severity fires and mortality of fire-resistant species. Fire-resistant tree species that may have had their natural fire frequencies suppressed include giant sequoias (*Sequoia giganteum*) in Yosemite, Sequoia, and Kings Canyon National Parks; ponderosa pine (*Pinus ponderosa*) in Grand Canyon and other southwestern parks; and southwestern white pine (*Pinus strobiformis*) in Guadalupe Mountains National Park. In other areas, such as Yellowstone or the subalpine forests of Rocky Mountain National Park (see Case Study Summary 4.1), fires are driven almost completely by historically infrequent weather events and post-fire forest regrowth (Romme and Despain, 1989). Recent land use or fire suppression have had little effect on fire regimes in the latter parks.

Coast and Soil Erosion

Coasts are naturally dynamic systems that respond to changes in sea level, storms, wind patterns, sediment inputs from river systems, and offshore bathymetry. Barrier islands, which provide protection to coasts, migrate in response to storms and currents and are replenished by winds, waves, currents, and tides. When sea level rise is gradual, ecosystems and landforms can adjust via accretion of sediments, and thus

CASE STUDY SUMMARY 4.1

Rocky Mountain National Park, Colorado
Western United States

Why this case study was chosen

Rocky Mountain National Park:

- Serves as a good example of the state in which most parks find themselves as they confront resource management in the face of climate change: regardless of the greater apparent urgency in some parks, all of them will have to initiate adaptation actions in order to meet the National Park Service mission and goals;
- Contains biomes that are vulnerable to climate change such that the distribution, condition, and abundance of ecological resources could be drastically altered;
- Is staffed with personnel who are already engaged in early stages of adaptation planning.
- Is a major destination for more than three million visitors per year from Colorado, the United States, and abroad, who come to experience the unique high-elevation environment and escape summer heat;
- Is a crucial component of the greater Southern Rockies Ecosystem, and nearly surrounded by other public lands, including wilderness.

Management context

Located in the Front Range of the Rocky Mountains, the 415-square-mile Rocky Mountain National Park (RMNP) was established in 1915 as a public park for the benefit and enjoyment of the people of the United States, with regulations primarily aimed at the freest use of the park and the preservation of natural conditions and scenic beauties. A primary management goal is to maintain the park in its natural condition. RMNP's wide elevation gradient—from 8,000 to more than 14,000 feet—includes montane forests and grasslands, old-growth subalpine forests, and the largest expanse of alpine tundra in the lower 48 states. More than 150 lakes and 450 miles of streams form the headwaters of the Colorado River to the west and the South Platte River to the east. Rich wetlands and riparian areas are regional hotspots of native biodiversity. Several small glaciers and rock glaciers persist in east-facing cirque basins along the Continental Divide. The park is home to populations of migratory elk, mule deer, and bighorn sheep; alpine plant and animal species such as white-tailed ptarmigan, pika, and yellow-bellied marmot; and several endangered species such as the boreal toad and the greenback cutthroat trout.

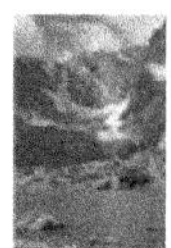

Key climate change impacts

- Projected biome shifts, fragmentation, and losses as temperatures warm and major habitats shift upward in elevation;
- Projected ecosystem disruptions due to increased risks of fire, insect pest outbreaks, invasion by non-native species, and population changes in native species (e.g., grazers and browsers);
- Projected reduction of snowpack;
- Projected warming of water bodies with resulting impacts to aquatic life;
- Projected species losses (e.g., white-tailed ptarmigan and other tundra obligates);
- Projected population increases in organisms that can stress the system (e.g., elk);
- Observed increases in summer temperatures (average increase of 3°C from 1991–2001) as well as increases in extreme heat events;
- Observed earlier melting of winter snowpack;
- Observed early emergence of animals from hibernation and early arrival of migratory species;
- Observed thinning of nearby Arapahoe Glacier by more than 40 m since 1960.

Opportunities for adaptation

- RMNP has benefited from long-term research and monitoring projects and climate change assessments that will be vital to ongoing adaptation planning.
- Park managers have been proactive in removing or preventing invasive species, managing fire through controlled burns and thinning, reducing regional air pollution through partnerships with regulatory agencies, purchasing water rights, restoring streams and lakes to free-flowing status, and preparing a plan to reduce elk populations to appropriate numbers.

CASE STUDY SUMMARY 4.1 (CONTINUED)

- Managers have identified a strategy for increasing their ability to adapt to climate change built on their current activities, what they know, and what they do not know about upcoming challenges related to climate change.
- Regular workshops with scientific experts offer opportunities to develop planning scenarios, propose adaptive experiments and management options, learn from high resolution models of species and process responses to possible climates and management activities, and keep abreast of the state of knowledge regarding climate change and its effects.
- An RMNP Science Advisory Board has been proposed to contribute strategic thinking to enable park managers to anticipate climate-related events.
- By developing a regional-scale approach toward adaptation with neighboring and regional resource managers, the park keeps its options open for allowing species to migrate in and out of the park and protects an important part of the greater Southern Rockies Ecosystem.
- Managers have recognized the need for learning activities and opportunities for all park employees to increase their knowledge of climate change-related natural resource issues within RMNP.

Conclusions

RMNP is home to a wide diversity of valued ecosystems and species. As such, it attracts large numbers of visitors. RMNP is also potentially highly vulnerable to climate change. Adaptation planning is vital if the health of RMNP biomes and the greater Southern Rockies Ecosystem is to be protected, and such planning has already begun. However, much remains to be accomplished. Complex climate change issues require flexible ways of thinking, and enough time and systems-level training to approach them with broad, strategic vision. Expanded monitoring programs within the park could ensure that early signs of impacts are detected in all biomes. Forums for identifying problems and solutions are already being initiated between park managers and regional scientists. Acceleration of these dialogues would speed identification of specific and realistic adaptation options for each of the major resources within the park.

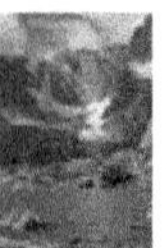

keep pace with the changes. Coastal responses may be nonlinear in response to abrupt natural disturbances; freshwater and salt marshes, mangroves, or beach regeneration may take years to decades to recover after severe storms, and irreversible changes can occur if there is salt-water intrusion or a lack of sediment source for replenishment (IPCC, 2007). Direct human activities have had significant impacts on coastlines and coastal zones, and a trend toward increasing coastal development is projected to occur through the next century (IPCC, 2007). Drainage of coastal wetlands, deforestation and reclamation, and discharge of pollutants of all kinds are examples of direct alterations of coasts. Extraction of oil and natural gas can lead to subsidence. Structures such as seawalls and dams harden the coast, impede natural regeneration of sediments, and prevent natural inland migration of sand and vegetation after disturbances. Channelization of marshes and waterways alters freshwater, sediment, and nutrient delivery patterns (IPCC, 2007).

Soils provide a critical foundation for ecosystems, and soil development occurs in geologic time. Natural soil erosion can also occur slowly, over eons, but rapid soil loss can happen in response to extreme physical and climatic events. Many of the changes in soil erosion rates in the parks are a legacy of human land use. Soil erosion rates are also influenced by interacting stressors, such as fire and climate change. Historic land uses such as grazing by domestic livestock have accelerated water and wind erosion in some semiarid national parks when overgrazing has occurred. This erosion has had long-term effects on ecosystem productivity and sustainability (Sydoriak, Allen, and Jacobs, 2000). In Canyonlands National Park, soils at sites grazed from the late 1800s until the 1970s have lost much of their vegetative cover. These soils have lower soil fertility than soils that never were exposed to livestock grazing (Belnap, 2003). Erosion after fires also can lead to soil loss, which reduces options for revegetation, and contributes sediment loads to streams and lakes. Excessive sediment loading degrades aquatic habitat. Long-term erosion in a humid environment like that in Redwood National Park is a direct legacy of intensive logging and road development.[10]

Altered Flow Regimes

Freshwater ecosystems are already among the most imperiled of natural environments worldwide, due to human appropriation of freshwater (Gleick, 2006). Few natural area national parks have rivers that are unaltered or unaffected by upstream manipulations. Reservoirs in several national parks have flooded valleys where rivers once existed. Examples of large impoundments include Hetch Hetchy Reservoir in Yosemite National Park, Lakes Powell and Mead on the Colorado River of Glen Canyon and Lake Mead National Recreation Areas, and Lake Fontana in Great Smoky Mountains National Park. There are many smaller dams and reservoirs in other national parks. Parks below dams and diversions, such as Big Bend National Park, are subject to flow regulation from many miles upstream. Irrigation structures, such as the Grand Ditch in Rocky Mountain National Park, divert annual runoff away from the Colorado River headwaters each year.[11] Volume, flow dynamics, temperature, and water quality are often highly altered below dams and diversions (Poff *et al.*, 2007). Everglades National Park now receives much less water than it did before upstream drainage canals and diversions were constructed to divert water for agriculture. Natural hydrologic cycles have been disrupted, and the water that Everglades now receives is of lower quality due to agricultural runoff. Altered hydrologic regimes promote shifts in vegetation; facilitate the invasion of non-native species such as tamarisk, Russian olive, and watermilfoil; and promote colonization by native species such as cattail.

[10] **National Park Service**, 2006: Redwood National and State Parks. National Park Service, http://www.nps.gov/redw/naturescience/environmentalfactors.htm, accessed on 5-15-2007.

[11] **National Park Service**, 2007: Rocky Mountain National Park - hydrologic activity. National Park Service, http://www.us-parks.com/rocky/hydrologic_activity.html, accessed on 4-6-2007.

Groundwater depletion, which influences replenishment of springs, has been suggested as a cause of decreased artesian flows at Chickasaw National Recreation Area and in desert parks such as Organ Pipe Cactus and Death Valley (*e.g.*, Knowles, 2003). Groundwater depletion also directly affects phreatophytes, or water-loving riparian and wetland species. Groundwater depletion increasingly is occurring throughout the United States, even in the southeastern parks such as Chattahoochee National River National Recreation Area (Lettenmaier *et al.*, 1999). Caves, such as Jewel Cave National Monument, and the processes that maintain them are at special risk from groundwater depletion. Impacts include drying of cave streams and pools, drying of speleothems (stalactites and other carbonate formations) so they do not continue to grow, and loss of habitat for aquatic cave fauna (Ford and Williams, 1989).

Land use, particularly urbanization, alters flow regimes through creation of impervious surfaces. Water that previously percolated through soils and was assimilated by native vegetation runs rapidly off paved surfaces, increasing the probability that streams and rivers will flood in response to storms. Flooding is a management concern in urban parks, such as Rock Creek Park in Washington, DC. When Rock Creek was established in 1890, it was at the edge of the city; its watershed is now wholly urbanized.

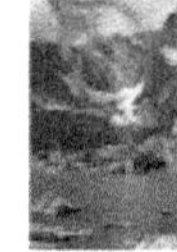

4.3.2.2 Habitat Alteration: Fragmentation and Homogenization

"Wild life" is identified specifically in the NPS enabling legislation, and regardless of whether the framers of the Organic Act intended the words to mean only birds and mammals, or all wild living things, large mammals have long been a central focus of NPS management and public discourse. Many wildlife challenges within parks stem from past extirpation of predators and overexploitation of game species, such as elk, and furbearers, such as beaver and wolverine. Restoration of species that were extirpated, and control of species that in the absence of predators have greatly expanded their populations, are important issues in many of the 270 natural area parks (Tomback and Kendall, 2002).

National parks may be affected by landscape alterations occurring either within or beyond their boundaries. Both fragmentation and landscape homogenization pose serious challenges to maintaining biodiversity. Roads, trails, campsites and recreational use can lead to fragmentation of habitat for various species. Fragmentation can directly or indirectly deter or prevent animal species from accessing food sources or accessing mating or birthing grounds (*e.g.*, some species of birds will not return to their nests when humans are present nearby, *e.g.*, Rodgers, Jr. and Smith, 1995). Moreover, fragmentation can impede dispersal of plant seeds or other propagules and migration of plant and animal populations that live along boundaries of national parks. However, fragmentation can also increase the amount and quality of habitat for some species, such as white-tailed deer, which, while native, are now considered a nuisance because of high numbers in many parts of the eastern United States.

Causes of fragmentation include road building and resource extraction such as timber harvest, mines, oil and gas wells, water wells, power lines, and pipelines. Coastal wetland ecosystems can be constrained by structures that starve them of sediments or prevent landward migration. In lands adjacent to parks, fragmentation increasingly is driven by exurban development—low-density rural home development within a landscape still dominated by native vegetation. Since 1950, exurban development has rapidly outpaced suburban and urban development in the conterminous United States (Brown *et al.*, 2005).[12] The effects of fragmentation are highly dependent on the spatial scale of disturbance and the particular taxonomic group being affected. And while there have been many studies on the effects of fragmentation on biodiversity, results of empirical studies are often difficult to interpret because they were conducted at patch scales rather than landscape scales, and did not distinguish between fragmentation and habitat loss (Fahrig, 2003). However, some known ecological effects include shifts in the distribution and composition of species, altered mosaics of land cover, modified disturbance

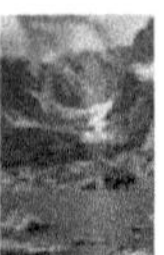

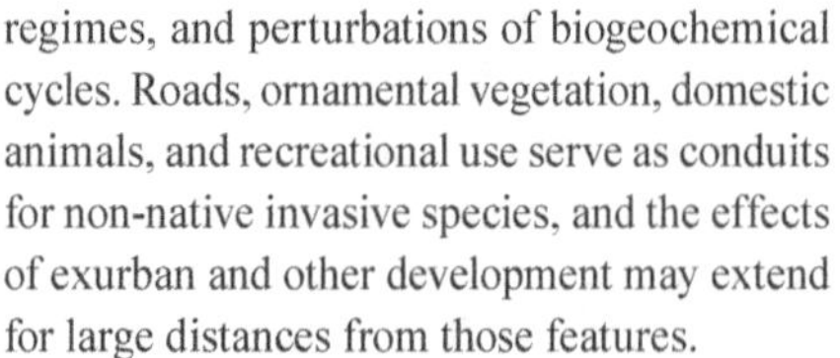

regimes, and perturbations of biogeochemical cycles. Roads, ornamental vegetation, domestic animals, and recreational use serve as conduits for non-native invasive species, and the effects of exurban and other development may extend for large distances from those features.

Management activities that homogenize landscapes have also contributed to changes in species composition and ecological processes. Landscape homogenization can select against local adaptation, reducing the ability of species to evolve in response to environmental change. For example, reductions in the naturally variable rates of freshwater inflows and increases in nutrients have converted much of the vegetation of Florida Bay in Everglades National Park from sea grasses to algae (Unger, 1999). Fire exclusion has created large tracts of even-aged forest and woodland in many western and midwestern parks, reducing heterogeneity of land cover and species richness (Keane *et al.*, 2002).

4.3.2.3 Invasive Species

The deliberate or inadvertent introduction of species with the capability to become nuisances or invaders is a major challenge to management throughout the national park system, and is likely to be exacerbated by climate change. These types of organisms are defined as invasive, whether or not they are non-native. Invasive species are those that threaten native species or impede current ecosystem function. Invasive plants are present across some 2.6 million acres in the national parks. Invasive animals are present in 243 parks.[13] The NPS has identified control of invasive species as one of its most significant land management issues, and has established a highly coordinated and aggressive invasive plant management program. Efforts to restore native plants also occur, but at much lower levels than control of invasive plants.

4.3.2.4 Air and Water Pollution

Air Pollution

Atmospheric processes link park ecosystems to sources of air and water pollution that may be hundreds of miles away. These pollutants

[12] **Hansen**, L.J., J.L. Biringer, and J.R. Hoffman, 2003: *Buying Time: a User's Manual for Building Resistance and Resilience to Climate Change in Natural Systems*. World Wildlife Foundation, Washington, DC.

[13] **National Park Service**, 2004: Invasive species management. National Park Service, http://www.nature.nps.gov/biology/invasivespecies/, accessed on 5-15-2007.

diminish both the recreational experience for park visitors and the ecological status of many park and wilderness ecosystems.

Ozone pollution from airsheds upwind of parks compromises the productivity and viability of trees and other vegetation. Because not all species are equally affected, competitive relationships are changed, leading to winners as well as losers. Ozone is also a human health hazard: during 2006, ozone health advisories were posted once each in Acadia and Great Smoky Mountains National Parks; and multiple times each in Sequoia, Kings Canyon, and Rocky Mountain National Parks.[14] Ozone concentrations are increasing in Congaree Swamp and 10 western park units, including Canyonlands, North Cascades, and Craters of the Moon.[15]

Acid precipitation is still a concern in many eastern parks. While sulfur dioxide emissions have decreased significantly in response to the Clean Air Act Amendments of 1990, the legacy of soil, lake, and stream acidification persists (Driscoll *et al.*, 2001). Acadia, Great Smoky Mountains, and Shenandoah National Parks have active monitoring programs that track stream acidity and biological responses. Acidic waters from air pollution in Shenandoah are responsible for the loss of native trout populations and decline in fish species richness (MacAvoy and Bulger, 1995; Bulger, Cosby, and Webb, 2000). Warmer future climate conditions, economic growth, and increasing populations will create more requirements for energy, and if the energy is derived from fossil fuels there is the potential for increasing acid rain.

Atmospheric nitrogen deposition, which is attributable to motor vehicles, energy production, industrial activities, and agriculture, contributes to acidification and also to fertilization of ecosystems, because nitrogen is an essential nutrient whose supply is often limited. Nitrogen saturation, or unnaturally high concentrations of nitrogen in lakes and streams, is of great concern to many national parks. Although nitrogen oxide emissions are decreasing in the eastern United States, nitrogen emissions and deposition are increasing in many western parks as human density increases. Gila Cliff Dwellings, Grand Canyon, Yellowstone, and Denali National Parks reported increased nitrogen deposition over the period 1995–2004.[14] Some classes of plants, especially many weedy herbs, may benefit from N-fertilization (Stohlgren *et al.*, 2002). Effects of excess nitrogen in Rocky Mountain National Park include changes in the composition of alpine tundra plant communities, increases in nutrient cycling and the nitrogen content of forests, and increased algal productivity and changes to species assemblages in lakes (Baron *et al.*, 2000; Bowman *et al.*, 2006).

The heavy metal mercury impairs streams and lakes in parks across the United States. Mercury is a byproduct of coal-fired energy production, incineration, mining, and other industrial activities. Mercury concentrations in fish are so high that many national parks are under fish advisories that limit or prohibit fish consumption. Parks in which levels of mercury in fish are dangerous to human health include Everglades, Big Cypress, Acadia, Isle Royale, and Voyageurs. Managers at many other parks, including Shenandoah, Great Smoky Mountains, and Mammoth Cave, have found significant bioaccumulation of mercury in taxonomic groups other than fish, including amphibians, bats, raptors, and songbirds. In Everglades, elevated mercury has been linked to mortality of endangered Florida panthers (Barron, Duvall, and Barron, 2004).

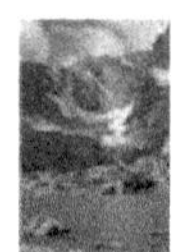

Water Pollution

Water quality in national parks is influenced not only by air pollution, but also by current or past land use activities and pollution sources within the watersheds in which national parks are located. Currently, agricultural runoff that includes nutrients, manure and coliform bacteria, pesticides, and herbicides affects waters in nearly every park downstream from where agriculture or grazing is located. Discharges from other non-point sources of pollution—such as landfills, septic systems, and golf courses—also cause problems for park resources, as they have for Cape Cod National Seashore, which now has degraded surface and groundwater quality.

---

[14] **National Park Service**, 2006: Ozone health advisory program yearly summaries. National Park Service, http://www2.nature.nps.gov/air/data/O3AdvisSum.cfm, accessed on 5-15-2007.

[15] **National Park Service**, 2006: Performance measures. National Park Service, http://www2.nature.nps.gov/air/who/npsPerfMeasures.cfm, accessed on 5-15-2007.

At least 10 parks, mostly in Alaska, are affected by past land-use activities and are designated as EPA Superfund sites. Severely polluted waters in Cuyahoga Valley National Park, in which surface oil and debris ignited in 1969, were an impetus for the Clean Water Act of 1972. Although the Cuyahoga River has become cleaner in the past three decades, it still receives discharges of storm water combined-sewer overflows, and partially treated wastewater from urban areas upstream of the park. Beaches of lakes and seashores, such as Indiana Dunes National Lakeshore, are sometimes affected by high levels of bacteria from urban runoff and wastewater after heavy rainfall events.

4.3.2.5 Direct Impacts of Climate Change

There will be some direct effects of climate change on national parks, as well as many interactive effects of climate change with the other major disruptions of natural processes described above. In addition to warming trends, climate change will influence the timing and rate of precipitation events. Both storms and droughts are expected to become less predictable and more intense. There will be direct effects on glaciers and hydrologic processes. Worldwide, glaciers are retreating rapidly, and glacier attrition is apparent in Glacier and North Cascades National Parks (Hall and Fagre, 2003; Granshaw and Fountain, 2006). The retreating Van Trump glacier on Mount Rainier has produced four debris flows between 2001-2006, filling the Nisqually River with sediment and raising the river bed at least six feet. Future high flow events will spread farther from the river banks because of the raised bed.[16] Data already show that climate change is modifying hydrologic patterns in seasonally snow-dominated systems (Mote, 2006). Snowmelt now occurs earlier throughout much of the United States (Huntington *et al.*, 2004; Stewart, Cayan, and Dettinger, 2005; Hodgkins and Dudley, 2006). Sea level rise has great potential to disturb coastal ecosystems, by intrusion of saltwater into freshwater marshes and by inundating coastal wetlands faster than they can compensate. Although coastlines are highly dynamic though geologic time, structural

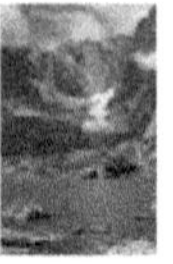

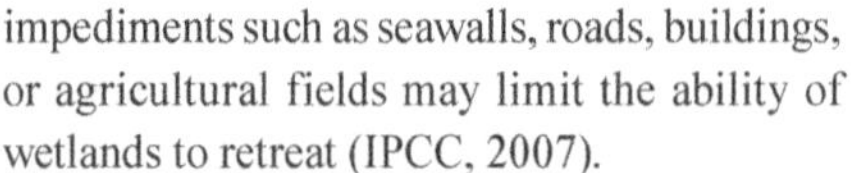

impediments such as seawalls, roads, buildings, or agricultural fields may limit the ability of wetlands to retreat (IPCC, 2007).

Climatic changes will have both direct and indirect effects on vegetation. With rapidly warming temperatures, more productive species from lower elevations that are currently limited by short growing seasons and heavy snowpack may eventually replace upper-elevation tree species (Hessl and Baker, 1997). Similarly, alpine meadows will be subject to invasion by native tree species (Fagre, Peterson, and Hessl, 2003). Subalpine fir is already invading the Paradise flower fields at Mt. Rainier National Park, taking advantage of mild years to establish, and forming tree islands that buffer individual trees against cold and snow. In Tuolumne Meadows, at 2,900 m in Yosemite National Park, lodgepole pine is rapidly establishing, and indeed is colonizing other more remote meadows above 3,000 m.[17] Vegetation will be redistributed along north-south gradients, as well as along elevation gradients, facilitated by dieback in southern ranges and possible expansion to cooler latitudes. Piñon pine forests of the Southwest are illustrative of how severe drought and unusual warmth exceeded species-specific physiological thresholds, causing piñon mortality across millions of hectares in recent years (Allen, 2007). Piñon pines are not dying in their northern range, according to the Forest Inventory Analysis (Shaw, Steed, and DeBlander, 2005), and model results suggest that their range could expand in Colorado over the next 100 years.[18] Where vegetation dieback occurs, it can interact with wildfire activity, and both fires and plant mortality can enhance erosion (Allen, 2007).

Climate change will influence fire regimes throughout the country. Extended fire seasons and increased fire intensity have already been observed to correlate directly with climate in the western United States, and these effects are

[16] **Halmon**, S., P. Kennard, S. Beason, E. Beaulieu, and L. Mitchell, 2006: River bed elevation changes and increasing flood hazards in the Nisqually River at Mount Rainier National Park, Washington. *American Geophysical Union, Fall Meeting 2006.*

[17] **Yosemite National Park**, 2006: Tuolumne Meadows lodgepole pine removal. National Park Service, www.nps.gov/archive/yose/planning/projects/tmtrees.pdf, accessed on 4-13-2007.

[18] **Ironside**, K., K.L. Cole, N. Cobb, J.D. Shaw, and P. Duffy, 2007: Modeling the future redistribution of pinyon-juniper woodland species. In: *Climate-Induced Forest Dieback As an Emergent Global Phenomenon: Patterns, Mechanisms, and Projections.* Proceedings of the ESA/SER Joint Meeting, 5, August 2007.

projected to continue (Westerling *et al.*, 2006). Air quality is likely to be adversely affected by warmer climates, brought about by increased smoke from fires and ozone, whose production is enhanced with rising temperature (Langner, Bergström, and Foltescu, 2005; McKenzie *et al.*, 2006). Water quality is likely to decrease with climate change. Post-fire erosion will introduce sediment to rivers, lakes, and reservoirs; warmer temperatures will increase anoxia of eutrophic waters and enhance the bioaccumulation of contaminants and toxins (Murdoch, Baron, and Miller, 2000). Reduced flows, either from increased evapotranspiration or increased human consumptive uses, will reduce the dilution of pollutants in rivers and streams (Murdoch, Baron, and Miller, 2000).

4.3.3 Current Approaches to NPS Natural Resource Management

To date, only a few individual parks address climate change in their General Management Plans, Resource Management Plans, Strategic Plans, or Wilderness Plans. Dry Tortugas' General Management Plan lists climate change as an external force that is degrading park coral reefs and seagrass meadows, but considers climate change beyond the scope of park management authority. Sequoia and Kings Canyon National Park's Resource Management Plan specifically references climate change as a restraint to achieving desired future conditions, and notes the need for inventory and monitoring to enable decision making.

NPS has made significant progress in recent years in gathering basic information, developing a rigorous structure for monitoring changes, and raising natural resource management to the highest level of importance. Decisions about the extent and degree of management actions that are taken to protect or restore park ecosystems are increasingly supported by management objectives and credible science (National Park Service, 2006). NPS management approaches to altered disturbance regimes, habitat fragmentation, invasive species, and pollution are described below.

Fire management in the NPS, while conducted in close coordination with other agencies, is driven by five-year prescribed burn plans in individual parks and suppression responses to fire seasons that have become increasingly severe. While NPS makes extensive use of fire as an ecological management tool, the decision to let naturally ignited fires burn is highly constrained by human settlements and infrastructure. Park managers apply preemptive approaches, including mechanical thinning and prescribed burns, to reduce the risk of anomalously severe crown fires in forest ecosystems in which fires historically have been frequent low-severity events. These treatments appear to work in some systems, including the Rincon Wilderness in Saguaro National Park (Allen *et al.*, 2002; Finney, McHugh, and Grenfell, 2004).

Erosion is prevented or repaired by necessity on a site-by-site basis. Terrestrial ecosystem restoration often uses heavy machinery in an effort to repair severely damaged wetlands, stream banks, and coastal dunes, and to restore landforms and connectivity among landscapes disturbed by roads. Restoration treatments after severe fire can increase herbaceous ground cover and thus resistance to accelerated runoff and erosion, as exemplified by work at Bandelier National Monument in New Mexico (Sydoriak, Allen, and Jacobs, 2000).

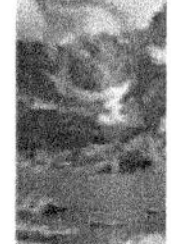

There are no national summaries of the extent of hydrologic alteration in national parks. Technical assistance and research on flow regimes are supplied by the NPS Water Resource Division and the U.S. Geological Survey to individual parks. For downstream parks that have extensive upstream watershed development, there is no management of altered hydrology (*e.g.*, Cuyahoga Valley NRA, Big Bend National Park). In other locations, research is being conducted on hydrologic alterations and management options. For example, at Organ Pipe Cactus National Monument, scientists and managers are identifying groundwater source areas. Upper Delaware Scenic and Recreational River is quantifying minimum flows necessary for protecting endangered dwarf wedgemussels. Adaptive management using experimental flows in Grand Canyon National Park below Glen Canyon Dam is helping to develop a flow regime that supports endangered fish, sediment, recreation, and hydropower generation. Some park units are actively removing dams (*e.g.*, Glines Canyon and Elwha Dams in Olympic National Park), purchasing water rights from previous owners in order to protect water flows (*e.g.*, Zion National Park,

Cedar Breaks National Monument, Craters of the Moon National Monument), and restoring wetlands, stream banks, and wildlife habitat in areas affected by logging (*e.g.,* Redwoods National Park, St Croix National Scenic Riverway) or road construction (*e.g.,* Klondike Gold Rush NHP).

Current wildlife management policies in national parks have been shaped by a combination of strong criticism of past wildlife management practices in Yellowstone and Rocky Mountain National Parks (Chase, 1987; Sellars, 1999) and by scientific research that has highlighted the role of parks as refuges for native wildlife. Individual parks manage their wildlife differently on the basis of history, current land use adjacent to the park, ecological feasibility, public sentiment, and legal directives. Large ungulates and carnivores attract much management attention, and there have been many studies on carrying capacity and the feasibility of reintroducing certain species in national parks. Reintroduction of gray wolves into Yellowstone National Park was accomplished in 1995 and 1996 after extensive study and environmental assessment. The number of packs and reproduction of individual wolves has increased substantially since the reintroductions. There have been remarkable effects on the entire trophic cascade and Yellowstone ecosystem as a result of the wolves' hunting tactics and behavioral changes among ungulates. Changes have occurred in vegetation and habitat for many other species, including songbirds, beaver, and willows in response to restructuring the Yellowstone food chain (Ripple and Beschta, 2005).

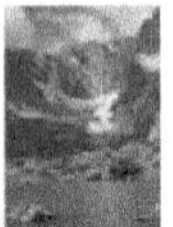

Restoration of bighorn sheep illustrates another successful application of contemporary wildlife ecology to park management. A geospatial assessment of the existence and quality of habitat for bighorn sheep within 14 western national parks from which bighorn sheep had been extirpated found that only 32% of the available area could support reintroduced populations (Singer, Bleich, and Gudorf, 2000). By reintroducing bighorn sheep only to areas with adequate habitat quality and quantity, managers have facilitated establishment of stable reproducing populations.

Many other examples, from restoring nesting populations of Kemp's Ridley sea turtles at Padre Island National Seashore, to directing more NPS funding toward protecting listed species whose need is most immediate, illustrate species-specific management activities that occur within park boundaries (Fig. 4.7). Management summaries have been completed for almost all of the 284 threatened and endangered species that occur in the national parks. The summaries that relate basic biological information to recovery goals for species are posted on a Web site in a form that is accessible to resource managers.[19]

[19] **National Park Service**, 2004: Threatened and endangered species. National Park Service, http://www.nature.nps.gov/biology/endangeredspecies/database/search.cfm, accessed on 5-15-2007.

Figure 4.7. Kemp's Ridley hatchlings heading for the water at a hatchling release. Photo courtesy National Park Service, Padre Island National Seashore.

At least two parks, Great Smoky Mountains and Point Reyes National Seashore, have embarked on All-Taxa Biodiversity Inventories (ATBIs) to catalog all living species of plants, vertebrates, invertebrates, bacteria, and fungi. Inventories are a critical first step toward tracking and understanding changes in species richness and composition. Through the Natural Resource Challenge, more than 1,750 park inventory data sets have recently been compiled. For all natural national parks, these sets of data include natural resource bibliographics, vertebrate and vascular plant species lists, base cartography, air and water quality measures, the location and type of water bodies, and meteorology. Additional inventories of geologic and vegetation maps, soils, land cover types, geographic distributions and status of vertebrates and vascular plants, and location of air quality monitoring stations are in progress.

Efforts to address regional landscape and hydrologic alteration occur in some park areas, and have been initiated either by individual parks or their regional partners. A pilot project to understand the role of NPS units in the fragmented landscape was conducted from 2004–2006. NPS and its partners used geospatial datasets and regional conservation frameworks to develop over 40 partnership proposals. The Greater Yellowstone Coordinating Committee (Box 4.5), and the Comprehensive Everglades Restoration Plan—which includes Everglades, Big Cypress National Preserve, and Biscayne National Parks—are two examples of large multi-agency efforts targeting landscape and hydrologic rehabilitation or protection. Some management within park units has also attempted to alleviate fragmentation. For example, road underpasses have been constructed for desert tortoises in Joshua Tree National Monument.

As part of the NPS commitments within the National Invasive Species Management Plan, 17 Exotic Plant Management Teams operating under the principles of adaptive management serve more than 200 park units (National Invasive Species Council, 2001). Exotic Plant Management Teams identify, develop, conduct, and evaluate invasive species removal projects. Modeled after rapid response fire management teams, crews aggressively control unwanted plants. Mechanical, chemical, and cultural management methods and biological control

BOX 4.5. The Greater Yellowstone Coordinating Committee.[20]

The Greater Yellowstone Coordinating Committee, established in 1964, has been highly effective at working on public lands issues for the nearly 14 million acres of public lands that include Yellowstone and Grand Teton National Parks, John D. Rockefeller, Jr. Memorial Parkway, five national forests, and two national wildlife refuges (see map below). Subcommittees of managers from federal agencies as well as state and private entities work on a wide variety of cross-boundary issues, including land cover and land use patterns and fragmentation, watershed management, invasive species, conservation of whitebark pine and cutthroat trout, threatened and endangered species, recreation, and air quality. Shared data, information, and equipment have been effective in coordinating specific activities including acquiring and protecting private lands through deeds and conservation easements, raising public awareness, providing tools such as a vehicle washer, and increasing purchasing power. These activities have helped combat the spread of invasive plants, restore fish passageways, conserve energy, reduce waste streams, educate the public, and develop a collective capacity for sustainability across the federal agencies.

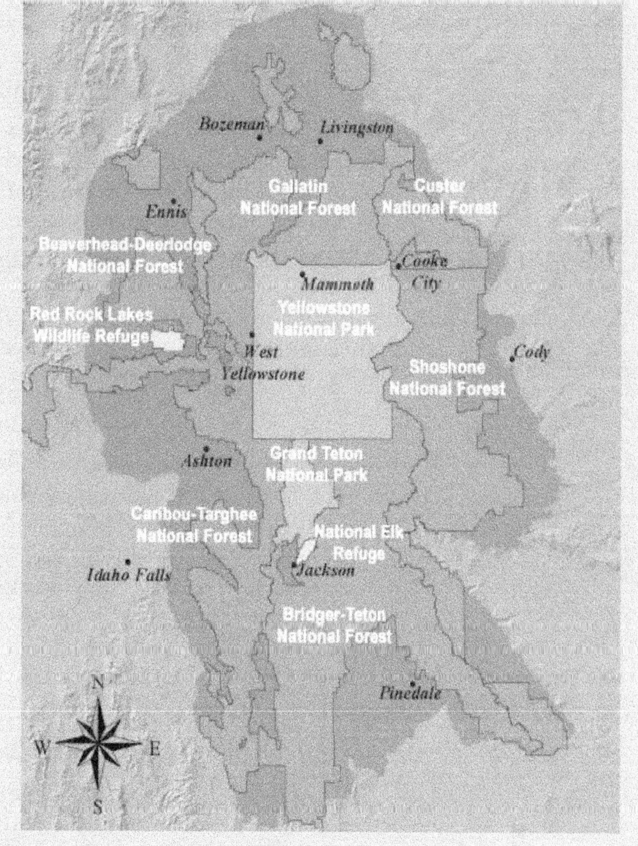

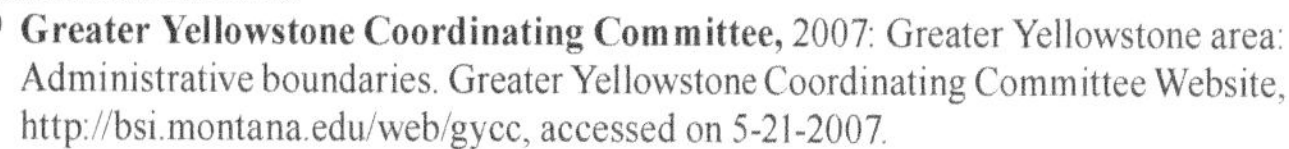
[20] **Greater Yellowstone Coordinating Committee,** 2007: Greater Yellowstone area: Administrative boundaries. Greater Yellowstone Coordinating Committee Website, http://bsi.montana.edu/web/gycc, accessed on 5-21-2007.

techniques are all used in the effort to rapidly remove unwanted plant species. Exotic plant management teams work collaboratively with the U.S. Department of Agriculture, other bureaus in the Department of the Interior, state and local governments, and non-governmental organizations such as the Rocky Mountain Elk Foundation to control invasive plants, many of which are common across extensive areas. In 2004, 6,782 acres with invasive plants were treated in national park units, and 387 were restored (National Park Service, 2004b).

If invasive insects, either native or alien, are considered a threat to structures or the survival of valued flora, they may be treated aggressively. Direct management interventions include use of biocides, biological control, and plant removal in "frontcountry" areas where safety and visitor perception are paramount. Non-native diseases are another major threat to native plants and animals. White pine blister rust (*Cronartium ribicola*), for instance, has caused die-offs of five-needled pines in western and Midwestern parks.

Several national parks either actively manage visitor use or are proposing to do so in order to control the spread of invasive species. Voyageurs National Park proposes to prohibit use of natural bait, privately owned watercraft, and float plane landings in all interior waters in order to limit the spread of the spiny water flea.[21] Glen Canyon National Recreation Area requires all boaters to display a certificate on their dashboard stating their boat is free of zebra or quagga mussels, or have their boats decontaminated.[22]

Because most sources of pollution are outside national park boundaries, NPS air and water managers work with state and federal regulatory agencies that have the authority to implement pollution control by requiring best management practices and adhering to air and water quality standards. Unlike many resource management programs that operate in individual parks, there is national oversight of air quality issues for all national parks. The Clean Air Act and the Wilderness Act set stringent standards for air quality in all 48 Class I Parks (those parks with the highest level of air quality protection), and the NPS Air Quality Program actively monitors and evaluates air quality in these parks, notifying the states and EPA when impairment or declining trends in air quality are observed.

Rocky Mountain National Park provides an example of a successful program to reduce nitrogen deposition. A synthesis of published research found many environmental changes in the park caused by increasing atmospheric nitrogen deposition. NPS used the information to convince the state of Colorado to take action, and NPS, Colorado, and EPA now have a plan in place to reverse deposition trends at the park. The Air Quality Program recently completed a risk assessment of the effects of increasing ozone concentrations to plants for all 270 natural resource parks (Kohut, 2007), and has planned a similar risk assessment of the potential for damage from atmospheric nitrogen deposition.

A baseline water quality inventory and assessment for all natural resource national parks is scheduled for completion in 2007, and 235 of 270 park reports were completed as of 2006. Reports are accessible online,[23] and electronic data are provided to individual parks for planning purposes. Measurement, evaluation of sources of water pollution, and assessment of biological effects currently are carried out by individual parks, with support from the NPS and USGS Water Resources Divisions. Most routine water quality monitoring is related to human health considerations.

A number of low-lying coastal areas and islands are at high risk of inundation as climate changes. The NPS Geologic Resources Division, in partnership with the USGS, conducted assessments of potential future changes in sea level. The two agencies used results of the assessments to create vulnerability maps to assist NPS in managing its nearly 7,500 miles of

21 **National Park Service**, 2007: Voyageurs National Park draft spiny water flea spread prevention plan. National Park Service, http://www.nps.gov/voya/parkmgmt/upload/FinalDraft%20SWFT%20Spread%20Prevention%20Planl%203-28-07%20.pdf, accessed on 11-20-2007.

22 **National Park Service**, 2007: Glen Canyon national recreation area. National Park Service, http://www.nps.gov/glca/parknews/advisories.htm, accessed on 11-21-2007.

23 **National Park Service**, 2004: Baseline water quality data inventory & analysis reports. National Park Service, http://www.nature.nps.gov/water/horizon.cfm, accessed on 4-6-2007.

shoreline along oceans and lakes. Vulnerability was based on risk of inundation. For example, the USGS coastal vulnerability index has rated six of seven barrier islands at Gulf Islands National Seashore highly vulnerable to sea level rise; the seventh island was rated moderately vulnerable.[24]

4.3.4 Sensitivity of NPS Goals to Climate Change

The features and ecosystems that define national parks were shaped by climate in the past, and they will be re-shaped in the future by climate change. Efforts to increase resilience through thoughtful reduction of non-natural disturbances, protection of refugia, and relocation of valued species to more favorable climates may help NPS meet its enabling language conservation goals. Even so, management applications that aim to increase the resilience of physical and biological resources in their current form to climate change will likely succeed only for the next few decades. As climate change continues, thresholds of resilience will be overcome. Science-based management principles will be even more important as park managers begin to manage for change rather than existing resources (Parsons, 2004).

One of the biggest challenges to the national parks revolves around protection and restoration of native species. The Natural Resource Challenge distinguishes between native and non-native plants, animals, and other organisms, and recommends that non-natives be controlled where they jeopardize natural communities in parks. However, species distributions will change, and indeed are already changing, as the climate warms. Changing distributions are evident in observations of gradual migrations (*e.g.*, northward and higher elevation observations of many species; Edwards *et al.*, 2005; Parmesan, 2006) and in massive diebacks (*e.g.*, piñon mortality in Bandelier National Monument; Allen, 2007). A recent study suggests that by 2100, between 4% and 39% of the world's land areas will experience combinations of climate variables that do not currently exist anywhere on Earth, eliciting a biological response unprecedented in human history (Williams, Jackson, and Kutzbach, 2007). Individual species, constrained by different environmental factors, will respond differently, with the result that some species may vanish, others stay in place, and new arrivals appear (Saxon *et al.*, 2005). This type of ecosystem reshuffling will occur in national parks as well as other places, and may confound the abilities of NPS to restore species assemblages to past (or even existing) conditions that may no longer be tenable. If, however, NPS accepts the inevitability of change, it and other collaborating agencies can anticipate, and even aid, the establishment of desirable climate-appropriate species that will take advantage of favorable conditions. By using species suited to anticipated future climates after disturbance or during restoration, for instance, managers may prevent establishment of less desirable species.

NPS goals of providing visitor services such as interpretation and protection will not be directly altered by climate change, although programs will need to adapt. National parks will remain highly desirable places for people to visit, but climate change may cause visitation patterns to shift in season or location. Parks may consider managing visitor use practices or patterns differently in order to prevent people from inadvertently contributing to climate-change enhanced damage. Climate change will alter the length of visitor seasons in many parks; coastal and mountain parks may see increased visitation, while desert parks may see decreased visitation during summer months. Extreme heat and heavy precipitation events, projected as being very likely by IPCC (2007), may strain visitor safety services. Interpretation efforts can play an important role in educating park visitors about changes occurring in national parks and what the park is doing to manage or reduce the impacts of those changes. Interpretation may also be a good way to engage the public in meaningful discussions about good environmental stewardship, and what climate change means for ecosystems and valued species within them.

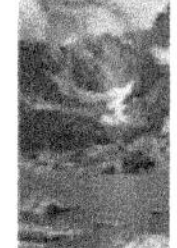

[24] **Pendleton**, E.A., E.S. Hammar-Klose, E.R. Thieler, and S.J. Williams, 2007: Relative coastal vulnerability assessment of Gulf Islands National Seashore (GUIS) to sea-level rise. U.S. Geological Survey, http://woodshole.er.usgs.gov/project-pages/nps-cvi/parks/GUIS.htm, accessed on 4-6-2007.

4.4 ADAPTING TO CLIMATE CHANGE

4.4.1 Coming to Terms with Uncertainty

Predicting climate change and its effects poses a variety of challenges to park managers. What is likely to happen? What potentially could happen? Do we have any control over what happens? The answers to these questions are associated with substantial uncertainties, including uncertainties particular to management of natural resources (Rittel and Webber, 1973; Lee, 1993; Regan, Colyvan, and Burgman, 2002). Resource uncertainties can be separated into two categories (Lee, 1993): the first type, *technical and scientific* uncertainty, centers on what we do and do not know about future climate change effects and our ability to ameliorate them. The second type, *social uncertainty*, focuses on our cultural and organizational capability to respond.

There is considerable uncertainty in predictions, understanding, and interpretation of climate change and its effects. Managers must consider at least three different categories of climate change impacts, each associated with a different level of uncertainty: foreseeable or tractable changes, imagined or surprising changes, and unknown changes.

Predictions of climate change are generally accepted if changes are foreseeable and evidence already exists that many of these predictions are accurate. For instance, we can predict with high confidence that atmospheric carbon dioxide concentrations will increase, sea levels will rise, snow packs across most of North America will shrink, global temperature will increase, fire seasons will become longer and more severe, and the severity of storms will increase (IPCC, 2007). We refer to a given change as foreseeable if there is a fairly robust model (or models) describing relationships between system components and drivers, and sufficient theory, data, and understanding to develop credible projections over the appropriate scales. We cannot project precisely the magnitude of foreseeable changes, but we can quantify the distribution of probable outcomes. For example, a 40-year record shows that snow is melting increasingly earlier in the spring in the Sierra Nevada, Cascade Range, and New England (Stewart, Cayan, and Dettinger, 2005; Hodgkins and Dudley, 2006). We also have understanding from the physical sciences of why the timing of snowmelt is likely to change in regions with winter and spring temperatures between -3 and 0°C as the climate warms (Knowles, Dettinger, and Cayan, 2006). Foreseeable changes are sufficiently certain that park managers can begin planning now for effects of earlier snowmelt on river flow, fishes and other aquatic species, and fire potential. Such plans for aquatic organisms could include establishing refugia for valued species at risk, removing barriers to natural species migrations, replicating populations as a bet-hedging strategy to reduce overall risk, restoring riparian vegetation to shade river reaches, or even conducting assisted migrations. As the risk of fire increases, planners might consider moving infrastructure out of fire-prone areas and restricting visitor access to fire-prone areas during fire seasons for safety reasons. Planners may also need to consider how to manage for increased smoke-related health alerts and possibly increased respiratory emergencies in parks. Many parks, such as Yosemite, have been managing fuels and fire ecology for decades, and have extensive prescriptive documents that describe where and how to manage in specific locations, complete with numbers of acres to treat each year and a targeted natural fire frequency return interval (National Park Service, 2004a). Methods that may have been effective in the past, however, should be regularly reviewed for their applicability, since historic ranges of variability in natural disturbance cycles may be less appropriate targets in a warmer climate.

The second category of climate change and its related effects includes changes that are known or imaginable, but difficult to predict with high certainty. These may include changes with which we have little or no past experience or history, or effects of changes in systems for which there is a great deal of experience. For example, nonlinear interactions among system components and drivers could reduce the certainty of predictions and generate unexpected or surprising dynamics. Surprises may present crises when the ecological system abruptly crosses a threshold into a qualitatively different state. For example, a November 2006 storm that caused severe flooding and damage in Mount Rainier National Park was surprising, because a storm of this magnitude had not been observed previously. An example of change that is known but difficult to project is rapid and extensive dieback of forests and woodlands from climate-induced physiological stress, and in some cases, associated insect outbreaks. Forest mortality in the Jemez Mountains of northern New Mexico had occurred before; the lower extent of the ponderosa pine zone in Bandelier National Monument retreated upslope by as much as 2 km in less than five years in response to severe drought and an associated outbreak of bark beetles in the 1950s (Allen and Breshears, 1998; Allen, 2007). Planning for these rare but major events requires that mechanisms be put in place to reduce the damage caused by those events. In some instances, minimizing the ecological effects of sudden changes in system state might require removing infrastructure or maintaining corridors for species migration.

The third category of climate change and related effects is unknown or unknowable changes. This group includes changes and associated effects that have not previously been experienced by humans. Perhaps the greatest uncertainties in projecting climate change and its effects are associated with the interaction of climate change and other human activities. The synergistic and cumulative interactions among multiple system components and stressors, such as new barriers or pathways to species movement, disruption of nutrient cycles, or the emergence of new diseases, may create emerging ecosystems unlike any ever seen before.

4.4.2 Approaches to Management Given Uncertainty

When confronting a complex issue, it is tempting to postpone action until more information or understanding is gained. Continuing studies and evaluations almost always are warranted, but not all actions can or should be deferred until there is unequivocal scientific information. Scenario planning and knowledge gained from research and adaptive management practices can help with decision-making, and can point toward implementation of actions to manage natural resources in the face of substantial uncertainty. Ideally, actions should be taken that are robust to acknowledged uncertainty. So-called "no-regrets" strategies that improve the environment increase resilience regardless of climate change, and thus are robust to uncertainty. It is critical to develop and implement frameworks that allow the NPS to learn from implementation of policies, regulations, and actions.

National parks are complex systems within a complex landscape. John Muir wrote "When we try to pick out anything by itself, we find it hitched to everything else in the universe" (Muir, 1911). Species co-occur, influenced by physical, chemical, and biological conditions. Parks are surrounded by lands that are managed with different goals and objectives. Although few problems can be solved easily, the adoption of a systems approach to management and a shared environmental protection vision with adjacent landowners increases the probability of achieving park objectives. The two major factors that influence selection of strategies for managing complex resource systems are the degree (and type) of uncertainty and the extent to which key ecological processes can be controlled (Fig. 4.8). Uncertainty can be qualitatively evaluated on a scale of low to high. Ability to control an ecological process depends on the process itself, the responsible management organization or institution, and the available technology. For example, supply of surface water can be manipulated upstream from some national parks, such as Everglades or Grand Canyon.

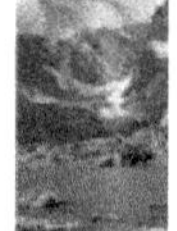

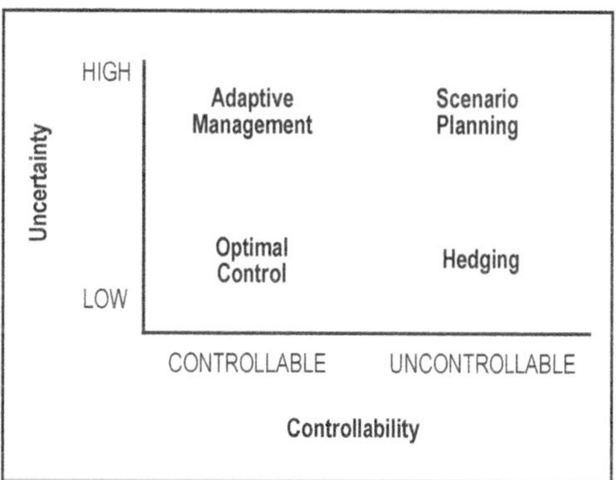

Figure 4.8. Scenario planning is appropriate for systems in which there is a lot of uncertainty that is not controllable. In other cases optimal control, hedging, or adaptive management may be appropriate responses. Reprinted from Peterson, Cumming, and Carpenter (2003).

Optimal Control and Hedging

The strategic approaches in Fig. 4.8 provide a broad set of tools for resource management. Each tool is appropriate for certain types of management, and, while not interchangeable, the lessons learned from application of one can and should inform the decisions on whether and how to employ the others. Most approaches toward current resource management in the NPS are appropriate when uncertainty is low. That is, most management is based on either an optimal control approach or a hedging approach. However, the attributes and effects of climate change present sufficient uncertainties to NPS managers that adaptive management or scenario development are much more appropriate than optimal control or hedging.

Fire and wildlife management as currently practiced are examples of optimal control. Many fire management plans are developed and implemented by controlling the timing—and hence the probable impact—of fire to achieve an optimal set of resource conditions. Control of wildlife populations through culling, birth control, or reintroduction of top predators is based on concepts about limits such as carrying capacity. Physical removal of invasive plants exemplifies optimal control. Hedging strategies involve management that may improve fitness or survival of species. For example, placing large woody debris in a stream to improve fish habitat is essentially a hedging strategy.

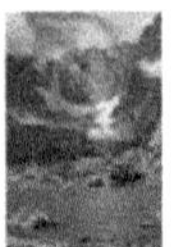

Scenario-Based Planning

Scenario-based planning is a qualitative, or sometimes quantitative process that involves exploration and articulation of a wide set of possible or alternative futures (Carpenter, 2002; Peterson, Cumming, and Carpenter, 2003; Raskin, 2005). Each of these alternative scenarios is developed through a discourse among knowledgeable persons, and is informed by data and either conceptual or simulation models. Scenarios are plausible—yet uncertain—stories or narratives about what might happen in the future. Scenario development is used routinely to assess a variety of environmental resource issues (National Research Council, 1999). Park Service managers, along with subject-matter experts, apply existing knowledge to conduct scenario planning related to climate change and resources of interest. A finite number of scenarios (*e.g.*, three to five) that depict the range of possible futures can be extremely useful for helping managers develop and implement plans, and also minimize the anxiety of frustration that comes from having to deal with uncertainty. Research into the rate, extent, or permanence of climate change-induced impacts on species and ecosystems of interest can inform the scenarios. Either passive or active contingency plans can be deployed for both (1) trends that are observed and have a high probability of continuing, and (2) events with low probability but high risk that result from any combination of climate change and other stressors.

Scenario planning and development of contingency plans can lead to several levels of preparedness. For example, plans can be constructed to trigger action if a threshold is crossed, similar to current air quality regulations for ozone. Mandatory reductions in ozone precursor emissions are imposed on ozone-producing regions by EPA when allowable ozone levels are exceeded. Plans could include management "drills" to prepare for low, but real, probabilities of an extreme event (fire drills are an example we are all familiar with). Scenarios should be built around consideration of how climate change will affect current resource management issues. If current habitat recovery plans for endangered species, for instance, do not take future climate change into account, recovery goals may not be met.

Scenarios provide the opportunity to explore and attempt to resolve the inevitable problems that will arise when management for one goal conflicts with laws or other management goals. Tradeoffs between air quality and the use of fire for ecosystem restoration and maintenance already need to be made, for instance. The prudent decision-maker will conduct planning exercises to identify where potential collisions may occur under various climate change and management scenarios, and address the balance between short-term costs and long-term benefits. Management responses to scenarios should consider the degree of uncertainty attached to impacts, the probable magnitude and character of impact, the resources available, and legal mandates as well as social and economic consequences.

Triage is an extreme form of tradeoff. In a resource- and staff-limited world, there will be a need to prioritize. Scenarios that evaluate the feasibility of adaptation against ecological, social, or economic returns will be valuable in making difficult decisions, and importantly, in conveying results of these decisions to the public. Public involvement in scenario building at all levels, from individual park or region up to national, will not only prepare people for the inevitable, but will help build support if goals need to modified.

Adaptive Environmental Assessment and Management

Adaptive environmental assessment and management refers to a set of processes to integrate learning with management actions (Holling, 1978; Walters, 1986; Lee, 1993). The processes focus on developing hypotheses or explanations to describe (1) how specific ecological dynamics operate and (2) how human interventions may affect the ecosystem. Adaptive environmental assessment is substantially different from environmental assessments routinely conducted within frameworks such as NEPA. The NEPA process presumes certainty of impacts and outcomes, and generally minimizes or ignores uncertainties. Adaptive environmental assessment and management, by contrast, highlights uncertainty. Managers design actions that specifically test uncertainties about ecosystem dynamics and outcomes of proposed interventions. The objectives of management actions explicitly include learning (hence reduction of uncertainty). Adaptive management views policies as hypotheses and management actions as treatments that are structured to "test" desired outcomes.

Adaptive management can be either active or passive. Active adaptive management involves direct manipulation of key ecological processes to test understanding of relationships among system components and drivers and to examine the effects of policies or decisions, such as the flood release experiments of 1996 and 2004 in the Grand Canyon (Walters *et al.*, 2000). Passive adaptive management, instead of direct hypothesis-testing, relies on historical information to construct a "best guess" conceptual model of how a system works and how it will respond to changing conditions. Management choices are made on the assumption that the ecosystem will respond according to the model (National Research Council, 2003). Whether active or passive, information gathered throughout the iterative adaptive management cycle is used to increase ecological understanding, and adjust and refine management (Walters and Holling, 1990).

Adaptive management has been successful in large-scale systems that meet both ecological and social criteria: sufficient ecological resilience to deterministic and stochastic change, and a willingness to experiment and participate in a formal structure for learning. Ecological resilience, or the capacity for renewal in a dynamic environment, buffers the system from the potential failure of management actions that unavoidably were based upon incomplete understanding. Resilience allows managers the latitude to learn and change. Trust, cooperation, and other forms of social capital are necessary for implementing management actions that are designed to meet learning and other social objectives.

Safe-to-Fail Strategies

Because the uncertainties associated with predictions of climate change and its effects are substantial, expected outcomes or targets of agency policies and actions have some probability of being incorrect. Accordingly, NPS could take the robust approach of designing actions that are "safe to fail." That is, even though managers intend to implement a "correct" action, they and their supervisors recognize that failure may occur.

Safe-to-fail policies apply to both natural resources and to human resources. For natural resource management, a safe-to-fail experiment or action is undertaken only where there is confidence the system can recover without irreversible damage to the targeted resource. This type of approach is employed in other fields, such as engineering systems (*e.g.*, air traffic control, or electric power distribution) where uncertainty is actively managed through flexible designs that adjust to changing conditions (Neufville, 2003). One low-tech example of where safe-to-fail strategies are already used in NPS resource management is in attempting to control invasive feral hogs. Feral hogs are common to many parks in the southeastern United States, California, the Virgin Islands, and Hawaii. The hogs are opportunistic omnivores whose rooting profoundly disrupts natural communities and individual populations, and facilitates establishment of invasive plants. Hogs compete directly with native wildlife for mast, prey on nests of ground-nesting birds and sea turtles, and serve as reservoirs for a variety of serious wildlife diseases and parasites. Fencing, hunting, and trapping efforts to eliminate feral hog populations in national parks often fail; either removal operations are unsuccessful or native plant and animal populations do not recover. Yet control tactics and restoration activities can be modified and managed adaptively as information accrues on probabilities of success associated with different sets of ecological conditions and interventions.

Safe-to-fail policies for human resources (*e.g.*, careers and livelihoods) empower managers to take reasoned management risks without concern for retribution. Although not desired, failures provide tremendous opportunities for learning. Learning from mistakes and successes is a critical part of adaptation to climate change. As climate changes, even the most well-reasoned actions have some potential to go awry. The wisdom, experience, and empirical data of front line managers, resource management personnel, and scientific staff need to be protected, preserved, and expanded. Public education about the complexity of resource management, transparency in the decision-making process, frequent public updates on progress or setbacks, and internal agency efforts that promote trust and respect for professionals within the agency are all important methods for promoting more nuanced and potentially unsuccessful management efforts.

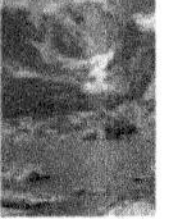

Acceptance of a gradient between success and failure might foster greater creativity in resource management and remove the need to assign blame. Shifting attitudes about failure increases institutional capacity to capture and expand learning. Punishing managers whose proactive management efforts fail may create an environment in which managers are risk-averse and act only on the basis of what is known with certainty.

4.4.3 Incorporating Climate Change Considerations into Natural Resource Management

Given that recent climate changes and climate variations are already beginning to have effects on natural systems, and warming trends are projected into the next century (IPCC, 2007), it is prudent to begin to implement adaptation strategies as soon as possible. Note that the kinds of management actions that increase resilience will be most effective in the near term, but will need to be re-evaluated as the climate, and environmental response, move into realms for which there is no historical analog. Clearly, methods manuals and handbooks of adaptation strategies should be used with caution and reviewed regularly to determine if they are still appropriate, since analogs from the past may not be effective for managing future environments.

The importance of action in national parks extends well beyond the parks themselves. The value of national parks as minimally disturbed refugia for natural processes and biodiversity becomes more important with increasing alteration of other lands and waters. Many parks have received international recognition as Biosphere Reserves or World Heritage sites because of their transcendent value worldwide. If protection of natural resources and processes is to be achieved during the

BOX 4.6. Process for Adaptations of Parks and the Park Service to Climate Change.

- Identify resources and processes at risk from climate change.
  - Characterize potential future climate changes, including inherent uncertainty and possible ranges.
  - Identify which resources are susceptible to change under future climates.
- Develop monitoring and assessment programs for resources and processes at risk from climate change.
- Define baselines or reference conditions for protection or restoration.
- Develop and implement management strategies for adaptation.
  - Consider whether current management practices will be effective under future climates.
  - Diversify the portfolio of management approaches.
  - Accelerate the capacity for learning.
  - Assess, plan, and manage at multiple scales.
- Let the issues define appropriate scales of time and space.
- Form partnerships with other resource management entities.
  - Reduce other human-caused stressors to park ecosystems.
  - Nurture and cultivate human and natural capital.

BOX 4.7. Examples of Adaptation Options for Resource Managers.

- Remove structures that harden the coastlines, impede natural regeneration of sediments, and prevent natural inland migration of sand and vegetation after disturbances.
- Move or remove human infrastructure from floodplains to protect against extreme events.
- Remove barriers to upstream migration in rivers and streams.
- Reduce or eliminate water pollution by working with watershed coalitions to reduce non-point sources and with local, state and federal agencies to reduce atmospheric deposition.
- Reduce fragmentation and maintain or restore species migration corridors to facilitate natural flow of genes, species and populations.
- Use wildland fire, mechanical thinning, or prescribed burns where it is documented to reduce risk of anomalously severe fires.
- Minimize alteration of natural disturbance regimes, for example through protection of natural flow regimes in rivers or removal of infrastructure that prohibits the allowance of wildland fire
- Minimize soil loss after fire or vegetation dieback with native vegetation and debris.
- Aggressively prevent establishment of invasive non-native species where they are documented to threaten native species or current ecosystem function.
- Allow the establishment of species that are non-native locally, but maintain native biodiversity or enhance ecosystem function in the overall region.
- Actively plant or introduce desired species after disturbances or in anticipation of the loss of some species.
- Manage Park Service and visitor use practices to prevent people from inadvertently contributing to climate change.
- Practice bet-hedging by replicating populations and gene pools of desired species.
- Restore vegetation where it confers biophysical protection to increase resilience, including riparian areas that shade streams and coastal wetland vegetation that buffers shorelines.
- Create or protect refugia for valued aquatic species at risk to the effects of early snowmelt on river flow.
- Assist in species migrations.

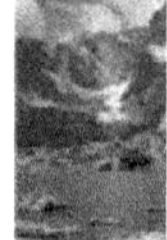

coming decades of climate change, NPS managers need to first identify what is at risk; define the baselines, or reference conditions, that constitute "unimpaired" in a changing world; monitor and evaluate changes over time; decide the appropriate scales at which to manage the processes and resources of national parks; and finally set measurable targets of protection by which to measure success or failure over time (Box 4.6). All of these actions require intimate and iterative connections among scientists, resource managers, other resource management partners, and the public. Dialog on management goals and resources at risk should include members of the public, adjacent land and resource managers, and state and local authorities. Moreover, efforts should be made to engage the full diversity of public opinion, rather than a selected set of public interests. Continuous dialog between scientists, managers, and the interested public will build the greatest possible understanding of the threats, consequences, and possible actions related to climate change (Box 4.7). Climate change literacy at all levels is a worthy goal, and one that is currently actively pursued by NPS. Climate change literacy will become even more important in the future in order to manage public expectations, since even the best management practices will not be able to prevent change.

While resource management is implemented at individual parks, planning and support can and should be provided at all management levels, with better integration between planners and resource management staff. A revision of NPS Management Policies to incorporate climate change considerations would help to codify the importance of the issue. Park General

Management Plans and resource management plans also should be amended to include the understanding, goals, and plans that address climate change issues. Climate change education and coordination efforts at the national level will be helpful for offering consistent guidance and access to information. Regional and network level workshops and planning exercises will be important for addressing issues at appropriate scales, as will interagency activities that address climate change impacts to physical and natural resources regardless of political boundaries.

Identify Resources and Processes at Risk from Climate Change

The first activity is to identify the important park processes and resources that are likely to change as a result of climate change and from the interactions of climate change with existing causes of stress. This should take place within each park, but the exercise should occur at the network, regional, and national scales as well, in order to prioritize which resources will respond most rapidly, thus warranting immediate attention. The process begins with characterizing potential future climate changes and systematically considering resources, as well as their current stressors, susceptible to change under future climates. This can be accomplished through summaries of the literature, guided research, gatherings of experts, and workshops where scientists and managers engage in discussing risks to resources. Some of these activities may have already been done during the process of identifying vital signs for the Inventory and Monitoring Program. Park managers may wish to rank resources and processes according to how susceptible they are to changes in climate, based on the rapidity of expected response, the potential for adaptation opportunities (or conversely, the threat of endangerment), the "keystone" effect (*i.e.*, species or processes that have disproportionate effects on other resources), and the importance of the species or resources to meeting the park's management goals. The direct and indirect influence of climate change itself on specific resources will vary in comparison with other resource management issues, but this exercise will ensure the potential effects are not ignored.

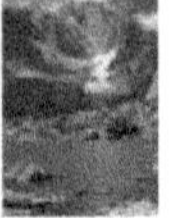

Develop Monitoring and Assessment Programs for Resources at Risk from Climate Change

In periods of accelerated change, it is critical to understand and evaluate the nature of change. As part of the NPS Inventory and Monitoring Program, every national park has established a number of vital signs for monitoring change over time; these vital signs lists should be reviewed in order to ensure they are adequate to capture climate-caused changes. If they are not, the list of vital signs and the frequency with which they are measured may need to be amended. Increasingly, ground-based monitoring can and should be augmented with new technologies and remote sensing. NPS maintains 64 sites as part of the Global Fiducial Program, which collects high-resolution geospatial data for predetermined sites over a period of years to decades.[25] Global Fiducial is an example of an important, and underutilized, type of information that has much to offer to national parks. Collaborations with universities and other agencies can accelerate the ability of NPS to obtain useful data that can be incorporated into adaptive management. Collaborations with other information gathering and assessment programs—such as programs of the USGS and National Science Foundation, including the National Ecological Observatory Network (NEON) and the Long-Term Ecological Research (LTER) networks—present benefits to all partners by developing broad integrated analyses.

Assessment involves tracking the vital signs and their major drivers of change to evaluate the presence of trends or thresholds. While it is important to look at the data that show what happened in the past, it is critically important to use monitored information to anticipate potential future trends or events. Projections of possible futures allow management intervention in advance of some undesired change, and can be conducted with simple extrapolations of monitored data. Simulation and statistical models are invaluable tools for projecting future events, but they need to be parameterized with physical and biological information, and validated against existing records. The data requirements for models, therefore, need to be

[25] **National Park Service**, 2007: OCIO factsheets, Global Fiducial Program. National Park Service, http://www.nps.gov/gis/factsheets/fiducial.html, accessed on 5-16-2007.

considered when choosing which environmental attributes to monitor.

Define Baselines or Reference Conditions for Protection or Restoration

As the change in biological assemblages and physical processes plays out in our national parks, certain common sense actions should be undertaken, among them establishment of quantifiable and measurable baseline conditions that describe unimpaired or current (not necessarily the same thing) conditions, and routine monitoring of select indicators that can be used to measure change. Management goals should be used to establish baselines for species, communities, or processes. Much can be learned from surveys of the literature on past conditions (including the geologic past as determined by paleoenvironmental records; Willis and Birks, 2006). Historic or prehistoric baselines may be unattainable, however, if the climates that produced them will not occur again, so caution needs to be employed in extrapolating from a past baseline condition to a management goal. Shifting baselines, or the circumstance by which a reference condition changes according to the perspective of the manager, can lead to acceptance of degraded conditions and loss of resource integrity (Pauly, 1995). Careful monitoring and clear resource protection goals are necessary for incorporating climate change into management.

Philosophical discussions will need to take place regarding the legitimacy of novel ecosystems made up of previously unrepresented species (Hobbs *et al.*, 2006). Natural migrations of plants and animals from outside park boundaries will occur, indeed will need to occur, as individual species seek favorable climatic conditions. Because of this, the definition of invasive may need to be relaxed so that natural species assemblages can develop in response to new climates. National park boundaries are porous, and corridors for naturally migrating species, either in or out of a national park, should be protected or restored. The dispersal of species does not only occur through migration to adjacent lands or waters, of course, and there are many dispersal mechanisms that species will employ to locate favorable new habitats. A more nuanced understanding of the constraints and selective pressures on dispersal will be important for deciding which new residents are unwelcome (Kokko and López-Sepulcre, 2006).

As part of this exercise, national park managers may need to address whether protecting or recovering certain processes or resources will be possible and what the ramifications are if such ends are not attainable. Individual species, such as the pika—a small-bodied mammal related to rabbits and hares that lives on isolated mountains in the Great Basin, Rocky Mountains, and Sierra Nevada—or features, such as glaciers in Glacier National Park, are extremely vulnerable to climate change (Beever, Brussard, and Berger, 2003; Hall and Fagre, 2003; Grayson, 2005). Establishment or protection of refugia for vulnerable species, or actively translocating them to new favorable habitats, may enable some highly vulnerable species to persist. Ramifications are economic as well as ecological. With limited resources, NPS will have hard decisions in the coming years over how to manage most effectively.

Develop and Implement Management Strategies for Adaptation

Developing and implementing strategies for adaptation to climate change will require NPS managers to adopt a broad array of tools well beyond control and hedging strategies. Current management practices may not be effective under future climates. Some strategies include:

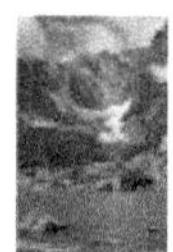

- *Diversify the portfolio of management approaches.* Because climate change is complex and predictions often have high levels of uncertainty, diverse management strategies and actions will be needed. It is important to think broadly about potential environmental changes and management responses and not be constrained by history, existing policies and their interpretation, current practices, and traditions. Initial assessments of effective approaches in general or specific environmental circumstances can be informed by the degree of uncertainty in management outcomes and the potential for control through human intervention. Managers can hedge bets and optimize practices in situations where system dynamics and responses are fairly certain. In situations with greater uncertainty, adaptive management can be undertaken if key ecosystem processes can be manipulated. In all situations, capacity to project changes and manage adaptively will be enhanced by scenario development, planning, and clear goals. Scenario development can rely primarily on qualitative conceptual models, but is more likely to be effective when data are available to characterize key system components, drivers, and mechanisms of responses.

- *Plan, and manage, for inevitable changes.* Sea level will rise, and the removal of barriers to landward migration of coastal wetlands may offer the chance that wetlands may persist. New climate conditions and assemblages are likely to favor opportunistic species, pests, and diseases in marine, freshwater, and terrestrial environments.[26] It is possible that invasive species cannot be controlled before native species are extirpated (Box 4.8). Potential responses may include aggressive efforts to prevent invasion of non-native species in specific locations at which they currently are absent and where future conditions may remain favorable for native species. Managers might relocate individuals or populations, or even consider conceding the loss of the species.

Although in many cases restoration and maintenance of historic communities may become impossible, useful efforts might be directed toward maintenance of ecosystem function. The protection of ecosystem services that supply food and habitat for wildlife, preserve beaches or soil, and regulate hydrologic processes is critically important to the NPS mission of conservation.

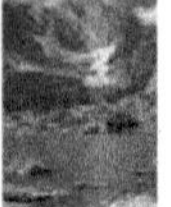

[26] **Lovejoy**, T.E., 2007: Testimony to congressional hearing on climate change and wildlife. United States Senate Committee on Environment and Public Works.

BOX 4.8. Examples of Invasive Species Impacts.

Buffelgrass (*Pennisetum ciliare*), an African bunchgrass, is spreading rapidly across the Sonoran Desert in southern and central Arizona. The Mojave Desert and Great Basin counterparts to buffelgrass, the brome grasses (*Bromus* spp.) and Arabian Schismus (*Schismus* spp.), cover millions of acres. Brome and Schismus grasses are highly flammable and spread rapidly after fires; their invasion into deserts that evolved with infrequent, low-intensity fires is hastening loss of native species. Among the many charismatic species at risk are saguaro cactuses, Joshua trees, and desert tortoises. Buffelgrass and the Mediterranean annual grasses thrive under most temperature regimes so they are likely to continue expanding (Weiss and Overpeck, 2005).

- *Accelerate the capacity for learning.* Given the magnitude of potential climate changes and the degree of uncertainties about specific changes and their effects on national parks, park managers, decision makers, scientists, and the public will need to learn quickly. Some amount of uncertainty should not be an excuse for inaction, since inaction can sometimes lead to greater harm than actions based on incomplete knowledge. Adaptive management—the integration of ongoing research, monitoring, and management in a framework of testing and evaluation—will facilitate that learning. Scenario planning exercises are effective ways of synthesizing much information for learning. Bringing together experts at issue-specific workshops can rapidly build understanding. Application of safe-to-fail approaches also will increase capacity for learning and effective management.
- *Assess, plan, and manage at multiple scales.* Complex ecological systems in national parks operate and change at multiple spatial and temporal scales. As climate changes, it will be important to match the management or intervention effort with the appropriate scale where environmental changes occur. The scales at which ecological processes operate often will dictate the scales at which management institutions must be developed. Migratory bird management, for instance, requires international collaboration; large ungulates and carnivores require regional collaboration; marine preserves require cooperation among many stakeholders; all are examples of cases in which park managers cannot be effective working solely within park boundaries. Similarly, preparation for rapid events such as floods will be managed very differently than responses to climate impacts that occur over decades. Species may be able to move to favorable climates and habitats over time if there is appropriate habitat and connectivity. There are several examples of management of park resources within larger regional or ecosystem contexts. The Greater Yellowstone Coordinating Committee, and the Southern Appalachian Man and the Biosphere (SAMAB) Program are building relationships across jurisdictional boundaries that will allow effective planning for species and processes to adapt to climate change. Olympic, Channel Islands,

BOX 4.9. Southern Appalachian Man and the Biosphere Program.[27]

The Southern Appalachian Man and the Biosphere (SAMAB) Program is a public/private partnership that focuses on the Southern Appalachian Biosphere Reserve. The program encourages the use of ecosystem and adaptive management principles. SAMAB's vision is to foster a harmonious relationship between people and the Southern Appalachian environment. Its mission is to promote the environmental health and stewardship of natural, economic, and cultural resources in the Southern Appalachians. It encourages community-based solutions to critical regional issues through cooperation among partners, information-gathering and sharing, integrated assessments, and demonstration projects. The SAMAB Reserve was designated by the United Nations Educational, Scientific, and Cultural Organization (UNESCO) in 1988 as a multi-unit regional biosphere reserve. Its "zone of cooperation" covers the Appalachian parts of six states: Tennessee, North Carolina, South Carolina, Georgia, Alabama, and Virginia, and includes Great Smoky Mountains National Park.

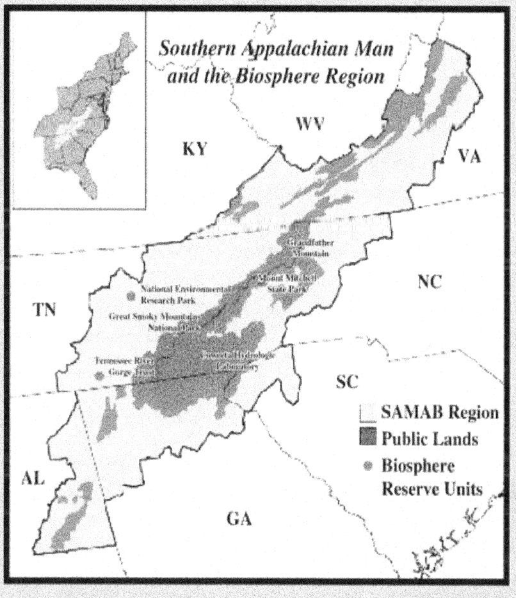

American Samoa, Everglades, Point Reyes, and other coastal parks cooperate with many other state and federal agencies in advising and managing national marine sanctuaries. These ecoregional consortia should serve as models for other park areas as they begin to address the multiple challenges that emanate from outside park boundaries (Box 4.9).

- *Reduce other human-caused stressors to park ecosystems.* In addition to the direct consequences of climate change to park resources, we know that interactions of climate with other stressors will have major influences on national park resources (McKenzie *et al.*, 2006). Therefore, one of the most basic actions park managers can take to slow or mitigate some effects of climatic change is to reduce the magnitude of other disturbances to park ecosystems.[28] Minimizing sources of pollution, competition between non-native and native species, spread of disease, and alteration of natural disturbance regimes should increase ecosystem resilience to

[27] **Southern Appalachian Man and the Biosphere,** 2007: SAMAB home page. Southern Appalachian Man and the Biosphere Website, http://samab.org/, accessed on 5-21-2007.

[28] *E.g.*, **Hansen**, L.J., J.L. Biringer, and J.R. Hoffman, 2003: Buying Time: a User's Manual for Building Resistance and Resilience to Climate Change in Natural Systems. World Wildlife Foundation, Washington, DC. and
Welch, D., 2005: What should protected areas managers do in the face of climate change? The George Wright Forum, 22(1), 75-93.

changing climate. Some combination of these stressors affects every one of the 270 natural national parks either directly or indirectly. Reducing threats and repairing damage to natural resources is the major purpose of the Natural Resource Challenge, among other NPS programs; the synergistic effect of other disturbances with climate change increases the urgency for getting other threats under control. The interactions between these drivers and climate change can lead to nonlinear ecological dynamics, sometimes causing unexpected or undesired changes in populations or processes (Burkett *et al.*, 2005). Once an ecosystem shifts from one state to another, it may be difficult, if not impossible, to return it to its prior desirable state (Gunderson and Holling, 2002). While it may be tempting to promote a return to some range of natural variability, this option must be considered very carefully. Ecosystems change in many ways as a result of management, and unexpected results may occur if management is focused on restoring only one kind of process. A historic flow and temperature regime for the Colorado River below Glen Canyon Dam, for instance, will allow non-native warm water fishes that are now established to move upstream to compete with endangered fishes (U.S. Geological Survey, 2005).

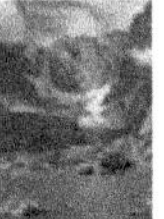

- *Nurture and cultivate human resources.* NPS is endowed with a wealth of human resources in terms of the wisdom, experience, dedication and understanding of its staff and affiliated personnel (such as advisory groups, research scientists, and volunteers). That human capital should be protected and preserved concurrent with natural resources. NPS can accomplish this by promoting training, continuous inquiry, an atmosphere of respect, allowance for periodic failure, and personal initiative. NPS could also allow time for managers and resource practitioners to step back from their daily routines once or twice a year to take in broad strategic views of national park resources, their stressors, and management approaches.

Use Parks to Demonstrate Responses to Climate Change

The goodwill of Americans toward national parks means that they can be used as examples for appropriate behavior, including mitigation strategies, education, and adaptive natural resource management. The NPS is well aware of its ability to serve as an example, and is rapidly becoming a "green" leader through its Climate Friendly Parks program, a partnership between NPS and EPA (Box 4.10). There is an initial cost to change operations in response to climate change, but the tradeoff between that cost and

BOX 4.10. Climate Friendly Parks.

With support from EPA, the National Park Service began the Climate Friendly Parks initiative in 2002.[29] The Climate Friendly Parks program provides tools for parks to mitigate their own contributions to climate change and increase energy efficiency. The program also aims to provide park visitors with examples of environmental excellence and leadership that can be emulated in communities, organizations, and corporations across the country. Parks begin with a baseline inventory of their own greenhouse gas emissions, using inventories and models developed by EPA. The baseline assessment is used to set management goals, prioritize activities, and demonstrate how to reduce emissions, both at the level of individual parks and service-wide. Solid waste reduction, environmental purchasing, management of transportation demands (e.g., increasing vehicle efficiency, reducing motorized vehicle use and total miles traveled), and alternative energy and energy conservation measures are considered in developing action plans for emissions reductions by individual parks. In addition, the NPS will extend these efforts to air pollutants regulated under the Clean Air Act, including hydrocarbons, carbon monoxide, sulfur dioxide, nitrogen dioxide, and particulate matter. Education and outreach are strong components of the Climate Friendly Parks program.

[29] **National Park Service**, 2007: Climate Friendly Parks. National Park Service, http://www.nps.gov/climatefriendlyparks/, accessed on 7-12-2007.

a high certainty of long-term tangible benefits makes decisions easier to make and implement. It is also fairly easy to incorporate information about the causes and effects of climate change into park education and interpretation activities. National parks offer tremendous opportunities for increasing ecological literacy, and park staff rely on sound science in their public education efforts.

No-regrets activities for national park operations, education, and outreach have already begun. The Climate Friendly Parks program is visionary in its efforts to inventory greenhouse gas emissions from parks, provide park-specific suggestions to reduce greenhouse gas emissions, and help parks set realistic emissions reduction goals. Education and outreach are addressed in the Climate Friendly Parks program with materials for educating staff and visitors about climate change. NPS's Pacific West Regional Office has been proactive in educating western park managers on issues related to climate change, as well as promoting messages for communication to the public and actions for addressing the challenge of climate change. Expansion of this type of proactive leadership is needed.

4.5 CONCLUSIONS

The National Park System contains some of the least degraded ecosystems in the United States. Protecting national parks for their naturally functioning ecosystems becomes increasingly important as these systems become more rare (Baron, 2004). However, all ecosystems are changing due to climate change and other human-caused disturbances, including those in national parks. Climate changes that have already been documented, and coupled with existing threats to national parks—including invasive species, habitat fragmentation, pollution, and alteration of natural disturbance regimes—constitute true global change. Climate change will overlay and influence all current resources and how they are managed. Rather than simply adding and ranking the importance of climate change against a host of pressing issues, managers need to begin to include climate change considerations into all activities. Natural resource managers are challenged to evaluate the possible ramifications, both desirable and undesirable, to the resources under their protection, and to develop strategies for minimizing harm under changing environmental conditions.

The definition of what is "unimpaired" may need to be reviewed in a future for which there is no past analog. Managing for resilience through protection, restoration, and reducing risks may be effective for protecting valued ecosystems in the short term. These efforts might buy some time for developing new methods and strategies for addressing longer-term ecosystem and environmental responses of continued climate change.

Within NPS, adaptation may involve prioritizing which resources, and possibly which parks, should receive immediate attention, while recognizing that the physical and biological changes that will accompany warming trends and increasing occurrences of extreme events will affect every one of the 270 natural national parks in the coming century. NPS can be a catalyst for regional collaboration with other land and resource management entities. Regional partnerships together can evaluate alternative scenarios of change and plausible collective responses. Uncertainties about how ecosystems will change, as well as the organizational responses to climate change, will need to be confronted, acknowledged, and incorporated into decision-making processes. Adaptation will be facilitated by the use of adaptive management, where management actions generate data that are used to evaluate the effects of alternative, feasible, management interventions. Flexibility, and institutionalizing trust in resource managers that can, and must, take some risks, will need to become more common than traditional management methods that emphasize control over nature.

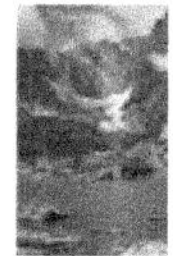

This chapter has addressed how climate change challenges both the natural resources within parks and the social system linked to those parks. Effective adaptations require that agencies, scientists, and the public think differently about how to manage natural resources. There are many strategies available to confront the uncertainties and complexities of climate change, but with climate change upon us, there is precious little time to wait.

National Wildlife Refuges

Lead Authors: J. Michael Scott, U.S. Geological Survey and University of Idaho; Brad Griffith, U.S. Geological Survey and University of Alaska Fairbanks

Contributing Authors: Robert S. Adamcik, U.S. Fish and Wildlife Service; Daniel M. Ashe, U.S. Fish and Wildlife Service; Brian Czech, U.S. Fish and Wildlife Service; Robert L. Fischman, Indiana University School of Law; Patrick Gonzalez, The Nature Conservancy; Joshua J. Lawler, University of Washington; A. David McGuire, U.S. Geological Survey; Anna Pidgorna, University of Idaho

5.1 SUMMARY

The U.S. National Wildlife Refuge System (NWRS) is the largest system of protected areas in the world. It encompasses more than 93 million acres (37.6 M ha) and is composed of 584 refuge units plus 37 wetland management districts that include waterfowl production areas in 193 counties. Compared with other federal conservation areas, the units are relatively small, typically embedded in a matrix of developed lands, and situated at low elevations on productive soils. The key mandate of the NWRS Improvement Act of 1997 is to maintain the integrity, diversity, and health of trust species and populations of wildlife, fish and plants. This species mandate provides the system with substantial legal latitude to respond to conservation challenges. The system has emerged and evolved in response to crises that have included market hunting at the beginning of the 20th century, dust-bowl drought during the 1930s, and recognition of dramatic reductions in biodiversity in the 1970s. Ongoing conservation challenges include habitat conversion and fragmentation, invasive species, pollution, and competition for water. The most recent pervasive and complex conservation challenge is climate change.

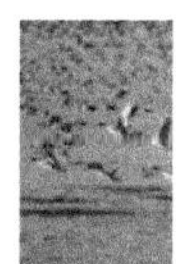

KEY FINDINGS

Climate change will have NWRS-wide effects on species and their habitats. Mean global temperature has risen rapidly during the past 50 years and is projected to continue increasing throughout the 21st century. Changes in precipitation, diurnal temperature extremes, and cloudiness—as well as sea level rise—are some of the factors that are projected to accompany the warming. A coherent pattern of poleward and upward (elevation) shifts in species distributions, advances in phenology of plants, and changes in the timing of arrival of migrants on seasonal ranges in concert with recent climate warming has been well documented and is expected to have NWRS-wide effects.

The effects of most concern are those that may occur on NWRS trust species that have limited dispersal abilities. Climate related changes in the distribution and timing of resource availability may cause species to become decoupled from their resource requirements. For example, the projected drying of the Prairie Pothole Region—the single most important duck production area in North America—will significantly affect the NWRS's ability to maintain migratory species in general and waterfowl in particular. Maintaining endangered aquatic species, such as the Devil's Hole pupfish, which occurs naturally in a single cave in Ash Meadows NWR in Nevada, will present even more challenges because, unlike waterfowl that can shift their breeding range northward, most threatened and endangered species have limited dispersal abilities and opportunities. Projected sea level rise has substantial negative implications for 161 coastal refuges, particularly those surrounded by human developments or steep topography. Projected climate-related changes in plant communities are likely to alter habitat value for trust species on most refuges; *e.g.*, grasslands and shrublands may become forested. Habitats for trust species at the southern limits of ecoregions and in the Arctic, as well as rare habitats of threatened or endangered species, are most likely to show climate-related changes.

Managing the "typical" challenges to the NWRS requires accounting for the interaction of climate change with other stressors in the midst of substantial uncertainties about how stressors will interact and systems will respond. Many NWRS trust species are migratory. Breeding, staging, and wintering habitats are typically dispersed throughout the system and on non-NWRS lands. The superimposition of spatially and temporally variable warming on spatially separated life history events will add substantial complexity to understanding and responding to ongoing conservation challenges. Climate change will act synergistically with other system stressors, and is likely to impose complex non-linear system responses to the "typical" challenges. It will be extremely difficult to clearly understand the influence of non-climate stressors on habitats, populations, and management actions without accounting for the effects of climate change. Local- to national-scale managers will face the dilemma of managing dynamic systems without fully understanding what, where, or when the climate related changes will occur, or how they might best be addressed. The actions suggested below will increase the chances of effectively resolving this dilemma.

Actions taken now may help avoid irreversible losses. Lost opportunities cannot be regained. The system is changing, and delaying action could result in irreversible losses to the integrity, diversity, and health of the NWRS. Heterogeneity in climate change effects will require diverse and innovative adaptations, increased emphasis on rigorous modeling projections at multiple scales, effective application of the experimental concepts fundamental to adaptive management, and enhanced collaboration with public and private stakeholders. However, expert opinion will need to be used in the initial response stages, and mistakes will be made while adaptation capabilities are being developed. Waiting for improved climate effect projections before acting would be inappropriate in view of the pervasive and immediate nature of the problem; developing a culture that rewards risk taking would enhance the speed of adaptation to climate change challenges. Expected decadal persistence of climate change effects suggests that a revision of contemporary planning and budgeting horizons will be necessary.

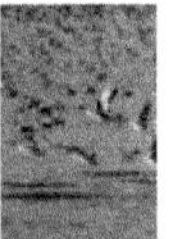

Knowing which species will be affected positively and negatively will allow NWRS managers to take advantage of positive outcomes and prepare for the management challenges of negative outcomes. If the near-term historical record is an accurate indicator, there will be substantial spatial heterogeneity in temperature and precipitation trends across the NWRS accompanying the system-wide increase in mean temperatures. As a result of this heterogeneity in regional- and local-scale climate change effects, some species will be "winners" and others will be "losers." Opportunities to capitalize on positive effects of climate change should be exploited. However, the scientific literature primarily documents negative effects. These negative effects of climate change present the NWRS with the most difficult management challenges. Once lost, conservation opportunities are extremely difficult to regain.

Responding to ecological effects may also be improved by projecting the possible futures of trust species, their NWRS habitats, and management options at all relevant management scales using the most rigorous scientific modeling tools, climate change scenarios, and suite of expected non-climate stressors. This activity would have several components: (1) clearly identifying conservation targets for the coming decades, and implementing effective and efficient monitoring programs to detect climate-related system changes; (2) identifying the species and systems most vulnerable to climate change, in the context of other system stressors, at the refuge, regional, and national scales, and prioritizing planning, budgeting, and management accordingly; (3) evaluating scale-specific (refuge > region > NWRS) suites of management and policy responses to alternative climate change scenarios; (4) developing objective criteria for choosing among these responses; and (5) proactively developing, comparing, executing, and evaluating multi-scale plans to mitigate vulnerability to climate change using adaptive management principles. Climate change can serve as a catalyst to develop an increased understanding of the ecological mechanisms affecting trust species and to improve the rigor of adaptive management programs.

A key requirement for adaptation to climate change is recognition that management for static conservation targets is impractical. The historical concept of refuges as fixed islands of safe haven for species is no longer viable. Even in special situations, such as the sole remaining habitat for a threatened or endangered species, management for the status quo will not be appropriate to the challenge of climate change. Managers and researchers will need to define and focus on a dynamic system "state" that provides representative, redundant, and resilient populations of trust species that fulfill the key legal mandate to maintain the integrity, diversity, and health of NWRS conservation targets. Managing for a dynamic system "state" that provides representative, redundant, and resilient populations of trust species provides the best opportunity to fulfill NWRS legal mandates in an environment that allows for evolutionary response to the effects of climate change and other selective forces.

The effective conservation footprint of the NWRS may be increased by using all available tools and partnerships. Maintaining and enhancing connectivity of system units is critical and may be accomplished by increasing the effective conservation footprint of NWRS. Approaches for increasing this footprint include new institutional partnerships; management responses that transcend traditional political, cultural, and ecological boundaries; greater emphasis on trans-refuge and trans-agency management and research; strong political leadership; and re-energized collaborations between the NWRS and its research partners at multiple spatial scales. Increasing the conservation footprint may bring about greater resilience of the NWRS to the challenge of climate change.

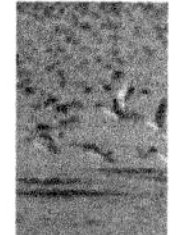

Actions that will enable more effective responses to climate change include initiating multi-scale communication, education, research, and training programs, and strengthening collaborations between USFWS and all conservation management and research partners. Effectively responding to climate-related complexity will be aided by substantial education and training, along with multi-scale, coordinated, and focused efforts by all NWRS partners (management, research, and other public and private land managers). Stronger management-research collaborations will help identify management- and policy-relevant climate-related ecological changes and responses, will keep decision makers informed, and will thus increase the likelihood that an effective response to climate change will be made. All levels and jurisdictions of management and research need to be integrated and empowered to meet the challenge of climate change. Climate change ignores administrative boundaries. Therefore it will be important to explore means of facilitating collaboration and communication among government and private land managers, such as an inter-agency climate information center that serves as a clearing house for documented climate change effects and available management tools.

A clearly elucidated vision of the desired state of the NWRS on the 150th anniversary of the system in 2053 would enhance the development of a framework for adaptation. This vision needs to explicitly incorporate the expected challenges of climate change and define the management philosophy necessary to meet this challenge. The complexity of expected climate effects and necessary management responses offers an opportunity to re-energize a focus on the interconnection of spatially separated units of the NWRS and to foster an integrated refuge-to-NWRS vision for managing climate change effects on system trust species.

Because climate change is a global phenomenon with national, regional, and local effects, it may be the largest challenge faced by the NWRS. Climate change adds a known forcing trend in temperature to all other stressors, and likely creates complex non-linear challenges that will be exceptionally difficult to understand and mitigate. New tools, new partnerships, and new ways of thinking will be required to maintain the integrity, diversity, and health of the refuges in the face of this complexity. The historic vision of refuges as fixed islands of safe haven for species met existing needs at a time when the population of the United States was less than half its current size and construction of the first interstate highway was a decade away. At that time, climates and habitats were perceived to be in dynamic equilibrium, and species were able to move freely among refuges. Today, the landscape is highly fragmented, much of the wildlife habitat present in the 1930s and 1940s has been lost, and climate-related trends in ecological systems are well documented. While Congress' aspiration for the refuges to serve as a national network for the support of biological diversity remains sound, the challenge now is to make the refuge network more resilient and adaptive to a changing environment.

5.2 BACKGROUND AND HISTORY

5.2.1 Introduction

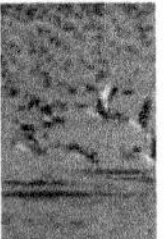

The National Wildlife Refuge System (NWRS)—the largest system of protected areas in the world established primarily to manage and protect wildlife—was born in and has evolved in crises. The first crisis was the threat to egrets, herons, and other colonial nesting waterbirds caused by hunting for feathers and plumes for the millinery trade; the second was the loss of wildlife habitat, accelerated by the Great Depression, drought, and agricultural practices in the dust bowl era. The third—still ongoing—is species extinction triggered by a growing human population and its demand on natural resources. The first two crises were largely regional in their influence and effect. Although the third crisis—extinction—is international, the response to it is local. The influence of the fourth crisis—climate change—is global and covers the full breadth and depth of the NWRS. It will require national to local responses.

In response to the first challenge, President Theodore Roosevelt established America's first national wildlife refuge (NWR), Pelican Island, Florida. Nearly three decades later, in response to depression-era challenges, Ira Gabrielson and Ding Darling had a vision for a system of refuges that would ensure the survival of recreationally viable populations of waterfowl for future generations of Americans. Whereas the first response resulted in an *ad hoc* collection of refuges, the second was the birth of the NWRS as the vision of Gabrielson and Darling, carried forward by three generations of wildlife biologists and managers. The U.S. Fish and Wildlife Service (USFWS), which manages the NWRS, has responded to the current extinction crisis in a number of ways, including the establishment and management of 61 refuges to recover threatened and endangered species. That response has been insufficient to meet the challenge of biodiversity loss, which will only progress as it is exacerbated by climate change.

Now, more than a century after Theodore Roosevelt established Pelican Island NWR, 584 refuges and nearly 30,000 waterfowl production areas encompassing 93 million acres and spanning habitats as diverse as tundra, tropical rainforests, and coral reefs, dot the American landscape (Figs. 5.1 and 5.2). However, rapidly increasing mean global temperature during the past 100 years, which is predicted to continue throughout the coming century (*i.e.*, climate change, IPCC, 2007a), challenges not only the existence of species and ecosystems on individual refuges, but also across the entire U.S. landscape—and thus the diversity, integrity, and health of the NWRS itself. If the historical record is an indicator

NATIONAL WILDLIFE REFUGE SYSTEM

"...various categories of areas that are administered...for the conservation of fish and wildlife, including species that are threatened with extinction, all lands, waters, and interests therein administered...as wildlife refuges, areas for the protection and conservation of fish and wildlife that are threatened with extinction, wildlife ranges, game ranges, wildlife management areas, or waterfowl production areas..."

National Wildlife Refuge

"...any area of the National Wildlife Refuge System, except coordination areas..."

Other Named Refuges

584 units with 16 types of names

523 – National Wildlife Refuges
37 – Farm Service Administration (FSA)
9 – Wildlife Management Areas
2 – Fish and Wildlife Refuge
1 – Antelope Refuge
1 – Bison range
1 – Conservation Area
1 – Elk Refuge
1 – Game Preserve
1 – International Wildlife Refuge
1 – Key Deer Refuge
1 – Migratory Bird Refuge
1 – Refuge for Columbian White-tail Deer
1 – Research Refuge
1 – Wildlife and Fish Refuge
1 – Wildlife Range
1 – Wildlife Refuge

Waterfowl Production Areas

"...any wetland or pothole area acquired pursuant to section 4(c) of the amended Migratory Bird Hunting Stamp Act

Over 27,655 individual units consisting of waterfowl production areas, wetland easements, wildlife management areas, and easements from Farm Service Administration. The units are grouped into counties, which are further grouped into wetland management districts.

37 Wetland Management Districts

193 Waterfowl Production Area Counties

Coordination Area

"...a wildlife management area...made available to a State by cooperative agreement..."

49 units with 17 types of names

21- Wildlife Management Areas
5 – Game Ranges
3 – Elk Winter Pastures
3 – Public Fishing Areas
3 – Waterfowl Management Areas
2 – Elk Refuges
2 – Winter Range and Wildlife Refuges
1 – Deer-Elk Range
1 – Deer Refuge and Winter Pasture
1 – Deer Winter Pasture
1 – Game and Fish Management Unit
1 – Game Management Area
1 – Migratory Bird Management Area
1 – Migratory Waterfowl and game Management Area
1 – State Game Range
1 – Waterfowl Project
1 – Wildlife Conservation Area

Figure 5.1. Structure of the NWRS. Adapted from Fischman (2003), Refuge Administration Act,[1] and FWS Regulations.[2]

(Figs. 5.3a; 5.3b), there will be substantial heterogeneity in future trends for temperature and precipitation across the NWRS. These refuges—conservation lands—support many activities, especially wildlife-dependent outdoor recreation, which attracts more than 35 million visitors a year (Caudill and Henderson, 2003), and other economic activities where compatible with refuge purposes.

Direct uses of the NWRS, such as wildlife-dependent outdoor recreation and farming, are the most readily valued in monetary terms. Ecological functions of the refuges that provide services to humans include water filtration in wetlands and aquifers, buffering

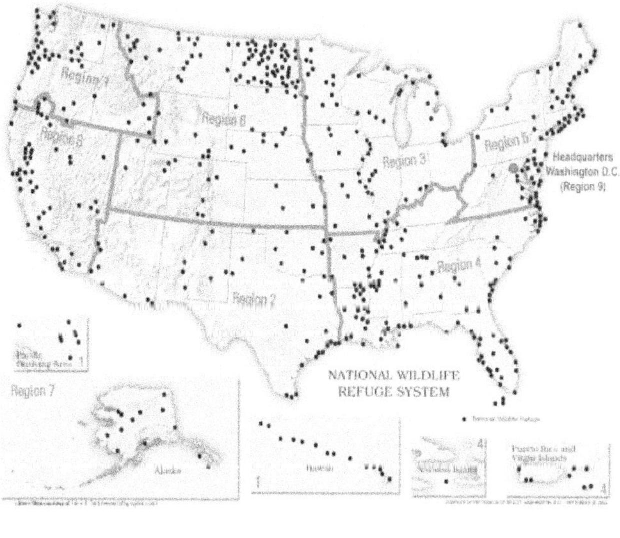

NATIONAL WILDLIFE REFUGE SYSTEM

Figure 5.2. The National Wildlife Refuge System. Adapted from Pidgorna (2007).

[1] P. L. No. 89-669, 16 U.S.C. '668dd
[2] FWS Regulations – CFR 50

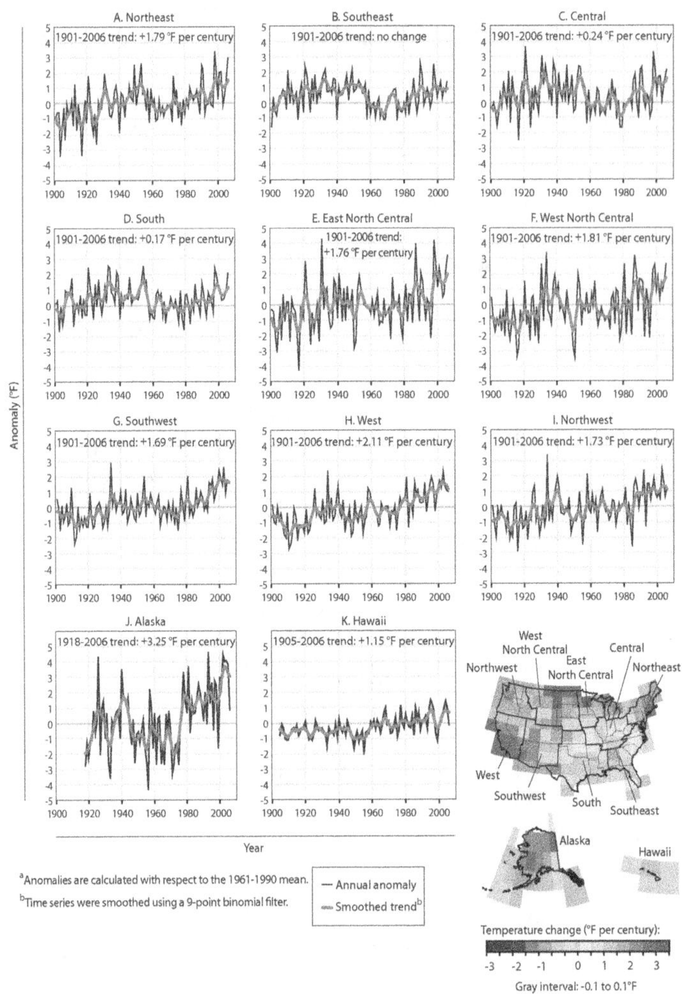

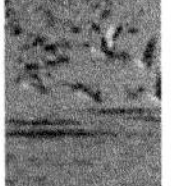

Figure 5.3a. Observed annual trends in temperature, 1901-2006, for the coterminous United States and Alaska. Data and mapping courtesy of NOAA's National Climate Data Center.

from hurricanes by coastal wetlands, and maintenance of pollinator species that pollinate agricultural plants off the NWRS. A recent estimate of the value of ecosystem services provided by the NWRS was $26.9 billion/year.[3]

Refuges were established as fixed protected areas, conservation fortresses, set aside to conserve fish, wildlife, and plant resources and their habitats. The NWRS design principles assumed an environment that varied but did not shift. Populations and ecosystems were thought to be in dynamic equilibrium, where species could move freely among the refuges and challenges could be dealt with through local management actions. Much has changed since then. The population of the United States in 1903, when the first refuge was established, was 76 million, and gross domestic product (GDP) was $300 billion[4] with no interstate highways. On the 100th anniversary of Pelican Island NWR, America's population reached 290 million, its GDP increased by a factor of 36, and more than 46,000 miles of interstate highways both linked and fragmented America's landscape. The assumption of plant and animal populations moving freely

[3] **Ingraham**, M.W., and S.G. Foster, in press: The indirect use value of ecosystem services provided by the U.S. National Wildlife Refuge System. *Ecological Economics*.

[4] In 1992 dollars.

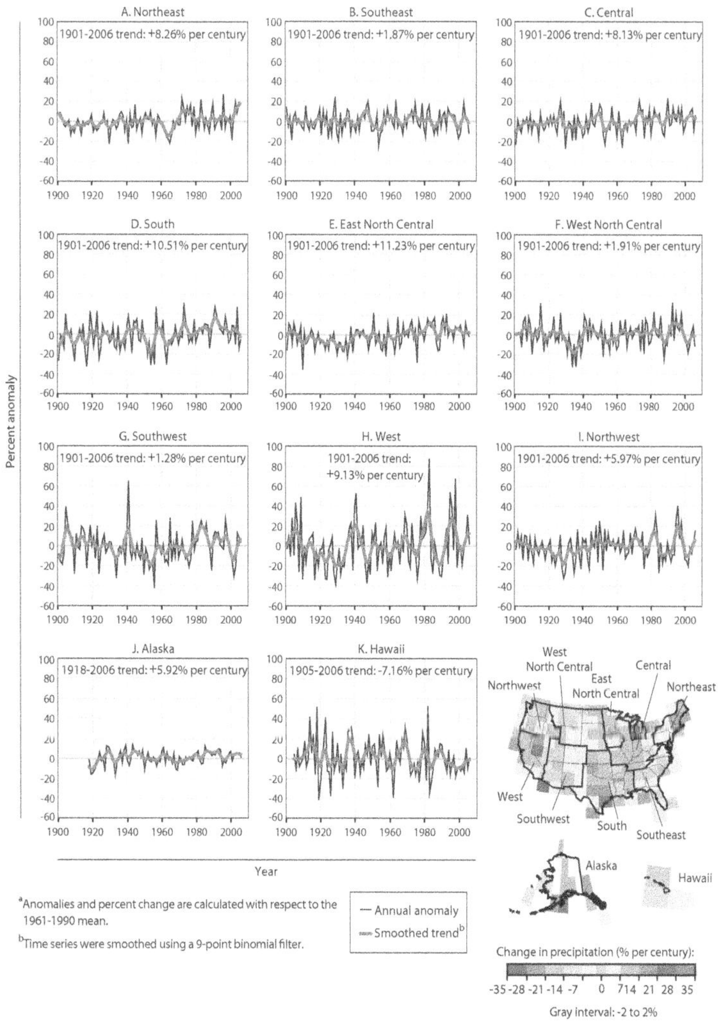

Figure 5.3b. Observed annual trends in precipitation, 1901-2006, for the coterminous United States and Alaska. Data and mapping courtesy of NOAA's National Climate Data Center.

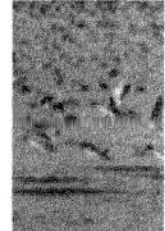

among refuges could no longer be made. Yet with climate change, the need for such free movement is greater. It is now apparent that species' ranges are dynamic, varying in space and time, but showing a globally coherent response to climate change (Parmesan and Yohe, 2003). Climate change may exacerbate the misfits between the existing NWRS and ecological realities. Coastal refuges are likely to become inundated, migrations supported by refuges may become asynchronous with the changing seasons, native and non-native invasive species will likely extend their ranges into new refuges, and vegetation types may shift to plant communities that are inappropriate for refuge trust species.

Today, a system established to respond to local challenges is faced with a global challenge, but also—as with the first three crises—with an opportunity. The NWRS is only beginning to consider how to address projected climate change effects through management activities; however, using enhanced understanding of ecological mechanisms and the administrative mandates of the NWRS Improvement Act of 1997, the USFWS is better equipped to take on this new crisis. Success will demand new tools, new ways of thinking, new institutions, new conservation partnerships, and renewed commitment for maintaining the biological integrity, diversity, and health of America's

BOX 5.1. USFWS Goals for the NWRS (601 FW1).[5]

1. Conserve a diversity of fish, wildlife, and plants and their habitats, including species that are endangered or threatened with becoming endangered.
2. Develop and maintain a network of habitats for migratory birds, anadromous and interjurisdictional fish, and marine mammal populations that is strategically distributed and carefully managed to meet important life history needs of these species across their ranges.
3. Conserve those ecosystems, plant communities, wetlands of national or international significance, and landscapes and seascapes that are unique, rare, declining, or underrepresented in existing protection efforts.
4. Provide and enhance opportunities to participate in compatible wildlife-dependent recreation (hunting, fishing, wildlife observation and photography, and environmental education and interpretation).
5. Foster understanding and instill appreciation of the diversity and interconnectedness of fish, wildlife, and plants and their habitats.

wildlife resources on the world's largest system of dedicated nature reserves. No longer can refuges be managed as independent conservation units. Decisions require placing individual refuges in the context of the NWRS. The response must be system-wide as well as local to match the scale and effects of the challenge. Such a response is unprecedented in the history of conservation biology.

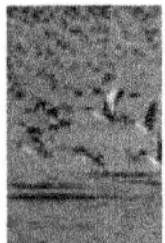

The ability of individual refuges and the entire NWRS to respond to the challenge of climate change is a function of the system's distribution, unit size, and ecological context. Familiarity with the legal, ecological, geographical and political nature of the NWRS is necessary for understanding both challenges and opportunities to adapting to climate change on the NWRS. It is equally important to understand that existing legal and policy guidelines direct refuge managers to manage for a set of predetermined conservation targets (trust species). Meeting legal and policy guidelines for maintaining biological integrity, diversity, and environmental health of the NWRS will require careful evaluation of the continuing role of individual refuges in the face of climate change.

With climate change there is a renewed realization that species' distributions are dynamic. This requires the NWRS to manage for change in the face of uncertainty. Climate change effects will be enduring, but existing models and projections typically span decades to a century. Unless otherwise specified, we focus on the decadal time frame for adaptation measures described in this chapter. The scientific literature is dominated by reports of negative effects of climate change, and this dominance is reflected in our treatment of effects on refuges because the negative effects of climate change will present the greatest challenges to managers and policy makers.

In the pages that follow we focus on regional and national scales, and: (1) describe the institutional capacity of the NWRS to respond to the challenge of climate change; (2) document challenges to integrity, diversity, and health of species, refuges, and the NWRS; (3) describe projected effects of climate change on components of the NWRS; (4) identify research themes and priorities, most vulnerable species and regions, and important needs; and (5) suggest new partnerships for conservation success.

5.2.2 Mission, Establishing Authorities, and Goals

The NWRS is managed by the USFWS (Fig. 5.4) under two sets of "purposes" (Fischman, 2003). The first is the generic (or System) purpose, technically called the "mission," defined in the NWRS Improvement Act of 1997: "The mission of the NWRS is to administer a national network of lands and waters for the conservation, management, and where appropriate, restoration of the fish, wildlife, and plant resources and their habitats within the United States for the benefit of present and future generations of Americans." The Act goes on to define the two most flexible terms of the mission, conservation and management, as a means "to sustain and, where appropriate, restore and enhance, healthy populations" of animals and plants using methods associated with "modern scientific resource programs."[6] In 2006, the USFWS interpreted this first congressional purpose in a policy (601 FW1),[7] which lists five goals that derive from the mission and other objectives stated in statute (see Box 5.1). The USFWS policy gives top

[5] U.S. Fish and Wildlife Service manual 601 FW 1 - FW 6.

[6] 16 USC 668dd P. L. 105–57

[7] U.S. Fish and Wildlife Service manual 601 FW 1

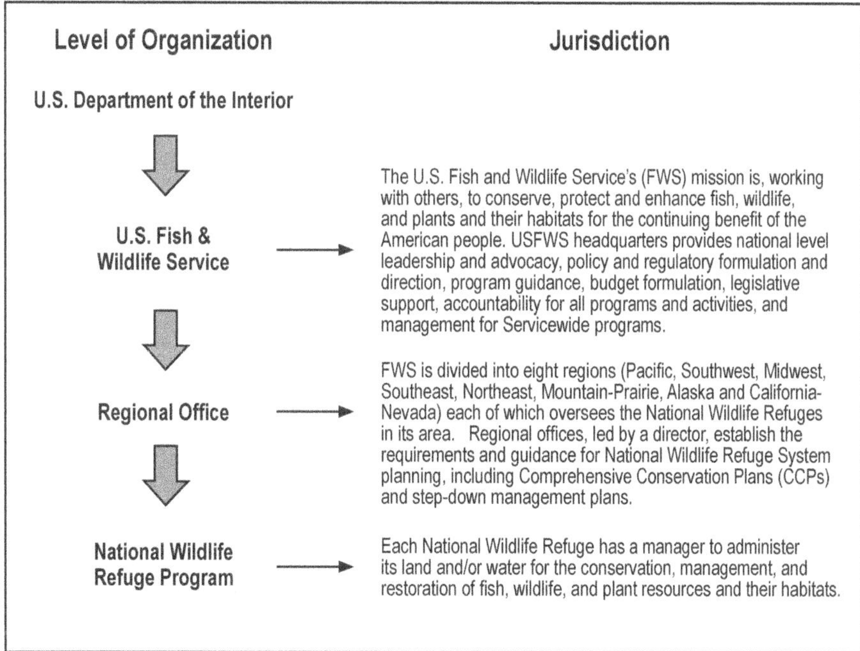

Figure 5.4. Organizational chart.[8]

priority to the first three goals listed in Box 5.1, which focus most directly on the ecological concerns that impel adaptation to climate change.

The second set of purposes is individual purposes specific to individual refuges or specific tracts or units within a refuge that may have been acquired under different authorities (Fig. 5.1). These are the authorities under which the refuge was originally created, as well as possibly additional ones under which individual later acquisitions may have been made. While it is difficult to conceive of a conflict between the NWRS mission and individual refuge purposes, in such an event the latter, or more specific, refuge purpose takes precedence. Furthermore, where designated wilderness (or some other overlay system, such as a segment of a wild and scenic river) occurs within a refuge boundary, the purposes of the wilderness (or any other applicable overlay statute) are additional purposes of that portion of the refuge.

Establishing authorities for a specific refuge may derive from one of three categories: presidential, congressional, and administrative (Fischman, 2003). Refuges established by presidential proclamation have very specific purposes, such as that for the first refuge, Pelican Island (a "preserve and breeding ground for native birds"). Congressional authorities stem from one or more of 15 different statutes providing generally for new refuges, such as the Migratory Bird Conservation Act ("for use as an inviolate sanctuary or for any other management purpose for migratory birds").[9] Or, they may be specific to a single refuge, such as the Upper Mississippi River NWR (as a refuge for birds, game, fur-bearing animals, fish, other aquatic animal life, wildflowers and aquatic plants).[10] The third source of refuge purposes are administrative documents such as public land orders, donation documents, and administrative memoranda (Fischman, 2003). These, however, are less clearly understood and documented, and are not addressed further in this document.

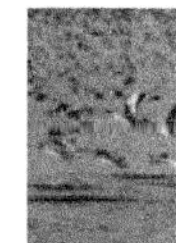

5.2.3 Origins of the NWRS

The first significant legislative innovation to systematically assemble protected areas

[8] **U.S. Fish and Wildlife Service**, 2007: America's national wildlife refuge system. FWS Website, http://www.fws.gov/refuges, accessed on 7-18-2007.

[9] 16 U.S.C. 715-715r; 45 Stat. 1222

[10] 16 USC § 721

was the Migratory Bird Conservation Act of 1929,[11] which authorized acquisition of lands to serve as "inviolate sanctuaries" for migratory birds (Fig. 5.5). But funds to purchase refuges were scarce. In the early 1930s, waterfowl populations declined precipitously. Congress responded with the Migratory Bird Hunting Stamp Act of 1934.[12] It created a dedicated fund for acquiring waterfowl conservation refuges from the sales of federal stamps that all waterfowl hunters would be required to affix to their state hunting licenses. This funding mechanism remains the major source of money for purchasing expansions to the NWRS. A quick glance at a map of today's NWRS (Fig. 5.2) confirms the legacy of the research findings and funding mechanism of the 1930s: refuges are concentrated in four corridors. The geometry of the NWRS conservation shifted from the enclave points on the map to the flyway lines across the country (Gabrielson, 1943; Fischman, 2005; Pidgorna, 2007).

After the push for protecting habitat of migratory waterfowl, the next impetus for NWRS growth came in the 1960s as Congress recognized that a larger variety of species other than just birds, big game, and fish needed protection from extinction. The Endangered Species Preservation Act of 1966 sought to protect species, regardless of their popularity or evident value, principally through habitat acquisition and reservation. In doing so, the law provided the first statutory charter for the NWRS as a whole. Indeed, the part of the 1966 law dealing with the refuges is often called the Refuge Administration Act.[13]

The 1966 statute consolidated the conservation land holdings of the USFWS: it was the first statute to refer to this hodgepodge as the "NWRS" and it prohibited all uses not compatible with the purpose of the refuge. The compatibility criterion, established by statute in 1966, but practiced by the USFWS for decades before that, would become a byword of international sustainable development in the

[11] 16 U.S.C. 715-715r; 45 Stat. 1222

[12] 16 U.S.C. § 718-718h

[13] P. L. No. 89-669, 16 U.S.C. § 668dd

[14] **U.S. Fish and Wildlife Service**, 2007: History of the national wildlife refuge system. U.S. Fish and Wildlife Service Website, http://www.fws.gov/refuges/history/index.html, accessed on 7-10-2007.

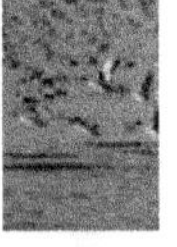

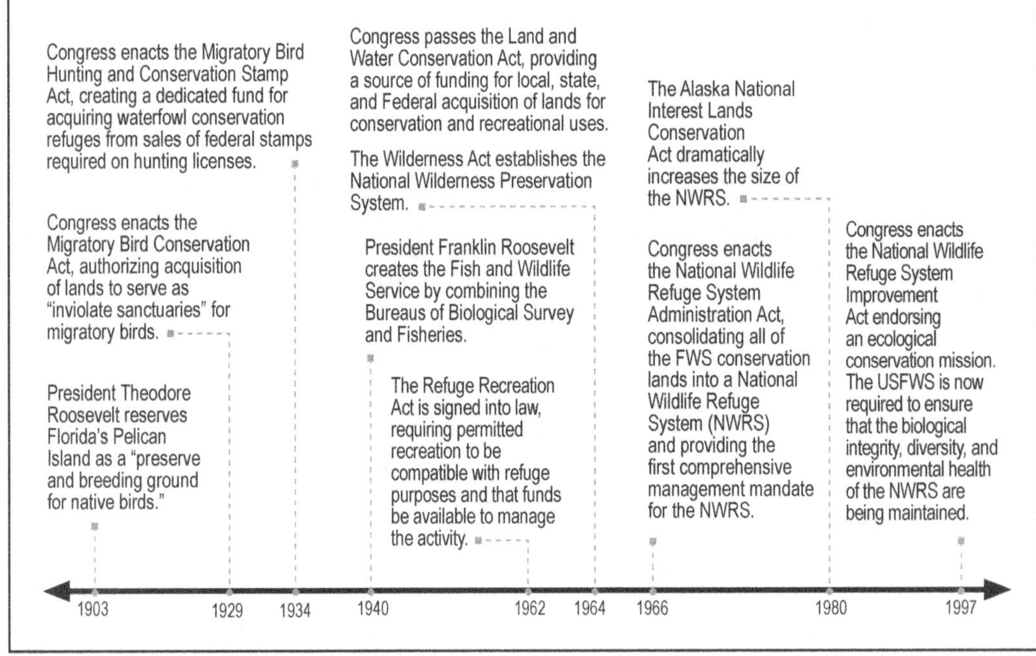

Figure 5.5. Timeline of milestone events of the NWRS.[14]

1980s. In 1973 the Endangered Species Act[15] replaced the portion of the 1966 law dealing with imperiled species, and succeeded it as an important source of refuge establishment authority. The ESA also provides a broad mandate for the Interior Department to review the NWRS and other programs and use them in furtherance of imperiled species recovery.

In 1980 Congress enacted the Alaska National Interest Lands Conservation Act. This added over 54 million acres to the NWRS.

5.2.4 The 1997 NWRS Improvement Act

The NWRS Improvement Act (NWRSIA) of 1997[16] marked the first comprehensive overhaul of the statutory charter for the NWRS since 1966. It is also the only significant public land "organic legislation" since the 1970s (Fischman, 2003). The term "organic legislation" describes a fundamental piece of legislation that either signifies the organization of an agency and/or provides a charter for a network of public lands. The key elements of the NWRSIA are described below.

The NWRSIA sets a goal of conservation, defined in ecological terms (*e.g.*, sustaining, restoring, and enhancing populations). The 1997 statute envisions the NWRS as a national network of lands and waters to sustain plants and animals. This realigns the geometry of refuge conservation from linear flyways to a more complex web of relationships. The NWRSIA requires each refuge to achieve the dual system-wide and individual refuge purposes, with the individual establishment purpose receiving priority in the event of a conflict with the NWRS mission.

5.2.4.1 Designated Uses

The NWRSIA constructs a dominant use regime, where most activities must either contribute to the NWRS goal or at least avoid impairing it. The primary goals that dominate the NWRS are individual refuge purposes and the conservation mission. The next level of the hierarchy are the "priority public uses" of wildlife-dependent recreation, which the statute defines as "hunting, fishing, wildlife observation, and photography, or environmental education and interpretation."[17] These uses may be permitted where they are compatible with primary goals. The statute affirmatively encourages the USFWS to promote priority public uses on refuges.

5.2.4.2 Comprehensive Conservation Plans (CCPs)

The NWRSIA requires comprehensive conservation plans ("CCP") for each refuge unit (usually a single refuge or cluster of them). The CCPs zone refuges into various areas suitable for different purposes and set out desired future conditions. The NWRSIA requires the USFWS to prepare a CCP for each non-Alaskan unit within 15 years and to update each plan every 15 years, or sooner if conditions change significantly. Planning focuses on habitat management and visitor services. The planning policy models its procedure on adaptive management.[18] Once approved, the CCP becomes a source of management requirements that bind the USFWS, though judicial enforcement may not be available.[19]

The majority of refuges are still in the process of completing their CCPs. In a review of 100 completed refuge CCPs available online as of February 1, 2007, only 27 CCPs included terms such as "climate change," "climate variability," "global change," or "global warming." None of these CCPs have identified explicit adaptation management strategies that are currently being implemented. This suggests that the perception of climate variability and change as a challenge is just emerging in the refuge management community. Much of the information needed to implement an effective response to climate change is unavailable to refuge managers. Furthermore, the system-wide nature of the climate change challenge will require system-wide responses. The magnitude of the challenge posed by climate change is unprecedented in scale and intensity, and the challenges exceed the capabilities of individual refuges. National coordination and guidance are needed. The CCPs provide a vehicle for engaging refuges in planning for response to climate change within the context of the NWRS.

15 P. L. 93-205, 16 U.S.C. § 1531-1544, 87 Stat. 884
16 P.L. 105-57, 16 USC § 668dd

17 P.L. 105-57, 16 USC § 668dd
18 U.S. Fish and Wildlife Service manual 602
19 Norton v. Southern Utah Wilderness Alliance, 2004. 542 U.S. 55.

5.2.4.3 Cross-Jurisdictional Cooperation

Like all of the modern public land organic laws, the NWRSIA calls for coordination with states, each of which has a wildlife protection program. This partnership with states is, of course, limited by federal preemption of state law that conflicts with USFWS management control on refuges. For instance, a state may not impose its own management programs or property law restrictions on the NWRS under circumstances where they would frustrate decisions made by the USFWS or Congress.[20] USFWS policy emphasizes state participation in most refuge decision-making, especially for comprehensive conservation planning and for determination of appropriate uses.

5.2.4.4 Substantive Management Criteria

The NWRSIA imposed many substantive management criteria, some of which are unprecedented in public land law. First, the Act expanded the compatibility criterion as a basic tool for determining what uses are allowed on refuges. The USFWS may not permit uses to occur where they are incompatible with either the conservation mission or individual refuge purposes. The Act defines "compatible use" to mean "a wildlife-dependent recreational use or any other use of a refuge that, in the sound professional judgment of the Director, will not materially interfere with or detract from the fulfillment of the mission of the NWRS or the purposes of the refuge."[21] The USFWS compatibility policy promises to assure that "densities of endangered or otherwise rare species are sufficient for maintaining viable populations."[22] The USFWS interprets its policy to prohibit uses that reasonably may be anticipated to fragment habitats.[23] Second, the NWRSIA requires that the USFWS maintain "biological integrity, diversity, and environmental health" on the refuges.[24] This element of the 1997 Act, discussed in more detail directly below, is the closest Congress has ever come to requiring a land system to ensure ecological sustainability, and creates a mandate unique to federal land systems in the United States.

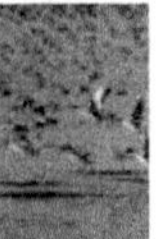

5.2.4.5 New Emphasis on Biological Integrity, Diversity, and Environmental Health

The Policy on Biological Integrity, Diversity, and Environmental Health[25] presents the process by which the NWRS fulfills the NWRSIA mandate to "...ensure that the biological integrity, diversity, and environmental health of the System are maintained..." The 2001 USFWS policy correspondingly focuses on the three distinct yet largely overlapping concepts of biological integrity, diversity, and environmental health. The core idea of the policy is maintaining composition and function of ecosystems (Fischman, 2004). Though climate change may make that impossible within the boundary of some refuges, it remains an appropriate guiding principle for the system as a whole. The policy's guidance on the biological integrity, diversity, and environmental health mandate is the single most important legal foundation for leadership in shifting NWRS management toward needed adaptations. There are other path-breaking criteria especially relevant to adaptation, but the USFWS has yet to implement them through new policies or other major initiatives. However, as climate change increases in importance to the public and refuge managers, the USFWS will find itself increasingly challenged by its 1997 duty to: (1) acquire water rights needed for refuge purposes; (2) engage in biological monitoring; and (3) implement its stewardship responsibility.[26] While the 2001 policy provides a basis for ecological sustainability, climate change presents new challenges at unprecedented scales for maintaining biological integrity, diversity, and environmental health of refuges and the refuge system. Explicit performance goals and objectives tied to biological integrity, diversity and environmental health of refuges and the services conservation targets will be needed to assess the degree and effectiveness of NWRS response to the challenges of climate change.

Rather than compare refuge conditions with existing reference sites, the USFWS policy encourages managers to use "historic conditions"

[20] North Dakota v. United States, 1983. 460 U.S. 300. and State of Wyoming v. United States, 2002. D.C. No. 98-CV-37-B, 61 F. Supp. 2d 1209-1225.

[21] 16 USC § 668dd

[22] U.S. Fish and Wildlife Service manual 601 FW 1 - FW 6.

[23] U.S. Fish and Wildlife Service manual 603, 65 Federal Register 62486

[24] 16 USC § 668dd

[25] U.S. Fish and Wildlife Service manual 601 FW 3

[26] 16 USC § 668dd

(for integrity and health, but not diversity) as a benchmark for success. "Historic conditions" are those present before significant European intervention. This policy assumes a range of variation that is constant. That assumption is not consistent with projected environmental changes that may result from climate change. Rather, historical benchmarks and their variability may provide long-term perspective for developing strategies for the management of self-sustaining native populations and ecosystems in the face of change and uncertainty.

With climate change, the future species composition of the community may be quite different from that of the time when the refuge was established. However, the opportunity to manage biological integrity, diversity, and environmental health of refuges and the NWRS, regardless of changes in species composition, remains. The policy on biological integrity, diversity, and environmental health does not insist on a return to conditions no longer climatically appropriate. Instead, it views historical conditions as a frame of reference from which to understand the successional shifts that occur within ecological communities as a result of climate change. The policy also implies that we can use the knowledge and insights gained from such analysis to develop viable site-specific management targets for biological integrity, diversity, and environmental health despite the changing climate.

In addition to addressing ecosystems or ecological communities, the policy also governs target fauna and flora, stressing that native populations in historic sex and age ratios are generally preferable over artificial ones, and that invasive or non-indigenous species or genotypes are discouraged. In general, except for species deemed beneficial (*e.g.*, pheasants), managers would consistently work to remove or suppress invasive and exotic species of both plants and animals. The policy directs special attention to target densities on refuges for rare species (viable densities) and migratory birds (higher-than-natural densities to accommodate loss of surrounding habitat). These targets, where extended to a broader spatial scale, provide good starting points for NWRS adaptation to climate change.

Meeting the NWRS's statutory and policy mandates will require an approach and philosophy that sees the "natural" condition of a given community as a moving target. A refuge manager must plan for the future in the context of past and present conditions and the likelihood of an altered community within the bounds of a new climate regime.

5.3 CURRENT STATUS OF THE NWRS

5.3.1 Key Ecosystem Characteristics on Which Goals Depend

One of the primary goals of the NWRS—to conserve the diversity of fish, wildlife, plants, and their habitats—is reflected in the design of the NWRS, which is the largest system of protected areas in the world primarily designated to manage and protect wildlife (Curtin, 1993). The NWRS includes 584 refuge units and nearly 30,000 waterfowl production areas[27] (Fig. 5.1) that encompass an area of over 93 million acres, distributed across the United States (Fischman, 2003; Scott *et al.*, 2004). The NWRS contains a diverse array of wildlife, with more than 220 species of mammals, 250 species of amphibians and reptiles, more than 700 species of birds, and 200 species of fish reported.

Another important goal of the NWRS is to maintain its trust species, which include threatened and endangered species, marine mammals, anadromous and interjurisdictional fish, and migratory birds. Of these, the latter remain the NWRS's largest beneficiary, with over 200 refuges established for the conservation of migratory birds (Gergely, Scott, and Goble, 2000). Shorebirds and waterfowl are better represented on refuges compared with landbirds and waterbirds (Pidgorna, 2007).

Twenty percent of refuges were established in the decade immediately following the enactment of the Migratory Bird Treaty Act (1930–1940). The NWRS captures the distribution of 43 waterfowl species in the continental United States at a variety of geographic, ecological, and temporal scales (Pidgorna, 2007).

The fact that many refuges were established in areas important to migratory birds, and especially waterfowl, can account for the

[27] Grouped into 37 wetland management districts.

abundance of wetland habitat found in the NWRS today and for the fact that refuges are found at lower elevations and on more productive soils compared with other protected areas in the United States (Scott *et al.*, 2004). Besides wetlands, other commonly occurring landcover types include shrublands and grasslands (Scott *et al.*, 2004).

The NWRS is characterized by an uneven geographic and size distribution. Larger refuge units are found in Alaska, with Alaskan refuges contributing 82.5% of the total area in the NWRS and average sizes more than two orders of magnitude greater than the average size of refuges found in the lower 48 states. Nearly 20% of the refuges are less than 1,000 acres in size, and effectively even smaller because more than half of the refuges in the system consist of two or more parcels. Median refuge area is 5,550 acres and the mean area is 20,186 acres (Scott *et al.*, 2004). In contrast, the median area of Alaskan refuges is 2.7 million acres.

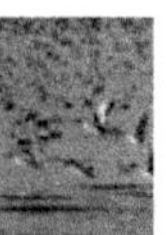

Approximately one sixth of the nation's threatened and endangered species are found on refuges. More than 50% of all listed mammals, birds, and reptiles are found on refuges (Davison *et al.*, 2006), while the percentage of listed invertebrates and plants is much lower. These, and the 10% of the threatened and endangered species for which refuges have been established, realize a conservation advantage over species not found on refuges (Blades, 2007). The NWRS plays an important role in the conservation of threatened and endangered species, providing core habitat, protection, and management. However, as most refuges are small, fragmented, and surrounded by anthropogenic habitats (Scott et al. 2004 and Pidgorna 2007), it may prove difficult for the NWRS to support and restore a diverse range of taxonomic groups and to maintain viable populations of some larger threatened and endangered species (Czech, 2005; Blades, 2007).

The distribution of refuges in geographical and geophysical space has given Americans a network of protected areas that function differently from other protected areas in the United States. In a nutshell, most refuges, with the exception of those in Alaska, are small islands of habitat located in a predominantly and increasingly anthropogenic landscape. Refuges contain lower-elevation habitat types important to the survival of a large number of species that are not included in other protected areas. Their small size and close proximity to anthropogenic disturbance sites (such as roads and cities) makes refuges vulnerable to external challenges and highly susceptible to a wide array of stressors. The lands surrounding individual refuge units (matrix lands) in the lower 48 states and Hawaii also decrease the ability of species to move from refuge to refuge; the barriers are far greater for species that cannot fly than for those that can. The positive side is that their proximity to population centers provides them with an opportunity to serve as educational centers for the public to learn more about the diversity of fish, wildlife, plants, and their habitats, as well as ecological processes and the effects of climate change. They also provide sites for researchers to develop new understanding of the ecology and management of conservation landscapes.

However, the ability of individual refuges to meet the first three of the USFWS goals, as well as the biological integrity, diversity, and environmental health clause of the NWRSIA, will depend upon the ability of refuge managers to increase habitat viability through restoration and through reduction of non-climate stressors. Other tools include integrating inholdings into refuge holdings, strategically increasing refuge habitat through CCPs, increased incentive programs, establishment of conservation easements with surrounding landowners, and, when desired by all parties, fee-title acquisitions of adjacent lands. These actions would in turn provide species with increased opportunities to adapt to a changing environment.

At the level of the NWRS, the integration of the USFWS's five goals and the biological integrity, diversity, and environmental health of species, ecosystems, and plant and animal communities may be achieved through increased representation and redundancy of target species and populations on refuge lands through strategic growth of the NWRS. The need for any such strategic growth has to be carefully evaluated in the context of maintaining the biological integrity, diversity, and environmental health of the NWRS trust species today and the uncertain effects of climate change. A national plan should be developed to assess the projected shifts in biomes and develop optimal placement of refuge lands on a landscape that is likely to exist 100 or more years into the future. Waterfowl species provide exemplars of what might be achieved for other trust species. Robust populations of ducks and geese have been achieved through seven decades of strategic acquisitions and cooperative conservation (Pidgorna, 2007), and a vision of a NWRS that conserved recreationally viable populations of North American waterfowl—a vision that was shared with many others (U.S. Fish and Wildlife Service and Canadian Wildlife Service, 1986). However, the ability to meet the objectives of the USFWS's five goals and the mandate of the NWSRIA necessitates strategic growth of the effective conservation footprint of the NWRS to increase the biological integrity, diversity, and environmental health of threatened and endangered species and at-risk ecosystems and plant communities.

5.3.2 Challenges to the NWRS

5.3.2.1 2002 Survey of Challenges to NWRS

In an effort to quantify challenges to the refuges, the NWRS surveyed all refuges and wetland management districts in 2002 with an extensive questionnaire. The result was a large database of challenges and management conflicts experienced by the NWRS. It contains 2,844 records, each representing a different challenge to a refuge or a conflict with its operations.

The most common challenges to refuges that could be exacerbated by climate change are ranked by frequency of reporting in Table 5.1. Each record covers a specific challenge, so a single refuge could have reported multiple records for the same category (*e.g.*, invasive species or wildlife disease), which are grouped for discussion purposes. The responses from the survey regarding challenges generally fall into four themes: off-refuge activities, on-refuge activities, flora and fauna imbalances, and uncontrollable natural events.

Off-refuge activities such as mining, timber harvest, industrial manufacturing, urban development, and farming often produce products or altered ecological processes that influence numbers and health of refuge species. The off-refuge activities often result in a range of environmental damage that affects the refuge, including erosion; degraded air and water quality; contaminants; habitat fragmentation; competition for water; expansion of the wildland-urban interface that creates conflicts over burning and animal control; noise and light pollution; and fragmentation of airspace with communication towers, wind turbines, and power lines.

Other activities that challenge refuges occur within refuge boundaries but are beyond USFWS jurisdiction. These activities include military activities on overlay refuges; development of mineral rights not owned by the USFWS; commercial boat traffic in navigable waters not controlled by USFWS; off-road vehicles; some recreational activities beyond USFWS jurisdiction; illegal activities such as poaching, trespassing, dumping, illegal immigration, and drug trafficking; and other concerns.

Imbalances in flora and fauna on and around the refuge also challenge refuges and the NWRS. Such concerns take the form of invasive non-native species, disease vectors such as mosquitoes, or unnaturally high populations of larger animals, usually mammals. The latter group includes small predators that take waterfowl or endangered species, beaver and muskrat that damage impoundments, and white-tailed deer that reduce forest understory (Garrott, White, and White, 1993; Russell, Zippin, and Fowler, 2001). Invasive plant species are far and away of the most concern, both within this category and within the NWRS overall (Table 5.1).

Table 5.1. The most common challenges to national wildlife refuges that could be exacerbated by climate change.[28]

Challenge	Number of Records	%
Invasive, exotic, and native pest species	902	32
Urbanization	213	7
Agricultural conflicts	170	6
Natural disasters	165	6
Rights-of-way	153	5
Industrial/commercial interface	145	5
Predator-prey imbalances	93	3
Wildlife disease	93	3

[28] **U.S. Fish and Wildlife Service**, 2002: USFWS unpublished data.

Extreme events such as hurricanes, floods, earthquakes, and volcanic eruptions also challenge refuges. While far less common than other challenges, the ecological and economic damage wrought by such events can be significant. For example, hurricanes can affect large coastal areas and multiple refuges, and cause habitat change (*e.g.*, from forest blowdowns), saline intrusion into freshwater wetlands, and loss of coastal wetlands and barrier islands. Equipment and infrastructure damage and loss can be significant and costly to repair or replace. The increasing ecological isolation of refuges and the species that reside on them decreases the ability of refuge managers to respond to effects of climate change and other stressors. However, tools and strategies used to respond to past stressors and challenges are many of the same tools that can be used to mitigate projected effects of global climate change.

5.3.2.2 Interactions of Climate Change with Other Stressors of Concern

Over the last 100 years, average annual temperatures in the United States have risen 0.8°C, with even greater increases in Alaska over the same period (2–4°C) (Houghton *et al.*, 2001). Global average surface temperatures are projected to rise an additional 1.1–6.4°C by 2100 (IPCC, 2007b). Most areas in the United States are projected to experience greater-than-average warming, with exceptional warming projected for Alaska (Houghton *et al.*, 2001). Coastal areas have experienced sea level rise as global average sea level has risen by 10–25 cm over the last 100 years (Watson, Zinyowera, and Moss, 1996). Global average sea level is projected to increase by 18–59 cm by 2100 (IPCC, 2007b). Due to thermal expansion of the oceans, even if greenhouse gas emissions were stabilized at year-2000 levels, the committed sea level rise would still likely be 6–10 cm by 2100, and sea level would continue to rise for four more centuries (Meehl *et al.*, 2005).

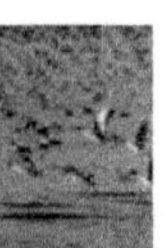

Other effects of climate change include altered hydrological systems and processes, affecting the inland hydrology of streams, lakes, and wetlands (Frederick and Gleick, 1999; Poff, Brinson, and Day, Jr., 2002). Warmer temperatures will mean reduced snowpack and earlier spring melts (Barnett, Adam, and Lettenmaier, 2005; Milly, Dunne, and Vecchia, 2005), changes in flood magnitudes (Knox, 1993), and redistribution of lakes and wetlands across the landscape (Poff, Brinson, and Day, Jr., 2002). Climate change is also likely to affect other physical factors, such as fire and storm intensity (Westerling *et al.*, 2006; IPCC, 2007b).

Climate changes may have cascading effects on ecological systems (Walther *et al.*, 2002; Parmesan and Yohe, 2003; Root *et al.*, 2003; Parmesan, 2006). These include changes in species' phenologies, distributions, and physiologies.

Climate change is likely to magnify the influences of other challenges—including habitat loss and fragmentation, changes in water quality and quantity, increased transportation corridors, etc.—on the NWRS. Climate change will also introduce new challenges or variations on existing ones, primarily by accelerating a convergence of issues (*e.g.*, water scarcity, non-native invasive species, off-refuge land-use change, and energy development), or creating such convergences where none existed before. Current and projected challenges have the potential to undermine the mission of the NWRS and the achievement of its goals.

The following pages of this section summarize the main challenges to the NWRS that could be exacerbated by climate change (see also Section 5.8, the Appendix). There is, however, a great deal of uncertainty associated with these projections, making it possible to show the overall trend but not the specific effect on an individual refuge. For example, IPCC (2007a) projects future increases in wind speeds of tropical cyclones, but does not yet offer detailed spatial data on projected terrestrial surface wind patterns. Changes in wind patterns may affect long-distance migration of species dependent on tailwinds.

Invasive Non-Native Species

Invasive non-native species are currently one of the most common challenges to the NWRS and could become even more serious with climate changes (Table 5.1) (Sutherst, 2000). Since species are projected to experience range shifts as a result of climate change and naturally expand and contract their historic ranges, it is important to distinguish between non-native species and native species. There is distinction in state and federal law between native and non-native species.[29] The text of this report reflects those differences. We consider non-native species to be those species that have been introduced to an area as a result of human intervention, whether accidental or purposeful. Native species moving into new areas as a result of climate-change-induced range expansions continue to be native. Both native and non-native species can be considered to be invasive. It is, however, the non-native invasive species that present the greatest challenge and are discussed here and elsewhere in this chapter.

An increase in the number and spread of non-native invasive species could undermine the NWRS's goal of maintaining wildlife diversity and preserving rare ecosystems and plant communities. By replacing native organisms, non-native invasive species often alter the ecological structure of natural systems by modifying predator-prey, parasite, and competitive relationships of species. Shifting distribution of native species in response to climate change will further increase the rate of change in species' composition, structure, and function on refuges.

Range shifts that result in range contractions and range expansions are the best-studied effects of climate change on invasive non-native species. Range expansions refer to the expansion of established invasive non-native species into previously unoccupied habitats. A rise in temperatures could allow invasive non-native species to expand their ranges into habitats that previously were inaccessible to them. For example, Westbrooks (2001) describes the expansion of the balsam wooly aphid (*Adelges piceae*) into stands of subalpine fir (*Abies amabilis*). Currently the aphid is restricted to areas of low and middle elevation because of its temperature requirements; however, an increase of 2.5°C would allow the aphid to expand its range to higher elevations where it would affect native subalpine fir. Species that are considered tropical today may also expand their ranges into more northern latitudes if the climate grows warmer. When temperatures become suitable, non-native invasive species could spread into new habitats and compete with stressed native species (Westbrooks, 2001).

[29] P.L. 101-646, 104 Stat. 4761; 16 U.S.C. 4701; and P.L. 104-332, 16 USC 4701.

Although climate change might not benefit non-native invasive species over native species in all cases, it is likely that non-native invasive species will benefit from a transitional climate (Dukes and Mooney, 1999). Non-native invasive species are highly adaptable and spread quickly. Many such non-native invasive species may extirpate native plants or even lead to complete regime shifts within vegetative communities. All of these traits make non-native invasive species much more likely to survive projected climate change effects compared to many of the native species.

Disease

Climate change has the potential to affect the prevalence and intensity of both plant and animal diseases in several ways. First, changes in temperature and moisture may shift the distribution of disease vectors and of the pathogens themselves (Harvell et al., 2002; Logan, Regniere, and Powell, 2003; Pounds et al., 2006). For example, Hakalau Forest NWR, now largely free of avian malaria, harbors one of the few remaining population centers of endangered Hawaiian forest birds. Climate change may eliminate this and other such refugia by changing conditions to favor avian malaria (LaPointe, Benning, and Atkinson, 2005). Second, climate-induced changes in hydrology can alter the spread and intensity of diseases in two key ways. First, in wetlands or other water bodies with reduced water levels and higher water temperatures, diseases may be able to spread much more quickly and effectively within a population. Increased temperatures have been demonstrated to speed pathogen and/or vector development (Rueda *et al.*, 1990). Second, increases in precipitation may result in increased connectivity among aquatic systems in some areas, potentially facilitating the spread of diseases among populations. Finally, climate change may also indirectly increase the prevalence and the magnitude of disease effects by affecting host susceptibility. Many organisms that are stressed due to changes in temperature or hydrology will be more susceptible to diseases. Corals are an excellent example of increased temperatures leading to increased disease susceptibility (Harvell *et al.*, 2001).

Urbanization and Increased Economic Pressure

Urbanization has the potential to further isolate refuges by altering the surrounding matrix, increasing habitat loss and fragmentation, and introducing additional barriers to dispersal. Roads and human-built environments pose significant barriers to the movement of many species. Poor dispersers (*e.g.*, many amphibians, non-flying invertebrates, small mammals, and reptiles) and animals that avoid humans (*e.g.*, lynx) will be more isolated by increased urbanization than more mobile or more human-tolerant species. This increased isolation of wildlife populations on refuges will prevent many species from successfully shifting their distributions in response to climate change.

Urbanization has the potential to interact with climate change in two additional ways. First, increased urbanization creates more impervious surfaces, increasing runoff and potentially confounding the effects of climate-altered hydrological regimes. Second, urbanization has the potential to affect local climatic conditions by creating heat islands, further exacerbating the increases in temperature and increased evaporation.

Refuges are highly susceptible to the effects of management activities on surrounding landscapes. More pressure will likely be put on the U.S. economy with rising energy demands, which will result in a push for increased oil and gas development in the western states. This will also increase habitat loss and fragmentation on lands surrounding refuges and could result in extraction activities within refuges themselves. Economic and social pressure for alternative energy sources may increase efforts to establish wind plants near refuges, or promote agricultural expansion or conversions to produce bio-fuels, including nearby biofuel production and transport facilities.

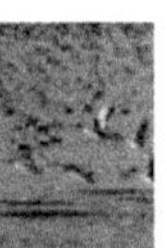

Although habitat loss and fragmentation will likely have a negative effect on the NWRS's biodiversity conservation goals, it could provide additional recreational and educational opportunities for people who will become attracted to the NWRS as open space becomes scarce. This could increase the number of visitors to the NWRS, which would raise public visibility of the refuges. Management of visitors and their activities to minimize effect on refuges and refuge species will be a challenge.

Altered Hydrological Regimes

Water is the lifeblood of the NWRS (Satchell, 2003) because much of the management of fish, migratory waterfowl, and other wildlife depends upon a reliable source of clean freshwater. Climate change is likely to result in significant changes to water resources at local, regional, and national scales, with varying effects on economies and ecosystems at all levels. The primary effects on water resources within the NWRS from climate change can be placed into two broad categories: changes in the amount and seasonality of precipitation and surface water flows.

While climate change models vary in projecting changes to precipitation to any given geographical area, at least some parts of the United States are projected to experience reduced precipitation (*e.g.*, Milly, Dunne, and Vecchia, 2005). Parts of the country where current water supplies are barely meeting demand—in particular, portions of the western United States—are especially vulnerable to any reduction in the amount, or change in timing, of precipitation. In 1995, central and southern California and western Washington experienced some of the largest water-withdrawal deficits in the United States (Roy *et al.*, 2005). Future projected increases in deficits are not just limited to the western United States, but are spread across much of the eastern part of the country as well (Roy *et al.*, 2005). Less precipitation would mean less water available for ecosystem and wildlife management, even at refuges with senior water rights. Refuges possessing junior water rights would be particularly susceptible to losing use of water as demand exceeds supply.

The other major consequence of climate change to water resources is a seasonal shift in the availability of water. Mountain snowpacks act as natural reservoirs, accumulating vast amounts of snow in the winter and releasing this stored precipitation in the spring as high flows in streams. Many wildlife life histories and agricultural economies are closely tied to this predictable high volume of water. Warmer temperatures would result in earlier snowmelt at higher elevations as well as more precipitation falling in the form of rain rather than snow in these areas. The result would be both high and low flows occurring earlier in the year, and an insufficient amount of water when it is needed.

This effect is most likely to affect the western United States (Barnett, Adam, and Lettenmaier, 2005).

Water quality is also likely to decline with climate change as contaminants become more concentrated in areas with reduced precipitation and lower stream flows. In addition, warmer surface water temperatures would result in lower dissolved oxygen concentrations and could jeopardize some aquatic species. In the far north, current thawing of permafrost has resulted in an increase in microbial activity within the active soil layer. This has reduced the amount of dissolved organic carbon reaching estuaries, lowering productivity (Striegl *et al.*, 2005).

Climate change will offer a challenge for the NWRS to maintain adequate supplies of water to achieve wildlife management objectives. Although it is not currently possible to project precisely where the greatest effects to water resources will occur, refuges in areas where demand already exceeds supply—as well as those in areas highly dependent upon seasonal flows from snowmelt—appear to be especially vulnerable.

Waterfowl occurring on refuges in areas such as the Prairie Pothole Region (PPR), for which warmer and drier conditions are projected (Poiani and Johnson, 1991; Sorenson *et al.*, 1998), may be expected to face more stressful conditions than those in areas that are projected to be warmer and wetter, such as the Northeast. The projected drying of the PPR—the single most important duck production area in North America—will significantly affect the NWRS's ability to maintain migratory species in general and waterfowl in particular. Maintaining endangered aquatic species, such as the desert hole pupfish, which occurs naturally in a single location in Ash Meadows NWR in Nevada, will present even more challenges because, unlike waterfowl that can shift their breeding range northward, most threatened and endangered species have limited dispersal abilities and opportunities.

Sea Level Rise

The NWRS includes 161 coastal refuges. Approximately 1 million acres of coastal wetlands occur on refuges in the lower 48 states. Sea level rise is the result of several factors, including land subsidence, thermal expansion of the oceans, and ice melt (IPCC, 2007a). The sea-level rise at any given location depends on the local rate of land subsidence or uplift relative to the other drivers of sea level rise. On a given refuge, the extent of coastal inundation resulting from sea level rise will be influenced by hydrology, geomorphology, vertical land movements, atmospheric pressure, and ocean currents (Small, Gornitz, and Cohen, 2000).

Historically, accretions of sediments and organic matter have allowed coastal wetlands to "migrate" to adjacent higher ground as sea levels have risen. However, wetland migration may not keep pace with accelerating rates of sea level rise because of upstream impoundments and bulkheaded boundaries. Also, in many cases topography or the structures and infrastructure of economically developed areas (essentially bulkheaded refuges) impede migration (Titus and Richman, 2001). In both scenarios, coastal wetlands will be lost, along with the habitat features that make them valuable to species the NWRS is intended to conserve, *e.g.*, waterfowl.

Along the mid-Atlantic coast, the highest rate of wetland loss is in the middle of the Chesapeake Bay region of Maryland. One example is Blackwater NWR, part of the Chesapeake Marshlands NWR Complex. This refuge has been affected by sea level rise for the past 60 years. Models project that in 50 years, continued sea level rise in conjunction with climate change will completely inundate existing marshes (Fig. 5.6) (Larsen *et al.*, 2004b; see also U.S. Climate Change Science Program, 2007). Along the Gulf Coast, substantial wetland loss is also occurring. For example, in Louisiana, the combination of sea level rise, high rates of subsidence, economic growth, and hurricanes has contributed to an annual loss of nearly 25,000 acres of wetlands, even prior to Hurricane Katrina (2005) (Erwin, Sanders, and Prosser, 2004). Sea level rise challenges a lesser extent of NWRS wetlands along the Pacific coast because few refuges there have extensive coastal wetlands, in part due to steep topography. Conversely, a higher proportion of these wetlands have limited potential for migration for the same topographical reasons. Additionally, up-elevation movements of plant and animal species among these refuges are prevented by presence of highways, industrial and urban

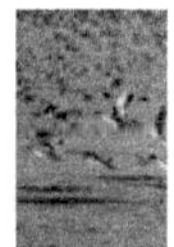

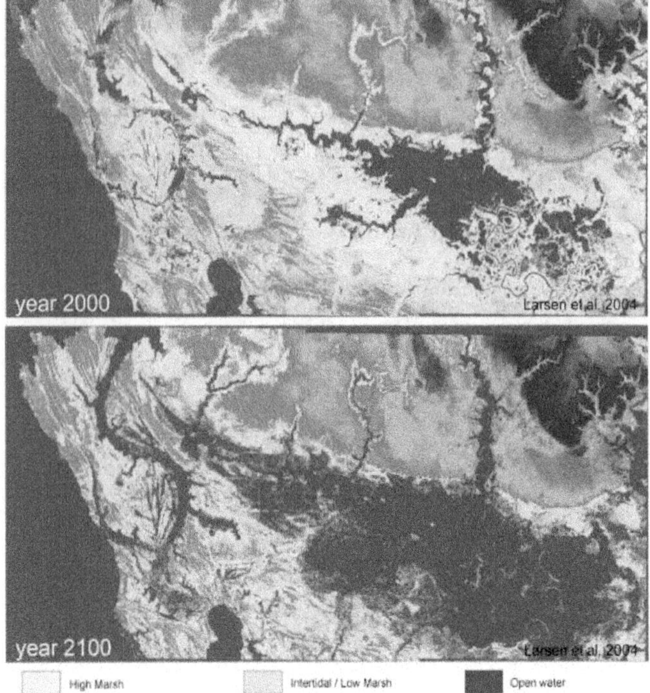

Figure 5.6. Blackwater National Wildlife Refuge, Chesapeake Bay, Maryland. Current land areas and potential inundation due to climate change (Larsen *et al.*, 2004b).

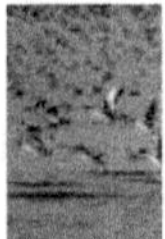

areas, and other products of development. They are, in effect, "bulkheaded." Alaskan refuge wetlands appear to be least at risk of sea level rise effects because of countervailing forces, most notably isostatic uplift (Larsen *et al.*, 2005), which has accelerated as a function of climate change and melting of glaciers (Larsen *et al.*, 2004a). In Alaska, permafrost thawing and resulting drainage of many of the lakes is a greater challenge to wetlands, both coastal and non-coastal. In Florida, Pelican Island NWR, the system's first refuge, is among the 161 coastal refuges challenged by sea level rise.

Recent studies have attempted to quantitatively project the potential effect of sea level rise on NWRS wetlands. For example, the Sea Level Affecting Marshes Model (SLAMM) was used to project coastal wetland losses for four refuges in Florida: Ding Darling (Fig. 5.7), Egmont Key, Pine Island, and Pelican Island. Significant wetland losses are projected at each refuge, but the types and extent of changes to wetlands may vary considerably. SLAMM was also used to model sea level rise at San Francisco Bay NWR (Galbraith *et al.*, 2002). The projections suggested that the refuge will be inundated in

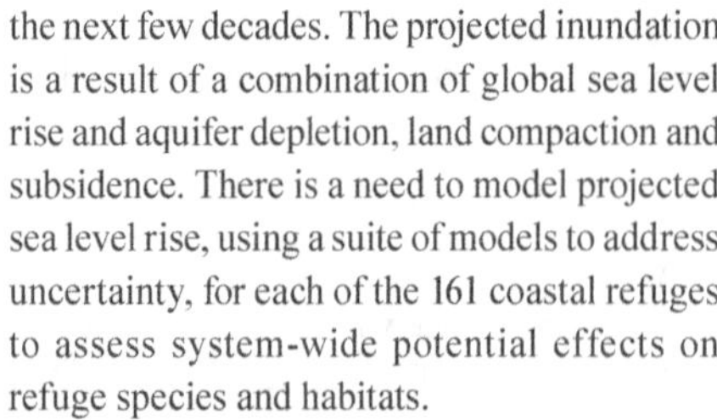

the next few decades. The projected inundation is a result of a combination of global sea level rise and aquifer depletion, land compaction and subsidence. There is a need to model projected sea level rise, using a suite of models to address uncertainty, for each of the 161 coastal refuges to assess system-wide potential effects on refuge species and habitats.

The effects of climate change on wetlands will not be uniform. For example, sea level rise could create new wetlands along the coast. However, changes in hydrological regimes and precipitation patterns will cause some existing wetlands to dry out and change the geomorphology and sedimentation of wetlands.

Extreme Weather Events

Increased frequency of extreme weather events, such as hurricanes, floods, or unusually high tides, could significantly alter coastal and other habitats. Observed and projected effects include loss of barrier islands and coastal marshes; damage or loss of storm- and tide-dampening mechanisms and other refuge equipment and infrastructure; and pollution of refuge habitats from storm-borne pollutants from nearby urban centers and industrial sites, increasing the strain on tight budgets. The loss of equipment and property damage could hinder both recreational and educational activities on refuges, thus affecting the ability of the NWRS to fulfill its relevant mandates as well as cutting individual refuges' income.

The potential effects of hurricanes and other extreme weather events on the NWRS's conservation target species and their habitats are complex and difficult to prevent and mitigate. Threatened and endangered species are likely to be the most affected. Documented negative effects of extreme weather events on threatened and endangered species and their habitats include the loss of 95% of breeding habitat of the red-cockaded woodpecker, loss of habitat for five red wolves in South Carolina, and diminished food supply for the Puerto Rican parrot as a result of hurricane Hugo (U.S. Fish and Wildlife Service, 1989).

The effects of storms and hurricanes are not limited to terrestrial species. Aquatic species managed by the USFWS on the NWRS could also be affected by some of the side

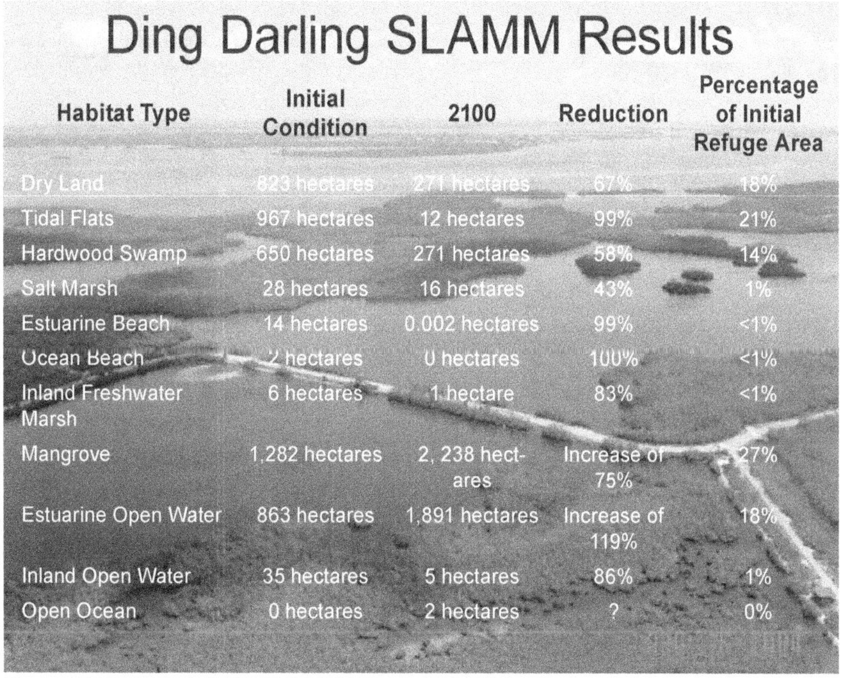

Ding Darling SLAMM Results

Habitat Type	Initial Condition	2100	Reduction	Percentage of Initial Refuge Area
Dry Land	823 hectares	271 hectares	67%	18%
Tidal Flats	967 hectares	12 hectares	99%	21%
Hardwood Swamp	650 hectares	271 hectares	58%	14%
Salt Marsh	28 hectares	16 hectares	43%	1%
Estuarine Beach	14 hectares	0.002 hectares	99%	<1%
Ocean Beach	2 hectares	0 hectares	100%	<1%
Inland Freshwater Marsh	6 hectares	1 hectare	83%	<1%
Mangrove	1,282 hectares	2, 238 hect-ares	Increase of 75%	27%
Estuarine Open Water	863 hectares	1,891 hectares	Increase of 119%	18%
Inland Open Water	35 hectares	5 hectares	86%	1%
Open Ocean	0 hectares	2 hectares	?	0%

Figure 5.7. Results of the Sea Level Affecting Marshes Model (SLAMM) for Ding Darling National Wildlife Refuge. Source: USFWS unpublished data.[30] Photo: Susan White.

effects of storms and hurricanes, such as oxygen depletion, changes in salinity, mud suffocation, and turbulence (Tabb and Jones, 1962). Such effects could also severely damage recreational fishing opportunities on affected refuges. Projected effects of tropical storms on southeastern wetlands (Michener *et al.*, 1997) could pose additional challenges to other NWRS trust species, such as migratory birds, that use those wetlands. Hurricane Hugo caused soil erosion on Sandy Point NWR, which had an adverse affect on nesting leatherback turtles (U.S. Fish and Wildlife Service, 1989).

5.3.2.3 Regime Shifts

Much of the NWRS lies in areas that could experience vegetation shifts by 2100 (Gonzalez, Neilson, and Drapek, 2005). Species may respond to climate change in several ways: ecologically (by shifting distributions), evolutionarily/genetically, behaviorally, and/or demographically. One of the more profound effects of climate change is total "regime shift," where entire ecological communities are transformed from their "historical" conditions. Such shifts are even now being witnessed in the black spruce forests of southern Alaska due to northern expansion of the spruce bark beetle, and the coastal shrublands of central and southern California, due to increased frequency of wildfires. Similar changes, though difficult to project, will likely occur with changing rainfall patterns. Increased moisture may create wetlands where none existed before, whereas declining rainfall may eliminate prairie potholes or other significant wetlands, especially in marginally wet habitats such as vernal pools and near-deserts.

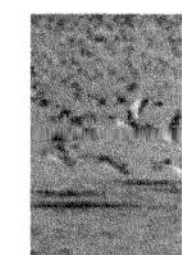

Where such regime shifts occur, even on smaller scales, it may become impossible to meet specific refuge purposes. For example, the habitats of a highly specialized refuge (such as one established for an endangered species) might shift away from the habitat occupied by the species for which the refuge was established; *e.g.*, Kirtland's Warbler Wildlife Management Area (Botkin, 1990). Likewise, shifts in migratory bird habitats in the prairie potholes of the Midwest might diminish available breeding habitat for waterfowl (Sorenson *et al.*, 1998; Johnson *et al.*, 2005). Less obviously, increasing competition for water in areas such as California's Central Valley, southern New Mexico, or Arizona may restrict a refuge's access to that critical resource, thus making attainment of its purposes virtually impossible.

[30] **McMahon**, S., Undated: USFWS unpublished data.

As suggested by emerging research, there will be winners and losers among the species and habitats currently found on the NWRS (Peterson and Vieglais, 2001; Peterson, Ball, and Cohoon, 2002; Parmesan and Yohe, 2003; Peterson *et al.*, 2005; Parmesan, 2006). Existing species' compositions in refuges may change; however, it will be possible to maintain the integrity, diversity, and environmental health of the NWRS, albeit with a focus on the composition, structure, and function of the habitat supported by the refuges, rather than any particular species or group of species that uses that habitat.

The prospect of regime shifts makes it more crucial that the USFWS train and educate refuge managers in methods of ascertaining how specific refuges can assess changing climate and their role in support of the system-wide response. Without such guidance it will be increasingly challenging to define what a refuge should "conserve and manage," and impossible in most cases to "restore" a habitat in an ecological milieu that no longer supports key species. This raises the question of what refuge managers are actually managing for: single species occurrences or maintenance of capacity for evolutionary and ecological change in self-sustaining ecosystems.

5.3.3 Ecoregional Implications of Climate Change for the NWRS

The NWRS is characterized by an uneven geographic and ecological distribution (Scott *et al.*, 2004). There are 84 ecoregions in North America (Omernik, 1987), ranging from temperate rainforests to the Sonoran desert. Eleven of these ecoregions host almost half of all refuges (Scott *et al.*, 2004). Over all the ecoregions, Alaskan ecoregions dominate; however, the Southern Florida Coastal Plain ecoregion has the largest area representation within the NWRS in the lower 48 states: 3.7%.

This section describes some of the implications of climate change on an ecoregion-by-ecoregion basis, based on a hierarchical agglomeration of the 84 ecoregions mentioned above (Omernik, 1987; level 1 ecoregions) (Fig. 5.8).

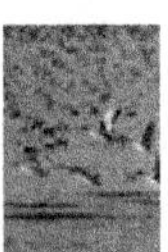

Figure 5.8. Ecoregions of North America (Level I).[31]

[31] **U.S. Environmental Protection Agency**, 2007: Ecoregions of North America. Environmental Protection Agency Website, http://www.epa.gov/wed/pages/ecoregions/na_eco.htm#Level%20I, accessed on 7-12-2007.

5.3.3.1 Arctic Cordillera, Tundra, Taiga, and the Hudson Plain (16 NWRs)

Although there are only 16 refuges in this ecoregion, they capture more than 80% of the area of the NWRS, provide important breeding habitat for waterfowl, and offer key habitat for many high-latitude species. The high latitudes have experienced some of the most dramatic recent climatic changes in the world. Arctic land masses have warmed over the last century by at least 5°C (IPCC, 2001). In North America, the most warming has occurred in the western Arctic region, including Alaska, and has been concentrated in the winter and spring (Serreze *et al.*, 2000). This warming has resulted in a decrease in permafrost (IPCC, 2001). Melting permafrost has implications for vegetation, hydrology, and ecosystem functioning. The thawing permafrost also releases carbon, which results in a positive feedback loop generating further warming (Zimov, Schuur, and Chapin, III, 2006). Furthermore, the melting of permafrost may connect shallow lakes and wetlands to groundwater, resulting in draining and the loss of many shallow-water systems (Marsh and Neumann, 2001).

Due to the rugged coast and lack of low-lying coastal areas, sea level rise is not projected to strongly affect Alaska except where sea ice affects the shoreline. The extent of Arctic sea ice has been decreasing at a rate of 2.7 % per decade from 1980 to 2005 (Lemke *et al.*, 2007). Loss of Arctic ice in areas near NWRs will decrease and eliminate foraging opportunities for those seabirds and mammals that congregate at the sea-ice interface.

Climate change will likely have large effects on the composition of ecological communities on many refuges in the northern ecoregions. As temperatures increase, many species will continue to shift their ranges to the north. For example, the boreal forest is projected to expand significantly into the tundra (Payette, Fortin, and Gamache, 2001). In the tundra itself, mosses and lichens will likely be replaced by denser vascular vegetation, resulting in increased transpiration and further altering hydrology (Rouse *et al.*, 1997). There will also be changes in animal communities as range shifts introduce new species. Some native species will likely be affected by new predators and new competitors. For example, red foxes have expanded their range to the north (Hersteinsson and Macdonald, 1992), potentially increasing competition with Arctic foxes for resources. This range expansion is likely to continue (MacPherson, 1964; Pamperin, Follmann, and Petersen, 2006).

Climate change also will amplify a number of the factors that already affect refuges in these ecoregions. The large projected increases in temperature may result in the introduction of new diseases and an increase in the effects of diseases already present on the refuges. For example, recent warming has already led to a shortening of the lifecycle of a specific nematode parasite, resulting in decreased fecundity and survival in musk oxen (Kutz *et al.*, 2005). Higher temperatures will potentially increase the role that fire plays in northern ecoregions and increase the frequency of ignition by dry lightning. Fires in the boreal forest are, for example, projected to increase in frequency with further warming (Rupp, Chapin, and Starfield, 2000). Finally, the combination of warming and acidification of streams and lakes in the boreal forest will have combined negative effects on freshwater fauna (Schindler, 1998).

Because the refuges of the northernmost ecoregions cover more than 80% of the area of the NWRS, and because the high latitudes are expected to undergo some of the most dramatic changes in climate, climate-driven effects to these refuges will greatly affect the ability of the NWRS to meet many of its mandated goals to maintain existing species assemblages. As a result of range shifts, recreational and conservation targets may change. This yet again raises the question of where conservation and management activities should be directed—at species, ecosystem, or conservation landscape scales.

5.3.3.2 Northern Forests and Eastern Temperate Forests (207 NWRs)

These two ecoregions cover almost all of the eastern United States (Fig. 5.8). In the northeastern United States, recent documented seasonal warming patterns, extended growing seasons, high spring stream flow, and decreases in snow depth are projected to continue; new trends such as increased drought frequency, decreased snow cover, and extended periods of

low summer stream flow are projected for the coming century (Hayhoe *et al.*, 2007). Changes in stream flow, drought frequency and duration, snow cover, and snow depth have significant implications for precipitation-fed wetlands on many northeastern refuges. Decreases in water availability will affect breeding habitat for amphibians, and feeding and nesting habitat for wading birds, ducks, and some migratory songbirds (Inkley *et al.*, 2004).

In both the northern forests and the eastern temperate forests, climate change will likely result in shifts in forest composition and structure (Iverson and Prasad, 1998). In addition, global vegetation models project the conversion of many southeastern forests to grasslands and open woodlands in response to changes in atmospheric CO_2 and climate (Bachelet *et al.*, 2001). Shifts of this magnitude will greatly change the availability of habitat for many species on national wildlife refuges. Shifts in the dominant vegetation type or even small changes in the understory composition may result in significant changes in animal communities. In addition, climatic changes in these regions will have implications for both terrestrial and aquatic ecosystem functioning (Allan, Palmer, and Poff, 2005) which, in turn, will affect wildlife. For example, increases in temperature will affect dissolved oxygen levels in the many lakes of this region, resulting in changes in lake biota (Magnuson *et al.*, 1997).

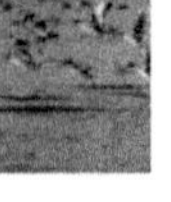

Urbanization continues across much of the eastern United States, and most significantly across the East Coast states. Urbanization and residential development have the potential to further isolate refuges and reduce the ability of organisms to move from one protected area to another. Concurrent warming, reduced stream flow, and increased urbanization may lead to increased bioaccumulation and potentially biomagnifications of organic and inorganic contaminants from agriculture, industry, and urban areas (Moore *et al.*, 1997). Finally, climate change will likely accelerate the spread of some exotic invasive species and shift the ranges of others (Alward, Detling, and Milchunas, 1999).

5.3.3.3 Great Plains (139 NWRs)

Changes in hydrology likely present the largest threat to refuges in the Great Plains. Several of these refuges encompass portions of the PPR, which is the most productive waterfowl habitat in the world. Population numbers for many waterfowl species in the area are positively correlated with the number of May ponds available in the PPR in the beginning of the breeding season (Batt *et al.*, 1989). For example, the number of May ponds in the PPR dropped from approximately 7 million in 1975 to a little over 3 million in 1990, and then rose again to roughly 7 million by 1997. Mallard duck numbers tracked this trend, dropping from roughly 5 million in 1975 to a little under 3 million in 1990 and rising to roughly 6 million in 1997.[32] Hydrological models have been used to accurately simulate the effect of changing climate on wetland stage (Johnson *et al.*, 2005). The projected continued rise in temperatures will likely cause severe drought in the central part of the PPR and a significant drop in waterfowl population numbers (Johnson *et al.*, 2005). Increased temperatures will result in increased evaporation, and lead to decreased soil moisture and the likely shrinkage and drying of many wetlands in the region (Sorenson *et al.*, 1998). More specifically, these changes have been projected to result in fewer wetlands (Larson, 1995), along with changes in hydroperiod, water temperature, salinity, dissolved oxygen levels, and aquatic food webs (Poiani and Johnson, 1991; Inkley *et al.*, 2004). The likely cascading effects on waterfowl in refuges across the region include reduced clutch sizes, fewer renesting attempts, and lower brood survival (Inkley *et al.*, 2004). Earlier projections of potential population declines for waterfowl have ranged from 9–69% by 2080 (Sorenson *et al.*, 1998). In addition, stresses from agricultural lands surrounding refuges in the Great Plains will likely be exacerbated by future climatic changes. In particular, decreases in precipitation and increases in evaporation have the potential to increase demands for water for agriculture and for refuges. In contrast, increases in precipitation have the potential to increase agricultural runoff.

[32] **U.S. Fish and Wildlife Service**, 2007: Migratory Bird Data Center. U.S. Fish and Wildlife Service Website, http://mbdcapps.fws.gov/, accessed on 11-20-2007.

In addition, stresses from agricultural lands surrounding refuges in the Great Plains will likely be exacerbated by future climatic changes. In particular, decreases in precipitation and increases in evaporation have the potential to increase demands for water for agriculture and for refuges. In contrast, increases in precipitation have the potential to increase agricultural runoff.

5.3.3.4 Northwestern Forested Mountains and Marine West Coast Forest (59 NWRs)

Together, these two ecoregions account for most of the mountainous areas in the western United States (Fig. 5.8). The Marine West Coast Forest ecoregion is generally relatively wet, with temperate ocean-influenced climates. The Northwestern Forest Mountains ecoregion is generally drier. Future projections for the region are for intermediate temperature increases and increased precipitation.

Some of the largest effects to this region are likely to come from changes in hydrological regimes resulting from reduced snowpack and earlier snowmelt. The resulting changes in stream flow and temperature will negatively affect salmon and other coldwater fish (Mote *et al.*, 2003). In addition, competition among different users for scarce summer water supplies will be intensified as snowpack is reduced and spring melts come earlier (Mote *et al.*, 2003). Water-use conflicts are already a major issue (National Research Council, 2007) in dry summers following winters with minimal snowpack (*e.g.*, Klamath Basin NWR Complex).

Climate change is also likely to affect fire regimes in the mountains of the western United States (Westerling *et al.*, 2006). Larger and more intense fires have implications for refuges at lower elevations that receive much of their water from the forested mountains. These fires will alter stream flows and sediment loads, changing the hydrology and vegetation in downstream wetlands. Changes in wetland habitats in the western mountains, whether driven by changing hydrology, fire regimes, or shifting vegetation patterns, have the potential to affect the ability of the NWRS to protect habitat and provide viable populations of species on refuges.

5.3.3.5 Mediterranean California (28 NWRs)

In the Sierra Mountains (as in the Northwest Forested Mountains ecoregion), the competition for water for agricultural, residential, industrial, and natural resource use will intensify (Hayhoe *et al.*, 2004). At the same time, changes in snowpack in the Sierra Mountains will also have the potential to affect the hydrology and habitat of refuges in the central valley and on the coast of California. Based on projections from two general circulation models, under the lower SRES B1 greenhouse gas emissions scenario, the Sierra Mountains will experience 30–70% less snowpack. Under the higher SRES A1FI emissions scenario, the Sierras are projected to have 73–90% less snowpack (Hayhoe *et al.*, 2004). The snow-fed streams draining the Sierras into the Central Valley of California will have lower summer flows and earlier spring flows, significantly changing the hydrology of the valley. Reduced stream flows and higher temperatures may result in increased salinity in bays and estuaries such as San Francisco Bay, significantly affecting the biological integrity, diversity, and health of species and populations in the San Francisco Bay NWR Complex. Sea level rise will compound these effects for refuges in low-lying estuaries and bays along the California coast.

5.3.3.6 North American Deserts and Southern Semiarid Highlands (53 NWRs)

Like most of the rest of the United States, the arid Southwest has been warming over the last century. Parts of southern Utah and Arizona have had greater than average increases in temperature (*e.g.*, 2–3°C) (Figure 5.3a). The southwestern United States has experienced the smallest increase in precipitation in the last 100 years of any region in the coterminous United States (Figure 5.3b).

Climate models project drying and continued warming in the arid ecoregions of the United States, which could have significant effects on many refuges. These projected climate trends could lead to changes in hydrology that, in turn, may have large effects on wetlands and other shallow water bodies. Although precipitation-fed systems are most at risk, groundwater-fed systems in which aquifer recharge is largely driven by snowmelt may also be heavily

affected (Winter, 2000; Burkett and Kusler, 2000). Reductions in water levels and increases in water temperatures will potentially lead to reduced water quality, in terms of increased turbidity and decreases in dissolved oxygen concentrations (Poff, Brinson, and Day, Jr., 2002). Increased productivity, driven by increased temperature, may lead to increases in algal blooms and more frequent anoxic conditions (Allan, Palmer, and Poff, 2005).

More so than in the other ecoregions, water resources in the arid portions of the western United States are already in high demand. Decreases in available water will exacerbate the competition for water for agriculture, urban centers, and wildlife (Hurd *et al.*, 1999). Competition for water already challenges the Moapa dace on the Desert NWR Complex in the Moapa Valley of Nevada and the wildlife of the Sonny Bono Salton Sea NWR in southern California.

Dams and other small water diversions, combined with the prevalence of east-west flowing rivers, will hinder migration of aquatic species to cooler waters (Allan, Palmer, and Poff, 2005). In addition, many endemic fish in arid ecoregions are highly adapted to local conditions and quite limited in distribution. Many of these species are projected to go extinct in response to temperature increases of just a few degrees (Matthews and Zimmerman, 1990). Reduced water levels and increased water temperatures may also lead to increases in disease outbreaks.

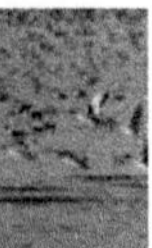

Grazing by cattle on refuges in the arid ecoregions will likely exacerbate the effects of drought stress and aid in the spread of non-native species. Furthermore, refuges may be sources of scarce water resources in the future, making them even more attractive to cattle. Grazing will also likely interact with climate-driven vegetation changes to further alter plant communities and wildlife habitat on refuges in arid regions (Donahue, 1999).

Although reduced precipitation and increased temperatures may reduce productivity in some arid regions, global vegetation models have projected an expansion of grasslands, shrublands, and woodlands into arid regions in response to increased water-use efficiency driven by increased atmospheric CO_2 concentrations. Increased abundance of invasive non-native grasses has altered fire regimes, increasing the frequency, intensity, and extent of fires in the American Southwest (D'Antonio and Vitousek, 1992; Brooks *et al.*, 2004).[33] These shifts could result in dramatic changes in wildlife communities in the affected areas. Overall, we would see a reduction in the number of desert species and an increase in species that inhabit dry grasslands, shrublands, and woodlands.

5.3.3.7 Sub-Tropical and Tropical Ecosystems (7 NWRs)

In the continental United States, the tropical wet forest ecoregion occurs only in southern Florida. The largest climate-driven challenge to the refuges in this ecoregion is sea level rise. With its extensive low-lying coastal areas, much of this region will be underwater or inundated with salt water in the coming century. The several refuges in the Florida Keys, Florida Panther NWR, and Key Deer NWR are all particularly at risk.

Invasive native and non-native species are also a major challenge in this ecoregion. As temperatures rise, South Florida will likely be the entry point of many new tropical species into the United States. Five new species of tropical dragonfly had established themselves in the country by 2000—each suspected to be the result of a northward range shift from populations in the Caribbean. Loss of land due to sea level rise in southern Florida will increase development pressure inland and in the north, potentially accelerating urbanization and exacerbating the isolating and fragmenting effects of development.

5.3.3.8 Coastal and Marine Systems: Marine Protected Areas (161 NWRs)

Low-lying coastal refuges face several climate-driven challenges. Sea level rise will likely be the largest challenge to refuges in the southeastern United States (Daniels, White, and Chapman, 1993; Ross, O'Brien, and Sternberg, 1994). Low-

[33] **Brooks**, M.L. and D.A. Pyke, 2002: Invasive plants and fire in the deserts of North America. In: *Proceedings of the Invasive Species Workshop: the Role of Fire in the Control and Spread of Invasive Species* [Gallery, K.E.M. and T.P. Wilson (eds.)]. Proceedings of the Fire Conference 2000: The First National Congress on Fire Ecology, Prevention, and Management, Tall Timbers Research Station, pp. 1-14.

lying coastal areas on the East and Gulf Coasts are some of the most vulnerable in the country. Some of the most vulnerable refuges include the Chincoteague NWR, on the Delmarva Peninsula; the Alligator River NWR, on the Albemarle Peninsula of North Carolina; San Francisco Bay NWR in California; and Merritt Island NWR in Florida. In fact, many of the refuges in New England, the Middle Atlantic states, North Carolina, South Carolina, and Florida are coastal and susceptible to sea level rise (Daniels, White, and Chapman, 1993; Titus and Richman, 2001). For many of these refuges, sea level rise will dramatically alter habitats by inundating estuaries and marshes and converting forests to marshes. Beach nesting birds such as the piping plover, migratory birds using the refuges as stopovers, and species using low-lying habitats such as the red wolf and Florida panther will likely lose habitat to sea level rise.[34] In addition, sea level rise may eliminate coastal stopover sites used by birds migrating up and down the East Coast (Galbraith *et al.*, 2002; Huntley *et al.*, 2006).

Warming ocean temperatures also challenge coastal and marine refuges. In fact, warming ocean temperatures are already having significant effects on many marine organisms. For example, increased water temperatures have resulted in increases in the frequency of toxic algal blooms (Harvell *et al.*, 1999), and future climate changes are projected to result in more intense tropical storms, resulting in increased disturbance for many coastal refuges (IPCC, 2007b). Coral bleaching is another effect of increased ocean temperatures, and has had profound effects on reefs in the Caribbean. Increased ocean acidity (from the accumulation of carbonic acid in the water—a direct result of more CO_2 entering the ocean from the atmosphere and combining with water) will dissolve calcium-rich shells, dramatically changing the species composition of zooplankton and having cascading effects on entire marine ecosystems (Guinotte *et al.*, 2006).

Over-fishing, eutrophication, and increasing temperatures may lead to toxic algal and jellyfish blooms (Jackson *et al.*, 2001). Temperature-stressed corals will be more susceptible to disease. Invasive species are likely to expand their ranges as water temperatures rise. And finally, pathogens and disease vectors may move with climate change. An example of this latter challenge is given by the expansion of an oyster parasite, *Perkinsus marinus,* up the East Coast of the United States in response to warmer waters (Ford, 1996).

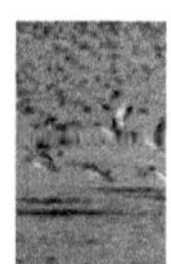

5.4 ADAPTING TO CLIMATE CHANGE

Adaptation measures aim to increase the resilience of species, communities, and ecosystems to climate change (Turner, II *et al.*, 2003; Tompkins and Adger, 2004). The law governing management of the NWRS affords the USFWS great latitude in deciding what is best for the system. Especially in dealing with the scientific uncertainty associated with the effects of climate change, the USFWS can act assertively within the broad power Congress delegated to make judgments about how best to achieve the system's objectives. Maintaining biological integrity, diversity, and environmental health, and sustaining healthy populations of species, two of the chief goals for the NWRS, provide ample

[34] **Schlyer**, K., 2006: *Refuges at Risk: the Threat of Global Warming and America's Ten Most Endangered National Wildlife Refuges.* Defenders of Wildlife, Washington, DC.

bases to support adaptation.[35] The uncertainty associated with climate change influences on refuges, the NWRS, and ecosystems, along with the complexity of conservation targets and their interactions, requires a structured and integrative approach to decision-making and management actions. The scale of the effects of climate change is global, and the scale of desired conservation responses—flyways, entire species' ranges—requires that management actions be implemented and conservation target responses be measured in areas unprecedented in their size and in their area of extent (Anderson *et al.*, 1987; Nichols, Johnson, and Williams, 1995; Johnson, Kendall, and Dubovsky, 2002).

National wildlife refuges are not yet implementing adaptation strategies to explicitly address climate change. However, various management approaches (*e.g.*, riparian reforestation, assisted dispersal) currently used to address other stresses could also be used to address climate change stresses within individual refuges. More importantly, beyond the scale of individual refuges, climate change warrants system-wide adaptive management.

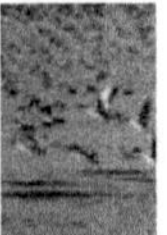

Representation, redundancy, and resilience are key conservation principles that could be used to strengthen the NWRS in the face of climate change, both within and beyond existing refuge boundaries (Shaffer and Stein, 2000). The resilience/viability of populations and ecosystems on an individual refuge level may be increased through habitat augmentation, restoration, reduction/elimination of environmental stressors, acquisition of inholdings, and by enhancing the surrounding matrix through conservation partnerships, conservation easements, fee-title acquisitions, etc. At the NWRS scale, opportunities for refuge species to respond and adapt to climate change effects can be enhanced by capturing the full geographical, geophysical, and ecological ranges of a species on as many refuges as possible. The goal of these management responses is not to create artificial habitats for species, but to restore and increase habitat availability and reduce stressors to provide species maximum opportunity to respond and adapt to climate change.

[35] 16 USC § 668dd

Most of the adaptation measures presented in the following sections will most effectively facilitate ecosystem adaptation to climate change when implemented within the framework of adaptive management.

5.4.1 Adaptive Management as a Framework for Adaptation Actions

Response to climate change challenges must occur at multiple integrated scales within the NWRS and among partner entities. Individual symptomatic challenges of climate change must be addressed at the refuge level, while NWRS planning is the most appropriate level for addressing systemic challenges to the system.

Adaptive management lends itself well to the adaptation of natural resource management actions to climate change. Adaptive management is an iterative approach that seeks to improve natural resource management by testing management hypotheses and learning from the results (Holling, 1978; Walters, 1986; Salafsky, Margoluis, and Redford, 2001). A management action can have the desired effect on the distribution and abundance of the target species. However, depending on the type of management action, there can also be a number of unintended consequences. Adaptive management provides a research/management tool to asses the frequency and intensity of unintended effects. It is an approach that is useful in situations where uncertainty about ecological responses is high, such as climate change.

Adaptive management proceeds generally through seven steps: (1) Establish a clear and common purpose; (2) Design an explicit model of the system; (3) Develop a management plan that maximizes results and learning; (4) Develop a monitoring plan to test the assumptions; (5) Implement management and monitoring plans; (6) Analyze data and communicate results; and (7) Iteratively use results to adapt and learn (Salafsky, Margoluis, and Redford, 2001). Public participation, scientific monitoring, and management actions based on field results form the core principles of adaptive management.

Adaptive management also incorporates a research agenda into plans and actions, so that they may yield useful information for future decision-making. For instance, the planning process for refuges and the NWRS does not end when a plan is adopted. It continues into a phase of implementation and evaluation.[36] Under adaptive management, each step of plan implementation is an experiment requiring review and adjustment.

In general, the law provides authority to USFWS for adaptive management. The general principles of administrative law give the USFWS wide latitude for tailoring adaptive management to the circumstances of the refuges. One element of adaptive management, monitoring, is affirmatively required by the NWRSIA of 1997.[37] The only legal hurdle for adaptive management is the need for final agency action in adopting CCPs and making certain kinds of decisions involving findings of no significant effect under the National Environmental Policy Act (NEPA).

Although the USFWS policy implementing its planning mandate makes a strong effort to employ adaptive management through modeling, experimentation, and monitoring, legal hurdles remain for the insertion of truly adaptive strategies into CCPs. These hurdles are acknowledged in DOI policy on adaptive management (Williams, Szaro, and Shapiro, 2007). Not only do the Administrative Procedure Act, NEPA, and the NWRSIA all emphasize finality in approval of a document, but the relative formality of the development of an administrative record, the preparation of an environmental impact statement for proposals significantly affecting the environment, and the need to prepare initial plans for all refuges by the statutory deadline of 2012 all tend to front-load resources in planning. Once the USFWS adopts an initial CCP for a refuge, adaptive management would call for much of the hard work to come in subsequent implementation. However, from a legal, budgetary, and performance-monitoring standpoint, few resources are available to support post-adoption implementation, including monitoring, experimentation, and iterative revisions. Despite these drawbacks, adaptive management remains the most promising management strategy for the NWRS in the face of climate change. The research and management objectives described below are thought out within the framework of adaptive management.

[36] U.S. Fish and Wildlife Service manual 602

[37] 16 USC § 668dd

5.4.2 Adaptation Strategies within Refuge Borders

One of the most important comparative advantages of the NWRS for adaptation (compared with other federal agencies) is its long experience with intensive management techniques to improve wildlife habitat and populations. The NWRSIA of 1997 provides for vast discretion in refuge management activities designed to achieve the conservation mission. Some regulatory constraints, such as the duty not to jeopardize the continued existence of listed species under the ESA, occasionally limit this latitude. Generally, intensive management occurs within the boundaries of an existing refuge, but ambitious adaptation projects may highlight certain locations as high priority targets for acquisition, easement, or partnerships. Also, programs such as animal translocations will require cooperation with all the involved parties within the organism's range (McLachlan, Hellmann, and Schwartz, 2007).

The chief legal limitation in using intensive management to adapt to climate change is the limited jurisdiction of many refuges over their water. Both the timing of water flows as well as the quantity of water flowing through the refuge are often subject to state permitting and control by other federal agencies, as discussed above. But, in general, the USFWS has ample proprietary authority to engage in transplantation-relocation, habitat engineering (including irrigation-hydrologic management), and captive breeding.

Because government agencies and private organizations already protect a network of remarkable landscapes across the United States, resource managers will need to develop specific land management actions that will help species adapt to changes associated with sea level rise, changes in water availability, increased air and water temperatures, etc. These measures may provide time for populations to adapt and evolve, as observed in select plant and animal species in the past few decades of increasing temperatures (Berteaux *et al.*, 2004; Davis, Shaw, and Etterson, 2005; Jump and Peñuelas, 2005). Strategic growth of the NWRS to capture the full ecological, genetic, geographical, behavioral, and morphological variation in species will increase the ability of refuge managers and the NWRS to meet legal mandates of maintaining biological integrity, diversity, and environmental health of biological systems on NWRS lands. These habitats will increase chances that species will be more resilient to the challenges posed by climate change (Scott *et al.*, 1993).

The tools available to the NWRS to confront and adapt to climate change are those it has historically used so successfully to address past crises: prescribed burning, water management, land acquisition, inventory and monitoring, research, in some cases grazing and haying, etc. Critically, however, the NWRS needs to regroup and reassess in a collective way the value of these tools—as well as where and how to apply them—in the context of the current dynamic environmental conditions. For example, 2007 has presented a dramatic shift in historic wildfire patterns in the contiguous United States, as the "fire season" and fire risk areas have expanded to the East Coast in addition to the traditionally notorious West. As of June, 2007, the Big Turnaround Complex Fire burning on and around Okefenokee NWR in southeastern Georgia had surpassed 600,000 acres, and was the largest wildfire in history within the lower 48 states. This suggests that the application of fire to habitat management fuel reduction on refuges throughout the eastern United States may need reconsideration. Some potential climate adaptation measures that could be used by the NWRS include:

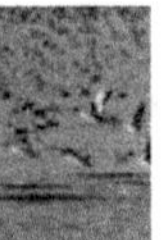

- *Prescribed burning to reduce risk of catastrophic wildfire.* Climate change is already increasing fire frequency and extent by altering the key factors that control fire, temperature, precipitation, wind, biomass, vegetation species composition and structure, and soil moisture (IPCC, 2001; IPCC, 2007a). In the western United States, increasing spring and summer temperatures of 1°C since 1970 have been correlated to increased fire frequency of 400% and burned area of 650% (Westerling *et al.*, 2006). Analyses project that climate change may increase future fire frequencies in North America (Flannigan *et al.*, 2005). Wildfires may also create a positive feedback for climate change through significant emissions of greenhouse gases (Randerson *et al.*, 2006). Prescribed burns could prevent catastrophic effects of stand-replacement fires in ecosystems characterized by less intense fire regimes. Fire

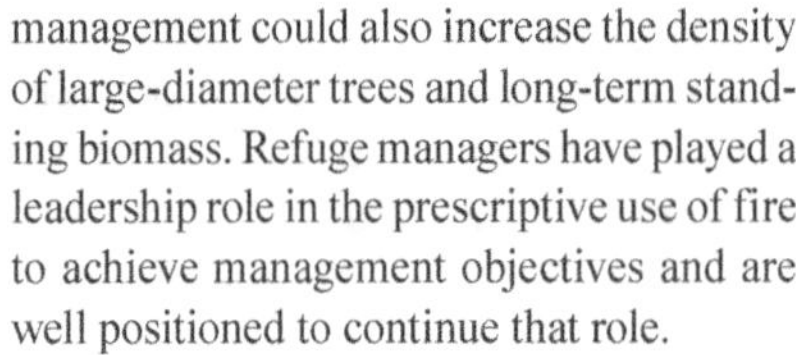

management could also increase the density of large-diameter trees and long-term standing biomass. Refuge managers have played a leadership role in the prescriptive use of fire to achieve management objectives and are well positioned to continue that role.

- *Facilitate the growth of plant species more adapted to future climate conditions.* Future conditions may favor certain types of species; for example, broadleaved trees over conifers. Favoring the natural regeneration of species better adapted to projected future conditions could facilitate the development of functional ecosystems. Nevertheless, high genetic diversity of species at the low-latitude edge of their range may require special protection in those areas (Hampe and Petit, 2005). Additional research is needed to better understand the long-term effects that such regeneration might have on natural communities.

- *Assisted dispersal.* Endemic species that occur in a limited area challenged with complete conversion by climate change may face extinction. Assisted dispersal is the deliberate long-distance transport by people of plants or animals in their historically occupied range and introduction into new geographic areas. Assisted dispersal offers an extreme measure to save such species (Hulme, 2005; McLachlan, Hellmann, and Schwartz, 2007). It risks, however, the release of non-native species into new areas and may not be as effective in altered environments. It also raises social and ethical issues, and should be viewed only as a last resort and considered on a case-by-case basis.

- *Interim food propagation for mistimed migrants.* The decline of long-distance migratory birds in Europe and the United States may originate in mistiming of breeding and food abundance due to differences in phenological shifts in response to climate change (Sauer, Pendleton, and Peterjohn, 1996; Both *et al.*, 2006). To compensate for the resource, it may become necessary to propagate food sources in the interim. The USFWS has provided food for waterfowl wintering on various refuges. For example, at Wheeler NWR, water levels are regulated in order to promote additional vegetation growth on the refuge. Parts of Columbia NWR are devoted to crop production, which is then available for waterfowl and other birds. Although a common practice on many refuges, it is important to remember that food propagation does not promote the biological integrity, diversity, and health of the refuges and the NWRS, nor the ability of the species to adjust to a changing landscape.

- *Riparian reforestation.* Reforestation of native willows, alders, and other native riparian tree species along river and stream banks will provide shade to keep water temperatures from warming excessively during summer months, while providing dispersal corridors for many species. This will create thermal refugia for fish and other aquatic species while also providing habitat for many terrestrial species. This adaptation strategy will only be sustainable if the riparian species are tolerant to the effects of climate change.

- *Propagation and transplantation of heat-resistant coral.* Climate change has increased sea surface temperatures that, in turn, have caused bleaching and death of coral reefs. The Nature Conservancy leads a consortium of 11 government and private organizations in the Florida Reef Resilience Program, a program to survey coral bleaching and test adaptation measures in the Florida Keys, an area that includes four refuges. The program has identified heat-resistant reefs and established nurseries to propagate live coral from those reefs. The program plans to transplant the heat-resistant coral to bleached and dead reefs.

On many refuges, external challenges are controlled principally by federal agencies other than the USFWS. Water flows may be dependent on decisions of sister federal agencies, such as the Federal Energy Regulatory Commission (for hydropower dams), the U.S. Army Corps of Engineers (for navigational and impoundment operations), and the Bureau of Reclamation (dam and water supply projects). Adaptation to climate change will require increased cooperation of these agencies with the USFWS if refuge goals are to be met.

Other possible management actions that could be applied to address climate change effects include building predator-free nest boxes, predator control programs, nest parasite control programs, translocation to augment genetics or demographics, prescribed burns to maintain preferred habitat types, creation of dispersal bridges, removal of migration barriers, habitat restoration, etc. Caution should be observed when any actions that assist one species over another are taken. There is always the risk of unintended consequences. The degree of assistance has to be evaluated on a case-by-case basis.

5.4.3 Adaptation Strategies Outside Refuge Borders

Adaptation to climate change requires the USFWS to consider lands and waters outside of refuge boundaries. In some instances acquisition of property for refuge expansion will best serve the conservation mission of the NWRS. In most cases, however, coordination with other land managers and governmental agencies (*e.g.*, voluntary land exchanges and conservation easements) will be more practical than acquisition. Coordination, like acquisition, can both reduce an external challenge generated by a particular land or water use and increase the effective conservation area through cooperative habitat management. Though the NWRSIA does little to compel neighbors to work with the USFWS on conservation matters external to the NWRS boundary, there are some regulatory hooks that USFWS managers can leverage. There are also several partnership incentive programs that could be used to create collaborative conservation partnerships (such as the Partners for Fish and Wildlife Program,[38] Refuge Partnership Programs,[39] Safe Harbor agreements,[40] Habitat Conservation Plans,[41] Candidate Conservation Agreements,[42] Natural

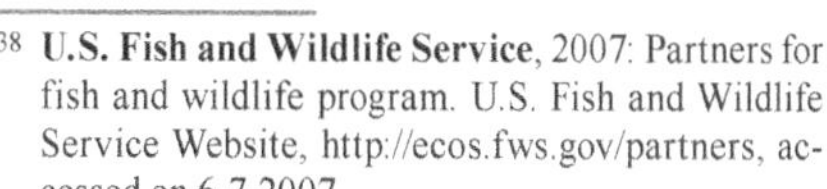

[38] **U.S. Fish and Wildlife Service**, 2007: Partners for fish and wildlife program. U.S. Fish and Wildlife Service Website, http://ecos.fws.gov/partners, accessed on 6-7-2007.

[39] **U.S. Fish and Wildlife Service**, 2007: Refuge partnership programs. U.S. Fish and Wildlife Service Website, http://www.fws.gov/refuges/generalInterest/partnerships.html, accessed on 6-7-2007.

[40] **U.S. Fish and Wildlife Service**, 2007: Safe harbor agreements. U.S. Fish and Wildlife Service Website, http://www.fws.gov/ncsandhills/safeharbor.htm, accessed on 6-7-2007.

[41] **U.S. Fish and Wildlife Service**, 2007: Endangered species habitat conservation planning. U.S. Fish and Wildlife Service Website, http://www.fws.gov/Endangered/hcp/, accessed on 6-7-2007.

[42] **U.S. Fish and Wildlife Service**, 2002: Candidate conservation agreements with assurances for non-federal property owners. U.S. Fish and Wildlife Service Website, U.S. Fish and Wildlife Service, http://www.fws.gov/endangered/listing/cca.pdf, accessed on 6-7-2007.

Resources Conservation Service incentive programs,[43] etc.) Increased partnerships of refuges with other service programs—the Endangered Species programs, in particular—could result in cost savings and increased achievement of the USFWS's five goals that they could not achieve acting individually.

Abating External Challenges through Increased Coordination. The 2001 USFWS biological integrity, diversity, and environmental health policy tells refuge managers to seek redress before local planning and zoning boards, and state administrative and regulatory agencies, if voluntary or collaborative attempts to forge solutions do not work.[44] In 2004, USFWS officials helped stop development of a 19,250-seat concert amphitheater on a tract of land adjacent to the Minnesota Valley NWR by testifying before the local county commissioners in opposition to a permit application. NWRS leaders may take such actions to achieve conservation as climate changes.

Abating External Challenges through the Regulatory Process. In addition to land use planning, other state legal procedures can offer refuge managers opportunities to address external challenges. The Clean Water Act requires states to revise water quality standards every three years.[45] The USFWS participation in this process could work to ensure that water quality does not limit adaptation to climate change. Designation of "outstanding national resource waters" in refuges, strengthening of water quality criteria, and establishment of total maximum daily loads of key stressors are three state tasks that can enhance the NWRS's adaptive capacity (see water quality standards, antidegradation policy[46]). Also, some states establish minimum stream flows or acquire instream water rights. Federal law requires the Secretary of the Interior to acquire water rights needed for refuge purposes.[47]

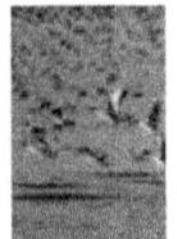

The ESA regulates private activities that may harm listed species and may be an important tool, particularly for listed species on refuges that suffer from external challenges.[48] Over the past 15 years, the ESA prohibitions have induced private cooperation to enhance conservation of species through tools such as habitat conservation plans and safe harbor agreements. The USFWS can encourage incorporation of adaptation terms into these tools.

5.4.3.1 Building Buffers, Corridors, and Improving the Matrix

Resilience is the capacity of an ecosystem to tolerate disturbance without changing into a different state controlled by a different set of processes (Holling, 1973). Fundamental ecosystem functions, including nutrient cycling, natural fire processes, maintenance of food webs, and the provision of habitat for animal species, often require land areas of thousands of square kilometers (Soulé, 1987; Millennium Ecosystem Assessment, 2006). Consequently, the relatively small size of most refuges and other conservation areas in the United States; their location in landscapes often altered by human activity; incomplete representation of imperiled species across the full range of their geographical, ecological, and geophysical range; and incomplete life history support on those refuges where it occurs; raise fundamental obstacles to achieving resilience on individual refuges and the NWRS (Grumbine, 1990). Indeed, the existing NWRS cannot fully support even genetically viable populations for a majority of threatened and endangered species (Czech, 2005). For those threatened

[43] **U.S. Department of Agriculture**, 2007: Natural resources conservation service. U.S. Department of Agriculture Website, U.S. Department of Agriculture, http://www.nrcs.usda.gov/, accessed on 6-7-2007.

[44] U.S. Fish and Wildlife Service manual 601 FW 1

[45] 33 U.S.C. § 1251-1376

[46] 40 C.F.R. § 131.12, Parts 87-135

[47] 16 USC § 668dd

[48] 16 U.S.C. § 1531-1544, 87 Stat. 884

and endangered species for which refuges were specifically established, the numbers are similar (Blades, 2007).

In response to the obstacle of small reserve size, the USFWS and other organizations engage in landscape-scale natural resource and conservation planning. A bolder strategic initiative to increase the effective conservation footprint of the NWRS may be needed to mitigate the projected effect of climate change on refuge species if the biological integrity, diversity, and health of the NWRS are all to be maintained. For example, the biological integrity, diversity, and environmental health of the least Bell's vireo (*Vireo bellii*) could be enhanced through restoration of riparian habitats on those refuges where it is found. Conservation partnerships with adjacent land managers and owners to increase the area and quality of least Bell's vireo habitat would include conservation easement and fee simple acquisition, where appropriate, and strategic acquisition of new refuges within the least Bell's vireo habitat range. The potential applications of these approaches to facilitate ecosystem adaptation to climate change concentrate on the optimum size and configuration of new and existing conservation areas at a landscape scale. State Wildlife Action Plans also provide an opportunity to create more favorable environment adjacent to refuges through which species disperse, by identifying strategic habitat parcels within the range of the least Bell's vireo.

The USFWS already engages in planning to prioritize land acquisition (U.S. Fish and Wildlife Service, 1996). Acquisition of easements often represents an attractive option for building a support network around refuges to facilitate adaptation. The USFWS has great flexibility in crafting easements to address the particular dynamic circumstances of climate uncertainty. Federal courts have consistently upheld federal easements, even in the face of state laws that imposed term limitations or contravened negotiated property restrictions.[49] However, given the projected increases in the American population and its demands on natural resources, options for easements may be fewer and pressure to remove existing easement restrictions may increase in the future. This potential currently is playing out as the U.S. Department of Agriculture considers policy proposals to reduce enrollment in the Conservation Reserve Program in order to stimulate crop production for biofuels. These factors attest to the necessity of creating a strategically planned conservation network today capable of meeting the challenges posed by climate change tomorrow.

Opportunities for maintaining the viability of refuge species, ecosystems, and ecosystem processes may be achieved through conservation partnerships, incentive programs, conservation easements, and fee simple acquisitions with willing sellers on refuge inholdings and adjacent properties. The USFWS already plays a leadership role in these best practices for conserving wildlife within watersheds and regions. The aspirational goals of refuge law along with the expertise of USFWS personnel are consistent with these outreach efforts, which may be informal or memorialized in memoranda or agreement among local landowners and jurisdictions surrounding refuges.

The alteration of habitat from climate change vegetation shifts produces one of the most significant challenges to conservation, because it reduces the viability of existing conservation areas. The targeted acquisition of new conservation areas, together with a structured configuration of the network of new and existing conservation areas across the landscape, offers an important approach to facilitating ecosystem adaptation. Landscape-scale adaptation strategies and tools—drawn from the literature and expert opinion—could include:

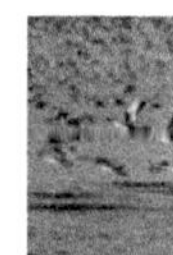

- *Establish and maintain wildlife corridors.* Connectivity among habitat patches is a fundamental component of ecosystem management and refuge design (Harris, 1984; Noss, 1987). Corridors provide connectivity and improve habitat viability in the face of conventional challenges such as deforestation, urbanization, fragmentation from roads, and invasive species. Because dispersal and migration become critical as vegetation shifts in response to climate changes, corridors offer a key adaptation tool (*e.g.*, highway over- and underpasses, Yellowstone to Yukon corridor) and help

[49] See North Dakota v. United States, 1983. 460 U.S. 300.

maintain genetic diversity and higher populations size (Hannah *et al.*, 2002). In many areas, riparian corridors provide connectivity among conservation units.

- *Expand the effective conservation footprint to include projected climate change refugia.* Climate change refugia are locations more resistant to vegetation shifts, due to wide climate tolerances of individual species, to the presence of resilient assemblages of species or to local topographic and environmental factors. Because of the lower probability of significant change, these refugia will likely require less-intense management interventions to maintain viable habitat, and should cost less to manage than vulnerable areas outside refugia. Acquisition of new land in potential climate change refugia will likely change past priorities for new conservation areas. This will require integration of climate change data from tools identified below into the USFWS Land Acquisition Priority System. Currently, The Nature Conservancy is analyzing effects of climate change in the seven ecoregions that cross the State of New Mexico in order to identify climate change refugia and to guide the development of new conservation areas under ecoregional plans developed in collaboration with government and private partners. Identification of refugia requires field surveys of refugia from past climate change events, or spatial analytical tools that include dynamic global vegetation models (DGVMs), bioclimatic models of individual species, and sea level rise models; each of these are described in more detail below.

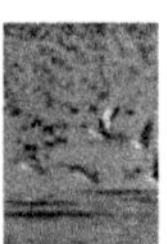

- *Eliminate dispersal barriers and create dispersal bridges.* This topic was addressed to some extent previously, but additional opportunities exist, including removal of dispersal barriers in and near refuges, establishing dispersal bridges by eliminating hanging culverts, building highway under- and overpasses, modification of land use practices on adjacent lands through incentive programs, habitat restoration, enhancement, and conservation partnerships with other public land managers.

- *Improve compatibility of matrix lands.* Strict preservation of a core reserve, and multiple-use management reflecting decreasing degrees of preservation in concentric buffer zones around the core, constitutes another climate change adaptation tool. These land use changes may be achieved through new acquisitions, conservation partnerships, or conservation incentives programs, all focused on meeting the needs of NWRS species subject to climate change stresses. In the United States, a national park, wilderness area, or national wildlife refuge often serves as the core area, with national forests serving as an immediate buffer zone, and non-urbanized state and private lands forming the outermost buffer zone. A conservation easement is a legal agreement that restricts building on open land in exchange for lower taxes for the landowner. It offers a mechanism for habitat conservation without the great expense and governmental processes required to purchase additional land for federal agencies through fee title acquisitions. As climate change shifts vegetation and animal ranges, conservation easements offer an adaptation tool to provide room for dispersal of species and maintenance of ecosystem function. If the ecosystem(s) maintained within a core conservation area and on lands adjacent to it is resilient, then—even if climate changes cause a shift in species composition—that core conservation area will remain an important part of a conservation network because new species will be able to expand their ranges into it.

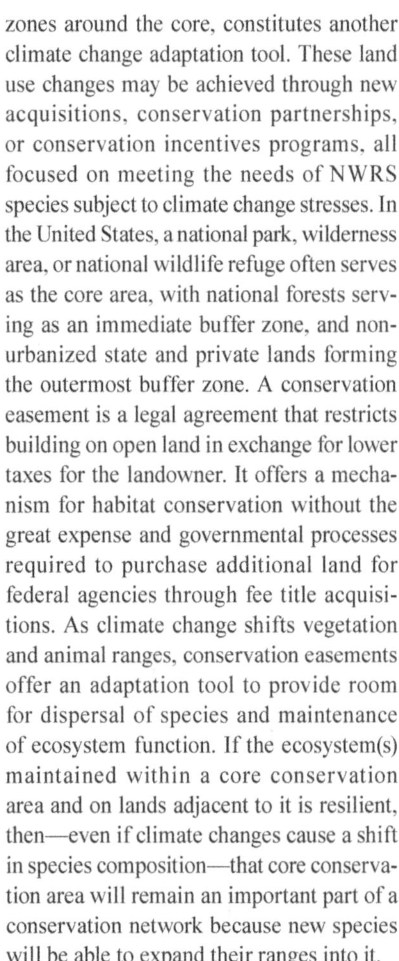

- *Restore existing and establish new marshland vegetation as sea level rise inundates coastal land.* The Nature Conservancy and USFWS are collaborating on a project in Alligator River NWR and on adjacent private land on the Albemarle Peninsula, North Carolina, to establish saltwater tidal marsh as the ocean inundates coastal land. The Nature Conservancy also plans to establish dune shrub vegetation in upland areas as coastal dunes move inland. In the Blackwater NWR in Chesapeake Bay, Maryland, the USFWS may be restoring marshland that oceans have recently inundated, by using clean dredging material from ship channels to recreate land areas.

- *Establish other marshland vegetation where freshwater lake levels fall.* Decreasing summer precipitation and increasing evapotranspiration may decrease water levels in the Great Lakes by 0.2–1.5 m (Chao, 1999). Depending on the slope of shoreline areas, the drop in lake level could translate into shore extensions 3 m wide or more. Managers of the Ottawa NWR at Lake Erie, Ohio, and other refuges on the Great Lakes may need to preemptively establish freshwater marshes as shoreline areas become shallower.

- *Reduce human water withdrawals to restore natural hydrologic regimes.* Water conservation in agricultural or urban areas may free up enough water to compensate for projected decreases in runoff due to climate change. NWR managers could work with water managers to change the timing of water flows as climate change alters fish behavior. For example, a half-day earlier migration of adult Atlantic salmon over the course of 23 years was associated with climate change (Juanes, Gephard, and Beland, 2004).

- *Install levees and other engineering works.* Levees, dikes, and other engineering works have been used widely to alter water availability and flows to the benefit of refuge species. Their use to hold back the changes brought by sea level rise and increases in storm intensity remains largely untested.

5.4.3.2 Reducing the Rate of Change

In addition to the adaptation options described in this chapter, there are a number of actions that could be taken to mitigate climate change. These actions are primarily about reducing greenhouse gases. Refuges can participate by: being educational centers for solutions to climate change; developing and showcasing energy-saving practices on refuges, such as using fuel-efficient vehicles (Eastern Neck NWR) or electrical vehicles; using solar energy (Imperial NWR, Mississquoi NWR), wind energy (Eastern Neck NWR, Mississquoi NWR), and geothermal heating and cooling (The John Heinz NWR at Tinicum, Chincoteague NWR); and, sequestering carbon through reforestation actions when consistent with refuge objectives, although this strategy needs to be further researched.

5.4.3.3 Managing to Accommodate Change

Rather than managing in order to retain species currently on refuges, refuges could manage to provide trust species the opportunity to respond to and evolve in response to emerging selective forces. Managing for change in the face of uncertainty is about buying time while planning for change. It also means working with other conservation land managers to increase linkages between protected areas, and with conservation partners on matrix lands, to increase suitability of these lands for the services to conservation targets. The scientific literature and expert opinion suggest the following possible management actions to improve the surrounding matrix:

- Creating artificial water bodies;
- Gaining access to new water rights;
- Reducing or eliminating stressors on conservation targets, *e.g.*, predator control, nest parasite control, control of non-native competitors;
- Introducing temperature-tolerant individuals, *e.g.*, resistant corals (see previous discussion) (Urban, Cole, and Overpeck, 2000);
- Eliminating barriers to dispersal;
- Building bridges for dispersal; and
- Increasing food availability.

Additional measures to help mitigate the effect of climate change on refuges could include expanding access to water and enhancing the quality of existing terrestrial and aquatic habitats, creating habitat islands near sea-ice foraging sites for seabirds, adding drip irrigation to increase humidity and moisture levels in amphibian microhabitats, etc. The possible unintended effects and side effects of these and other management actions need to be further studied.

Management/conservation partnerships with adjacent landowners to establish more refuge-compatible land are another useful tool for dealing with the effects of climate change on the NWRS. For example, refuges could enter into partnerships with organizations such as the Natural Resources Conservation Service in the USDA,[50] which offers an extensive list of programs and opportunities to manage and improve the landscape and to better meet challenges of climate change. Also, refuges could use existing general

[50] **U.S. Department of Agriculture**, 2007: NRCS conservation programs. U.S. Department of Agriculture Website, U.S. Department of Agriculture, http://www.nrcs.usda.gov/Programs/, accessed on 6-7-2007.

statutory (programmatic) authorities to manage collaboratively with federal, state, tribal, and local governments to meet the challenges of climate change. The NWRS has approximately six such resource-related (non-administrative) programs. Each program has one or more statutes that guide or govern its activities, and some of these statutes overlap among programs. Examples include the Migratory Birds and State Programs (guided by the Migratory Bird Treaty Act, Pittman-Robertston, Dingell-Johnson) and the Endangered Species program (Endangered Species Act of 1973, Marine Mammals Act, etc.).

It is probable that the stress from climate change will continue to increase over time, forcing national wildlife refuge managers and scientists to communicate, collaborate, manage, and plan together with managers and scientists from adjacent lands. One possible mechanism that the Department of the Interior could consider to enhance such collaboration is establishing national coordination entities for both management and informational aspects of responding to climate change. The National Interagency Fire Center, in Boise, Idaho,[51] is a potential model to consider. Establishing entities such as a national interagency climate change council and a national interagency climate change information network could help ensure that refuges are managed as a system, which will be a key element in climate change adaptation, as the scale of climate change effects are such that refuges must be managed in concert with all public lands, not in isolation. A cabinet-level interagency committee on climate change science and technology integration has already been created by the current administration.[52] This committee, co-chaired by the secretaries of commerce and energy, oversees subcabinet interagency climate change programs.

A coordinated information network could assemble information on successful and unsuccessful management actions and adaptations, and provide extensive literature information and overviews of all climate-change related research. It could also offer technical assistance in the use of all available climate change projection models, as well as support for geographic information systems, databases, and remote sensing for managers within each of the participating agencies.

The scale of the challenge presented by climate change and its intersection with land-use changes and expanding human populations necessitates new research and management partnerships. Building on existing partnerships between USGS and the USFWS, agencies could convene a national research and management conference bringing together managers and researchers to identify research priorities that are management-relevant and conducted at scales that are ecologically relevant (Box 5.2). The biannual Colorado Plateau Research conference provides a model to emulate (van Riper, III and Mattson, 2005).

BOX 5.2. Research Priorities for NWRS.

1. Identify
    - Conservation targets;
    - Vulnerable species.
2. Monitor and predict responses.
3. Select best management strategies.
4. Game alternative climate change scenarios.

The relatively small size and disjunct distribution of refuges presents a challenge to maintaining biological integrity, diversity, and environmental health. Yet, the NWRS has a great deal of experience with land- and water-intensive management, habitat restoration, and working across jurisdictional boundaries to achieve population objectives. These skills are critical to effective climate change adaptation. External challenges to refuge goals have forced refuge managers to deal with transboundary issues more than most other land managers. Also, because refuge land management is often similar to private land management in a surrounding ecoregion, refuges can demonstrate practices that private landowners might adopt in responding to climate change.

[51] **National Interagency Fire Center**, 2007: Welcome, National Interagency Fire Center. National Interagency Fire Center Website, National Interagency Fire Center, Boise, Idaho, www.nifc.gov, accessed on 6-7-2007.

[52] **The White House**, 2007: Addressing global climate change. The White House Website, http://www.whitehouse.gov/ceq/global-change.html, accessed on 6-7-2007.

In order to be efficient in managing refuges in the face of changing climate, the NWRS should produce a strategic plan for adaptation to global climate change. This plan would include research priorities, management strategies, and adaptation scenarios that will guide the USFWS in its task of managing refuges.

The collaborative science paradigm must guide the management-science relationship in order to meet the challenge of global climate change. A beginning would be a small (8–12 individuals) workshop of service managers and scientists to flesh out the dimensions of the challenge, using this report and those prepared for other public land managers. Further collaboration could be facilitated by a national conference of managers and researchers on challenges of climate change to conservation areas. A central piece of the conference would be the use of alternative refuge scenarios, documenting the past and current characteristics of the refuge (including their ecological content and context) and what they might become, under three alternative climate change scenarios and perhaps two to three different management scenarios. The fundamental questions throughout this conference would be: what are we managing toward? What do we expect the NWRS to be 100 years from now? Which will be the target species and where will they be? What will be the optimal configuration of refuges under such a climate shift and large scale changes in vegetation? This national conference could be followed by regional conferences hosted by each of the USFWS regions. A manager/researcher conference would need to include thematic breakout sessions to frame management-relevant questions, identify possible funding sources, and develop collaborative relationships. Ultimately these conferences would be focused on building bridges between research and management. To be successful, they would be convened every two years. The highly successful manager/researcher partnership on the Colorado Plateau (van Riper, III and Mattson, 2005) and the recent (February 2007) joint USGS-USFWS Alaska Climate Change Forum offer models for such efforts.

5.4.4 Steps for Determining Research and Management Actions

Modeling efforts are one tool that researchers and managers may use to project the effects of climate change on conservation target species and ecosystems. The following section describes the different tasks that can be accomplished using modeling tools, highlights research and management priorities in the face of climate change, and provides examples of the successful application of these tools (Box 5.3).

BOX 5.3. National Wildlife Refuges: Adaptation Options for Resource Managers.

- Manage risk of catastrophic fires through prescribed burns.
- Reduce or eliminate stressors on conservation target species.
- Improve the matrix surrounding the refuge by partnering with adjacent owners to improve existing habitats or build new habitats.
- Install levees and other engineering works to alter water flows to benefit refuge species.
- Remove dispersal barriers and establish dispersal bridges for species.
- Use conservation easements around the refuge to provide room for species dispersal and maintenance of ecosystem function.
- Facilitate migration through the establishment and maintenance of wildlife corridors.
- Reduce human water withdrawals to restore natural hydrologic regimes.
- Reforest riparian areas with native species to create shaded thermal refugia for fish species in rivers and streams.
- Identify climate change refugia and acquire necessary land.
- Facilitate long-distance transport of threatened and endangered endemic species.
- Strategically expand the boundaries of NWRs to increase ecological, genetic, geographical, behavioral, and morphological variation in species.
- Facilitate the growth of plant species more adapted to future climate conditions.
- Provide redundant refuge types to reduce risk to trust species.
- Restore and increase habitat availability, and reduce stressors, in order to capture the full geographical, geophysical, and ecological ranges of species on as many refuges as possible.
- Facilitate interim propagation and sheltering or feeding of mistimed migrants, holding them until suitable habitat becomes available.

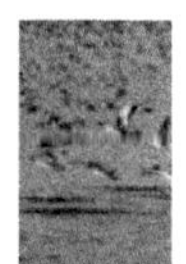

5.4.4.1 Modeling and Experimentation

In general, federal law encourages public agencies to employ science in meeting their mandates. The USFWS has a stronger mandate than most. Indicative of the congressional encouragement to partner with scientists and use refuges as testing grounds for models is the statutory definition of key terms in the NWRS mission:

> *The terms "conserving," "conservation," "manage," "managing," and "management," mean to sustain and, where appropriate, restore and enhance, healthy populations of fish, wildlife, and plants utilizing ... methods and procedures associated with modern scientific resource programs. Such methods and procedures include, ... research, census, ... habitat management, propagation, live trapping and transplantation, and regulated taking.*[53]

This definition provides ample authority and encouragement for modeling and experimentation.

Inventorying and Monitoring

The NWRS is unique among federal public lands in having a legislative mandate for monitoring. Congress requires the USFWS to "monitor the status and trends of fish, wildlife, and plants in each refuge."[54] However, as with other federal land management agencies, budgets have not prioritized the implementation of monitoring. Enlisting outside researchers can leverage resources and help achieve mutual goals for monitoring, but this cannot substitute for a systematic effort to monitor key indicators identified in unit plans and consistent with a national (or international) system of data collection. The USFWS policy guiding comprehensive refuge planning is rife with monitoring mandates, including exhortations to establish objectives that can be measured,[55] to create monitoring strategies (ibid. at 3.4C(4)(e)), and to perform the monitoring (ibid. at 3.4C(7)). The National Park Service has developed an extensive survey monitoring program as well as one suitable for adaptive management (Oakley, Thomas, and Fancy, 2003). Information from monitoring efforts may be used to document how species respond to alternative management actions and thus inform adaptive management decisions for the next generation of management actions. Thus, well-designed and -implemented monitoring programs are absolutely necessary to conducting rigorous adaptive management efforts.

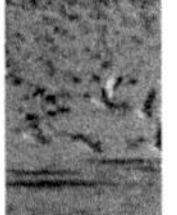

Understanding and Modeling Interactions between Populations and Habitat

As climate change drives habitat transformation, the abundance and distribution of wildlife populations will shift—often in unanticipated ways. Therefore, it will become increasingly important to support adaptive management efforts with greater understanding of the relationships between habitat and focal species or groups of focal species. By modeling these relationships at management-relevant scales, the work to protect and restore additional habitat, promote connectivity, and manipulate habitat through intensive management can be evaluated against population objectives.

There will be winners and losers among the species currently found on the NWRS. The challenge is to project possible shifts in species distributions, phenologies, and interspecific relationships, and shifts in ecological and hydrological regimes, and then to manage toward these new assemblages and distributions. Essential to that process will be a comprehensive review of the literature. The NWRS is operating in a data-deficit environment. It does not have an all-taxa survey of refuges; while 80% of refuges have presence/absence information for birds, many of those have no information on abundance or seasonal occurrence (Pidgorna, 2007). It is the rare refuge that has even presence/absence data for lesser-known vertebrates. Checklists for plants and invertebrates are almost unknown. The initial survey effort should be directed at refuges in which the greatest change is anticipated, and at those species that are identified as most vulnerable to the effects of climate change, *e.g.*, species occurring on a refuge that is at the southernmost extreme of a species' range. More explicitly, the NWRS could carry out the following tasks to target adaptation efforts:

---

[53] 16 USC § 668dd

[54] 16 USC § 668dd

[55] U.S. Fish and Wildlife Service manual 601 FW 1 - FW 6

- *Task*: Facilitate identification of species that occur on refuges.

  Tools: Different tools are available to help facilitate the identification of species that occur on refuges (Pidgorna, 2007). The Cornell Lab of Ornithology and Audubon have created an interactive database called "eBird."[56] It allows birders from North America to add their observations to existing data on bird occurrences across the continent. The data can then be queried to reveal information on birds sighted at specific locations, *e.g.*, the NWRS. Refuge employees could also be engaged in providing species occurrence information for refuges, and this database could later be expanded to include other taxonomic groups.

- *Task*: Develop detailed inventory of species, communities, and unique ecological features. Few, if any, detailed inventories of the species, communities, and unique ecological features on refuges have been conducted. The exceptions, *e.g.*, waterfowl numbers and reproductive success, provide valuable information by which refuge managers may measure the effects of climate change on this group of species. Without these data it will be impossible to monitor changes and to determine how to allocate resources to protect the biota of the different refuges.

  Tools: Traditional inventory and monitoring methods (Anderson *et al.*, 1987; Nichols, Johnson, and Williams, 1995) could be used to develop information (in a database) on sensitivity of all management targets to climate change. These sensitivities are described in the previous section. Additional information may be derived from literature searches and existing digital databases. The species monitoring program used by the National Park Service and the eBird database (described above) could also be used to facilitate this effort. This will also help fulfill the USFWS mandate to determine the biological integrity, diversity, and environmental health of the NWRS, another important research priority.

[56] **National Audubon Society and Cornell Lab of Ornithology**, 2007: North America's destination for birding on the web. eBird Website, www.eBird.org, accessed on 10-20-2006.

- *Task*: Develop more detailed coastal elevation maps. Addressing sea level rise will require more detailed maps of coastal elevations and accurate, easily applied models to integrate these maps with projected sea level increases. These maps and models are also needed to translate projected habitat changes into population changes and remedies for conservation targets. Expansion of sea water as climate change raised sea temperatures, along with increases in ocean water volume as terrestrial ice melted, increased global mean sea level by 17 ± 5 cm in the 20th century and may raise sea level another 18–59 cm by 2100 (IPCC, 2007a). As a first approximation, reserve managers can use topographic maps and local surveys of high tide levels and add 18–59 cm to estimate areas subject to inundation from climate change.

  Tools: Coastal geomorphology and other factors determine local patterns of sea level rise. The U.S. Geological Survey has analyzed sea level rise projections, geomorphology, shoreline erosion and accretion, coastal slope, mean tidal range, and mean wave height to generate a coastal vulnerability index for the entire coast of the lower 48 states (Thieler and Hammar-Klose, 1999; 2000a; 2000b). The GIS data are available online.[57]

  Because local topography determines actual inundation patterns, only detailed elevation surveys can identify exact areas subject to flooding from climate change. USGS has flown light detection and ranging (LIDAR) surveys and produced a topographic data layer with a 30 cm contour interval for the Blackwater NWR on Chesapeake Bay, Maryland, which lies entirely below 1 meter above sea level and has lost land area since at least 1938 (Larsen *et al.*, 2004b). The Blackwater inundation model identifies the land areas that may be submerged by 2100 (Fig. 5.6), providing USFWS staff with the information needed to plan potential new fee title acquisitions or conservation easements in contiguous upland areas and potential restoration of inundated wetlands using clean dredging material from ship channels.

  In order to estimate local effects of subsidence, isostatic adjustment, sedimentation, and hydrologic structures on sea level rise in the Ding Darling, Egmont Key, Pelican

[57] http://woodshole.er.usgs.gov/project-pages/cvi

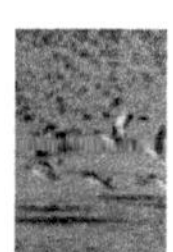

Island, and Pine Island refuges in Florida, the USFWS, the National Wildlife Federation, and Virginia Polytechnic Institute and State University used the Sea Level Affecting Marshes Model (SLAMM) (Park *et al.*, 1989). The output of this and similar models include maps that provide "before and after" images of coastal habitats and tables that provide data on habitat transformations corresponding to a specific period of time. However, SLAMM requires considerable skill with GIS and is expensive to use.

- *Task*: Provide estimates of uncertainty and model concurrence for climate projections.

  Tools: This task can be accomplished with comprehensive analyses of the variability across different climate model projections. Specifically, maps of model agreement and disagreement can be produced using recently derived methods (*e.g.*, Dettinger, 2005; Araújo and New, 2007). Both maps and concise summaries of the future projections written for managers and field biologists need to be made readily available on an easily accessed website and easily downloaded for any given region.

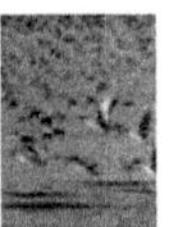

- *Task*: Obtain projections of future climate at management-relevant scales. Projected trends in climate must be summarized and made available to refuge managers at scales and in forms that are useful to them. The USFWS raw climate projections from climate models are at a coarse spatial resolution (on the order of thousands of km^2). Finer resolution projections of future climate for all of the most recent model outputs are needed. All downscaled climate data will require peer review and validation against actual observations.

  Tools: Finer-resolution projections could be generated from downscaled climate model output using statistical downscaling approaches (*e.g.*, Wilby *et al.*, 1998), but more preferably would be generated using regional climate models (*e.g.*, Giorgi, 1990) capable of running off of boundary conditions generated by one or more global climate models.

- *Task*: Project climate-induced shifts in vegetation, individual species ranges, and ranges of invasive and exotic species and summarize data for managers and field biologists. These projections of climate-induced shifts will aid mangers in determining how specific species or communities on refuges are likely to change in response to climate change. The projections should quantify uncertainty in order to account for the variability among future scenarios of climate change. The challenge of climate change to biotic interactions has been a focus of attention for over a decade (Kareiva, Kingsolver, and Huey, 1993; Peters and Lovejoy, 1994; Parmesan and Yohe, 2003; Parmesan, 2006; Lovejoy and Hannah, 2006). These types of projections for both plants (Bachelet *et al.*, 2001; Shafer, Bartlein, and Thompson, 2001) and animals (Price and Glick, 2002) in North America are now becoming available, but more projections at management-relevant resolutions are needed. As with the climate data, these data need to be summarized and made available to managers and field biologists. In addition to projecting shifts in the distributions of species that are currently protected on the refuges, models can be used to project the expansion of ranges of invasive and exotic species (*e.g.*, Peterson and Vieglais, 2001; Scott *et al.*, 2002).

  Tools: Dynamic global vegetation models (DGVMs) simulate the spatial distribution of vegetation types, biomass, nutrient flows, and wildfire by iterative analysis of climate and soil characteristics against observed characteristics of plant functional types and of biogeochemical, hydrologic, and fire processes. The LPJ DGVM (Sitch *et al.*, 2003) and the MC1 DGVM (Daly *et al.*, 2000) are the two most extensively tested and applied DGVMs (Neilson *et al.*, 1998; Bachelet *et al.*, 2003; Lenihan *et al.*, 2003; Scholze *et al.*, 2006). The Nature Conservancy, the USDA Forest Service, and Oregon State University are currently engaged in a collaborative research effort to run MC1 globally at a spatial resolution of 0.5 geographic degrees, approximately 50 km at the Equator, in order to estimate spatial probabilities of climate change vegetation shifts and to identify climate change refugia (Gonzalez, Neilson, and Drapek, 2005). The Nature Conservancy is using these data in order to help set global ecoregional priorities for site-based conservation, based on climate change and other challenges to habitat (Hoekstra *et al.*, 2005).

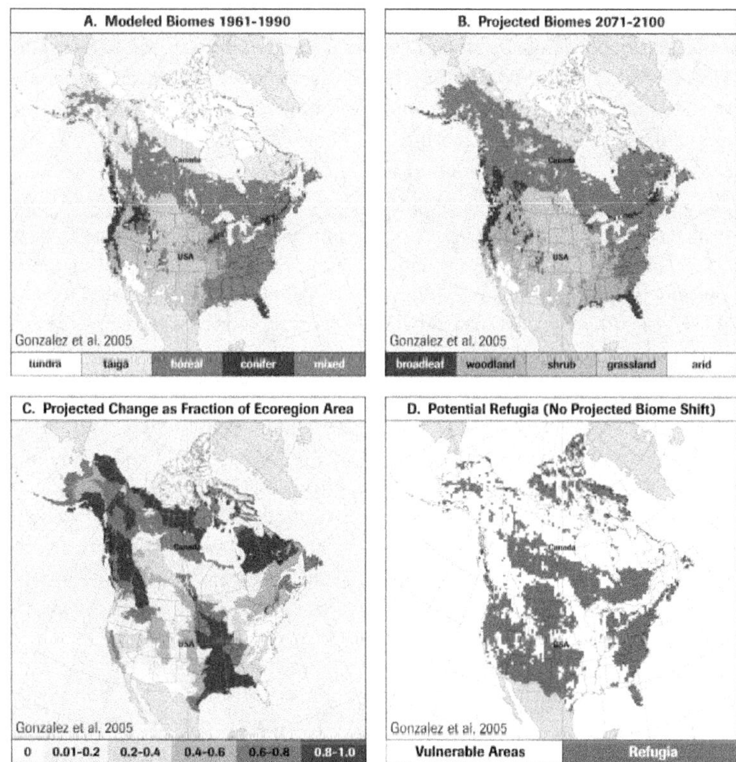

Figure 5.9. Potential climate change vegetation shifts across North America. A. Vegetation 1990. B. Projected vegetation 2100, HadCM3 general circulation model, IPCC (2000) SRES A2 emissions scenario. C. Projected change as fraction of ecoregion area. D. Potential refugia (Gonzalez, Neilson, and Drapek, 2005).

The Nature Conservancy-USDA Forest Service-Oregon State University project is analyzing potential effects from a set of general circulation models of the atmosphere and Intergovernmental Panel on Climate Change (2000) greenhouse gas emissions scenarios. This analysis is producing four spatial indicators of climate change: temperature change, precipitation change, estimated probability of vegetation shift at the biome level, and refugia, defined as areas that all emission scenarios project as stable (Fig. 5.9). Many of the refuges in the NWRS are projected to experience a biome shift and thus be outside refugia by 2100, and there is substantial heterogeneity among administrative regions. Even vegetation changes that do not constitute a biome shift may have substantial implications for trust species populations as well.

Several other modeling tools and mapping efforts will be required to address the challenges posed by climate change. An easily applied hydrological model is needed to assess the relative vulnerability of all refuges to changes in temperature and precipitation. Several hydrological models exist and could be applied to individual refuges. This would be a major, but important, undertaking. It will also be critical to assess the current and projected future level of connectivity among refuges and among all protected lands in general. Maps of current land-cover can be used to derive estimates of which refuges are most isolated from other protected lands, and where potential future corridors should be located to connect protected lands. These maps can be integrated with projections of future development to determine where additional reductions in connectivity will likely occur. Land-cover analyses can also be used to identify areas where there will likely be increased conflicts over water-use for agriculture, residences, and refuges.

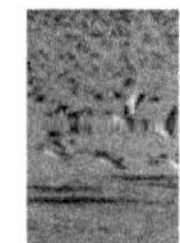

While DGVMs model the biogeography of vegetation types, bioclimatic models for individual species simulate the range of single species (Pearson *et al.*, 2002; Thomas *et al.*, 2004b; Thuiller, Lavorel, and Araujo, 2005). These models generally identify areas that fall within the climate tolerance, or envelope, of a species. Alternatively, some bioclimatic models define species-specific climate envelopes by correlating field occurrence and climate data. Like DGVMs, bioclimatic models generally do not simulate dispersal, interspecific interactions, or evolutionary change (Pearson and Dawson, 2003). Analysis of climate envelopes for 1,103 plant and animal species and the effect of climate change on habitat areas defined by species-area relationships indicates that climate change places 15–37 % of the world's species at risk of extinction (Thomas *et al.*, 2004a). The USDA Forest Service has analyzed climate envelopes and projected potential range shifts for 80 North American tree species (Iverson, Schwartz, and Prasad, 2004) and has posted all of the spatial data.[58] These data are available for anyone proficient in GIS. Natural resource managers could use these species-specific data to locate refugia or to anticipate migration of new species into an area.

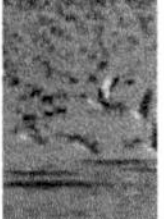

Intercomparisons of bioclimatic models for animal and plant species (Lawler *et al.*, 2006; Elith *et al.*, 2006) show variation among models, although MARS-COMM (Elith *et al.*, 2006) and random forests estimators (Breiman, 2001) have demonstrated abilities to correctly simulate current species occurrences. Moreover, ensemble forecasting of species distributions can reduce the uncertainty of future projections (Araújo and New, 2007). Nevertheless, research has not adequately tested the ability of bioclimatic models to simulate the new and unforeseen distributions and assemblages of species that climate change may generate (Araújo and Rahbek, 2006). The computer-intense and specialized nature of bioclimatic models has restricted them to academic research.

Documenting species' responses to climate change will be crucial for developing models to project responses in abundance, migration arrival and departure dates, and distribution for those species that have not yet responded to climate change (Root *et al.*, 2003). Once the projected responses are available, it will be possible to identify relevant management options and strategies. It may also be important to project responses of competitors, parasites, and host species of conservation targets in order to better manage conservation targets and also prevent invasions of refuges by non-native weedy species. Quantification of the uncertainty of projections of climate change, biome shifts, and changes in species ranges will allow natural resource managers to appropriately weight the results of modeling efforts that currently show moderate skill and will increase in skill over time. Validation against field observations will allow objective assessment of climate, biome, and species data.

Paleoclimatic and paleobiological information may be used to estimate the range of historical changes in species and ecosystem distributions, as well as rates of past change and their possible implications for future management. However, past rates of change, and the conditions that caused them, may not be indicative of future conditions or rates of change. The future will be uncertain. Thus we suggest that, rather than managing for historical range of variation, or against historical benchmarks, refuges and the refuge system be managed to maintain self-sustaining native populations and ecosystems. Refuge managers can increase their options at the refuge level by reducing non-climatic stressors and increasing habitat quality and quantity. At the systems level, chances of species surviving on the refuge system are increased by insuring that the full range of a species' ecological, geographical, genetic and behavioral variation is found on refuges, and that it occurs in more than one refuge. For example occurrence of mallard ducks on a single refuge in the central flyway would be insufficient to insure the integrity, diversity, and health of mallards in the refuge system.

- *Task*: Identify those species and ecosystems most vulnerable to effects of climate change in the context of other pressures on the

[58] http://www.fs.fed.us/ne/delaware/atlas

system(s). Strategic decisions for refuges and the NWRS regarding the biological integrity, diversity, and health of refuge species require understanding which occurrences of a species on NWRS lands are most or least likely to be affected by climate change.

Tools: Species/populations that will be most vulnerable can be identified through reviews of the literature to identify species that have already shown shifts in phenology, distribution, or abundance consistent with climate change, and through vulnerability assessment to identify the species likely to be most vulnerable to climate change, *i.e.*, species with poor dispersal capabilities; those that occur at the extremes of their ecological, geophysical, or geographical ranges; narrowly distributed species; species with small populations and/or fragmented distributions; and species susceptible to predation or crowding out by invasive non-native species.

- *Task*: Identify those regions and refuges within the NWRS that are most vulnerable to climate change in the context of other pressures on the system(s).

  Tools: In considering system-wide responses to the challenge of global climate change, managers need to think about management actions necessary to maintain the integrity, diversity, and health of the NWRS as well as that of individual refuges. This will require identifying those refuges that are most vulnerable to climate change through a system-wide vulnerability assessment. A quick review of work to date suggests that the 161 refuges that are characterized as Marine Protected Areas, the 16 refuges in Alaska that account for 82% of the total area in refuges, and the 70 refuges in the Prairie Pothole Region—thus nearly 250 refuges and perhaps 90% of the area of refuges—occur in areas subject to significant climate changes.

- *Task*: Use designated wilderness areas to track environmental changes that result from climate change.

  Tools: The larger, more intact wilderness tracts would be key elements in our ability to track environmental changes due to climate change. The larger wilderness tracts are predominantly free of the "environmental noise" of more developed areas; therefore, observed changes in ecosystems within wilderness areas could more easily and reliably be attributed to climate change rather than some other factor. Selected wilderness areas should be considered as priority locations to institute baseline inventory work and long-term monitoring.

- *Task*: Weigh projected losses of waterfowl, other conservation targets, and their habitat with possible acquisition of new refuges, and establish new conservation partnerships outside refuge lands as future conditions dictate.

  Tools: If and when refuges are managed as part of a larger conservation landscape, gains and losses will have to be weighed in terms of the refuges' conservation partners' activities (*e.g.*, the Bureau of Land Management, U.S. Forest Service, The Nature Conservancy, National Park Service), the continental or ecoregion system of public and private reserves, as well as land-use practices on matrix lands.

- *Task*: Develop renewed and enhanced management/science partnerships between USFWS, USGS, other state and federal agencies, and academia.

  Tools: Collaborative relationships could be fostered through host researcher/manager conferences locally, regionally, nationally, and internationally that would allow researchers/managers working together to frame management-relevant research questions. The answers to such questions would increase the ability of refuges and the NWRS to meet the legal mandate of maintaining biological integrity, diversity, and environmental health in the face of the change and uncertainty projected to occur with climate change.

Because the ecological needs of many refuge species are more complex than what is supported by the current NWRS design, their biological integrity, diversity, and environmental health can only be managed through partnerships with the National Park Service, U.S. Forest Service, and other public and private managers with stewardship responsibilities for America's publicly held conservation lands. For example, the harlequin duck breeds in clear and sparkling mountain stream habitats of Olympic National Park and in the U.S. Forest Service's Frank Church Wilderness, and it may be found wintering in the marine waters of Willapa NWR and Oregon Islands NWR. As another example, the State of California has taken account of climate change in its latest state wildlife action plan (Bunn *et al.*, 2007), which identifies management opportunities for natural habitat that crosses state, federal, and private land boundaries.

- *Task*: Develop a vision for the NWRS on its 150th anniversary in 2053.

  Tools: What will the conservation targets be: those species that currently occur on the NWRS, those species for which refuges were established, or threatened and endangered species for which refuges were established? Or, possibly, some subset of one of those categories, *e.g.*, waterfowl of North America? Threatened and endangered species? Invertebrates? Once target species are selected, what level of abundance will be targeted: minimally viable, ecologically viable, evolutionarily viable populations, recreationally viable, or something else? It is important to also consider species that are currently absent from the NWRS, but that could expand their ranges into the NWRS and become conservation targets in the future, *e.g.*, Mexican songbirds and hummingbirds. Much of the success of the NWRS's efforts to conserve waterfowl species can be attributed to the clearly articulated vision of Ira Gabrielson and Ding Darling for a system of refuges that would provide habitat for recreationally viable populations of ducks and geese for the enjoyment of the American public.

  Due to the uncertainty associated with climate change, it is essential that conservation targets not be static. Stopgap targets eventually will contribute to failure of the adaptation process. Ambiguity and conflict among targets are potential problems. Regulations and statutes may need to be assessed and amended in some cases. Refuges with broad mission statements, such as those created as a result of the Alaska National Interest Lands Conservation Act (ANILCA), will have the greatest flexibility to accommodate future change in species composition. Non-ANILCA refuges will be required to emphasize species identified in refuge creation mission statements.

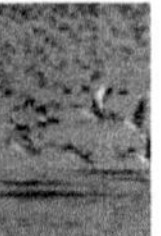

There are four other key research priorities that will likely involve a combination of modeling and empirical studies. First, managers need information on how climate change will affect the prevalence and the intensity of wildlife and plant diseases and pathogens that pose challenges to refuge species. Are outbreaks of certain diseases mediated by changes in temperature and moisture? How will a given disease respond to a change in temperature? How will the geographic ranges of diseases change with climate?

A second research need is projections of how the disturbance regimes on refuges will change. For example, how sensitive to an increase in temperature is the current fire regime or drought cycle at a given refuge?

A third priority is to investigate the implications of key translocations or "assisted dispersals." For species that will likely need to be moved to new sites or other refuges, where are these new sites, and what are the ecological implications of introducing the new species?

Finally, research priorities that include developing and enhancing methods and tools to identify and select the best possible management actions under alternative climate change scenarios would provide managers with badly needed information. The use of rigorously tested models, and enhanced species occurrence information for assessing the costs and benefits of alternative climate change scenarios, would enhance the ability to anticipate and proactively respond to changes projected under different climate scenarios at both the refuge and NWRS scales. One could also project species and ecosystem effects with current or alternate management practices, strategic growth of the refuge, strategic growth of the NWRS, or establishment of coastal barriers. Developing these and other research questions in collaborative workshops of managers and researchers will likely increase chances that results of research will be relevant to managers and increase chances that the information will be used to make a difference on refuges.

5.5 CONCLUSIONS

Climate change may be the largest challenge ever faced by the NWRS. It is a global phenomenon with national, regional, and local effects. It adds a known forcing trend in temperature to all other stressors and likely creates complex non-linear challenges that will be exceptionally difficult to understand and to mitigate. New tools, new partnerships and new ways of thinking will be required to maintain the integrity, diversity, and health of the refuges in the face of this complexity. The historic vision of refuges as fixed islands of safe haven for species met existing needs at a time when the population of the United States was less than half its current size and construction of the first interstate highway was a decade away. At that time, climates and habitats were perceived to be in dynamic equilibrium, and species were able to move freely among refuges. Today, the landscape is highly fragmented, much of the wildlife habitat present in the 1930s and 1940s has been lost, and the dynamic nature of ecological systems is well known. While Congress' aspiration for the refuges to serve as a national network for the support of biological diversity remains sound, the challenge now is to make the refuge network more resilient and adaptive to a changing environment. Changes have already occurred that are consistent with those projected under climate change, thus increasing confidence that future changes in species distribution and behavior will occur with increasing frequency. Refuge managers are faced with the dilemma of managing for a future challenge without fully understanding where and when the changes will occur and how they might best be addressed. How can USFWS fulfill the key legal mandate to maintain the integrity, diversity, and health of conservation targets in an environment that allows for evolutionary response to the effects of climate change and other selective forces?

In this chapter we have identified research initiatives, management/research partnerships, and efforts that may be used to meet the challenges of climate change. Alaskan refuges, where effects of climate change are already apparent, have been used to illustrate some of the challenges facing researchers and managers locally, regionally, and nationally (see Case Study Summary 5.1). While there is uncertainty about the scale of the projected effects of climate change on sea level rise, species distributions, phenologies, regime shifts, precipitation, and temperature, most of these changes have already begun and will most likely significantly influence the biological integrity, diversity, and health of the NWRS. These changes will require management actions on individual refuges to restore habitat; build dispersal bridges for species; eliminate dispersal barriers; increase available habitat for species through strategic fee title acquisitions, easements or other tools; and increase cooperative, consultative conservation partnerships if biological integrity, diversity, and environmental health of refuge populations and systems is to be maintained. National wildlife refuges, especially those near urban centers, could increase public awareness of the challenges facing wildlife by developing educational kiosks that provide information on the effects of climate change, habitat loss and fragmentation on refuge species.

However, actions on individual refuges will be insufficient. NWRS-wide challenges require system-wide responses. The USFWS's response to the three previous challenges faced by the NWRS (overhunting in the late 1800s, dust bowl era effects, and the ongoing loss of biodiversity that began in the second half of the 20th century) helped shape the current system, which is viewed worldwide as a model of what a natural areas system can be. Climate change, the fourth crisis facing the NWRS, offers us the opportunity to build on past successes and to do so with a more complete understanding of ecological systems. While the scale of climate change is unprecedented, so are the opportunities to make a difference for the future of wildlife and the ecosystems on which they depend. A response sufficient to the challenge will require new institutional partnerships; management responses that transcend traditional political, cultural, and ecological boundaries; greater emphasis on trans-refuge and trans-agency management and research; strong political leadership and reenergized collaborations between the USFWS and its research partners in USGS, other federal, state, tribal, and private organizations, and academic institutions. The scope and magnitude of expected changes—inundation of coastal refuges, regime shifts, shifts in species distributions and phenologies—challenges the viability of populations on single

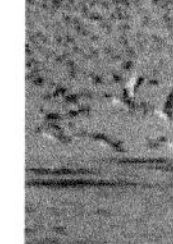

CASE STUDY SUMMARY 5.1

Alaska and the Central Flyway
Alaska and Central United States

Why this case study was chosen

Alaska and the Central Flyway:

- Together produce 50–80% of North American ducks, as well as a variety of other migratory waterfowl that are National Wildlife Refuge System (NWRS) trust species;
- Support migratory species that have an energetically costly and complex life history strategy, with separate breeding, migratory stopover, and wintering habitats dispersed throughout the system;
- Show strong historical and projected warming in migratory species breeding areas (most of Alaska and the Prairie Pothole Region of the Central Flyway);
- Demonstrate heterogeneity in non-climate stressors that creates substantial complexity in both documenting and developing an understanding of the potential effects of climate warming on major trust species;
- Differ in the expected relative magnitude of climate and non-climate stressors as drivers of populations; climate is expected to be the dominant driver of migratory trust species performance in Alaska, whereas pervasive non-climate stressors such as habitat conversion and fragmentation, invasive species, pollution, and competition for water are expected to complicate estimation of the net effects of climate change on migrants in the Central Flyway.

Management context

The first unit of the NWRS was established in 1903, and the system has since grown to encompass 584 units distributed throughout the continental United States, Alaska, Hawaii, and the Trust Territories. These refuges provide the seasonal habitats necessary for migratory waterfowl to complete their annual life cycles, and conditions on one seasonal habitat may affect waterfowl performance in subsequent life history stages at remote locations within the NWRS. The key mandate of the NWRS is to maintain the integrity, diversity, and health of trust species and populations of wildlife, fish and plants, and this species mandate provides the system with substantial legal and cooperative latitude to respond to conservation challenges. Individual symptomatic challenges of climate change can be addressed at the refuge level, while NWRS planning is the more appropriate level for addressing systemic challenges to the system using all legal and partnership tools that are available.

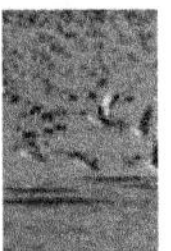

Key climate change effects

- Observed warming that is more pronounced in Alaska than in southerly regions of the United States;
- Observed earlier thaw in Alaska that increases the length of the ice-free season;
- Observed increases in summer water deficits in Alaska;
- Observed lake drying in Alaska;
- Observed shifts to later freeze-up and longer growing seasons in the Central Flyway in Canada and in the Northern United States;
- Observed increases in temperatures that account for 60% of the variation in the number of wet basins in the Prairie Pothole Region of the Central Flyway;
- Projected further increases in temperature for much of the Central Flyway, with northerly regions expected to warm more than southern regions;
- Projected drying of the Prairie Pothole Region in the Central Flyway, the single most important duck production area in North America, which may significantly affect the NWRS's ability to maintain migratory species in general and waterfowl in particular;
- Projected sea level rise and increased urbanization in southern regions of the Central Flyway, which are expected to cause reductions in refuge area and increased insularity of remaining fragments, respectively;

CASE STUDY SUMMARY 5.1 (CONTINUED)

- Projected changes in vegetation , which suggest that most of the Central Flyway will experience a biome shift by the latter part of the 21st century while interior Alaska will remain relatively stable.

Opportunities for adaptation

- Increased emphasis on design of inventory and monitoring programs could enhance early detection of climate change effects;
- A focus on climate change in Comprehensive Plans and Biological Reviews could allow early identification of potential mechanisms for adaptation;
- Enhanced education, training, and long-term research-management partnerships could increase the likelihood that adaptive management responses to climate change will be implemented and be successful;
- Emphasis on multiple integrated-scale responses to climate change and developing enhanced formal mechanisms to increase inter- and intra- agency communication may be particularly effective for migratory species.

Conclusions

The integrity, diversity, and health of NWRS migratory trust species populations are affected by habitat conditions throughout the system. The value of seasonal refuges can be evaluated only in the context of their relative contribution to trust species populations. Breeding areas in Alaska contribute birds to all four flyways from the Pacific to the Atlantic, but the status of staging and wintering habitats throughout these flyways also influences the number and condition of birds returning to Alaska to breed. Climate change adds substantial uncertainty to the problems associated with accessing resources necessary to meet energy requirements for migration and reproduction, and this climate challenge may interact synergistically in unexpected ways with non-climate stressors. For example, depending on the migratory species, lengthened access to migratory stopover areas that is caused by climate change combined with changing agricultural crop mixes that are driven by market forces may eventually result in either reduced or increased reproduction on breeding areas. The primary climate challenge to migratory waterfowl is that resource availability may become spatially or temporally decoupled from need, and, in a warming climate, individual refuges may no longer meet the purposes for which they were established. An emphasis on the contribution of all conservation lands to the NWRS mission and strategic system growth, using all available tools, will likely provide the greatest latitude for migratory trust species and the NWRS to adapt to climate change. The unresolved complexity of understanding the net effects of variable climate and non-climate stressors throughout the NWRS represents an opportunity to focus on the importance of strong interconnections among system units, and to foster a national vision for accommodating net climate warming effects on system trust species.

refuges as well as the existence of trust species (threatened and endangered species, migratory birds, marine mammals, and anadromous and interjurisdictional fish) in the refuge system. The most important tools available are the species themselves and their abilities to evolve genetic, physiological, morphological, and behavioral responses to changing climates, site-specific relationships, and environments. The opportunities for species to evolve in response to changing environments can be enhanced by ensuring that the full range of the target species' biogeographical, ecological, geophysical, morphological, behavioral, and genetic expression is captured in the NWRS (Scott *et al.*, 1993; Shaffer and Stein, 2000).

A national interagency climate change council, a national interagency climate change information network, researcher/manager conferences, research themes and management strategies, and the species inventories and monitoring programs identified in this chapter represent some of the initial tools that could enable the USFWS to best meet the challenge of global climate change. In particular, there is a need for in-depth studies of the projected effects of climate change on refuges in different ecoregions. Comparing and contrasting effects in different ecoregional setting may provide insights to future management, partnership and research opportunities.[59] The most important take-away messages about the management of the NWRS in the face of climate change are summarized below.

Response to climate change challenges must occur at multiple integrated scales. This must occur both within the NWRS and among partner entities. Individual symptomatic challenges of climate change must be addressed at the refuge level, while NWRS planning is the most appropriate level for addressing systemic challenges to the system. Both top-down and bottom-up approaches must be integrated. Due to the heterogeneous nature of observed (Figs. 5.3a and 5.b) and predicted changes in temperature and precipitation, a "one-size-fits-all" solution will not be appropriate.

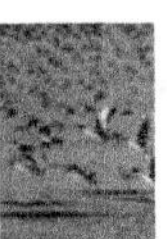

Immediately convene a national research-management workshop. At this workshop, researches and managers could identify and discuss the challenges presented by projected effects of climate change and collectively identify, frame, and prioritize management-relevant research questions. Similar workshops could be convened regionally.

Establish coordinating bodies, such as a national interagency climate change information network, to provide information and advice on the management of ecosystems and resources. The scale of climate change is such that public lands (including refuges) and private lands may be best managed in concert rather than in isolation. Management and information mechanisms could be established to support this new level of cooperation. Adaptation to climate change will likely require an entirely new level of coordination among public lands at multiple spatial scales. Such coordination could involve national and regional councils that bring together federal, state, county, and private land owners to share information, and resources to develop cooperative management/research responses to climate change. Essential to this effort would be a center that would serve as a clearinghouse for information on climate change, its effects, and available management tools. Increased international cooperation will also be necessary, since climate change does not respect political borders. Lessons could be learned from the work done by the intergovernmental Arctic Council and its six working groups.

Conduct vulnerability assessments and identify conservation targets. Peer reviewed and validated national and regional assessments could be carried out to identify ecosystems, species, and protected areas facing the greatest risks; this information then could be used to develop shared conservation targets and objectives. The most vulnerable species on refuges include those with restricted ranges, limited dispersal capabilities, and those that occur on a refuge that is at the geographical, ecological, or geophysical extreme of a species range and/or on a refuge that provides incomplete life history support.

Conduct a series of workshops that compare the costs and benefits of alternative management scenarios. A series of workshops that evaluate alternative management scenarios in the face of climate change would provide refuge managers with a portfolio of tools, solutions, and actions to both proactively and reactively respond to the effects of climate change.

Manage lands as dynamic systems. It may not be possible to manage for static conservation targets. Species ranges will shift, disturbance regimes will change, and ecological processes will be altered. Management actions to decrease non-climate stressors and enhance the biological integrity, diversity, and health of refuge species, ecosystems, and ecological processes could include water impoundment; control of water flow; control of predators, competitors, and nest parasites on conservation targets; and enhancement of food resources and breeding habitat (*e.g.*, red-cockaded woodpecker).

[59] **U.S. Global Change Research Program**, 1997: Impact of land use and climate change in the southwestern United States. U.S. Geological Survey Website, http://geochange.er.usgs.gov/sw/, accessed on 11-17-2007.

Ensure that conservation targets provide a representative, resilient, and redundant sample of trust species and communities. If the conservation targets are managed through adequate and well-coordinated interagency efforts, their evolutionary capabilities will be enhanced, viable populations will be maintained, and the potential for recreational and subsistence uses will be maximized.

Strategically increase the effective conservation footprint of the NWRS. Adaptation to climate change may require strategic growth of individual refuges and the NWRS, to increase resilience of populations and the conservation value of the NWRS through increased representation and redundancy of conservation target populations in the NWRS. Increased emphasis on providing connectivity and dispersal corridors among units, especially for trust species that cannot fly, will be critical. A refuge that has "lost" its establishment and/or acquisition purpose could still be valuable to the NWRS, if it provides connectivity or is resilient enough to support different species and processes. The strategic growth of the NWRS and successful adaptation to climate change will require refuge managers, scientists, government officials and other stakeholders to look beyond any one species and any single refuge purpose. The mandate of the NWRS—to maintain biological integrity, diversity, and environmental health of the Refuge System—is so complex and broad that it would be difficult if not impossible to state that a refuge has lost its larger purpose and will no longer contribute to the fulfillment of this mandate. The size and distribution of refuges in the NWRS, and the question of whether individual refuges continue to be capable of contributing to maintenance of biological integrity, diversity, and environmental health of various conservation targets need to be vigorously assessed before any decisions regarding divestiture of existing refuge lands can be made.

The NWRS was designed principally as a migratory bird network. The widely dispersed units provide for the seasonally variable life history requirements for trust species. Because many birds make use of different parts of the NWRS throughout the year, the performance of birds on any one component of the NWRS will be affected by climate-induced changes throughout the NWRS. Thus, innovative inter- and intra-flyway, inter- and intra-agency, and inter-regional communication and coordination are needed to understand and adapt to climate change.

The policy of managing toward pre-settlement biological integrity, diversity, and environmental health will be more problematic under projected future climate conditions. Historical benchmarks and their variability may provide long-term perspective for managers, but historical conditions (species composition, abundance, distribution, and their variability) are unlikely to be reasonable management goals in the face of climate change. Pursuing such goals would force managers to attempt to sustain species in areas where environmental conditions were no longer suitable. However management for self-sustaining native populations and ecosystems in the face of change and uncertainty as the standard would be consistent with maintaining integrity diversity and health of native species and ecosystems.

The NWRS has extensive experience working with private landowners and can be a model for private landowner responses to climate change. With 4 million acres in easements, the NWRS has developed valuable experience working with landowners to develop collaborative conservation projects, conservation incentive programs, and agreements that support system-wide objectives. Because refuge lands are more productive and at lower elevation than other protected areas, they are more similar in these characteristics to private lands and thus better suited to demonstrate practices that private landowners might adopt in responding to climate change. All public lands should be models for other landowners, but the refuges may be the most relevant models in many parts of the country.

Refuges are more disturbed and fragmented than other public land units. These characteristics may exacerbate the challenges presented by climate-induced habitat changes. However, the NWRS has substantial experience with intensive management, a wide range of habitat restoration methods, and cross-jurisdictional partnerships that should enhance the refuges' ability to achieve objectives compared with other federal land management systems.

Education and training of NWRS staff, at all levels, regarding potential implications of climate change for NWRS planning and sustainability is critical. To facilitate inclusion of climate change considerations into CCPs we suggest that workshops be held that instruct national, regional, and refuge staff on ways to identify options for responding to effects of climate change and means to incorporate this information in planning documents.

The challenge today is to manage to accommodate change in the face of uncertainty. If responses to projected climate change effects fail to match the scale of the challenges, it may not be possible to meet the legal mandate of managing refuges and the NWRS to maintain their biological integrity, diversity, and environmental health. The USGS and USFWS cross-programmatic, strategic, habitat conservation initiative illustrates the type of thinking and planning that will be needed to tackle climate change within the NWRS, across the USFWS, and in collaboration with other agencies (National Ecological Assessment Team, 2006). The integrity and functioning of ecological systems will be maintained only if USFWS manages to accommodate change and reintegrates refuges into the American mind and the American landscape. Our challenge is no different than that faced by Ira Gabrielson, Ding Darling, and other professionals in the 1930s. Isolated conservation fortresses managed to resist change will not fulfill the promise (U.S. Fish and Wildlife Service, 1999) of the NWRSIA, nor will they meet the needs of American wildlife. We must articulate a vision of the NWRS that focuses on system status in 2053, the 150th anniversary of establishment of the first refuge. What will the NWRS contain, how healthy will it be, and what must we do to fulfill that vision?

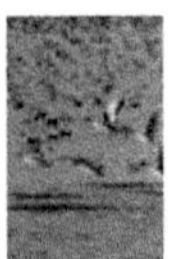

APPENDIX

Actions to Assist Managers in Meeting the Challenges Posed by Climate Change[60]

Climate-related Stressor	Ecological Impacts	Information Needed	Would It Require a Change in Management/Can It Be addressed?	Management Approach/ Activity	Opportunities	Barriers or Constraints
Changes in invasive species (increases or shifts in the types)	New invasive species may affect refuges; warming temperatures may enable the survival of exotic species that previously were controlled by cold winter temperatures.	Need better models and projections of non-native terrestrial and aquatic species distributions.	Can be addressed in small areas; large areas would be more challenging.	Remove exotics; prevent and control invasive pests.[61]	Expand collaboration with other federal agencies, state agencies, private organizations to increase/share knowledge.	Need better monitoring systems. Managers need better management tools to implement at ecologically relevant scales.
Sea level rise	Loss of high and intertidal marsh; species affected: migratory waterfowl, shorebirds, threatened and endangered species, anadromous fish.	Need better models and projections of sea level rise; more extensive use of SLAMM (Sea Level and Marsh Migration Model).	Refuge boundaries may need to be established in a different way (e.g., Arctic refuge has ambulatory boundaries that are going to shift with sea level rise—meaning that the islands and lagoon will be lost); dikes and impoundments are temporary, so longer term solutions need to be sought.	Avoid acquiring additional bunkered/ coastal lands; do acquire land further inland in areas where sea level projected to rise; avoid maladaptive activities such as moving wetland grasses/removing peat content.	Expand collaboration with other federal agencies, state agencies, private organizations to increase/share knowledge.	Need better monitoring system. Managers need adaptation tools.

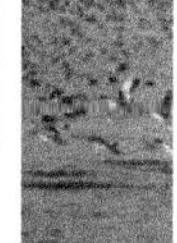

[60] The content of this table was taken from the ideas that emerged during the stakeholder workshop.

[61] **Combes**, S., 2003: Protecting freshwater ecosystems in the face of global climate change, In: *Buying Time: a User's Manual for Building Resistance and Resilience to Climate Change in Natural Systems*, [Hansen, L.J., J.L. Biringer, and J.R. Hoffman (eds.)]. World Wildlife Foundation, Washington, DC, pp. 1-244 as cited in: **Matson**, N., 2006: *Letter From Defenders of Wildlife to Beth Goldstein, Refuge Planner at the U.S. Fish and Wildlife Service: Comments on the Silvio O. Conte National Fish and Wildlife Refuge Comprehensive Conservation Plan.* Noah Matson, director of Defenders of Wildlife, provided this letter at the SAP 4.4 NWR Stakeholder Workshop, January 10–11, 2006.

Climate-related Stressor	Ecological Impacts	Information Needed	Would It Require a Change in Management/Can It Be addressed?	Management Approach/ Activity	Opportunities	Barriers or Constraints
Salt water intrusion	Flooding of coastal marshes and other low-lying lands and loss of species that rely on marsh habitat, beach erosion, increases in the salinity of rivers and groundwater.[62]	Better models and projections of sea level rise at the scale of individual refuges.	Yes, but will need to decide if managers should manage for original conditions or regime shift.	Restoration of saltmarshes may be facilitated by removal of existing coastal armoring structures such as dikes and seawalls, which may create new coastal habitat in the face of sea level rise. Presence of seawalls at one site in Texas increased the rate of habitat loss by about 20% (Galbraith *et al.*, 2002).	Cooperative agreements with adjacent landowners.	Bulkheaded refuges and expense.
Hydrologic changes	See Cinq-Mars and Diamond (1991) for discussion of how changes in precipitation may affect fish and wildlife resources. See Larson (1995) for a discussion on the effects of changes in precipitation on northern prairie wetland basins. Van Riper III, Sogge, and Willey discuss the effects of lower precipitation on bird communities in the southwestern United States.[63]	Need better models and projections of hydrological changes.	May require accessibility to new sources of water.	Use projected changes in hydrology to help manage impacts caused by hydrologic changes. Cinq-Mars and Diamond (1991) recommend that "monitoring programs must be established for fish and wildlife resources; migration corridors must be identified and protected; and new concepts must be developed for habitat conservation."	Increased cooperation with upstream land managers.	Increasing demands on water resources.

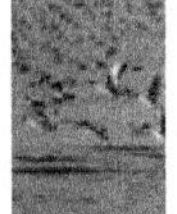

[62] **Matson**, N., 2006: *Letter From Defenders of Wildlife to Beth Goldstein, Refuge Planner at the U.S. Fish and Wildlife Service: Comments on the Silvio O. Conte National Fish and Wildlife Refuge Comprehensive Conservation Plan*. Noah Matson, director of Defenders of Wildlife, provided this letter at the SAP 4.4 NWR Stakeholder Workshop, January 10–11, 2006.

[63] **van Riper**, C., III, M.K. Sogge, and D.W. Willey, 1997: Potential impacts of global climate change on bird communities of the Southwest. In: *Proceedings of the U.S. Global Change Research Program Conference hosted by US DOI and USGS*: Impact of Climate Change and Land Use in the Southwestern United States.

Climate-related Stressor	Ecological Impacts	Information Needed	Would It Require a Change in Management/Can It Be addressed?	Management Approach/ Activity	Opportunities	Barriers or Constraints
Melting ice and snow	Polar bears are increasingly using coastal areas as habitat changes due to sea ice melting; there also have been changes in wintering patterns for waterfowl due to food availability. Bildstein (1998) describes observations about how timing of cold fronts affects raptor migration. Changes in snowpack in the West will result in reduced summer streamflow, which could affect habitat.	More detailed life history information on polar bear movements and use of sea ice.	May require significant changes in management, including development of artificial foraging platforms.	Provide artificial foraging platforms, mitigate effects of climate change globally.	Increase cooperation with other Arctic nations where polar bears occur.	Lack of global commitment to mitigate climate change.
Diseases	Diseases may move around or enter new areas (e.g., avian malaria in Hawaii may move upslope as climate changes). Diseases would seem to be a major concern considering shift in migration ranges, the changes in endemic disease patterns (northern shifts of traditionally "tropical" diseases, for example), and the ability for certain diseases to be spread rapidly through migratory bird populations.	More detailed information on phenology of diseases and their effects on species' vital rates.	Control of vectors at unprecedented scales.	Control vectors, increase habitat beyond projected range shifts of diseases and disease vectors.	Expand collaboration with other federal agencies, state agencies, private organizations to increase/share knowledge.	Lack of information and lack of funding.

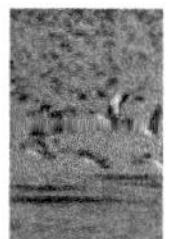

Climate-related Stressor	Ecological Impacts	Information Needed	Would It Require a Change in Management/Can It Be addressed?	Management Approach/ Activity	Opportunities	Barriers or Constraints
Warming temperatures	Species range shifts/ phenology: loss of keystone species (*e.g.*, polar bears and seals, salmon, beaver); 90% decline in population of sooty shearwater; habitat loss for cold water fishes. Breeding range of songbirds may migrate north, which could negatively affect forests (the birds eat gypsy moths and other pests).[64] Trees will become sterile, and dying trees will become more susceptible to invasive pathogens.[65] Native species will be affected by the change in tree species.[66] Warmer conditions can lead to food spoiling prematurely for species that rely on freezing winter temperatures to keep food fresh until spring.[67] Prolonged autumns can also delay breeding, which can lead to lower reproductive success. See also Hannah *et al.* (2005).	Need better models and projections of species shifts.	Yes; if species that are the purpose of a refuge shift out of the refuge area, management must be changed either to focus on management of different species or thinking about the refuge boundaries.	(1) Baseline inventorying: need to determine what species are where; an available tool for doing this is eBIRD; (2) monitoring along gradient such as latitude, longitude, distance to sea; GLORIA: mountain top assessments of species shifts; GIS layers on land prices, LIDAR data (3) build redundancy into system (4) establish new refuges for single species (5) build connectivity into the conservation landscape (change where agriculture is located and what crops are planted to allow migratory corridors to exist); (6) acquire land to north when projected species shifts northward; (7) identify indicator species that will help detect changes in ambient temperatures.	Expand collaboration with other federal agencies, state agencies, private organizations to increase/share knowledge.	Need better monitoring system. Fifteen-year planning cycle may limit ability to think about long-term implications. Managers need adaptation tools. Cannot deal with this issue in a piecemeal fashion because will likely be a great deal of spatial redistribution in and out of refuge system.

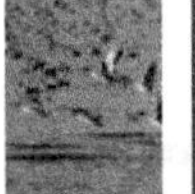

64 **Matson**, N., 2006: *Letter From Defenders of Wildlife to Beth Goldstein, Refuge Planner at the U.S. Fish and Wildlife Service: Comments on the Silvio O. Conte National Fish and Wildlife Refuge Comprehensive Conservation Plan.* Noah Matson, director of Defenders of Wildlife, provided this letter at the SAP 4.4 NWR Stakeholder Workshop, January 10-11, 2006.

65 **Matson**, N., 2006: *Letter From Defenders of Wildlife to Beth Goldstein, Refuge Planner at the U.S. Fish and Wildlife Service: Comments on the Silvio O. Conte National Fish and Wildlife Refuge Comprehensive Conservation Plan.* Noah Matson, director of Defenders of Wildlife, provided this letter at the SAP 4.4 NWR Stakeholder Workshop, January 10-11, 2006.

66 **Matson**, N., 2006: *Letter From Defenders of Wildlife to Beth Goldstein, Refuge Planner at the U.S. Fish and Wildlife Service: Comments on the Silvio O. Conte National Fish and Wildlife Refuge Comprehensive Conservation Plan.* Noah Matson, director of Defenders of Wildlife, provided this letter at the SAP 4.4 NWR Stakeholder Workshop, January 10-11, 2006.

67 **Waite**, T. and D. Strickland, 2006: Climate change and the demographic demise of a hoarding bird living on the edge. In: *Proceedings of the Royal Society B: Biological Sciences*, **273(1603)**, 2809-2813 as cited in:**Matson**, N., 2006: *Letter From Defenders of Wildlife to Beth Goldstein, Refuge Planner at the U.S. Fish and Wildlife Service: Comments on the Silvio O. Conte National Fish and Wildlife Refuge Comprehensive Conservation Plan.* Noah Matson, director of Defenders of Wildlife, provided this letter at the SAP 4.4 NWR Stakeholder Workshop, January 10-11, 2006.

Climate-related Stressor	Ecological Impacts	Information Needed	Would It Require a Change in Management/Can It Be addressed?	Management Approach/ Activity	Opportunities	Barriers or Constraints
Wildfires	Fires are becoming more intense and longer in Alaska and elsewhere. Schoennagel, Veblen, and Romme (2004) discuss the interaction of fires, fuels, and climate in the Rocky Mountains.	It is known that fires are becoming more intense and longer, but managers are not sure what to do about it.	Increased collaborative fire management practices and response. Increased fuel management activities over larger areas.	Pre-emptive fire management: use prescribed burning to mimic typical fires (increase fire frequency cycle to prevent more catastrophic fire later).	Increased interagency cooperation.	Need to tie into wildlife management goals, but managers are not sure how to do that.
More frequent and extreme storm events	Debris from human settlements may be blown in or washed into refuges, and may include hazardous substances. Eutrophication due to excess nutrients coming in from flood events could stimulate excessive plant growth and negatively affect habitats.[68] Soils could be affected through erosion, changes in nutrient concentrations, seed losses, etc. Hydrology could be affected through stream downcutting, changes in bedload dynamics, loss of bank stability, changes in thermal dynamics, etc.	It is uncertain what the refuge system can do to manage for this issue.	Harden infrastructures.	Space populations widely apart; if a catastrophic weather event occurs, population loss may be less.[69]	Cooperative agreements with up-elevation landowners and managers. Restoration of wetland habitats.	Limited resources. Large scale of the problem. Hulme (2005): Species translocation can lead to unpredictable consequences, so should only be used in extreme situations.
Alaska central flyway (see Case Study Summary 5.1): stressors include early thaw/ late freeze, sea level rise, storm events, warming temperatures	Early thaw/late freeze: resource access; increased rearing season length, crop mix, early spring migration, delayed fall migration, short-stopping, northward-shifted harvest, redistribution; warming: habitat access, disease.	Refined estimates of projected climate-warming-related changes in nesting and rearing lake number and area in Alaska and the Prairie Pothole Region; projections of climate-change related changes in agricultural crop mixes and distribution in the Central Flyway.	Management may expect different distributions of waterfowl as a result of climate change, may be addressed through directional emphasis on partnerships (e.g., emphasize collaborative projects in areas where net gain under projected climate change is the greatest.	Recognition and monitoring; establish secure network of protected areas.	Enhance educational outreach, in-agency training, and focused monitoring.	Lack of a national vision; uncertainty; resources/ political climate; non-climate stressors: agricultural disturbances, urbanization, fragmentation, pollution.

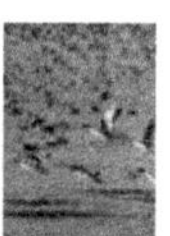

[68] **Matson**, N., 2006: *Letter From Defenders of Wildlife to Beth Goldstein, Refuge Planner at the U.S. Fish and Wildlife Service: Comments on the Silvio O. Conte National Fish and Wildlife Refuge Comprehensive Conservation Plan.* Noah Matson, director of Defenders of Wildlife, provided this letter at the SAP 4.4 NWR Stakeholder Workshop, January 10-11, 2006.

[69] **Matson**, N., 2006: *Letter From Defenders of Wildlife to Beth Goldstein, Refuge Planner at the U.S. Fish and Wildlife Service: Comments on the Silvio O. Conte National Fish and Wildlife Refuge Comprehensive Conservation Plan.* Noah Matson, director of Defenders of Wildlife, provided this letter at the SAP 4.4 NWR Stakeholder Workshop, January 10-11, 2006.

Wild and Scenic Rivers

Lead Authors: Margaret A. Palmer, University of Maryland

Contributing Authors: Dennis Lettenmaier, University of Washington; N. LeRoy Poff, Colorado State University; Sandra Postel, Global Water Policy Project; Brian Richter, The Nature Conservancy; Richard Warner, Kinni Consulting

6.1 SUMMARY

Wild and Scenic Rivers (WSRs) provide a special suite of goods and services, valued highly by the public, that are inextricably linked to their flow dynamics and the interaction of flow with the landscape. The WSR System was created to protect and preserve the biological, ecological, historic, scenic, and other "outstandingly remarkable values" for which they have been selected. The management goals for WSRs center on the preservation and protection of these conditions and values. Currently there are 165 WSRs across the country, representing more than 11,000 stream miles. Most states have at least one designated river or river segment, but 100 of the WSRs fall within just four states (Oregon, Alaska, Michigan, and California with 46, 25, 16, and 13 WSRs respectively). With the exception of the state of Alaska, most WSRs are within watersheds affected by human activities, including development (agricultural, urban, or suburban land use) or dams. In fact, many WSR segments lie downstream of these impacts, meaning their management for scenic or free-flowing condition is difficult.

Climate change adds to and magnifies risks that are already present in many watersheds with WSRs through its potential to alter rainfall, temperature, and runoff patterns, as well as to disrupt biological communities and sever ecological linkages in any given locale. Thus, the anticipation of climate change effects requires both reactive and proactive management responses if the nation's valuable river assets are to be protected.

KEY FINDINGS

The context of WSRs within their watershed and the ability to manage the many stressors that interact with climate change exert a large influence on their future. Anticipating the future condition of a river in the face of climate change requires explicit consideration not only of the current climatic, hydrogeologic, and ecological conditions, but also of how it is managed and how human behavior will affect the river (the human context). Even if impacts are small at present, consideration of the human context is critical because so many WSRs are not within a fully protected basin. This means that in addition to climate change, impacts associated with activities such as development and water withdrawals are likely to become issues in the future. Thus, stress associated with the *future human context* will interact with climate change, often exacerbating problems and intensifying management challenges. To the extent that managers are able to control aspects of this "context," they are better placed to manage for adaptation to climate change.

Impacts of climate change on WSRs will vary by region and human context, and will be manifest through changes in hydrology, geomorphology, and ecology. Climate change is expected to have a significant impact on running waters throughout the world, including WSRs. Impacts are not only in terms of changes in flow magnitude and timing, but in terms of thermal regimes and the flora and fauna that currently inhabit these waters. For a given change in temperature, rainfall, and CO_2 relative to the natural range of variability, WSRs in highly developed watersheds are expected to experience the most significant changes. Changes outside the natural range of flow or temperature variability may have drastic consequences for ecosystem structure and function, and thus the values for which the river was designated as wild and scenic. Species may be locally extirpated or shift their distributions. Changes in flow regimes also may affect recreational opportunities, and could affect valued cultural resources.

Management approaches for many WSRs will require collaborations with federal and non-federal partners in the respective river basins. WSR managers could strengthen collaborative relationships among federal, state, and local resource agencies and stakeholders to ease the implementation of adaptive river management strategies. Options to protect WSRs and river segments are diverse and most of them require cooperation and collaboration with other groups, including local landowners, reservoir and dam managers, as well as city, county or state agencies. Options presented assume WSR managers/administering agencies will actively seek cooperative arrangements with the needed parties to ensure WSR ecosystems are protected. Land acquisition is an option that may provide the most security for WSRs that are in watersheds with some non-federal land.

Managers may forge partnerships and develop mechanisms to ensure environmental flows for WSRs in basins that experience water stress, work with land use planners to minimize additional development in WSR watersheds, or ensure that land adjacent to a WSR is in protected status. Methods to manage and store surface and groundwater will be important for WSRs in developed or dammed watersheds that are in regions expected to experience more floods or droughts. With more than 270 dams located within 100 miles (upstream or downstream) of a designated WSR, collaborative arrangements with dam managers offer great potential to secure beneficial flows for WSRs under various climate change scenarios. Similarly, working to develop agreements to limit water extractions, purchase additional water rights or dry-year agreements with willing parties, and working with land use planners to minimize additional development may be very important in regions of the country that are expected to experience water stress.

In the face of climate change, management of WSRs will require both proactive approaches as well as reactive actions to be taken if impacts occur. The ability of a WSR to provide the ecosystem goods and services in the future that originally prompted its designation will depend largely on how it is managed. Without deliberate management actions that react to stress already occurring or that anticipate future stress, the provision of ecosystem services will not be guaranteed. Some actions are far more desirable to undertake proactively (*e.g.*, acquire land to protect floodplains), and others may be done proactively *or* reactively (*e.g.*, restore riparian habitat). Those actions that are more desirable to undertake reactively occur where the costs of acting before an event are high and the uncertainty of an event occurring is high (*e.g.*, severe damage occurs from an extreme event that requires channel reconfiguration). Among the most important proactive measures is expanding the technical capacity of WSR managers so they have the needed tools and expertise to prepare for and implement new management.

Priority management strategies that include a focus on increased monitoring and the development of tools to project future impacts will better enable river managers to prioritize actions and evaluate effectiveness. A task critical to prioritizing actions and evaluating effectiveness is to monitor and develop regional-scale (preferably WSR basin-specific) tools for projecting the likely impacts of climate change in concert with other stressors. Monitoring efforts may begin by providing adequate baseline information on water flows and water quality. Then management plans for WSRs may be designed with flexibility built in so that they may be updated regularly to reflect new information and scientific understanding, based on monitoring and modeling efforts.

6.2 BACKGROUND AND HISTORY

In the late summer of 1958, the greatest anadromous fish disaster in history was unfolding on the Snake River near the small town of Oxbow, Idaho. Once known for its booming copper mines and rowdy saloons, this small town would soon be known as the site of the "Oxbow Incident." Chinook salmon and steelhead had started their fall spawning run but became stranded in stagnant, un-aerated pools of water just below the 205-foot Oxbow Dam. Plans to trap the fish and transport them around the dam were failing. By the end of the season, 10,000 fish had perished before spawning.[1]

Oxbow is situated just below Hell's Canyon—North America's deepest river gorge—which was carved by the Snake River and remains one of the largest wilderness areas in the West. In the 1950s, this gorge contained one of the last free-flowing stretches of the Snake River (Fig. 6.1) and became the focus of a major fight that spanned two decades. Idaho Senator Frank Church played a pivotal role in deciding who would build dams and where they would be built (Ewert, 2001). As a New Deal Democrat, Church had supported development and dam construction that he felt were keys to the growth and prosperity of Idaho. However, the Oxbow Incident had a profound effect on Church. He witnessed the severe effect of dams on fisheries, and even began to ponder the value of riverine corridors to wildlife and their growing value to tourism and recreation.

Frank Church's efforts in the U.S. Senate resulted in passage of the national Wild and Scenic Rivers Act in 1968. While it was not until 1975 that the Hell's Canyon of the Snake River was designated as wild and scenic, two of the eight rivers originally designated as wild and scenic were in Idaho.

Fundamental to the Act was the desire to preserve select rivers with "outstandingly remarkable values" in a "free-flowing condition." The Act defines free-flowing as "any river or section of a river existing or flowing in natural condition without impoundment, diversion, straightening, rip-rapping, or other modification of the waterway."[2] One should note, however, that low dams or other minor structures do not preclude a river from being considered for designation. The "outstandingly remarkable values" encompass a range of scenic, biological, and

Figure 6.1. Photo of Snake River below Hell's Canyon Dam. Photograph courtesy of Marshall McComb, Fox Creek Land Trust.

1 **Barker**, R., 1999: Saving fall Chinook could be costly. The Idaho Statesman, http://www.bluefish.org/saving.htm, accessed on 2-9-2006.

2 Section 16(b) of the Wild and Scenic Rivers Act, 16 U.S.C. 1271-1287 P.L. 90-542.

cultural characteristics that are valued by society. The management goals for Wild and Scenic Rivers (WSRs) center on the preservation and protection of these conditions and values (Box 6.1), including attempting to keep them in a free-flowing condition with high water quality and protected cultural and recreational values.

BOX 6.1. Management Goals for Wild and Scenic Rivers.

1. Preserve "free flowing condition":
   - with natural flow
   - with high water quality
   - without impoundment
2. Protect "outstandingly remarkable values":
   - scenic
   - recreational
   - geologic
   - fish and wildlife
   - historic
   - cultural

There are currently 165 WSRs across the country, representing more than 11,000 stream miles (Fig. 6.2). Oregon ranks highest with 46 designations, most of which were designated in 1988 when a large number of forest management plans were developed to deal with concerns over salmonids. Alaska follows with 25 WSRs that became designated as a result of the Alaska National Interests Land Conservation Act in 1980. This act created nearly 80 million acres of wildlife refuge land in Alaska, much of which is wilderness. Michigan and California are the only other states with a significant number of rivers that have the wild and scenic designation (16 and 13, respectively); however, most states have at least one designated river or river segment. Selected milestones in the evolution of the Wild and Scenic Rivers system are shown in Fig. 6.3.

As severe as the dam effects were on fisheries in Oxbow, Idaho, there is equal or greater concern today about the potential future impacts of climate change on WSRs. Climate

Figure 6.2. Wild and Scenic Rivers in the United States. Data from USGS, National Atlas of the United States.[3] *Note: this map is missing three Wild and Scenic Rivers updated through 2006. The Missouri River in Nebraska, White Clay Creek in Delaware and Pennsylvania, and Wilson Creek in North Carolina will be included in the final version.*

[3] **U.S. Geological Survey**, 2005: Federal land features of the United States - parkways and scenic rivers. *Federal Land Features of the United States*. http://www-atlas.usgs.gov/mld/fedlanl.html. Available from nationalatlas.gov.
U.S. Geological Survey, 2006: Major dams of the United States. *Federal Land Features of the United States*. http://www-atlas.usgs.gov/mld/dams00x.html. Available from nationalatlas.gov.

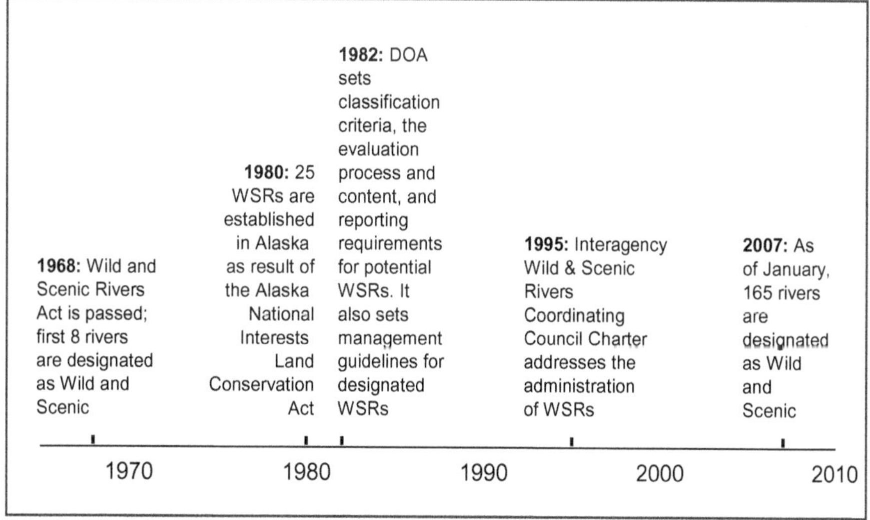

Figure 6.3. Selected milestones in the evolution of the Wild and Scenic Rivers system. Adapted from National Wild and Scenic Rivers System website.[4]

change is expected to alter regional patterns in precipitation and temperature, and this has the potential to change natural flow regimes at regional scales. The ecological consequences of climate change and the required management responses for any given river will depend on how extensively the magnitude, frequency, timing, and duration of key runoff events change relative to the historical pattern of the natural flow regime for that river, and how adaptable the aquatic and riparian species are to different degrees of alteration.

6.3 CURRENT STATUS OF MANAGEMENT SYSTEM

With the exception of the state of Alaska, most WSRs are within watersheds affected by human activities, including development (agricultural, urban, or suburban land use) or dams. In fact, many WSR segments lie downstream of these impacts, meaning their management for scenic or free-flowing condition is difficult. Thus in many ways, WSRs are like rivers all over the United States—they are not fully protected from human impacts. They are distinctive because river-specific outstanding values have been identified and river-administrating agencies have been directed to monitor and protect them as much as possible. More specifically, it is the responsibility of the relevant federal agency—the Forest Service, the National Park Service, the Bureau of Land Management, or the Fish and Wildlife Service—in conjunction with some state and local authorities, to manage them in ways to best protect and enhance the values that led to the designation as wild and scenic. This makes WSRs ideal for implementing and monitoring the results of management strategies to minimize the impacts of climate change—the responsible manager (*e.g.*, the river-administering agency) is specified and the ecosystem values in need of protection have been identified.

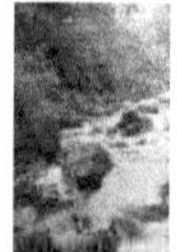

6.3.1 Framework for Assessing Present and Future Status

Climate change is expected to have a significant impact on running waters throughout the world, not only in terms of changes in flow magnitude and timing, but in terms of thermal regimes and the flora and fauna that currently inhabit these waters (Sala *et al.*, 2000). The focus in this chapter is not only on identifying the likely impacts of climate change, but also identifying management options for protecting riverine ecosystems and their values against these impacts. However, rivers across the United

[4] **National Wild and Scenic Rivers System**, 2007: Homepage: National Wild and Scenic Rivers System. National Wild and Scenic Rivers System Website, http://www.rivers.gov, accessed on 5-30-2007.

States have been designated as wild and scenic for diverse reasons, and they exist in diverse settings. Thus climate change is not the only risk they face.

Anticipating the future condition of a river in the face of climate change requires explicit consideration not only of the current climatic, hydrogeologic, and ecological conditions (the *hydrogeomorphic context*), but also of how it is currently managed and how human behavior will affect the river (the *human context*) (Fig. 6.4). Even if impacts are small at present, consideration of the human context is critical to a river's future unless it is within a fully protected basin. If it is not, then impacts associated with activities such as development and water withdrawals are likely to become issues in the future. Stress associated with the *future human context* will interact with climate change, often exacerbating problems and intensifying management challenges (Fig. 6.4)

The ability of a WSR to provide the ecosystem goods and services in the future that originally prompted its designation will largely depend on how it is managed. Without deliberate management actions that anticipate future stress, managers will be left "reacting" to problems (*reactive management)* that come along, and the provision of ecosystem services will not be guaranteed.

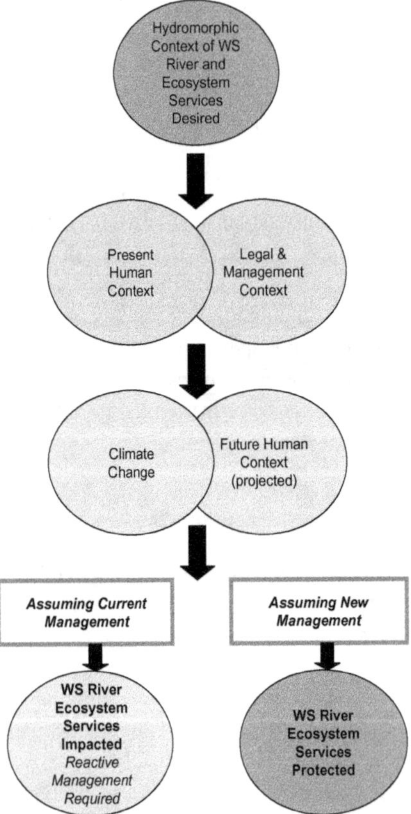

Figure 6.4. Conditions and factors affecting the future conditions of Wild and Scenic Rivers.

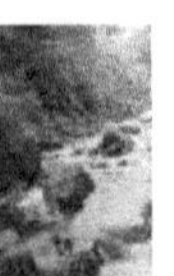

6.3.2 Hydrogeomorphic Context

6.3.2.1 Ecosystem Goods and Services

WSRs provide a special suite of goods and services valued highly by the public (Box 6.2) that are inextricably linked to their flow dynamics and the interaction of flow with the landscape. The ecological processes that support these goods and services are fueled by the movement of water as it crosses riparian corridors, floodplains, and the streambed transporting nutrients, sediment, organic matter, and organisms. Thus, water purification, biological productivity and diversity, as well as temperature and flood control, are all mediated by interactions between the local hydrology and geologic setting. For this reason, the particular goods and services offered by WSRs vary greatly across the nation, reflecting the great variety of landscape settings and climates in which WSRs occur.

The Rogue River in Oregon supports whitewater rafting through dramatic gorges, while the Loxahatchee River in Florida supports highly productive cypress swamp. The goods and services provided by any river depend in no small measure on how "healthy" it is, *i.e.*, the degree to which the fundamental riverine processes that define and maintain the river's normal ecological functioning are working properly. One of the main threats of climate change to WSRs is that it may modify these critical underlying riverine processes and thus diminish the health of the system, with potentially great ecological consequences. Of particular concern is the possibility that climate-induced changes can exacerbate human-caused stresses, such as depletion of water flows, already affecting these rivers. The likelihood of this happening will depend on the current conditions in the river and the extent to which future changes in precipitation and temperature differ from present conditions.

Although every river is arguably unique in terms of the specific values it provides and the

BOX 6.2. Rivers provide a number of goods and services, referred to here as ecosystem functions, that are critical to their health and provide benefits to society. The major functions are outlined below along with the ecological processes that support the function, how it is measured, and why it is important (information synthesized from Palmer et al., 1997; Baron et al., 2002; Naiman, Décamps, and McClain, 2005)

Ecosystem Function	Supporting Ecological Process	Measurements Required	Potential Impacts if Impaired
Water Purification (a) Nutrient Processing	Biological uptake and transformation of nitrogen, phosphorus, and other elements.	Direct measures of rates of transformation of nutrients; for example: microbial denitrification.	Excess nutrients can build up in the water, making it unsuitable for drinking or supporting life.
Water Purification (b) Processing of Contaminants	Biological removal by plants and microbes of materials such as excess sediments, heavy metals, contaminants, etc.	Direct measures of contaminant uptake or changes in contaminant flux.	Toxic contaminants kill biota; excess sediments smother invertebrates, foul the gills of fish, etc; water not potable.
Decomposition of Organic Matter	The biological (mostly by microbes and fungi) degradation of organic matter such as leaf material or organic wastes.	Decomposition is measured as the rate of loss in weight of organic matter over time.	Without this, excess organic material builds up in streams, which can lead to low oxygen and thus death of invertebrates and fish; water may not be drinkable.
Primary Production Secondary Production	Photosynthesis; chemosynthesis; consumption (e.g., herbivory, predation).	For primary production, measure the rate of photosynthesis in the stream; for secondary, measure growth rate of organisms or annual biomass.	Primary production supports the food web; secondary production support fish and wildlife and humans.
Temperature Regulation	Infiltration and vegetative shading: temperature is "buffered" if there is sufficient infiltration in the watershed & riparian zone AND shading by riparian vegetation.	Measure the rate of temperature change in the water and air (in riparian zone immediately next to river) as surrounding air temperature changes or as increases in discharge occur.	If infiltration or shading are reduced (due to clearing of vegetation along stream), stream water and riparian air heat up beyond what biota are capable of tolerating.
Flood Control	Slowing of water flow from the land to streams or rivers so that flood frequency and magnitude are reduced; intact floodplains and riparian vegetation help buffer increases in discharge.	Measure the rate of infiltration of water into soils OR discharge in stream in response to rain events.	Without the benefits of floodplains, healthy stream corridor, and watershed vegetation, floods become more frequent and higher in magnitude.
Biodiversity Maintenance	Maintenance of intact food web and genetic resources that together provide other ecosystem goods. Local genetic adaptation contributes to landscape-scale resilience of river ecosystems.	Enumeration of genotypes, species, or species guilds.	Impoverishment of genetic diversity at broader spatial scales. Reduced capacity for resilience and sustainability of many ecosystem goods and services.

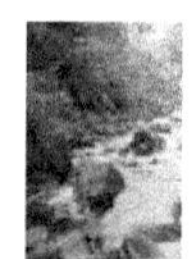

wildlife it supports, an important scientific perspective is to identify the general underlying processes that dictate how a river functions, so that researchers may consider the vulnerabilities of these systems to climate change. This report uses the phrase "hydrogeomorphic context" to mean the combination of fundamental riverine processes that interact with the particular landscape setting of a river to define its fundamental character and potential for ecological resilience in the face of natural variation and future climate change.

From a physical perspective, rivers function to move water and sediment off the landscape and downhill toward the sea. The regime of rainfall and the geology of a river's watershed control landscape soil erosion rates and influence how fast precipitation falling on a watershed is moved to the river channel, as well as the likelihood that the channel will develop an active floodplain (Knighton, 1998). Thus, a river's hydrogeomorphic context is largely defined by the nature of the flow regime and the river's channel features. For example, rivers flowing through steep mountains with bedrock canyons and boulder-strewn beds, such as Colorado's Cache la Poudre River, represent very different environments than rivers flowing slowly across flat land where channels can be wide and meandering due to sandy banks, such as Mississippi's Black Creek. Likewise, rivers draining watersheds with porous soils and high groundwater levels respond very sluggishly to rainfall storm events, compared with those that drain impervious soils and show a rapid flood response to heavy rains (Paul and Meyer, 2001). Such differences exert strong control over the temporal dynamics of critical low and high flow events and thus directly influence many ecological processes and populations of aquatic and riparian species (Poff *et al.*, 1997; Bunn and Arthington, 2002).

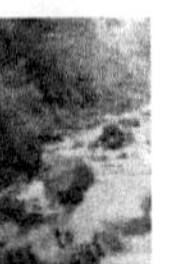

But the hydrogeomorphic context can also be extended beyond precipitation and geology. Specifically, the thermal regime of a river is also a critical component of its fundamental nature, because water temperature directly controls animal and plant metabolism and thus influences the kinds of species that can flourish in a particular environment and the rates of biogeochemical processes within the river ecosystem (Ward, 1992; Allan, 1995). This thermal response explains the categorization of fishes as being either cold-water species (*e.g.*, trout, salmon) or warm-water species (*e.g.*, largemouth bass) (Eaton and Scheller, 1996; Beitinger, Bennett, and McCauley, 2000). Regional climate largely determines air temperature, and hence water temperature (Nelson and Palmer, 2007), and this factor also influences whether precipitation falls as rain or snow. When it falls as snow, regional climate also influences the time and rate of melt to provide the receiving river with a prolonged pulse of runoff.

At a broad, national scale, it is important to appreciate the differences in hydrogeomorphic context of WSRs. Not only do these differences influence the kind and quality of human interactions with WSRs, they also serve to generate and maintain ecological variation. For example, the cold and steep mountain rivers of the West, such as Montana's Flathead River, support different species of fish and wildlife than the warmer rivers in the South, such as the Lumber River in the south-central coastal plains of North Carolina. Aquatic and riparian species are adapted to these local and regional differences (Lytle and Poff, 2004; Naiman, Décamps, and McClain, 2005), thereby generating great biodiversity across the full range of river types across the United States. The wide geographic distribution of WSRs is important not only in ensuring large-scale biodiversity, but also the concomitant ecosystem processes associated with different river systems. This is particularly true for "wild" rivers, *i.e.*, those that are not dammed or heavily modified by human activities and that are protected over the long term due to their WSR status. Thus, wild rivers across the United States can serve as a valuable natural repository of the nation's biological heritage (*e.g.*, Poff *et al.*, 2007; Moyle and Mount, 2007), and the threats of climate change to this ecological potential is of great national concern.

6.3.2.2 What it Means to be Wild

WSRs include headwaters with undisturbed watersheds as well as river segments that have only modest watershed impacts. The term "wild river" in its strictest sense would include a river with no human impacts in its entire watershed. One of the key features of these truly wild rivers is their natural flow regime; *i.e.*, the day-to-day and year-to-year variation in the amount of water flowing through the channel. Research over the last 10 years has clearly demonstrated that human modification of the natural flow regime of streams and rivers degrades the ecological integrity and health of streams and rivers in the United States and around the world (Poff *et al.*, 1997; Richter *et al.*, 1997; Bunn and Arthington, 2002; Postel and Richter, 2003; Poff *et al.*, 2007).

From an ecological perspective, some of the key features of a natural flow regime are the occurrence of high flood flows and natural drought flows. These flows act as natural disturbances that exert strong forces of natural selection on species, which have adapted to

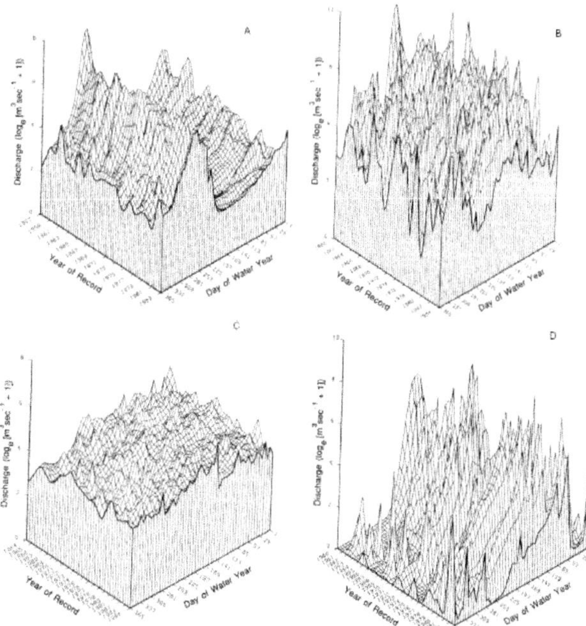

Figure 6.5. Illustration of natural flow regimes from four unregulated streams in the United States: (A) the upper Colorado River (CO), (B) Satilla Creek (GA), (C) Augusta Creek (MI), and (D) Sycamore Creek (AZ). For each the year of record is given on the x-axis, the day of the water year (October 1–September 30) on the y-axis, and the 24-hour average daily streamflow on the z-axis (Poff and Ward, 1990).

these critical events over time (Lytle and Poff, 2004). But it's not just the magnitude of these critical flows that is ecologically important; it's also their frequency, duration, timing, seasonal predictability, and year-to-year variation (Poff *et al.*, 1997; Richter *et al.*, 1997; Lytle and Poff, 2004), because various combinations of these features can dictate the success or failure of aquatic and riparian species in riverine ecosystems. Thus, for example, a river that has frequent high flows that occur unpredictably at any time of the year provides a very different natural environment than one that typically has only one high flow event predictably year-in and year-out.

Across the United States there are large differences in climate and geology, and thus there is a geographic pattern to the kinds of natural flow regimes across the nation. This is illustrated in Fig. 6.5 from Poff and Ward (1990). For example, in the Rocky Mountain states and in the northern tier of states, most annual precipitation falls in the winter in the form of snow, which is stored on the land until the spring, when it melts and enters the rivers as an annual pulse (Fig. 6.5a). In more southerly regions where there is frequent rainfall, floods can occur unpredictably and flow regimes are much more variable over days to weeks (Fig. 6.5b). In watersheds with highly permeable soils, such as those in Michigan, falling rain infiltrates into the ground and is delivered slowly to the stream as groundwater (Fig. 6.5c). The frequency of floods and river low flows depends on precipitation patterns and specific hydrologic conditions within a given watershed. Yet other streams may be seasonally predictable but present harsh environments because they cease to flow in some seasons (Fig. 6.5d).

These different flow regime types result in very different hydrogeomorphic contexts, which in turn support very different ecological communities. For example, Montana's Upper Missouri River supports extensive stands of native cottonwood trees along the riverbanks. These trees become established during annual peak flows that jump the banks and create favorable establishment conditions during the annual snowmelt runoff event. Arkansas' Buffalo River is nestled in the Ozark Mountains and supports a tremendous diversity of fish and other aquatic life such as native mussels,

CASE STUDY SUMMARY 6.1

Wekiva River Basin, Florida
Southeast United States

Why this case study was chosen

The Wekiva River Basin:
- Is a spring-fed system that requires management of surface and sub-surface water resources;
- Is a sub-tropical, coastal ecosystem and thus faces potential impacts from tropical storms and sea level rise;
- Is dealing directly with large and expanding urban and suburban populations, and associated water and land use changes.

Management context

The Wekiva River basin is a complex system of streams, springs, lakes, and swamps that are generally in superb ecological condition and harbor an impressive list of endangered species, including the West Indian Manatee and endemic invertebrates. The springs that feed the river are affected by pumping of groundwater and by proximity to the expanding population of Orlando. Agricultural and urban expansion is affecting groundwater and surface water systems critical to the ecological balance of the WSR. Other management issues include urban and agricultural pollution, and invasive exotic species. The National Park Service has overall coordinating responsibility for the Wekiva WSR, while land, water, and natural resources management in the basin is provided through cooperation among state agencies, local governments, and private landowners. Even without climate change considerations, the basin is expected to reach maximum sustained yields of water use by 2013. Agencies in the basin are monitoring water quantity and quality, ecosystem health, and native and invasive species populations, and are taking an increasingly proactive approach to water management.

Key climate change impacts

- Projected increase in average temperatures (2.2–2.8°C in Central Florida by 2100);
- Projected increase in the frequency of tropical storms and hurricanes;
- Projected sea level rise of 0.18–0.59 m by 2099;
- Projected decline of water availability due to increased evaporation and transpiration.

Opportunities for adaptation

- Monitoring programs could support more robust modeling to project management needs in a climate change scenario, including how rising sea level might affect saltwater intrusion into the groundwater.
- The possible shift to longer droughts, punctuated by more intense rain events, could be addressed through aggressive practices to maintain water quality and availability, e.g., by maximizing recharge of the aquifer during rain events and minimizing withdrawals during droughts through water conservation programs;
- Additional measures could be pursued to reduce pollution of surface and groundwater reaching the Wekiva River; management changes should be informed by more research into how pollutants in reclaimed water are transported through the porous karst geology to the aquifer and springs.
- There is considerable public interest in the importance of water; therefore, management programs have the opportunity to provide education to the public and other stakeholder groups on conserving water and reducing pollution, including limiting runoff of nitrate-based fertilizers and encouraging the use of central sewage treatment facilities instead of septic tanks.

Conclusions

The preservation of ecological conditions in the Wekiva WSR will require integrated management of the complex interactions between surface and ground water in the watershed. Expanded water monitoring and advanced modeling programs will be keys to maintaining water quantity and quality in the Floridian aquifer, and for regulating runoff to maximize reuse for urban and rural uses while ensuring optimal water reaching the river.

as well as diverse riparian tree species. This near-pristine river is seasonally very dynamic, due to the steep mountain topography and rapid runoff from frequent rainfall events. Florida's Wekiva River is a flatwater system that is heavily influenced by groundwater and streamside wetlands that store and release water to the river over the year (see Case Study Summary 6.1). This creates a highly stable flow regime and stable wetland complexes that support a great diversity of plant species and community types.

These natural flow regime types occur across the nation and reflect the interaction of precipitation, temperature, soils, geology, and land cover. For every region of the country there can be a natural flow regime representative of the unaltered landscape; *i.e.*, with native vegetation and minimally altered by human activities such as point- or non-point source pollution (Poff *et al.*, 2006).

6.3.3 Present Human Context

To the American public, the designation of a river as "Wild and Scenic" conjures an image of a river protected in pristine condition, largely unchanged by human development. However, as mentioned above, in reality many of the rivers in the WSR system have experienced some ecological degradation from a variety of human activities.

Due to their vulnerable position as the lowermost features of landscapes, rivers are the recipients of myriad pollutants that flush from the land, the bearers of sediment loads washed from disturbed areas of their watersheds, and the accumulators of changes in the hydrologic cycle that modify the volume and timing of surface runoff and groundwater discharge. As Aldo Leopold once said, "It is now generally understood that when soil loses fertility, or washes away faster than it forms, and when water systems exhibit abnormal floods and shortages, the land is sick" (Leopold, 1978). Because rivers are integrators of changes in a watershed, they are also often indicators of ecological degradation beyond their banks.

WSR managers have limited authority or control over human activities occurring outside of federally owned WSR corridors. The vulnerability of rivers generally increases in relation to the area of contributing watershed in nonfederal control; the protection of these areas depends on coordinated management with local landowners and governments. In general, designated headwater reaches are considerably less vulnerable to human impacts than reaches situated downstream of cities and agricultural areas. This reality makes the Middle Fork of the Salmon River in Idaho, a headwater river embedded in a federal wilderness area, far less susceptible to human influences than the Rio Grande in Texas (see Case Study Summary 6.2). Protection of headwaters is especially important, since they support critical (keystone) ecosystem processes and often support sensitive species.

To prepare a foundation for understanding the potential consequences of climate change, this report summarizes current influences and historic trends in water use and dam operations that affect the ecological condition of WSRs.

6.3.3.1 Water Use

Excessive withdrawals of water from rivers can cause great ecological harm. The nature and extent of this ecological damage will depend upon the manner in which water is being withdrawn. The hydrologic and ecological effects of surface water withdrawals may differ considerably from the impact of the same amount of water being withdrawn through groundwater extraction. When on-channel reservoirs are used to store water for later use, the placement and operation of dams can have considerably greater ecological impact than direct withdrawal of water using surface water intakes, as discussed below.

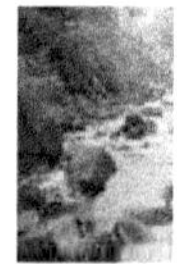

The depletion of river flows fundamentally alters aquatic habitats because it reduces the quantity of habitat available (Poff *et al.*, 1997; Richter *et al.*, 1997; Bunn and Arthington, 2002). Adequate water flows can also be important in maintaining proper water temperature and chemistry, particularly during low-flow periods. The depth of water can strongly influence the mobility of aquatic animals such as fish, and river levels can also influence water table levels in adjacent riparian areas, particularly in rivers with high degrees of hydraulic connectivity between the rivers and alluvial floodplain aquifers.

During the latter half of the 20th century, water withdrawals in the United States more than

CASE STUDY SUMMARY 6.2

Rio Grande River
Southeast United States

Why this case study was chosen

The Rio Grande River:

- Is the second largest river in the Southwest, and provides an important water resource for hydropower and agricultural and municipal needs in the United States and Mexico;
- Exemplifies the complex domestic and international water rights issues typical of the American West;
- Is an example of a WSR managed by federal agencies, as is typical for many WSR in the West;
- Provides so much water to diversions and extraction in Colorado and New Mexico that the riverbed is dry for about 80 miles south of El Paso, Texas, resulting in two distinct hydrologic systems: the northern segment of the WSR is strongly influenced by spring snowmelt, while the segment forming the border between Texas and Mexico receives most of its water from summer rains in Mexico.

Management context

Management responsibilities for the Rio Grande WSR corridor rest with the Bureau of Land Management, the Forest Service, the National Park Service, and state and local agencies, while water in the river basin is largely controlled through complex water rights agreements and international treaties. Ecological management goals in the upper and lower WSR address similar priorities: preserving the natural flow regime, maintaining and improving water quality, conserving plant and animal species, and addressing invasive species. Impoundments and water extractions have reduced stream flow by over 50%, and invasive species have significantly altered ecosystems, particularly in the lower segment of the WSR. Water rights were established before the river was designated as a WSR, so they have priority over management goals of the WSR. Extraction of groundwater exceeds recharge in parts of the basin, and existing international agreements to provide the river with water have not been met in recent drought years, leaving the river as a series of pools in segments of the WSR along the border with Mexico.

Key climate change impacts

- Projected increase in average temperatures;
- Projected reductions in snowpack and earlier spring melts;
- Projected 5% decrease in annual precipitation by 2010, leading to recurring droughts;
- Projected increases in population and development, leading to greater water demands;
- Projected decline in water availability due to increased evaporation and runoff;
- Projected increase in invasive species due to warming of water and irregularity of the flow regime.

Opportunities for adaptation

- Scenario-based forecasting could be used by water managers to better anticipate trends and address their ramifications.
- Management of water releases, diversions, and extractions could be adapted to store water from early snowmelt and summer rains, and release water to the river to mimic the natural flow regime.
- Economic incentives can bring flexibility to water rights, including purchasing or leasing of water rights for the river and incentives that promote water efficiency and reduce pollution.
- Improving efficiency of agricultural and urban water use through conservation and reuse of water could reduce demand and improve water quality.

Conclusions

Meeting the management goals for the Rio Grande WSR is challenging even today, and will be more so as historic problems of water availability and international water rights are complicated by climate change. Even so, the WSR may be maintained through improved water use forecasting, water conservation, and reduced water demand, combined with economic incentives to ensure that enough water is provided to the WSR on a schedule that mimics the natural flow regime.

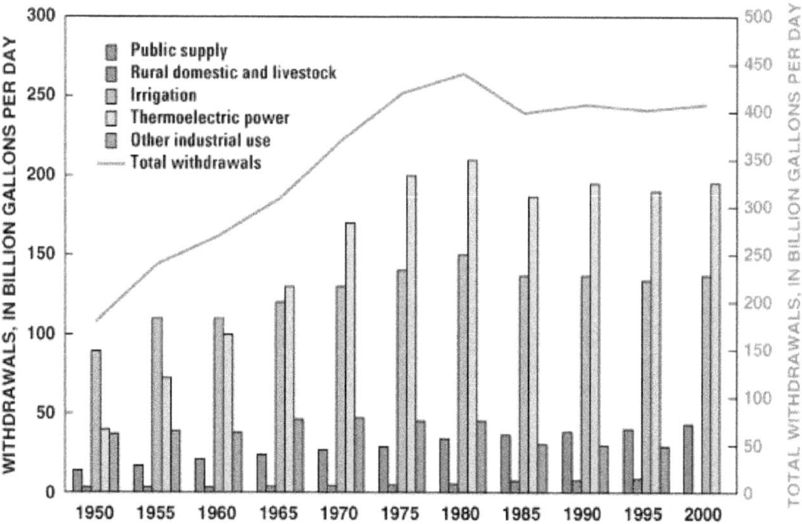

Figure 6.6. Trends in water withdrawals by water-use category. As the population has grown, water has been increasingly withdrawn for public use since 1950 as indicated by total withdrawals (blue line). Water withdrawn for power production and water for irrigation represent the largest use, followed by water for industrial uses, then public supply.[5]

doubled (Fig 6.6).[5] Virtually all of this increase occurred during 1950–1980, and withdrawals leveled off in 1980–2000 even while the U.S. population grew by 24%. This flattening of water withdrawals resulted primarily from lessened demand for thermoelectric power and irrigation. Thermoelectric-power water withdrawals primarily were affected by federal legislation that required stricter water quality standards for return flow, and by limited water supplies in some areas of the United States.[5] Consequently, since the 1970s, power plants increasingly were built with or converted to closed-loop cooling systems or air-cooled systems, instead of using once-through cooling systems. Declines in irrigation withdrawals are due to changes in climate, shifts in crop types, advances in irrigation efficiency, and higher energy costs that have made it more expensive to pump water from ground- and surface-water sources.

An important exception to the recent nationwide declines in total water withdrawals has been a continuous increase in public water supply withdrawals (withdrawals for urban use) during the past 50+ years; withdrawals for public water supplies more than tripled during 1950–2000 (Fig 6.6).[5] These rises in urban water demand have been driven by overall population growth as well as the higher rate of urban population growth relative to rural population growth. Fifty U.S. cities with populations greater than 100,000 experienced growth rates of at least 25% during recent decades.[6]

Water withdrawals for urban and agricultural water supplies are having substantial impacts on the natural flow regimes of rivers across the United States, including WSRs. For example, upstream withdrawals for New York City's water supply have depleted average annual flows in the Upper Delaware Scenic and Recreational River by 20%, with flows in some months lowered by as much as 40% (Fig. 6.7 and Case Study Summary 6.3) (Fitzhugh and Richter, 2004). Heavy agricultural and municipal withdrawals along the Rio Grande in Colorado, New Mexico, Texas, and Mexico have increasingly depleted river flows during the past century (Collier, Webb, and Schmidt, 1996).

[5] **Hutson**, S.S., N.L. Barber, J.F. Kenny, K.S. Linsey, D.S. Lumia, and M.A. Maupin, 2004: Estimated use of water in the United States in 2000. *U. S. Geological Survey Circular 1268*. http://water.usgs.gov/pubs/circ/2004/circ1268/.

[6] **Gibson**, C., 1998: *Population of the 100 Largest Cities and Other Urban Places in the United States: 1790–1990*. Population Division, U.S. Bureau of the Census, Washington, DC.

CASE STUDY SUMMARY 6.3

Upper Delaware River, New York, and Pennsylvania
Northeast United States

Why this case study was chosen

The Upper Delaware River:

- Has recently been affected by unusually frequent and severe flooding, including three separate hundred-year flood events in less than two years;
- Serves as the major water source to New York City and surrounding areas;
- Exemplifies a largely natural river on the Atlantic coast;
- Represents a WSR "Partnership River," with little public ownership of the WSR corridor.

Management context

Predominately private ownership of the WSR corridor requires that the National Park Service, along with local and state government agencies, work with private interests to develop and implement the river management plan. The goals of the plan include maintaining and improving water quality and ecosystems, providing opportunities for recreation, and maintaining scenic and historic values of the river. The rights of private landowners are especially emphasized in the management plan. In addition to providing water to New York City (the city takes about 50% of the available water) and flood control, the reservoirs in the upper tributaries strategically release water downstream to the keep the salt front in the tidal zone from reaching upstream infrastructure that would be damaged by the salt water. The timing and quantity of these water releases do not match natural flow regimes of the river, and occasional low water levels tend to concentrate pollutants and increase water temperature in some river segments. Water conservation in the Delaware Basin and New York City has significantly helped address drought-related water shortages.

Key climate change impacts

- Observed and projected increase in mean temperature and annual precipitation, changes in amount and timing of precipitation;
- Observed and projected increase in severe flood events;
- Projected decrease in snowpack and earlier spring melts;
- Projected periodic droughts;
- Projected rise in sea level that will push the salt front further upstream.

Opportunities for adaptation

- Modeling tools can be used to project climate change impacts on the water system, and to determine the reservoir levels and water releases that can best establish an optimal water flow regime and offset river water warming in the WSR.
- Incentives and ordinances could be used to improve water quality by reducing agricultural pollutants reaching the river, reducing storm water runoff, and improving flood and erosion control through restoration of wetlands and riparian buffers.
- Support for water-efficient measures could further improve efficiency of water use in New York City and throughout the basin, thereby reducing per-capita demand for household water.
- Reservoir management could be adapted to store water from early snowmelt and release water to the river, in order to mimic the natural flow regime.

Conclusions

The Upper Delaware River currently has good water quality and provides natural and scenic resources for residents of nearby urban areas. However, recent acute climatic events and projected climate change strongly suggest that new management programs must be considered by the Delaware River Basin Commission, local communities, and private interests that manage land and water resources in the basin and Upper Delaware WSR corridor. Reservoir and landscape management to reduce impacts of floods, to manage flow regime and water temperature, and to expand water conservation programs will become increasingly important as the population continues to grow and impacts of climate change increase.

While national trends in water use provide insight into large-scale factors influencing river flows in WSRs, the impact of water withdrawals on hydrologic systems varies greatly across the United States. Ultimately, the consequences of water withdrawals on a specific WSR can best be understood by developing hydrologic simulation models for the local region of interest, or by examining changes or trends in river flows such as those presented in Fig. 6.7.

6.3.3.2 Dam Operations

Nearly 80,000 dams are listed in the National Inventory of Dams for the United States.[7] Approximately one-third of these dams are publicly owned, with ownership divided among federal, state, local, and public utility entities. An estimated 272 of these dams are located within 100 miles upstream or downstream of WSRs (Fig. 6.8).

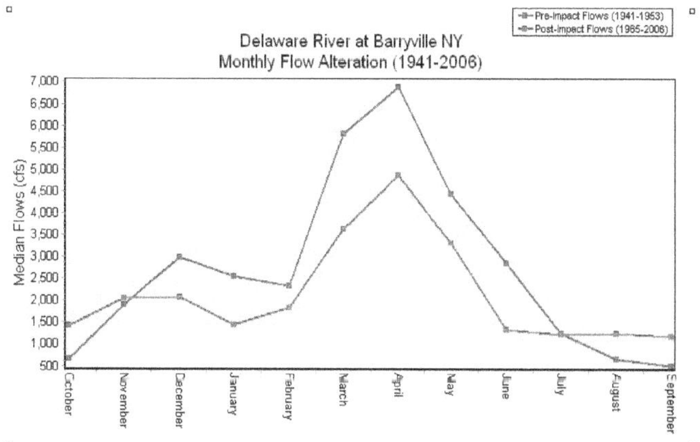

Figure 6.7. Changes in monthly average river flows on the Delaware River, in the Upper Delaware Scenic and Recreational River segment. Lowered flows in December–July result from upstream depletions for New York City water supply. Increased flows result from upstream reservoir releases during summer months for the purpose of controlling salinity levels in the lower Delaware. Figure based on data provided by USGS.[8]

[7] **U.S. Army Corps of Engineers**, 2000: National inventory of dams. http://crunch.tec.army.mil/nid/webpages/nid.cfm, Federal Emergency Management Agency. CD-ROM.

[8] **U.S. Geological Survey**, 2007: USGS surface water data for the nation. USGS Website, http://waterdata.usgs.gov/nwis/sw, accessed on 7-26-2007.

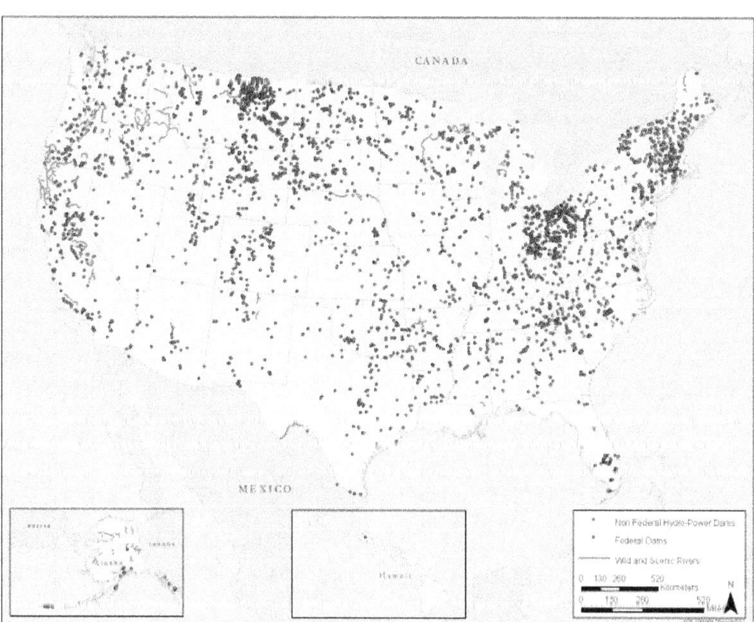

Figure 6.8. Location of dams and WSRs in the United States. Data from USGS, National Atlas of the United States.[3] *Note: map is missing three Wild and Scenic Rivers updated through 2006. The Missouri River in Nebraska, White Clay Creek in Delaware and Pennsylvania, and Wilson Creek in North Carolina will be included in the final version.*

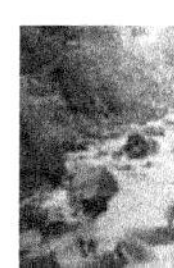

Most dams provide substantial benefits to local or regional economies (World Commission on Dams, 2000). Hydroelectric power dams currently provide 7% of the U.S. electricity supply. By capturing and storing river flows for later use, dams and reservoirs have contributed to the national supply of water for urban, industrial, and agricultural uses. Storage of water in reservoirs helped to meet the steep growth in water use in the United States during the 20th century, particularly for agricultural water supply. Nearly 9,000 (12%) of the U.S. dams were built solely or primarily for irrigation.

However, damming of the country's rivers has come at great cost to their ecological health and ecosystem services valued by society (Ligon, Dietrich, and Trush, 1995; World Commission on Dams, 2000; Postel and Richter, 2003; Poff *et al.*, 2007). The most obvious change in river character results from the conversion of a flowing river into an impounded reservoir. Also obvious is the fact that dams create barriers for upstream-downstream movements of mobile aquatic species such as fish. A dam can artificially divide or isolate species populations, and prevent some species from completing anadromous or diadromous life cycles, such as by blocking access to upriver spawning areas (Silk and Ciruna, 2005). For example, Pacific salmon migrations through WSR segments on the Salmon and Snake rivers in Idaho and pallid sturgeon migrations on the Missouri River are impeded by dams. The consequences of such population fragmentation have been documented for many fish species, including many local extirpations following damming. Hence, dams located downstream of WSRs likely have consequences for movements of aquatic animals, particularly widely ranging fish.

Dams have considerable influence on downstream river ecosystems as well, in some cases extending for hundreds of miles below a dam (Collier, Webb, and Schmidt, 1996; McCully, 1996; Willis and Griggs, 2003). Dam-induced changes affect water temperature (Clarkson and Childs, 2000; Todd *et al.*, 2005) and chemistry (Ahearn, Sheibley, and Dahlgren, 2005); sediment transport (Williams and Wolman, 1984; Vörösmarty *et al.*, 2003); floodplain vegetation communities (Shafroth, Stromberg, and Patten, 2002; Tockner and Stanford, 2002; Magilligan, Nislow, and Graber, 2003). Dams may even affect downstream estuaries, deltas, and coastal zones by modifying salinity patterns, nutrient delivery, disturbance regimes, and the transport of sediment that builds deltas, beaches, and sandbars.[9] Of all the environmental changes wrought by dam construction and operation, the alteration of natural water flow regimes (Fig. 6.5) has had the most pervasive and damaging effects on river ecosystems (Poff *et al.*, 1997; Postel and Richter, 2003). Dams can heavily modify the magnitude (amount) of water flowing downstream, change the timing, frequency, and duration of high and low flows, and alter the natural rates at which rivers rise and fall during runoff events.

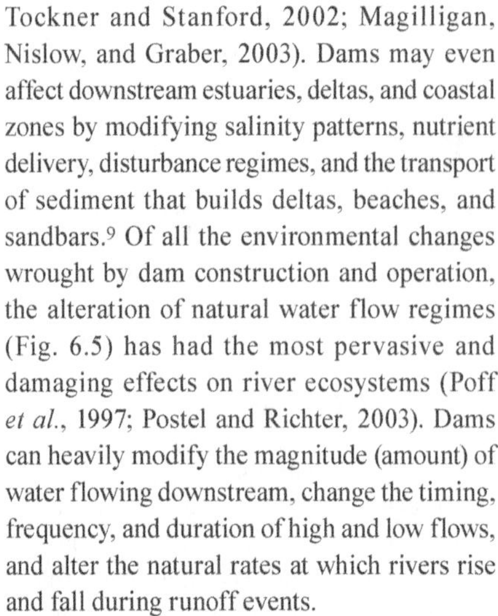

The location of a WSR relative to upstream dams can have great influence on the ecological health of the WSR. As a general rule, ecological conditions improve with distance downstream of dams due to the influence of tributaries, which moderate dam-induced changes in water flow, sediment transport, water temperature, and chemistry. For example, flow alterations associated with hydropower dams in the Skagit River are most pronounced immediately downstream of the dams, but lessen considerably by the time the river reaches its estuary. It is quite difficult to assess the dam-induced biophysical changes that have transpired in WSRs, because long-term measurements of sediment, temperature, water quality, and biological conditions are rarely available. However, for many rivers, dam-related changes to hydrologic regimes can be evaluated by examining streamflow changes before and after dams were built (see Fig. 6.7 for example).

6.3.3.3 Land-Use Changes

As humans have transformed natural landscapes into cities and farms, and increasingly utilized resources such as timber and metals, the consequences to river ecosystems have been quite severe. Beyond the impacts on water quantity and timing of river flows discussed above, landscape conversion has had substantial influence on water quality (Silk

[9] **Olsen**, S.B., T.V. Padma, and B.D. Richter, Undated: Managing freshwater inflows to estuaries: a methods guide. U.S. Agency for International Development, Washington, DC.

and Ciruna, 2005).[10] The potential impact of land use on WSRs depends upon a number of factors, including proximity of the WSR to various land uses and the proportion of the contributing watershed that has been converted to high-intensity uses such as agriculture or urbanization.

Nearly half of the billion hectares of land in the United States has been cultivated for crops or grazed by livestock. As described above, agriculture accounts for approximately 70% of water withdrawals in the United States. While most of this water is consumed through evapotranspiration, the portion of irrigation water that returns to streams and rivers is commonly tainted with chemicals or laden with sediment (National Research Council, 1993).[11] Because much of the land converted to agricultural use in recent decades has been wetlands and riparian areas, this conversion has severely affected the natural abilities of landscapes to absorb and filter water flows. Major pollutants in freshwater ecosystems include excessive sediment, fertilizers, herbicides, and pesticides (Silk and Ciruna, 2005). Agriculture is the source of 60% of all pollution in U.S. lakes and rivers; nitrogen is the leading pollution problem for lakes and the third most important pollution source for rivers in the United States (U.S. Environmental Protection Agency, 2000). The U.S. Geological Survey National Water Quality Assessment (NAWQA) found that most of the rivers sampled in agricultural areas contained at least five different pesticides,[11] including DDT, dieldrin, and chlordane. Intensive agriculture often leads to the eutrophication of freshwater ecosystems, resulting in deoxygenation of water, production of toxins, and a general decline in freshwater biodiversity. Agriculture is a major source of sedimentation problems as well, resulting from large-scale mechanical cultivation, channelization of streams, riparian clearing, and accentuated flood runoff.

After agriculture, the next three top sources of river ecosystem degradation include hydromodification, urban runoff/storm sewers, and municipal point sources—all associated with urban environments (Silk and Ciruna, 2005). Although urban areas occupy only a small fraction of the U.S. land base, the intensity of their impacts on local rivers can exceed that of agriculture (see Fig. 6.9 for an example). More than 85% of the U.S. population lives in cities, potentially concentrating the impacts from urban activities and exacerbating conditions affected by rainfall runoff events, such as water use, wastewater discharge, polluted surface runoff, and impervious surfaces. Industrial activities located in cities pose several threats to river ecosystems, including effluent discharge and risk of chemical spills, in addition to water withdrawals. The NAWQA program reports the highest levels of phosphorus in urban rivers. Other highly problematic forms of pollution in urban areas include heavy metals, hormones and pharmaceutical chemicals, and synthetic organic chemicals from household uses.[11] Excellent reviews on the effects of urbanization on streams have been published (Paul and Meyer, 2001; Walsh *et al.*, 2005), but in brief the most obvious impacts are increases in

Figure 6.9. Photo of scientists standing on the bed of an urban stream whose channel has been incised more than 5 m due to inadequate storm water control. Incision occurred on the time scale of a decade, but the bank sediments exposed near the bed are marine deposits laid down during the Miocene epoch. Photograph courtesy of Margaret Palmer.

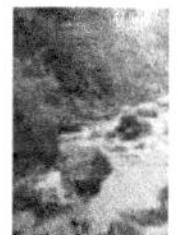

[10] See also **U.S. Geological Survey**, 2006: *Rates, Trends, Causes and Consequences of Urban Land-Use Change in the United States*. USGS Professional Paper 1726.

[11] See also **U.S. Geological Survey**, 2001: Hydrological simulation program—Fortran. http://water.usgs.gov/software/hspf.html. U.S. Geological Survey, Reston, VA.

impervious surface area resulting in increased runoff, higher peak discharges, higher sediment loads, and reduced invertebrate and fish biodiversity (Dunne and Leopold, 1978; Arnold, Jr. and Gibbons, 1986; McMahon and Cuffney, 2000; Walsh, Fletcher, and Ladson, 2005).

6.3.4 The Policy Context: Present Management Framework Legal and Management Context

The creation of the National System of Wild and Scenic Rivers (the WSR System) under the Wild and Scenic Rivers Act of 1968 (Box 6.3) was an attempt by the U.S. Congress to proactively rebalance the nation's river management toward greater protection of its river assets. Every river or river segment included within the WSR System must be managed according to goals associated with preserving and protecting the values for which the river was designated for inclusion in the system (see Box 6.1). The degree of protection and enhancement afforded each river or river segment is a prerogative of the agency responsible for a particular river's management, but the values that made the river suitable for inclusion in the WSR System must be protected. (Throughout the rest of this chapter, the term "river," in the context of a WSR, refers to the segment of river designated under the Act.)

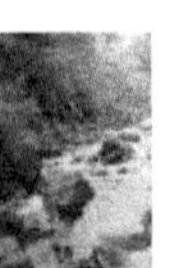

When a river is admitted into the WSR System, it is designated under one of three categories: "wild," "scenic," or "recreational." These categories are defined largely by the intensity

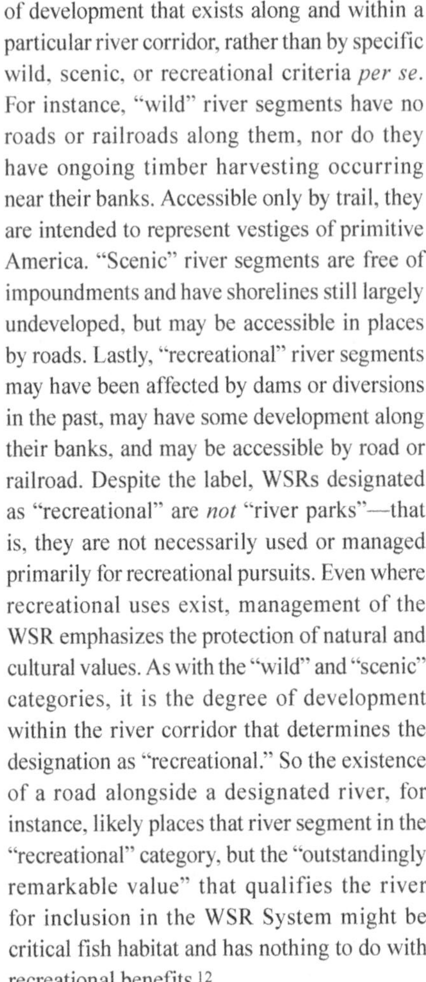

of development that exists along and within a particular river corridor, rather than by specific wild, scenic, or recreational criteria *per se*. For instance, "wild" river segments have no roads or railroads along them, nor do they have ongoing timber harvesting occurring near their banks. Accessible only by trail, they are intended to represent vestiges of primitive America. "Scenic" river segments are free of impoundments and have shorelines still largely undeveloped, but may be accessible in places by roads. Lastly, "recreational" river segments may have been affected by dams or diversions in the past, may have some development along their banks, and may be accessible by road or railroad. Despite the label, WSRs designated as "recreational" are *not* "river parks"—that is, they are not necessarily used or managed primarily for recreational pursuits. Even where recreational uses exist, management of the WSR emphasizes the protection of natural and cultural values. As with the "wild" and "scenic" categories, it is the degree of development within the river corridor that determines the designation as "recreational." So the existence of a road alongside a designated river, for instance, likely places that river segment in the "recreational" category, but the "outstandingly remarkable value" that qualifies the river for inclusion in the WSR System might be critical fish habitat and has nothing to do with recreational benefits.[12]

Regardless of how a WSR is classified—wild, scenic, or recreational—administering agencies must seek to protect existing river-related values and, to the greatest extent possible, enhance those values. Once placed under one of the three classifications, the river must be managed to maintain the standards of that classification. A river classified as wild, for instance, cannot be permitted to drop to the less-strict criteria of scenic. A non-degradation principle therefore guides river management. So, for example while many WSRs had dams in place prior to the river segment being designated as wild and scenic (Fig. 6.8), the Wild and Scenic Rivers Act charges the administering agency with reviewing any new federally assisted water

BOX 6.3. Wild and Scenic Rivers Act of 1968.

It is hereby declared to be the policy of the United States that certain selected rivers of the Nation which, with their immediate environments, possess outstandingly remarkable scenic, recreational, geologic, fish and wildlife, historic, cultural, or other similar values, shall be preserved in free-flowing condition, and that they and their immediate environments shall be protected for the benefit and enjoyment of present and future generations. The Congress declares that the established national policy of dam and other construction at appropriate sections of the rivers of the United States needs to be complemented by a policy that would preserve other selected rivers or sections thereof in their free-flowing condition to protect the water quality of such rivers and to fulfill other vital conservation purposes.

12 **Interagency Wild and Scenic Rivers Coordinating Council**, 2002: *Wild & Scenic River Management Responsibilities*. National Wild and Scenic Rivers System.

resource projects (such as dams) to ensure they will not degrade river values.

6.3.4.1 Administering Agencies and Authorities

The management of WSRs is complex due to the overlapping and at times conflicting federal and state authorities that are responsible for managing these rivers, as well as to the mix of public and private ownership of lands within or adjacent to WSR corridors. The four federal agencies administering WSRs are the Bureau of Land Management (BLM), the National Park Service (NPS), the U.S. Forest Service (USFS), and the U.S. Fish and Wildlife Service (USFWS) (Fig. 6.10). WSRs administered by the NPS and the USFWS are managed as part of the National Park System or the National Wildlife Refuge System, respectively. If a conflict arises between laws and regulations governing national parks or refuges and the WSR Act, the stricter of them—that is, the laws and regulations affording the greatest protection to the river applies.

In addition to ensuring that the management of lands within the river corridor sufficiently protects WSR values, the administering agency must work to ensure that activities on lands adjacent to the river corridor do not degrade WSR values. Other (non-administering) federal agencies must also protect WSR values when exercising their oversight of activities within and adjacent to a WSR corridor. For rivers designated by states and added to the WSR System under Section 2 (a)(ii) of the Act, authorized state agencies have primary responsibility for river management. In all cases, a partnership among federal, state, and local entities is encouraged.

A number of environmental laws that are applicable to all federal resource agencies—including the Clean Water Act, the National Environmental Policy Act, the Endangered Species Act, and the National Historic Preservation Act—come into play in the management of WSRs. The four primary administering agencies therefore work collaboratively with agencies that administer these "cross-cutting acts," such as the Army Corps of Engineers and the Environmental Protection Agency. The Act also encourages river-administering federal agencies to enter into cooperative agreements with state and local political entities where necessary or beneficial to protect river values. For example, state and local authorities implement zoning restrictions and pollution control measures that may be critical to protecting the river's water quality or specific outstandingly remarkable values. Finally, where private landholdings

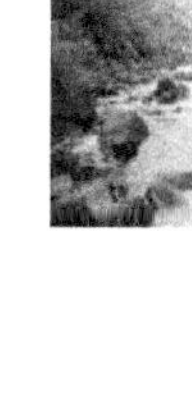

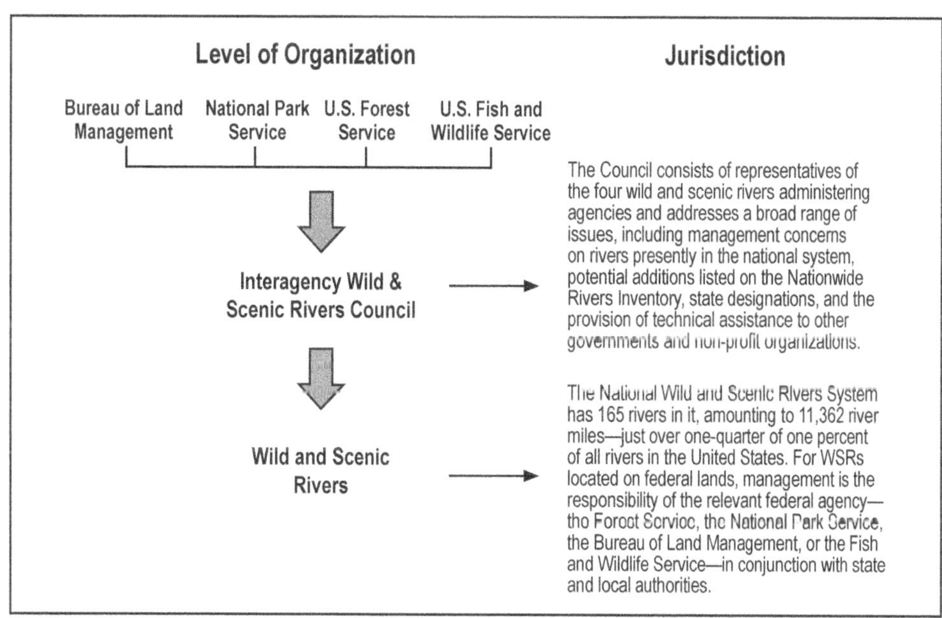

Figure 6.10. Organization of the WSR system. Adapted from National Wild and Scenic Rivers System website.[4]

abut WSRs, the administering agencies may need to negotiate arrangements with private landowners to ensure adequate protection of the river's values.[12]

6.3.4.2 Management Plans

For all WSRs designated by Congress, a Comprehensive River Management Plan (CRMP) must be developed within three full fiscal years of the river's addition to the WSR System. CRMPs essentially amend the broader land management plans of the agency administering the river (the BLM, for example, would amend its Resource Management Plans) in order to ensure that the designated river corridor's values are protected or enhanced. For rivers designated at the request of a state, a CRMP is not required, but the state's application for a river's inclusion in the WSR System must include a strategy to ensure that the river will be managed so as to meet the goals (see Box 6.1) associated with the purposes of the Act. In developing CRMPs, federal agencies will typically consult with state and local agencies and solicit intensive public involvement. Over the years, various parties have challenged the allowance of certain activities (*i.e.*, timber harvesting, livestock grazing, road-building) when a CRMP has not been prepared and the effects of the potentially harmful activities in question cannot be adequately assessed. CRMPs are an important vehicle for establishing the flow and quality objectives that will sustain the values for which the river was designated. They are also vehicles for setting forth adaptive strategies to mitigate the effects of future human stressors on WSRs, including potential climate change impacts.

The Interagency Wild and Scenic Rivers Coordinating Council, a government body established to coordinate management of WSRs among the responsible agencies, has identified six steps to identify the water quantity and quality that are needed to ensure river values are protected: (1) clearly define the water-related values to be protected, (2) document baseline conditions against which to assess future changes or threats, (3) identify potential threats and protection opportunities, (4) identify an array of protection options in the management plan, (5) vet the plan through legal counsel, and (6) decide upon and implement the best protection strategies for achieving the management objectives for the river.[13]

In order to fulfill the Act's intent to "protect and enhance" WSR values, the collection and documentation of adequate baseline information for each WSR, along with a detailed narrative description of the characteristics and values that qualified the river for the WSR designation, is critical to both river managers and stakeholders. For example, a long-term record of river flows is invaluable for developing a water rights claim (see water rights discussion below), and background data on water quality are often essential for pursuing action to stop some proposed activity that threatens a river's ecological services and outstandingly remarkable values. In a case decided in 1997, for instance, the Oregon Natural Desert Association claimed that the BLM's river management plan was failing to protect the riparian vegetation and aquatic habitat of the Donner and Blitzen WSR, which studies had shown were adversely affected by livestock grazing. The court ultimately determined that grazing could continue, but only in a manner that fulfilled BLM's obligation to "protect and enhance" the values that qualified the river as a WSR. Without adequate baseline information, it is difficult, if not impossible to implement a "protect and enhance" policy.

Since passage of the Act, scientific understanding of the ecological importance of the natural variability of a river's historic flow regime has expanded markedly (Poff *et al.*, 1997; Postel and Richter, 2003; Richter *et al.*, 2003). In particular, a prior emphasis on the maintenance of "minimum flows"—ensuring that some water flows in the channel—has been succeeded by the more sophisticated and scientifically based "natural flow paradigm," which calls on river managers to mimic, to some degree, the variable natural flows that created the habitats and ecological conditions that sustain the river's biodiversity and valuable goods and services. Especially in the face of climate change and the resulting likelihood of altered river flow patterns, an understanding of the importance of a river's historical natural flow pattern to the maintenance of its ecological services will be

[13] **Interagency Wild and Scenic Rivers Coordinating Council**, 2003: *Water Quantity and Quality As Related to the Management of Wild & Scenic Rivers*. National Wild and Scenic Rivers System.

critical to the development of effective climate adaptation strategies.

6.3.4.3 Legal and Management Tools

The federal and state agencies administering WSRs have a number of tools and measures at their disposal to fulfill their obligations to "protect and enhance" the water flows, water quality, and outstandingly remarkable values that qualify a particular river for inclusion in the WSR System. This section describes a few of these tools. Later sections suggest how these and other tools can be used to more effectively adapt the management of WSRs to climate change impacts and related human stressors.

Water Rights Claims and Purchases

By virtue of two U.S. Supreme Court rulings, one in 1908 (Winters *v.* United States) and another in 1963 (Arizona *v.* California), national parks, forests, wildlife refuges, and other federal land reservations, as well as Indian reservations, may claim federal "reserved" water rights to the extent those rights are necessary to carry out the purposes for which the reservation was established. The WSR Act makes clear that such reserved rights also apply to designated WSRs.[12] The quantity of the right cannot exceed that necessary to protect the specific river values that qualified the river for inclusion in the WSR System. To date, there are approximately 15 WSRs with water rights adjudications completed or in progress.

Because most WSR designations are less than 30 years old, WSRs typically have very junior rights in the western system of "first-in-time, first-in-right" water allocations. In over-allocated western rivers, another way of ensuring flows for a WSR segment is often to purchase water rights from private entities willing to sell them. In any effort to secure more flow for a WSR, the CRMP developed for the river must demonstrate how the river's outstandingly remarkable values depend on a particular volume or pattern of flow, and include a strategy for protecting flow-dependent river values.

Environmental Flow Protections

An environmental flow study can assist river managers in establishing scientifically based limits on flow alterations that are needed to protect a WSR's habitat, biodiversity, fishery, and other values (Richter *et al.*, 1997; Postel and Richter, 2003). Where allowed by state laws, state agencies (often working in partnership with federal and local authorities) may secure more flows for designated rivers by legislating environmental flows, using permit systems to enforce limits on flow modifications, transferring water rights for in-stream purposes, and implementing water conservation and demand-management strategies to keep more water in-stream (Postel and Richter, 2003; Postel, 2007). The WSR study for Connecticut's Farmington River (pictured in Fig. 6.11), for example, resulted in state water allocation authorities and a water utility committing themselves to the protection of flows needed to safeguard fisheries and other flow-dependent outstandingly remarkable values.[14]

Figure 6.11. Farmington WSR. Photo courtesy of the Farmington River Watershed Association.

Land Protection Agreements with Landowners Adjacent to WSR Corridors

Protection of the land included in the designated river corridor is critical to the protection of the habitat, scenic, scientific, and other values of a WSR. The boundary of a WSR includes up to 320 acres per river mile (twice this for Alaskan rivers), measured from the ordinary high water mark.[14] Under the WSR Act, the federal government may acquire non-federal lands, if necessary, to achieve adequate river protection, but only if less than 50% of the entire acreage within the WSR boundary is in public ownership. However, other options for land protection, besides acquisition, exist.[14] For

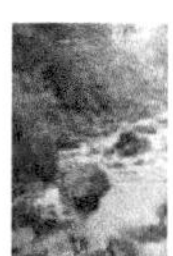

[14] **Interagency Wild and Scenic Rivers Coordinating Council**, 1996: *Protecting Resource Values on Non-Federal Lands*. National Wild and Scenic Rivers System.

instance, the administering agency can work cooperatively with landowners and establish binding agreements that offer them technical assistance with measures to alleviate potentially adverse impacts on the river resulting from their land-use activities. The National Park Service proposes such cooperative agreements, for instance, in its management plan for the Rio Grande WSR in Texas (National Park Service, 2004). In addition, landowners may voluntarily donate or sell lands, or interests in lands (*i.e.*, easements) as part of a cooperative agreement. Local floodplain zoning and wetlands protection regulations can also be part of a land-protection strategy.[14]

Limitations on Impacts of Federally Assisted Water Projects on WSRs

The WSR Act is clear that no dams, diversions, hydropower facilities, or other major infrastructure may be constructed within a designated WSR corridor. In addition, the Act states that no government agency may assist (through loans, grants, or licenses) in the construction of a water project that would have a "direct and adverse effect" on the river's values. A grayer area exists, however, when projects upstream or downstream of a designated WSR would "invade" or "unreasonably diminish" the designated river's outstandingly remarkable values. Legal decisions in a number of WSR cases suggest that proposed water projects above or below a designated stream segment, or on a tributary to a WSR, should be evaluated for their potential to "unreasonably diminish" the scenic, recreational, fish, or wildlife values of the designated river. For example, when the U.S. Army Corps of Engineers proposed to complete the Elk Creek Dam, located 57 miles upstream of the Rogue WSR, the two administering agencies— BLM and the USFS—issued a determination that the dam would result in "unreasonable diminishment to the anadromous fisheries resource [within the designated area] because of impediments to migration and some loss of spawning and rearing habitat." While it was left to Congress to decide whether the dam should be built, the Rogue WSR's administering agencies weighed in to protect the river's values.[12]

Cooperative Arrangements with Other Agencies to Mitigate Impacts on WSRs

The WSR administering agencies can work proactively with other federal or state agencies to secure their cooperation in protecting the natural flows and outstandingly remarkable values of designated rivers. For example, the NPS could establish an agreement with an upstream dam operator, such as the Army Corps of Engineers, to help ensure flows adequate to protect the WSR's habitat and other values. In addition, working with local governments and communities to secure zoning restrictions that protect a WSR's water quality or other values can be effective. For example, cooperative work on WSR studies for the Sudbury, Assabet, and Concord Rivers in Massachusetts (which received WSR designation in 1999) led to a "nutrient trading" program designed to reduce pollution loads and eutrophication problems within the river systems.[13]

Establishment of Effective Baseline Information

Although there is sufficient authority for the administering agencies to acquire land interests and water rights, information is often lacking to answer the important detailed questions about where to acquire these interests and water rights, when to do so, for how much, and for what purposes. Baseline data that are needed to adequately implement authorities under the Act are often skimpy or lacking altogether. It is very difficult for a river manager to propose a change when it cannot be demonstrated what that change will do to the river's protection. Without baseline data as a reference point, it will also be impossible to detect climate-induced changes in flow regimes. Thus, it is critical to begin to develop baseline data.

Technical Assistance

The spirit of the WSR Act is one of cooperation and collaboration among all the entities involved—whether public or private, and including local, state, regional, and national political divisions. The provision of technical assistance to communities within or near a designated or potential WSR can be a powerful tool for implementing the Act. In some cases, for example, communities may see the value of zoning restrictions only when given assistance with GIS mapping that shows the potential for harmful flooding in the future.

6.4 ADAPTING TO CLIMATE CHANGE

Climate change arises from human activity and, unlike climate variation resulting from natural forces operating at historical time scales, the rate of climate change expected over the next 100 years is extremely high (IPCC, 2007a). The magnitude and form of the changes will be variable across the United States—some regions may experience more frequent and intense droughts, while others may have fewer or less severe dry periods. This regional variability will be pronounced among the WSRs because they already vary dramatically in terms of their local climates and in terms of the extent to which their watersheds are influenced by human activities that exacerbate climate change impacts. Because impacts due to human activities (*e.g.*, land use change, water extraction) will persist or grow in the future, this discussion focuses on climate change impacts and the interactive effects of climate change with other stressors on ecosystems and their services. This section finishes by presenting adaptation options for WSRs.

6.4.1 Climate Change Impacts

Output from climate change models indicate that global temperature will increase, with the direction and magnitude varying regionally. Projections of changes in precipitation are less certain but include change in the amount or timing of rainfall as well as the frequency and magnitude of extreme rainfall events. The latest IPCC (2007b) assessment report states: [We are] "*virtually certain* to experience warmer and fewer cold days over most land areas as well as warmer and more frequent hot days; we are *very likely* to experience heat waves and heavy rainfall events more frequently; and we are *likely* to experience more drought in some regions." Thus, much of the world can expect warmer conditions and many watersheds will experience more severe weather events.

6.4.1.1 Temperature

During the 21st century, the average global surface temperature is projected to increase with the best estimate across six IPCC (2007a) scenarios being 1.8–4.0°C during the 21st century. Increases will vary geographically and seasonally. For instance, in summer, rivers in Nevada, Utah, and Idaho will be most strongly affected (Fig. 6.12). In the past, for snowmelt-dominated rivers in the western United States, temperature increases have affected the onset of the spring pulse and the timing of the center of mass for flow (Stewart, Cayan, and Dettinger, 2005) (Fig. 6.12). Because streams and rivers are generally well mixed and turbulent, they respond to changes in atmospheric conditions fairly easily and thus they would become warmer under projected climate change (Eaton and Scheller, 1996). Rivers that are fed by groundwater, such as Michigan's Au Sable and Florida's Wekiva, should be somewhat buffered from atmospheric heating (Allan, 2004). Those that do warm could experience reductions in water quality due to increased growth of nuisance algae and to lower oxygen levels (Murdoch, Baron, and Miller, 2000).

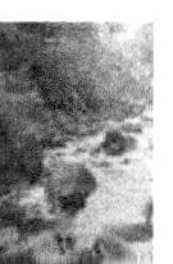

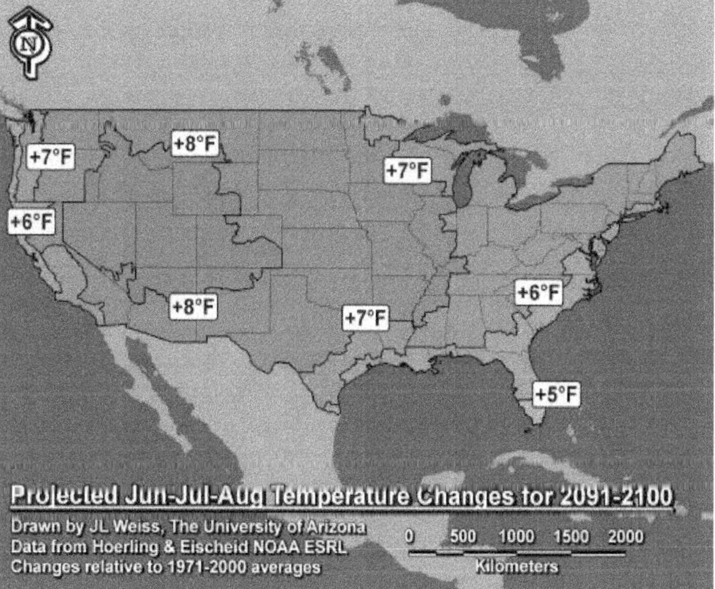

(www.geo.arizona.edu/dgesl/research/regional/projected_US_climate_change.htm)

Figure 6.12. Projected temperature changes for 2091-2100.[15]

[15] **University of Arizona**, Environmental Studies Laboratory, 2007: Climate change projections for the United States. University of Arizona, http://www.geo.arizona.edu/dgesl/, accessed on 5-17-2007.

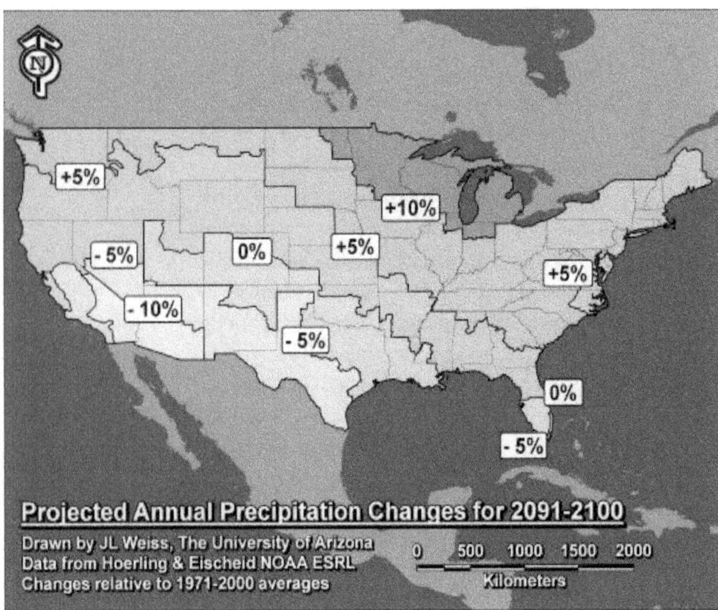

Figure 6.13. Projected annual precipitation changes for 2091-2100.[15]

6.4.1.2 Precipitation

Little to no change in precipitation is projected in southern Utah, southern Colorado, northeastern New Mexico, eastern Texas, and Louisiana, where only a few WSRs are designated (the Saline Bayou, Louisiana; Upper Rio Grande and Pecos, New Mexico) (Fig. 6.13). Up to a 10% increase in rainfall may occur around the Great Lakes region, where there are a number of designated rivers including the Indian, Sturgeon, Presque Isle, and St. Croix. As much as a 10% decrease in precipitation may occur in southern Arizona and southeastern California, where the Verde, Kern, Tuolumne, and Merced rivers are designated as Wild and Scenic.

In regions that receive most of their precipitation as snow, the increased temperatures may result in a shift from winter snow to rain or rain plus snow. A recent analysis of long-term USGS discharge gauge records showed that most rivers north of 44° North latitude—roughly from southern Minnesota and Michigan through northern New York and southern Maine—have had progressively earlier winter-spring streamflows over the last 50–90 years (Hodgkins and Dudley, 2006). Rivers in mountainous regions also may experience earlier snowmelt, and in some regions, less snowpack (Stewart, Cayan, and Dettinger, 2005; McCabe and Clark, 2005). Many parts of Oregon and southern Washington, which are states notable for their large number of WSRs, may experience earlier snowmelt and thus higher winter-spring discharges.

6.4.1.3 Discharge

Because of the projected changes in temperature, precipitation, and CO_2 concentrations, river discharges are expected to change in many regions (Lettenmaier, Wood, and Wallis, 1994; Vörösmarty *et al.*, 2000; Alcamo *et al.*, 2003). The total volume of river runoff and the timing of peak flows and low flows are expected to shift significantly in some regions. In humic, vegetated regions of the world, the majority of runoff follows subsurface pathways and the majority of precipitation returns to the atmosphere as evapotranspiration (Allan, Palmer, and Poff, 2005). Since climate change will affect the distribution of vegetation (Bachelet *et al.*, 2001), the dominant flow paths to some rivers may shift, resulting in higher or flashier discharge regimes (Alcamo, Flörke, and Märker, 2007).

Milly, Dunne, and Vecchia (2005) evaluated relative (*i.e.*, percent) change in runoff from a 1900–1970 baseline (2006 IPCC 20C3M model runs) to a 2041–2060 period (2006 IPCC A1B model runs). They averaged the relative change across 24 pairs of model runs, obtained from 12 different models, some of which performed replicate runs. Fig. 4 in Milly, Dunne, and Vecchia (2005) shows projected changes in runoff in two ways: (1) as the mean, across 24 pairs of runs, of the relative changes in runoff, and (2) as the difference between the number of pairs of runs showing increases in runoff minus the number showing decreases in runoff. Fig. 6.14 shows similar results from the same analysis, but with (1) central estimates of change based on the more stable median instead of the mean, (2) equal weighting of the 12 models instead of the 24 pairs of model runs, and (3) relative changes of areal-averages of runoff over United States water regions instead of relative changes of point values of runoff.

The median projections are for increased runoff over the United States Midwest and Middle-Atlantic, through slightly decreased

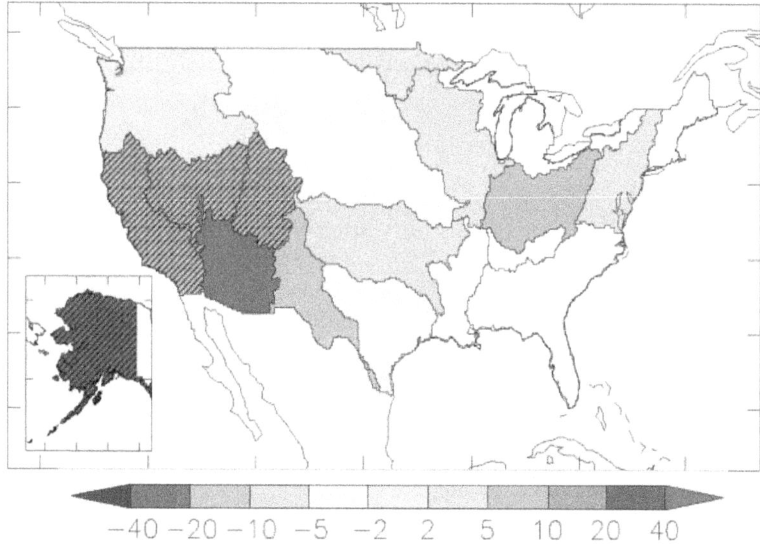

Figure 6.14. Median, over 12 climate models, of the percent changes in runoff from United States water resources regions for 2041–2060 relative to 1901–1970. More than 66% of models agree on the sign of change for areas shown in color; diagonal hatching indicates greater than 90% agreement. Recomputed from data of Milly, Dunne, and Vecchia (2005) by Dr. P.C.D. Milly, USGS.

runoff in the Missouri River Basin and the Texas Gulf drainage, to substantial change (median decreases in annual runoff approaching 20%) in the Southwest (Colorado River Basin, California, and Great Basin). Median estimates of runoff changes in the Pacific Northwest are small. Large (greater than 20%) increases in runoff are projected for Alaska.

Fig. 6.14 also contains information on the degree of agreement among models. Uncolored regions in the Southeast, New England, and around the Great Lakes indicate that fewer than two thirds of the models agreed on the direction of change in those regions. Elsewhere, the presence of color indicates that at least two thirds of the models agreed on the direction of change. Diagonal stippling in Alaska and the Southwest indicate that more than 90% (*i.e.*, 11 or 12) of the 12 models agree on the direction of change.

It is important to note that and some of the regions in Fig. 6.14 are small and are not well resolved by the climate models, so important spatial characteristics—such as mountain ranges in the western United States—are only very approximately represented in these results. However, these regions are generally larger than many of the river basins for which Milly, Dunne, and Vecchia (2005) demonstrated substantial model skill in reproducing historical observations.

In regions in which snowmelt occurs earlier due to warmer temperatures, stream flows will increase early in the season and flooding may be pronounced (see Fig. 6.15 for a picture of river flooding) if high flows coincide with heavy rainfall events ("rain on snow events"). As evidenced by increases in discharge, a shift in the timing of springtime snowmelt toward

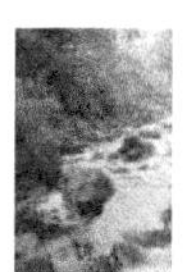

Figure 6.15. Photo of snowmelt in WSR during winter-spring flows. Photo courtesy of National Park Service, Lake Clark National Park & Preserve.

earlier in the year is already being observed (1948–2000) in many western rivers (Fig. 6.16), particularly in the Pacific Northwest, Sierra Nevada, Rockies, and parts of Alaska (Stewart, Cayan, and Dettinger, 2004).

6.4.1.4 Channel and Network Morphology

Large changes in discharge that are not accompanied by changes in sediment inputs that offset the flow changes will have dramatic impacts on river geomorphology (Wolman, 1967). Rivers with increases in discharge will experience more mobilization of bed sediments (Pizzuto *et al.*, 2008), which may result in changes in the river's width and depth (Bledsoe and Watson, 2001). Regions that lose vegetation under future climate may have increased runoff and erosion when it does rain (Poff, Brinson, and Day, Jr., 2002). The drier conditions for extended periods of time may result in some perennial streams becoming intermittent and many intermittent or ephemeral streams potentially disappearing entirely, thus simplifying the network.

6.4.2 Future Human Context: Interactive Effects of Multiple Stressors

The effects of multiple environmental stressors on ecosystems are still poorly understood, yet their impacts can be enormous. Any consideration of climate change is by definition a consideration of future conditions; *i.e.*, a look at what is expected over the next century. Many factors other than climate influence the health of ecosystems, and these factors certainly will not remain static while climate changes (see Box 6.4 for examples). The stressors most likely to intensify the negative effects of climate change include land use change—particularly the clearing of native vegetation for urban and suburban developments—and excessive extractions of river water or groundwater that feed WSRs (Allan, 2004; Nelson and Palmer, 2007).

WSRs in watersheds with a significant amount of urban development are expected to not only experience the greatest changes in temperature under future climates, but also to experience temperature spikes during and immediately following rain storms (Nelson and Palmer, 2007) (Fig 6.17). Such changes may result in the extirpation of cool water species.[16]

The number of extreme flow events would also increase more in WSRs in urbanized basins compared with those that are mostly wild. Large amounts of impervious cover are well known to cause an increase in flashiness in streams—both higher peak flows during the rainy season and lower base flows in the summer (Walsh *et al.*, 2005). Thus, flooding may be a very serious problem in regions of the United States that are expected to have more rainfall and more urbanization in the future (*e.g.*, the Northeast and portions of the mid-Atlantic) (Nowak and Walton, 2005) (see Fig. 6.13). Areas of the United States that will experience the greatest increase in population size are the South and West, with increases of more than 40% between the year 2000 and 2030.[17] More specifically, significant growth is occurring in the following regions that have rivers designated as wild and scenic: most of Florida; central and southern

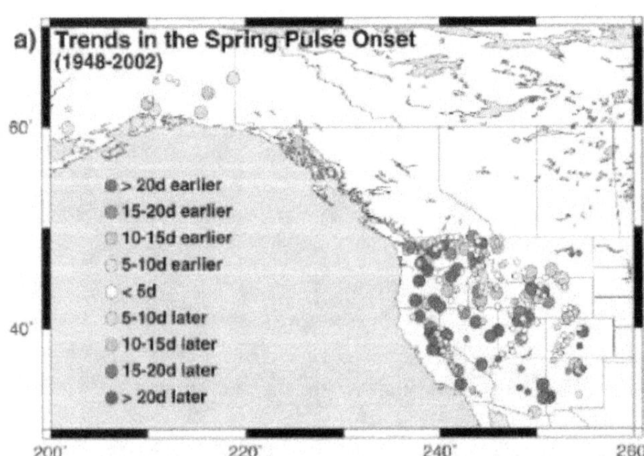

Figure 6.16. Earlier onset of spring snowmelt pulse in river runoff from 1948–2000. Shading indicates magnitude of the trend expressed as the change (days) in timing over the period. Larger symbols indicate statistically significant trends at the 90% confidence level. From Stewart, Cayan, and Dettinger (2005).

[16] **Nelson**, K., M.A. Palmer, J.E. Pizzuto, G.E. Moglen, P.L. Angermeier, R. Hilderbrand, M. Dettinger, and K. Hayhoe, submitted: Forecasting the combined effects of urbanization and climate change on stream ecosystems: from impacts to management options. *Journal of Applied Ecology.*

[17] **U.S. Census Bureau**, 2004: State interim population projections by age and sex: 2004–2030. U.S. Census Bureau Projection Website, www.census.gov/population/www/projections/projectionsagesex.html, accessed on 4-1-2007.

BOX 6.4. Climate Change and WSRs in Alaska.

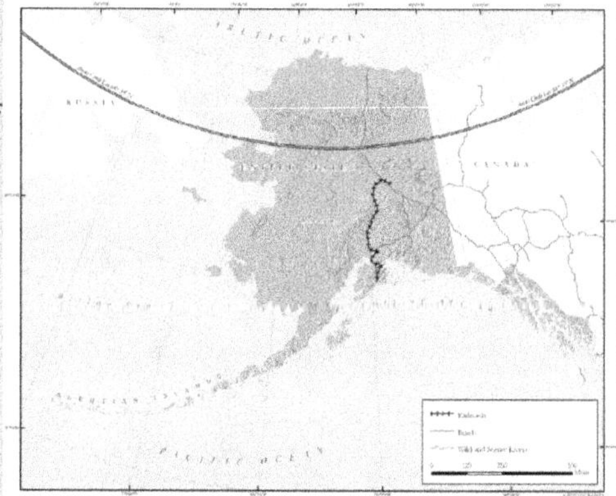

Approximately 28% of the designated WSR river miles in the nation are in Alaska, including 55% of those designated as wild. In Alaska there are 3,210 WSR miles, of which 2,955 are wild, 227 scenic, and 28 recreational. About half of Alaska's 25 WSRs are located north of the Arctic Circle. The federal government owns much of the designated river corridors and in many cases controls most or all of the upstream watersheds. None of the WSRs in Alaska are dammed above or below the designated segments.

Potential Effects of Climate Change on Ecosystems and Current Management

Climate change is happening faster in the Arctic than at lower latitudes and is the predominant stressor of WSR ecosystems in Alaska today. The annual average Arctic temperature has risen almost twice as fast as in temperate and equatorial zones, precipitation has increased, glaciers are melting, winter snows and river ice are melting earlier, and permafrost is vanishing (Hassol, 2004). Research in Siberia has shown large lakes permanently lost and attributes the loss to thawing of permafrost, which allows the lakes and wetlands to drain (Smith *et al.*, 2005). Major impacts of climate change on the rivers include earlier ice breakup in spring, earlier floods with higher flows, more erosion, and greater sediment loads. These trends are projected to accelerate as warming continues.

Major shifts in ecological assemblages may occur. For example, where permafrost thaws, new wetlands will form—although these may be temporary and in turn may be displaced by forest. In currently forested areas, insect outbreaks and fires are very likely to increase and may facilitate invasions of non-native species (Hassol, 2004). Invasive plants have also begun to colonize gravel bars near roads, railway and put-ins; although this is not attributed to climate change, climatic changes may favor these species to displace some native species.

Shifts in flow regime (from earlier snowmelt), increased sedimentation, and warmer water, combined with climate change impacts on marine and estuarine systems, may negatively affect anadromous fish populations with far-reaching ecological and human impacts. Higher water temperatures in rivers are thought to be associated with outbreaks of fish diseases such as *Ichthyophonus*, a fungal parasite suspected of killing some salmon before they spawn and degrading the quality of dried salmon. Salmonid runs are an important component of many WSRs, providing a critical food source for other wildlife and for Alaska Natives. Increased erosion along riverbanks results in loss of archeological sites and cultural resources, since there is a long history of seasonal human settlement on many Alaskan rivers.

Potential for Altering or Supplementing Current Management Practices to Enable Adaptation to Climate Change

Managing these large rivers in extremely remote regions of Alaska can not be compared to managing WSRs in the lower 48 states, where river managers are dealing with urban centers, intensive rural land use, dams, diversions, and water extraction infrastructure—all of which can potentially be manipulated. Most of the WSRs in Alaska are truly *wild* rivers.

Even in these remote regions, there are opportunities to manage WSRs affected by climate change. For example, invasive species might be minimized by educating people to avoid introducing problematic species. Archeological and cultural resources of Alaska Natives and their ancestors are abundant along the rivers that have been the transportation corridors for millennia. In consultation with Alaska Natives, these sites should be inventoried, studied, and, where possible, saved from negative impacts of permafrost thaw and erosion resulting from climate change.

Finally, the wild rivers of Alaska are a laboratory for researching climate change impacts on riverine ecosystems and species, and for informing managers farther south years before they face similar changes.

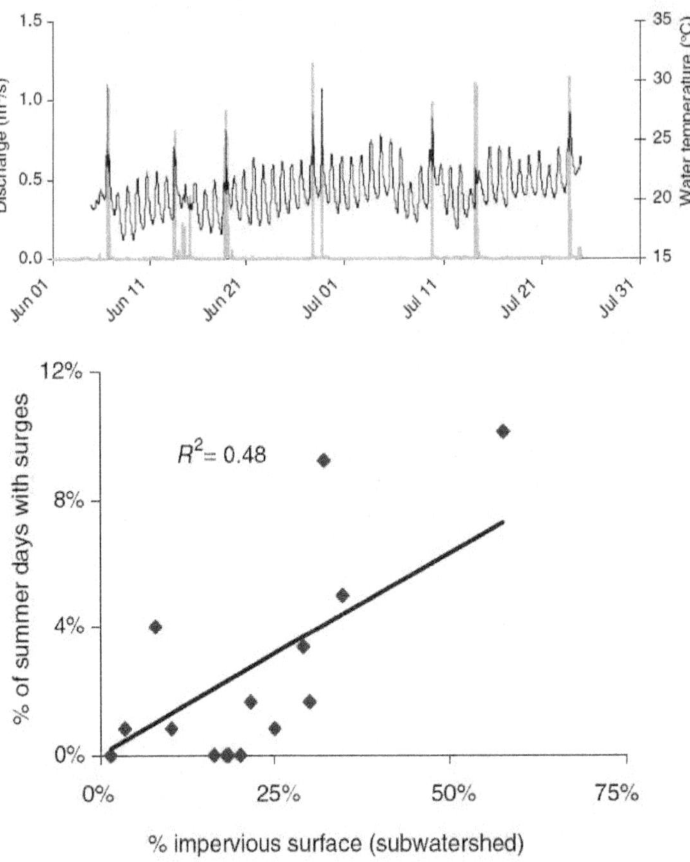

Figure 6.17. Very rapid increases (1–4 hours) in water temperature (temperature "spikes") in urban streams north of Washington D.C. have been found to follow local rain storms. *Top graph:* dark line shows stream discharge that spikes just after a rainfall in watersheds with large amounts of impervious cover; gray line shows temperature surges that increase 2–7°C above pre-rain levels and above streams in undeveloped watersheds in the region. There is no temperature buffering effect that is typical in wildlands where rain soaks into soil, moves into groundwater, and laterally into streams. *Bottom graph:* shows that the number of temperature surges into a stream increases with the amount of impervious cover. From Nelson and Palmer (2007).

California; western Arizona; around Portland, Oregon; much of the mid-Atlantic; and parts of Wisconsin, northern Illinois, and Michigan.[18]

Excessive water extractions are already affecting some WSRs (*e.g.*, the Rio Grande) and this impact will be exacerbated in regions of the country expected to experience even more water stress under future climates. Alcamo, Flörke, and Märker (2007) used a global water model to analyze the combined impacts of climate change and future water stress due to socioeconomic driving forces (income, electricity production, water-use efficiency, etc.) that influence water extractions. Their models indicate that for the 2050s, areas under severe water stress will include not only parts of Africa, Central Asia, and the Middle East, but also the western United States. (Fig. 6.18)

Water managers will need to adjust operating plans for storing, diverting, and releasing water as the timing and intensity of runoff change due to climate change (Bergkamp, Orlando, and Burton, 2003). If these water management adjustments do not keep pace with climate change, water managers will face increasingly severe water and energy shortages due to lessened efficiency in capturing and storing

[18] **Auch**, R., J. Taylor, and W. Acevedo, 2004: *Urban Growth in American Cities*. U.S. Geological Survey Circular 1252, US Geological Survey, EROS Data Center, Reston, VA.

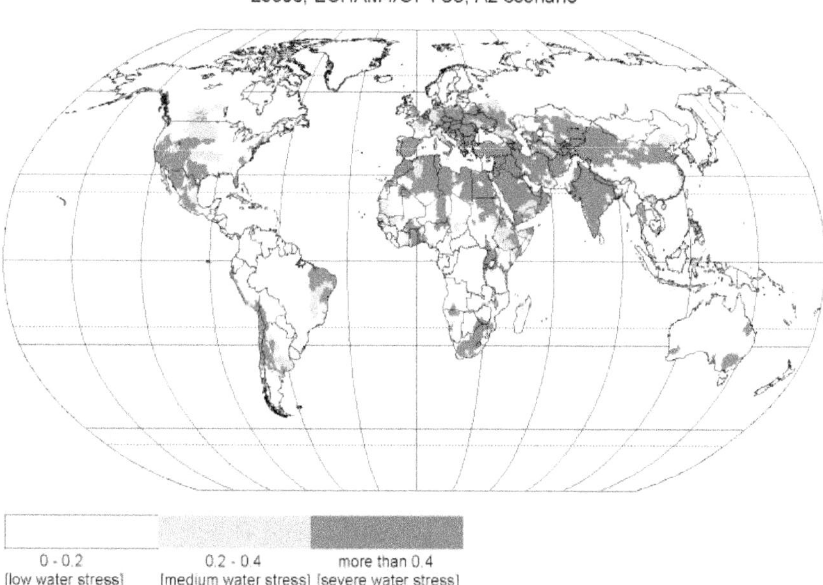

Figure 6.18. Water stress projected for the 2050s based on withdrawals-to-availability ratio, where availability corresponds to annual river discharge (combined surface runoff and groundwater recharge). From Alcamo, Flörke, and Märker (2007).

water to supply cities and farms, or to generate electricity.

Dam building in the United States has slowed considerably relative to the past century, so river impacts related to the interactive effects of dams and climate change will result primarily from changes in management of the dams, particularly as water withdrawals for irrigation or urban water supplies increase in response to a changing climate. For basins expected to experience high water stress in the future (*e.g.*, in the southwestern United States), drawdown of reservoirs is expected, with less water available to sustain environmental flows in the downstream rivers. In regions expected to experience increased precipitation, such as the Great Lakes, flooding problems may increase—particularly if climate change brings greater intensity of rainfall. Shifts in the timing of snowmelt runoff or ice break-up will force dam managers to adjust their operating plans to avoid catastrophic high releases of water into downstream areas. In general, WSRs in basins that are affected by dams or are highly developed will require more changes in management than free-flowing rivers in basins that are mostly wild (Palmer *et al.*, 2008). Ideally this will be done proactively to minimize the need to repair and restore damaged infrastructure and ecosystems.

6.4.3 Ecosystem Goods and Services Assuming Present Management

This chapter has outlined expectations given future climate projections that include warmer water temperatures for most rivers and changes in flow regimes, with extreme events (floods and droughts) increasing in frequency for many rivers. While the impacts will vary among the WSRs depending on their location, their ability to absorb change—which is largely related to the "wildness" of their watershed—also depends on the management response. If proactive measures to buffer ecosystems (such as those discussed in the next section) are taken, then the consequences may be reduced. The need for these proactive measures should be least for WSRs that are classified as "wild," followed by those that are designated "scenic." Presumably wild rivers are the least affected by human activities that may exacerbate the impacts of climate change (Palmer *et al.*, 2008). However, as noted earlier, because many WSRs are in reality river *segments* within watersheds that may be affected by development or even dams, each designated river must be evaluated to determine the management needs.

This section describes the impacts to ecosystems assuming "business as usual" in management—*i.e.*, no changes from current practices. The discussion focuses on species and ecological processes, because these two factors influence most of the attributes valued in WSRs: clean water and healthy ecosystems, with flow regimes that support diverse plant and animal assemblages. Even though recreational use of some WSRs is focused primarily on water sports, it may be that other users still have a strong preference for the other attributes listed above. Clean and beautiful waterways are only possible if materials entering that water—*e.g.*, nutrients, excess organic matter, etc.—do not interfere with natural biophysical processes or the health of flora and fauna.

For a given level of "wilderness," the impacts of climate change on WSRs will depend on how much the changes in thermal and flow regimes deviate from historical and recent regimes (Fig. 6.5). Changes outside the natural range of flow or temperature variability may have drastic consequences for ecosystem structure and function (Richter *et al.*, 1997; Poff, Brinson, and Day, Jr., 2002). The impacts will also depend on the rate of change in temperature or discharge relative to the adaptive capacity of species (amount of genetic diversity). Finally, the impacts will depend on the number and severity of other stressors. Thus, the warmer temperatures and drier conditions expected in southwestern rivers may lead to severe degradation of river ecosystems, which will be exacerbated if water withdrawals for consumptive uses increase (Xenopoulos *et al.*, 2005). For example, the Verde River north of Phoenix, Arizona is in a region of the United States that is experiencing increases in population size, and is expected to have reduced rainfall as well as higher winter and summer temperatures under future climates. The Verde is one of the few perennial rivers within Arizona, but its headwaters are an artificial reservoir (Sullivan Lake) and its flows are affected by groundwater pumping and diversions despite being largely in national forest land.

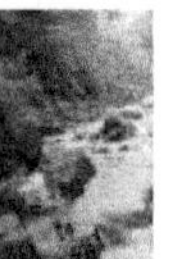

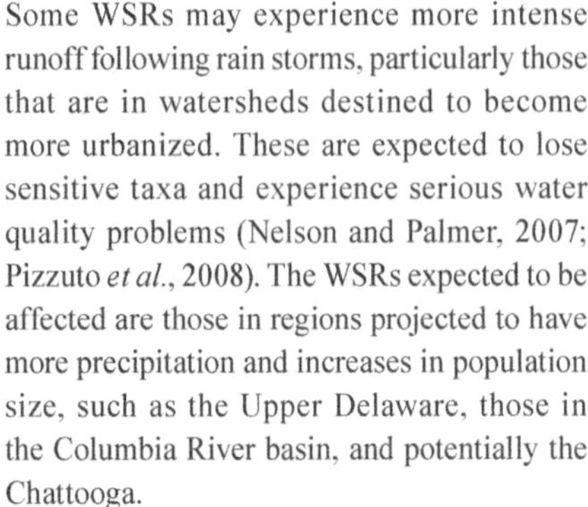

Some WSRs may experience more intense runoff following rain storms, particularly those that are in watersheds destined to become more urbanized. These are expected to lose sensitive taxa and experience serious water quality problems (Nelson and Palmer, 2007; Pizzuto *et al.*, 2008). The WSRs expected to be affected are those in regions projected to have more precipitation and increases in population size, such as the Upper Delaware, those in the Columbia River basin, and potentially the Chattooga.

6.4.3.1 Species-Level Impacts

As the water warms, individual growth and reproductive rates of fish are expected to increase so long as thermal tolerances of any life history stage are not exceeded; typically, eggs and young juveniles are the most sensitive to temperature extremes (Van der Kraak and Pankhurst, 1997; Beitinger, Bennett, and McCauley, 2000). Faster growth rates and time to maturation typically result in smaller adult size and, because size is closely related to reproductive output in many aquatic invertebrates (Vannote and Sweeney, 1980), population sizes may decline over time. The spawning time of fish may also shift earlier if river waters begin to warm earlier in the spring (Hilborn *et al.*, 2003). Further, some aquatic species require prolonged periods of low temperatures (Lehmkuhl, 1974); these species may move northward, with local extirpations. However, dispersal to more northern rivers may be restricted by habitat loss, and riverine insects with adult flying stages that depend on vegetated corridors for dispersal may not survive (Allan and Flecker, 1993). For fish, amphibians, and water-dispersed plants, habitat fragmentation due to dams or the isolation of tributaries due to drought conditions may result in local extirpations (Dynesius *et al.*, 2004; Palmer *et al.*, 2008).

Depending on their severity, climate-induced decreases in river discharge may reduce freshwater biodiversity, particularly if other stressors are at play. Xenopoulos *et al.* (2005) predict that up to 75% of local fish biodiversity could be headed toward extinction by 2070 due to the combined effects of decreasing discharge and increasing water extractions. Even if streams do not dry up in the summer, those that experience reductions in baseflow (*e.g.*, in the

Southwest) may have stressed biota and riparian vegetation (Allan, 2004). Dissolved oxygen levels may decline, as may critical habitat for current-dependent (rheophilic) species (Poff, 2002). Physiological stress and increased predation resulting from crowding (less depth means less habitat), combined with habitat fragmentation in stream networks (isolated pools), may dramatically reduce survival and constrain dispersal (Poff, 2002).

Rivers in which future discharge exceeds historical bounds will also experience a loss of species unless they are capable of moving to less-affected regions. Since species life histories are closely tied to flow regime, some species may not be able to find suitable flow environments for feeding, reproducing, or surviving major flood events. Further, with higher flows come higher suspended sediment and bedload transport, which may interfere with feeding. If sediment deposition fills interstitial spaces, this will reduce hyporheic habitat availability for insects and spawning areas for lithophilic fish (Pizzuto *et al.*, 2008). Whether deposition or net export of these sediments occurs depends on the size of the sediment moving into channels in concert with peak flows (*i.e.*, the stream competency). Particle size and hydraulic forces are major determinants of stream biodiversity (both the numbers and composition of algae, invertebrates, and fish) and excessive bottom erosion is well known to decrease abundances and lead to dominance by a few taxa (Allan, 1995).

6.4.3.2 Impacts on Ecological Processes

Many of the ecological processes that ensure clean water for drinking and for supporting wildlife will be influenced by higher water temperatures and altered flows. Primary production in streams is very sensitive to temperature and flow levels (Lowe and Pan, 1996; Hill, 1996); climate change may thus result in an increase in food availability to herbivorous biota that could support higher abundances and also shift species composition. If riparian plants also grow at faster rates, inputs of leaves and other allochthonous material to rivers may increase. While this could be expected to provide more food for detritivores, this may not be the case if the rate of breakdown of those leaves is higher under future climates. This may occur with higher water temperatures and thus increased microbial growth, or with higher flows that contribute to the physical abrasion of leaves (Webster and Benfield, 1986). Further, allochthonous inputs may represent lower-quality food since plants growing under elevated CO_2 levels may have higher carbon-to-nitrogen ratios, and compounds such as lignin (Tuchman *et al.*, 2002) that reduce microbial productivity (Rier *et al.*, 2002). They also may experience higher leaf decay rates (Tuchman *et al.*, 2003) and detritivore growth rates in streams (Tuchman *et al.*, 2002).

There is a great deal of uncertainty about how rates of nutrient processing in streams will be influenced by climate change. Dissolved inorganic nitrogen (as NO_3) levels may decrease if rates of denitrification are increased (*e.g.*, by higher temperatures and lower oxygen), which could be important given increasing levels of nitrogen deposition (Baron *et al.*, 2000). On the other hand, if discharge and sediment transport increase, then the downstream movement of nitrogen (as NH_4) and phosphorus (as PO_4) may increase. In short, there is a high degree of uncertainty with respect to how climate change will affect ecological processes. This means that our present ability to predict changes in water quality and food availability for aquatic biota is limited. To date, few studies have been conducted to simultaneously examine the many interacting factors that are both subject to change in the future and known to influence ecological processes.

6.4.4 Options for Protection Assuming New Management

Options to protect WSRs and river segments are diverse, and most of them require cooperation and collaboration with other groups. Depending on the specific watershed and the level of human use (development, agriculture, forestry, etc.), these groups could include local landowners, reservoir and dam managers, as well as city, county or state agencies. As pointed out several times in this chapter, WSRs are distinctive—as are some other ecosystems on federally owned land—because rivers are affected by *all* activities in their watershed whether the land is federal or not. Thus the options we discuss below extend well beyond federal boundaries and assume WSR managers/administering agencies will be proactive in seeking cooperative arrangements with the needed parties to ensure WSR ecosystems are protected.

Rivers are inherently dynamic systems—in their native state they are constantly "adjusting" to changes in sediment and water inputs by laterally migrating across the landscape and by changing the depth, width, and sinuosity of their channels. These changes are part of a healthy river's response to changes in the landscape and the climate regime. However, the new temperature and precipitation regimes expected as a result of global climate change would occur much more quickly than historical climate shifts did (IPCC, 2007a). Further, many WSRs are affected by development in their watershed, dams, and excessive water extractions. Thus, the ability to adjust to changes in the flux of water and material, particularly on rapid time scales, is impeded in many watersheds.

In general, WSRs that are in fairly pristine watersheds with no development and few human impacts will fare the best under future climates because their natural capacity to adjust is intact. Even in the face of climate change impacts, rivers surrounded by uninhabited and undeveloped land may experience shifts in channels—perhaps even a deepening and widening of those channels—but their provision of ecosystem services may remain intact. The access points for wildlife or river enthusiasts may need to be shifted and existing trails moved, but largely these rivers are expected to remain beautiful and healthy. In contrast, rivers in Illinois, which will also experience increased discharge, may experience serious problems because flooding and erosion may be exacerbated by development. That said, even some pristine rivers may be negatively affected. For example, the Noatak River in Alaska is already experiencing very large temperature shifts because of its fairly high latitude. This could have serious consequences for migrating salmon and other highly valued species (National Research Council, 2004) (Box 6.4).

The question becomes, what is the appropriate management response? Following Palmer *et al.*, (2008) we distinguish between *proactive* and *reactive* responses. The former includes management actions such as restoration, land purchases, and measures that can be taken now to maintain or increase the resilience of WSRs (*i.e.*, the ability of a WSR to return to its initial state and functioning despite major disturbances). Reactive measures involve responding to problems as they arise by repairing damage or mitigating ongoing impacts. Some actions are far more desirable to undertake proactively (*e.g.*, acquire land to protect floodplains), others may be done proactively *or* reactively (*e.g.*, riparian restoration), and some are more desirable to undertake reactively, such as where the costs of acting before an event are high and the uncertainty of an event occurring is high (*e.g.*, severe damage occurs from an extreme event that requires channel reconfiguration). (Boxes 6.5 and 6.6).

6.4.4.1 Reactive Management

Reactive management basically refers to what managers will be forced to do once impacts are felt if they have not prepared for them. When it comes to rivers, examples of reactive measures include responding to events such as floods, droughts, erosion, and species loss as they occur. Extreme flow events in areas expected to have later snowmelt with the potential for rain-on-snow events may lead to substantial erosion of river banks, not only placing sensitive riparian ecosystems at risk but potentially causing water quality problems downstream due to higher suspended sediment loads.[16] At the other extreme, arid regions that experience more droughts may find populations of valued species isolated due to dropping water levels. For these examples, reactive management efforts may be needed to stem future degradation of ecosystems or extirpation of a species.

The most expensive and serious reactive measures will be needed for WSRs in basins that are heavily developed or whose water is managed for multiple uses. In areas with higher discharge, reactive measures may include river restoration projects to stabilize eroding banks or projects to repair in-stream habitat. To reduce future occurrences of severe erosion, more stormwater infrastructure may be needed. Other measures, such as creating wetlands or off-channel storage basins, may be a way to absorb high flow energy and provide refugia for fauna during droughts or floods. Removing sediment from the bottom of reservoirs could be a short-term solution to allow for more water storage, perhaps averting dam breaches that

BOX 6.5. WSR Adaptation Options.

- Maintain the natural flow regime through managing dam flow releases upstream of the WSR (through option agreements with willing partners) to protect flora and fauna in drier downstream river reaches, or to prevent losses from extreme flooding.
- Use drought-tolerant plant varieties to help protect riparian buffers.
- Create wetlands or off-channel storage basins to reduce erosion during high flow periods.
- Actively remove invasive species that threaten key native species.
- Purchase or lease water rights to enhance flow management options.
- Manage water storage and withdrawals to smooth the supply of available water throughout the year.
- Develop more effective stormwater infrastructure to reduce future occurrences of severe erosion.
- Consider shifting access points or moving existing trails for wildlife or river enthusiasts.
- Increase genetic diversity through plantings or by stocking fish.
- Increase physical habitat heterogeneity in channels to support diverse biotic assemblages.
- Establish special protection for multiple headwater reaches that support keystone processes or sensitive species.
- Conduct river restoration projects to stabilize eroding banks, repair in-stream habitat, or promote fish passages from areas with high temperatures and less precipitation.
- Restore the natural capacity of rivers to buffer climate-change impacts (e.g., through land acquisition around rivers, levee setbacks to free the floodplain of infrastructure, riparian buffer repairs).
- Plant riparian vegetation to provide fish and other organisms with refugia.
- Acquire additional river reaches for the WSR where they contain naturally occurring refugia from climate change stressors.
- Create side-channels and adjacent wetlands to provide refugia for species during droughts and floods.
- Establish programs to move isolated populations of species of interest that become stranded when water levels drop.

could be disastrous. Water quality problems due to high sediment loads or contaminants may appear in WSR reaches downstream of developed (urbanized or agricultural) regions, and these problems are very difficult to cope with in a reactive manner.

In regions with higher temperatures and less precipitation, reactive projects might include fish passage projects to allow stranded fish to move between isolated river reaches during drought times, replanting of native riparian vegetation with drought-resistant vegetation, or removal of undesirable non-native species that take hold. If dams are present upstream of the WSR, flow releases during the summer could be used to save flora and fauna in downstream river reaches that are drying up, and accentuated floods can be managed to avert potentially disastrous ecological consequences of extreme floods.

These are simply examples of reactive management that are discussed more fully in Palmer *et al.*, (2008) but the most important point is that a reactive approach is not the most desirable response strategy to climate change, because a high degree of ecosystem and infrastructure damage is likely to occur before reactive measures are taken. The best approach for reactive management is to continuously evaluate river health over time with rigorous monitoring and scientific research, so that management begins as soon as problems are detected; *i.e.*, before problems are severe. Further, this monitoring and research should help identify proactive needs, thus minimizing costs of repair and loss of ecological services.

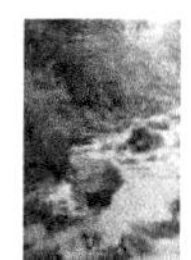

6.4.4.2 Proactive Management

Many of the management actions that are needed to respond to the risks of climate change arise directly from changes in the frequency and magnitude of extreme events, in addition to changes in average conditions or baseflow. Anticipating how climate impacts will interact with other ongoing stressors is critical to developing strategies to protect the values of WSRs. Proactive measures that restore the natural capacity of rivers to buffer climate-change impacts are the most

BOX 6.6. Examples of potential river management and restoration actions. Actions may be taken proactively to prepare for and minimize the impacts of climate change on ecosystems and people, or could be required reactively at the time of or after impact. The type and extent of these actions will vary among rivers and river segments that experience an increase in available water (increased discharge and/or groundwater storage) vs. those that experience water stress. WSRs that are free-flowing throughout their watersheds are expected to require fewer management interventions than river segments in watersheds with dams (as outlined in Palmer et al., 2008); however, the need for intervention will also vary depending on if and how much a watershed containing a WSR segment is in developed use (e.g., agriculture, urban) and the magnitude of climate change for the region.

Type of Management Action	Context and Purpose
Improve environmental monitoring and develop WSR-scale climate forecasts	To facilitate planning and better understand local effects of climate change.
Build capacity to offer technical assistance	National or regional enhancement of technical capacity can provide assistance to WSR managers who may not have the resources to do this on their own.
Designate more WSRs and/or acquire land around existing WSRs	May raise awareness of value of WSRs, potentially leading to additional protection; land acquisition may enhance floodplain extent and buffer river segments from impacts in surrounding watershed, and could provide "replication" in space of at-risk habitats and refugia for species.
Conjunctive Groundwater/ Surface Water Management	Purchasing more water rights may be needed for WSRs under water stress due to droughts or extractions. If dams are present, develop reservoir release options with dam managers and/or design structures for temporary storage of flood waters before they reach reservoir; remove dams in areas with high evaporation, and consider methods to divert water to groundwater storage to provide for later use; adjust outlet height on dam to release high quality water to downstream rivers.
Restoration Projects	Needed particularly for rivers in watersheds with some level of development: riparian management to revegetate damaged areas to slow runoff in the event of more floods, OR to remove drought-tolerant exotic species in drier regions; stormwater management projects and wetland creation to reduce runoff and sediment flux to river or to store flood water; channel reconfiguration and/or stream bank stabilization—some configurations may help channel withstand peak flow releases, OR in drier regions stream bank may need to be re-graded to reconnect floodplain to channel to enhance water storage and habitat.

desirable actions since they may also lead to other environmental benefits such as higher water quality and restored fish populations. Examples of such measures might include stormwater management in developed basins or, even better, land acquisition around the river or setting back existing levees to free the floodplain of infrastructure, absorb floods, and allow regrowth of riparian vegetation.

For WSR segments fed by non-designated headwaters that are not protected in some way from human impacts, efforts should be made to extend the designation to these small tributaries through land acquisition or partnerships with landowners. Indeed, since headwaters often support rare and sensitive species, protecting multiple small headwaters will provide a sort of "insurance" against regional species loss if losses occur in one or a few tributaries.

While shifting climate regimes may result in local shifts in species assemblage (Thuiller, 2004), if there are flora and fauna of special value associated with a WSR then proactive responses to ensure the persistence of these species are needed. These responses will require detailed understanding of their life histories and ecology. For rivers in regions expected to experience hot, dry periods, planting or natural establishment of drought-tolerant varieties of plants may help protect the riparian corridor from erosion. A focus on increasing genetic diversity and population size through plantings or via stocking fish may increase the adaptive capacity of species. Aquatic fauna may benefit from an increase in physical habitat heterogeneity in the channel (Brown, 2003), and replanting or widening any degraded riparian buffers may protect river fauna by providing more shade and maintaining sources of allochthonous input (Palmer *et al.*, 2005).

Incorporating the potential impacts of climate change into water management strategies inevitably involves dealing constructively with uncertainty. Enough is now known about the likelihood of certain impacts of climate change on water availability and use that it is possible to design proactive management responses to reduce future risks and to protect important river assets. At the core of these strategies is the ability to *anticipate* change and to *adapt* river management to those changing circumstances. Water managers need to know, for example, when to take specific actions to ensure the maintenance of adequate flows to sustain river species. It is important that this adaptive capacity be built at the watershed scale, incorporating factors such as grazing, farming, forestry, and other land-uses; reservoir management; water withdrawals; and other features. A new layer of cooperation and coordination among land and water managers will thus be essential to the successful implementation of these adaptive strategies for the management of WSRs.

Legal and institutional barriers exist in many river systems, and will need to be overcome for the adoption of effective management strategies. Water rights, interstate water compacts, property rights, and zoning patterns may all present constraints to effective adaptation strategies. Studies of the Colorado River basin, for example, have found that much of the potential economic damage that may result from climate change is attributable to the inflexibility of the Colorado River Compact (Loomis, Koteen, and Hurd, 2003). The new stressor of climate change, on top of the existing pressures of population growth, rising water demand, land-use intensification, and other stressors, may demand a re-evaluation of the institutional mechanisms governing water use and management, with an eye toward increasing flexibility.

Along with the management tools described above, a number of other categories of actions and measures can enhance the WSR System's ability to protect the nation's rivers under changing climatic regimes, as described below. Box 6.5 presents a summary list of specific actions WSR managers can take to promote adaptation.

Improve Water Monitoring Capabilities and Apply Climate Forecasting

It is critical that river flow monitoring be supported adequately to detect and adapt to flow alterations due to climate change and other stressors. However, many stream gauges maintained by USGS have been discontinued due to resource limitations. Without sufficient monitoring capabilities, river managers simply cannot do their jobs adequately and researchers cannot gather the data needed to elucidate trends. For instance, adequate monitoring to detect trends in flow is needed to show that flooding is increasing as a consequence of more rapid melting in spring. River managers may use the monitoring data to determine where to pursue additional land conservation easements or where to encourage local zoning that limits development on floodplains.

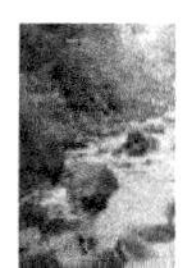

Climate forecasts can enable water managers to minimize risk and avoid damage to WSR values. The development of scenarios that capture the spectrum of possible outcomes is an invaluable tool for anticipating the ramifications of climate-related hydrological and land-use changes, including reduced snowpacks, greater spring flooding, lower summer flows, and warmer stream temperatures. The utility of forecasting tools, however, depends on the ability to apply their results to water management planning.

For instance, the possibility of severe drought occurring in three out of five years indicates that river flows may be affected not only by lack of rainfall and runoff, but by increased evapotranspiration from vegetative regrowth after forest fires. Anticipating such flow depletion, and its potential magnitude, is critical to devising plans that mitigate the impacts. For example, warming trends across the Southwest exceed global averages by 50%, providing ample evidence of the importance of planning for reduced water availability and streamflows in the Rio Grande and other southwestern rivers.[19]

Build Capacity to Offer Technical Assistance

The ability to demonstrate to communities the importance of certain zoning restrictions, land conservation measures, land-use modifications, or floodplain restrictions may require user-friendly models or tools that exhibit potential climate change impacts within specific watersheds. While sophisticated tools may be feasible to use in reaches with ample resources to support management activities, there is a need for affordable tools that enable managers to offer technical assistance in areas with fewer resources.

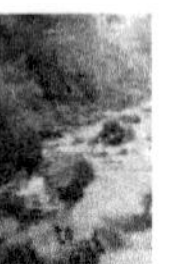

Designate More River Corridors as Wild and Scenic and Acquire Land Adjacent to WSRs

Rivers may be designated as Wild and Scenic by acts of Congress or by the Secretary of Interior upon a state's request. Designation of additional rivers to the WSR program may raise visibility and expand protection to river assets at a time when they are coming under increased human and climatic pressures. Possible candidates for designation include rivers in the Nationwide Rivers Inventory (NRI). The NRI, which is maintained by the National Park Service (updated last in the 1980s), includes more than 3,400 free-flowing river segments that are believed to possess at least one outstandingly remarkable value of national significance. By virtue of a 1979 Presidential directive, all federal agencies must seek to avoid or mitigate actions that would affect NRI segments. The WSR System would also benefit from hastening the review of rivers that have already been submitted for designation, but about which no decision has yet been made. For new designations, there is an opportunity to think strategically about climate change impacts when identifying and prioritizing rivers for designation. Climate change may affect the priority order and rationale for designation.

A second reason for increasing the number of designated rivers in some regions is that if there is a high risk of species extinctions, due for example to a high drought probability, spreading that risk among rivers within the same ecoregions may provide protection (across space) for species. At any given time, there may be rivers within the ecoregion that are not as affected by drought. Land acquisition around existing WSRs may also reduce extinction if the land helps buffer the river segment from nearby development pressures or the land allows for floodplain expansion.

Consider Conjunctive Groundwater/Surface Water Management

The protection of river health and natural flows under a changing climatic regime will require more concerted efforts to secure environmental flows, namely flows that will support the ecosystem, for rivers. With more than 270 dams located within 100 miles (upstream or downstream) of a designated WSR, collaborative arrangements with dam managers offer great potential to secure beneficial flows for WSRs under various climate change scenarios. For WSR segments in watersheds with dams, there may be a need to develop reservoir release options with dam managers and/or design structures for temporary storage of flood waters before they reach reservoirs. In regions with extremely high rates of evaporation, managers may wish to work with requisite authorities to consider removing dams below shallow, high-surface-area reservoirs. In such cases, alternative strategies for water storage will be needed. Finally, with large changes in reservoir water levels, the outlet height on dams may need adjusting to ensure high quality water to downstream WSRs.

Because the agencies administering WSRs have little or no authority over dam operations, a proactive collaboration among the agencies involved—at federal, state, and local levels—is critical. Additionally, the purchase or leasing of water rights to enhance flow management

[19] **New Mexico Office of State Engineer** and Interstate Stream Commission, 2006: *The Impact of Climate Change on New Mexico's Water Supply and Ability to Manage Water Resources.* New Mexico Office of State Engineer/Interstate Stream Commission.

options can be a valuable tool. For example, the establishment of dry-year option agreements with willing private partners can ensure that flows during droughts remain sufficient to protect critical habitats and maintain water quality. A strengthening of environmental flow programs and water use permit conditions to maintain natural flow conditions will also be critical.

Implement Restoration Projects

Restoration can be done either proactively to protect existing resources or, as in the examples provided in Section 6.4.4.1 above, projects may be required to repair damage associated with a changing climate. Since floodplains and riparian corridors are critical regions both for mitigating floods and for storing water, measures should be taken to ensure they are as healthy as possible. This could include removal of invasive plants that threaten native species, re-grading river banks to reconnect floodplains to the active channel, and a whole host of other measures that are more fully described elsewhere (Bernhardt *et al.*, 2007; Palmer *et al.*, 2008; Wohl, Palmer, and Kondolf, 2008).

Develop and Amend CRMPs to Allow for Adaptation to Climate Change

For river managers to fulfill their obligations to protect and enhance the values of WSRs, their management plans need to be evaluated and amended as appropriate to take into account changing stressors and circumstances due to shifting climate (Poff, Brinson, and Day, Jr., 2002). For example, the severe drought in Australia in recent years has not only had serious short-term impacts on river flows, but—due to the effects of fires—may have severe long-term flow effects as well. Studies of the Murray River system by researchers at the University of New South Wales have found that large-scale forest regeneration following extensive bush fires will deplete already low flows further due to the higher evapotranspiration rates of the younger trees compared with the mature forests they are replacing. The 2003 fires, for example, may reduce flows by more than 20% for the next two decades in one of the major tributaries to the Murray.[20] Similar flow alterations might be

anticipated in the American Southwest, which can expect a significant increase in temperature, reduction in snowpack, and recurring droughts that may cause more frequent fires and related vegetation changes. Management of the Rio Grande Wild and Scenic corridors in both New Mexico and Texas will need to take such scenarios into account.

Rebalance the Priority of Values used for Designation of WSRs

In light of climate change impacts and their anticipated effects on habitat, biodiversity, and other ecological assets, it may be useful to emphasize such natural values when designating new WSRs. In addition, where two outstandingly remarkable values are in conflict within the same designated river—as sometimes happens, for example, between habitat and recreational values—an open and fair process in which climate change impacts are considered needs to be used to evaluate the priorities. To protect ecosystem services, strong consideration should be given to prioritizing those natural assets *most at risk* from climate change.

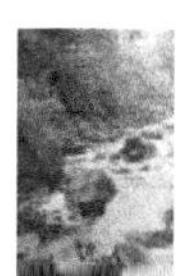

6.5 CONCLUSIONS

The WSR System was created to protect and preserve the biological, ecological, historic, scenic and other "remarkable" values of the nation's rivers. These assets are increasingly at risk due to land-use changes, population growth, pollution discharges, flow-altering dams and diversions, excessive groundwater pumping, and other pressures within watersheds and river systems. Climate change adds to and magnifies these risks through its potential to

[20] **University of New South Wales**, 2007: Fire in the snow: thirsty gum trees put alpine water yields at risk. University of New South Wales Website, http://www.science.unsw.edu.au/news/2007/bushfire.html, accessed on 1-20-2007.

alter rainfall, temperature, and runoff patterns, as well as to disrupt biological communities and sever ecological linkages in any given locale. Thus, the anticipation of climate change effects requires a proactive management response if the nation's valuable river assets are to be protected.

It is critical to recognize that only a subset of WSRs are headwater rivers in watersheds that are free of development, extractive uses, or dams. Since human activities on the land and those affecting ground waters have a very significant impact on rivers and will exert stress that could exacerbate any problems associated with climate change, WSR managers alone can not ensure the protection of many WSRs. Thus, forging partnerships with nonfederal water managers, land owners, towns, and states will be necessary to protect and to preserve the "outstandingly remarkable values" that are the basis for the designation of many rivers as wild and scenic.

In a world of limited budgets, it may not be possible to implement all of the measures identified in the previous section and summarized in Box 6.5. But given limited financial and human resources, the highest priorities for the protection of WSR assets under conditions of climatic change are the following:

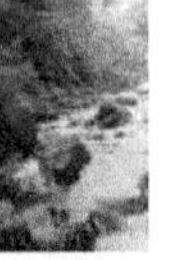

- Increase monitoring capabilities in order to acquire adequate baseline information on water flows and water quality, thus enabling river managers to prioritize actions and evaluate effectiveness.
- Increase forecasting capabilities and develop comprehensive scenarios so that the spectrum of possible impacts, and their magnitude, can reasonably be anticipated.
- Strengthen collaborative relationships among federal, state, and local resource agencies and stakeholders to facilitate the implementation of adaptive river management strategies.
- Forge partnerships and develop mechanisms to ensure environmental flows for WSRs in basins that experience water stress.
- Work with land use planners to minimize additional development on parcels of land adjacent to WSRs, and optimally to acquire floodplains and nearby lands that are not currently federally owned or ensure they are placed in protected status.
- Build flexibility and adaptive capacity into the CRMPs for WSRs, and update these plans regularly to reflect new information and scientific understanding.

National Estuaries

Lead Author: Charles H. Peterson, University of North Carolina

Contributing Authors: Richard T. Barber, Duke University; Robert R. Christian, East Carolina University; Kathryn L. Cottingham, Dartmouth College; Heike K. Lotze, Dalhousie University; Charles A. Simenstad, University of Washington; Michael F. Piehler, University of North Carolina; John Wilson, U.S. Environmental Protection Agency

7.1 SUMMARY

National estuaries comprise a group of 28 estuaries, distributed around the United States and its protectorates and territories, that form the U.S. Environmental Protection Agency's National Estuary Program (NEP). The NEP mandates and supports the grass-roots development of estuary-specific Comprehensive Conservation and Management Plans (CCMPs), which, because national estuaries have no regulatory authority, rely on voluntary commitments to targets and on a wide suite of existing federal, state, and local authorities for implementation. The CCMPs hold several management goals in common: maintaining water quality; sustaining fish and wildlife populations, preserving habitat, protecting human values, and fulfilling water quantity needs.

Maintaining the status quo of estuarine management would guarantee growing failures in meeting all of these management goals under progressive climate change. This chapter thus reviews the suite of management adaptations that might accommodate effects of climate change in ways that could preserve the ecosystem services of estuaries. On time scales of a few decades, management strategies exist that may build resilience sufficiently to minimize ecosystem service losses from estuaries. However, over longer time scales, despite these actions to enhance resilience, dramatic net losses in ecosystem services will arise, requiring trade-offs to be made among which services to preserve and which to sacrifice.

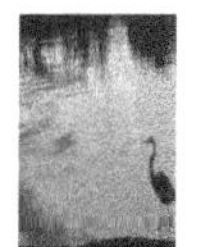

KEY FINDINGS

In the short time frame of a few decades, negative consequences of climate change may be avoided or minimized by enhanced efforts in managing traditional stressors of estuarine ecosystems through existing best management practices (BMPs). For example, climate change will enhance eutrophication in many estuaries by increasing stratification of the water column, elevating biological oxygen demand by increasing temperatures, elevating nutrient loading as wetland buffers are inundated and eroded with sea level rise, and increasing organic loading in runoff from more frequent intense storms. Thus, traditional BMPs to minimize eutrophication are appropriate to expand so as to protect against the climate change enhancement of eutrophication. Protection and restoration of wetland buffers along riverine and estuarine shores should emphasize those shorelines where no barriers exist to prevent wetland transgression to higher ground as sea level rises. This strategy may require modification of present priorities in policy for protection and restoration of riparian wetlands. BMPs that remove non-native invasive species, and maintain and restore native genetic, species, and landscape diversity in estuarine habitats may build resilience to changing climate, although this ecological concept needs further testing to confirm its practical value.

Many management adaptations to climate change can be achieved at modest expense by strategic shifts in existing practices. Reviews of federal, tribal, state, and local environmental programs could be used to assess the degree to which climate change is being addressed by management activities. Such reviews would identify barriers to and opportunities for management adaptation. One major form of adaptation involves recognition of the projected consequences of sea level rise and then application of policies that create buffers to anticipate them. An important example would be redefining riverine flood hazard zones to match the projected expansion of flooding frequency and extent. Other management adaptations could be designed to build resilience of ecological and social systems. These adaptations could include choosing only those sites for shoreline habitat restoration that allow natural recession landward, and thus provide resilience to sea level rise.

The appropriate time scale for both planning and implementing new management adaptations requires considering and balancing multiple factors. Management adaptations to climate change can occur on three different time scales: (a) reactive measures taken in response to observed negative impacts; (b) immediate development of plans for management adaptation to be implemented later, either when an indicator signals that delay can no longer occur without risking serious consequences, or in the wake of a disaster that provides a window of socially feasible opportunity; or (c) immediate implementation of proactive policies. The factors determining which of these time frames is appropriate for any given management adaptation include balancing expenditures associated with implementation against the magnitude of risks of injurious consequences under the status quo of management; the degree of reversibility of negative consequences of climate change; recognition and understanding of the problem by managers and the public; the uncertainty associated with the projected consequences of climate change; the time table on which change is anticipated; and the extent of political, institutional, and financial impediments.

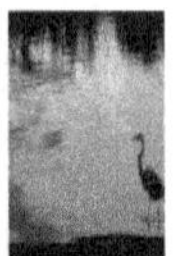

To minimize negative consequences of climate change beyond a few decades, planning for some future management adaptations and implementing other present management adaptations is necessary now. For estuaries, the most critical management challenge to sustain ecosystem services over longer time frames is to implement actions now that will allow orderly retreat of development from shorelines at high risk of erosion and flooding, or to preclude development of undeveloped shorelines at high risk. Such proactive management actions have been inhibited in the past by: (a) uncertainty over climate change and its implications; (b) failures to include true economic, social, and environmental costs of present policies allowing and subsidizing such risky development; and (c) legal tenets of private property rights. One possible proactive management option would be to establish and enforce "rolling easements" along largely undeveloped estuarine shorelines as sea level continues to rise, thereby sustaining the public ownership of tide lands yet allowing private property use to continue. Another proactive management action could be developing and implementing effective ecosystem-based management (EBM). This requires collaboration that crosses traditionally separate levels of management (e.g., state and federal) and management authorities (e.g., water quality and land-use planning) to coordinate and focus actions of all agencies with responsibilities to manage and influence stressors that affect estuarine organisms and ecosystems.

Even with sufficient long-term planning and enhancing short-term resilience by instituting BMPs, dramatic long-term losses in ecosystem services are inevitable and will require tradeoffs among services to protect and preserve. The most serious conflict arises between sustaining public trust values and private property. This is because current policies allowing shoreline armoring to protect private property from damaging erosion imply escalating losses of public tidewater lands, especially including tidal wetlands, as sea level continues to rise and the frequency of intense storms increases. In regions where relative sea level is rising most rapidly, coastal wetlands and other shoreline habitats that maintain water quality and support fish and wildlife production can be sustained only where transgression of tidal marshes and other shoreline habitats to higher ground can occur: such transgression is incompatible with bulkheading and other types of shoreline armoring that protect development from erosion. One possible management adaptation for maximizing natural ecosystem services of estuaries with minimal loss of shoreline development involves establishment of rolling easements to achieve orderly retreat, perhaps only politically feasible where estuarine shoreline development is slight.

Establishing baselines and monitoring ecosystem state and key processes related to climate change and other environmental stressors is an essential part of any adaptive approach to management. Going back into the past to identify baselines from historic environmental, agency, and ecological records, and from paleoecological reconstructions, is critical so as to enhance our understanding of estuarine responses to historic climate change and thereby improve our models of the future. A key goal of monitoring is to establish and follow indicators that signal an approach toward an ecosystem threshold that—once passed—implies passage of the system into an alternative state from which conversion back is difficult. Avoiding conversion into such alternative states, often maintained by positive feedbacks, is one major motivation for implementing proactive management adaptation. This is especially critical if the transition is irreversible, or very difficult and costly to reverse, and if the altered state delivers dramatically fewer ecosystem services. One example of such ecosystem conversions involves nitrogen-induced conversion from an estuary dominated by submersed benthic grasses to an alternative dominated by seaweeds and planktonic microalgae. Detecting ecosystem responses to climate change plays an integral role in management adaptation, because it can trigger implementation of planned but delayed management responses and because such monitoring serves to test the accuracy, and reduce the uncertainty, of the models that guide our management actions. This is the essence of agency learning and adapting management accordingly. Various federal programs for global and national observing systems are currently in development, but they need to include more focus on estuaries and more biological targets to accompany the physical parameters that dominate the current plans.

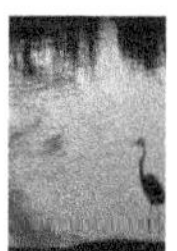

The nature and scope of many anticipated consequences of climate change are not widely recognized by policy makers, managers, and the public because they involve interactions among stressors. Consequently, an effective class of management adaptation involves reducing levels of those existing stressors to minimize the risks and magnitudes of interactive consequences of climate change. These interactions and their potential significance also imply a need for more substantive rather than superficial evaluations of interacting effects of climate change in environmental impact and environmental assessments conducted in response to the National Environmental Policy Act and its state analogs. Interactions of climate change with other stressors leads to a management priority for including consideration of climate change sensitivity, resilience, and adaptation responses in all relevant federal and state funding programs. In the absence of such actions, for example, climate impacts on estuarine wetlands will likely violate the "no net loss of wetlands" policy, which underlies the Clean Water Act, in two ways: (a) wetland losses resulting from sea level rise and increasing frequency of intense storms will compound the continuing loss of wetlands from small development projects with inadequate mitigation; and (b) measures used to protect human developments and infrastructure from climate change impacts will inhibit wetland adaptation to climate change. Management adaptations taken in response to the importance of potential interactions between climate change and existing stressors could include ending direct and indirect public subsidies that now support risky development on coastal barriers and estuarine shores at high risk of flooding and storm damage.

7.2 BACKGROUND AND HISTORY

7.2.1 Historical Context and Enabling Legislation

This chapter focuses on meeting the challenges of managing national estuaries and estuarine ecosystem services under influence of changing climate. Our contribution is distinguished from previous reviews of estuarine responses to climate change (*e.g.*, National Coastal Assessment Group, 2000; National Assessment Synthesis Team, 2000; Scavia *et al.*, 2002; Kennedy *et al.*, 2002; Harley and Hughes, 2006) by its focus on developing adaptive management options and analyzing the characteristics of human and ecological systems that facilitate or inhibit management adaptation. The chapter is thus written mostly for an audience of natural resource and environmental managers and policy makers.

A summary of federal legislation for the protection and restoration of estuaries is presented in the Appendix. There are 28 national estuaries in the U.S. National Estuarine Program, which is administered by the U.S. Environmental Protection Agency (Fig. 7.1). These estuaries span the full spectrum of estuarine ecosystem types and encompass the diversity of estuarine ecosystem services across the country.

Estuaries are sometimes defined as those places where fresh and salt water meet and mix, thereby potentially excluding some largely enclosed coastal features such as marine lagoons and including, for some vigorous rivers like the Mississippi, extensive excursions into the coastal ocean. So as to match common characteristics of the 28 national estuaries, we choose an alternative, geomorphologically based definition of an estuary as a semi-enclosed body of water on the seacoast in which fresh and salt water mix (adapted from Pritchard, 1967). Such a definition includes not only those water bodies that are largely perpendicular to the coastline where rivers approach the sea, but also marine lagoons, which are largely parallel to the shoreline and experience only occasional fresh water inflow, thereby retaining high salinities most of the time. In the landward direction, we include the intertidal and supratidal shore zone to be part of the estuary and thus include marshes, swamps and mangroves (*i.e.*, the coastal wetlands).

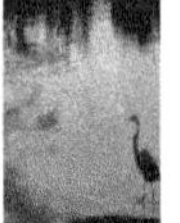

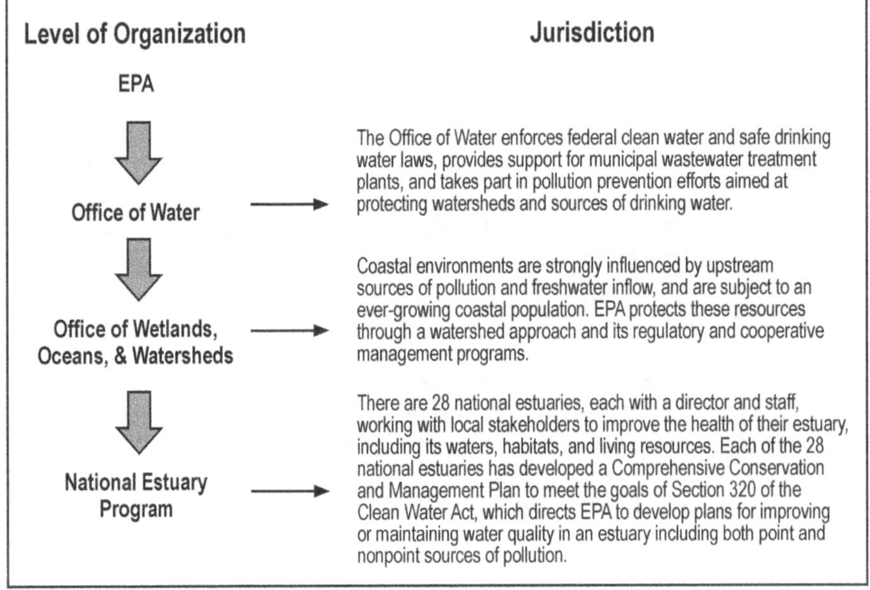

Figure 7.1. Organization of the NEP System.[1]

[1] **U.S. Environmental Protection Agency**, 2007: Office of Water organizational chart. EPA Website, http://www.epa.gov/water/org_chart/index.htm, accessed on 5-30-2007.

BOX 7.1. Ecosystem services provided by coastal wetlands, adapted from the Millennium Ecosystem Assessment (2005).

1. Habitat and food web support
   - High production at base of food chain
     - Vascular plants
     - Microphytobenthos
     - Microbial decomposers
     - Benthic and phytal invertebrates (herbivores and detritivores)
   - Refuge and foraging grounds for small fishes and crustaceans
   - Feeding grounds for larger crabs and fishes during high water
   - Habitat for wildlife (birds, mammals, reptiles)
2. Buffer against storm wave damage
3. Shoreline stabilization
4. Hydrologic processing
   - Flood water storage
5. Water quality
   - Sediment trapping
   - Nutrient cycling
   - Chemical and metal retention
   - Pathogen removal
6. Biodiversity preservation
7. Carbon storage
8. Socioeconomic services to humans
   - Aesthetics
   - Natural heritage
   - Ecotourism
   - Education
   - Psychological health

Estuaries are notoriously idiosyncratic because of intrinsic differences among them in physical, geological, chemical, and biological conditions (Wolfe, 1986). There can also be considerable variation within an estuary. This variation exists over wide spectra of time and space (Remane and Schlieper, 1971). This high level of environmental variability in estuaries places physiological constraints on the organisms that can occupy them, generally requiring broad tolerances for varying salinity but also for temperature and other factors. Consequently, the organisms of estuaries represent a biota that may have unusually high intrinsic capability for species-level physiological adaptation to changing salinity, temperature, and other naturally varying aspects of historic climate change. The challenge is to predict how these species will respond to accelerated rates of change and how species interactions will alter communities and ecosystems.

Estuaries possess several features that render them unusually valuable for their ecosystem services, both to nature and to humans. The biological productivity of estuaries is generally high, with substantial contributions from vascular plants of historically extensive tidal marshes and coastal wetlands as well as from sea grasses and other submerged aquatic vegetation. A large fraction of the fisheries of the coastal ocean depend on estuaries to provide nursery or even adult habitat necessary to complete the life cycle of the fish or shellfish. Similarly, many species of coastal wildlife, including terrestrial and marine mammals and coastal birds, depend on estuaries as essential feeding and breeding grounds. Although depicting the ecosystem services of only one estuarine habitat, the wetlands and marshes, the Millennium Ecosystem Assessment (2005) provides a table of ecosystem services that helps indicate the types and range of natural and human values that are vested in estuarine ecosystems more broadly (Box 7.1). Partly in recognition of the value of estuaries and the threats to their health, the National Estuary Program (NEP) was established by Congress in 1987 and housed within EPA (Fig. 7.1).[2] After the establishment of this program, the 28 national estuaries were added over a 10-year period (Fig. 7.2).

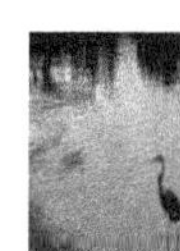

Estuaries represent the collection point past which runoff from the entire watershed must flow. The health and functioning of estuaries are at risk from pollutants that are discharged and released over the entire catchment area and reach these collection points. Degradation of estuarine habitats, water quality, and function is traceable to human modification of watersheds, with substantial cumulative consequences worldwide (Jackson *et al.*, 2001; Worm *et al.*, 2006; Lotze *et al.*, 2006). More recently, threats

[2] 33 U.S.C. 1251-1387 P.L. 100-4

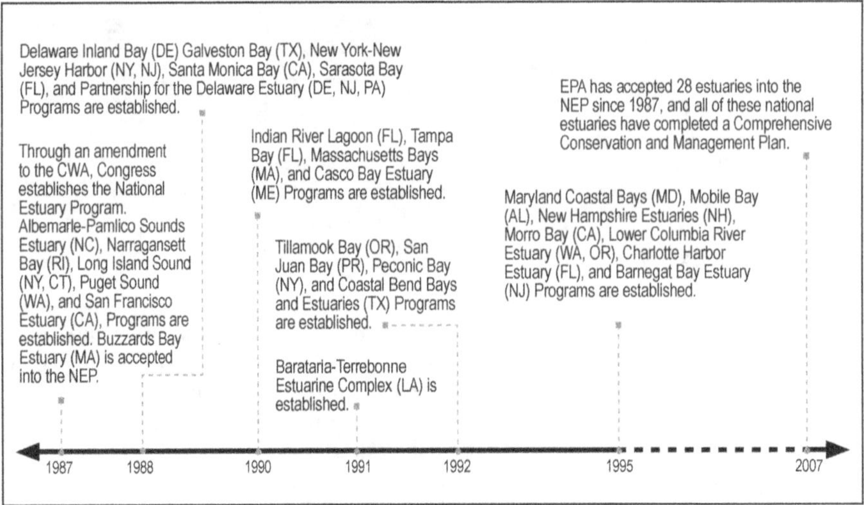

Figure 7.2. Timeline of National Estuaries Program Formation.[3]

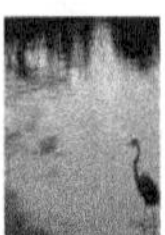

to estuaries have arisen from sources even closer to estuarine waters as human population migration and growth have targeted the coasts, especially waterfront property. Although more than half of the U.S. population now lives on the 17% of lands considered coastal, within the next 25 years human populations on the coast are expected to increase by 25% (National Coastal Assessment Group, 2000). Thus, the threats to estuarine ecosystems are not only widespread, requiring a basin-wide scope for management, but increasingly local as more people choose to occupy habitats of higher risk. The growing human occupation of estuarine shores increases the challenge of managing for climate change, because estuarine services are placed at growing risk from both direct impacts of changing climate as well as indirect consequences of human responses to personal and property risks from climate change.

7.2.2 Interpretation of National Estuary Program Goals

Under the goals of Section 320 of the Clean Water Act, each national estuary[4] is required to develop a Comprehensive Conservation and Management Plan (CCMP). Many national estuaries have watersheds found within a single state, and therefore their CCMP is contained within one state. Other estuarine watersheds are trans-boundary and more than one state participates. Emphasis is on "integrated, watershed-based, stakeholder-oriented water resource management."[5] These plans are produced by a full range of stakeholders within each national estuary through a process involving (1) assessments of trends in water quality, natural resources, and uses of the estuary; (2) evaluation of appropriate data; and (3) development of pollutant loading relationships to watershed and estuarine condition. The final CCMP is approved by the governors of the states in the study area and the EPA administrator. The programs are then obligated to implement the CCMPs and monitor effectiveness of actions.[6] Each national estuary prepares an annual plan, approved by EPA, to guide implementation of its CCMP.

The national estuaries represent a wide variety of sizes, geomorphologies, and watershed characteristics. For example Santa Monica Bay is a relatively small, open embayment or coastal lagoon; the Maryland Coastal Bays are a group

[3] **U.S. Environmental Protection Agency**, 2007: National Estuary Program: program profiles. EPA Website, http://www.epa.gov/owow/estuaries/list.htm, accessed on 5-30-2007.

[4] In the National Estuary Program, individual national estuaries are referred to as National Estuary Programs. To avoid confusion between individual estuary programs and the umbrella program, this chapter uses the term "national estuaries" to refer to the individual programs.

[5] **U.S. Environmental Protection Agency**, 2006: The National Estuary Program: a Ten Year Perspective. U.S. Environmental Protection Agency Website, http://www.epa.gov/owow/estuaries/aniv.htm, accessed on 4-6-2007.

[6] 33 U.S.C. 1251-1387 § 320

of more closed lagoons; and the Albemarle-Pamlico Sound is a complex of drowned river valleys emptying into largely closed coastal lagoons. The Columbia River Estuary and the Delaware Estuary are the more traditional drowned river valleys. This diversity has largely prevented classification, grouping, and synthetic assessment of the constituent national estuaries. The NEP separates national estuaries into four geographic regions: West Coast (six sites), Gulf of Mexico (seven sites), South Atlantic (six sites, including San Juan Bay, Puerto Rico), and Northeast (nine sites). Although the estuaries do not share easily identified geomorphic characteristics, they are recognized to share common stressors (Bricker *et al.*, 1999; Worm *et al.*, 2006; Lotze *et al.*, 2006). These stressors include "eutrophication, contamination from toxic substances and pathogens, habitat loss, altered freshwater inflows, and endangered and invasive species" (Bearden, 2001). This particular list ignores direct and indirect fishing impacts, which are important and included in many CCMPs. Even more importantly, this list fails to include the direct and indirect effects of climate change, particularly the threats posed by sea level rise.

A hallmark of the NEP is that it is largely a local program with federal support. While federal grants provide a critical source of base funding, most national estuaries have successfully raised significant local and state support, primarily to finance specific projects or activities. The individual national estuaries lack regulatory authority; thus they depend on voluntary cooperation using various incentives, plus existing federal, state, tribal, and local legislation and regulation. Their purpose is to coordinate these local efforts and promote the mechanisms to develop, implement, and monitor the CCMPs. The NEP was designed to provide funding and guidance for the 28 estuaries around the country to work in a bottom-up science-based way within the complex policy-making landscape of federal, state, and local regulations. Non-regulatory strategies must complement the limited federal and even state authority or regulations. Lessons learned about how monitoring, research, communication, education, coordination, and advocacy work to achieve goals are transferable to all estuaries, not just NEP members.

The overarching areas of concern in national estuaries can be classified as water quality, fisheries, habitat, wildlife, introduced species, biodiversity, human values, and freshwater quantity. More specifically the goals include "protection of public water supplies and the protection and propagation of a balanced, indigenous population of shellfish, fish, and wildlife, and [allowing] recreational activities, in and on water, [and requiring]...control of point and nonpoint sources of pollution to supplement existing controls of pollution."[2] Thus, overwhelmingly, the interest has been on anthropogenic impacts and their management (Kennish, 1999).

Within recent years, each national estuary has developed or begun to develop system-specific ecosystem status indicators. These indicators allow ongoing assessments of the success of management activities resulting from the CCMPs. However, almost none of the CCMPs mention climate change, and only one national estuary (Puget Sound) has completed a planning process to assess implications of climate change for the perpetuation of ecosystem services in its system (Snover *et al.*, 2005). Managers may fail to account for the effects of climate change on the estuaries if the choices of indicators are not reconsidered in the context of changing climate. Perhaps more importantly, climate change may confound the interpretation of the indicator trend results and thus the interpretation of the effectiveness of the CCMPs.

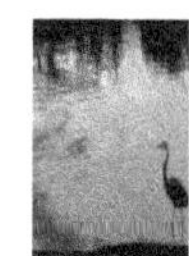

7.3 CURRENT STATUS OF MANAGEMENT SYSTEMS

7.3.1 Key Ecosystem Characteristics on Which Goals Depend

To understand how climate drivers might affect individual national estuaries, it is useful to identify the susceptibility of characteristics of the entire management system. At a large scale, the location of the estuary on Earth (*i.e.*, its latitude and longitude) determines its susceptibility. Climate varies over the globe, and expectations for change likewise differ geographically on a global scale. Expected temperature and precipitation changes and range shifts can be estimated from global-scale geographic position quite well, whereas local variation of these and other variables (*e.g.*, winds) of climate change are less predictable.

Next in scale is the airshed. This is the area capable of influencing the estuary through the contribution of quantitatively significant pollutants, especially nitrogen oxides (NO_x). For the Chesapeake Bay, this area includes Midwestern states, the source of nutrients from industrial and transportation activities. Estuaries on the Gulf and East coasts are likely to have different dependencies on their airsheds for nutrient enrichment than their western counterparts. Western estuaries are affected more by fog banks emanating from coastal waters. Climate drivers that change wind, ultraviolet radiation, and precipitation patterns are particularly important at this scale.

Next in hierarchical context is the watershed. The NEP takes a watershed perspective to management. Land and watershed use, population density, and regulatory effectiveness combine to determine the potential loading of pollutants, extraction of freshwater and resources, and transformation of habitat and coastline. Climate change can influence each of these factors. Changes in temperature, sea level, storminess, precipitation, and evapotranspiration patterns can alter human settlement and migration, agricultural and fisheries practices, and energy and resource use. These responses are likely to be long-term and large-scale, although their influence on estuarine dynamics may be exhibited on shorter time scales. For example, seasonal nutrient loading varies as a result of changes in tourism or crop choice. These factors largely affect the concentration of nutrients, while changes in runoff and river flow affect the discharge component of loading.

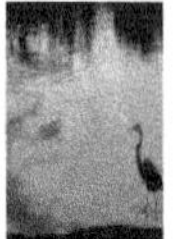

At the opposite end of the estuary is the marine environment, which also serves as an intermixing boundary susceptible to climate change. The oceans and coastal marine waters have responded—or are expected to respond—to climate change by changes in sea level, circulation patterns, storm intensity, salinity, temperature, and pH. Some of these factors may change little over the large scale, but may be altered locally outside the mouths of estuaries. All of these factors influence the biota, with all but pH exerting additional indirect effects by modifying estuarine hydrodynamics.

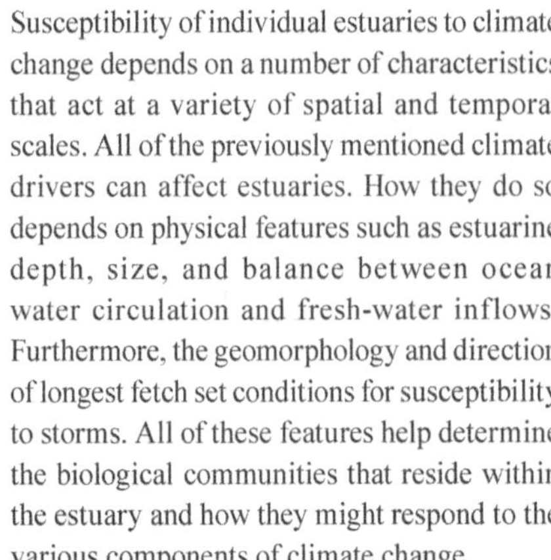

Susceptibility of individual estuaries to climate change depends on a number of characteristics that act at a variety of spatial and temporal scales. All of the previously mentioned climate drivers can affect estuaries. How they do so depends on physical features such as estuarine depth, size, and balance between ocean water circulation and fresh-water inflows. Furthermore, the geomorphology and direction of longest fetch set conditions for susceptibility to storms. All of these features help determine the biological communities that reside within the estuary and how they might respond to the various components of climate change.

The way in which a specific estuary responds to climate change depends on the anthropogenic stressors acting on it. These stressors include those that pollute and contaminate the system, as well as those that remove or disrupt estuarine resources. Pollutants include nutrients, metals, pathogens, sediments, and organic toxicants. Extractions include uses of fresh and brackish water, sediments, and living resources within the ecosystem. Disruption of a variety of biological communities occurs through overfishing, introduction of invasive species, habitat destruction, damming, boat traffic, and shoreline conversion and stabilization activities.

Finally, there are the social, political, and economic contexts for susceptibility. Some of these contexts play out in ways already mentioned. But it is clear that stakeholder attitudes about estuaries and their perceptions about climate change are critical to wise management for climate change. Each stakeholder group, indeed each individual, uses estuaries in different ways and places different importance on specific ecosystem services. One aim of this report is to provide a common body of knowledge to stakeholders and to managers at higher levels (local, state, tribal, and federal governments) to inform their choices.

7.3.2 Current Stressors of Concern

Estuaries are generally stressful environments because of their strong and naturally variable gradients of salinity, temperature, and other parameters. However, estuaries are also essential feeding and reproduction grounds, and provide refuge for a wide variety of seasonal and permanent inhabitants. Throughout history, estuaries have been focal points of human settlement and resource use, and humans have added multiple stressors to estuarine ecosystems (Lotze *et al.*, 2006). A stressor is any physical, chemical, or biological entity that can induce an adverse response (U.S. Environmental Protection Agency, 2000). This document focuses specifically on those stressors that significantly affect the services that estuaries are managed to provide. The major stressors currently imposed on estuaries are listed in Table 7.1. Almost all current efforts to manage estuarine resources are focused on these stressors (Kennish, 1999 and the various CCMPs).

Several stressors result from modified rates of loading of naturally occurring energy and materials. Nutrient loading is perhaps the most studied and important material addition. Although essential to the primary production of any open ecosystem, too much nutrient loading can cause eutrophication, the subject of considerable concern for estuaries and the target for much management action (Nixon, 1995; Bricker *et al.*, 1999). Nutrient (especially nitrogen) loading comes from diverse point- and non-point sources, including agriculture, aquaculture, and industrial and municipal discharges, and can lead to harmful and nuisance algal blooms, loss of perennial vegetation, bottom-water hypoxia, and fish kills.

Sediment delivery has also been altered by human activities. Again, sediments are important to estuarine ecosystems as a material source for the geomorphological balance in the face of sea level rise, and for nutrients (especially phosphorus) for primary production. However, land clearing, agriculture, and urban land use can increase sediment load (Howarth, Fruci, and Sherman, 1991; Cooper and Brush, 1993; Syvitski *et al.*, 2005), while dams may greatly restrict delivery and promote deltaic erosion (Syvitski *et al.*, 2005). Historically, sediment loading has increased on average 25-fold, and nitrogen and phosphorus loading almost 10-fold, in estuaries since 1700 (Lotze

Table 7.1. The major stressors currently acting on estuaries, and their expected impacts on management goals, as determined by consensus opinion of the contributing authors. Evidence is mounting that sea level rise is already having direct and indirect impacts on estuaries (e.g., Galbraith *et al.*, 2002), but because this factor has not yet been widely integrated into management, we do not list it here despite its dominating significance in future decades.

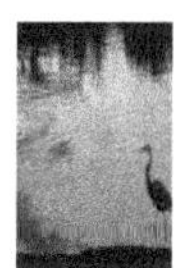

Stressor	Water Quality	Fisheries	Habitat	Human Value & Welfare	Water Quantity
Excess Nutrients	negative	positive then negative	positive then negative	positive then negative	
Sediments	negative	positive or negative	positive or negative	negative	
Pathogens	negative	negative		negative	
Oyster Loss & Habitat Destruction	negative	negative	negative	negative	
Benthic Habitat Disturbance	negative	positive or negative	positive then negative	negative	
Wetland Habitat Loss from Development	negative	negative	negative	positive or negative	positive or negative
Toxics	negative	negative	negative	negative	
Invasive Species	positive or negative	positive or negative	positive or negative	positive or negative	
Thermal Pollution	positive then negative or down	positive then negative	positive then negative or down	positive then negative	
Biological Oxygen Demand (BOD)	negative	negative	negative	negative	

et al., 2006). Because riverine loading of both nutrients and sediments depends on their concentration and river flow, modifications of river flow will further alter the amount and timing of material delivery. River flow also contributes to the energy budget through mechanical energy. River flow may be a major determinant of flushing times, salinity regime, and stratification, and thus determine community structure and resource use patterns. Modifications in river flow come from dam management decisions, land development, loss of riparian wetlands, extraction of freshwater, and surface and ground water consumption. Thermal pollution, largely from power plants, is a direct enhancement of energy with resultant local changes in metabolic rates, community structure, and species interactions.

Human activities also cause or enhance the delivery of materials and organisms that are not normally part of the natural systems. Pathogen loading compromises the use of estuarine resources, causing shellfish bed closures and beach closures (*e.g.*, Health Ecological and Economic Dimensions of Global Change Program, 1998), human health advisories, and diseases to estuarine organisms themselves. Other anthropogenic contributions include the discharge and ongoing legacy of organic wastes and persistent organic pollutants (*e.g.*, DDT, dioxin, PCBs, petroleum) (Kennish, 1999). The toxicity of some of the persistent organic pollutants has been recognized for decades, dating to the publication of *Silent Spring* by Rachel Carson (1962). More recently, the potential importance of other endocrine-disrupting chemicals is causing concern (Cropper, 2005). Added to these organic pollutants are metals entering estuaries from direct dumping, riverine waters, sediments, and atmospheric deposition. Moreover, biodegradable organic wastes contribute to eutrophication and dissolved oxygen deficits (Nixon, 1995). Finally, the introduction and spread of non-indigenous species are enhanced by globalization and shipping, intentional decisions for commerce or other human use, and unintentional actions (Mooney and Hobbs, 2000). For those locations that have been surveyed, the known number of resident non-indigenous species ranges from about 60 to about 200 species per estuary in the United States (Ruiz *et al.*, 1997; Lotze *et al.*, 2006), likely the result of an increasing rate of invasions over the last 300 years (Lotze *et al.*, 2006).

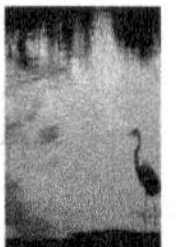

Human use and development in and around estuaries alter wetland and subtidal habitats directly. Wetland destruction has occurred during much of human history as a result of the perceptions of wetlands as wastelands and the value of waterfront land. For example, 12 estuaries around the world have lost an average of more than 65% of their wetland area (with a range of 20–95%) over the last 300 years (Lotze *et al.*, 2006). Wetland habitat loss from development continues, despite changes in perceptions about wetland value and regulations intended to protect wetlands. Coastal wetlands represent a diverse assortment of hydrogeomorphic classes (Brinson, 1993; Christian *et al.*, 2000), both sea-level controlled (*e.g.*, marshes and mangroves), non-sea-level controlled (*e.g.*, swamps, fens, bogs, and pocosins) and subtidal (*e.g.*, submerged aquatic vegetation (SAV), seagrass, and macroalgal) habitats. Supratidal and intertidal wetlands are subject to land use change, dredging and filling, and changes in water quality. Subtidal habitats are particularly susceptible to not only these impacts but also activities within the water. For example, SAV loss also occurs from bottom-disturbing fishing practices and eutrophication. Oyster reef habitat destruction occurs from direct exploitation and bottom disturbance from fishing practices (*e.g.*, trawling). For 12 study sites around the world, both seagrass meadows and oyster reefs have experienced substantial losses over the last 300 years (>65% and about 80%, respectively) (Lotze *et al.*, 2006). Together with the loss of wetlands, these changes have resulted in great reductions of essential nursery habitats, important filtering functions (nutrient cycling and storage), and coastal protection (barriers and floodplains) in estuaries (Worm *et al.*, 2006; Lotze *et al.*, 2006).

Another important anthropogenic stressor in estuaries is the extraction of living and non-living material that alters estuarine ecosystem structure and functioning. Historically, estuaries provided a wide variety of resources used and valued by humans as sources of food, fur, feathers, fertilizer, and other materials (Lotze *et al.*, 2006). Since the 19th century, however, the ecological service of estuaries receiving greatest management attention has been their support of fisheries. Pollution, damming, and habitat destruction affect fisheries. Recently, more emphasis has been placed on overfishing as a negative impact, not only on target species but also on the community and food web structure (*e.g.*, Dayton, Thrush, and Coleman, 2002). Large apex predators have been greatly reduced from many, if not most, estuarine and coastal ecosystems (Lotze *et al.*, 2006). The absence of these large consumers (including marine mammals, birds, reptiles, and larger fish) translates through the food web, creating ecosystem states that are distinct from those of the past (*e.g.*, Jackson *et al.*, 2001; Lotze *et al.*, 2006; Myers *et al.*, 2007). Ongoing fishing pressure targets species lower and lower in the food chain, affecting detritivorous and herbivorous invertebrates and marine plants; consequences can include further alteration of ecosystem structure and functioning and negative effects on habitat integrity and filtering functions (Pauly *et al.*, 1998; Worm *et al.*, 2006; Lotze *et al.*, 2006). Management goals to stabilize current or restore former ecosystem states are jeopardized if large consumers are not also recovered (Jackson *et al.*, 2001).

It is rare that an estuary is subject to only one of these stressors. Management decisions must consider not only stressors acting independently but also interacting with each other (Breitburg, Seitzinger, and Sanders, 1999; Lotze *et al.*, 2006). Multiple stressors can interact and cause responses that cannot be anticipated from our understanding of each one separately. For example, Lenihan and Peterson (1998) demonstrate that habitat damage from oyster dredging and the stress of bottom-water hypoxia interact to affect oyster survival. Tall oyster reefs, both those that remain and those that have been rebuilt, project above hypoxic bottom waters and therefore allow oyster survival in the upper wind-mixed layers even as water quality further deteriorates. Unfortunately, management of fisheries and water quality is done by different agencies, inhibiting the integrated approach that such interacting stressors demand.

Interactive effects of multiple stressors are likely to be common and important because of both the interdependence of physiological rate processes within individuals and the interdependence of ecological interactions within communities and ecosystems (Breitburg and Riedel, 2005). Individual stressors fundamentally change the playing field upon which additional stressors act, by selecting for tolerant species while also changing the abundance, distribution, or interactions of predators, prey, parasites, hosts, and structural foundation species (*e.g.*, organisms such as bivalves and corals that create physical structures upon which other species depend). These direct and indirect effects can be common when stressors occur simultaneously, but they also occur from exposure to stressors in sequence. Across hierarchical levels from individuals through ecosystems, the recovery period from a particular stressor can extend beyond the period of exposure, thus influencing responses to subsequent stressors. For example, Peterson and Black (1988) demonstrated that bivalves that were already stressed from living under crowded conditions exhibited higher mortality rates after experimental application of the stress of sedimentation. Moreover, effects of stressors on indirect interactions within populations and communities can extend the spatial scale of stressor effects and delay recovery (Peterson *et al.*, 2003), increasing the potential for interactions with additional stressors. For example, years after the Exxon Valdez oil spill, female harlequin ducks exposed to lingering oil during feeding on benthic invertebrates in contaminated sediments, and exhibiting activation of detoxification enzymes, suffered lower survivorship over winter. Winter is a period of energetic stress to these small-bodied ducks (Peterson *et al.*, 2003). On longer time scales, heritable adaptations that increase tolerance to one class of stressors may enhance susceptibility to others (Meyer and Di Giulio, 2003).

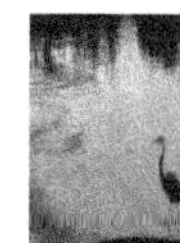

One hallmark of the NEP is the recognition that management actions need to take account of the complexity of the larger watershed and the potentially diverse socioeconomic demands and objectives within them. The NEP tracks habitat restoration and protection efforts with annual updates from the component estuaries.[7] The reality of interacting stressors has important implications for estuarine management. Specifically, because climate change affects some pre-existing stressors,

[7] **U.S. Environmental Protection Agency**, 2007: Performance indicators visualization and outreach tool introduction. EPA Website, www.epa.gov/owow/estuaries/habitat/index.html, accessed on 7-25-2007.

and the magnitude of such interactive effects typically increases with the intensity of each stressor, more effective management of the pre-existing stressor can help reduce climate change consequences.

7.3.3 Legislative Mandates Guiding Management of Stressors

Because of the intrinsically wide range of estuarine resources and diversity of human activities that influence them, management of estuarine services is achieved via numerous legislative acts at the federal level. Many of these acts possess state counterparts, and local laws—especially land use planning and zoning—also play roles in management of estuarine services. This web of legal authorities and guiding legislation is a historical legacy, reflective of prevailing management that compartmentalized responsibilities into multiple agencies and programs.

The presentation here of applicable federal legislative acts is long, yet incomplete, and does not attempt to list state and local laws. One motivation in providing this spectrum of applicable legislation is to illustrate the challenges involved for estuaries in the integration of management authorities that is urged under the umbrella of ecosystem-based management by the U.S. Commission on Ocean Policy.

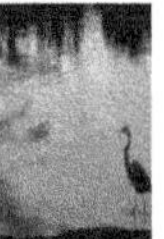

7.3.3.1 Basin-Wide Management of Water Quality

As one of the tools to meet the goal of "restoration and maintenance of the chemical, physical, and biological integrity of the Nation's waters" under §402 of the Federal Water Pollution Control Act, any entity that discharges pollutants into a navigable body of water must possess a National Pollutant Discharge Elimination System (NPDES) permit.[8] This requirement applies to public facilities such as wastewater treatment plants, public and private industrial facilities, and all other point sources. While EPA was the original administrator of the program, many states have now assumed the administrative function. All states have approved State NPDES Permit Programs except Alaska, the District of Columbia, Idaho, Massachusetts, New Hampshire, New Mexico,

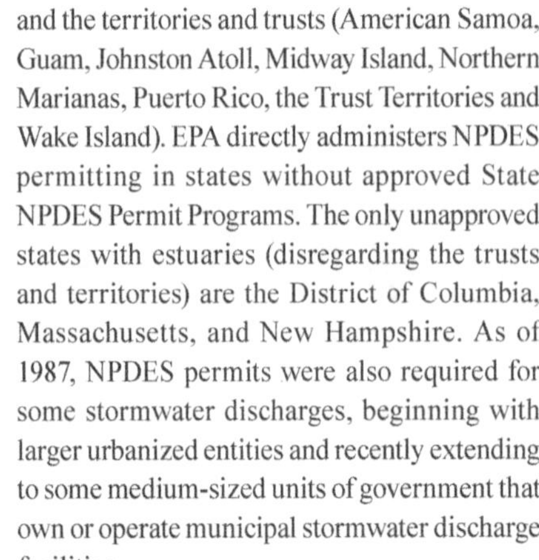

and the territories and trusts (American Samoa, Guam, Johnston Atoll, Midway Island, Northern Marianas, Puerto Rico, the Trust Territories and Wake Island). EPA directly administers NPDES permitting in states without approved State NPDES Permit Programs. The only unapproved states with estuaries (disregarding the trusts and territories) are the District of Columbia, Massachusetts, and New Hampshire. As of 1987, NPDES permits were also required for some stormwater discharges, beginning with larger urbanized entities and recently extending to some medium-sized units of government that own or operate municipal stormwater discharge facilities.

Although the content, style, and length of any given NPDES permit for point-source discharge will be slightly different depending on where and when it is written, all permits contain certain core components mandated by the Clean Water Act, including testing, monitoring, and self reporting. NPDES permits are renewed every five years, and monitoring and/or reporting requirements may change. These changes are determined by the local Regional Water Quality Control Boards or the State Water Resources Control Board through their research and monitoring efforts.

In addition to traditional NPDES permitting for point sources, states are required by the Clean Water Act of 1972 (modified in 1977, 1981, and 1987) to manage and protect water quality on a basin-wide scale. This involves assessing the assimilative capacity of the water body for wastes of various sorts and managing loads from all sources to prevent water quality violations in any of the key water quality standards used to indicate degradation. The inputs of most concern for estuaries are nutrient loading, sedimentation, BOD, and fecal coliform bacteria. EPA has developed several technical guidance manuals to assist the states in their basin-wide planning, including those for nutrients, sediments, and biocriteria of estuarine health. When chronic water quality violations persist, then TMDLs (total maximum daily loads) are mandated by EPA and must be developed to cap loading and restore water quality. TMDLs are also now triggered by inclusion of any water body on the 304(d) list of impaired waters, which the states are obligated to provide annually to EPA. In the 2000s, EPA has expanded the scope of

8 33 U.S.C. 1251-1387 § 420

the NPDES program to include permits for municipal stormwater discharges, thereby bringing a traditionally non-point source of water pollution under the NPDES permitting program. Non-point sources must also be considered in any basin-wide plans, including establishment of TMDLs and allocation of loads among constituent sources to achieve the necessary loading caps. Climate change has great potential to influence the success of basin-wide water quality management and the effectiveness of TMDLs through possible changes in rainfall amounts and patterns, flooding effects, stratification of waters, salt penetration and intrusion, and acidification.

7.3.3.2 Habitat Conservation under Federal (Essential Fish Habitat) and State Fishery Management Plans

As administered under NOAA, the Magnuson Fishery Conservation and Management Act of 1976 (amended as the Sustainable Fisheries Act (SFA) in 1996[9] and reauthorized as Magnuson-Stevens Fishery Conservation and Management Reauthorization Act (MSA) of 2006[10] established eight regional fishery management councils that are responsible for managing fishery resources within the federal 200-mile zone bordering coastal states. Management is implemented through the establishment and regulation of Fishery Management Plans (FMPs). In addition to "conservation and management of the fishery resources of the United States...to prevent overfishing, rebuild overfished stocks and insure conservation," the Act also mandates the facilitation of long-term protection of *essential fish habitats,* which are defined as "those waters and substrate necessary to fish for spawning, breeding, feeding, or growth to maturity." The Act states "One of the greatest long-term threats to the viability of commercial and recreational fisheries is the continuing loss of marine, estuarine, and other aquatic habitats." It emphasizes that habitat considerations "should receive increased attention for the conservation and management of fishery resources of the United States" and "to promote the protection of essential fish habitat in the review of projects conducted under Federal permits, licenses, or other authorities that affect or have the potential to affect such habitat."

[9] P.L. 94-265
[10] P.L. 109-479

FMPs prepared by the councils (or by the Secretary of Commerce/NOAA) must describe and identify essential fish habitat to minimize adverse effects on such habitat caused by fishing. In addition, they must identify other actions to encourage the conservation and enhancement of essential fish habitat, and include management measures in the plan to conserve habitats, "considering the variety of ecological factors affecting fishery populations."[2]

Because managed species use a variety of estuarine/coastal habitats throughout their life histories, few are considered to be "dependent" on a single, specific habitat type (except, for example, larger juvenile and adult snappers and groupers on ocean hard bottoms) or region. As a result, federal FMPs do not comprehensively cover species' habitats that are not specifically targeted within their region. In addition, the only estuarine-dependent fish stocks under federal management authority are migratory stocks, such as red drum and shrimp, so estuarine habitats are not a key focus for essential fish habitat. However, many states also have FMPs in place or in preparation for target fisheries under their jurisdiction (the non-migratory inshore species) and participate with the regional councils under the SFA/MSA.

Thus, threats to marshes and other estuarine systems that constitute essential fish habitat or state-protected fisheries habitat should include all potential stressors, whether natural or anthropogenic, such as climate change and sea level rise. Although essential fish habitats have been codified for many fisheries, and science and management studies have focused

Kevin Rosseel, EPA

on the status and trends of fisheries-habitat interactions, most management consideration has targeted stresses caused by different types of fishing gear. Because few fisheries take place in emergent marshes, the essential fish habitat efforts have not provided much protection to this important habitat. Seagrass and oyster reef habitats have been targeted for additional management concern because of the federal essential fish habitat provisions. State protections of fishery habitat vary, but generally include salt marsh and other habitats. Nearly two decades ago, EPA projected extensive loss of coastal marshes and wetlands from sea level rise by 2100, with an elimination of 6,441 square miles (65%) of marshes in the continental United States associated with a probable rise of 1m (Park *et al.*, 1989).

7.3.3.3 Estuarine Ecosystem Restoration Programs

While comprehensive planning of coastal restoration is inconsistent at the national level, a number of national, regional, and local programs are coordinated to the extent that stressors are either the target of restoration or addressed as constraints to restoration. These programs tend to be oriented toward rehabilitation of injuries done by individual stressors, such as eutrophication or contaminants, or toward restoration of ecosystems that have not been so extensively modified that their loss or degradation is not irreversible. Federal programs that authorize restoration of estuaries include:

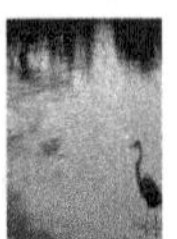

Estuary Restoration Act of 2000 (P.L. 106-457, Title I)

Probably the most prominent federal program that involves non-regulatory restoration in the nation's estuaries is the Estuary Restoration Act of 2000 (ERA). The ERA promotes estuarine habitat restoration through coordinating federal and non-federal restoration activities and more efficient financing of restoration projects. It authorizes a program under which the Secretary of the Army, through the Corps of Engineers (USACE), may carry out projects and provide technical assistance to meet the restoration goal. The purpose of the Act is to promote the restoration of estuarine habitat; to develop a national Estuary Habitat Restoration Strategy for creating and maintaining effective partnerships within the federal government and with the private sector; to provide federal assistance for and promote efficient financing of estuary habitat restoration projects; and to develop and enhance monitoring, data sharing, and research capabilities. Guidance provided by an Estuary Habitat Restoration Council, consisting of representatives of NOAA, EPA, USFWS, and USACE, includes soliciting, evaluating, reviewing, and recommending project proposals for funding; developing the national strategy; reviewing the effectiveness of the strategy; and providing advice on development of databases, monitoring standards, and reports required under the Act. The Interagency Council implementing the ERA published a strategy in December of 2002 with the goal of restoring one million acres of estuarine habitat by the year 2010. Progress toward the goal is being tracked via NOAA's National Estuaries Restoration Inventory.

Although the guiding principles that contributed to the development of this legislation argued for the "need to learn more about the effects of sea level rise, sedimentation, and a host of other variables to help set appropriate goals and success indicators for restoration projects in their dynamic natural environments," climate change is not explicitly addressed in the ERA. Similarly, the Council's Estuarine Habitat Restoration Strategy, published in 2002, neglects to explicitly mention climate change or sea level rise.

National Estuary Program and National Monitoring Program (EPA)

The National Estuary Program (NEP), administered under Section 320 of the 1987 amended Clean Water Act, focuses on point- and non-point source pollution in targeted, high-priority estuarine waters. Under the NEP, EPA assists state, regional, and local governments, landowners, and community organizations in developing a Comprehensive Conservation and Management Plan (CCMP) for each estuary. The CCMP characterizes the resources in the watershed and estuary and identifies specific actions to restore water quality, habitats, and other designated beneficial uses. Each of the 28 national estuaries has developed a CCMP to meet the goals of Section 320. Because the primary goal of the NEP is maintenance or restoration of water quality in estuaries, the CCMPs tend to focus on source control or treatment of pollution. NEP tracks estuarine habitat restoration and protection, with annual updates using information provided by the constituent national estuaries.[7] While climate change is not considered a direct stressor, it is gradually being addressed in individual CCMPs in the context of potential increased nutrient loading from watersheds under future increased precipitation. For instance, the Hudson River Estuary Program has initiated with other partners an ongoing dialogue about how climate change constitutes a future stressor of concern to the estuary and its communities.[11] The Puget Sound and Sarasota Bay Estuary Programs have been the most proactive relative to anticipating a range of climate change challenges, although their assessments have been completed only recently.

7.3.3.4 National Coastal Zone Management Act and Its Authorized State Programs

The federal Coastal Zone Management Act of 1972 (CZMA) provides grants to states to develop and implement federally approved coastal zone management plans. Approval of the state plan then allows that state to participate in reviews of federal actions and determine whether they are consistent with the approved state plan. In addition, CZMA authorized establishment of the National Estuarine Research Reserve System (NERRS). Individual states have responded by creating various governmental structures, legislation, commissions, and processes for developing and implementing the coastal planning process. Planning extends down to the local level as local communities take responsibility for local Land Use Plans, which are then reviewed for approval by the state authority. Thus, this process has substantial capacity for responding to and adapting to climate change. CZMA explicitly identifies planning for climate change as one of its mandates: "Because global warming may result in a substantial sea level rise with serious adverse effects in the coastal zone, coastal states must anticipate and plan for such an occurrence."[12] The act calls for balancing of the many uses of the coastal zone with protection of natural resources.

The Coastal States Organization, an organization established in 1970 to represent the governors of the 35 coastal states, commonwealths, and territories on policy issues related to management of coastal and ocean resources, released a recent report reviewing how the states are using their Coastal Program under the CZMA to anticipate climate change and practice adaptive management.[13] This report identifies the very same suite of climate change impacts that we emphasize and address here. The report used surveys, to which 18 state programs responded, to develop information on how the state Coastal Management Programs are currently addressing climate change and the new challenges posed by accelerating rates of sea level rise, enhanced frequencies of intense storms, and rainfall and flood risk changes.[13] Several states are actively examining climate change impacts to their coastal zone planning, often through interagency commissions. New policies are being considered and developed in response to rising rates of sea level rise and enhanced storm and flood risk to reconsider siting of public infrastructure, site-level project planning, wetland conservation and restoration, shoreline building setbacks, building elevations, and alternatives to shoreline "armoring" to counteract erosion.

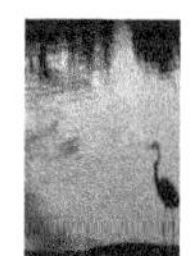

The NOAA NERRS Program authorized by CZMA now includes 27 constituent estuaries from around the country. This program uses a local grassroots process to help monitor and create public awareness of the resources, threats, and values of constituent estuaries. Clearly, the goals of NERRS are compatible with the goals of the National Estuary Program and CZMA, implying need for cross-agency and federal-state partnerships to develop integrated management adaptations to climate change.

---

11 **New York State Department of Environmental Conservation**, 2006: Hudson Valley climate change conference, December 4, 2006. New York State Department of Environmental Conservation, http://www.dec.state.ny.us/website/hudson/hvcc.html, accessed on 3-23-2007.

12 16 U.S.C. 1451-1456 P.L. 92-583

13 **CSO Climate Change Work Group**, 2007: *The Role of Coastal Zone Management Programs in Adaptation to Climate Change*. Coastal States Organization.

7.3.3.5 State Sedimentation and Erosion Control, Shoreline Buffers, and Other Shoreline Management Programs Involving Public Trust Management of Tidelands and Submerged Lands

Protection from shoreline erosion has a long legal history, as far back as the tenets of property law established under the court of Roman Emperor Justinian.[14] In general, property law protection of tidelands held in public trust (most of the U.S. coastline) is conveyed either as the *law of erosion* (public ownership migrates inland when shores erode) or the *public trust doctrine* (the state holds tidelands in trust for the people unless it decides otherwise). Shoreline planners in many states (*e.g.*, Texas, Rhode Island, South Carolina, and Massachusetts) use these laws to plan for natural shoreline dynamics, including policies and tools such as "rolling easements" (*i.e.*, as the sea rises, the public's easement "rolls" inland; owners are obligated to remove structures if and when they are threatened by an advancing shoreline), setbacks (*i.e.*, prohibitions against development of certain areas at a set distance from the shoreward property line), prohibition of future shoreline armoring, and direct purchase of land that will allow wetlands or beaches to shift naturally (IPCC, 2001).[14] Some states are beginning to prohibit new structures in areas likely to be eroded in the next 30–60 years (*e.g.*, North Carolina through its Coastal Resources Commission).

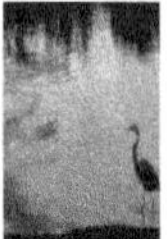

7.3.3.6 Species Recovery under Federal Endangered Species Act

Recovery plans for aquatic species that are threatened or endangered under the Endangered Species Act (ESA)[15] may be contingent on implicit assumptions about habitat conditions in the coastal zone. However, explicit accounting for impacts and strategic designing of recovery efforts to consider climate variability and change is rare. A recent analysis of current ESA recovery plans indicates that of 101 plans that mention climate change, global warming, or related terms, only 60 actually discuss these topics, and only 47 identify climate change or its effects as a threat, possible threat, or factor in the species' decline.[16] Strategies and approaches that specifically address climate include monitoring for metapopulation variability that could link climate variation to extinction/recolonization probabilities or to unpredictable changes in existing or proposed future habitat. For example, the NOAA recovery plan for the Hawaiian monk seal (*Monachus schauinslandi*) suggests that habitat loss that has already been observed could be exacerbated by "...sea level rise over the longer term [that] may threaten a large portion of the resting and pupping habitat..." (National Marine Fisheries Service, 2006).

Climate variability and change will undoubtedly involve an even more consequential response by diadromous fishes and macroinvertebrates that require extensive, high-quality juvenile or adult transitional habitats during migrations between ocean and estuarine or freshwater aquatic systems. For example, in the Pacific Northwest and Alaska, sea level rise and shifts in timing and magnitude of snowmelt-derived riverine runoff may be particularly exacerbated by climate variability and change. Consequently, the recovery plans for threatened or endangered Pacific salmon (*e.g.*, juvenile, "ocean-type" Chinook [*Oncorhynchus tshawytscha*] and summer chum [*O. keta*] salmon) may need to account for their extreme sensitivity to climate-induced changes in environmental conditions of their estuarine wetland habitats during different life stages of the fish.

7.3.3.7 Wetland Protection Rules Requiring Avoidance, Minimization, and Mitigation for Unavoidable Impacts

Federal jurisdiction of waters of the United States began in 1899 with the Rivers and Harbors Act of 1899, and wetlands were included in that definition with the passing of the Clean Water Act of 1977 (CWA). This jurisdiction does not extend beyond the wetland/upland boundary. However, many state environmental laws, such as those of New York[17] and New Jersey, require permits for alterations in adjacent upland areas in addition to protecting the wetland itself. While not originally intended for the purpose of

[14] Spyres, J., 1999: Rising tide: global warming accelerates coastal erosion. Erosion Control, http://www.forester.net/ec_9909_rising_tide.html, accessed on 3-22-2007.

[15] 16 U.S.C. 1531-1544, 87 Stat. 884

[16] **Jimerfield**, S., M. Waage, and W. Snape, 2007: Global Warming Threats and Conservation Actions in Endangered Species Recovery Plans: a Preliminary Analysis. Center for Biological Diversity.

[17] **New York State**, 1992: Tidal wetlands - land use regulations. **6 NYCRR Part 661**.

increasing climate change preparedness, many of these regulations could facilitate adaptation to sea level rise (Tartig *et al.*, 2000).

The U.S. Army Corps of Engineers regulates dredging, the discharge of dredged or fill material, and construction of structures in waterways and wetlands through Section 404 of the CWA,[18] the provisions of which have been amended progressively through 1987. Although not explicitly required within the language of the amended law, the CWA provides the Corps with the implicit authority to require that dredge or fill activities avoid or minimize wetland impacts (Committee on Mitigating Wetland Losses, National Research Council, 2001). The Corps and EPA developed criteria (Section 404(b)(1) guidelines) that over the years (latest, 1980) have defined mitigation as both minimization of wetland impacts and compensation for wetland losses. Thus, mitigation has been loosely interpreted to include a range of actions from wetland restoration and enhancement to creation of wetlands where they have never occurred. However, a 1990 memorandum of agreement between the Corps and EPA established that mitigation must be applied sequentially. In other words, an applicant must first avoid wetland impacts to the extent practicable, then minimize unavoidable impacts, and finally—only after these two options are reasonably rejected—compensate for any remaining impacts through restoration, enhancement, creation, or in exceptional cases, preservation (Committee on Mitigating Wetland Losses, National Research Council, 2001). The Corps now grants permits for shoreline development that include armoring of the present shoreline, which guarantees future loss of wetlands as sea level rises, thereby violating the requirement for mitigation in the application of this authority (Titus, 2000).

7.3.3.8 Compensatory Restoration Requirements for Habitat and Natural Resource Injuries from Oil Spills or Discharges of Pollutants

Federal legislation requires compensatory restoration of estuarine habitats and natural resources after environmental incidents such as spills of oil or other toxicants (*e.g.*, Fonseca, Julius, and Kenworthy, 2000). For example, the Oil Pollution Act of 1990 specifies the procedures that federal agencies are required to follow to assess injury from pollution events and to conduct quantitatively matching restoration actions so the responsible parties replace the lost ecosystem services. Similar federal legislation, such as the Comprehensive Environmental Response, Compensation, and Liability Act, also specifies formation of natural resource trustees composed equally of state and federal agencies to oversee the injury assessments, pursue funding from the responsible party(ies) sufficient to achieve restoration, and then design and implement the restoration. The process of restoration typically involves rehabilitation of biogenic habitats such as salt marshes, seagrass beds, or oyster reefs. The modeling done to insure that the restoration will provide ecosystem services equal to the injuries may need to be modified to reflect impacts of climate change, because services from habitat restorations are assumed to extend for years and even decades in these computations.

7.3.3.9 Federal Legislation Controlling Location of Ballast Water Release to Limit Introduction of Non-Indigenous Marine and Estuarine Species

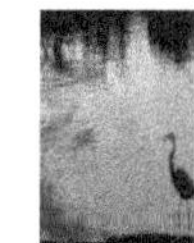

One of the more troubling implications of climate change for estuaries is the probability of expanded distributions of non-indigenous species with the potential of progressively warmer waters in temperate zones. Ballast water discharged from ships in harbors after transiting from foreign ports (and domestic estuaries with extensive species invasions, such as San Francisco Bay) is one of the major sources of aquatic nuisance species. The primary federal legislation regulating ballast water discharge of invasive species is the National Invasive Species Act of 1996, which required the Coast Guard to establish national voluntary ballast water management guidelines. Because of a lack of compliance under the initial nationwide self-policing program that began in 1998, the voluntary program became mandatory in 2004. All vessels equipped with ballast water tanks that enter or operate within U.S. waters must now adhere to a national mandatory ballast water management program and maintain a ballast water management plan. Ballast water

[18] Codified generally as 33 U.S.C. §1251; 1977.

BOX 7.2. Estuarine properties and the climate-driven processes that affect them. The order of the properties and processes is a subjective ranking of the importance of the property and the severity of the particular process.

Semi-enclosed geomorphology is affected by:
- sea level rise – (Rahmstorf, 2007)
- storm intensity – (Emanuel, 2005)
- storm frequency – (Emanuel, 2005)
- storm duration – (Emanuel, 2005)
- sediment delivery – (Cloern et al., 1983)

Fresh water inflow is affected by:
- watershed precipitation – (Arora, Chiew, and Grayson, 2000)
- system-wide evapotranspiration – (Arora, Chiew, and Grayson, 2000)
- timing of maximum runoff – (Ramus et al., 2003)
- groundwater delivery – (Wolock and McCabe, 1999)

Water column mixing is affected by:
- strength of temperature-driven stratification – (Li, Gargett, and Denman, 2000)
- strength of salinity-driven stratification – (Li, Gargett, and Denman, 2000)
- wind velocity – (Li, Gargett, and Denman, 2000)

Water temperature is affected by:
- air temperature via sensible heat flux – (Lyman, Willis, and Johnson, 2006)
- insolation via radiant heat flux – (Lyman, Willis, and Johnson, 2006)
- temperature of fresh water runoff – (Arora, Chiew, and Grayson, 2000)
- temperature of ocean seawater advected into the estuary – (Lyman, Willis, and Johnson, 2006)

Salinity is affected by:
- exchange with the ocean – (Griffin and LeBlond, 1990)
- evaporation from estuary or lagoon – (Titus, 1989)

discharge may fall under the scope of the Clean Water Act, which adjudication may resolve.

7.3.3.10 Flood Zone Regulations

Tidal flood surge plains will likely be the estuarine regions most susceptible to climate change forcings, with consequent effects on human infrastructure, especially as development pressures continue to increase along the nation's coastal zone. Before the more recent projections of (higher) sea level rise rates, the Federal Emergency Management Agency (Federal Emergency Management Agency, 1991) estimated that existing development in the U.S. Coastal Zone would experience a 36%–58% increase in annual damages for a 0.3-meter rise in sea level, and a 102%–200% percent increase for a 1-meter rise. While state and local governments regulate building and other human activities in existing flood hazard zones, FEMA provides planning assistance by designating Special Flood Hazard Areas and establishing federal flood insurance rates according to the risk level.

7.3.3.11 Native American Treaty Rights

More than 565 federally recognized governments of American Indian and other indigenous peoples of Alaska, Hawaii, and the Pacific and Caribbean islands carry unique status as "domestic dependent nations" through treaties, Executive Orders, tribal legislation, acts of Congress, and decisions of the federal courts (National Assessment Synthesis Team, 2000). While climate variability and change are likely to impinge on all of these tribal entities, the impacts will perhaps be most strongly felt on the large coastal Native reservations, which are integrally linked to tourism, human health, rights to water and other natural resources, subsistence economies, and cultural resources. While these Native peoples have persisted through thousands of years of changes in their local environment, including minor ice ages, externally driven climate change will likely be more disruptive of their long, intimate association with their environments. In some cases, climatic changes are already affecting Natives such as those in Alaska who are experiencing melting of permafrost and the dissolution of marginal sea ice, altering their traditional subsistence-based economies and culture.

Where climate variability and change intersect with resource management of shared natural resources, Natives' treaty status may provide them with additional responsibility and influence. For example, on the basis of the "Boldt II decision," treaty tribes in Washington State have treaty-based environmental rights that make them legal participants in natural resource and environmental decision making, including salmon and shellfish habitat protection and restoration (Brown, 1993; 1994).

7.3.4 Sensitivity of Management Goals to Climate Change

7.3.4.1 Climate Change and Changing Stressors of Estuarine Ecosystems

Many estuarine properties are expected to be altered by climate change. Global-scale modeling has rarely focused on explicit predictions for estuaries because realistic estuarine modeling would require very high spatial and temporal resolution. It is, however, reasonable to assume that estuaries will be affected by the same climate forcing that affects the coastal and marginal oceans. With increases in atmospheric CO_2, models project increases in oceanic temperature and stratification, decreases in convective overturning, decreases in salinity in mid- and high latitudes, longer growing seasons in mid- and high latitudes, and increases in cloud cover (Table 7.2). Such changes will necessarily force significant alterations in the physics, chemistry, and biology of estuaries. In particular, climate change may have significant impacts on those factors that are included in the definition of an estuary (Box 7.2). For example, climate-driven alterations to geomorphology will affect every physical, chemical, biological, and social function of estuaries.

The 2007 report of the Intergovernmental Panel on Climate Change (IPCC, 2007) summarizes the results of multiple credible models of climate change, providing various ranges of estimated change by year 2100. Whereas these projections carry varying degrees of uncertainty, and in some cases fail to include processes of likely significance in the modeling due to high scientific uncertainty, these projections of rates of change over the next century help ground our scenario building for consequences of

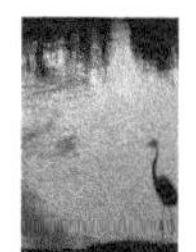

Table 7.2. Percentage change in oceanic properties or processes as a result of climate change forcing by 2050. This table is adapted from Sarmiento *et al.* (2004). Physical changes used as inputs to the biological model are the mean of six global Atmosphere-Ocean Coupled General Circulation Models (AOCGCMs) from various laboratories around the world. The AOCGCMs were all forced by the IPCC IS92a scenario, which has atmospheric CO_2 doubling by 2050.

Domain	Percentage Change by 2050 Due to Climate Change Forcing					
	Mixed layer	Upwelling volume	Vertical stratification	Growing season	Chlorophyll concentration	Primary productivity
Marginal ice zone	-41	-10	+17	-14	+11	+18
Subpolar gyre, seasonally stratified	-22	+1	+11	+6	+10	+14
Subtropical gyre, seasonally stratified	-12	-6	+13	+2	+5	+5
Subtropical gyre, permanently stratified	nd*	-7	+8	0	+3	-3
Low-latitude and equatorial upwelling	nd*	-6	+11	0	+6	+9

*No data

climate change on estuarine dynamics and on ability to attain management goals. The best estimates of average global temperature rise in the surface atmosphere vary from a low scenario of 1.1–2.9°C and a high scenario of 2.4–6.4°C by 2100. Scenarios of sea level rise range from a low projection of 0.18–0.38 meters to a high projection of 0.26–0.59 meters by 2100. The modeled sea level does not, however, include enhanced contributions from shifts of the Greenland and Antarctic ice shelves and could therefore be a serious underestimate. The future temperatures projected for Greenland reach levels inferred to have existed in the last interglacial period 125,000 years ago, when paleoclimate information suggests reductions of polar ice extent and a 4–6-meter rise in sea level. The IPCC projects growing acidification of the ocean, with reductions in pH of between 0.14 and 0.35 units over the next century. In our report, so as to standardize our framework for climate change across responses, we discuss a short term of two to three decades, and also project the consequences of a 1-meter rise in sea level. This increase may not occur within the next century, but if ice sheet shifts add to the present rate of sea level rise, a 1-meter increase may occur sooner than the IPCC projects.

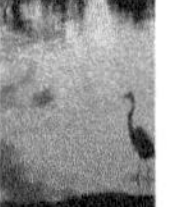

Climate change may also modify existing stressors (described in Section 7.2.2) and create new ones not discussed above. For example, the nutrient, sediment, pathogen, and contaminant stressors usually carried downstream with freshwater runoff will change in proportion to that runoff. If runoff increases, it can be expected to deliver more deleterious material to estuaries, leading to increased eutrophication via nutrients, smothering of benthic fauna via sediment loading, decreased photosynthesis via sediment turbidity, decreased health and reproductive success via a wide spectrum of toxins, and increased disease via pathogens. In contrast, "novel" stressors created by climate change include increased temperatures, shifts in the timing of seasonal warming and cooling, and the acidification caused by increased CO_2 (Box 7.3). The most important emerging and enhanced stressors related to climate change have largely negative consequences for the ecosystem services and management goals of the Nation's estuaries (Table 7.3).

Importantly, there are likely to be interactions among existing and novel stressors, between those factors that define estuaries and stressors, and between stressors and existing management strategies. As noted above (Section 7.2.2), interactions among the multiple stressors related to climate change are likely to pose considerable challenges. Nonetheless, it is important for successful natural resource management and conservation that managers, researchers, and policy makers consider the myriad stressors to which natural systems are exposed. Importantly, interactions among multiple stressors can change not only the magnitude of stressor effects, but also the

Table 7.3. Effects of emerging or enhanced stressors on estuaries arising from climate change.

Stressor	Water Quality	Fisheries & Wildlife	Habitat	Human Value & Welfare	Water Quantity
Sea Level Rise (shoreline armoring prevents transgression of habitats)	positive then negative	positive then negative	positive then negative	negative	negative
Increased Intensive Storms (shoreline erosion; pulsed floods and runoff)	negative	negative	negative	negative	
Temperature Increases (new species mix; disease and parasitism increase, phenology mismatch)	positive then negative	positive then negative	positive then negative	positive then negative	
Increased CO_2 and Acidification (CaCO3 deposition inhibited)	negative	negative	negative	negative	
Precipitation Change (stratification changes)	negative	positive or negative	positive or negative	positive or negative	positive or negative
Species Introduction (facilitated by disturbance)	unpredictable	positive or negative	positive or negative	positive or negative	

BOX 7.3. "Novel" stressors resulting from climate change, together with a listing of potential biological responses to these stressors. The most important of these changes are highlighted in the main text. Not included are increases in sea levels and modifications in geomorphology of estuarine basins (barrier island disintegration), which are of utmost importance but act through complex interactions with other factors, as explained in the text.

Temperature increases, acting through thermal physiology, may cause:

- Altered species (fauna and flora) distributions, including expanding ranges for tropical species currently limited by winter temperatures and contracting ranges due to increased mortality via summer temperatures
- Altered species interactions and metabolic activity
- Altered reproductive and migration timing
- Increased microbial metabolic rates driving increased hypoxia/anoxia
- Increased desiccation lethality to intertidal organisms
- Increased roles of disease and parasitism
- All of the above open niches for invasive species

Timing of seasonal temperature changes, acting through phenology, disrupts:

- Predator and prey availability
- Food and reproductive pulses
- Runoff cycle and upstream migration
- Temperature-driven behavior from photoperiod-driven behavior
- Biological ocean-estuary exchanges (especially of larvae and juveniles)

CO_2 increases drive acidification (lowered pH), forcing:

- Reduced carbonate deposition in marine taxa
- Greatly increased coral reef dieoff
- Reduced photosynthetic rates
- Increased trace metal toxicity
- Evaporation from estuary or lagoon – (Titus, 1989)

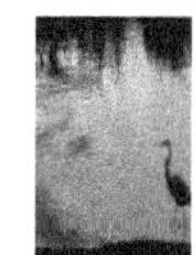

patterns of variability and predictability on which management strategies rely (Breitburg *et al.*, 1998; Breitburg *et al.*, 1999; Vinebrooke *et al.*, 2004; Worm *et al.*, 2006). Enhancing ecosystem resilience by establishing better controls on current stressors would limit the strength of interactions with climate change.

7.3.4.2 Impacts to and Responses of the Ecosystem

7.3.4.2.1 Temperature Effects on Species Distributions

Because species distributions are determined in part by physiological tolerances of climatic extremes, ecologists expect that species will respond to climate warming by shifting distributions towards the poles—so long as dispersal and resources allow such shifts (Walther *et al.*, 2002). In fact, a wide array of species is already responding to climate warming worldwide (Walther *et al.*, 2002; Parmesan and Yohe, 2003; Root *et al.*, 2003; Parmesan and Galbraith, 2004; Parmesan, 2006). Global meta-analyses of 99 species of birds, butterflies, and alpine herbs demonstrate that terrestrial species are migrating poleward at a rate of 6.1 km per decade (Parmesan and Yohe, 2003). Moreover, 81% of 920 species from a variety of habitats showed distributional changes consistent with recent climate warming (Parmesan and Yohe, 2003). In marine systems, warm water species of zooplankton, intertidal invertebrates, and fish have migrated into areas previously too "cool" to support growth (Barry *et al.*, 1995; Southward, Hawkins, and Burrows, 1995; Walther *et al.*, 2002; Southward *et al.*, 2004). Some copepod species have shifted hundreds to 1,000 kilometers northward (Beaugrand *et al.*, 2002), and the range of the oyster parasite *Perkinsus marinus* expands in warm years and contracts in response to cold winters (Mydlarz, Jones, and Harvell, 2006). Its range expanded 500 kilometers from Chesapeake Bay to Maine during one year—1991—in response to above-average winter temperatures (Ford, 1996).

It is important to keep in mind that each species responds individualistically to warming: ecological communities do not move poleward as a unit (Parmesan and Yohe, 2003; Parmesan, 2006). This pattern was first demonstrated by paleoecological studies tracking the poleward expansions of individual species of plants following Pleistocene glaciation (*e.g.*, Davis, 1983; Guenette, Lauck, and Clark, 1998) and has since been extended to animals in phylogeographic studies (*e.g.*, Turgeon *et al.*, 2005). Climate warming is therefore likely to create new mixes of foundation species, predators, prey, and competitors. For example, "invading" species may move poleward faster than "resident" species retreat, potentially creating short-term increases in species richness (Walther *et al.*, 2002). Competitive, plant-herbivore, predator-prey, and parasite-host interactions can be disrupted by shifts in the distribution, abundance, or phenology of one or more of the interacting species (Walther *et al.*, 2002; Parmesan, 2006). Not surprisingly, therefore, it is difficult, if not impossible, to predict how community dynamics and ecosystem functioning will change in response to species shifts (Walther *et al.*, 2002).

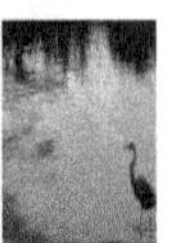

Evidence from studies that have monitored changes in marine biota over the last three decades has shown that in coastal waters, the response of annual temperature cycles to climate change is both seasonally and regionally asymmetric. Along the mid-Atlantic East Coast, maximal summer temperatures are close to 30°C. When greenhouse gas forcing provides more heat to the surface waters in summer, they do not get warmer; instead the additional heat increases evaporation and is transferred to the atmosphere as a latent heat flux. Consequently maximum summer temperatures have not changed in the mid-Atlantic regions, but the minimum winter temperatures are now dramatically higher, by as much as 1–6°C (Parker Jr. and Dixon, 1998). In the reef fish community off North Carolina, the reduction over 30 years in winter kill during the coldest months made it possible for two new (to the area) families and 29 new species of tropical fishes to become permanent residents on the reef (Parker Jr. and Dixon, 1998). In addition, the 28 species of tropical reef fishes that have been present on the site for the entire three decades increased in abundance. An increase in fish-cleaning symbiosis was especially noticeable. Over the 30-year study period, no new temperate species became permanent residents and, while no temperate species dropped out of the community, the temperate species that was most abundant at the start of the study decreased in abundance by a factor of 22. This kind of seasonal asymmetry in temperature change expands the range of tropical species to the north, but so far has not changed the southern limit of temperate species—although it has reduced the biomass of temperate species that were previously abundant.

On the West Coast, changes in the species composition of a rocky intertidal community showed that between the 1930s and 1990s most species' ranges shifted poleward (Barry *et al.*, 1995). The abundance of eight of nine southern species increased and the abundance of five of eight northern species decreased. Annual mean ocean temperatures at the central California coastal site increased by 0.75°C during the past 60 years, but more importantly the monthly mean maximum temperatures during the warmest month of year were 2.2°C warmer. On the West Coast, summer conditions are relatively cool and foggy due to strong coastal upwelling that produces water temperatures from 15–20°C. For intertidal organisms adapted to these relatively cool summer temperatures, a 2°C increase in monthly mean temperature during the warmest month of the year was enough to decrease survival of northern species and increase the survival of southern species. It is clear that climate change has already altered the species composition and abundance of marine fauna, but is equally clear that the physical and biological response of organisms to warming in marine waters is extremely complex.

These effects of temperature on species distributions have influenced and will continue to influence fish and wildlife populations, and will modify habitat provided by organisms such as mangroves, requiring many site-specific adaptive modifications in management.

7.3.4.2.2 Temperature Effects on Risks of Disease and Parasitism

Not only will species' distributions change, but scientists expect that higher temperatures are likely to lead to increased risks of parasitism and disease, due to changes in parasites and pathogens as well as host responses (Harvell *et al.*, 2002; Hakalahti, Karvonen, and Valtonen, 2006). For example, temperature has the potential to alter parasite survival and development rates (Harvell *et al.*, 2002), geographic ranges (Harvell *et al.*, 2002; Poulin, 2005; Parmesan, 2006), transmission among hosts (Harvell *et al.*, 2002; Poulin, 2005), and local abundances (Poulin, 2005). In particular, shortened or less-severe winters are expected to increase potential parasite population growth rates (Hakalahti, Karvonen, and Valtonen, 2006). On the host side, higher temperatures can alter host susceptibility (Harvell *et al.*, 2002) by compromising physiological functioning and host immunity (Mydlarz, Jones, and Harvell, 2006). Animals engaged in partnerships with obligate algal symbionts, such as anemones, sponges, and corals, are at particular risk for problems if temperatures alter the relationship between partners (Mydlarz, Jones, and Harvell, 2006).

Reports of marine diseases in corals, turtles, mollusks, marine mammals, and echinoderms have increased sharply over the past three decades, especially in the Caribbean (Harvell *et al.*, 2002; Ward and Lafferty, 2004). For example, temperature-dependent growth of opportunistic microbes has been documented in corals (Ritchie, 2006). Poulin and Mouritsen (2006) documented a striking increase in cercarial production by trematodes in response to increased temperature, with potentially large effects on the intertidal community (Poulin and Mouritsen, 2006). Geographic range expansion of pathogens with broad host ranges is of particular concern because of the potential to affect a broad array of host species (Dobson and Foufopoulos, 2001; Lafferty and Gerber, 2002).

Importantly, however, we cannot predict the effects of climate change on disease and parasitism based solely on temperature (Lafferty, Porter, and Ford, 2004). Temperature is likely to interact with a variety of other stressors to affect parasitism and disease rates (Lafferty, Porter, and Ford, 2004), including excess nutrients (Harvell *et al.*, 2004), chemical pollutants such as metals and organochlorines (Harvell *et al.*, 2004; Mydlarz, Jones, and Harvell, 2006), and hypoxia (Mydlarz, Jones, and Harvell, 2006). For example, the 2002 die-off of corals and sponges in Florida Bay co-occurred with a red tide (*Karenia brevis)* driven by high nutrient conditions (Harvell *et al.*, 2004). Moreover, not all parasites will respond positively to increased temperature; some may decline (Harvell *et al.*, 2002; Roy, Guesewell, and Harte, 2004) and others may be kept in check by other factors (Harvell *et al.*, 2002; Hall *et al.*, 2006). This suggests that generalizations may not always be possible; idiosyncratic species responses may require that we consider effects on a species-by-species, or place-by-place basis, as with the species distributions discussed earlier.

Such changes in risk of parasitism and disease will influence populations of fish and wildlife, and can affect habitat that is provided by organisms like corals, thereby affecting management.

7.3.4.2.3 Effects of Shoreline Stabilization on Estuaries and Their Services

Estuarine shorelines along much of the U.S. coast have been affected by human activities. These activities have exacerbated both water- and land-based stressors on the estuarine land-water interface. Real and perceived threats from global sea level rise, increased intensity of tropical storms, waves from boat wakes, and changes in delivery of and erosion by stream flows have contributed to greater numbers of actions taken to stabilize estuarine shorelines using a variety of techniques. Shoreline stabilization can affect the physical (bathymetry, wave environment, light regime, sediment dynamics) and ecological (habitat, primary production, food web support, filtration capacity) attributes of the land-water interface in estuaries. Collectively, these physical and ecological attributes determine the degree to which ecosystem services are delivered by these systems (Levin *et al.*, 2001). Shoreline stabilization on the estuarine shoreline has only recently begun to receive significant attention (Committee on Mitigating Shore Erosion along Sheltered Coasts, National Research Council, 2006).

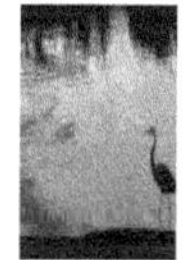

Surprisingly little is known about the effects of estuarine shoreline stabilization structures on adjacent habitats (Committee on Mitigating Shore Erosion along Sheltered Coasts, National Research Council, 2006). Marsh communities at similar elevations with and without bulkheads behind them were found

to be indistinguishable in a study in Great Bay Estuary in New Hampshire (Bozek and Burdick, 2005). However, this study also reported that bulkheads eliminated the up-slope vegetative transition zone. This loss is relevant for both current function of the marsh and also future ability of the marsh to respond to rising sea level. In several systems within Chesapeake Bay, Seitz and colleagues (2006) identified a link between the hardening of estuarine shorelines with bulkheads or rip-rap and the presence of infaunal prey and predators. This study illustrated the indirect effects that can result from shoreline stabilization, and found them to be on par with some of the obvious direct effects. Loss of ecological function in the estuarine land-water margin as a result of shoreline stabilization is a critical concern. However, the complete loss of the structured habitats (SAV, salt marsh) seaward of shoreline stabilization structures as sea level rises is a more dire threat. In addition, the intertidal sand and mud flats, which provide important foraging grounds for shorebirds and nektonic fishes and crustaceans, will be readily eliminated as sea level rises and bulkheads and other engineered shoreline stabilization structures prevent the landward migration of the shoreline habitats. Absent the ability to migrate landward, even habitats such as marshes, which can induce accretion by organic production and sediment trapping, appear to have reduced opportunity to sustain themselves as water level rises (Titus, 1998).

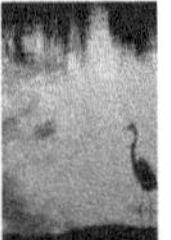

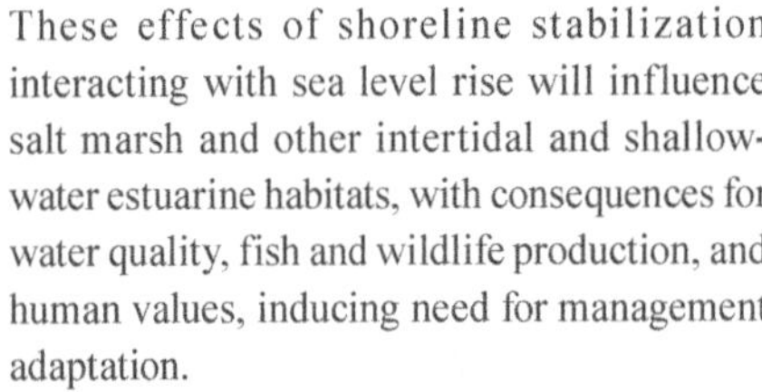

These effects of shoreline stabilization interacting with sea level rise will influence salt marsh and other intertidal and shallow-water estuarine habitats, with consequences for water quality, fish and wildlife production, and human values, inducing need for management adaptation.

7.3.4.2.4 Effects of Climate Change on Marsh Trapping of Sediments and Geomorphologic Resiliency

Coastal wetlands have been relatively sustained, and even expanded, under historic eustatic sea level rise. Marsh surfaces naturally subside due to soil compaction, other geologic (subsidence) processes, and anthropogenic extraction of fluids such as groundwater and oil. However, marsh surfaces (marsh plain) also build vertically due to the combined effect of surface sediment deposition and subsurface accumulation of live and dead plant roots and decaying plant roots and rhizomes. Both of these processes are controlled by tidal-fluvial hydrology that controls delivery of sediments, nutrients, and organic matter to the marsh, as well as the oxygen content of the soil. Local landscape setting (wave energy) and disturbance regime (storm frequency and intensity) are also factors over the long term. Thus, the relative sea level (the simultaneous effect of eustatic sea level rise and local marsh subsidence) can be relatively stable under a moderate rate of sea level rise, because marsh elevation increases at the same rate as the sea level is rising (*e.g.*, Reed, 1995; Callaway, Nyman, and DeLaune, 1996; Morris *et al.*, 2002). Whether a marsh can maintain this equilibrium with mean sea level and sustain characteristic vegetation and associated attributes and functions is uncertain. It will depend on the interaction of complex factors, including sediment pore space, mineral matter deposition, initial elevation, rate of sea level rise, delivery rates of sediments in stream and tidal flows, and the production rate of below-ground organic matter (U.S. Climate Change Science Program, in press).

Thus, changes in sediment and nutrient delivery and eustatic sea level rise are likely to be the key factors affecting geomorphic resiliency of coastal wetlands. Sediment delivery may be the critical factor: estuaries and coastal zones that currently have high rates of sediment loading, such as those on the southeast and northwest coasts, may be able to persist up to thresholds of 1.2 cm per year that are optimal for marsh primary production (Morris *et al.*, 2002). If sea level rise exceeds that rate, then marsh surface elevation decreases below the optimum for primary production. Increased precipitation and storm intensities commensurate with many future climate scenarios (*e.g.*, in the Pacific Northwest) would likely increase sediment delivery, but also would erode sediments where flows are intensified. The large-scale responses to changes in sediment delivery to estuarine and coastal marshes have not been effectively addressed by most hydrodynamic models incorporating sediment transport. SAP 4.1 elucidates potential impacts by providing maps depicting the wetland losses in the mid-Atlantic states that are anticipated under various rates of sea level rise (U.S. Climate Change Science Program, in press). Such changes in sediment and nutrient delivery to the estuary will threaten the geomorphologic resilience of salt marsh habitat, thereby altering water quality and fish and wildlife production; these changes imply the need for management adaptation.

7.3.4.2.5 Effects of Sea Level Rise and Storm Disturbance on Coastal Barrier Deconstruction

Two important consequences of climate change are accelerated sea level rise and increased frequency of high-intensity storms. Sea level rise and intense storms work alone and in combination to alter the hydrogeomorphology of coastal ecosystems and their resultant services. Furthermore, the extent to which they act on ecosystems is dependent on human alterations to these ecosystems. Perhaps the best known example of the current interaction of sea level rise, storm intensity, and human activity is the coast of the Gulf of Mexico around the Mississippi River. Relative sea level rise of the Louisiana coast is one of the highest in the world, in large part as a result of human activities, and this has caused significant losses of wetlands (Boesch *et al.*, 1994; González and Törnqvist, 2006; Day, Jr. *et al.*, 2007). The consequences of intense storms (*e.g.*, Hurricanes Katrina and Rita) on coastal ecosystems of the Gulf of Mexico, human-dominated and natural, are now legend (Kates *et al.*, 2006). New Orleans and other cities were devastated by these storms. Wetland loss was dramatic, with sharp alterations to community structure (Turner *et al.*, 2006).[19] Barrier islands were eroded, overwashed, and breached, with severe impacts to both human lives and infrastructure. The impacts of these storms are linked to the damaged conditions and decreased area of the wetlands and their historical loss (Day, Jr. *et al.*, 2007). Reconstruction of New Orleans and other affected cities has begun, and plans are being offered for the replenishment and protection of wetlands and barrier islands (U.S. Army Corps of Engineers, in press; Day, Jr. *et al.*, 2007; Coastal Protection and Restoration Authority of Louisiana, 2007).

Although the impacts of the hurricanes of 2005 and the influence of relative sea level rise on their impacts were the most costly to the United States, they are not the only examples of how storms and sea level rise influence hydrogeomorphology. Sea level rise and erosion, fostered by storms, have caused estuarine islands to disappear and led to significant changes in shorelines (Hayden *et al.*, 1995; Riggs and Ames, 2003). Barrier island shape and position are dynamic, dependent on these two processes. These processes are natural and have occurred throughout the Holocene; what is relatively new are the ways in which human values are in conflict with these processes and how humans either promote or inhibit them.

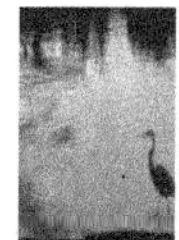

Wetlands can maintain themselves in the face of sea level rise by accretion. This accretion is supported by both sedimentation and organic matter accumulation (Chmura *et al.*, 2003). The ability to accrete makes it difficult to assess the true consequences of sea level rise on landscape pattern and resultant area of wetlands, especially over large areas (Titus and Richman, 2001). We do not know exactly the potential accretion and subsidence rates of most wetlands and the thresholds at which relative sea level rise exceeds net elevation change, causing increased inundation and ultimately wetland loss. Based on the experiences of Louisiana, we can estimate that the maximum accretion rate may be less than 10 mm per year, but applicability to other systems is undetermined. Two things are clear: First, the limits depend on the source of material for accretion (*i.e.*, sediment or organic matter) and hence the rates

[19] **U.S. Geological Survey**, 2007: Hurricanes Katrina and Rita. USGS, http://www.nwrc.usgs.gov/hurricane/katrina.htm, accessed on 3-23-2007.

of processes that introduce and remove the materials. Second, the rates of these processes will differ with location both locally within the coastal landscape and regionally due to climate, community, and hydrogeomorphic conditions.

Sea level rise and storm disturbance have not only severe consequences as described, but also are important drivers of the natural progression of coastal ecosystems. One can consider the coastal landscape as having a sequence of ecosystem states, each dependent upon a particular hydroperiod and tidal inundation regime (Brinson, Christian, and Blum, 1995; Hayden *et al.*, 1995; Christian *et al.*, 2000). For example in the mid-Atlantic states, coastal upland, which is rarely flooded, would be replaced by high salt marsh as sea level rises. High marsh is replaced by low marsh, and low marsh is replaced by intertidal flats. While sea level rise alone may effect these changes in state, they are promoted by disturbances that either kill vegetation (*e.g.*, salt intrusion from storms killing trees) or change elevation and hence hydroperiod (*e.g.*, erosion of sediment). It is unclear how accelerated sea level rise and frequency of severe storms will alter the balance of this sequence.

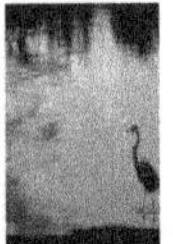

Normally one considers that disturbances would be local, such as salt water intrusion or wrack deposition. But these state changes can actually result from regional impacts of disturbance. For example, *Juncus roemerianus* is a rush species commonly found in high marshes along the mid-Atlantic, southern Atlantic, and Gulf of Mexico coasts of the United States. It is less common where astronomical tidal signals are strong (Woerner and Hackney, 1997; Brinson and Christian, 1999), and it is replaced by *Spartina alterniflora* or perhaps other species. Any disturbance that increases the strength of astronomical tides promotes this shift. Such a disturbance could be the breaching of barrier islands in which increased flow through new inlets may foster more dominant astronomical tides and the ecosystem state change. The projected disintegration of barrier islands as a consequence of intense storm damage acting from a higher base sea level has catastrophic implications (Riggs and Ames, 2003). Coastal barriers function to protect mainland shorelines from tidal energy, storm surge, and wave forces, such that loss of the protections implies catastrophic inundation, erosion, and loss of wetlands and other coastal habitats on mainland shores as well as back-barrier shores.

Sea level rise and increased frequency of intense storms will influence salt marsh and other wetland habitats by erosion and salt water intrusion, thereby influencing fish and wildlife production, available quantity of fresh water, and provision of human values, with consequences for management.

7.3.4.2.6 Joint Effects of Increasing Temperature and Carbon Dioxide

As a consequence of increasing global temperatures, the limits of climate-adapted habitats are expected to shift latitudinally. Temperate herbaceous species that dominate tidal wetlands throughout many southern U.S. estuaries may be replaced by more tropical species such as mangroves (Harris and Cropper Jr., 1992). Salt marshes and mangroves are not interchangeable, despite the fact that both provide structure to support productive ecosystems and perform many of the same ecosystem functions. Mangroves store up to 80% of their biomass in woody tissue, whereas salt marshes lose 100% of their aboveground biomass through litterfall each year (Mitsch and Gosselink, 2000). Production of litter facilitates detrital foodwebs and supports many ecological processes in wetlands, so this distinction has implications for materials cycling such as carbon sequestration (Chmura *et al.*, 2003). There are significant differences in structural complexity and biological diversity between these wetland systems. These differences will affect the capacity of the wetlands to assimilate upland runoff, maintain their vertical position, and provide flood control. Temperature-driven species redistribution will be further complicated as sea level increases and vegetation is forced landward.

Since pre-industrial times, the atmospheric concentration of carbon dioxide (CO_2) has risen by 35% to 379 ppm in 2005 (IPCC, 2007). Ice cores have proven that this concentration is significantly greater than the natural range over the last 650,000 years (180–300 ppm). In addition, the annual average growth rate in CO_2 concentrations over the last 10 years is larger than the average growth rate since the beginning of continuous direct atmospheric measurements: an average of 1.9 ppm per year from 1995–2005 compared with an average of 1.4 ppm per year from 1960–2005 (IPCC, 2007). Because CO_2 is required for photosynthesis, these changes may have implications for estuarine vegetation. Plants can be divided into two groups based on the way in which they assimilate CO_2. C3 plants include the vast majority of plants on earth (~95%) and C4 plants, which include crop plants and some grasses, comprise most of the rest. Early in the process of CO_2 assimilation, C3 plants form a pair of three carbon molecules whereas C4 plants form four carbon molecules. The distinction between C3 and C4 species at higher atmospheric CO_2 concentrations is that C3 species increase photosynthesis with higher CO_2 levels, while C4 species generally do not (Drake *et al.*, 1995). In wetland systems dominated by C3 plants (*e.g.*, mangroves, many tidal fresh marshes), elevated CO_2 will increase photosynthetic potential and may increase the related delivery of ecosystems services from these systems (Drake *et al.*, 2005). Ongoing research is examining the potential for shifts in wetland community composition driven by elevated CO_2. Data from one of these efforts indicate that despite the advantage afforded to C3 species at higher CO_2 levels, CO_2 increases alone are unlikely to cause black mangrove to replace cordgrass in Louisiana marshes.[20] However, many important estuarine ecosystem effects from elevated CO_2 levels have been documented, including increases in fluxes of CO_2 and methane (Marsh *et al.*, 2005), augmented nitrogen fixation by associated microbial communities (Dakora and Drake, 2000), increased methanogenesis (Dacey, Drake, and Klug, 1994) and changes in the quantity and composition of root material (Curtis *et al.*, 1990).

The joint effects of rising temperature and increased CO_2 concentrations will influence composition and production of shoreline plants that are critical habitat providers and contributors to detrital food chains, thereby also affecting fish and wildlife production and provision of human values, and inducing need for management adaptations.

7.3.4.2.7 Effects of Increased CO_2 on Acidification of Estuaries

Ocean acidification is the process of lowering the pH of the oceans by the uptake of CO_2 from the atmosphere. As atmospheric CO_2 increases, more CO_2 is partitioned into the surface layer of the ocean (Feely *et al.*, 2004). Since the industrial revolution began to increase atmospheric CO_2 significantly, the pH of ocean surface waters has deceased by about 0.1 units and it is estimated that it will decrease by another 0.3–0.4 units by 2100 as the atmospheric concentration continues to increase (Caldeira and Wickett, 2003). The resulting decrease in pH will affect all calcifying organisms because as pH decreases, the concentration of carbonate decreases, and when carbonate becomes under-saturated, structures made of calcium carbonate begin to dissolve. However, dissolution of existing biological calcium carbonate structures is only one aspect of the threat of acidification; another threat is that as pH falls and carbonate becomes undersaturated it requires more and more metabolic energy for an organism to deposit calcium carbonate. The present lowered pH is estimated to have reduced the growth of reef-building by about 20% (Raven, 2005). While corals get the most attention regarding acidification, a wide spectrum of ocean and estuarine organisms are affected, including coraline algae; echinoderms such as sea urchins, sand dollars, and starfish; as well as coccolithophores, foraminifera, crustaceans, and molluscan taxa with shells, of which pteropods are particularly important (Orr *et al.*, 2005). The full ecological consequences of the reduction in calcification by marine calcifiers are uncertain, but it is likely that the biological integrity of ocean and estuarine ecosystems will be seriously affected (Kleypas *et al.*, 2006).

Effects of climate change on estuarine acidification will influence water quality, provision of some biogenic habitat like coral reefs, fish and wildlife production, and human values, thus implying need for management adaptation.

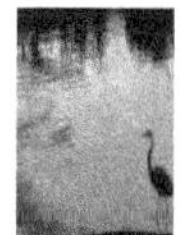

[20] **U.S. Geological Survey**, 2006: Potential effects of elevated atmospheric carbon dioxide (CO_2) on coastal wetlands. USGS, http://www.nwrc.usgs.gov/factshts/2006-3074/2006-3074.htm, accessed on 4-1-2006.

7.3.4.2.8 Effects of Climate Change on Hypoxia

Low dissolved oxygen (DO) is a problematic environmental condition observed in many U.S. estuaries (Bricker *et al.*, 1999). Although a natural summer feature in some systems, the frequency and extent of hypoxia have increased in Chesapeake Bay, Long Island Sound, the Neuse River Estuary, and the Gulf of Mexico over the past several decades (Cooper and Brush, 1993; Paerl *et al.*, 1998; Anderson and Taylor, 2001; Rabalais, Turner, and Scavia, 2002; Cooper *et al.*, 2004; Hagy *et al.*, 2004; Scavia, Kelly, and Hagy, 2006). Persistent bottom water hypoxia (*e.g.*, DO concentration < 2.0 mg per L) results from interactions among meteorology and climate, the amounts and temporal patterns of riverine inflows, estuarine circulation, and biogeochemical cycling of allochthonous and autochthonous organic matter (Kemp *et al.*, 1992; Boicourt, 1992; Buzzelli *et al.*, 2002; Conley *et al.*, 2002). Over time, the repeated bottom water hypoxia can alter biogeochemical cycling, trophic transfers, and estuarine production at higher trophic levels (Baird *et al.*, 2004). Ecological and economic consequences of fish kills, bottom habitat degradation, and reduced production at the highest trophic levels in response to low DO have provided significant motivation to understand and manage hypoxia (Tenore, 1970; Officer *et al.*, 1984; Turner, Schroeder, and Wiseman, 1987; Diaz and Rosenberg, 1995; Hagy *et al.*, 2004).

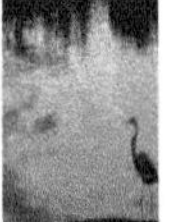

Various scenarios predict that climate change will influence the vulnerability of estuaries to hypoxia through changes in stratification caused by alterations in freshwater runoff, changes in water temperature, increases in sea level, and altered exchanges with the coastal ocean (Peterson *et al.*, 1995; Scavia *et al.*, 2002). Additionally, warmer temperatures should increase metabolism by the water-column and benthic microbial communities, whose activity drives the depletion of DO. Many of the factors that have been found to contribute to the formation of hypoxia (Borsuk *et al.*, 2001; Buzzelli *et al.*, 2002) will be affected by one or more predicted changes in climate (Table 7.4). Because hypoxia affects valued resources, such as fish and wildlife production, reductions in hypoxia are a management target for many estuaries, and adaptations will be required as a consequence of climate change.

7.3.4.2.9 Effects of Changing Freshwater Delivery

Climate change is predicted to affect the quality, rate, magnitude, and timing of the freshwater delivered to estuaries (Alber, 2002), potentially exacerbating existing human modifications of these flows, as described by Sklar and Browder (1998). However, the exact nature of these changes is difficult to predict for a particular estuary, in part because there is not clear agreement among general circulation models (GCMs) on precipitation changes over drainage basins (National Assessment Synthesis Team, 2000). There does seem to be agreement among models that increases in frequencies of extreme rainfall will occur (Scavia *et al.*, 2002), suggesting that there will be changes in potential freshwater inflow amounts and patterns (hydrographs). These inflows will then be subjected to human modifications that differ across estuaries. For example, where dams are used in flood regulation, there is reduced variability within and among seasons, damping, for example, normally peak flows at snowmelt in temperate regions (Poff *et al.*, 1997; Alber, 2002). In some watersheds, increased reuse of wastewater in agriculture, municipalities, and industry may offset changes in supply by reducing demand for "clean" freshwater.

The potential physical and chemical consequences of altered freshwater flows to estuaries include changes in salinity and stratification regimes, loadings of nutrients

Table 7.4. Factors that control the occurrence of estuarine hypoxia and the climate change-related impacts that are likely to affect them.

Factor	Climate-Related Forcing
Water temperature	Δ T
River discharge	Δ precipitation
N&P loading	Δ T, Δ precipitation
Stratification	Δ T, Δ precipitation, Δ RSL*
Wind	Δ weather patterns, Δ tropical storms
Organic carbon source	Δ T, Δ precipitation, Δ RSL*

*RSL = relative sea level

and sediments, water residence times, and tidal importance (reviewed in Alber, 2002). Potential biological consequences include changes in species composition, distribution, abundance, and primary and secondary productivity, all in response to the altered availability of light, nutrients, and organic matter (Cloern *et al.*, 1983; Howarth *et al.*, 2000; Alber, 2002).

Increases in the delivery of freshwater to estuaries may enhance estuarine circulation and salt wedge penetration up the estuary (Gedney *et al.*, 2006), resulting in stronger vertical stratification. For individual estuaries there is the potential for increased freshwater inflow to shift the degree of mixing along the gradient from the fully mixed toward the stratified state. Those estuaries that receive increased supplies of organic matter and nutrients and exhibit enhanced stratification may be particularly susceptible to enhanced hypoxia and the negative effects described in the previous section. However, at some level, increased freshwater delivery will reduce residence time and thus reduce the potential for hypoxia. This threshold will be specific to individual estuaries and difficult to predict in a generic sense.

In some estuaries, climate change may also lead to a reduction in freshwater inflow, which will generally increase salinity. This could lead to more salt-water intrusion upstream, negatively affecting species intolerant of marine conditions (Copeland, 1966; Alber, 2002) and/or lengthening the estuary by extending the distance along the freshwater-to-full-seawater gradient (Alber, 2002). Water residence times within the estuary will likely increase with reduced freshwater inflow, potentially creating a more stable system in which phytoplankton can grow and reproduce (Cloern *et al.*, 1983; Howarth *et al.*, 2000). Thus, one might expect a greater response to nutrients—*i.e.*, greater primary productivity and/or larger phytoplankton populations (Mallin *et al.*, 1993)—than under baseline rates of freshwater discharge. This may be especially true for estuaries that are currently somewhat "protected" from eutrophication symptoms by high freshwater flow, such as the Hudson River (Howarth *et al.*, 2000). However, reduced flushing times will also keep water in the estuary longer, potentially increasing the risks posed by pollutants and pathogens (Alber and Sheldon, 1999; Sheldon and Alber, 2002).

Other biological consequences of changing freshwater delivery include alterations in secondary productivity (the directions of which are difficult to predict), the distributions of plants and sessile invertebrates (Alber, 2002), and cues for mobile organisms such as fish, especially migratory taxa with complex life histories (Whitfield, 1994; Whitfield, 2005). Not surprisingly, therefore, a whole branch of management is developing around the need to determine the optimal freshwater flows required to maintain desired ecosystem services (*e.g.*, Robins *et al.*, 2005; Rozas *et al.*, 2005).

Changes in freshwater delivery to the estuary will affect freshwater quantity, water quality, stratification, bottom habitats, fish and wildlife production, and human values, inducing needs for management adaptation.

7.3.4.2.10 Phenology Modifications and Match/Mismatch

Estuaries are characterized by high temporal variability, on multiple time scales, and spatial variability, which includes sharp environmental gradients with distance upstream and vertically in the water column (Remane and Schlieper, 1971). One mode of adaptation that many free-living estuarine species use to exploit the many resources of estuaries is to move in and out of the estuary, as well as upstream and downstream within the estuary, on a complex temporal schedule. A study in North Carolina found that the most abundant fish species in small tributaries of the upper estuary differed in 10 of the 12 months of the year (Kuenzler *et al.*, 1977). Ten different species were dominant during the 12 months of the year. To accomplish such movements, many estuarine species have evolved behavior that uses various sensory cues to control the timing of their activities (Sims *et al.*, 2004). The timing of behavior cued by environment information is referred to as "phenology" (Mullins and Marks, 1987; Costello, Sullivan, and Gifford, 2006). The best understood type of phenology that occurs in estuaries involves matching critical feeding stages with the timing of primary productivity blooms (Scavia *et al.*, 2002). As many estuarine stressors are altered by climate change, we can expect that phenology will be one of the first biological processes to be seriously disrupted.

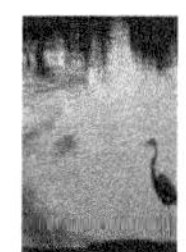

Changing phenology has large implications for fish and wildlife production because trophic coupling of important species in the food chain can be disrupted, thereby presenting a need for management adaptation.

7.3.4.2.11 River Discharge and Sea Level Impacts on Anadromous Fishes

Anadromous fishes, such as Pacific salmon, are an important economic and cultural resource that may be particularly vulnerable to significant shifts in coastal climates in the Pacific Northwest and Alaska. The combined effect of shifts in seasonal precipitation, storm events, riverine discharge, and snowmelt (Salathé, 2006; Mote, 2006) are likely to change a broad suite of environmental conditions in coastal wetlands upon which salmon depend at several periods in their life histories. The University of Washington's Climate Impacts Group (UW-CIG) has summarized current climate change in the Pacific Northwest to include region-wide warming of ~0.8°C in 100 years, increased precipitation, a decline in snowpack, especially at lower elevations, and an earlier spring.[21] The UW-CIG predictions for future climate change in the region include an increase in average temperatures on the order of 0.1–0.6°C (best estimate = 0.3°C) per decade throughout the coming century, with the warming occurring during all seasons but with the largest increases in the summer. Precipitation is also likely to increase in winter and decrease in summer, but with no net change in annual mean precipitation. As a consequence, the mountain snowpack will diminish and rivers that derive some of their flow from snowmelt will likely demonstrate reduced summer flow, increased winter flow, and earlier peak flow. Lower-elevation rivers that are fed mostly by rain may also experience increased wintertime flow due to increases in winter precipitation. Summer river flows in the Pacific Northwest are projected to decline by as much as 30% and droughts would become more common (Leung and Qian, 2003), implying significant changes in estuarine salinity distribution that has not yet been examined in any detail. Chapter 6, Wild and Scenic Rivers, provides an expanded discussion of these and other climate change effects on rivers in the United States.

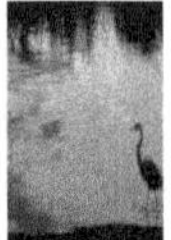

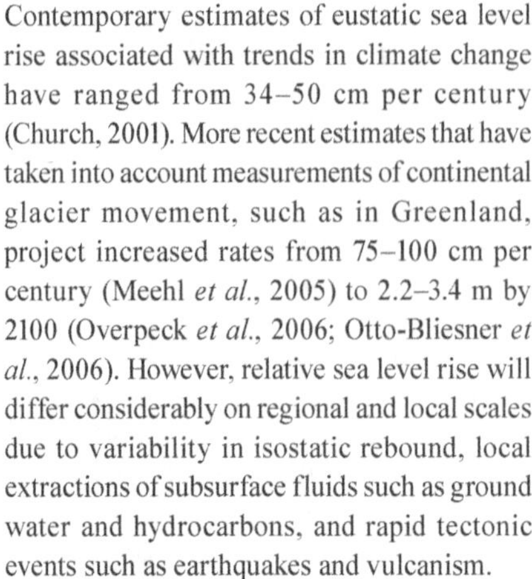

Contemporary estimates of eustatic sea level rise associated with trends in climate change have ranged from 34–50 cm per century (Church, 2001). More recent estimates that have taken into account measurements of continental glacier movement, such as in Greenland, project increased rates from 75–100 cm per century (Meehl *et al.*, 2005) to 2.2–3.4 m by 2100 (Overpeck *et al.*, 2006; Otto-Bliesner *et al.*, 2006). However, relative sea level rise will differ considerably on regional and local scales due to variability in isostatic rebound, local extractions of subsurface fluids such as ground water and hydrocarbons, and rapid tectonic events such as earthquakes and vulcanism.

Because different anadromous species occupy estuarine wetlands according to their divergent life history strategies, impacts of these climate changes vary among and within species. In the case of Pacific salmon, the "ocean-type" species and life history types would be the most vulnerable because they occupy transitional estuarine waters significantly longer than "stream-type" salmon. For instance, juvenile Chinook and chum salmon representing this "ocean-type" life history strategy may occupy estuarine wetlands for more than 90 days (Simenstad, Fresh, and Salo, 1982), seeking (1) refugia from predation at their small size, (2) time to achieve physiological adaptation from freshwater to marine salinities, and (3) high densities of appropriate prey organisms. Based on our knowledge of the habitat requirements and landscape transitions of migrating juvenile ocean-type salmon (Simenstad *et al.*, 2000; Parson *et al.*, 2001; Mote *et al.*, 2003), the present spatial coincidence of necessary physical habitats, such as marsh platforms and tidal creeks, will change with the appropriate salinity regime as sea water penetrates further up the estuary. This would have potentially large impacts on the ocean-type salmon performance.

In the Pacific Northwest, shifts from snowmelt runoff to more winter storm precipitation will potentially disrupt the migration timing and residence of juvenile salmon in estuarine wetlands. For example, juvenile Chinook salmon in many watersheds migrate to estuaries coincident with the spring freshet of snowmelt, and occupy the extensive brackish marshes available to them during that period. This opportunity often diminishes

[21] **Climate Impacts Group**, University of Washington, 2007: Climate Change. University of Washington, http://www.cses.washington.edu/cig/pnwc/cc.shtml, accessed on 3-23-2007.

as water temperatures increase and approach physiologically marginal limits (*e.g.*, 19–20°C) with the decline of snowmelt and flows in early summer. Under current climate change/ variability scenarios, much of the precipitation events will now be focused in the winter, providing less brackish habitat opportunities during the expected juvenile salmon migration and even more limiting temperatures during even lower summer flows. Whether migration and other life history patterns of salmon could adapt to these climate shifts are unknown.

The sustainability of estuarine wetlands under recent sea level rise scenarios is also of concern if estuarine habitat utilization by anadromous fish is density-dependent. Estuaries that are positioned in a physiographic setting allowing transgressive inundation, such as much of the coastal plain of the southeastern and Gulf of Mexico coasts, have a buffer that will potentially allow more inland development of estuarine wetlands. Other coasts, such as those of New England and the Pacific Northwest, have more limited opportunities for transgressive development of estuarine wetlands, and many estuaries are already confined by upland agricultural or urban development that would prevent further inland flooding (Brinson, Christian, and Blum, 1995). For one example, Hood[22] found that a 45-cm sea level rise over the next century would result in a 12% loss, and an 80-cm rise would eliminate 22%, of the tidal marshes in the Skagit River delta (Puget Sound, Washington), which could be translated to an estimated reduction in estuarine rearing capacity for juvenile Chinook salmon of 211,000–530,000 fish, respectively. These estimates are based entirely on the direct inundation effects on vegetation and do not incorporate the potential response of existing marshes to compensate for the increased rate of sea level rise, which can include increased sediment accretion and maintenance of marsh plain elevation or increased marsh progradation due to higher sediment loads from the river (see section 7.2.4.2.15 below). Nor do these estimates take into account increased marsh erosion from greater winter storm activity or changes in salinity distribution due to declining summer river flows. Court cases have already

overturned general permits for shoreline armoring where salmon (an endangered species under ESA) would be harmed. With projected rises in sea level, the needs of salmon may come even more often into conflict with management policies that generally permit bulkheads and other shoreline armoring to protect private property.

Salmon represent such an iconic fish of great importance to fisheries, wildlife, subsistence uses, and human culture that climate-related impacts on salmon populations would require management adaptation.

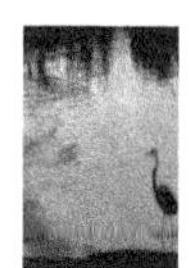

7.3.4.2.12 Effects of Climate Change on Estuarine State Changes

The many direct and indirect influences of climate change may combine to cause fundamental shifts in ecosystem structure and functioning. Some shifts, such as those associated with transgression of wetlands, can be considered part of the normal responses to sea-level rise (Brinson, Christian, and Blum, 1995; Christian *et al.*, 2000). Of particular concern is the potential for ecosystems to cross a threshold beyond which there is a rapid transition into a fundamentally different state that is not part of a natural progression. Ecosystems typically do not respond to gradual change in key forcing variables in a smooth, linear fashion. Instead, there are abrupt, discontinuous, non-linear shifts to a new state (or "regime") when a threshold is crossed (Scheffer *et al.*, 2001; Scheffer and Carpenter, 2003; Burkett *et al.*, 2005). Particularly relevant here is the hypothesis that gradual changes in "slow" variables that operate over long time

[22] **Hood**, W.G., Unpublished: Possible sea-level rise impacts on the Skagit River tidal marshes. Skagit River System Cooperative.

scales can cause threshold-crossing when they alter interactions among "fast" variables whose dynamics happen on short temporal scales (Carpenter, Ludwig, and Brock, 1999; Rinaldi and Scheffer, 2000). We anticipate that some climate changes will fall into this category, such as gradual increases in temperature. The diversity of additional stressors arising from consequences of climate change greatly enhances the likelihood of important stressor interactions. Thus, in estuaries, where so many stressors operate simultaneously, there is great potential for interactions among stressors to drive the system into an alternative state.

Regime shifts can sometimes be catastrophic and surprising (Holling, 1972; Scheffer and Carpenter, 2003; Foley *et al.*, 2005), and reversals of these changes may be difficult, expensive, or even impossible (Carpenter, Ludwig, and Brock, 1999). Moreover, the social and economic effects of discontinuous changes in ecosystem state can be devastating when accompanied by the interruption or cessation of essential ecosystem services (Scheffer *et al.*, 2001; *e.g.*, Foley *et al.*, 2005). Recognizing and understanding the drivers of regime change and the inherent nonlinearities of biological responses to such change is a fundamental challenge to effective ecosystem management in the face of global climate change (Burkett *et al.*, 2005; Groffman *et al.*, 2006).

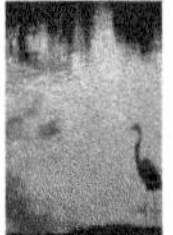

All the potential regime shifts described below have large implications for sustaining biogenic habitat, provision of fish and wildlife, and many human values, thereby implying need for management adaptation.

7.3.4.2.13 Climate Change Effects on Suspension-Feeding Grazers and Algal Blooms

The Eastern oyster (*Crassostrea virginica*) is a historically dominant species in estuaries along the Atlantic and Gulf of Mexico coasts of the United States. At high abundances, oysters play major roles in the filtration of particles from the water column, biodeposition of materials to the benthos, nutrient cycling, and the creation of hard substrate habitat in otherwise soft-bottom systems (Kennedy, 1996; Coen, Luckenbach, and Breitburg, 1999; Newell and Ott, 1999; Newell, Cornwell, and Owens, 2002). Dominant consumers (*e.g.*, the schyphomedusan sea nettle, *Chrysaora quinquecirrha*) are dependent on oysters for habitat for sessile stages, and large numbers of estuarine fish species benefit either directly or indirectly from habitat and secondary production of oyster reefs (Coen, Luckenbach, and Breitburg, 1999; Breitburg *et al.*, 2000). Oysters are structural as well as biological ecological engineers (Jones, Lawton, and Shachak, 1994), and have been shown to reduce shoreline erosion (Meyer, Townsend, and Thayer, 1997) and facilitate regrowth of submerged aquatic vegetation by reducing nearshore wave action.

Oyster abundances in Atlantic Coast estuaries have declined sharply during the past century, with a precipitous decline in some systems during the past two to three decades. The primary stressors causing the recent decline are likely overfishing and two pathogens: *Haplosporidium nelsoni*—the non-native protist that causes MSX—and *Perkinsus marinus*, a protistan that causes Dermo and is native to the United States but has undergone a recent range expansion and possible increase in virulence (Rothschild *et al.*, 1994; National Research Council, 2004). Both overfishing and disease cause responses in the relatively slow-responding (*i.e.,* years to decades) adult oysters and oyster reefs, making recovery to the oyster-dominant regime quite difficult. High sediment loading (Cooper and Brush, 1993), eutrophication (Boynton *et al.*, 1995), and blooms of ctenophores (Purcell *et al.*, 1991) may further contribute to oyster decline or prevent recovery to the high-oyster state. These factors—all of which are likely to increase with changes in climate—appear to act most strongly on the larval and newly settled juvenile stages, raising the possibility that this system will at best exhibit hysteretic recovery to the high-oyster state.

7.3.4.2.14 N-Driven Shift from Vascular Plants to Planktonic Micro- and Benthic Macroalgae

Seagrasses are believed to be in the midst of a global crisis in which human activities are leading to large scale losses (Orth *et al.*, 2006). Human and natural impacts have had demonstrable detrimental effects on SAV (Short and Wyllie-Echeverria, 1996). Enhanced loading of nutrients to coastal waters has been found to alter primary producer communities, through shifts toward species with faster growth-nutrient uptake rates (Duarte, 1991). The shift is often toward phytoplankton, which reduces light availability and can lead to losses of other benthic primary producers such as seagrasses. The disappearance of seagrass below critical light levels is dramatic (Duarte, 1991), and has been linked to nutrient loading in some systems (Short and Burdick, 1996). In Waquoit Bay, Massachusetts, replacement of SAV by macroalgae has also been observed and was primarily attributed to shading (Hauxwell *et al.*, 2001). Increases in macroalgal biomass, macroalgal canopy height and decreases in SAV biomass were linked to nitrogen loading rate using a space-for-time substitution (Hauxwell *et al.*, 2001). It is essential to understand the potential for thresholds in water quality parameters that may lead to loss of SAV through a state change. SAV is sensitive to environmental change, and thus may serve as a "coastal canary," providing an early warning of deteriorating conditions (Orth *et al.*, 2006). SAV also provides significant ecological services (Williams and Heck Jr., 2001) and its loss would have appreciable effects on overall estuarine function.

7.3.4.2.15 Non-linear Marsh Accretion with Sea Level Rise

Coastal inundation is projected to lead to land loss and expansion of the sub-tidal regions along estuarine shorelines (Riggs, 2002). Intertidal habitats that do not accrete or migrate landward proportionally to relative sea level rise are susceptible to inundation. Wetlands are often present in these areas, and have shown the ability to keep up with increases in sea level in some systems (Morris *et al.*, 2002). However, the ability to maintain their vertical position is uncertain, and depends on a suite of factors (Moorhead and Brinson, 1995). Recent work in the Venice Lagoon found a bimodal distribution of marsh (higher elevation) and flat (lower elevation) intertidal habitats, with few habitats at intermediate intertidal elevations (Fagherazzi *et al.*, 2006). The findings indicate that there may be an abrupt transition from one habitat type to another. Should this model hold true for a broad range of coastal systems, there are clearly significant implications for coastal geomorphology and the ecological services provided by the different habitat types.

7.4 ADAPTING TO CLIMATE CHANGE

Biologists have traditionally used the term "adaptation" to apply to intrinsic biological responses to physical or biological changes that may serve to perpetuate the species, with implications for the community and ecosystem. This definition includes behavioral, physiological, and evolutionary adaptation of species. This question therefore arises: Can biological adaptation be relied upon to sustain ecosystem services from national estuaries under conditions of present and future climate change? In the short term of a few decades, the capability of estuarine organisms to migrate farther toward the poles in response to warming temperatures and farther up the shore in response to rising water levels has potential to maintain estuarine ecosystem processes and functioning that do not differ greatly from today's conditions. However, over longer time frames, depending on the realized magnitude of climate changes, estuarine ecosystems may not be able to adapt biologically and thereby retain high similarity to present systems. The scope and pace of current and anticipated future climate change are too great to assume that management goals will be sustained by intrinsic biological adjustments, without also requiring management adaptation (Parmesan and Galbraith, 2004; Parmesan, 2006; Pielke *et al.*, 2007).

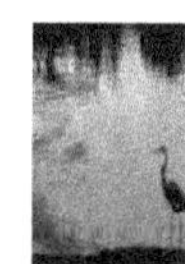

The extremely high natural variability of estuarine environments has already selected for organisms, communities, and ecosystems with high capacity for natural physiological, behavioral, and perhaps also evolutionary adaptation (Remane and Schlieper, 1971; Wolfe, 1986). Nevertheless, the current rapid rates of change in many variables, such as temperature, and the absolute levels of key environmental variables, such as CO_2 concentration, that ultimately may be reached, could fall outside the historical evolutionary experience of estuarine organisms. The historical experience with environmental variability may not help much to achieve effective biological adaptation under these novel rates of change and conditions. While behavioral (*e.g.*, migration, dispersal) adaptation of individual species may take place to some degree, the dramatic suite of projected changes in estuarine environments

BOX 7.4. Adaptation Options for Resource Managers

- Help protect tidal marshes from erosion with oyster breakwaters and rock sills, and thus preserve their water filtration and fisheries enhancement functions.
- Preserve and restore the structural complexity and biodiversity of vegetation in tidal marshes, seagrass meadows, and mangroves.
- Adapt protections of important biogeochemical zones and critical habitats as the locations of these areas change with climate.
- Prohibit bulkheads and other engineered structures on estuarine shores to preserve or delay the loss of important shallow-water habitats, by permitting their inland migration as sea levels rise.
- Connect landscapes with corridors to enable migrations to sustain wildlife biodiversity across the landscape.
- Conduct integrated management of nutrient sources and wetland treatment of nutrients to limit hypoxia and eutrophication.
- Manage water resources to ensure sustainable use in the face of changing recharge rates and saltwater infiltration.
- Maintain high genetic diversity through strategies such as the establishment of reserves specifically for this purpose.
- Maintain landscape complexity of salt marsh landscapes, especially preserving marsh edge environments.
- Support migrating shorebirds by ensuring protection of replicated estuaries along the flyway.
- Restore important native species and remove invasive non-natives to improve marsh characteristics that promote propagation and production of fish and wildlife.
- Direct estuarine habitat restoration projects to places where the restored ecosystem has room to retreat as sea level rises.
- Restore oyster reefs in replication along a depth gradient to provide shallow water refugia for mobile species, such as fish and crustaceans, to retreat to in response to climate-induced deep water hypoxia/anoxia, or to spread the risk of losses due to other climate-related environmental disturbances.
- Develop practical approaches to apply the principle of rolling easements, to prevent engineered barriers from blocking landward retreat of coastal marshes and other shoreline habitats as sea level rises.

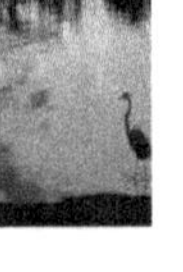

and stressors that we summarized earlier poses complex challenges to individual species, including those of estuaries, on a timetable that is inconsistent with the capacity for evolutionary change to keep up (Pielke *et al.*, 2007). Even if evolutionary change could proceed at a rapid pace, the diversity of environmental changes implies that conflicting demands may be placed on selection such that adaptation to the full suite of changes may be compromised. The success of individual species in adapting to climate change does not lead to intrinsic resilience at the community and ecosystems levels of organization. Because virtually all ecosystem processes involve some form of interaction between or among species, biological adaptation by individual species to climate-driven changes is not a process that will protect functioning estuarine ecosystems, because species adapt and migrate at differing rates (Sims *et al.*, 2004; Parmesan, 2006).

Among the most important estuarine species that dictate overall community composition and ecosystem dynamics are the structural foundation species, namely intertidal marsh plant and subtidal seagrass (SAV) vegetation. Donnelly and Bertness (2001) have assembled ecological evidence that, starting in the late 1990s, the low marsh plant *Spartina alterniflora* has begun to move upslope and invade the higher marshes of New England that are typically occupied by a more diverse mix of *Juncus gerardi*, *Distichlis spicata*, and *Spartina patens*. Their paleontological assessment revealed that in times of rapid sea level rise in the late 19th and early 20th centuries, *Spartina alterniflora* similarly grew upwards and dominated the high marsh. Such replacement of species and structural diversity of foundation species is likely to modify the functioning of the salt marsh ecosystem and affect its capacity to deliver traditional goods and services. Similarly, among SAV species, some like *Halodule wrightii* are known to be better colonizers with greater ability to colonize and spread into disturbed patches than other seagrasses like *Thalassia testudinum* (Stephan, Peuser, and Fonseca, 2001). In general, seagrasses that recolonize by seed set can move into newly opened areas more readily than those that largely employ vegetative spread. Analogous to the marsh changes, if storm disturbance and rising water levels favor more opportunistic seagrass species, then the new SAV community may differ from the present one and provide different ecosystem services. Vascular plants of both intertidal and shallow subtidal estuaries possess characteristically few species relative to terrestrial habitats (Day, Jr. *et al.*, 1989; Orth *et al.*, 2006), so these differences in behavior of important foundation species in the marsh and in SAV beds will have disproportionately large influences on function. Thus, the web of interactions among biotic and abiotic components of the estuarine ecosystem cannot be expected to be preserved through intrinsic biological adaptation alone, which cannot regulate the physical changes. Management adaptations must be considered to sustain ecosystem services of national estuaries. Examples of specific adaptation options are presented in Box 7.4 and elaborated further throughout the sections that follow.

7.4.1 Potential for Adjustment of Traditional Management Approaches to Achieve Adaptation to Climate Change

Three different time frames of management adaptation can be distinguished: (1) avoidance of any advance adaptation strategy (leading to *ad hoc* reactive responses); (2) planning only for management responses to climate change and its consequences (leading to coordinated, planned responses initiated either after indicators reveal the urgency or after emergence of impacts); and (3) taking proactive measures to preserve valuable services in anticipation of consequences of climate change. Rational grounds for choosing among these three options involve consideration of the risks and reversibility of predicted negative consequences, and the expenditures associated with planning and acting now as opposed to employing retroactive measures. Political impediments and lack of effective governance structures may lead to inaction, even if planning for intervention or initiating proactive intervention represents the optimal strategy. For example, the partitioning of authority for environmental and natural resource management in the United States among multiple federal and state agencies inhibits effective implementation of ecosystem-based management of our estuarine and ocean resources (Peterson and Estes, 2001; Pew Center on Global Climate Change, 2003; U.S. Commission on Ocean Policy, 2004; Titus, 2004). Even if governance structures were developed that allow cooperation among agencies and among levels of government, successful application of ecosystem-based management of estuaries may not be a realistic expectation for estuarine management because of the intrinsic conflicts of interest among stakeholders, which include land users across the entire watershed and airshed as well as coastal interests.

Planning for adaptation to climate change, without immediate implementation, may represent the most prudent response to uncertainty over timing and/or intensity of negative consequences of global change on estuarine ecosystem services, provided that advance actions are not required to avoid irreversible damage. Issues of expense also deserve attention in deciding whether to delay management actions. An ounce of prevention may be worth a pound of cure. For example, by postponing repairs and vertical extensions of levees around New Orleans, the estimated expenditures for retroactive repair and all necessary restorations of about $54 billion following Hurricanes Katrina and Rita greatly exceed what proactive levee reconstruction would have cost (Kates *et al.*, 2006). On the other hand, the protections provided against natural disasters are typically designed to handle more frequent events, such as storms and floods occurring more frequently than once a century, but inadequate to defend against major disasters like the direct hit by a category 5 hurricane. Such management protections even enhance losses and restoration costs by promoting development under the false sense

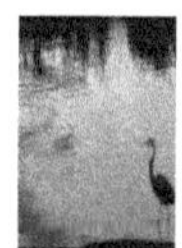

of security that is based on success in the face of more frequent, smaller storm events (Kates *et al.*, 2006). This example has direct relevance to adaptation management in estuaries, because there is broad consensus that climate change is increasing sea levels and increasing the frequency of intense hurricanes (IPCC, 2007). Engineered dikes for estuarine shorelines may represent one possible management adaptation, protective of some human values but injurious to natural resources. Thus, the need for understanding the effectiveness and consequences of alternative management policies relating to dikes, levees, and other such structural defenses makes the New Orleans experience relevant.

A decision to postpone implementation of adaptation actions may rely on continuing scientific monitoring of reliable indicators and modeling. Based on inputs from evolving ocean observing systems, model predictions could provide comfort that necessary actions, although delayed, may still be timely. Other important prospective management actions may be postponed because they are not politically feasible until an event alters public opinion sufficiently to allow their implementation. Such adaptations are best planned in advance to anticipate the moment when they could be successfully triggered. Other management actions may involve responding to events and therefore only have relevance in a retrospective context. Catastrophic events provide opportunities for changes that increase ecological and human community resilience, by addressing long-standing problems such as overbuilding in floodplains or degradation of coastal wetlands (Box 7.5).[23] However, pressures to expediently restore conditions to their familiar pre-disaster state often lead to the loss of these opportunities (Mileti, 1999). Therefore, decisions about whether and where to rebuild after damage from major floods and storms should be carefully examined and planned in advance in order to avoid making poorer judgments during chaotic conditions that follow these types of incidents. This strategy becomes more relevant as storm intensity and flood damages increase.

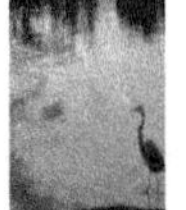

BOX 7.5. Storms as Opportunities for Management Change

Catastrophic events provide management opportunities that make difficult decisions more publicly acceptable for increasing ecological and human resilience to climate change. Comprehensive planning could be initiated at federal, tribal, state, and local levels before—and applied after—major storm events to avoid future loss of life and property, and at the same time protect many environmental assets and ecosystem services in the interest of the public trust. Examples of proactive management activities include:

- Planning to prevent rebuilding in hazardous areas of high flood risk and storm damage.
- Establishing setbacks, buffer widths, and rolling easements based on reliable projections of future erosion and sea level rise, and implementing them rapidly after natural disasters.
- Prohibiting development subsidies (e.g., federal flood insurance and infrastructure development grants) to estuarine and coastal shorelines at high risk.
- Modifying local land use plans to influence redevelopment after storms and direct it into less risky areas.
- Using funds from land trusts and programs designated to protect water quality, habitat, and fisheries, to purchase the most risky shorelines of high resource value.

Proactive intervention in anticipation of consequences of climate change represents rational management under several conditions. These conditions include irreversibility of undesirable ecosystem changes, substantially higher costs to repair damages than to prevent them, risk of losing important and significant ecosystem services, and high levels of scientific certainty about the anticipated change and its ecological consequences (Titus, 1998; 2000). Avoiding dramatic structural ("phase") shifts in estuarine ecosystem state may represent a compelling motivation for proactive management, because such shifts threaten continuing delivery of many traditional ecosystem services and are typically difficult or exceedingly expensive to reverse (Groffman *et al.*, 2006). Reversibility is especially at issue in cases of potential transitioning to an alternative stable state, because positive feedbacks maintain the new state and resist reversal (Petraitis and Dudgeon, 2004). For example, the loss of SAV removes a baffle to water flow, thus increasing near-bottom currents. The faster currents in turn mean that seagrass seeds are less likely to be deposited,

[23] **H. John Heinz III Center for Science, Economics, and the Environment**, 2002: *Human Links to Coastal Disasters*. Washington, DC.

BOX 7.6. Responding to the Risk of Coastal Property Loss

The practice of protecting coastal property and infrastructure with hard engineered structures, such as bulkheads, prevents marshes and beaches from migrating inland as the sea level rises. Ultimately, many marshes and beaches seaward of bulkheads will disappear as sea level rises (Titus, 1991).

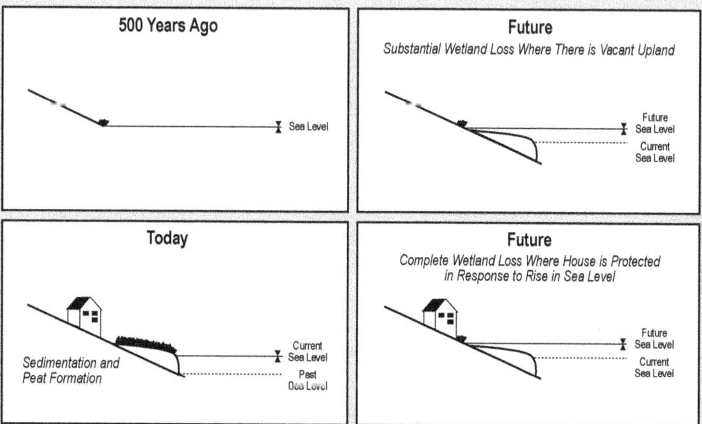

Coastal marshes have generally kept pace with the slow rate of sea level rise that has characterized the last several thousand years. Thus, the area of marsh has expanded over time as new lands have been inundated. If, in the future, sea level rises faster than the ability of the marsh to keep pace, the marsh area will contract. Construction of bulkheads to protect economic development may prevent new marsh from forming and result in a total loss of marsh in some areas.

Beach nourishment may also contribute to the loss of salt marsh on coastal barriers, because it prevents natural processes of coastal barrier migration through overwash. Overwash of sediments to the estuarine shoreline is a process that extends and revitalizes salt marsh on the protected side of coastal barriers.

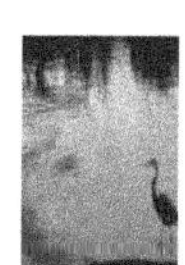

and seedlings are more likely to be uprooted by erosion; this feedback makes reestablishment of lost beds much more difficult.

With adequate knowledge of the critical tipping point and ongoing monitoring of telling indicators, proactive intervention could in some cases be postponed and still be completed in time to prevent climate change from pushing the system over the threshold into a new phase. Nevertheless, many processes involved in ecosystem change possess substantial inertia such that even after adjusting levels of drivers, a memory of past stress will continue to modify the system, making postponement of action inadvisable. Climate change itself falls into this class of processes, in that if greenhouse gas emissions were capped today, the Earth would continue to warm for decades (IPCC, 2007).

Financial costs of climate change may be minimized by some types of proactive management. For example, enacting legislation that prohibits bulkheads and other engineered structures and requires rolling easements along currently undeveloped estuarine shores could preserve or at least delay loss of important shallow-water habitats, such as salt marsh, by allowing them to migrate inland as sea level rises (Box 7.6) (Titus, 1998). A law to require rolling easements is not likely to be ruled as a taking, especially if enacted before property is developed, because "the law of erosion has long held that the public tidelands migrate inland as sea level rises, legislation saying that this law will apply in the future takes nothing" (Titus, 1998). However, absent such a law and this interpretation of it, the value of habitat and associated ecosystem services may exceed the value of property losses that would occur if property owners could not protect their investment. Some other proactive steps that enhance adaptation to climate change are likely to come at very little expense, and deserve immediate inclusion in policy and management plans. For example, the simple incorporation of climate change consequences in management plans for natural and environmental resources will trigger inclusion of forward-looking modifications that might provide resistance to climate change, build resiliency of ecological and socioeconomic systems, and avoid interventions incompatible with anticipated change and sustained ecosystem services (Titus, 2000). Principles for environmental planning could be adopted that (1) prohibit actions that will exacerbate negative consequences of climate change, (2) allow actions that are climate-change neutral, and (3) promote actions that provide enhanced ecosystem resilience to climate change. Such principles may lead to many low-cost modifications of existing management plans that could be initiated today.

The scientific basis for predicting climate change and its ecosystem consequences must be especially compelling in order to justify any costly decisions to take proactive steps to enhance adaptation to climate change. Willingness to take costly actions should vary with the magnitude of predicted consequences, the confidence associated with the predictions, and the timing of the effects. The scientific basis for the predictions must also be transparent, honest, and effectively communicated, not just to managers but also to the general public, who ultimately must support adaptation interventions. Thus, there is an urgent need to continue to refine the scientific research on climate change and its ecosystem consequences to reduce uncertainty over all processes that contribute to climate change and sea level rise, so that future projections and GCM scenarios are more complete and more precise. Because of the tremendous publicity associated with the release of each IPCC report, this process of periodic re-evaluation of the science and publication of the consensus report plays an integral role in public education. Scientific uncertainty about the magnitudes and timetables of potentially important processes, such as melting of the Greenland ice sheet (Dowdeswell, 2006; Rignot and Kanagaratnam, 2006), leads to their exclusion from IPCC projections. Further scientific research will allow inclusion of such now uncertain contributions to change.

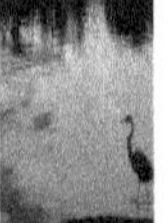

BOX 7.7. Estuarine Water Quality and Climate Change

Climate change may lead to changes in estuarine water quality, which in turn would affect many of the vital ecosystem services offered by estuaries.

- Changes in nutrient concentrations and light penetration into estuarine waters may affect productivity of submerged aquatic vegetation, which provides a range of services such as nursery habitat for fish species, sediment stabilization, and nutrient uptake.
- Changes in water quality may affect oxygen demand as well as directly affecting availability of dissolved oxygen. An increase in freshwater discharge to estuaries may lead to increased frequency, scope, and duration of bottom-water hypoxia arising from stronger stratification of the estuarine water column and greater microbial oxygen demand at higher temperatures.

7.4.2 Management Adaptations to Sustain Estuarine Services

7.4.2.1 Protecting Water Quality

All national estuaries, and estuaries more generally, include water quality as a priority management target. The federal Clean Water Act serves to identify explicit targets for estuarine water quality nationwide, but state and local programs can also include other numeric standards for explicit parameters. Some CCMPs specify explicit, sometimes numeric, targets for specific member estuaries. Parameters with federally mandated standards include chlorophyll concentration; turbidity; dissolved oxygen; fecal coliform bacteria; nutrient loading where TMDLs apply; and conditions for NPDES discharge permits that maintain balanced and indigenous communities of fish, shellfish, and wildlife. In addition, coastal marsh and other riparian wetland buffers serve to treat non-point-source storm waters before they enter the open waters of estuaries, so preserving marsh extent and functionality is an important management target relating to water quality (Mitsch and Day Jr, 2006).

Perhaps the greatest threat to estuarine water quality from climate change derives from the loss of water treatment of diffuse nutrient pollution by constricted tidal marsh and wetland buffers (Box 7.7). These vegetated buffers are threatened by the joint effects of sea level rise and increasingly intense storms interacting with hardening of estuarine shorelines through installation of bulkheads, dikes, and other engineered structures (Titus, 1998). Such structures are now readily permitted along estuarine shorelines to protect private property and public infrastructure from shoreline erosion; however, by preventing orderly retreat of intertidal and shallow subtidal habitats shoreward as sea level rises (Schwimmer and Pizzuto, 2000), marsh will be lost and its functions eliminated over extensive portions of estuarine shorelines (Titus, 2000; Reed, 2002; Committee on Mitigating Shore Erosion along Sheltered Coasts, National Research Council, 2006). The loss of salt marsh on coastal barriers is further facilitated by beach nourishment, which prevents natural processes of coastal barrier recession through overwash. Overwash of sediments to the estuarine shoreline is a process that extends and revitalizes salt marsh on the protected side of coastal barriers.

Estuarine shorelines differ in their susceptibility to erosion and recession under rising sea levels (U.S. Environmental Protection Agency, 1989). Relative sea level is rising at very different rates around the country and the globe. The subsiding shores of the Louisiana Gulf Coast are losing more salt marsh to sea level rise than any other region of the United States (U.S. Environmental Protection Agency, 1989). Marsh losses on the Mississippi River Delta are enhanced by modification of river flows in ways that inhibit sediment delivery to the marshes, and by extraction of subsurface fluids (oil and gas). Extraction of groundwater from shallow aquifers also induces subsidence and enhances relative sea level rise along the shores of some estuaries, such as San Francisco Bay. For many estuaries, salt marsh does not currently face increased flooding and erosion from rising sea levels, either because relative sea level is not rising rapidly in these regions or because the accumulation of organic peat, along with the trapping and deposition of largely inorganic sediments by emergent marsh plants, is elevating the land surface at a rate sufficient to keep up with sea level rise (Reed, 2002). Despite the capability of salt marsh to rise with sea level, this gradual process produces a marsh on an elevated platform where the estuarine shore is increasingly more steeply sloped. The consequently deeper water does not dissipate wave energy as readily as the previously shallow slope, leading to increased risk of shoreline and marsh erosion at the margin (Committee on Mitigating Shore Erosion along Sheltered Coasts, National Research Council, 2006). Therefore, even marsh shores that today are maintaining elevation and position as sea level rises are at risk of greater erosion at their seaward margin in the future. Nevertheless, substantial geographic variation exists in erosion risk and susceptibility to marsh loss (U.S. Environmental Protection Agency, 1989).

Maintaining present management policy allowing bulkheads will likely lead to the loss of marshes, and the development of walled estuaries composed only of subtidal habitats, wherever development exists on the shoreline. Only on undeveloped estuarine shorelines can marshes recede landward. But with the ongoing dramatic expansion of coastal human communities, little undeveloped estuarine shoreline is likely to remain except in public parks, reserves, and sanctuaries. Along estuarine salinity gradients, much more development takes place toward the ocean end and less up-estuary. Therefore, as sea level rises, an increasing fraction of remaining marsh habitat will be found along these undefended, up-estuary shores (see maps in SAP 4.1; U.S. Climate Change Science Program, in press). All specific water quality parameters for which standards exist will suffer under this scenario of current management without adaptation. Reactive management holds little promise of reversing impacts, because it would require dismantling or moving structures and infrastructure, which is expensive, unpopular, and increasingly infeasible as coastal land becomes increasingly developed. Reactive marsh restoration would require removals of at least some portion of the engineered walls protecting estuarine shoreline property, so as to allow flooding of the proper elevations supporting salt marsh restoration. Implementing any public policy that would lead directly to widespread private property loss represents a large challenge under the prevailing property rights laws, but one that should be decided in favor of retaining the estuarine habitats, if done in a way that can involve rolling easements to preserve the public tidelands (Titus, 1998).

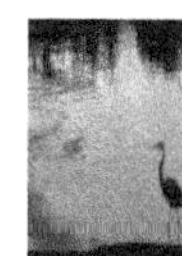

The process of retreat achieved by rolling easements or by some other administrative construct has been discussed in the United States for at least two decades. Retreat has an advantage over establishment of fixed buffer zones, because the abandonment need not be anticipated and shoreline use modified until sea level has risen enough to require action (Titus, 1998). An analogous proactive response to global climate change and sea level rise, known as "managed alignment," is being actively considered in the United Kingdom and European Union.[24] Managed alignment refers to deliberately realigning engineering structures affecting rivers, estuaries, and the coastline. The process could involve retreating to higher ground, constructing set-back levees, shortening the length of levees and seawalls, reducing levee heights, and widening river floodplains. The goals of managed realignment may be to:

- Reduce engineering costs by shortening the overall length of levees and seawalls that require maintenance;
- Increase the efficiency and long-term sustainability of flood and coastal levees by recreating river, estuary, or coastal wetlands, and using their flood and storm buffering capacity;
- Provide other environmental benefits through re-creation of natural wetlands; or
- Construct replacement coastal wetlands in or adjacent to a designated European site, to compensate for wetland losses resulting from reclamation or coastal squeeze.

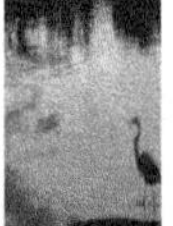

Under this UK/EU perspective, the goods and services provided by wetland coastal defenses against sea level rise appear to outweigh anticipated costs under some scenarios.

Locally in the United States, proactive management to protect tidal marshes, on which water quality of estuaries so strongly depends, may have some notable success in the short term of a few decades, although prospects of longer-term success are less promising. Only Rhode Island and parts of Massachusetts have regulations in place that recognize the need to allow wetlands the capacity to migrate inland as sea level rises, and thereby provide long-term protection (Titus, 2000).

An alternative to bulkheading is using natural breakwaters of native oysters, in quiescent waters of Atlantic and Gulf Coast estuaries, to dissipate wave action and thus help inhibit shoreline and marsh erosion inshore of the reef. Rock sills (so-called "living shorelines" as developed and permitted in Maryland)[13] can be installed in front of tidal marshes along more energetic estuarine shores, where oysters would not survive (Committee on Mitigating Shore Erosion along Sheltered Coasts, National Research Council, 2006). Such natural and artificial breakwaters can induce sediment deposition behind them, and thereby may help sediments rise and marshes persist with growing sea levels. As sea level rises, oyster reefs can also grow taller and rock sills can be artificially elevated, thereby keeping up protection by the breakwaters. Oysters are active suspension feeders and help reduce turbidity of estuarine waters. Rock breakwaters in the estuary are also often colonized by oysters and other suspension-feeding invertebrates. Restoration of oyster reefs as breakwaters, and even installation of rock breakwaters, contribute to water quality through the oysters' feeding and through protection of salt marshes by these alternatives to bulkheads and dikes. This proactive adaptation to sea level rise and risk of damaging storms will probably fail to be sustainable over longer time frames, because such breakwaters are not likely to provide reliable protection against shoreline erosion in major storms as sea level continues to rise. Ultimately, the owners of valuable estuarine shoreline may not be satisfied with breakwaters as their only defense against the rising waters, and may demand permission to install levees, bulkheads, or alternative forms of shoreline armoring. This could lead to erosion of all intertidal habitats along the shoreline and consequent loss of the tidal marsh in developed areas. Some of these losses of marsh acreage would be replaced by progressive drowning of river mouths and inundation of flood plains up-estuary as sea level rises, followed by transgression and spread of wetlands into those newly flooded areas. The most promising suite of management adaptations on highly developed shorelines down-estuary is likely a combination of rolling easements, setbacks,

[24] **Department for Environment, Food and Rural Affairs (DEFRA) and the UK Environment Agency**, 2002: *Managed Realignment Review - Project Report.* Policy Research Project FD 2008, DEFRA, Cambridge, UK.

density restrictions, and building codes (Titus, 1998). Political resistance may preclude local implementation of this adaptation, but financial costs of implementation are reasonable, if done before the shoreline is developed (Titus, 2000).

Given the political barriers to implementing these management adaptations to protect coastal wetlands, globally instituted mitigation of climate change may be the only means in the longer term (several decades to centuries) of avoiding large losses of tidal marsh and its water treatment functions. Losses will be nearly total along estuarine shorelines where development is most intense, especially in the zone of high hurricane risk from Texas to New York (see SAP 4.1; U.S. Climate Change Science Program, in press). Although rapid global capping of greenhouse gas emissions would still result in decades of rising global temperatures and consequent physical climatic changes (IPCC, 2007), it may be possible in the short term (years to a few decades) to partially alleviate damage to tidal marshes and diminution of their water treatment role on developed shores by local management adaptations, such as installation of natural and artificial breakwaters. On undeveloped estuarine shorelines, implementation of rolling easements is a critical need before development renders this approach too politically and financially costly. However, much public education will be necessary for this management adaptation to be accepted.

Estuarine water quality is also threatened by a combination of rising temperature, increased pulsing and, in many regions such as the East Coast, growing quantities of freshwater riverine discharge and more energetic upstream wedging of sea waters with rising sea level (Scavia *et al.*, 2002). Temperature increases drive faster biochemical rates, including greater rates of microbial decomposition and animal metabolism, which inflate oxygen demand. When increased fresh water discharges into the estuary, this less-dense fresh water at the surface, when combined with stronger salt water wedging on the bottom, will enhance water column stability because of greater density stratification. Such conditions are the physical precursor to development of estuarine bottom water hypoxia and anoxia in warm seasons, because oxygen-rich surface waters are too light to be readily mixed to depth (Paerl *et al.*, 1998). This water quality problem leads to persistent hypoxia and anoxia, creating dead zones on the bottoms of estuaries, one of the most serious symptoms of eutrophication (Paerl *et al.*, 1998; Bricker *et al.*, 1999). Under higher water temperatures and extended warm seasons, high oxygen demand is likely to extend for longer periods of the year while greater stratification further decreases dissolved oxygen in bottom waters. Erosion of riparian marshes from rising water levels also adds previously sequestered organic carbon to the estuary, further increasing oxygen demand for its microbial decomposition. In regions such as the Pacific Northwest, where summertime droughts are predicted rather than summer increases in storm-driven pulses of rain, this scenario of greater water-column stability and higher oxygen demand at elevated temperature will not apply. Nevertheless, negative consequences of summertime drought also are likely.

Failing to act in advance of increases in incidence, scope, and duration of bottom water hypoxia implies widespread climate-related modifications of many estuaries, inconsistent with maintaining a balanced indigenous population of fish, shellfish, and wildlife. Nutrient reduction in the watershed and airshed could limit algal blooms, and thereby reduce organic loading and oxygen demand (Conley *et al.*, 2002). However, discharge limits for point sources are already close to what is technically feasible in many rivers. From an economic standpoint, further limiting atmospheric nitrogen deposition would affect many activities, such as electric power generation, industrial operations, and automobile use. It is possible that wetland restoration over the drainage basin could be greatly enhanced to reduce the fraction of diffuse nutrient loading that reaches the estuary, and to help counteract the increased estuarine stratification and warming temperatures that drive higher microbial decomposition and oxygen demand (Mitsch and Day Jr, 2006). Thus, integrated management of nutrient sources and wetland treatment of nutrients can play a role in management to limit eutrophication and hypoxia.

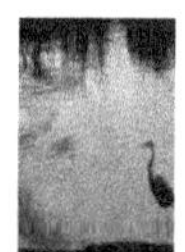

At state levels of management, recognition of the likelihood of climate change and anticipation of its consequences could lead to important proactive steps, some with potentially minimal financial costs. Regulatory change represents one major example of an institutional approach at this level. Rhode Island and Massachusetts deserve praise for appropriately responding to risk of wetland loss under sea level rise by instituting regulations to allow landward migration of these habitats (Titus, 2000). Examination of state laws, agency rules, and various management documents in North Carolina, on the other hand, suggests that climate change is rarely mentioned and almost never considered. One example of how changes in rules could provide proactive protection of water quality would be to anticipate changes in sea level rise and storm intensity by modifying riparian buffer zones to maintain water quality. Permitting rules that constrain locations for construction of landfills, hazardous waste dumps, mine tailings, and facilities that store toxic chemicals could be modified to insure that, even under anticipated future conditions of sea level rise, shoreline recession, and intense storms, these facilities would remain not only outside today's floodplains but also outside the likely floodplains of the future. Riverine floodplain maps and publicly run flood insurance coverage could be redrafted to reflect expectations of flooding frequency and extent under changing rainfall amounts and increasing flashiness of rainfall as it is delivered in more intense discrete storms. Such changes in floodplain maps would have numerous cascading impacts on development activities along the river edges in the entire watershed, many of which would help protect water quality during floods. Water quality degradation associated with consequences of floods from major storms such as hurricanes can persist for many months in estuaries (Paerl and Bales, 2001). Thus, if climate change leads to increases in storm intensity, proactive protection of riparian floodplains could help reduce the levels of pollutants that are delivered during those floods. Acting now to address this stressor helps enhance ecosystem resistance to impacts of climate change on eutrophication and pollution by toxicants. Floodplains may offer some of the last remaining undeveloped components of our coastal landscape over which transgressive expansion of sea level might occur with minimal human impact, so expanding protected areas of floodplains also helps build resilience of the socioeconomic system. Even during the past two decades, many estuarine watersheds have experienced multiple storms that exceeded standards for "100-year floods," implying that recomputation and remapping of those hazardous riverine floodplains is already necessary.

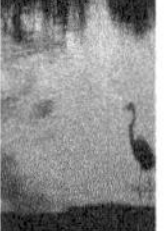

Kevin Rosseel, EPA

7.4.2.2 Sustaining Fisheries and Wildlife Populations

Sustaining fish production and wildlife populations represent important management goals of most national estuaries and essentially all estuaries nationwide. Fisheries are likely to suffer large declines from both of the major processes that affect water quality: (1) loss of tidal marshes associated with rising sea levels, and enhanced incidence of intense storms as these drivers interact with hardened shorelines; and (2) increased frequency, scope, and duration of bottom-water hypoxia arising from stronger stratification of the estuarine water column and greater microbial oxygen demand at higher temperatures.

Marshes and other wetlands perform many valuable ecosystem services (Box 7.1) (Millennium Ecosystem Assessment, 2005), several of which lead to enhanced fish production. Numerous studies have demonstrated the high use of salt marshes by killifish, grass shrimps, and crabs, which are important prey for larger commercially important fishes, and for wading birds at higher trophic levels. Salt marsh habitat supports several endemic species of birds, such as some rails, and small mammals, some of which are on federal or state threatened and endangered lists (Greenberg *et al.*, 2006). The combination of high primary production and structural protection makes the marsh significant as a contributor to important detrital-based food webs based on export of vascular plant detritus from the marsh, and also means that the marsh plays a valuable role as nursery habitat for small fishes and crustaceans. Zimmerman, Minello, and Rozas (2000) demonstrated that penaeid shrimp production in bays along the Gulf of Mexico varies directly with the surface area of the salt marsh within the bay. Maintaining complexity of salt marsh landscapes can also be an important determinant of fish, shellfish, and wildlife production, especially preserving marsh edge environments (*e.g.*, Peterson and Turner, 1994). Thus, marsh loss and modification in estuaries are expected to translate directly into lost production of fish and wildlife.

The climate-driven enhancement of bottom water hypoxia and anoxia will result in further killing of oysters and other sessile bottom invertebrates (Lenihan and Peterson, 1998), thereby affecting the oyster fishery directly and other fisheries for crabs, shrimp, and demersal fishes indirectly (Lenihan *et al.*, 2001). These demersal consumers prey upon the benthic invertebrates of the estuary during their nursery use of the system, in the warm season of the year. When the benthic invertebrates are killed by lack of oxygen and resulting deadly hydrogen sulfide, fish production declines as energy produced by phytoplankton enters microbial loops and is thereby diverted from passing up the food chain to higher tropic levels (Baird *et al.*, 2004). This enhanced diversion of energy away from pathways leading to higher trophic levels will not only affect demersal fish production, but also diminish populations of sea birds and marine mammals, such as bottle-nosed dolphins. Because estuaries contribute so greatly to production of coastal fisheries generally, such reductions in fish and wildlife transcend the boundaries of the estuary itself.

Fish and wildlife suffer additional risks from climate change, beyond those associated with loss of marsh and other shoreline habitats and those associated with enhanced hypoxia. Higher temperatures are already having and will likely have additional direct effects on estuarine species. Increased temperature is associated with lower bioenergetic efficiency, and greater risk of disease and parasitism. As temperatures increase, species will not move poleward at equal rates (Parmesan, 2006), so new combinations will emerge with likely community reorganization, elevating abundances of some fishes and crustaceans while suppressing others. Locally novel native species will appear through natural range expansion as water warms, adding to the potential for community reorganization. In addition, introductions of non-native species may occur at faster rates, because disturbed communities appear more susceptible to invasion. Finally, the changes in riverine flows—both amounts and temporal patterns—may change estuarine physical circulation in ways that affect transport of larval and juvenile life stages, altering recruitment of fish and valuable invertebrates.

The challenges of adapting management to address impacts of climate change on fish and wildlife thus include all those already presented for water quality, because the goals of preventing loss of tidal marsh and other shallow shoreline habitats and of avoiding expansion of hypoxic bottom areas are held in common. However, additional approaches may be available or necessary to respond to risks of declines in fish and wildlife. For example, fisheries management at federal and state levels is committed to the principle of sustainability, which is usually defined as maintaining harvest levels at some fixed amount or within some fixed range. With climate-driven changes in estuarine ecosystems, sustainable fisheries management will itself need to become an adaptive process as changes in estuarine carrying capacity for target stocks occur through direct responses to warming and other physical factors, and indirect responses to changes in biotic interactions. Independent of any fishing impacts, there will be a moving target for many fish, shellfish, and wildlife populations, necessitating adaptive definitions of what is sustainable. This goal calls for advance planning for management responses to climate change, but not implementation until the ecosystem changes have begun. Absent any advance planning, stasis of management could conceivably induce stock collapses by inadvertent overfishing of a stock in decline from climate modifications.

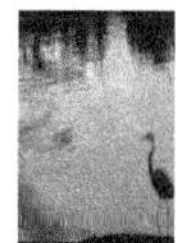

Extermination of injurious non-native species after their introduction into estuarine systems has not proved feasible. However, one proactive type of management adaptation in contemplation of possible enhancement of success of introduced species into climate-disrupted estuarine ecosystems may be to strengthen rules that prevent the introductions themselves. This action would be especially timely as applied to the aquarium fish trade, which is now a likely vector of non-native fish introductions.[25] Local removals of invasive non-natives, combined with restoration of the native species, may be a locally viable reactive management response to improve marsh characteristics that promote propagation and production of fish and wildlife. This type of action may best be applied to vascular plants of the salt marsh. Such actions taken now to reduce impacts of current stressors represent means of enhancing ecosystem resilience to impacts of climate change on fish and wildlife.

7.4.2.3 Preserving Habitat Extent and Functionality

All national estuaries and managers of estuarine assets nationwide identify preservation of habitat as a fundamental management goal. The greatest threat to estuarine habitat extent and function from climate change arises as sea level rise and enhanced incidence of intense storms interact with the presence of structural defenses against shoreline erosion. As explained earlier in the description of threats to water quality and fisheries, barriers that prevent horizontal migration of tidal marshes inland will result in loss of tidal marsh and other intertidal and then shallow subtidal habitats. This process will include losses to seagrass beds and other submerged aquatic vegetation down-shore of bulkheads, because if the grass cannot migrate upslope, the lower margin will die back from light limitation (Dennison *et al.*, 1993; Short and Wyllie-Echeverria, 1996) as water levels rise. The presence of bulkheads enhances the rate of erosion below them because wave energy is directed downwards after striking a hard wall, excavating and lowering the sediment elevation faster than if no bulkhead were present (Tait and Griggs, 1990). As shoreline erosion below bulkheads continues along with rising water levels, all currently intertidal

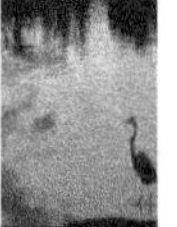

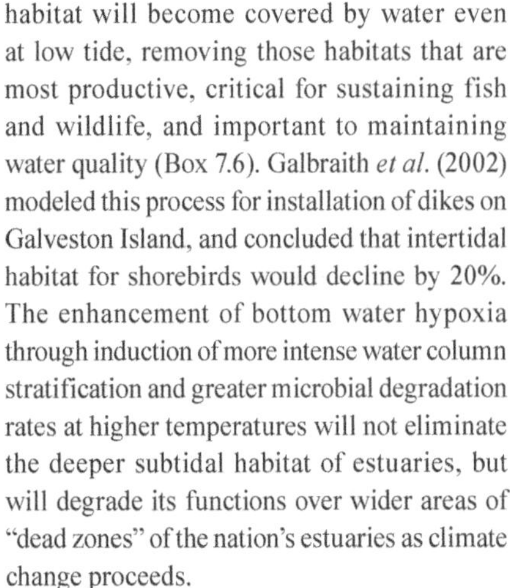

habitat will become covered by water even at low tide, removing those habitats that are most productive, critical for sustaining fish and wildlife, and important to maintaining water quality (Box 7.6). Galbraith *et al.* (2002) modeled this process for installation of dikes on Galveston Island, and concluded that intertidal habitat for shorebirds would decline by 20%. The enhancement of bottom water hypoxia through induction of more intense water column stratification and greater microbial degradation rates at higher temperatures will not eliminate the deeper subtidal habitat of estuaries, but will degrade its functions over wider areas of "dead zones" of the nation's estuaries as climate change proceeds.

Adaptations to address impacts of climate change on estuarine habitat extent and function face the same challenges as those already presented for water quality, due to common goals of preventing loss of marsh and other shallow shoreline habitats and avoiding expansion of hypoxic bottom areas. However, there may also be additional approaches available or necessary to respond to risks of areal and functional declines in estuarine habitats. At local levels, expanding the planning horizons of land use planning created in response to the federal Coastal Zone Management Act to incorporate the predictions of consequences of global change over at least a few decades would represent a rational proactive process. Such a longer view could inhibit risky development and simultaneously provide protections for important estuarine habitats, especially salt marshes and mangroves at risk from barriers that inhibit recession. Land use plans themselves rarely incorporate hard prohibitions against development close to sensitive habitats. They also have limited durability over time, as local political pressure for development and desires for protection of environmental assets wax and wane. Nevertheless, requiring planners to take a longer-term view could have only positive consequences in educating local decision makers about what lies ahead under alternative development scenarios. States run ecosystem restoration programs, largely targeted toward riparian wetlands and tidal marshes. The choice of sites for such restoration activities can be improved by strategically selecting only those where the restored wetland can move up-slope as sea level rises. Thus, planning and

[25] See, for example, **National Ocean Service**, 2005: Lionfish discovery story. NOAA Website, www.oceanservice.noaa.gov/education/stories/lionfish/lion03_blame.html, accessed on 7-25-2007.

decision-making for ecosystem restoration may require purchase of upland development rights or property to insure transgression potential, unless that upland is already publicly owned and managed to prevent construction of any impediment to orderly movement. This consideration of building in resilience to future climate change is necessary for compensatory habitat restorations that must mitigate for past losses for any restoration project that is projected to last long enough that recession would occur. In areas that are currently largely undeveloped, legislation requiring establishment of rolling easements represents a more far-reaching solution to preventing erection of permanent barriers to inland migration of tidelands. Rolling easements do not require predictions about the degree and rate of sea level rise and shoreline erosion. Purchasing development rights has the disadvantage that the uncertainty about rate of sea level rise injects uncertainty over whether enough property has been protected. In addition, rolling easements allow use of waterfront property until the water levels rise enough to require retreat, and thus represent a lower cost (Titus, 2000). Implementation of either solution should not be delayed, because delay will risk development of the very zone that requires protection.

At state and federal levels, environmental impact statements and assessments of consequences of beach nourishment do not sufficiently incorporate consideration of climate change and its impacts. Similarly, management policies at state and local levels for responding to the joint risks posed by sea level rise and increased frequencies or intensities of storms, including hurricanes, have not recognized the magnitude of growth in expenditures of present shoreline protection responses as climate change continues. Most state coastal management programs discourage hardening of shorelines, such as installation of sea walls, groins, and jetties, because they result in adverse effects on the extent of the public beach (Pilkey and Wright III, 1988). Beach nourishment, a practice involving repeated use of fill to temporarily elevate and extend the width of the intertidal beach, is the prevailing (Titus, 2000), rapidly escalating, and increasingly expensive alternative. On average, the fill sands last three to five years (Leonard, Clayton, and Pilkey, 1990) before eroding away, requiring ongoing nourishment activities indefinitely. As sea level rises, more sand is needed to restore the desired shoreline position, at escalating cost. The public debate over environmental impacts of and funding for beach nourishment will change as longer-term consequences are considered. Because beach nourishment on coastal barriers inhibits overwash of sediments during storms and the consequent landward retreat of the coastal barrier, erosion of the estuarine shoreline is intensified without this source of additional sediments. Continually elevating the shore of barrier land masses, above their natural level relative to depth on the continental shelf, implies that wave energy will not be as readily dissipated by bottom friction as the waves progress towards shore. This process brings more and more wave energy to the beach, and increases risk of storm erosion and substantial damage to the land mass in major storms.

Within less than a century, the rising sea may induce geomorphological changes historically typical of geological time scales (Riggs and Ames, 2003). These changes include predicted fragmentation of coastal barriers by new inlets, and even disintegration and loss of many coastal barriers (Riggs and Ames, 2003). Such changes would cause dramatic modifications of the estuaries lying now in protected waters behind the coastal barriers, and would shift inland the mixing zone of fresh and salt waters. As climate change progresses and sea level continues to rise, accompanied by more intense hurricanes and other storms, the beach nourishment widely practiced today on ocean beaches (Titus, 2000) may become too expensive to sustain nationwide (Titus *et al.*, 1991; Yohe *et al.*, 1996), especially if the federal government succeeds in withdrawing from current funding commitments. Miami Beach and other densely developed ocean beaches are likely to generate tax dollars sufficient to continue beach nourishment with state and local funding. Demand for groins, geotubes, sand bags, and other structural interventions will likely continue to grow as oceanfront property owners seek protection of their investment. These come at a price of loss of beach, which is the public trust resource that attracts most people to such areas. Retreat from and abandonment of coastal barriers affected by high relative rates of sea level rise and incidence

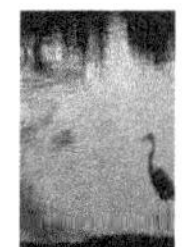

Kevin Rosseel, EPA

of intense storms does not seem to represent a politically viable management adaptation.

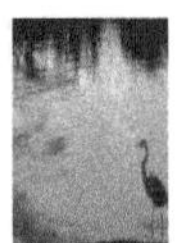

7.4.2.4 Preserving Human Values

All national estuaries recognize that estuaries provide diverse ecosystem services to people living in close proximity and to others who benefit from the estuaries' resources and functions, even passively. This category of human values relies on so many functions that the CCMPs vary widely in terms of the services they highlight and target for special management protection or restoration. Various consequences of climate change will modify these human values, and a complete assessment of how and by how much for each of the diverse values would be extensive. Nevertheless, it is clear that implications of many predictable climate-induced changes in the estuarine ecosystems are serious. Humans have a public trust stake in all other major management targets of the national estuaries, including water quality, fish and wildlife, and habitat, so to that extent we already address issues of perhaps the most importance to human interests in the estuary. However, other human values not expressly included deserve comment. Conflicts between private values of people living on estuarine shores and the public trust values are already evident, but will become increasingly prominent as sea level rises.

Probably the most serious effects of climate change on private human values associated with estuaries are those arising from climate-change-driven increases in shoreline erosion, flooding, and storm damage. Rising sea level and increased incidence of intense storms brings higher risk of extensive loss of real estate, houses, infrastructure, and even lives on estuarine shores. The houses and properties at greatest risk are those on coastal barriers lying between the ocean and outer estuary, because development on such coastal barriers is exposed during major storms to large waves in addition to storm surge and high winds. Economic and social costs of major storm events under conditions of elevated sea level may be staggeringly high, as illustrated by hurricane damage during the past decade. The management of such risks can already be considered proactive: on ocean beaches, nourishment is practiced to widen and elevate the beach, and bulkheads are widely installed on estuarine shorelines. However, each of these defenses is largely ineffective against major storms, and climate change models project more such storms developing on a continually warming Earth. Additional proactive management in the future may involve construction of dikes and levees, designed to withstand major storms and capable of vertical extension as sea level increases. Such intervention into natural processes on ocean and estuarine shores is technically feasible, but probably affordable only where development is intense enough to have created very high aggregate real estate values. It sacrifices public trust values for private values. Long-term sustainability of such barriers is questionable. In places experiencing rapid erosion but lacking dense and expensive development, shoreline erosion is likely to be accepted; retreat and abandonment will occur. Even before extensive further storm-related losses of houses, businesses, and infrastructure on ocean and estuarine shores, property values may deflate as sea level and risks of storm and flood damage increase. Many property insurers are already cancelling coverage and discontinuing underwriting activities along wide swaths of the coast in the areas most at risk to hurricanes, from Texas through New York. State governments are stepping into that void, but policy coverage is far more costly. Availability of mortgage loans may be the next economic blow to coastal development. As losses from storms mount further, the financial risks of home ownership on estuarine shorelines may create decreased demand for property and thus cause declines in real estate demand and values.

Comprehensive planning could be initiated now at federal, tribal, state, and local levels to act proactively, or opportunistically after major storm events, to modify rules or change policies to restructure development along coastal barrier and estuarine shorelines to avoid future loss of life and property, and at the same time protect many environmental assets and ecosystem services in the interest of the public trust. For example, up-front planning to prevent rebuilding in hazardous areas of high flood risk and storm damage may be feasible. Establishing setbacks from the water and buffer widths, based on the new realities of shoreline erosion and on reliable predictions of shoreline position into the future, may be possible if advance planning is complete so that rules or policies can be rapidly implemented after natural disasters. Many programs, such as federal flood insurance and infrastructure development grants, subsidize development. For undeveloped coastal barriers, such subsidies were prohibited by the Coastal Barriers Resources Act, and these prohibitions could be extended to other estuarine and coastal shorelines now at high and escalating risk. Local land use plans could be modified to influence redevelopment after storms and direct it into less risky areas. Nevertheless, such plans would result in financial losses to property owners who cannot make full use of their land. Land trusts and programs to protect water quality, habitat, and fisheries may provide funding to purchase the most risky shorelines of high resource value.

7.4.2.5 Water Quantity

Many national estuaries, especially those on the Pacific coast where snowmelt is a large determinant of the hydroperiod, identify water quantity issues among their management priorities. These issues will become growing concerns directly and indirectly for all estuaries as climate continues to change. Projected climate changes include modifications in rainfall amount and temporal patterns of delivery, in processes that influence how much of that rain falling over the watershed reaches the estuary, and in how much salt intrusion occurs from altered river flows and rising sea levels penetrating into the estuary. These climate changes interact strongly with human modifications of the land and waterways, as well as with patterns of water use and consumption. The models predicting effects of climate change on rainfall amount are not all in agreement, complicating adoption of proactive management measures. Thus, complex questions of adaptive management arise that would help smooth the transition into the predictably different rainfall future, whose direction of change is uncertain. Many of these questions will have site (basin)-specific conditions and solutions; however a generic overview is possible.

As freshwater delivery patterns change and salt water penetration increases in the estuaries, many processes that affect important biological and human values will be affected. Where annual freshwater delivery to the estuary is reduced, and in cases where only seasonal reductions occur, salt water intrusion into groundwater will influence the potable yield of aquifers. In the Pacific Northwest, predicted patterns of precipitation change imply that increased salt water penetration up-estuary will be a summertime phenomenon when droughts are likely. Fresh water is already a limiting resource globally (Postel, 1992), and is a growing issue in the United States even in the absence of climate change. Failure to develop proactive management responses will have serious consequences on human welfare and economic activity. Proaction includes establishing or broadening "use containment areas" (where withdrawal is allocated and capped) in the managed allocation of aquifer yields, so that uses are sustainable even under predicted climate-related changes in recharge rates and salt water infiltration. This may result in the need to develop reverse osmosis plants to produce potable water and replace ground water sources currently tapped to supply communities around estuaries. Further actions may be needed to modify permitting procedures for affected development, plan for growing salt water intrusion as sea level rises, and maintain aquifer productivities. Proactive planning

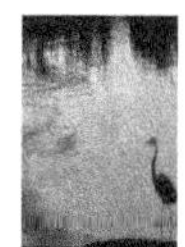

measures for water shortage can include much greater water reuse and conservation.

The enhanced flashiness of runoff from seasonal rainfall events, as they come in discrete, more intense storms, and fall upon more impervious surface area in the drainage basin, will have several consequences on human values and on natural resources of management priority. Greater pulsing of rain runoff reaching the rivers will lead to much higher frequency and extent of floods after intense storms. The resulting faster downstream flows will erode sediment from estuarine shorelines, and thus reduce the area of shallow habitats along the shores. In the Pacific Northwest, rain-on-snow events are major sources of flood waters (Marks *et al.*, 1998; Mote *et al.*, 2003) and are likely to become more frequent and intense under current climate change scenarios. These events have economic, health and safety, and social consequences for humans living or working in the newly enlarged flood plain. Bank stability and riparian habitats are threatened by increased water velocities in flood flows, which would affect water quality and ultimately fish and wildlife. When these pulses of water reach the estuary, they bring pollutants from land as well as nutrient and organic loading that have negative effects on estuarine functions for relatively long periods of time—on the order of a year or more. In estuaries where freshwater runoff is increased by global climate change, and in all estuaries where salt water has penetrated further upstream as sea level rises, the specific locations of important zones of biogeochemical processes and biotic use will shift in location. These shifts may have the effects of moving those zones, such as the turbidity maximum zone, which could influence the performance of anadromous fishes that make use of different portions of the rivers and estuaries for completing different life history stages and processes. Accurate modeling of such position changes in estuaries could allow proactive management to protect fish and wildlife habitats along the rivers and estuaries that will become critical for propagation of important fish stocks as positional shifts occur.

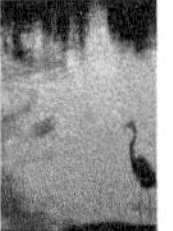

7.4.3 New Approaches to Management in the Context of Climate Change

Historically, little attention has been paid to preserving and enhancing ecosystem resilience in the management of estuaries and estuarine resources. Resilience refers to the amount of disturbance that can be tolerated by a socioecological system (*e.g.*, an estuary plus the social system interacting with it) before it undergoes a fundamental shift in its structure and functioning (Holling, 1972; Carpenter *et al.*, 2001; Gunderson *et al.*, 2002; Carpenter and Kinne, 2003). The ability of a system to maintain itself despite gradual changes in its controlling variables or its disturbance regimes is of particular concern for those interested in predicting responses to climate change. Importantly, resilience of a socioecological system results in part from appropriate management strategies. Human behaviors can reduce resilience in a variety of ways, including increasing flows of nutrients and pollutants; removing individual species, whole functional groups (*e.g.*, seagrasses, bivalves), or whole trophic levels (*e.g.*, top predators); and altering the magnitude, frequency, and duration of disturbance regimes (Carpenter *et al.*, 2001; Folke *et al.*, 2004). Importantly, climate change has the potential to exacerbate poor management and exploitation choices and cause undesirable regime shifts in ecosystems, as seen in the North Sea cod fishery and recent declines in coral reefs (Walther *et al.*, 2002). It is critical that we pursue wise and active adaptive management in order to prevent undesirable regime changes in response to climate change.

In recent years, basic research has dramatically improved our understanding of the ecosystem characteristics that help promote resilience. For example, the study of the roles of biodiversity in ecosystem dynamics has demonstrated several examples where productivity (Tilman and Downing, 1994; Naeem, 2002), biogeochemical functioning (Solan *et al.*, 2004), and community composition (Duffy, 2002; Bruno *et al.*, 2005) are stabilized under external stresses if biodiversity is high. Worm *et al.* (2006) likewise demonstrated that many services of marine ecosystems, including fisheries production, and ecosystem properties, such as resilience, are greater in more diverse systems. Some evidence exists to suggest that proliferation of non-native species can be suppressed by ecosystem biodiversity (*e.g.*, Stachowicz, Whitlatch, and Osman, 1999; but see Bruno *et al.*, 2004). These research results have not yet been directly translated into management of estuarine systems. This represents a potential approach to the goal of enhancing adaptation in contemplation of climate change. However, acting on the knowledge that higher biodiversity implies higher resilience represents a challenge for estuaries, where application of this concept is not necessarily appropriate and where any effectiveness may last only for a few decades given accelerating sea level rise.

Absent system-specific knowledge, some management actions are likely to preserve or enhance biodiversity (genetic, species, and landscape) and thus may support resilience, based upon current theory and some empirical evidence. Maintaining high genetic diversity provides high potential for evolutionary adaptation of species, and provides short-term resilience against fluctuating environmental conditions (Hughes and Stachowicz, 2004). This goal may be achieved by establishing diversity refuges, which in aggregate protect each of a suite of genotypes. Implementing this proactive management concept depends on knowledge of genetic diversity and spatial patterns of its genotypic distribution—a task most readily achieved for structural habitat providers, such as marsh and sea grasses and mangroves. Maintaining or restoring habitat and ecosystem diversity and spatial heterogeneity is another viable management goal, again most applicable to the important plants that provide habitat structure. Preserving or restoring landscapes of the full mix of different systems, and including structural corridors among landscape elements otherwise fragmented or isolated, can be predicted to enhance resilience by establishing replication of systems that can enable migrations to sustain biodiversity across the landscape (Micheli and Peterson, 1999). Structural complexity of vegetation has been related to its suitability for use of some (endangered) species (Zedler, 1993), so preserving or restoring the vegetational layering and structure of tidal marshes, seagrass meadows, and mangroves has potential to stabilize estuary function in the face of climate perturbations. In addition to salt marshes, oyster reefs have been the target of much active restoration. Success is mixed, with many reefs failing the test of sustainability because of insufficient oyster recruitment and early death of adult oysters from disease. Lenihan et al. (2001) demonstrated experimentally that the concept of representation applies well to enhance the resiliency of restored oyster reefs. They constructed more than 100 new oyster reefs along a depth gradient in the Neuse River Estuary, and showed that when persistent bottom-water hypoxia developed during summer, reef fishes were able to feed on reef-associated crustacean prey and survive the widespread mortality on reefs in deeper water by moving to shallow-water reefs, which were within the surface mixed layer. Thus, the creation of a system of reefs with representation in different environmental conditions protected against catastrophic loss of mobile fishes when eutrophication caused mass mortality of oysters and other benthic invertebrates in deeper waters.

Modifications of natural estuarine ecosystems, communities, and species populations through various forms of aquaculture represent human perturbations that may affect resilience of the estuarine ecosystem to climate change. For example, the modification and frequently the reduction in genetic diversity of cultured species can modify the gene pool of wild stocks, probably reducing their capacity for biological adaptation (Goldburg and Triplett, 1997). Flooding a system with unnaturally high densities of a cultured species such as salmon in Maine and Washington, or Pacific oysters in Oregon and Washington, carries risks of promoting disease and of simplifying the natural species composition of the fish and benthic communities respectively, thereby losing the biodiversity and natural balance of the system, which may reduce resilience. On the other hand, culturing species that are currently depleted relative to natural baselines, such as oysters and other suspension-feeding bivalve mollusks, can serve to restore missing ecosystem functions and build resilience to eutrophication (Jackson *et al.*, 2001). Similarly, culturing seaweeds can result in enhanced

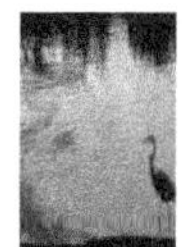

uptake of nutrients, thereby buffering against eutrophication (Goldburg and Triplett, 1997). Impacts of aquaculture in the estuaries have not been adequately considered in the context of emerging stresses of climate change, and deserve further integration into the ecosystem context (*e.g.*, Folke and Kautsky, 1989).

Analogous need exists for enhanced understanding of factors that contribute to resilience of human communities and of human institutions in the context of better preparation for consequences of changing climate. Both social science and natural science monitoring may require expansion to track possible fragility, and to look for signs of cracks in the system, as a prelude to instigating adaptive management to prevent institutional and ecological disintegration. For example, more attention should be paid to tracking coastal property values, human population movements, demography, insurance costs, employment, unemployment, attitudes, and other critical social and economic variables, in order to indicate need for proactive interventions as climate change stresses increase. An analogous enhancement of in-depth monitoring of the natural ecosystem also has merit; this likely would require changes in indicators now monitored to be able to enhance resilience through active intervention of management when the need becomes evident. Thus, monitoring in a context of greater understanding of organizational process in socioeconomic and natural systems is one means of enhancing resilience.

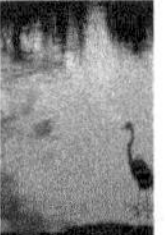

Both managers and the general public need better education to raise awareness of how important management adaptation will be if negative impacts of climate change are to be averted or minimized. Surely, managers undergo continuing education almost daily as they conduct their jobs, but targeted training on expected changes within the ecosystem they are responsible for managing is an emerging necessity. Careful articulation of uncertainties about the magnitudes, timelines, and consequences of climate change will also be important. Such education is vital to induce the broad conversations necessary for public stakeholders and managers to rethink in fundamental ways how we have previously treated and managed estuaries to provide goods and services of value.

Whereas we have used the term "management adaptation" to mean taking management actions that expressly respond to or anticipate climate change, and that are intended to counteract or minimize any of its negative implications, natural resource managers and academics have developed a different process termed "adaptive management" (Walters, 1986). Adaptive management in this context (see Chapter 9, Synthesis) refers to designing and implementing regulations or other management actions as an experiment, and employing rigorous methods of assessing the impacts of the actions. Monitoring the status of the response variables provides the data against which a management action's effectiveness can be judged. This blending of experimental design into management provides perhaps the most rigorous means of testing implications of management actions. Adaptive management has the valuable characteristic that it continuously re-evaluates the basis on which predictions are made, so that as more information becomes available to reduce the uncertainties over physical and biological changes associated with climate change, the framework of adaptive management is in place to incorporate that new knowledge. Use of this approach where feasible in testing management adaptations to global climate change can provide much-needed insight in reducing uncertainty about how to modify management to preserve delivery of ecosystem services. Unfortunately, this approach is very complex and difficult to implement, in large part because of the multiple and often conflicting interests of important stakeholders.

Because its holistic nature includes the full complexity of interactions among components, the most promising new approach to adapt estuarine management to global climate change is the further development and implementation of ecosystem-based management (EBM) of estuarine ecosystem services, in a way that incorporates climate change expectations (Peterson and Estes, 2001). The concept of EBM has its origins among land managers, where it is most completely developed (Grumbine, 1994; Christensen *et al.*, 1996). EBM is an approach to management that strives for a holistic understanding of the complex of interactions among species, abiotic components, and humans in the system and evaluates this complexity in pursuit of specific management goals (Lee,

1993; Christensen *et al.*, 1996). EBM explicitly considers different scales and thus may serve to meet the challenges of estuarine management, which ranges across scales from national and state planning and regulation to local implementation actions. Practical applications of the EBM approach are now evolving for ocean ecosystems (Pikitch *et al.*, 2004) and hold promise for achieving sustainability of ecosystem services. Both the Pew Oceans Commission (2003) and the U.S. Commission on Ocean Policy (2004) have identified EBM as our greatest hope and most urgent need for preserving ecosystem services from the oceans. The dramatic potential impacts of climate change on estuarine ecosystems imply many transformations that simply developing and applying EBM cannot reverse, but development of synthetic models for management may help optimize estuarine ecosystem services in a changing world. Ecosystems are sufficiently complex that no practical management model could include all components and processes, so the more simplified representations of the estuarine system might best be used to generate hypotheses about the effectiveness of alternative management actions that are then tested through rigorous protocols of adaptive management. One widely advocated approach to implementing EBM is the use of marine protected areas, which does not require an elaborate understanding of ecosystem structure and dynamics (Halpern, 2003; Roberts *et al.*, 2003; Micheli *et al.*, 2004). This approach may be applicable to solving important management challenges in estuaries, especially where fishery exploitation and collateral habitat injury exist; clearly, these issues apply to many estuarine systems.

Kevin Rosseel, EPA

7.4.4 Prioritization of Management Responses

Setting priorities is important to the development of management adaptations to respond to global climate change. Because responsibilities for managing estuaries are scattered among so many different levels of government and among so many different organizations within levels of government, building the requisite integrated plan of management responses will be difficult. EBM is designed to bring these disparate groups together to achieve the integration and coordination of efforts (Peterson and Estes, 2001). However, implementing EBM for national estuaries and other estuaries may require changes in governance structures and, even then, may prove politically impractical. The State of North Carolina has made progress in bringing together diverse state agencies with management authority for aspects of estuarine fisheries habitats in its Coastal Habitat Protection Plan, which approaches an EBM plan. However, this governance method is targeted toward producing fish, rather than the complete scope of critical estuarine functions and broad suite of estuarine goods and services. This model approach also lacks a mechanism to engage the relevant federal authorities. The national estuaries bring to the table a wider range of managers and stakeholders, including those from federal, tribal, state, and local levels, as are contemplated in the genesis of an EBM plan. However, the CCMPs that arise from the national estuaries do not carry any force of regulation and often lack explicit numerical targets, instead expressing wish lists and goals for improvements that are probably unattainable without substantially more resources and powers. Perhaps the national estuaries could provide the basis for a new integrative governance structure for estuaries that could

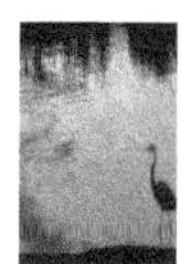

be charged with setting priorities among the many management challenges triggered by climate change.

Factors that probably would dictate priorities are numerous, including socioeconomic consequences of inaction, feasibility of effective management adaptations, the level of certainty about the projected consequence of climate change, the time frame in which action is best taken, the popular and political support for action, and the reversibility of changes that may occur in the absence of effective management response. Clearly, the processes that threaten to produce the greatest loss of both natural ecosystem services and human values are the rise of sea level and ascendancy of intense storms, with implications for land inundation, property loss, habitat loss, water quality degradation, declines in fisheries and in wildlife populations associated with shallow shoreline habitats, and salt water intrusion into aquifers. These issues attract the most attention in the media and from the public, but the global capping of greenhouse gases may not represent a feasible management response. Thus, removing and preventing engineered shoreline armoring such as bulkheads, levees, and dikes, combined with shoreline property acquisition, may be the focus of discussion if their costs are not an overwhelming impediment. Because the complexity of intermingled responsibilities for managing interacting components inhibits establishment of EBM, attention to modifying governance structures to meet this crisis would also rank high among priorities.

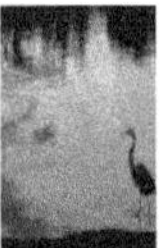

7.5 CONCLUSIONS

7.5.1 Management Response

(1) Maintaining the status quo in management of estuarine ecosystems would result in substantial losses of ecosystem services as climate change progresses.

(2) In the absence of effective management adaptation, climate-related failures will appear in all of the most important management goals identified in the CCMPs of national estuaries: maintaining water quality, sustaining fish and wildlife populations, preserving habitat, protecting human values and services, and fulfilling water quantity needs.

(3) Changes in the climate system would continue into the future even if global reductions in greenhouse gas emission were to be implemented today; thus, impacts of climate change and sea level rise, in particular, are inevitable. As an example, climate change impacts on sea level are already evident in the growing demand for and costs of beach nourishment.

(4) Many of the anticipated consequences of climate change occur via mechanisms involving interactions among stressors, and therefore may not be widely appreciated by policy makers, managers, stakeholders, and the public. The magnitude of such interactive effects typically declines as each stressor is better controlled, so enhanced management of traditional estuarine stressors has value as a management adaptation to climate change as well.

(5) Among the consequences of climate change that threaten estuarine ecosystem services, the most serious involve interactions between climate-dependent processes and human responses to climate change. In particular, conflicts arise between sustaining public trust values and private property, in that current policies protecting private shoreline property become increasingly injurious to public trust values as climate changes and sea level rises further.

(6) Many management adaptations to climate change to preserve estuarine services can be achieved at all levels of government at modest expense. One major form of adaptation involves recognizing the projected consequences of sea level rise and then applying policies that create buffers to anticipate associated consequences. An important example would be redefining riverine flood hazard zones to match the projected expansion of flooding frequency and extent.

(7) Other management adaptations can be designed to build resilience of ecological and social systems. These adaptations include choosing only those sites for habitat restoration that allow natural recession landward, thus providing resilience to sea level rise.

(8) Management adaptations to climate change can occur on three different time scales: (a) reactive measures taken in response to observed negative impacts; (b) immediate development

of plans for management adaptation to be implemented later, either when an indicator signals that delay can occur no longer, or in the wake of a disastrous consequence that provides a window of socially feasible opportunity; or (c) immediate implementation of proactive policies. The factors determining which of these time frames is appropriate for any given management adaptation include balancing costs of implementation with the magnitude of risks of injurious consequences under the status quo of management; the degree of reversibility of negative consequences of climate change; recognition and understanding of the problem by managers and the public; the uncertainty associated with the projected consequences of climate change; the timetable on which change is anticipated; and the extent of political, institutional, and financial impediments.

(9) A critical goal of monitoring is to establish and follow indicators that signal approach toward an ecosystem threshold that—once passed—implies passage of the system into an alternative state from which conversion back is difficult. One example of such ecosystem conversions involves nitrogen-induced conversion from an estuary dominated by submersed benthic grasses to an alternative dominated by seaweeds and planktonic microalgae. Avoiding conversion into such alternative states, often maintained by positive feedbacks, is one major motivation for implementing proactive management adaptation. This is especially critical if the transition is irreversible or very difficult and costly to reverse, and if the altered state delivers dramatically fewer ecosystem services. Work to establish environmental indicators is already being done in national estuaries, and can be used to monitor climate change impacts.

(10) One critically important management challenge is to implement actions to achieve orderly retreat of development from shorelines at high risk of erosion and flooding, or to preclude development of undeveloped shorelines at high risk. Such proactive management actions have been inhibited in the past by: (a) uncertainty over or denial of climate change and its implications; (b) failures to include true economic, social, and environmental costs of present policies allowing and subsidizing such risky development; and (c) legal tenets of private property rights. One possible proactive management option would be to establish and enforce "rolling easements" along estuarine shorelines as sea level continues to rise, thereby sustaining the public ownership of tide lands.

(11) Management adaptation to climate change may include ending public subsidies that now support risky development on coastal barrier and estuarine shores at high risk of flooding and storm damage as sea level rises further and intense storms are more common. Although the flood insurance system as a whole may be actuarially sound, current statutes provide people along the water's edge in eroding areas of highest risk with artificially low rates, subsidized by the flood insurance policies of people in relatively safe areas. Ending such subsidization of high-risk developments would represent a form of management adaptation to sea level rise. The federal Coastal Barriers Resources Act provides some guidance for eliminating such subsidies for public infrastructure and private development, although this act applies only to a list of undeveloped coastal barriers and would require extension to all barriers and to estuarine shorelines to enhance its effectiveness as an adaptation to climate change.

(12) Building upon ongoing efforts to operationalize ecosystem-based management (EBM) for oceans, analogous research is required for estuarine ecosystems. This research needs to address a major intrinsic impediment to EBM of estuarine services, which is the absence of a synthetic governance structure that unites now disparate management authorities, stakeholders, and the public. The U.S. Commission on Ocean Policy appealed for just this type of modification of governance structure to serve to implement EBM. EBM is necessary to facilitate management of interacting stressors, an almost ubiquitous condition for estuaries, because under present governance schemes management authority is partitioned among separate agencies or entities. Although national estuaries lack regulatory authority, they do unite most, if not all, stakeholders and could conceivably be reconstructed as quite different entities to develop and implement EBM. Such coordination among diverse management authorities must involve land managers in order to incorporate a major source of inputs to estuaries. Under changing climate, scales of management actions ultimately extend upward to include need for international

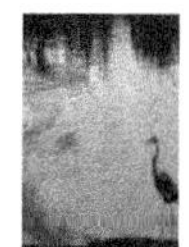

CASE STUDY SUMMARY 7.1

Albemarle-Pamlico National Estuary Program, North Carolina Southeast United States

Why this case study was chosen

The Albemarle-Pamlico National Estuary:

- Possesses more low-lying land within 1.5 m of sea level than any other national estuary;
- Is expected to lose large areas of wetlands and coastal lands to inundation, according to sea level rise projections;
- Faces projected disintegration of the protective coastal barrier of the Outer Banks of North Carolina and conversion to an oceanic bay, if the integrity of the banks is breached;
- Has a Coastal Habitat Protection Plan for fisheries enhancement (mandated under the state's Fisheries Reform Act in 1997), which provides a model opportunity for integrating climate change into an ecosystem-based plan for management adaptation.

Management context

The Albemarle-Pamlico system is a large complex of rivers, tributary estuaries, extensive wetlands, coastal lagoons, and barrier islands. It became part of the National Estuary Program in 1987. Initial efforts focused on assessments of the condition of the system through the Albemarle-Pamlico Estuarine Study. Assessment results were used in the stakeholder-based development of a Comprehensive Conservation and Management Plan (CCMP) in 1994. The CCMP presented objectives for plans in five areas: water quality, vital habitats, fisheries, stewardship, and implementation. Although long-term solutions to climate change are not specifically addressed in the Coastal Habitat Protection Plan, it does contemplate several anticipated impacts of climate change and human responses to threats.

Key climate change impacts

- Observed rise in mean sea level (current rate of relative sea level rise estimated at over 3 mm per year);
- Projected increase in interannual variability of precipitation;
- Projected increase in frequency of intense storms;
- Observed increase and projected future increase in water temperatures.

Opportunities for adaptation

- The Coastal Habitat Protection Plan ongoing process provides a means for adaptation planning across management authorities that can overcome historic constraints of compartmentalization.
- A recently established (2005) state commission on effects of climate change provides opportunity for education and participation of legislators, in a forward-looking planning process that can address issues with time frames that extend well beyond a single election cycle.
- Sparse human populations and low levels of development along much of the interior mainland shoreline of the Albemarle-Pamlico National Estuary provide openings for implementation of policies that protect the ability of the salt marsh and other shallow-water estuarine habitats to retreat as sea level rises. (Implementing the policies required to achieve this management adaptation would be extremely difficult in places where development and infrastructure are so dense that the economic and social costs of shoreline retreat are high.)
- Rolling easements and other management adaptations to climate change could be promoted by the Clean Water Management Trust Fund and the Ecosystem Enhancement Program of North Carolina.

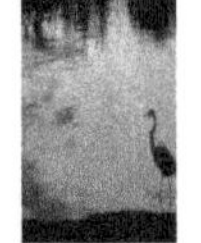

Conclusions

Community education and continuous dialogue with stakeholders are critically important in this situation, where the most economically valuable part of the ecosystem (the coast) is also the most vulnerable to climate. In estuaries, the human interest in protecting the shoreline from change is in direct conflict with the need for the shallow marshlands to transgress. Thus, the Albemarle-Pamlico National Estuary Program's stakeholder-driven process is well suited to catalyze necessary dialog on planning issues and thereby encourage legislative or regulatory actions to adapt to climate change.

The Coastal Habitat Protection Plan process provides a model on which to base further development and application of estuarine ecosystem-based management. Similarly, the North Carolina study commission established to report on the consequences of climate change and to make recommendations for management responses can serve as a model for other states and the National Estuary Program to synthesize information on climate change impacts and adaptation measures.

Finally, even the Albemarle-Pamlico National Estuary Program, which is among the most sensitive estuaries to climate change and is equipped with an active management planning process, does not explicitly include climate change adaptation measures in its Comprehensive Conservation and Management Plan. This highlights the need for increased attention to this issue by the National Estuary Program.

collaboration, placing even greater challenges to implementation of EBM.

(13) Using the Albemarle-Pamlico National Estuarine Program as a case study illustrates several management challenges posed by changing climate (see Case Study Summary 7.1). Risks of rising sea level, together with increases in intense storms, pose a serious threat to the integrity of the Outer Banks and thus to the character of the Albemarle and Pamlico Sounds, which are now sheltered and brackish, possessing little astronomical tide. A state analog to EBM, the Coastal Habitat Protection Plan, unifies state agencies to provide synthetic protection for fish habitats. This provides a model on which to base further development and application of estuarine EBM. The Legislature of the State of North Carolina established a study commission to report on the consequences of climate change and to make recommendations for management responses. This procedure too can form a model for other states and the federal government through the NEP. Although the Albemarle-Pamlico National Estuary is among the estuaries most sensitive to climate change, in large part because of the huge area of low-lying wetlands along the estuarine shorelines, and has an active management planning process in place, the absence of explicit adaptive management consideration in its CCMP reflects a need for attention to this issue by all national estuaries.

(14) Include climate change sensitivity, resilience, and adaptation responses as priorities on all relevant funding programs at state and federal levels. In the absence of such actions, for example, climate impacts on estuarine wetlands will likely violate the national "no-net-loss of wetlands" policy, which underwrites the current application of the Clean Water Act, in two ways: (a) wetland loss due to climate change will increasingly compound the continuing loss of wetlands due to development and inadequate mitigation; and; (b) measures used to protect human infrastructure from climate impacts will prevent wetland adaptation to climate change.

(15) Review all federal and state environmental programs to assess whether projected consequences of climate change have been considered adequately, and whether adaptive management needs to be inserted to achieve programmatic goals. For example, Jimerfield *et al.* conclude that "There clearly needs to be [a] comprehensive approach by federal agencies and cooperating scientists to address climate change in the endangered species recovery context. The current weak and piece-meal approach will waste precious resources and not solve the problem we are facing."[16]

7.5.2 Research Priorities

7.5.2.1 Conceptual Gaps in Understanding

(1) There is urgent need for further study of factors affecting sea level rise that may be significant, but now remain so uncertain that they cannot yet be included in IPCC projections. This especially includes enhancing our understanding of processes and rates of melting of Antarctic and Greenland ice sheets as a function of changing temperature and other coupled climatic conditions. Furthermore, it is important to resolve uncertainties about the fate of water in liquid phase released from the Greenland ice sheet, which involves the ability to project how land surface levels will respond to release from the weight of ice cover.

(2) Our understanding of processes affecting elevation change in land masses needs to be enhanced generally, so that risk of flooding, shoreline erosion, and storm damage can be better based upon geography-specific predictions of change in relative sea level, which combines rate of eustatic sea level change with land subsidence or emergence rate.

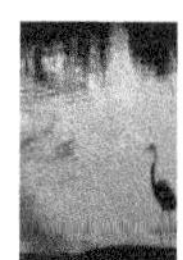

(3) Quantitative monitoring and research should be established in some model estuarine systems to develop mechanistic understanding of changes projected as consequences of climate change. Many climate change drivers (*e.g.*, CO_2 concentration, ocean temperature at the surface and with depth, sea level) are currently monitored. However, projected consequences (*e.g.*, shoreline erosion rates; estuarine physical circulation patterns; water column stratification and extent of hypoxia; species range extensions and subsequent consequences of interactions within these new combinations of predators, prey, and competitors; the incidence and impacts of disease and parasitism) require new targeted monitoring and research efforts to fill the many conceptual gaps in our understanding of these processes.

(4) Integrated, landscape-scale numerical modeling will have to become a fundamental tool to predict potential estuarine responses to the complex and often interacting stressors induced by climate change. For instance, in most cases significantly modified hydrology and sediment transport predictions will need to be linked at the estuarine interface to sea level and storm (wind/wave regime) predictions in order to evaluate the interactive effects on sediment accretion and erosion effects in estuarine marshes. Models will have to take into account complex aspects such as changes in contribution of snowmelt and rain-on-snow to timing, magnitude and hydroperiod of river discharges (*e.g.*, Mote, 2006), changes in storm tracks (*e.g.*, Salathé, 2006), changes in sediment loading to and circulation within estuaries, and how river management and regulation will be a factor (Sanchez-Arcilla and Jimenez, 1997) Ultimately, these models will need to be tied to coastal management models and other tools that allow assessment of both climate change and human response and infrastructure response.

(5) Research is needed on alternative implementation mechanisms, costs, and feasibility of achieving some form of coastal realignment, probably involving rolling easements. This would include legal, social, and cultural considerations in alternative methods of resolving or minimizing conflicts between public trust and private property values, in context of building resilience to climate change by requiring rolling easements for development in now largely undeveloped waterfront and riparian areas at risk of flooding, erosion, and storm damage.

7.5.2.2 Data Gaps

There is great need for socioeconomic research and monitoring on how social and economic variables and systems are changing, and likely to change further, in coastal regions as sea level rises. This includes developing better information on economic, social, and environmental costs of estuarine-relevant management policies under global climate change. Economic and social impacts of the growing abandonment of risky coastal areas by property insurers, and the possible future challenges in finding mortgage loans in such regions, may be important inputs into decisions on regulating development and redevelopment of such areas.

7.5.2.3 Governance Issues

(1) As stated in Management Response recommendation 12 above, a synthetic governance structure that unites now disparate management authorities, stakeholders and the public may be needed to address major impediments to EBM of estuarine services. Because of its reliance on stakeholder involvement, a restructured NEP could represent a vehicle for developing and implementing EBM.

(2) EBM of estuaries involves at minimum an approach that considers the entire drainage basin. Management plans to control estuarine water quality parameters sensitive to eutrophication, for example, must take a basin-wide approach to develop understanding of how nutrient loading at all positions along the watershed is transferred downstream to the estuary. Basin-scale management by its very nature thus prospers from uniting local governments across the entire watershed to develop partnerships that coordinate rule development and implementation strategies. Often trading programs (*e.g.*, non-point source pollution "credits") are available that allow economies to be realized in achieving management goals. To this end of facilitating management adaptation to climate change, new ecologically based partnerships of local governments could be promoted and supported.

7.5.2.4 Tool Needs

(1) New and enhanced research funds need to be invested in development and implementation of estuarine observing systems that are currently in a planning stage, such as NEON, ORION, US IOOS, and others. These observing systems need full integration with global coastal observing programs and the Global Earth Observation System of Systems. Whereas physical and chemical parameters lend themselves to automated monitoring by remote sensing and observing system platforms, more basic technological research is also necessary to allow monitoring of key biological variables as part of these observing systems. Furthermore, it is critical that current efforts to develop monitoring systems in coastal ocean waters be brought into estuaries and up into their watersheds, where

the largest human populations concentrate and where ecosystem values are most imperiled.

(2) New, more complete, interdisciplinary models are needed to project social, economic, and cultural consequences of alternative management scenarios under projected consequences of climate change. These models include decision tools that are accessible by and applicable to managers and policy makers at all levels of government.

(3) New tools are required to enhance local capacity for developing and implementing management adaptations in response to climate change, including especially the ability to use alternative scenarios to produce more effective local land-use planning.

(4) New tools are not enough: older, well-accepted tools must be used more effectively. Government agencies responsible for monitoring the environment have been reducing their commitment to this mission because of funding cuts. Extending historical records of environmental conditions is now even more urgent as a means of detecting climate change.

7.5.2.5 Education

(1) Urgent need exists to inform policy makers, managers, stakeholders, and the public about the specific evidence of climate change and its predicted consequences on estuaries. Education on the scale necessary will require new initiatives that make use of a variety of media tools, and that provide the public with accurate and unbiased information. Effective efforts must involve diverse suites of educational media including information delivery on evolving platforms such as the internet and cell phones. The information cannot reach far enough or rapidly enough if restricted to traditional delivery in school curricula and classes, but must propagate through churches, civic organizations, and entertainment media. Such education is particularly challenging and requires creative approaches.

(2) One goal of education about implications of climate change for estuaries is to build capacity for local citizen involvement in decision making. This is particularly important because of the dramatic changes required to move from management-as-usual to adaptive management. Especially challenging is the process of reconsideration of developing and redeveloping shorelines at risk of flooding, erosion, and storm damage.

(3) Some countries and states provide periodic assessments of the state of their environment. Monitoring data from many national estuaries often now serve this goal when placed in a sufficiently long time frame that extends back before establishment of the NEP. Similar scoreboards relating the status of stressors associated with climate change and of the consequences of climate change might be valuable additions to websites for all national estuaries and for our country's estuaries more broadly. To illustrate these aspects of climate change, longer-term records are required than those typically found in state of environment reports. One simple example would be provision of empirical data on sea level from local recording stations. Similarly, maps of historical shoreline movement would provide the public with a visual indication of site-specific risks. Historical hurricane tracks are similarly informative and compelling.

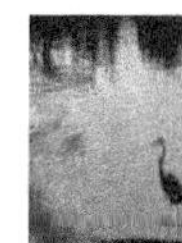

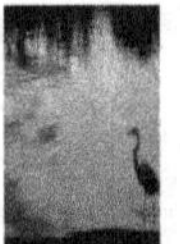

APPENDIX

Federal Legislation for Protection and Restoration of Estuaries

Legislation	As It Pertains to Estuaries	Link
Clean Water Act (1972, 1977, 1981, 1987)	Authorizes EPA to implement pollution control programs; established the basic structure for regulating discharges of pollutants and requirements to set water quality standards for all contaminants in surface waters.	http://www.epa.gov/region5/water/cwa.htm
Sec. 320 National Estuary Program (1987)	Authorizes EPA to develop plans for improving or maintaining water quality in estuaries of national significance including both point and nonpoint sources of pollution.	http://www.epa.gov/owow/estuaries/
Sec. 404 Permits for Dredged or Fill Materials (1987)	Authorizes the Corps of Engineers (U.S. Army) to issue permits for the discharge of dredged or fill material into the navigable waters at specified disposal sites.	http://www.epa.gov/owow/wetlands/
Sec. 601 State Water Pollution Control Revolving Funds (1987)	Authorizes EPA to capitalize state grants for water pollution control revolving funds for (1) for construction of public treatment facilities (2) for management program under section 319 (nonpoint source), and (3) for conservation and management plans under section 320 (NEP).	http://www.epa.gov/owm/cwfinance/
Coastal Zone Management Act (1972)	Provides grants to states that develop and implement federally approved coastal zone management plans; allows states with approved plans the right to review federal actions; authorizes the National Estuarine Research Reserve System.	http://www.legislative.noaa.gov/Legislation/czma.html
National Environmental Policy Act (NEPA) (1969)	Establishes national environmental policy for the protection, maintenance, and enhancement of the environment; integrates environmental values into decision making processes; requires federal agencies to integrate environmental values into their decision making processes by considering the environmental impacts of their proposed actions and reasonable alternatives to those actions.	http://www.epa.gov/compliance/nepa/
Magnuson-Stevens Fishery Conservation and Management Act (1996, amended)	Provides for the conservation and management of the fishery resources; ensures conservation; facilitates long-term protection of essential fish habitats; recognizes that one of the greatest long-term threats to the viability of fisheries is the continuing loss of marine, estuarine, and other aquatic habitats; promotes increased attention to habitat considerations.	http://www.nmfs.noaa.gov/sfa/
Endangered Species Act (1973)	Provides a means for ecosystems, upon which endangered species and threatened species depend, to be conserved; applicants for permits for activities that might harm endangered species must develop a Habitat Conservation Plan, designed to offset any harmful effects of the proposed activity.	http://www.fws.gov/Endangered/
National Flood Insurance Program (1968)	Component of FEMA that makes federally backed flood insurance available to homeowners, renters, and business owners in ~20,000 communities who voluntarily adopt floodplain management ordinances to restrict development in areas subject to flooding, storm surge or coastal erosion; identifies and maps the Nation's floodplains.	http://www.fema.gov/business/nfip/
Nonindigenous Aquatic Nuisance Prevention and Control Act (1990)	Provides means to prevent and control infestations of the coastal inland waters of the United States by nonindigenous aquatic nuisance species, control of ballast water, and allows for development of voluntary State Aquatic Nuisance Species Management Plans.	http://nas.er.usgs.gov/links/control.asp
Coastal Barrier Resources Act (CBRA) (1982)	Designates various undeveloped coastal barrier islands for inclusion in the Coastal Barrier Resources System. Areas so designated are made ineligible for direct or indirect federal financial assistance that might support development, including flood insurance, except for emergency life-saving activities.	http://www.fws.gov/habitatconservation/coastal_barrier.htm

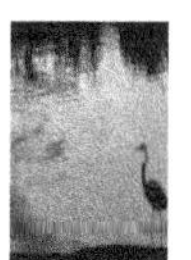

Marine Protected Areas

Lead Author: Brian D. Keller, National Oceanic and Atmospheric Administration

Contributing Authors: Satie Airamé, University of California, Santa Barbara; Billy Causey, National Oceanic and Atmospheric Administration; Alan Friedlander, National Oceanic and Atmospheric Administration; Daniel F. Gleason, Georgia Southern University; Rikki Grober-Dunsmore, National Oceanic and Atmospheric Administration; Johanna Johnson, Great Barrier Reef Marine Park Authority; Elizabeth McLeod, The Nature Conservancy; Steven L. Miller, University of North Carolina at Wilmington; Robert S. Steneck, University of Maine; Christa Woodley, University of California at Davis

8.1 SUMMARY

Marine protected areas (MPAs) such as national marine sanctuaries provide place-based management of marine ecosystems through various degrees and types of protective actions. A goal of national marine sanctuaries is to maintain natural biological communities by protecting habitats, populations, and ecological processes using community-based approaches. Biodiversity and habitat complexity are key ecosystem characteristics that must be protected to achieve sanctuary goals, and biologically structured habitats (such as coral reefs and kelp forests) are especially susceptible to degradation resulting from climate change. Marine ecosystems are susceptible to the effects of ocean acidification on carbonate chemistry, as well as to direct and indirect effects of increasing temperatures, changing circulation patterns, increasing severity of storms, and other factors.

KEY FINDINGS

Implementing networks of MPAs may help spread the risks posed by climate change by protecting multiple replicates of the full range of habitats and communities within an ecosystem. Recognizing that the science underlying our understanding of resilience is developing and that climate change will not affect marine habitats and species equally everywhere, an element of risk spreading is needed in MPA design. To help avoid the loss of a particular habitat type, managers can protect multiple examples of all habitats. In designing networks, managers can consider information on areas that may represent potential refugia from climate change impacts as well as information on connectivity (current patterns that support larval replenishment and recovery) among sites that vary in sensitivity to climate change. Larger MPAs may be necessary for networking to achieve goals such as protecting multiple refugia and addressing variability in connectivity.

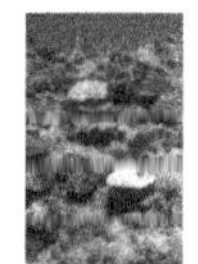

Managers can increase resilience to climate change by managing other anthropogenic stressors that also degrade ecosystems and by protecting key functional groups. Examples of anthropogenic stressors that can be managed at the site level include overfishing and overexploitation; excessive inputs of nutrients, sediments, and pollutants; and habitat damage and destruction. Reduction of these stressors may boost the ability of species, communities, and ecosystems to tolerate climate-related stresses or recover after impacts have occurred. Resilience is also affected by trophic linkages, which are key characteristics maintaining ecosystem integrity. Thus, a mechanism that has been identified to maintain resilience is the management of functional groups, specifically herbivores. In an experimental manipulation on the Great Barrier Reef, recovery from an algae-dominated to a coral-dominated state was driven by a single batfish species, not grazing by dominant parrotfishes or surgeonfishes that normally keep algae in check on Indo-Pacific reefs. This finding highlights the need to protect a diversity of species within functional groups, and the need for further research on key species and ecological processes that maintain resilience.

Overcoming the challenges of climate change will require creative collaboration among a variety of stakeholders. MPAs that reinforce social resilience can provide communities with opportunities to strengthen social relations and political stability, and diversify economic options. A variety of management actions that have been identified to reinforce social resilience include: (1) providing opportunities for shared leadership roles within government and management systems; (2) integrating MPAs and networks into broader coastal management initiatives to increase public awareness and support of management goals; (3) encouraging local economic diversification so communities are better able to deal with environmental, economic, and social changes; (4) encouraging stakeholder participation and incorporating stakeholders' ecological knowledge in a multi-governance system; and (5) making culturally appropriate conflict resolution mechanisms accessible to local communities.

A range of case studies highlight various ecological issues and management challenges found across MPAs. Three case studies are based on coral reef ecosystems, which have experienced coral bleaching events over the past two decades (see Case Study Summaries 8.1, 8.2, and 8.3). They span a range of levels of protection, from relatively low (Florida Keys) to moderate (Great Barrier Reef) to complete (Northwestern Hawaiian Islands). The Great Barrier Reef Marine Park is an example of an MPA with a relatively highly developed climate change program in place that can serve as an example to other MPAs. A Coral Bleaching Response Plan is part of its Climate Change Response Program, which is linked to a Representative Areas Program and a Water Quality Protection Plan in a comprehensive approach to enhance resilience of the coral reef ecosystem. In contrast, the Florida Keys National Marine Sanctuary is developing a bleaching response plan but does not have a climate change response program. The Florida Reef Resilience Program, led by The Nature Conservancy, is implementing a quantitative assessment of coral reefs before and after bleaching events. Finally, the recently established Papahānaumokuākea (Northwestern Hawaiian Islands) Marine National Monument is one of the largest MPAs in the world and provides a unique opportunity to examine the effects of climate change on a nearly intact large-scale marine ecosystem that will soon be a highly protected marine reserve.

A fourth case study (see Case Study Summary 8.4) examines the Channel Islands National Marine Sanctuary, located off the coast of southern California. The Sanctuary Management Plan for the Channel Islands National Marine Sanctuary mentions, but does not fully address, the issue of climate change. The plan describes a strategy to identify, assess, and respond to emerging issues through consultation with a stakeholder advisory committee and local, state, or federal agencies. Emerging issues that are not yet addressed by the management plan include ocean warming, sea level rise, shifts in ocean circulation, ocean acidification, spread of disease, and shifts in species ranges.

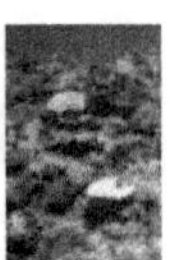

A number of opportunities exist for addressing barriers to implementation of adaptation options in MPAs. Barriers to implementation of adaptation options include lack of resources, varying degrees of interest in and concern about climate change impacts, and gaps in basic research on marine ecosystems and climate change effects. Opportunities include a growing public concern about the marine environment, recommendations of two ocean commissions, and an increasing dedication of marine scientists to conduct research that is relevant to MPA management. References to climate change as well as MPAs permeate both the Pew Oceans

Commission and U.S. Commission on Ocean Policy reports on the state of the oceans. Both commissions held extensive public meetings, and their findings reflect changing public attitudes about protecting marine resources and threats of climate change. The National Marine Sanctuary Program recently formed a Climate Change Working Group that is developing recommendations about adaptations to climate change to incorporate in site management plans. Concurrent with public and policy interests and needs, interests and involvement of the marine science community have evolved, with diminishing distinction between basic and applied research over recent decades. Although there is considerable research on physical impacts of climate change in marine systems, there are major opportunities for research on biological effects and ecological consequences of climate change. Attitudes of MPA managers have changed as well, with a growing recognition of the need to better understand ecological processes in order to implement science based adaptive management in the ocean. Managers also perceive the increasing need to consider regional- and global-scale issues in addition to traditional local-scale approaches.

The most effective configuration of MPAs may be a network of highly protected areas nested within a broader management framework. As part of this configuration, areas that are ecologically and physically significant and connected by currents, larval dispersal, and adult movements could be identified and included as a way of enhancing resilience to climate change. Connectivity is fundamental to ensuring larval exchange and the replenishment of populations in areas damaged by natural or human-related agents, and thus can enhance recovery following disturbance events. Critical areas to consider include nursery grounds, spawning grounds, areas of high species diversity, areas that contain a variety of habitat types in close proximity, and potential climate refugia. A high level of protection for these types of areas should help protect key ecological processes that enhance resilience such as larval production and recruitment, ecological interactions among full complements of species, and ontogenetic changes in habitat utilization. Management of the areas surrounding MPAs helps increase the likelihood of success of MPAs by creating a buffer zone between areas with high levels of protective actions and those with none.

8.2 BACKGROUND AND HISTORY

8.2.1 Introduction

Coastal oceans and marine ecosystems are central to the lives and livelihoods of a large and growing proportion of the U.S. population. They provide extensive areas for recreation and tourism, and support productive fisheries. Some areas produce significant quantities of oil and gas, and commercial shipping crosses coastal waters. In addition, coral reefs and barrier islands provide coastal communities with some protection from storm-generated waves. In their global analysis of the value of ecosystem services, Costanza *et al.* (1997) estimated that coastal marine ecosystem services were worth more than one-third the value of all terrestrial and marine ecosystem services combined ($12.5 of $33 trillion). Despite their value, coastal ecosystems and the services they provide are becoming increasingly vulnerable to human pressures, and management of coastal resources and human impacts generally is insufficient or ineffective (Millennium Ecosystem Assessment, 2005).

As a result of human activities, marine ecosystems are exposed to a long list of threats and stressors, including overexploitation of living marine resources, pollution, redistribution of sediments, and habitat damage and destruction. There is an equally long list of regulatory responses, including managing fisheries for sustainability, restricting ocean dumping, reducing loads of nutrients and contaminants, controlling dredge-and-fill operations, managing vessel traffic to reduce large-vessel groundings, and so on. These regulations are managed by coastal states and the federal government, with state jurisdiction extending three nautical miles (nm) offshore (9 nm in the Gulf of Mexico) and federal jurisdiction on out to 200 nm or the edge of the continental shelf (the U.S. Exclusive Economic

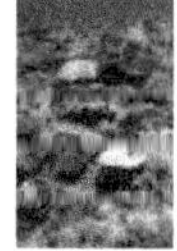

Zone, or U.S. EEZ). The total area of the U.S. EEZ exceeds the total landmass of the conterminous United States by about one-half (Pew Oceans Commission, 2003).

Broad-scale protections in the U.S. EEZ cover a wide range of types of marine ecosystems, from low to high latitudes and across the Atlantic and Pacific Oceans. Shallow areas of these systems share basic features in the form of biologically generated habitats: temperate kelp forests and salt marshes, tropical coral reefs and mangroves, and temperate and tropical seagrass beds. These biogenic habitats are fundamental to ecosystem structure and function, and support a range of different community types (Bertness, Gaines, and Hay, 2001). In addition, there are significant deep-water coral formations about which we are just starting to increase our understanding (Rogers, 1999; Watling and Risk, 2002).

Embedded within the general protections of the U.S. EEZ are hundreds of federal marine protected areas (MPAs) that are designed to provide place-based management at "special" places (Barr, 2004) and other areas that have been identified as meriting particular regulations. The term "marine protected area" has been used in many ways (*e.g.*, Kelleher, Bleakley, and Wells, 1995; Agardy, 1997; Palumbi, 2001; National Research Council, 2001; Agardy *et al.*, 2003). We use the following definition: "Marine protected area" means any area of the marine environment that has been reserved by federal, state, territorial, tribal, or local laws or regulations to provide lasting protection for part or all of the natural and cultural resources therein.[1] It is important to emphasize at the onset that MPAs are managed across a wide range of approaches and degrees of protection (Wooninck and Bertrand, 2004). At the most protective end of the spectrum are highly protected (no-take) marine reserves (Sobel and Dahlgren, 2004). These reserves eliminate fishing and other forms of resource extraction, and enable some degree of recovery of exploited populations and restoration of ecosystem structure and function, generally within relatively small areas. It is also important to highlight at the onset that management of waters surrounding MPAs is critically important both to the effectiveness of the MPAs themselves as well as to the overall resilience of larger marine systems. By "resilience" we refer to the amount of change or disturbance that can be absorbed by a system before the system is redefined by a different set of processes and structures (*i.e.*, the ecosystem recovers from the disturbance without a major phase shift; see Glossary).

Federal MPAs have been established by the Department of the Interior (National Park Service and U.S. Fish and Wildlife Service) and the Department of Commerce, National Oceanic and Atmospheric Administration (National Marine Fisheries Service, National Estuarine Research Reserve System, and National Marine Sanctuary Program) (Table 8.1). A 2000 executive order established the National Center

Table 8.1. Types of federal marine protected and marine managed areas, administration, and legislative mandates. MPAs are intended primarily to protect or conserve marine life and habitat, and are a subset of marine managed areas (MMAs), which protect, conserve, or otherwise manage a variety of resources and uses including living marine resources, cultural and historical resources, and recreational opportunities.[2]

Type of MPA/MMA	Number of Sites	Administration	Mandate
National Marine Sanctuary	13	NOAA/National Marine Sanctuary Program	National Marine Sanctuaries Act
Fishery Management Areas	216	NOAA/National Marine Fisheries Service	Magnuson-Stevens Act, Endangered Species Act, Marine Mammal Protection Act
National Estuarine Research Reserve[3]	27	NOAA/Office of Ocean and Coastal Resource Management	Coastal Zone Management Act
National Park	42	National Park Service	NPS Organic Act
National Monument[4]	3	National Park Service	NPS Organic Act
National Wildlife Refuge	109	U.S. Fish and Wildlife Service	National Wildlife Refuge System Administration Act

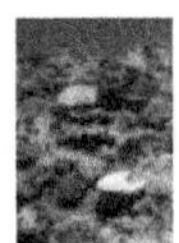

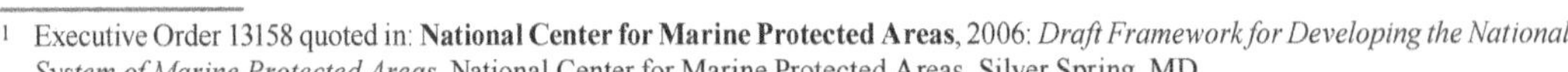

1 Executive Order 13158 quoted in: **National Center for Marine Protected Areas**, 2006: *Draft Framework for Developing the National System of Marine Protected Areas*. National Center for Marine Protected Areas, Silver Spring, MD.

2 **California Department of Fish and Game**, 2007: Marine life protection act initiatives. California Department of Fish and Game Website, http://www.dfg.ca.gov/mrd/mlpa/defs.html#mma, accessed on 7-27-2007.

3 The National Estuarine Research Reserve System is a state partnership program.

4 The Papahānaumokuākea Marine National Monument is included here. It is co-managed by NOAA/National Marine Sanctuary Program and National Marine Fisheries Service, the U.S. Fish and Wildlife Service, and the State of Hawaii and was established by Presidential Proclamation 8031.

Table 8.2. Type, number, area, and no-take area of federal marine managed areas (MMAs) and areas of Exclusive Economic Zones (EEZs) by region in U.S. waters.[5] *NP = National Parks, NWR = National Wildlife Refuges, NMS = National Marine Sanctuaries, FMA = Fishery Management Areas, NERR = National Estuarine Research Reserves, and NM = National Monuments.*

Region	Type of MMA	Number	Total Area (km^2)[6]	Total Area No Take (km^2)	% Area No Take	Area of EEZ in Region (km^2)
New England (Maine to Connecticut)						197,227
	NP	0	0	0	0%	
	NWR	1	30	0	0%	
	NMS	1	2,190	0	0%	
	FMA	30	212,930	0	0%	
	NERR[7]	1	27	0	0%	
Mid Atlantic (New York to Virginia)						218,151
	NP	3	36,472	0	0%	
	NWR	22	15	0	0%	
	NMS	0	0	0	0%	
	FMA	9	686,379	0	0%	
	NERR	1	27	0	0%	
South Atlantic (North Carolina to Florida)						525,627
	NP	8	1,421	119	8%	
	NWR	19	3,705	564	15%	
	NMS	3	9,853	591	6%	
	FMA	11	974,243	349	<0.1 %	
	NERR	5	928	0	0%	
Caribbean						212,371
	NP	2	27	1	2%	
	NWR	0	0	0	0%	
	NM[8]	2	128	76	59%	
	NMS	0	0	0	0%	
	FMA	6	168	55	33%	
	NERR	1	7	0	0%	
Gulf of Mexico						695,381
	NP	4	4,612	0	0%	
	NWR	24	2,375	2	<0.1%	
	NMS	1	146	0	0%	
	FMA	7	368,446	0	0%	
	NERR	5	2,195	0	0%	
West Coast						823,866
	NP	6	595	0	0%	
	NWR	15	226	16	7%	
	NMS	5	30,519	257	1%	
	FMA	56	386,869	0	0%	
	NERR	5	57	0	0%	
Alaska						3,710,774
	NP	3	29,795	0	0%	
	NWR	3	212,620	0	0%	
	NMS	0	0	0	0%	
	FMA	17	1,326,177	0	0%	
	NERR	1	931	0	0%	
Pacific Islands						3,869,806
	NP	4	21	< 1	<1%	
	NWR	10	281	158	56%	
	NM[8]	1	352,754	352,754	100%	
	NMS	3	3,556	1	<1%	
	FMA	6	1,467,614	0	0%	
	NERR	0	0	0	0%	
National Total						10,413,230
	NP	42	72,943	120	0.16%	
	NWR	109	219,252	740	0.34%	
	NM	3	352,882	352,882	100%	
	NMS	13	46,264	591	1.3%	
	FMA	216	5,422,826	488	0.01%	
	NERR	27	4,606	0	0.00%	
	TOTAL ALL FEDERAL MMAS[9]	410	6,118,773	354,820	5.8%	

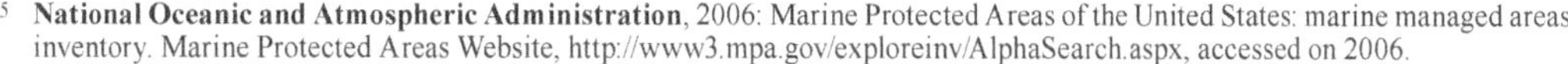

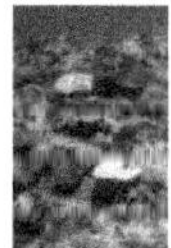

[5] **National Oceanic and Atmospheric Administration**, 2006: Marine Protected Areas of the United States: marine managed areas inventory. Marine Protected Areas Website, http://www3.mpa.gov/exploreinv/AlphaSearch.aspx, accessed on 2006.

[6] Total area includes only those sites for which data are available.

[7] NERRs are state/federal partnership sites.

[8] The Papahānaumokuākea Marine National Monument is scheduled to become a no-take area in five years when all fishing is phased out. This site has been included in the no-take category and will be the largest no-take MPA in the United States.

[9] This total is corrected for overlapping jurisdictions of Federal MMAs.

for Marine Protected Areas[10] to strengthen and expand a national system of MPAs. The total area of MMAs within the U.S. EEZ is considerable, but only a small area lies within highly protected marine reserves (Table 8.2). Only 3.4% of the U.S. EEZ lies within highly protected marine reserves, and most of this area is in the Papahānaumokuākea (Northwestern Hawaiian Islands) Marine National Monument; excluding the Monument reduces the percentage to 0.05%.

Manifestations of climate change are strengthening (IPCC, 2007c) against a background of long-standing alterations to ecological structure and function of marine ecosystems caused by fisheries exploitation, pollution, habitat degradation and destruction, and other factors (Pauly *et al.*, 1998; Jackson *et al.*, 2001; Pew Ocean Commission, 2003; U.S. Commission on Ocean Policy, 2004). Nowhere is the stress of elevated sea surface temperatures more dramatically expressed than in coral reefs, where local-scale coral bleaching has occurred in the Eastern Pacific and Florida for more than two decades (Glynn, 1991; Obura, Causey, and Church, 2006)[11] prior to the first global mass bleaching event in 1998. Impacts of climate variability and change in temperate ecosystems have not been as dramatic as coral bleaching. Interestingly, the combined effects of climate change, regime shifts, and El Niño-Southern Oscillation events (ENSOs) can strongly affect kelp forests (Paine, Tegner, and Johnson, 1998; Steneck *et al.*, 2002), but apparently not associated communities (Halpern and Cottenie, 2007).

The purpose of this chapter is to examine adaptation options for MPAs in the context of climate change. We will focus on the 14 MPAs that comprise the National Marine Sanctuary System (Table 8.3, Fig. 8.1) because they encompass a range of ecosystem types and are the only U.S. MPAs managed under specific enabling legislation. The National Marine Sanctuary Program has explicit approaches to and goals for MPA management, which simplify discussion of existing MPA management and how it may be adapted to climate change. Further, a goal of the program is to support ecosystem-based management (EBM) and, as will be discussed, EBM will become increasingly important in the context of climate change.

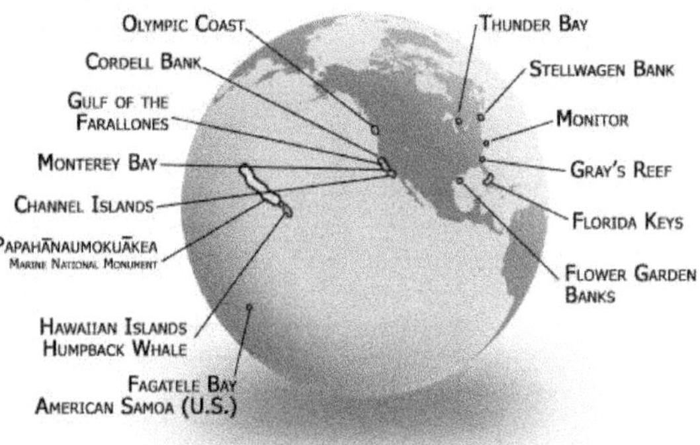

Figure 8.1. Locations of the 14 MPAs that comprise the National Marine Sanctuary System.[12]

The chapter provides background information about the historical context and origins of MPAs, with National Marine Sanctuaries highlighted as an example of effectively managed MPAs (Kelleher, Bleakley, and Wells, 1995; Agardy, 1997). MPAs are managed by several federal organizations other than the National Oceanic and Atmospheric Administration (NOAA) (Table 8.1), but it is beyond the scope of this chapter to cover all entities. National Marine Sanctuaries were selected to illustrate adaptation options for MPAs that apply broadly with respect to major anthropogenic and climate change stressors.

It is also beyond the scope of this chapter to cover issues concerning marine ecosystems from tropical to polar climates. This chapter highlights coral reef ecosystems, which have

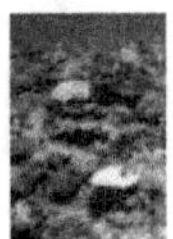

[10] http://mpa.gov/

[11] See also **Causey**, B.D., 2001: Lessons learned from the intensification of coral bleaching from 1980-2000 in the Florida Keys, USA. In: *Coral Bleaching and Marine Protected Areas. Proceedings of the Workshop on Mitigating Coral Bleaching Impact Through MPA Design, Bishop Museum, Honolulu, HI, 29-31 May 2001* [Salm, R.V. and S.L. Coles (eds.)]. Asia Pacific Coastal Marine Program Report # 0102, The Nature Conservancy, Honolulu, HI, pp. 60-66.

[12] **National Marine Sanctuary Program**, 2006: National Marine Sanctuary system and field sites. National Marine Sanctuaries Program Webpage, http://www.sanctuaries.nos.noaa.gov/visit/welcome.html, accessed on 5-18-2007.

already shown widespread and dramatic responses to oceanic warming and additional global and local stressors. Mass coral reef bleaching events became worldwide in 1998, and have resulted in extensive mortality of reef-building corals (Wilkinson, 1998; 2000; 2002; Turgeon *et al.*, 2002; Wilkinson, 2004; Wadell, 2005). There now exists a substantial and rapidly growing body of research on impacts of climate change on corals (such as bleaching) and coral reef ecosystems (*e.g.*, Smith and Buddemeier, 1992; Glynn, 1993; Hoegh-Guldberg, 1999; Wilkinson, 2004; Buddemeier, Kleypas, and Aronson, 2004; Donner *et al.*, 2005; Phinney *et al.*, 2006; Berkelmans and van Oppen, 2006). Climate change stressors, including effects of ocean acidification on carbonate chemistry (Kleypas *et al.*, 1999; Soto, 2001; The Royal Society, 2005; Caldeira and Wickett, 2005), will be reviewed later in this chapter. Management approaches to coral reef ecosystems in response to mass bleaching and/or climate change have also received some attention (Hughes *et al.*, 2003; Hansen, Biringer, and Hoffman, 2003; West and Salm, 2003; Bellwood *et al.*, 2004; Wooldridge *et al.*, 2005; Marshall and Schuttenberg, 2006).[13]

Table 8.3. Sites in the National Marine Sanctuary System.[12]
Regions: PC = Pacific Coast, PI = Pacific Islands, SE = Southeast, Gulf of Mexico, and Caribbean, NE = Northeast.

Site	Location	Region	Year Designated	Size (km²)	Yr of First Mgt Plan	Status of Mgt Plan Revision
Channel Islands	CA	PC	1980	4,263	1983	2008 planned publication
Cordell Bank	CA	PC	1989	1,362	1989	Central CA Joint Mgt Plan Review[14]
Fagatele Bay	Amer. Samoa	PI	1986	0.66	1984	Ongoing
Florida Keys	FL	SE	1990	9,844	1996	Published 2007
Flower Garden Banks	TX	SE	1992	2.0	In preparation	
Gray's Reef	GA	SE	1981	58	1983	Published 2006
Gulf of the Farallones	CA	PC	1981	3,252	1983	Central CA Joint Mgt Plan Review
Hawaiian Islands HW[15]	HI	PI	1992	3,548	1997	Published 2002
Monitor[16]	NC	NE	1975	4.1	1997[17]	
Monterey Bay	CA	PC	1992	13,784	1992	Central CA Joint Mgt Plan Review
Olympic Coast	WA	PC	1994	8,573	1994	Ongoing
Papahānaumokuākea MNM[18]	HI	PI	2006	~360,000	In preparation	
Stellwagen Bank	MA	NE	1992	2,188	1993	2009 planned publication
Thunder Bay[16]	MI	NE	2000	1,160	1999	Ongoing
Key Largo[19]	FL		1975	353		
Looe Key[19]	FL		1981	18		

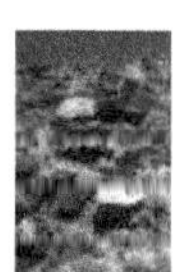

[13] See also Salm, R.V. and S.L. Coles (eds.), 2001. *Coral Bleaching and Marine Protected Areas. Proceedings of the Workshop on Mitigating Coral Bleaching Impact Through MPA Design, Bishop Museum, Honolulu, HI, 29-31 May 2001.* Asia Pacific Coastal Marine Program Report # 0102, The Nature Conservancy, Honolulu, HI, 118 pp.
Marshall, P. and H. Schuttenberg, 2006: A Reef Manager's Guide to Coral Bleaching. Great Barrier Reef Marine Park Authority, http://www.coris.noaa.gov/activities/reef_managers_guide/, pp.1-178.

[14] The Central California Joint Management Plan Review is a coordinated process to obtain public comments on draft management plans, proposed rules, and draft environmental impact statements for the three Central California Sanctuaries.

[15] HW = humpback whale.

[16] The Monitor (http://monitor.noaa.gov/) and Thunder Bay (http://thunderbay.noaa.gov/) NMSs were designated for protection of maritime heritage resources.

[17] This plan is a comprehensive, long-range preservation plan for the Civil War ironclad U.S.S. *Monitor*.

[18] The Papahānaumokuākea Marine National Monument is co-managed by NOAA/National Marine Sanctuary Program and National Marine Fisheries Service, U.S. Fish and Wildlife Service, and the State of Hawaii.

[19] The Key Largo and Looe Key NMSs were subsumed within the Florida Keys NMS as Existing Management Areas.

Climate-change stressors in and ecological responses of colder-water marine ecosystems only partially overlap those of warmer-water and tropical marine ecosystems (IPCC, 2001; Kennedy *et al.*, 2002). The Channel Islands National Marine Sanctuary is included as a temperate-zone case study (see Case Study Summary 8.4) to contrast with case studies of tropical coral reef ecosystems from the Florida Keys to Hawaii to Australia (Case Study Summaries 8.1–8.3), which differ in extent of no-take protection.

8.2.2 Historical Context and Origins of National Marine Sanctuaries and Other Types of Marine Protected Areas

8.2.2.1 Mounting Environmental Concerns and Congressional Actions

In 1972 the United States acknowledged the dangers and threats of uncontrolled industrial and urban growth and their impacts on coastal and marine habitats through the passage of a number of Congressional acts that focused on conservation of threatened coastal and ocean resources. The Water Pollution Control Act addressed the nation's threatened water supply and coastal pollution. The Marine Mammal Protection Act imposed a five-year ban on killing whales, seals, sea otters, manatees, and other marine mammals. The Coastal Zone Management Act provided a framework for federal funding of state coastal zone management plans that created a nationwide system of estuarine research reserves. A final environmental bill that focused on ocean health, the Marine Protection, Research and Sanctuaries Act of 1972, established a system of marine protected areas—national marine sanctuaries (NMS)—administered by NOAA (Fig. 8.2).

8.2.2.2. Types of Federal MPAs and Focus on National Marine Sanctuaries

In addition to the 13 national marine sanctuaries and one marine national monument, there are hundreds of marine managed areas (MMAs) under other, sometimes overlapping jurisdictions (Table 8.2) (National Research

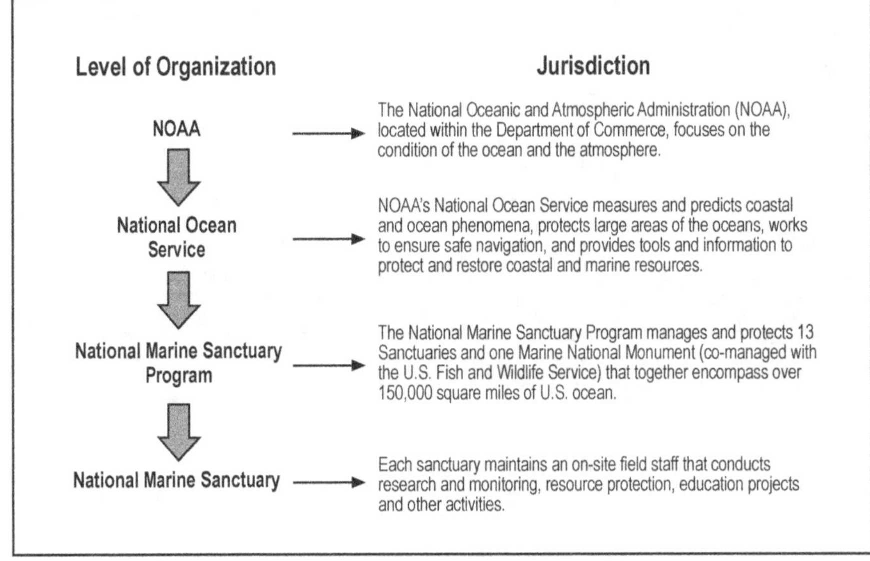

Figure 8.2. Organizational chart of the National Marine Sanctuary Program.[20]

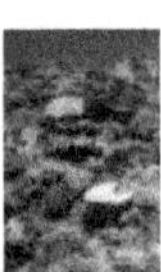

[20] **NOAA National Ocean Service**, 2006: NOAA's National Ocean Service: program offices. NOAA Website, http://www.oceanservice.noaa.gov/programs/, accessed on 7-29-2007.

CASE STUDY SUMMARY 8.1

Florida Keys National Marine Sanctuary
Southeast United States

Why this case study was chosen

The Florida Keys National Marine Sanctuary:

- Surrounds the Florida Reef Tract, the only system of bank-barrier coral reefs in the conterminous United States and one of the most biodiverse areas in North America;
- Draws millions of visitors each year due to its ready access to a unique environment, a burgeoning population in southern Florida, and its status as a destination for cruise ships at Key West;
- Is a relatively data-rich environment, with an existing baseline of information for detecting presumptive climate change effects;
- Is an example of a marine protected area with a relatively low level of protection using no-take marine reserves (6% of total area).

Management context

The Florida Keys National Marine Sanctuary encompasses multiple areas with different degrees of protection and management histories, some going back to 1960. It was designated as a national marine sanctuary in 1990, but management regulations did not go into effect until 1997, once the final management plan was approved. There are five types of management zones, with varying degrees of restrictions, including "no-take," limits on specific types of fishing or vessel access, and research only access. In addition, a water quality protection program is administered through the U.S. Environmental Protection Agency, working with the State of Florida and the National Oceanic and Atmospheric Administration. Enforcement efforts complement education and outreach programs; mooring buoys and waterway markers help minimize physical impacts from anchoring and vessel groundings.

Key climate change impacts

- Projected increase in water temperatures by several degrees in the next 100 years;
- Projected reduction in rates of calcification associated with increased ocean acidification;
- Projected increase in intensity of storms;
- Expected exacerbation of coral bleaching events;
- Potential increased prevalence of diseases;
- Potential changes in ocean circulation patterns;
- Potential geographic range shifts of individual species, and changes in reef community composition, in response to temperature increases.

Opportunities for adaptation

- Bleaching-resistant sites could be targeted for priority protection as refugia and as larval sources for recovery; the National Oceanic and Atmospheric Administration's Coral Reef Watch program to predict mass bleaching events presents an opportunity for designing before-during-after sampling around bleaching events, which will be crucial for site identification.
- The Florida Reef Resilience Program, led by The Nature Conservancy, is conducting surveys to identify resilient areas and is promoting public awareness and education.
- In the short time since their establishment, no-take zones have been shown to enhance heavily fished populations, which in turn may support resilience through re-establishment of key predators. (Much additional research is needed on the effects of community structure on resilience.)
- Protecting habitats similar to those that thrived during the middle Holocene, when coral reefs flourished north of their current distribution, could allow for northward range migration. (This would be contingent on mitigation of existing stressors that may otherwise limit the ability of corals to migrate.)
- Mangrove restoration not only provides habitat and shoreline protection, but is also a source of dissolved organic compounds that have been shown to provide protection from photo-oxidative stress in corals.

Conclusions

Environmental problems that spurred the creation of the Florida Keys National Marine Sanctuary are already being exacerbated by climate change, in particular coral bleaching and disease. Some of the management protections to reduce other anthropogenic stressors may also increase coral reef resilience and allow range expansion northward in response to climate change. Monitoring and research can identify bleaching resistant and resilient sites, so that protection efforts can be adjusted for future climate conditions.

CASE STUDY SUMMARY 8.2

Great Barrier Reef Marine Park
Northeastern Australia

Why this case study was chosen

The Great Barrier Reef Marine Park:

- Is at the forefront of climate change adaptation planning for marine protected areas (MPAs) and is thus an excellent model for U.S. MPAs;
- Has exhibited signs of climate change effects, with increases in coral bleaching events and seabird nesting failures correlated with increases in sea and air temperatures;
- Has a high conservation value as a World Heritage Area and as the largest coral reef ecosystem in the world;
- Is an example of an MPA with a moderate level of no-take protection.

Management context

The Great Barrier Reef (GBR) Marine Park has been under a management regime since 1975. Marine park zoning was revised in 2003 to increase no-take zones to 33% of the total area, with at least 20% protected in each habitat bioregion. Also in 2003, the Reef Water Quality Protection Plan was implemented to manage diffuse sources of pollution entering the GBR from the adjacent large catchment area. Tourism and fishing industries are highly regulated through the GBR Marine Park Authority and the Queensland Government, respectively. The GBR coast is one of the fastest growing regions in Australia, with different aspects of coastal development regulated at the local, state, and federal levels. The GBR Climate Change Response Program developed a Climate Change Action Plan in 2007 to facilitate: 1) targeted science; 2) a resilient GBR ecosystem; 3) adaptation of GBR industries and communities; and 4) reduced climate footprints.

Key climate change impacts

- Observed increase in regional sea surface temperatures (0.4°C since 1850) and projected further increase of 1–3°C by 2100, which will increase coral bleaching and disease, and will have implications for primary productivity;
- Projected decrease in ocean pH of 0.4–0.5 units by 2100, which will limit calcification rates of corals, forams, some plankton and molluscs;
- Projected rise in sea level of 30–60 cm by 2100, which will affect seabird and turtle nesting, island and coastal habitats, light penetration, and connectivity;
- Projected increase in tropical cyclone intensities, with potentially greater damage to coastal and shallow habitats including coral reefs;
- Projected changes in rainfall, river flow, and El Niño Southern Oscillation regimes;
- Expected losses of coral reef habitat, with associated decreases in ecosystem diversity and changes in community composition.

Opportunities for adaptation

- Areas with high resilience factors (water quality, coral cover, community composition, larval supply, recruitment success, herbivory, disease, and effective management) are being identified as priority areas to protect from other stresses; areas with low resilience are also being identified as candidates for more active management to improve their condition.
- Landward areas could be conserved through land acquisition and removal of barrier structures to allow migration of mangroves and wetlands as sea level rises.
- Sites of specific importance could be protected from coral bleaching through artificial shading or water mixing in summer months;
- Through partnerships with stakeholders to identify impacts on tourism, options for how the industry can respond, and strategies for becoming climate ready, the GBR has developed a Marine Tourism and Climate Change Action Strategy.
- By having a variety of management tools ready as new information becomes available, it may be possible to manage flexibly and respond rapidly to ongoing climatic changes.

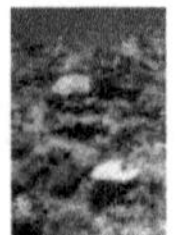

Conclusions

The GBR Climate Change Response Program has already documented observed climate change effects, identified likely vulnerabilities, and developed a Climate Change Action Plan. The combination of dramatic potential ecosystem effects and a strong national and international constituency for protection has made the GBR Marine Park an international leader in addressing climate change impacts on coral reefs. Management examples for other MPAs include initiatives that support local industries and communities in adapting to climate change, management plans that are flexible in the face of uncertainty, and resilience-based management strategies.

CASE STUDY SUMMARY 8.3

Papahānaumokuākea (Northwestern Hawaiian Islands) Marine National Monument Pacific United States

Why this case study was chosen

The Papahānaumokuākea (Northwestern Hawaiian Islands) Marine National Monument:

- Provides an opportunity to assess how a nearly intact, large-scale coral reef ecosystem responds to climate change;
- Has a high conservation value due to high levels of endemism, a unique apex-predator-dominated ecosystem, and the occurrence of a number of protected and endangered species;
- Is an example of a large Marine Protected Area with a high level of no-take protection.

Management context

The Northwestern Hawaiian Islands (NWHI) are an isolated, low lying, primarily uninhabited archipelago that is relatively free from human impacts due to its remoteness. Eight of the 10 NWHI have been protected since 1909 as part of what is now the Hawaiian Islands National Wildlife Refuge. The Papahānaumokuākea Marine National Monument was designated in 2006 as the largest marine protected area in the world, managed jointly by the State of Hawaii, the U.S. Fish and Wildlife Service, and the National Oceanic and Atmospheric Administration. The new protections will phase out commercial fishing over five years, and already ban other types of resource extraction and waste dumping. The dominant stressors are natural ones, including large inter- and intra-annual water temperature variations, seasonally high wave energy, and inter-annual and inter-decadal variability in ocean productivity. Marine debris is the largest anthropogenic stressor; a debris removal program between 1999 and 2003 resulted in a removal of historical debris accumulation, but the current level of effort is not sufficient to keep up with the annual rate of accumulation. The draft Monument Management Plan does not address climate and ocean change management actions specifically, but many of the research, monitoring, and education plans focus on climate, which will provide managers with tools for addressing climate change.

Key climate change impacts

- Projected increase in the intensity of storm events, which will in turn intensify wave impacts on habitat;
- Projected decreases in important habitat for sea turtles, endangered monk seals, and seabirds as sea level rise inundates low-lying emergent areas;
- Expected increase in temperature-related coral bleaching events like those observed in 2002 and 2004;
- Projected increases in ocean temperature that could lead to shifts in the distribution of corals and other organisms; shallow-water species that are adapted to cooler water may see habitat loss, while those adapted to warmer water might extend their range.

Opportunities for adaptation

- Monitoring and research provide an opportunity to evaluate the hypothesis that large, intact predator-dominated ecosystems are more resistant and resilient to stressors, including climate change, and expanded efforts will help better understand how climate change affects an ecosystem in the absence of localized human stressors.
- The Coral Reef Ecosystem Integrated Observing System (CREIOS) serves to alert resouce managers and researchers to environmental events considered significant to the health of the surrounding coral reef ecosystem, allowing managers to implement response measures in a timely manner and allowing researchers to increase spatial or temporal sampling resolution, if warranted; with supplementary sensors, CREIOS can help to capture climate change impacts at finer spatial and temporal scales than currently exist.
- The draft monument science plan includes several specific climate change research activities, including determining habitat changes due to sea level rise; mapping areas that will be most affected by extreme wave events; and determining how specific habitat, communities, and populations will be affected by climate change effects.
- Beach nourishment could counter the effects of sea level rise on the habitats of critical endemic and protected species.

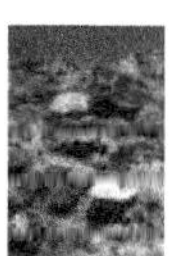

Conclusions

The high level of protection, the healthy intact predator-dominated ecosystem, the limited human impact, and the current ocean observing system present a unique research opportunity for studying adaptation to climate change in the Papahānaumokuākea Marine National Monument (PMNM). An increased understanding of natural resistance and resilience in this system will inform management planning in other marine protected areas. To date, management goals for adapting the PMNM to climate change have not looked beyond efforts to understand the system, but as endangered species habitat becomes affected, more active management efforts will be necessary.

CASE STUDY SUMMARY 8.4

Channel Islands National Marine Sanctuary
Western United States

Why this case study was chosen

The Channel Islands National Marine Sanctuary:

- Supports a diverse community based around the dominant, habitat-forming, giant kelp forests;
- Is sensitive to natural variability and has exhibited large responses to El Niño Southern Oscillation events, in particular;
- Encompasses a biogeographic boundary between the warm waters of the Davidson Current and the cool, nutrient-rich waters of the California Current.

Management context

The Channel Islands National Marine Sanctuary was designated in 1980 and was managed through overlapping state and federal jurisdictions. In 2003, 10 new fully protected marine reserves and two conservation areas that allow limited take were established to protect marine habitats and species of interest. The network of marine protected areas, which was designed with input from a broad array of stakeholders, offers additional protection to 10% of sanctuary waters. In 2007, the sanctuary implemented a second phase of the network of marine protected areas, by extending seven reserves and one conservation area into federal waters and adding a reserve to form a network of marine protected areas that includes 21% of sanctuary waters. The Sanctuary Management Plan includes a mechanism for addressing emerging issues; climate change has not yet been, but could be, explicitly identified as an emerging issue.

Key climate change impacts

- Projected increases in storm intensity that may increase damage to kelp stocks and rip kelp holdfasts from rocky substrata;
- Projected increase in frequency of El Niño-like conditions, which may suppress kelp growth by lowering nutrient levels due to associated relaxation of coastal winds;
- Projected increase in water temperature, which will affect metabolism, growth, reproduction, rates of larval development, spread of non-native species, and outbreaks of marine disease;
- Projected changes in currents and upwelling that may affect the location of biogeographic boundaries, and change primary productivity and species assemblages.

Opportunities for adaptation

- Marine reserves can be used as a management tool to increase resilience of kelp forest communities; in a marine reserve where fishing has been prohibited since 1978, kelp forests were less vulnerable to storms, ocean warming, overgrazing, lower nutrient concentrations, and disease compared with other areas of the sanctuary.
- With a slight adjustment, monitoring and research can be refocused to capture important information about climate and ocean change; observed changes associated with climate could be used to trigger more intensive observations.
- Outreach mechanisms such as the Sanctuary Naturalist Corps, Ocean Etiquette program, and sanctuary publications are well positioned to communicate information to the public on climate change impacts, mitigation, and adaptation options.
- Protection in reserves and more hands-on techniques, such as removal of non-indigenous species, could preserve the integrity of marine communities in the sanctuary.

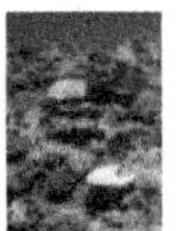

Conclusions

The high degree of natural environmental variability in the Channel Islands National Marine Sanctuary supports remarkable biological diversity. Climate change, in concert with anthropogenic stressors, will likely intensify the range of variability of the system. A marine reserve within the sanctuary has allowed kelp forests to flourish and increased their resilience to environmental shifts, such as those associated with El Niño events. Similarly, marine reserves are likely to be effective tools for minimizing the negative ecological impacts of climate change. The Sanctuary Management Plan is an appropriate mechanism for identifying climate change as an emerging issue and developing a strategic plan for management of climate change impacts, and for research, education, and outreach about climate change.

Council, 2001).[21] The National Park System, administered by the National Park Service of the Department of the Interior, includes more than 70 ocean sites (Davis, 2004). Certain national parks such as Everglades (founded in 1947), Biscayne (founded in 1968 as Biscayne National Monument), and Dry Tortugas National Parks (founded in 1935 as Fort Jefferson National Monument) have much longer histories of functioning as MPAs than the 35-year history of National Marine Sanctuaries. The National Marine Sanctuary Program and National Park Service have collaborated on ocean stewardship for a number of years (Barr, 2004). The U.S. Fish and Wildlife Service, also under the Department of the Interior, manages more than 100 national wildlife refuges that include marine ecosystems (Table 8.2). In some cases, jurisdictions overlap. For example, there are four national wildlife refuges within the Florida Keys National Marine Sanctuary (Keller and Causey, 2005), three of which cover large areas of nearshore waters (Fig. 8.3).

NOAA's National Marine Fisheries Service (NMFS) has jurisdiction over a large number of fishery management areas (Table 8.2). Collectively, these areas are more than an order of magnitude greater in size than all the other MMAs combined, but with a very small area under no-take protection (Table 8.2). NOAA also administers the National Estuarine Research Reserve System, which is a partnership program with coastal states that includes 27 sites.

This chapter is focused on NOAA's National Marine Sanctuary Program (NMSP), because it is dedicated to place-based protection and management of marine resources at nationally significant locations and has gained international

[21] See also **National Center for Marine Protected Areas**, 2006: *Draft Framework for Developing the National System of Marine Protected Areas.* National Center for Marine Protected Areas, Silver Spring, MD.

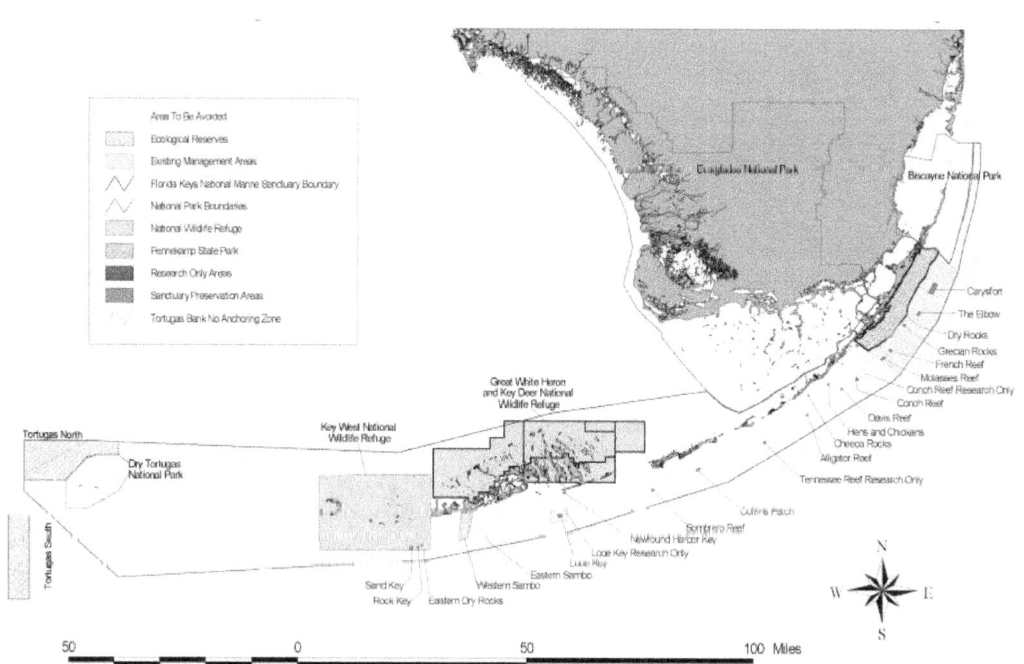

Figure 8.3. Map of the Florida Keys National Marine Sanctuary. The 1990 designation did not include the Tortugas Ecological Reserve, located at the western end of the sanctuary, which was implemented in 2001. The Key Largo NMS corresponded to the Existing Management Area (EMA) just offshore of the John Pennekamp Coral Reef State Park; the Looe Key NMS corresponded to the EMA surrounding the Looe Key Sanctuary Preservation Area and Research Only Area.[8]

recognition over the years (Barr, 2004). The principles of adaptation of MPA management to climate change (*i.e.*, institutional responses) that are identified will be broadly applicable to MPAs under other jurisdictions and forms of management, such as national parks, national wildlife refuges, and MMAs established by the NMFS, although institutional responses to adaptation likely will differ among the agencies responsible for resource management (Holling, 1995; McClanahan, Polunin, and Done, 2002). As the only federal program specifically mandated to manage MPAs, the NMSP is in a unique position to respond to challenges and recommendations in reports by the U.S. Commission on Ocean Policy (U.S. Commission on Ocean Policy, 2004) and Pew Oceans Commission (Pew Ocean Commission, 2003). Both reports encourage the use of ecosystem-based management, which is one of the hallmarks of the NMSP.

8.2.2.3 The National Marine Sanctuary Program

The NMSP was established to identify, designate, and manage ocean, coastal, and Great Lakes resources of special national significance to protect their ecological and cultural integrity for the use and enjoyment of current and future generations. In addition to natural resources within national marine sanctuaries, NOAA's Maritime Heritage Program is committed to preserving historical, cultural, and archaeological resources.[22]

The inclusion of consumptive human activities as a major part of the management programs in national marine sanctuaries distinguishes them from other federal or state resource protection programs. Sanctuaries are established for the long-term public benefit, use, and enjoyment, both recreationally and commercially. However, it is critical that sanctuary management policies, practices, and initiatives ensure that human activities in sanctuaries are compatible with long-term protection of sanctuary resources.

Thirteen national marine sanctuaries and one marine national monument, representing a wide variety of ocean environments as well as one cultural heritage site in the Great Lakes, have been established since 1975 (Table 8.3; Fig. 8.1; Fig 8.4). The national marine sanctuaries encompass a wide range of temperate and tropical environments: moderately deep banks, coral reef-seagrass-mangrove systems, whale migration corridors, deep sea canyons, and underwater archaeological sites. The sites range in size from 0.66 km^2 in Fagatele Bay,

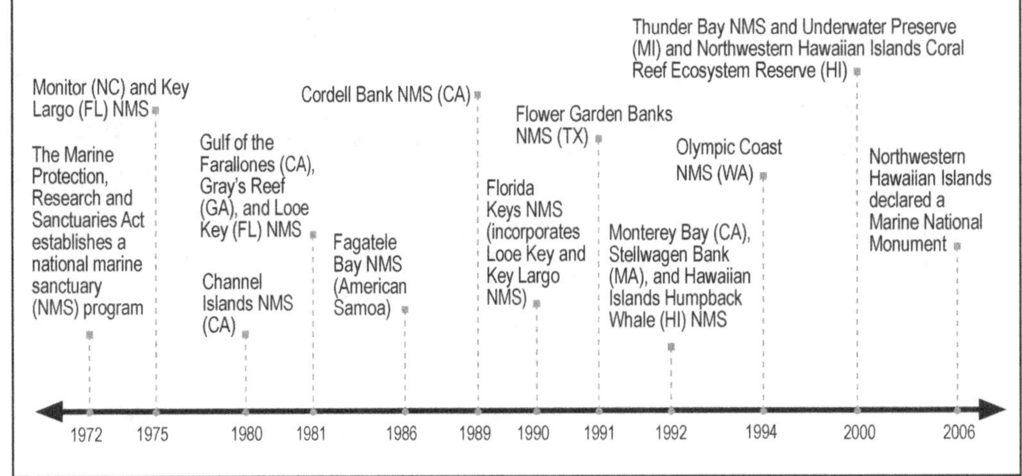

Figure 8.4. Timeline of the designation of the national marine sanctuaries in the National Marine Sanctuary Program.[23]

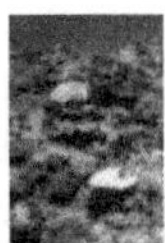

[22] **National Marine Sanctuary Program**, 2006: Maritime heritage program. National Marine Sanctuaries Program Webpage, http://www.sanctuaries.nos.noaa.gov/maritime/welcome.html, accessed on 5-18-2007.

[23] **National Marine Sanctuary Program**, 2006: History of the national marine sanctuaries. NOAA's National Marine Sanctuaries Website, http://sanctuaries.noaa.gov/about/history, accessed on 7-29-2007.

American Samoa, to more than 360,000 km^2 in the Northwestern Hawaiian Islands (Table 8.3), the largest marine protected area in the world.

The NMSP has implemented a regional approach to managing the system of sanctuaries[12] (Table 8.3). Four regions have been established to improve support for the sites and to enhance an integrated ecosystem-based approach to management of sanctuaries. An important function of the regions is to provide value-added services to the sites, while taking a broader integrated approach to management. The four regions are the Pacific Islands; West Coast; Northeast-Great Lakes; and the Southeast, Gulf of Mexico, and Caribbean. Boundaries for these regions are focused on physical and biological connectivity among sites, rather than political boundaries.

8.2.3 Enabling Legislation

8.2.3.1 Enabling Legislation for Different Types of MPAs

The U.S. National Park System Organic Act established the National Parks System in 1916. Several parks and national monuments have marine waters within their boundaries or are primarily marine; they were the earliest federal MPAs. Similarly, a large number of national wildlife refuges function as MPAs (Table 8.1) under the authority of the U.S. Fish and Wildlife Service. The 1966 National Wildlife Refuge System Administration Act was the first comprehensive legislation after decades of designations of federal wildlife reservations and refuges.[24]

NOAA's National Marine Fisheries Service implements and manages more than 200 fishery management areas (Table 8.1) under several different statutory authorities, with four major categories: Federal Fisheries Management Zones, Federal Fisheries Habitat Conservation Zones, Federal Threatened and Endangered Species Protected Areas, and Federal Marine Mammal Protected Areas. The purposes of these fishery management areas include rebuilding and maintaining sustainable fisheries, conserving and restoring marine habitats, and promoting the recovery of protected species. NOAA's National Estuarine Research Reserve System was established by the Coastal Zone Management Act of 1972.[25] This system consists of partnerships between NOAA and coastal states to protect habitat, offer educational opportunities, and provide areas for research. This same year, Congress also established a system of national marine sanctuaries.

8.2.3.2 The Marine Protection, Research, and Sanctuaries Act

The Marine Protection, Research, and Sanctuaries Act[26] established both the NMSP and a regulatory framework for ocean dumping, which was a major issue at the time. In Title III of the Act, later to be known as the National Marine Sanctuaries Act (NMSA),[27] the Secretary of Commerce received the authority to designate national marine sanctuaries for the purpose of preserving or restoring nationally significant areas for their conservation, recreational, ecological, or esthetic values. The NMSA is reauthorized every four to five years, allowing for updating and adaptation as necessary.

8.2.3.3 Legislation Designating Particular National Marine Sanctuaries

On November 16, 1990, the Florida Keys National Marine Sanctuary and Protection Act (FKNMS Act), P.L. 101-605, set out as a note to 16 U.S.C. 1433, became law. The FKNMS Act designated an area of waters and submerged lands, including the living and nonliving resources within those waters, surrounding most of the Florida Keys (Fig. 8.3). This was the first national marine sanctuary to be designated by an act of Congress.

The FKNMS Act immediately addressed two major concerns of the residents of the Florida Keys. First, it placed an instant prohibition on oil drilling, including mineral and hydrocarbon leasing, exploration, development, or production, within the sanctuary. Second, the Act created an internationally recognized area to be avoided (ATBA) for ships greater than 50 m in length, with special designated access corridors into ports (Fig. 8.3). The ATBA provides a buffer

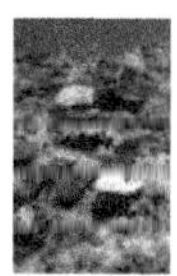

[24] **U.S. Fish and Wildlife Service**, 2007: Origins of the U.S. Fish and Wildlife Service. U.S. Fish and Wildlife Service Website, http://training.fws.gov/history/origins.html, accessed on 5-18-2007.

[25] 16 U.S.C. 1451-1456 P.L. 92-583

[26] 33 U.S.C. 1401-1445, 16 U.S.C. 1431-1445 P. L. 92-532

[27] 16 U.S.C. 1431-1445 P.L. 106-513

zone along the coral reef tract to protect it from oil spills and groundings by large vessels.

The FKNMS Act also called for a comprehensive, long-term strategy to protect and preserve the Florida Keys marine environment. The sanctuary seeks to protect marine resources by educating and interpreting for the public the Florida Keys marine environment, and by managing those uses that result in resource degradation. At the time it was thought that the greatest challenge to protecting the natural resources of the Keys and the economy they support was to improve water quality. To address this challenge, the FKNMS Act brought together various agencies to develop a comprehensive Water Quality Protection Program (WQPP). The U.S. Environmental Protection Agency (EPA) is the lead agency in developing and implementing the WQPP, the purpose of which is to "recommend priority corrective actions and compliance schedules addressing point and nonpoint sources of pollution to restore and maintain the chemical, physical, and biological integrity of the sanctuary, including restoration and maintenance of a balanced, indigenous population of corals, shellfish, fish, and wildlife, and recreational activities in and on the water" (U.S. Department of Commerce, 1996).

The FKNMS Act called for an Interagency Core Group to be established to compile management issues confronting the sanctuary as identified by the public at scoping meetings, from written comments, and from surveys distributed by NOAA. The Core Group consisted of representatives from several divisions of NOAA, National Park Service, U.S. Fish and Wildlife Service, EPA, U.S Coast Guard, Florida Governor's Office, Florida Department of Environmental Protection, Florida Department of Community Affairs, South Florida Water Management District, and Monroe County.

The FKNMS Act also called for the public to be a part of the planning process using a Sanctuary Advisory Council (SAC) to aid in the development of a comprehensive management plan. A 22-member SAC was selected by the Governor of Florida and the Secretary of Commerce. The council consisted of members of various user groups; local, state, and federal agencies; scientists; educators; environmental groups; and private citizens.

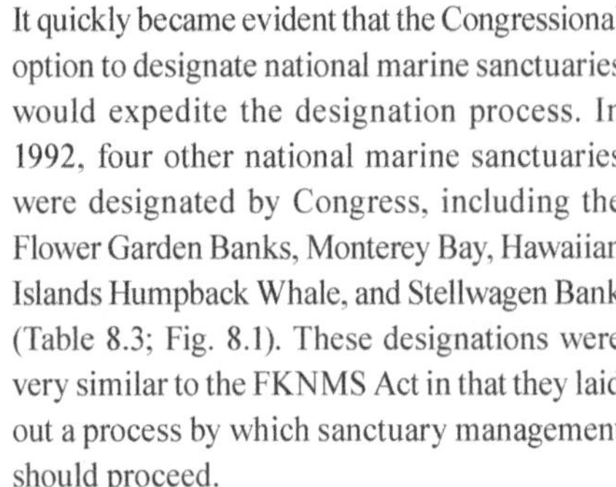

It quickly became evident that the Congressional option to designate national marine sanctuaries would expedite the designation process. In 1992, four other national marine sanctuaries were designated by Congress, including the Flower Garden Banks, Monterey Bay, Hawaiian Islands Humpback Whale, and Stellwagen Bank (Table 8.3; Fig. 8.1). These designations were very similar to the FKNMS Act in that they laid out a process by which sanctuary management should proceed.

8.2.3.4 Recent Proclamation of the Papahãnaumokuãkea (Northwestern Hawaiian Islands) Marine National Monument

In 2000 President William J. Clinton signed Executive Orders that created the Northwestern Hawaiian Islands (NWHI) Coral Reef Ecosystem Reserve. The orders also initiated a process to designate the waters of the NWHI as a national marine sanctuary. Scoping meetings for the proposed sanctuary were held in 2002. In 2005 Hawaii Governor Linda Lingle signed regulations establishing a state marine refuge in the nearshore waters of the NWHI (out to 3 nautical miles, except Midway Atoll) that excluded all extractive uses of the region, except those permitted for research or other purposes that benefited management. In 2006, after substantial public comment in support of strong protections for the area, President George W. Bush issued Presidential Proclamation 8031, creating the Northwestern Hawaiian Islands Marine National Monument. The President's actions followed Governor Lingle's lead and immediately afforded the NWHI the highest form of marine environmental protection as the world's largest MPA (360,000 km^2). Administrative jurisdiction over the islands and marine waters is shared by NOAA/NMSP and NMFS, U.S. Fish and Wildlife Service, and the State of Hawaii.

8.2.4 Interpretation of Goals

The mission of the NMSP is to identify, protect, conserve, and enhance natural and cultural resources, values, and qualities. The NMSP has developed a draft strategic plan with a set of goals (Box 8.1) to provide a bridge between the broad mandates of the NMSA and daily operations at the site level.

At the site level, management and annual operating plans for each national marine sanctuary and the marine national monument identify specific plans and tasks for day-to-day management of the 14 sites. Sanctuaries work closely with their stakeholder Sanctuary Advisory Councils in the processes of developing and revising management plans. Sanctuary staff support SAC members in forming working groups to analyze each of the action plans that comprise a management plan. There are public scoping meetings to ensure the opportunity for participation by the public. The NMSA stipulates that plans should be reviewed and revised on a five-year time frame, and various sanctuaries are at different phases of this process (Table 8.3). Three Central California sanctuaries are undergoing a joint management plan review, some revisions have been completed, and some are nearing completion. Sanctuary management plans are available via the internet (http://sanctuaries.noaa.gov).

8.3 CURRENT STATUS OF MANAGEMENT SYSTEM

8.3.1 Key Ecosystem Characteristics on Which Goals Depend

In keeping with the goals of the National Marine Sanctuary Program (Box 8.1), sanctuaries within U.S. waters generally are set aside for the preservation of natural or maritime heritage resources. Sites such as the Florida Keys and Channel Islands NMS are of the former, while the Monitor NMS is of the latter. Sites designated to protect marine biological resources have their primary focus on maintaining biodiversity or preserving key species, and are therefore directly related to NMSP Goals 1 and 4. These sites are in particular need of management in response to climate change, yet have management plans that were designed to address local stressors, not to protect flora and fauna from climate change. Management options in the context of climate change will be discussed below (section 8.4).

8.3.1.1 Biodiversity

The extraordinary biodiversity of tropical and subtropical coral reef sites is well recognized (see Case Study Summaries 8.1–8.3), but recent findings underscore the fact that high biodiversity is also characteristic of many temperate sanctuaries. For example, the recent discovery of deep, temperate corals in the Olympic Coast NMS raises the possibility that benthic invertebrate and associated fish diversity is significantly higher than previously thought. Though receiving substantially less attention from the scientific community than their tropical counterparts, subtidal temperate reefs may be no less important in promoting species diversity and enhancing production (Jonsson *et al.*, 2004, Roberts and Hirshfield, 2004). In the past, these reefs have been overlooked and under-studied primarily because of limited accessibility; they often occur in deeper or lower-visibility waters than those of tropical reefs. Recently, and primarily because of greater accessibility to deep-water ecosystems, the importance of temperate reefs as critical habitat has begun to be fully

BOX 8.1. Draft Goals of the National Marine Sanctuary Program, 2005–2015.

Goal 1. Identify, designate, and manage sanctuaries to maintain the natural biological communities in sanctuaries and to protect and, where appropriate, restore and enhance natural habitats, populations, and ecological processes, through innovative, coordinated and community-based measures and techniques.

Goal 2. Build and strengthen the nation-wide system of marine sanctuaries, maintain and enhance the role of the NMSP's system in larger MPA networks and help provide both national and international leadership for MPA management and marine resource stewardship.

Goal 3. Enhance nation-wide public awareness, understanding, and appreciation of marine and Great Lakes ecosystems and maritime heritage resources through outreach, education, and interpretation efforts.

Goal 4. Investigate and enhance the understanding of ecosystem processes through continued scientific research, monitoring, and characterization to support ecosystem-based management in sanctuaries and throughout U.S. waters.

Goal 5. Facilitate human use in sanctuaries to the extent such uses are compatible with the primary mandate of resource protection, through innovative public participation and interagency cooperative arrangements.

Goal 6. Work with the international community to strengthen global protection of marine resources, investigate and employ appropriate new management approaches, and disseminate NMSP experience and techniques.

Goal 7. Build, maintain, and enhance an operational capability and infrastructure that efficiently and effectively support the attainment of the NMSP's mission and goals.

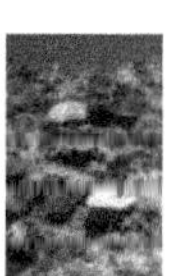

recognized (*e.g.*, Reed, 2002; Jonsson *et al.*, 2004; Roberts and Hirshfield, 2004; Roberts, Wheeler, and Freiwald, 2006). These reefs may host an array of undescribed species, including endemic gorgonians, corals, hydroids, and sponges (Koslow *et al.*, 2001; Jonsson *et al.*, 2004). Furthermore, the value of these offshore reefs to fisheries has long been recognized by commercial and recreational fishers. Fish tend to aggregate on deep-sea reefs (Husebø *et al.*, 2002), and scientific evidence supports the contention by commercial fishers that damage to temperate reefs affects both the abundance and distribution of fish (Fosså, Mortensen, and Furevik, 2002; Krieger and Wing, 2002).

8.3.1.2 Key Species

Key species within sanctuary boundaries may be resident as well as migratory, and may or may not represent species that are extracted by fishing (*i.e.*, NMSP Goal 5; Box 8.1). For example, three adjacent sanctuaries off the California coast—Cordell Bank, Gulf of the Farallones, and Monterey Bay—are frequented by protected species of blue (*Balaenoptera musculus*) and humpback (*Megaptera novaeangliae*) whales. In contrast, during the spring of each year king mackerel (*Scomberomorus cavalla*) migrate through Gray's Reef NMS off the coast of Georgia, representing a vibrant and sought-after recreational fishery. Under various climate change scenarios, management strategies employed to protect these key species may differ. For example, marine zones with dynamic boundaries reflecting shifting areas for feeding or reproduction may need to be considered by MPA managers.

Key species within sanctuaries may not be limited to subtidal marine organisms but, depending on the sanctuary, may also include intertidal species (*e.g.*, *Mytilus californianus* in Monterey Bay NMS) or sea- and shorebirds. It has been suggested that intertidal species are more likely to be stressed by climate change and may serve as a bellwether for change in other ecosystems (Helmuth, 2002).

8.3.1.3 Habitat Complexity

National marine sanctuary sites, especially subtidally, are characterized by habitat complexity that is either biologically or geologically structured; this complexity is an invaluable resource supporting biodiversity.

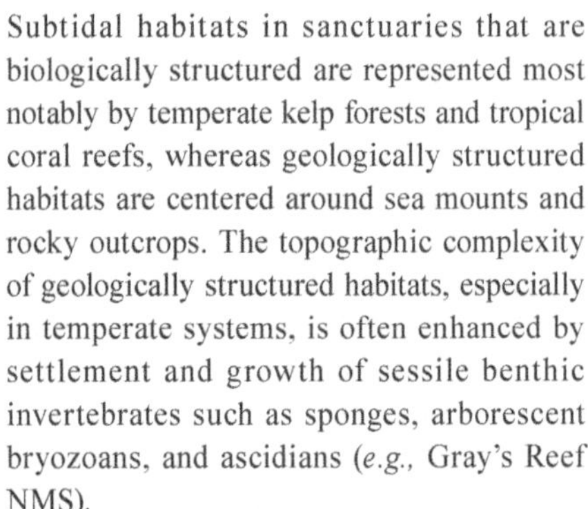

Subtidal habitats in sanctuaries that are biologically structured are represented most notably by temperate kelp forests and tropical coral reefs, whereas geologically structured habitats are centered around sea mounts and rocky outcrops. The topographic complexity of geologically structured habitats, especially in temperate systems, is often enhanced by settlement and growth of sessile benthic invertebrates such as sponges, arborescent bryozoans, and ascidians (*e.g.*, Gray's Reef NMS).

Habitat complexity is a key ecosystem characteristic that must be protected in order to achieve NMSP Goals 1 and 4 (Box 8.1). Biologically structured habitats are more susceptible to degradation resulting from climate change than geologically structured habitats. When habitat-building organisms such as corals are killed by climate change and other sources of mortality, skeletal material increases in susceptibility to bioerosion that may lead to reduced habitat complexity. As indicated in section 8.3.2 (Stressors of Concern), excess CO_2 absorbed by sea water lowers pH and results in reduced calcification rates in organisms that provide complex structure, such as arborescent bryozoans, bivalves, coralline algae, and temperate and tropical corals (Hoegh-Guldberg, 1999; Kleypas *et al.*, 1999; Kleypas and Langdon, 2006). Non-calcifying biological structures, such as kelp, as well as all shallow-water structures, are also at risk primarily from changes in storm intensity, ocean warming, and reduced upwelling associated with climate change (see Case Study Summary 8.4: Channel Islands National Marine Sanctuary).

8.3.1.4 Trophic Cascades

In addition to biodiversity and habitat complexity, trophic links between the benthos and water column help maintain ecosystem integrity within sanctuaries. In keeping with NMSP Goal 5 (Box 8.1) regarding human use, the strength of these benthic-pelagic linkages must be considered when designating fishing restrictions (Grober-Dunsmore, Wooninck, and Wahle, forthcoming).[28] Fishing regulations often involve removal of top predators and

[28] See also **Wahle**, C., R. Grober-Dunsmore, and L. Wooninck, 2006: Managing recreational fishing in MPAs through vertical zoning: the importance of understanding benthic-pelagic linkages. *MPA News*, **7(8)**, 5.

have direct impacts on trophic cascades that are defined as: (1) having top-down control of community structure, and (2) having conspicuous indirect effects on two or more links distant from the primary one (Frank *et al.*, 2005). The consequences of ignoring past experiences regarding these trophic cascades could be deleterious to sanctuary goals (Hughes *et al.*, 2005). As highlighted in a recent workshop sponsored by the MPA Science Institute, however, knowledge in this critical area is lacking.[28] Facilitating a better understanding of trophic cascades by supporting scientific inquiry into this topic would do much to enhance understanding of ecosystem processes in marine sanctuaries (NMSP Goal 4, Box 8.1). Further research may also provide insight into how these processes might be affected by climate change.

8.3.1.5 Connectivity

The open nature of marine ecosystems means that they do not function, and likewise should not be managed, in isolation (Palumbi, 2003). Connectivity among marine ecosystems and across biological communities contributes to maintaining the biological integrity of all marine environments (Kaufman *et al.*, 2004). While NMS boundaries are well defined, the separation between ecosystems and communities is blurred because of export and import of resources. At the broadest scale these linkages are manifested as sources and sinks of nutrients and recruits (*e.g.*, Crowder *et al.*, 2000).

8.3.1.6 Nutrient Fluxes

While excess nutrients can lead to degradation of offshore ecosystems (Rabalais, Turner, and Wiseman Jr, 2002), it is also hypothesized that the function of offshore ecosystems is dependent on nutrients that have their origins in upland productivity. Estuaries are thought to represent the conduit through which dissolved and particulate material from the continent passes to offshore areas through rivers (Gattuso, Frankignoulle, and Wollast, 1998). This "outwelling" characteristic was first proposed by Odum[29] and has since been applied to mangroves and seagrasses (Lee, 1995). The direct and indirect trophic links that exist between these ecosystems are thought to be critical to ecosystem function, and highlight the importance of assessing the downstream effects that upland and nearshore activities have on increasing and decreasing nutrient availability offshore. In areas where climate change alters historical rainfall patterns, concomitant alteration of the supply of nutrients to offshore ecosystems might also occur.

8.3.1.7 Larval Dispersal and Recruitment

One of the strengths of the NMSP is protection of entire ecosystems rather than management of single species. As such, a key characteristic of these ecosystems rests in their ability to serve as sources of recruits for both fish and invertebrate species and as foci for fish aggregations. Most benthic marine invertebrates and fish species have a planktonic larval stage that results from spawned gametes (Pechenik, 1999). Successful recruitment of planktonic larvae to the benthos depends on processes that function at multiple spatial scales in contrast to non-planktonic larvae, which generally recruit at a small spatial scale. At the broadest scale, hydrodynamic forces may disperse passive larvae long distances, potentially delivering them to suitable settlement sites far from the source population (Williams, Wolanski, and Andrews, 1984; Lee *et al.*, 1992). Alternatively, complex, three-dimensional secondary flows resulting from barriers, such as headlands, islands, and reefs, as well as cyclonic motion can retain passive larvae within estuaries, around islands, or within ocean basins, resulting in more settlement to natal populations (Black, Moran, and Hammond, 1991; Lee *et al.*, 1992; Black *et al.*, 1995; Lugo-Fernandez *et al.*, 2001).

Because of their small size and limited swimming ability, invertebrate larvae may be passively dispersed at a broad spatial scale (Denny, 1988, Mullineaux and Butman, 1991). Yet larvae of many marine invertebrates, including coral planulae, use swimming behavior, stimulated by chemical or physical cues, to control their position within the water column—thereby increasing the probability that they will be transported to suitable settlement substrata (Scheltema, 1986; Raimondi and Morse, 2000; Gleason, Edmunds, and Gates, 2006; Levin, 2006). In contrast, researchers

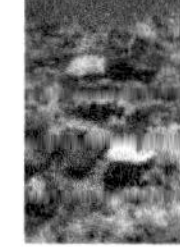

[29] **Odum**, E.P., 1969: A research challenge: evaluating the productivity of coastal and estuarine water. In: *Proceedings of the Second Sea Grant Conference.* University of Rhode Island, Kingston, Rhode Island, pp. 63-64.

continue to be surprised by the swimming and sensory capabilities of fish larvae (Stobutzki and Bellwood, 1997; Tolimieri, Jeffs, and Montgomery, 2000; Leis and McCormick, 2002; Leis, Carson-Ewart, and Webley, 2002; Lecchini *et al.*, 2005; Lecchini, Planes, and Galzin, 2005). That these larvae orient in the water column and swim directionally either at hatching or soon thereafter may explain recent evidence for localized recruitment (Jones *et al.*, 1999; Swearer *et al.*, 1999; Taylor and Hellberg, 2003; Cowen, Paris, and Srinivasan, 2006).

While connectivity among ecosystems and among biological communities in terms of both nutrients and recruits is an important feature of marine sanctuaries, boundaries of protected areas rarely encompass the continuum of habitats (*e.g.*, rivers to estuaries to mangroves to seagrasses to reefs) or the maximum dispersal distances of critical species. Recent information obtained for dispersal of fish and invertebrates suggests that sanctuaries must be managed for both self-recruitment and larval subsidies from upstream (Roberts, 1997b; Hughes *et al.*, 2005; Cowen, Paris, and Srinivasan, 2006; Steneck, 2006). Effective exchange of offspring is facilitated by MPA networks that are in close proximity [10–50 km apart according to Roberts *et al.* (2001)]. This would allow larval exchange among populations and also buffer these populations from climate-driven changes in current regimes. The NMSP should be a critical player in the development of such an MPA network. NMSP Goal 2 (Box 8.1) provides for the expansion of the nationwide system of MPAs and encourages cooperation among MPAs administered under a range of programs.

8.3.2 Stressors of Concern

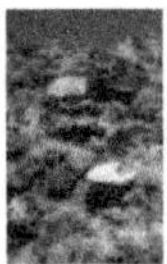

Population growth and coastal development increasingly affect U.S. MPAs; an estimated 153 million people (53% of the U.S. population) lived in coastal counties in 2003, and that number continues to rise (World Resources Institute, 1996; National Safety Council, 1998; U.S. Census Bureau, 2001; Crossett *et al.*, 2004).[30] Growing human impacts are

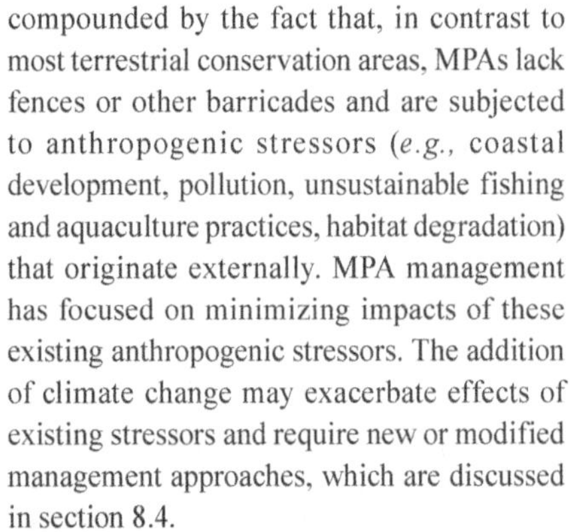

compounded by the fact that, in contrast to most terrestrial conservation areas, MPAs lack fences or other barricades and are subjected to anthropogenic stressors (*e.g.*, coastal development, pollution, unsustainable fishing and aquaculture practices, habitat degradation) that originate externally. MPA management has focused on minimizing impacts of these existing anthropogenic stressors. The addition of climate change may exacerbate effects of existing stressors and require new or modified management approaches, which are discussed in section 8.4.

The purpose of this section is: (1) to outline major stressors on marine organisms and communities resulting from climate change and (2) to introduce ways in which major "traditional" stressors may interact with climate change stressors.

There are excellent, extensive reviews of impacts of climate change on marine organisms and communities (*e.g.*, Scavia *et al.*, 2002; Walther *et al.*, 2002; Goldberg and Wilkinson, 2004; Harley *et al.*, 2006). By contrast, the scientific knowledge required to reach general conclusions related to the impact of multiple stressors at community and ecosystem levels is for the most part absent for marine systems. Thus, information concerning interactions among stressors is limited and MPA managers are faced with even higher levels of uncertainty about likely outcomes of management actions as climate change impacts have increasingly strong interactions with existing stressors.

8.3.2.1 Direct Climate Change Stressors

Ocean Warming

According to Bindoff *et al.* (2007), there is high confidence that an average warming of 0.1°C has occurred in the 0–700 m depth layer of the ocean between 1961 and 2003. Increasing ocean temperatures, especially near the surface, affect physiological processes in organisms ranging from enzyme reactions to reproductive timing (Fields *et al.*, 1993;

[30] See also **National Ocean Service**, 2000: Spatial patterns of socioeconomic data from 1970 to 2000: a national research dataset aggregated by watershed and political boundaries. http://cads.nos.noaa.gov/.
Hinrichsen, D., B. Robey, and U.D. Upadhyay, 1998: *Solutions for a Water-Short World.* Population Report, Series M, No. 14, Population Information Program, Center for Communication Programs, the Johns Hopkins University School of Public Health, Baltimore, MD, pp.1-60.
World Resources Institute, 2000: *Gridded Population of the World.* Version 2, Center for International Earth Science Information Network, Columbia University, Palisades, NY.

Roessig *et al.*, 2004; Harley *et al.*, 2006). The historical stability of ocean temperatures makes many marine species sensitive to thermal perturbations just a few degrees higher than those experienced over evolutionary time (Wainwright, 1994). However, it is not always intuitive which species might be most intolerant of temperature increases. For example, studies on porcelain crabs (*Petrolisthes*) and intertidal snails (*Tegula*) show that individuals in the mid-intertidal are closer to upper temperature limits and have less capacity to acclimate to temperature perturbations than subtidal congeners in temperature-stable conditions (Tomanek and Somero, 1999; Stillman, 2003; Harley *et al.*, 2006).

What is clear is that increasing sea temperatures will continue to influence processes such as foraging, growth, and larval duration and dispersal, with ultimate impacts on the geographic ranges of species. In fact, poleward latitudinal shifts in some zooplankton, fish, and intertidal invertebrate communities have already been observed along the California coast and in the North Atlantic (reviewed in Walther *et al.*, 2002). Within marine communities, these temperature changes and range shifts may result in new species assemblages and biological interactions that affect ecological processes such as larval dispersal, competitive interactions, and trophic interactions and webs (Barry *et al.*, 1995; Roessig *et al.*, 2004; Precht and Aronson, 2004; O'Connor *et al.*, 2007). Species that are unable to shift geographic ranges (perhaps due to physical barriers) or compete with other species for resources may face local—and potentially global—extinction. Conversely, some species may find open niches and dominate regions because of release from competition or predation.

Impacts at the ecosystem or community level are even more difficult to predict. For example, warmer waters stimulate increases in population sizes of the mid-intertidal sea star, *Pisaster ochraceus*, and its per capita consumption rates of mussels (Sanford, 1999). Continued warming may enable *P. ochraceus* to clear large sections of mussel beds, indirectly affecting hundreds of species associated with these formations (Harley *et al.*, 2006). How such an outcome affects trophic links and other biological processes within this community is not clear.

The latest reports from the IPCC (2007b; 2007c) state that temperature increases over the last 50 years are nearly twice those for the last 100 years, with projections that temperature will rise 2–4.5°C, largely caused by a doubling of atmospheric carbon dioxide emissions. Increases in seawater surface temperature of about 1–3°C are likely to cause more frequent coral bleaching events that cause widespread mortality, unless thermal adaptation or acclimatization by corals occurs (IPCC, 2007c). However, the ability of corals to adapt or acclimatize to increasing seawater temperature is largely unknown (Berkelmans and van Oppen, 2006) and remains a research topic of paramount importance.

Consequences of coral bleaching, during which corals lose their symbiotic algae, depend on the severity and duration of the bleaching event. They range from minimal affects on growth and reproduction to widespread mortality. Coral bleaching at the ecosystem level is a relatively recent phenomenon, first receiving widespread attention in 1987 when abnormally high summer seawater surface temperatures throughout the Caribbean resulted in a mass bleaching event (Williams, Goenaga, and Vicente, 1987; Ogden and Wicklund, 1988; Williams and Bunkley-Williams, 1990). Soon after, coral reef scientists identified climate change as a major long-term threat to coral reefs (Glynn, 1991; Smith and Buddemeier, 1992) and determined that irradiance interacts with temperature to cause bleaching (Gleason and Wellington, 1993; see also Hoegh-Guldberg, 1999; and Hoegh-Guldberg *et al.*, 2007). Reciprocity between these two parameters may provide MPA managers with options to alleviate stress during bleaching events (see section 8.4.2).

In 1997–1998, a mass bleaching event in association with an ENSO event caused worldwide bleaching and coral mortality (Wilkinson, 1998; 2000), and in 2005 the most devastating Caribbean-wide coral bleaching event to date occurred that, based on modeling, is highly unlikely to have occurred without anthropogenic forcing (Donner, Knutson, and Oppenheimer, 2007). Over the last 20 years, an extensive body of literature has conclusively identified anomalously high summer surface seawater temperatures as the major cause of coral bleaching (Wilkinson, 1998; 2000; Fitt *et*

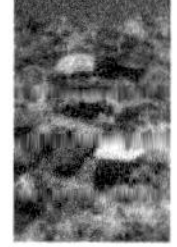

al., 2001; Wilkinson, 2002; U.S. Climate Change Science Program and Subcommittee on Global Change Research, 2003; Donner *et al.*, 2005; Donner, Knutson, and Oppenheimer, 2007), with widespread agreement that continued warming—as little as 1°C warmer than the average summer maxima is sufficient—will increase the severity and frequency of mass bleaching events (Smith and Buddemeier, 1992; Hoegh-Guldberg, 1999; Hughes *et al.*, 2003; Douglas, 2003; Done and Jones, 2006).

Effects of coral reef bleaching are both biological, including lost biodiversity and other ecosystem services, and economic, resulting in the decline of fisheries and tourism (Buddemeier, Kleypas, and Aronson, 2004). Coral reefs affected by mass bleaching typically take decades or longer to recover and sometimes may not recover at all. In general, coral reef decline throughout the Caribbean region has been caused by a combination of bleaching, disease, die-off of the sea urchin *Diadema antillarum*, overfishing, pollution, hurricanes, and other factors (Gardner *et al.*, 2003; Gardner *et al.*, 2005).

Ocean Acidification

Increased CO_2 concentrations lower oceanic pH, making it more acidic. According to the most recent IPCC report, the total inorganic carbon content of the ocean increased by 118 (±19) billion metric tons of carbon from 1750–1994, and continues to increase through absorption of excess CO_2 (Bindoff *et al.*, 2007). Furthermore, time series data for the last 20 years show a trend of decreasing pH of 0.02 pH units per decade (Bindoff *et al.*, 2007). Long-term exposures to low pH (-0.7 unit) have been shown to reduce metabolic rates, growth, and survivorship of both invertebrates and fishes (Michaelidis *et al.*, 2005; Shirayama and Thornton, 2005; Pane and Barry, 2007), but by far the greatest threat of reducing pH is to organisms that build their external skeletal material out of calcium carbonate ($CaCO_3$). Calcifying organisms such as sea urchins, cold-water corals, coralline algae, and various plankton that reside in cooler temperate waters appear to be the most threatened by acidification, because CO_2 has greater solubility in cooler waters (Hoegh-Guldberg, 1999; Kleypas *et al.*, 1999; Hughes *et al.*, 2003; Feely *et al.*, 2004; Kleypas and Langdon, 2006).

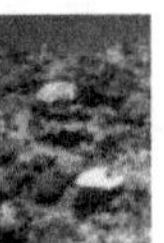

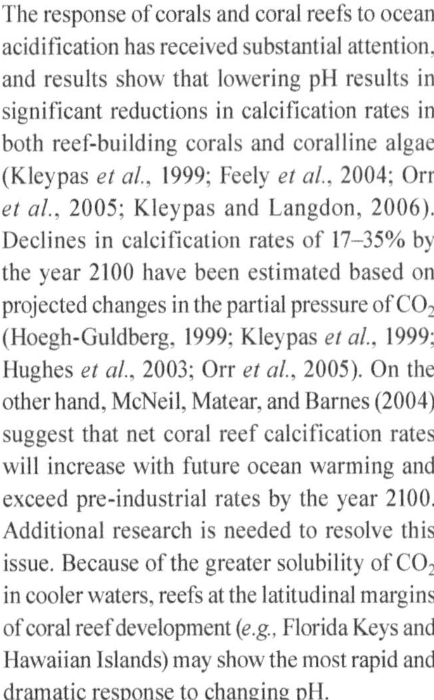

The response of corals and coral reefs to ocean acidification has received substantial attention, and results show that lowering pH results in significant reductions in calcification rates in both reef-building corals and coralline algae (Kleypas *et al.*, 1999; Feely *et al.*, 2004; Orr *et al.*, 2005; Kleypas and Langdon, 2006). Declines in calcification rates of 17–35% by the year 2100 have been estimated based on projected changes in the partial pressure of CO_2 (Hoegh-Guldberg, 1999; Kleypas *et al.*, 1999; Hughes *et al.*, 2003; Orr *et al.*, 2005). On the other hand, McNeil, Matear, and Barnes (2004) suggest that net coral reef calcification rates will increase with future ocean warming and exceed pre-industrial rates by the year 2100. Additional research is needed to resolve this issue. Because of the greater solubility of CO_2 in cooler waters, reefs at the latitudinal margins of coral reef development (*e.g.*, Florida Keys and Hawaiian Islands) may show the most rapid and dramatic response to changing pH.

Rising Sea Level

During the last 100 years, global average sea level has risen an estimated 1–2 mm per year and is expected to accelerate due to thermal expansion of the oceans and melting ice-sheets and glaciers (Cabanes, Cazenave, and Le Provost, 2001; Albritton and Filho, 2001; Rignot and Kanagaratnam, 2006; Chen, Wilson, and Tapley, 2006; Shepherd and Wingham, 2007; Bell *et al.*, 2007; IPCC, 2007c). Rates of sea level rise at a local scale vary from -2 to 10 mm per year along U.S. coastlines (Nicholls and Leatherman, 1996; Zervas, 2001; Scavia *et al.*, 2002). Low-lying areas, especially intertidal zones, along the eastern and Gulf coasts are at the greatest risk of damage from rising sea level (Scavia *et al.*, 2002). The consequences of sea level rise include inundation of coastal areas, erosion of vulnerable shorelines, and landward shifts in species distributions.

On undeveloped coasts with relatively gentle slopes, it is thought that plant communities such as mangroves and *Spartina* salt marshes will move inland as sea level rises (Scavia *et al.*, 2002; Harley *et al.*, 2006). In contrast, coastline development will interfere with these plant migrations (see the National Estuaries chapter, section 7.3.2, for further discussion). As a result, wetlands may become submerged and soils may become waterlogged, resulting in plant physiological stress due to chronic and intolerable elevated salinity. Marshes, mangroves and dune plants are critical to coastal environments because they produce and add nutrients to coastal systems, stabilize substrata, and serve as refuges and nurseries for many species. Their depletion or loss would therefore affect nutrient flux, energy flow, and essential habitat for a multitude of species, with ultimate long-term impacts on biodiversity (Scavia *et al.*, 2002; Galbraith *et al.*, 2002; Harley *et al.*, 2006). The projected 35–70% loss of barrier islands and intertidal and sandy beach habitat over the next 100 years could also drastically reduce nesting grounds for key species such as sea turtles and birds as these critical habitats disappear (Scavia *et al.*, 2002).

Climatic Variability and Ocean Circulation

Natural climatic variability resulting from ocean-atmosphere interactions such as the El Niño Southern Oscillation (ENSO), the Pacific Decadal Oscillation (PDO), and the North Atlantic Oscillation/Northern Hemisphere Annular Mode result in changes in open ocean productivity, shifts in the distribution of organisms, and modifications in food webs that foreshadow potential consequences of accelerated climate change (*e.g.,* Mantua *et al.*, 1997; McGowan *et al.*, 1998). These recurring patterns of ocean-atmosphere variability have very different behaviors in time. For example, whereas ENSO events persist for 6–18 months and have their major impact in the tropics, the PDO occurs over a much longer time frame of 20–30 years and has primary effects in the northern Pacific (Mantua *et al.*, 1997). Regardless of the temporal scale and region of impact, however, these natural modes of climate variability have existed historically, independent of anthropogenically driven climate change. These climate phenomena may act in tandem with (or in opposition to) human-induced alterations, with consequences that are difficult to predict (Philip and Van Oldenborgh, 2006).

Ocean-atmosphere interactions on a warming planet may also result in long-term alterations in the prevailing current and upwelling patterns (Bakun, 1990; McPhaden and Zhang, 2002; Snyder *et al.*, 2003; McGregor *et al.*, 2007). While at present there is no clear indication that ocean circulation patterns have changed (Bindoff *et al.*, 2007), modifications could have large effects within and among ecosystems through impacts on ecosystem and community connectivity in terms of both nutrients and recruits (see section 8.3.1., Key Ecosystem Characteristics on Which Goals Depend). Considering that there is evidence for warming of Southern Ocean mode waters and Upper Circumpolar Deep Waters from 1960–2000, changes in oceanic current and upwelling patterns are likely in the future (Bindoff *et al.*, 2007). The direction that these changes will take, however, is not evident. For example, it has been hypothesized that the greater temperature differential between the land mass and ocean that will occur with climate warming will increase upwelling because of stronger alongshore winds (Bakun, 1990). In contrast, Gucinski, Lackey, and Spence (1990) proposed that warming at higher latitudes will reduce latitudinal temperature gradients, resulting in decreased wind strength and less upwelling; some models show potential for Atlantic thermohaline circulation to end abruptly if high-latitude waters are no longer able to sink (Stocker and Marchal, 2000).

Storm Intensity

Whether or not storm frequency has changed over time is not clear, due to large natural variability resulting from such climate drivers as ENSO (IPCC, 2007c). However, since the mid 1970s there has been a trend toward longer storm duration and greater storm intensity (IPCC, 2007c). An increase in storm intensity generally has impacts on two fronts. First, it may increase pulses of fresh water to coastal and nearshore habitats (see below). Second, increasing storm intensity may cause physical damage to coastal ecosystems, especially those in shallow water (IPCC, 2007c).

Recent hurricanes in the southern United States have caused extensive destruction to homes and businesses; altered nearshore water quality; scoured the ocean bottom; over-washed beaches; produced immense amounts of marine debris (wood, metals, plastics) and pollution (household hazardous wastes, pesticides, metals, oils and other toxic chemicals) from floodwaters; and damaged many mangrove, marsh, and coral reef areas (Davis *et al.*, 1994; Tilmant *et al.*, 1994; McCoy *et al.*, 1996; Lovelace and MacPherson, 1998; Baldwin *et al.*, 2001).[31] Even 30–60 days after

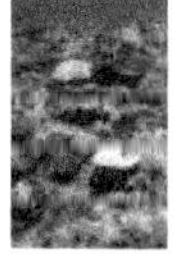

[31] See also **U.S. Fish and Wildlife Service**, 2005: U.S. Fish and Wildlife Service conducting initial damage assessments to wildlife and National Wildlife Refuges. http://www.fws.gov/southeast/news/2005/r05-088.html.

storms, some areas still experienced increased turbidity, breakdown of mangrove peat soils, and elevated concentrations of ammonia, dissolved phosphate, and dissolved organic carbon (Davis *et al.*, 1994; Tilmant *et al.*, 1994; Lovelace and MacPherson, 1998). In some instances, algal blooms from high nutrients further increased turbidity while driving down dissolved-oxygen concentrations (*i.e.*, caused eutrophication), resulting in mortalities in fish and invertebrate populations (Tilmant *et al.*, 1994; Lovelace and MacPherson, 1998). Given that most climate change models project increasing storm intensity as well as higher sea levels in many areas, it is evident that low-lying and shallow marine ecosystems such as mangroves, salt marshes, seagrasses, and coral reefs are at greatest risk of long-term damage.

Freshwater Influx

Observations indicate that changes in the amount, intensity, frequency, and type of precipitation are occurring worldwide (IPCC, 2007c). Consistent with observed changes in precipitation and water transport in the atmosphere, large-scale trends in oceanic salinity have become evident for the period 1955–1998 (Bindoff *et al.*, 2007). These trends are manifested as lowered salinities at subpolar latitudes and increased salinities in shallower parts of the tropical and subtropical oceans.

In addition to altering salinity in major oceanic water masses, changes in precipitation patterns can have significant impacts in estuarine and other nearshore environments (see the National Estuaries chapter, section 7.3.4.2.9). For instance, in regions where climate change results in elevated rainfall, increased runoff may cause greater stratification of water layers within estuaries as fresh water floats out on top of higher salinity layers (Scavia *et al.*, 2002). One consequence of this stratification may be less water column mixing and thus lower rates of nutrient exchange among water layers. Combining this stratification effect with shorter water residence times stemming from higher inflow (Moore *et al.*, 1997) may result in significantly reduced productivity because phytoplankton populations may be flushed from the system at a rate faster than they can grow and reproduce. On the other hand, estuaries that are located in regions with lower rainfall may also show decreased productivity due to lower nutrient influx. Thus, the relationship between precipitation and marine ecosystem health is complex and difficult to predict.

Another source of fresh water is melting of polar ice (IPCC, 2007c). In the Atlantic Ocean, accelerated melting of Arctic ice and the Greenland ice sheet are predicted to continue producing more freshwater inputs that may alter oceanic circulation patterns (Dickson *et al.*, 2002; Curry, Dickson, and Yashayaev, 2003; Curry and Mauritzen, 2005; Peterson *et al.*, 2006; Greene and Pershing, 2007; Boessenkool *et al.*, 2007).

8.3.2.2 Climate Change Interactions with "Traditional" Stressors of Concern

Pollution

Marine water quality degradation and pollution stem primarily from land-based sources, with major contributions to coastal watershed and water quality deterioration falling into two broad categories: point-source pollution and non-point-source pollution. Point-source pollution from factories, sewage treatment plants, and farms often flows into nearby waters. In contrast, marine non-point-source pollution originates from coastal urban runoff where the bulk of the land is paved or covered with buildings. These impervious surfaces prevent soils from capturing runoff, resulting in the input of untreated pollutants (*e.g.*, fuels, oils, plastics, metals, insecticides, antibiotics) to coastal waters. Increased terrestrial runoff due to more intense storm events associated with climate change may increase land-based water pollution from both of these sources. In some areas, increased groundwater outflows may also contribute to coastal pollution.

Deterioration and pollution of coastal watersheds can have far-reaching effects on marine ecosystems. As an example, the Gulf of Mexico "dead zone" that occurs each summer and extends from the Mississippi River bird-foot delta across the Louisiana shelf and onto the upper Texas coast can range from 1–125 km offshore (Rabalais, Turner, and Wiseman Jr, 2002). This mass of hypoxic (low-oxygen) water has its origins in the increased nitrate flux coincident with the exponential growth of fertilizer use that has occurred since the 1950s in the Mississippi River basin. This hypoxia results in changes in species diversity and community structure of the benthos and has impacts on trophic links that include higher-order consumers in the pelagic zone (Rabalais, Turner, and Wiseman Jr, 2002).

Until recently, pollution has been the major driver of decreases in the health of marine ecosystems such as coral reefs, seagrass beds, and kelp forests (Jackson *et al.*, 2001; Hughes *et al.*, 2003; Pandolfi *et al.*, 2003). Because pollution is usually more local in scope, it historically could be managed within individual MPAs; however, the addition of climate change stressors such as increased oceanic temperature, decreased pH, and greater fluctuations in salinity present greater challenges with regard to potentially deleterious effects of pollution (Coe and Rogers, 1997; Carpenter *et al.*, 1998; Khamer, Bouya, and Ronneau, 2000; Burton, Jr. and Pitt, 2001; Sobel and Dahlgren, 2004; Orr *et al.*, 2005; Breitburg and Riedel, 2005; O'Connor *et al.*, 2007; IPCC, 2007c). Also, in regions where climate change causes precipitation and freshwater influxes to increase, MPA managers may need to expand the scale at which they attempt to address issues of water quality, for example by forging stronger partnerships with organizations involved in watershed management both nearby and at more-distant locations.

For example, coral bleaching from the combined stresses of climate change and local pollution (*e.g.*, high temperature and sedimentation) have already been observed (Jackson *et al.*, 2001; Hughes *et al.*, 2003; Pandolfi *et al.*, 2003). Identifying those stressors with the greatest effect is not trivial. Research in coral genomics may provide diagnostic tools for identifying stressors in coral reefs and other marine communities (*e.g.*, Edge *et al.*, 2005).

Commercial Fishing and Aquaculture

Commercial fishing has ecosystem effects on three fronts: through physical impacts of fishing gear on habitat, overfishing of commercial stocks, and incidental take of non targeted species. The use of trawls, seines, mollusk dredges, and other fishing gear can cause damage to living seafloor structures and alterations to geologic structures, reducing habitat complexity (Engel and Kvitek, 1998; Thrush and Dayton, 2002; Dayton, Thrush, and Coleman, 2002; Hixon and Tissot, 2007). Overfishing is also common in the United States, with a conservative estimate of 26% of fisheries overexploited (Pauly *et al.*, 1998; National Research Council, 1999; Jackson *et al.*, 2001; Pew Oceans Commission, 2003; National Marine Fisheries Service, 2005; Lotze *et al.*, 2006). Meanwhile, non-specific fishing gear (*e.g.*, trawls, seines, dredges) causes considerable mortality of by-catch that includes invertebrates, fishes, sea turtles, marine mammals, birds, and early life stages of commercially targeted species (Condrey and Fuller, 1992; Norse, 1993; Sobel and Dahlgren, 2004; Hiddink, Jennings, and Kaiser, 2006).

Aquaculture has sometimes been introduced to augment fisheries production. Unfortunately, experiences in countries such as Southeast Asia show that aquaculture can have negative environmental impacts, including extensive mangrove and coastal wetland conversion to ponds, changes in hydrologic regimes, and discharge of high levels of organic matter and pollutants into coastal waters (Eng, Paw, and Guarin, 1989; Iwama, 1991; Naylor *et al.*, 2000). Furthermore, many aquacultural practices are not sustainable because farmed species consume natural resources at high rates and the intense culture environment (*e.g.*, overcrowding) creates conditions for disease outbreaks (Eng, Paw, and Guarin, 1989; Iwama, 1991; Pauly *et al.*, 2002; 2003).

Fishery populations that are overstressed and overfished exhibit greater sensitivity to climate change and other anthropogenically derived stressors than do healthy populations (Hughes *et al.*, 2005). Overfishing can reduce mean life span as well as lifetime reproductive success and larval quality, making fished species more susceptible to both short- and long-term perturbations (such as changes in prevailing current patterns) that affect recruitment success (Pauly *et al.*, 1998; Jackson *et al.*, 2001; Dayton, Thrush, and Coleman, 2002; Pauly *et al.*, 2003; Sobel and Dahlgren, 2004; Estes, 2005; Law and Stokes, 2005; Steneck and Sala, 2005; O'Connor *et al.*, 2007). Changing climatic regimes can also influence species' distributions, which are set by physiological tolerances to temperature, precipitation, dissolved oxygen, pH, and salinity. Because rates of climate change appear to exceed the capacity of many commercial species to adapt, species will shift their ranges in accordance with their physiological thresholds and may ultimately be forced to extend past the boundaries of their "known" native range,

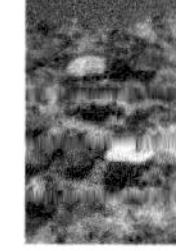

becoming invasive elements (Murawski, 1993; Walther *et al.*, 2002; Roessig *et al.*, 2004; Perry *et al.*, 2005; Harley *et al.*, 2006).

Commercial exploitation of even a single keystone species, such as a top consumer, can destabilize ecosystems by decreasing redundancy and making them more susceptible to climate change stressors (Hughes *et al.*, 2005). Examples of such ecosystem destabilization through overfishing abound, including the formerly cod-dominated system of the western North Atlantic (see Box 8.2), and the fish-grazing community on Caribbean coral reefs (*e.g.*, Frank *et al.*, 2005; Mumby *et al.*, 2006; 2007).

Interestingly, the theoretical framework that links protection against overfishing using no-take marine reserves to improved coral condition is hotly debated (Jackson *et al.*, 2001; Grigg *et al.*, 2005; Pandolfi *et al.*, 2005; Aronson and Precht, 2006). This framework hinges on an increase in herbivory [directly, through reduced fishing pressure on herbivores, or indirectly, through cascading effects (Mumby *et al.*, 2006, 2007)] that then reduces algal growth that can compete with coral colonies or inhibit coral recruitment. The heat of the debate is perhaps surprising because of the strong intuitive sense such arguments make. However, reserves also protect predators, so declines in herbivorous

BOX 8.2. The Western North Atlantic Food Web.

Marine carnivores of the western North Atlantic were both more abundant and larger in the past. In Maine, archaeological evidence indicates that coastal people subsisted on Atlantic cod for at least 4,000 years (Jackson *et al.*, 2001).[32] Prey species such as lobsters and crabs were absent from excavated middens in the region, perhaps because large predators had eaten them (Steneck, Vavrinec, and Leland, 2004; Lotze *et al.*, 2006).

Today cod are ecologically extinct from western North Atlantic coastal zones due to overfishing. The abundant lobsters and sea urchins that had formerly been the prey of apex predators became the primary target of local fisheries. By 1993, the value of sea urchins harvested in Maine for their roe was second only to that of lobsters. As sea urchin populations declined, so too did communitywide rates of herbivory.[32] In less than a decade, sea urchins became so rare that they could no longer be found over large areas of the coast (Andrew *et al.*, 2002; Steneck, Vavrinec, and Leland, 2004).

These and other instances of "fishing down food webs" in the Gulf of Maine have resulted in hundreds of kilometers of coast now having dangerously low biological and economic diversity. Today, bloodworms used for bait are worth more to Maine's economy than cod (see figure). The trophic level dysfunction (*sensu* Steneck, Vavrinec, and Leland, 2004) of both apex predators and herbivores leave a coastal zone suited for crabs and especially lobsters—the latter attaining staggering population densities exceeding one per square meter along much of the coast of Maine (Steneck and Wilson, 2001). The economic value of lobsters is high, accounting for nearly 80% of the total value of Maine's fisheries as of 2004 (see figure). The remaining 42 harvested species account for the remaining 20%. If a disease such as the one that recently decimated Rhode Island's lobster stocks (Glenn and Pugh, 2006) infects lobsters in the Gulf of Maine, there will be serious socioeconomic implications for the fishing industry. Prospects for such a disease outbreak may increase because of climate-induced changes in the environment such as temperature increases that favor pathogen growth (Harvell *et al.*, 1999; 2002). The figure is adapted from Steneck and Carlton (2001).

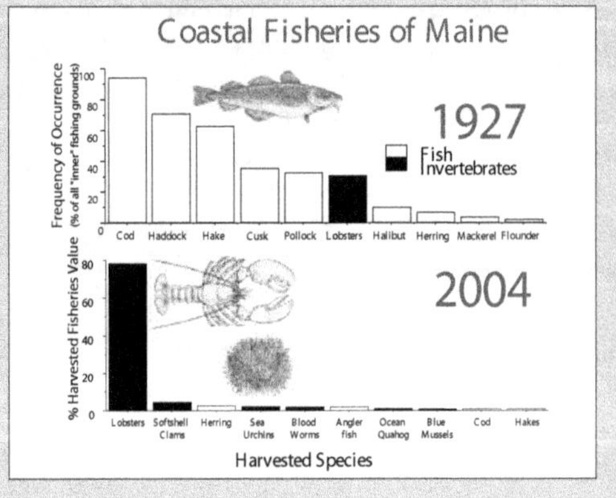

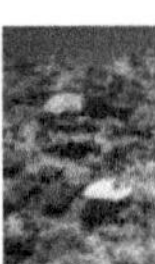

32 See also **Steneck**, R.S., 1997: Fisheries-induced biological changes to the structure and function of the Gulf of Maine ecosystem. In: Proceedings of the Gulf of Maine Ecosystem Dynamics Scientific Symposium and Workshop, RARGOM Report 91-1, Regional Association for Research in the Gulf of Maine, Hanover, NH, pp. 151-165.

fish and/or herbivory might occur, as opposed to increases, unless an escape in size from predation occurs for herbivores as was observed in the in the Bahamas (Mumby *et al.*, 2006). Also, data from field studies provide conflicting results on the role of herbivores. Mumby *et al.* (2006) showed that increased fish herbivory in a marine reserve reduced algal growth after mass bleaching caused extensive coral mortality. However, such herbivore densities (and presumably herbivory rates) do not always increase after protection is provided (Mosquera *et al.*, 2000; Graham, Evans, and Russ, 2003; Micheli *et al.*, 2004; Robertson *et al.*, 2005). Further, there is widespread belief that the mass mortality of the sea urchin *Diadema antillarum*—a major grazer on reefs—in 1983–1984 was a significant proximal cause of coral reef decline throughout the Caribbean. However, as reported in Aronson and Precht (2006), half the coral reef decline throughout the Caribbean reported by Gardner *et al.* (2003) occurred before the die-off of *D. antillarum*, and immediately after the die-off coral cover remained unchanged (Fig. 8.5) (Gardner *et al.*, 2003). Subsequent declines in cover throughout the region were due to coral bleaching (1987, 1997–1998) and disease. It is important to highlight this complexity, because it emphasizes how much is unknown about basic ecological processes on coral reefs and consequently how much needs to be learned about whether no-take marine reserves work effectively to enhance resilience when disease and bleaching remain significant sources of coral mortality (Aronson and Precht, 2006).

Nonindigenous/Invasive Species

Invasive species threaten all marine and estuarine communities. Currently, an estimated 2% of extinctions in marine ecosystems are related to invasive species while 6% are the result of other factors, including climate change, pollution, and disease (Dulvy, Sadovy, and Reynolds, 2003). Principal mechanisms of introduction vary and have occurred via both accidental and intentional release (Ruiz *et al.*, 2000; Carlton, 2000).[33] Invasive species are often opportunistic and can force shifts in the relative abundance and distribution of native species, and cause significant changes in species richness and community structure (Sousa, 1984; Moyle, 1986; Mills, Soulé, and Doak, 1993; Baltz and Moyle, 1993; Carlton, 1996; Carlton, 2000; Marchetti, Moyle, and Levine, 2004).

Some native species, particularly rare and endangered ones with small population sizes and gene pools, are unlikely to be able to adapt quickly enough or shift their ranges rapidly enough to compensate for the changing climatic regimes proposed by current climate change models (IPCC, 2007c). These native species will likely have their competitive abilities compromised and be more susceptible to displacement by invasive species, and therefore should be considered for stronger protective

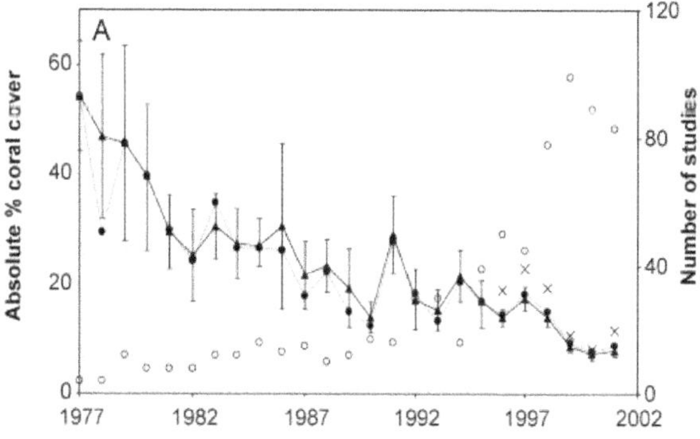

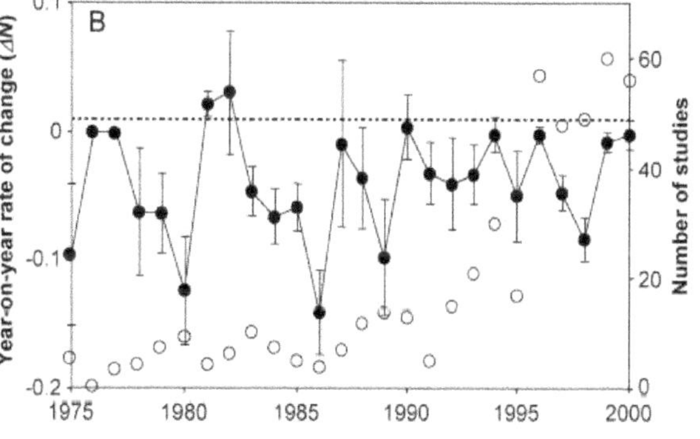

Figure 8.5. Total observed change in coral cover (%) across the Caribbean basin over the past 25 years (From Gardner *et al.*, 2003. Reprinted with permission from AAAS). A. Coral cover (%) 1977-2001. Annual estimates (▲) are weighted means with 95% bootstrap confidence intervals. Also shown are unweighted estimates (•), unweighted mean coral cover with the Florida Keys Coral Reef Monitoring Project (1996-2001) omitted (x), and the number of studies each year (○). B. Year-on-year rate of change (mean ΔN ± SE) in coral cover (%) for all sites reporting two consecutive years of data 1975-2000 (•) and the number of studies for each two-year period (○).

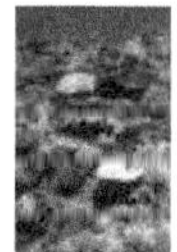

[33] See also **Hare**, J.A. and P.E. Whitfield, 2003: *An Integrated Assessment of the Introduction of Lionfish (Pterois Volitans/Miles Complex) to the Western Atlantic Ocean.* NOAA Technical Memorandum NOS NCCOS 2, pp.1-21.

measures by MPA managers. Increased seawater temperatures resulting from climate change may also allow introduced species to spawn earlier and for longer periods of the year, thus increasing their population growth rates relative to natives while simultaneously expanding their range (Carlton, 2000; McCarty, 2001; Stachowicz *et al.*, 2002; Marchetti, Moyle, and Levine, 2004). Furthermore, the same characteristics that make species successful invaders may also pre-adapt them to respond to, and capitalize on, climate change. As one example, Indo-Pacific lionfish (*Pterois volitans* and *P. miles*) are now widely distributed off the southeastern coast of the United States and in the Bahamas less than 10 years after being first observed off Florida (Whitfield *et al.*, 2007; Snyder and Burgess, 2007). One of the few factors limiting their spread is intolerance to minimum water temperatures during winter (Kimball *et al.*, 2004). Ocean warming could facilitate depth and range expansion in these species.

Diseases

Disease outbreaks alter the structure and function of marine ecosystems by affecting the abundance and diversity of vertebrates (*e.g.*, mammals, turtles, fish), invertebrates (*e.g.*, corals, crustaceans, echinoderms, oysters), and plants (*e.g.*, seagrasses, kelps). Pathogen outbreaks or epidemics spread rapidly, due to the lack of dispersal barriers in some parts of the ocean and the potential for long-term survival of pathogens outside the host (Harvell *et al.*, 1999; Harvell *et al.*, 2002). Many pathogens of marine taxa such as coral viruses, bacteria, and fungi are positively responsive to temperature increases within their physiological thresholds (Porter *et al.*, 2001; Kim and Harvell, 2004; Munn, 2006; Mydlarz, Jones, and Harvell, 2006; Boyett, Bourne, and Willis, 2007). However, it is noteworthy that white-band disease was the primary cause (though not the only cause) of reduced coral cover on Caribbean reefs from the late 1970s through the early 1990s (Aronson and Precht, 2006). That outbreak did not correspond to a period of particularly elevated temperature (Lesser *et al.*, 2007).

Exposure to disease compromises the ability of species to resist other anthropogenic stressors, and exposure to other stressors compromises species' ability to resist disease (Harvell *et al.*, 1999; Harvell *et al.*, 2002). For example, in 1998, the most geographically extensive and severe coral bleaching ever recorded was associated with the high sea surface temperature anomalies facilitated by an ENSO event (Hoegh-Guldberg, 1999; Wilkinson *et al.*, 1999; Mydlarz, Jones, and Harvell, 2006). In some species of reef-building corals and gorgonians, this bleaching event was thought to be accelerated by opportunistic infections (Harvell *et al.*, 1999; Harvell *et al.*, 2001). Several pathogens—such as bacteria, viruses, and fungi that infect such diverse hosts as seals, abalone, and starfish—show possible onset with warmer temperatures (reviewed in Harvell *et al.*, 2002), and some coral species may become more susceptible to disease after bleaching events (Whelan *et al.*, 2007). The mechanisms for pathogenesis, however, are largely unknown. Given that exposure to multiple stressors may compromise the ability of marine species to resist infection, the most effective means of reducing disease incidence under climate change may be to minimize impacts of stressors such as pollution and overfishing.

8.3.3 Management Approaches and Sensitivity of Management Goals to Climate Change

Marine protected area programs have been identified as a critical mechanism for protecting marine biodiversity and associated ecosystem services (National Research Council, 2001; Palumbi, 2002; Roberts *et al.*, 2003a; Sobel and Dahlgren, 2004; Palumbi, 2004; Roberts, 2005; Salm, Done, and McLeod, 2006).[34] MPA networks are being implemented globally to address multiple threats to the marine environment, and are generally accepted as an improvement over individual MPAs (Salm, Clark, and Siirila, 2000; Allison *et al.*, 2003; Roberts *et al.*, 2003a; Mora *et al.*, 2006).[21] Networks are more effective than single MPAs at protecting the full range of habitat and community types because they spread the risk of losing a habitat or community type following a disturbance such as a climate-change impact across a larger area. Networks are better able than individual MPAs to protect both short- and long-distance dispersers, and thus have more

[34] See also **Ballantine**, B., 1997: Design principles for systems of no-take marine reserves. Proceedings of the the design and monitoring of marine reserves, Fisheries Center, University of British Colombia, Vancouver.

potential to achieve conservation and fishery objectives (Roberts, 1997a). Networks provide enhanced larval recruitment among adjacent MPAs that are linked by local and regional dispersal patterns, enhanced protection of critical life stages, and enhanced protection of critical processes and functions, *e.g.*, migration corridors (Gerber and Heppell, 2004). Finally, networks allow for protection of marine ecosystems at an appropriate scale. A network of MPAs could cover a large gradient of biogeographic and oceanographic conditions without the need to establish one extremely large reserve, and can provide more inclusive representation of stakeholders (National Research Council, 2001; Hansen, Biringer, and Hoffman, 2003).

While MPA networks are considered a critical management tool for conserving marine biodiversity, they must be established in conjunction with other management strategies to be effective (Hughes *et al.*, 2003). MPAs are vulnerable to activities beyond their boundaries. For example, uncontrolled pollution and unsustainable fishing outside protected areas can adversely affect species and ecosystem function within the protected area (Kaiser, 2005). Therefore, MPA networks should be established considering other forms of fisheries management (*e.g.*, catch limits and gear restrictions) (Allison, Lubchenco, and Carr, 1998; Beger, Jones, and Munday, 2003; Kaiser, 2005), as well as coastal management to control land-based threats such as pollution and sedimentation (Cho, 2005). In the long term, the most effective configuration would be a network of highly protected areas nested within a broader management framework (Salm, Done, and McLeod, 2006). Such a framework might include an extensive multiple-use area managed for sustainable fisheries as well as protection of biodiversity, integrated with coastal management regimes, where appropriate, to enable effective control of threats originating upstream and to maintain high water quality (*e.g.*, Done and Reichelt, 1998).

The National Marine Sanctuary Program has developed a set of goals (Box 8.1) to help clarify the relationship between operations at individual sanctuaries and the broad directives of the National Marine Sanctuaries Act. A subset of these goals (Goals 1, 4, 5, and 6) are relevant to resource protection and climate change. Box 8.3 expands upon Goals 1, 4, 5, and 6 to display their attendant objectives, which provide guidance for management plans that are developed by sanctuary sites (see Table 8.3). Sanctuary management plans are developed and subsequently reviewed and revised on a five-year cycle as a collaboration between sanctuary staff and local communities. After threats and stressors to resources are identified, action plans are prepared that identify activities to address them. Threats and stressors may include such things as overexploitation of natural resources, degraded water quality, and habitat damage and destruction. Sanctuary management plans are designed to address additional issues raised by local communities, such as user conflicts, needs for education and outreach, and interest in volunteer programs.

Highly protected marine reserves within national marine sanctuaries have been implemented at some sites (*e.g.*, Channel Islands and the Florida Keys; Keller and Causey, 2005) to reduce fishing pressure, restore ecosystem structure and function, and protect biodiversity; the entire area of the Papahānaumokuākea Marine National Monument will become no-take within five years. These additional protective actions complement existing fishery regulations. Some sites, such as Monterey Bay and the Florida Keys, have Water Quality Protection Programs to address issues such as watershed pollution, vessel discharges, and, in the case of the Florida Keys, wastewater and stormwater treatment systems. Habitat damage may be addressed using waterway marking programs to reduce vessel groundings and mooring buoys to minimize anchor damage. Many of these activities are supported through education and outreach programs to inform the public, volunteer programs to help distribute information (*e.g.*, Team OCEAN[35]), and law enforcement.

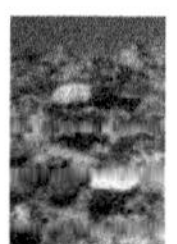

Sanctuary management plans are intended to be comprehensive, and may take years of community involvement to develop. For example, it took more than five years to develop the management plan for the Florida Keys

[35] **Florida Keys National Marine Sanctuary**, 2003: Florida Keys NMS Team OCEAN. Florida Keys National Marine Sanctuary Webpage, http://floridakeys.noaa.gov/edu/ocean.html, accessed on 5-21-2007.

BOX 8.3. Draft Objectives of the Goals of the National Marine Sanctuary Program That Are Relevant to Resource Protection and Climate Change (Goals 1, 4, 5, and 6 from Box 8.1).*

Goal 1: Protect Resources.

Objective 1. Prepare sanctuary-specific management plans and regional and national programs and policies that utilize all program capacities to protect and manage resources.

Objective 2. Conduct and maintain routine contingency planning, emergency response, damage assessment, and restoration activities to preserve and restore the integrity of sanctuary ecosystems.

Objective 3. Develop and maintain enforcement programs and partnerships to maximize protection of sanctuary resources.

Objective 4. Review and evaluate the NMSP's effectiveness at site, regional, and national levels, through both internal and external mechanisms.

Objective 5. Anticipate, characterize, and mitigate threats to resources.

Objective 6. Assess and predict changes in the NMSP's operating, natural, and social environments, and evolve sanctuary management strategies to address them through management plan reviews, reauthorizations, and program regulatory review.

Objective 7. Designate new sanctuaries, as appropriate, to ensure the nation's marine ecosystems and networks achieve national expectations for sustainability.

Goal 4: Improve Sanctuary Science.

Objective 1. Expand observing systems and monitoring efforts within and near national marine sanctuaries to fill important gaps in the knowledge and understanding of the ocean and Great Lakes ecosystems.

Objective 2. Support directed research activities that support management decision making on challenges and opportunities facing sanctuary ecosystems, processes, and resources.

Objective 3. Develop comprehensive characterization products of ocean and Great Lakes ecosystems, processes, and resources.

Goal 5: Facilitate Compatible Use.

Objective 1. Work closely with partners, interested parties, community members, stakeholders, and government agencies to assess and manage human use of sanctuary resources.

Objective 2. Create, operate, and support community-based sanctuary advisory councils to assist and advise sites and the overall program in the management of their resources, and to serve as liaisons to the community.

Objective 3. Consult and coordinate with federal agencies and other partners conducting activities in or near sanctuaries.

Objective 4. Use other tools such as policy development, permitting, and regulatory review and improvement to help guide human use of sanctuary resources.

Goal 6: Improve International Work.

Objective 1. Develop multilateral program relationships to interact with, share knowledge and experience with, and learn from international partners to improve the NMSP's management capacity, and bring new experiences to MPA management in the United States.

Objective 2. Investigate the use of international legal conventions and other instruments to help protect sanctuary resources, including those that are transboundary or shared.

Objective 3. Cooperate to the extent possible with global research initiatives in order to improve the overall understanding of the ocean.

Objective 4. Make NMSP education and awareness programs accessible through international efforts to increase the global population's awareness of ocean issues.

* Additional goals of the NMSP are in Box 8.1.

National Marine Sanctuary (Keller and Causey, 2005), and an additional three years were required to prepare a supplemental plan for the Tortugas Ecological Reserve (Cowie-Haskell and Delaney, 2003; Delaney, 2003). However, the focus of sanctuary management plans has been on local stressors and not on additional impacts of climate change. As suggested below, climate change will need to be included in MPA planning, management, and evaluation.

Effective management and preservation of ecosystem characteristics in the face of climate change projections is relevant to achieving NMSP Goals 1, 2, 4, and 5 (Box 8.1). The NMSP is a leader in the use of stakeholders in the development of new management approaches (Sanctuary Advisory Councils and public scoping meetings at the site level). This model of public involvement should serve well as management strategies adapt under the stresses of climate change. Exporting lessons learned to the general public, managers of other MPAs, and the international community will further address NMSP Goals 2, 3, and 6.

An additional approach of the NMSP that should further efforts toward adaptive management in the context of climate change is the development of performance measures to help evaluate the success of the program (Box 8.4). Although climate change stressors are not yet explicitly addressed in these performance measures, attainment of a number of these measures clearly will be increasingly affected by climate change. The performance-measure approach should encourage sanctuary managers to address climate change impacts using the public processes of Sanctuary Advisory Councils and public scoping meetings. In addition, national marine sanctuaries are preparing Condition Reports,[36] which provide summaries of resources, pressures on resources, current condition and trends, and management responses to pressures that threaten the integrity of the marine environment. These reports will provide opportunities for sanctuaries to evaluate climate change as a pressure, and identify management responses on a site-by-site basis as well as across the system of national marine sanctuaries.

8.4 ADAPTING TO CLIMATE CHANGE

MPA managers can respond to challenges of climate change at two scales: actions at individual sites and implementing MPA networks. At particular MPAs, managers can increase efforts to ameliorate existing anthropogenic stressors with a goal of reducing the overall load of multiple stressors (Breitburg and Riedel, 2005). For example, the concept of protecting or enhancing coral reef resilience has been proposed to help ameliorate negative consequences of coral bleaching (Hughes *et al.*, 2003; Hughes *et al.*, 2005; Marshall and Shuttenberg, 2006). Under this approach, resilience is an ecosystem property that can be managed and is defined as the ability of an ecosystem to resist or absorb disturbance without significantly degrading processes that determine community structure, or if

[36] **National Marine Sanctuary Program**, 5-21-2007: National Marine Sanctuaries condition reports. NOAA Website, http://sanctuaries.noaa.gov/science/condition/, accessed on 7-27-2007.

BOX 8.4. Draft Natural Resource Performance Measures of the National Marine Sanctuary Program.

2015: 12 sites with water quality being maintained or improved.

2015: 12 sites with habitat being maintained or improved.

2015: 12 sites with living marine resources being maintained or improved.

2010: 100% of the System is adequately characterized.

2010: 6 sites are achieving or maintaining an optimal management rating on the NMSP Report Card.

2007: 100% of NMSP permits are handled in a timely fashion and correctly.

2010: 100% of sites with zones in place are assessing them for effectiveness.

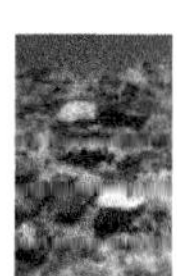

alterations occur, recovery is *not* to an alternate community state (Gunderson, 2000; Nyström, Folke, and Moberg, 2000; Hughes *et al.*, 2003). In short, managing for resilience includes dealing with causes of coral reef disturbance and decline that managers can address at local and regional levels, such as overfishing and pollution. These are the things that managers would want to do anyway, even if climate change were not a threat, because these activities help to maintain the ecological and economic value of ecosystems.

In addition to the approach of ameliorating existing stressors, MPA managers can protect putatively resistant and potentially resilient areas, develop networks of MPAs, and integrate climate change into planning efforts. Specific examples of adaptation options from across these approaches are presented in Box 8.5 and elaborated upon further in the sections that follow.

It is important to emphasize that variable and complex effects of climate on oceanographic processes and production (Soto, 2001; Mann and Lazier, 2006) present MPA managers with major uncertainties about climate change impacts and effective management approaches. An excellent discussion of uncertainty and scenario-based planning is provided in the National Parks chapter, sections 4.4.1 and 4.4.2.

BOX 8.5. Marine Protected Areas: Adaptation Options for Resource Managers.

- Manage human stressors such as overfishing and excessive inputs of nutrients, sediments, and pollutants within MPAs.
- Improve water quality by raising awareness of adverse effects of land-based activities on marine environments, implementing integrated coastal and watershed management, and developing options for advanced wastewater treatment.
- Manage functional species groups necessary to maintaining the health of reefs and other ecosystems.
- Identify and protect areas that appear to be resistant to climate change effects or to recover from climate-induced disturbances.
- Identify and protect ecologically significant ("critical") areas such as nursery grounds, spawning grounds, and areas of high species diversity.
- Identify ecological connections among ecosystems and use them to inform the design of MPAs and management decisions such as protecting resistant areas to ensure sources of recruitment for recovery of populations in damaged areas.
- Design MPAs with dynamic boundaries and buffers to protect breeding and foraging habits of highly migratory and pelagic species.
- Establish dynamic MPAs defined by large-scale oceanographic features, such as oceanic fronts, where changes in types and abundances of organisms often occur.
- Maximize habitat heterogeneity within MPAs and consider protecting larger areas to preserve biodiversity, ecological connections among habitats, and ecological functions.
- Include entire ecological units (e.g., coral reefs with their associated mangroves and seagrasses) in MPA design to help maintain ecosystem function and resilience.
- Ensure that the full breadth of habitat types is protected (e.g., fringing reef, fore reef, back reef, patch reef).
- Replicate habitat types in multiple areas to spread risks associated with climate change.
- Monitor ecosystems and have rapid-response strategies prepared to assess ecological effects of extreme events as they occur.
- Following extreme events, consider whether actions should be taken to enhance natural recovery processes through active restoration.
- Consider mangrove restoration for potential benefits including shoreline protection, expansion of nursery habitat, and release of tannins and other dissolved organic compounds that may reduce photo-oxidative stress in corals.

8.4.1 Ameliorate Existing Stressors in Coastal Waters

Managers may be able to increase resilience to climate change within MPAs by reducing impacts of local- and regional-scale stressors, such as overfishing; excessive inputs of nutrients, sediments, and pollutants; and degraded water quality. While this concept is logical and has considerable appeal, evidence in support of this approach is weak at best, which provides an excellent opportunity for adaptive-management research. Kelp forest ecosystems in marine reserves, where no fishing is allowed, are more resilient to ocean warming than those in areas where overfishing occurs (Behrens and Lafferty, 2004). This ecological response is a result of changes in trophic structure of communities in and around the reserves. When top predators such as spiny lobster are fished, their prey, herbivorous sea urchins, increase in abundance and consume giant kelp and other algae. When kelp forests are subjected to intense grazing by these herbivores, the density of kelp is reduced, sometimes becoming an "urchin barren," particularly during ocean warming events such as ENSO cycles. In reserves where fishing is prohibited, lobster populations were larger, urchin populations were diminished, and kelp forests persisted over a period of 20 years—including four ENSO cycles (Behrens and Lafferty, 2004).

Managing water quality has been identified as a key strategy for maintaining ecological resilience (Salm, Done, and McLeod, 2006).[37] In the Florida Keys National Marine Sanctuary and the Great Barrier Reef Marine Park, water quality protection is recognized as an essential component of management (U.S. Department of Commerce, 1996; The State of Queensland and Commonwealth of Australia, 2003; Grigg *et al.*, 2005, also see the Monterey Bay National Marine Sanctuary's water quality agreements with land-based agencies).[37] Strong circumstantial evidence exists linking poor water quality to increased macroalgal abundances, internal bioerosion, and susceptibility to some diseases in corals and octocorals (Fabricius and De'ath, 2004). Addressing sources of pollution—especially nutrient enrichment, which can lead to increased algal growth and reduced coral settlement—is critical to maintaining ecosystem health. In addition to controlling point-source pollution within an MPA, managers must also link their MPAs into the governance system of adjacent areas to control sources of pollution beyond the MPA boundaries (*e.g.*, Crowder *et al.*, 2006). Further actions necessary to improve water quality include raising awareness of how land-based activities can adversely affect adjacent marine environments, implementing programs for integrated coastal and watershed management, and developing options for advanced wastewater treatment (The Group of Experts on Scientific Aspects of Marine Environmental Protection, 2001).

Managers may be able to build resilience to climate change into MPA management strategies by protecting marine habitats such as coral reefs and mangroves from direct threats such as pollution, sedimentation, destructive fishing, and overfishing. Therefore, managers should continue to develop and implement strategies to reduce land-based pollution, decrease nutrient and sediment runoff, eliminate the use of persistent pesticides, and increase filtration of effluent to improve water quality. As noted above, the efficacy of these measures needs research in an adaptive-management context.

Another mechanism that may maintain resilience is the management of functional groups, specifically herbivores (Hughes *et al.*, 2003; Bellwood *et al.*, 2004). Bellwood *et al.* (2004) identified three functional groups of herbivores that assist in maintaining coral reef resilience: bioeroders, grazers, and scrapers. These groups work together to break down dead coral, providing sites for recruitment, graze macroalgae, and reduce the development of algal turfs to generate relatively bare substratum for coral settlement. Algal biomass must be kept low to maintain healthy coral reefs (Sammarco, 1980; Hatcher and Larkum, 1983; Steneck and Dethier, 1994). Bellwood, Hughes, and Hoey (2006) identify the need to protect both the species that prevent phase shifts from coral-dominated to algal-dominated reefs and the species that help reefs recover from algal dominance. They suggest that while parrotfishes and surgeonfishes appear to play a critical role in preventing phase shifts to macroalgae, their ability to remove algae may be limited if a

[37] **Monterey Bay National Marine Sanctuary**, 2007: Water quality protection program for the MBNMS. Monterey Bay National Marine Sanctuary Website, http://www.mbnms.nos.noaa.gov/resourcepro/waterpro.html, accessed on 5-23-2007.

phase shift to macroalgae has already occurred (Bellwood, Hughes, and Hoey, 2006). In their study on the Great Barrier Reef, the phase shift reversal from macroalgal-dominated to a coral- and epilithic algal-dominated state was driven by a single batfish species (*Platax pinnatus*), not grazing by dominant parrotfishes or surgeonfishes (Bellwood, Hughes, and Hoey, 2006). This finding highlights the need to protect the full range of species to maintain resilience, at least in some systems. For example, Ledlie *et al.* (2007) found that a shift from coral to algal dominance occurred at a marine reserve in the Seychelles after the 1998 mass coral bleaching event, despite the presence of abundant herbivorous fishes. Many herbivorous fishes avoid macroalgae, and more research on functional groups is needed.

Although protecting functional groups may be a component of MPA management to enhance resilience, understanding which groups should be protected requires a detailed knowledge of species and interactions that is not often available for all species. Therefore, managers should strive to maintain the maximum number of species in the absence of detailed data on ecological and species interactions. For example, for managing coral reefs, regional guidelines identifying key herbivores that reduce macroalgae and encourage coral reef settlement should be developed. For kelp forests, the opposite approach may apply—managers may need to identify key predators on herbivores and limit fishing on those predators to reduce herbivory and promote growth of healthy kelp forests. These guidelines should be field tested at different locations to verify the recommendations.

8.4.2 Protect Apparently Resistant and Potentially Resilient Areas

Marine ecosystems that contain biologically generated habitats face potential loss of habitat structure as climate change progresses (*e.g.*, coral reefs, seagrass beds, kelp forests, and deep coral communities) (see Hoegh-Guldberg, 1999; Steneck *et al.*, 2002; Roberts, Wheeler, and Freiwald, 2006; Orth *et al.*, 2006). As discussed earlier in this chapter, it is likely that climate change contributes to mass coral bleaching events (Reaser, Pomerance, and Thomas, 2000), which became recognized globally in 1997–1998 (Wilkinson, 1998; 2000) and have affected large regions in subsequent years (Wilkinson, 2002; 2004; Whelan *et al.*, 2007). The amount of live coral has declined dramatically in the Caribbean region over the past 30 years as a result of bleaching, diseases, and hurricanes (Gardner *et al.*, 2003; 2005). In the Florida Keys, fore-reef environments that formerly supported dense growths of coral are now nearly depauperate, and the highest coral cover is in patch reef environments (Porter *et al.*, 2002; Lirman and Fong, 2007). Irrespective of the mechanism—resistance, resilience, or exposure to relatively low levels of past environmental stress— these patch-reef environments might be good candidates for additional protective measures because they may have high potential to survive climate stress.

Done[38] (see also Marshall and Schuttenberg, 2006) presented a decision tree for identifying areas that would be suitable for MPAs under a climate change scenario. Two types of favorable outcomes included reefs that survived bleaching (*i.e.*, were resilient) and reefs that were not exposed to elevated sea surface temperatures (*e.g.*, may be located within refugia such as areas exposed to upwelling or cooler currents). This type of decision tree has already been adapted to guide site selection for mangroves (McLeod and Salm, 2006), and it could be extended further for other habitat types such as seagrass beds and kelp forests.

In addition, thermally stressed corals exhibit less bleaching and higher survival if they are shaded during periods of elevated temperatures (Hoegh-Guldberg *et al.*, 2007). On a small scale, MPA managers may be able to shade areas during bleaching events to reduce overall stress. On a larger scale, managers should protect mangrove shorelines and support restoration of areas where mangroves have been damaged or destroyed, because tannins and dissolved organic compounds from decaying mangrove vegetation contribute to absorbing light and reducing stress (Hallock, 2005) (see

[38] **Done**, T., 2001: Scientific principles for establishing MPAs to alleviate coral bleaching and promote recovery. In: *Coral Bleaching and Marine Protected Areas. Proceedings of the Workshop on Mitigating Coral Bleaching Impact Through MPA Design, Bishop Museum, Honolulu, HI, 29-31 May 2001* [Salm, R.V. and S.L. Coles (eds.)]. Asia Pacific Coastal Marine Program Report # 0102, The Nature Conservancy, Honolulu, HI, pp. 60-66.

also section 8.4.3.1). Extensive discussions of coral bleaching and management responses are provided in Marshall and Schuttenberg (2006) and Johnson and Marshall (2007).

Because climate change impacts on marine systems are patchy (with reefs that avoid bleaching one year potentially bleaching the following year), it is essential that areas that appear to be resistant or resilient to climate change impacts be monitored and tested to ensure that they continue to provide benefits (see section 8.4.4.1 for more on monitoring and research). This allows managers to target potential refugia for MPA design now, while also monitoring these areas over time so that management can be modified as circumstances and habitats change.

8.4.3 Develop Networks of MPAs

The concept of systems or networks of MPAs has considerable appeal because of emergent properties (*i.e.*, representation, replication, sustainability, connectivity) (National Research Council, 2001; Roberts *et al.*, 2003a),[21] spreading the risk of catastrophic habitat loss (Palumbi, 2002; Allison *et al.*, 2003), and the provision of functional wilderness areas sufficient to resist fundamental changes to entire ecosystems (Kaufman *et al.*, 2004). While MPA networks have been recognized as a valuable tool to conserve marine resources in the face of climate change, there have been a number of challenges to implementation (Pandolfi *et al.*, 2005; Mora *et al.*, 2006); nevertheless, a number of principles have been developed and are gradually being applied to aid MPA network design and implementation. These principles are described below.

8.4.3.1 Protect Critical Areas

Critical areas—areas that are biologically or ecologically significant—should be identified and included in MPAs. These critical areas include nursery grounds, spawning grounds, areas of high species diversity, areas that contain a variety of habitat types in close proximity to each other, and climate refugia (Allison, Lubchenco, and Carr, 1998; Sale *et al.*, 2005).[39] Coral assemblages that demonstrate resistance or resilience to climate change may be identified and provided additional protection to ensure a secure source of recruitment to support recovery in damaged areas. Managers can analyze how assemblages have responded to past bleaching events to assess possible resilience to climate change impacts. For example, some coral reefs resist bleaching due to genetic characteristics or avoid bleaching due to environmental factors. Managers can protect those that either resist or recover quickly from mass bleaching events, as well as those that are located in areas where physical conditions (*e.g.*, currents, shading) afford them some protection from temperature anomalies. Reefs that are resistant and reefs that are located in refugia from climate extremes may play a critical role in reef survival by providing a source of larvae for dispersal to and recovery of affected areas.[40] For coral reefs, indicators of potential refugia include a ratio of live to dead coral and a range of colony sizes and ages suggesting persistence over time. Refugia must be large enough to support high species richness to maximize their effectiveness as sources of recruits to replenish areas that have been damaged (Palumbi *et al.*, 1997; Bellwood and Hughes, 2001; Salm, Done, and McLeod, 2006).

Following extreme events, MPA managers should consider whether actions should be taken to enhance natural recovery processes through active restoration of biologically structured habitats. For example, damaged areas in seagrass beds may recover more rapidly if steps are taken to stabilize sediments (Whitfield *et al.*, 2002). Due to the loss of mangroves from many areas, mangrove restoration is another option for MPA managers that may have multiple benefits, including shoreline protection, expansion of nursery habitat (Nagelkerken, 2007), and release of tannins and other dissolved organic compounds that may reduce photo-oxidative stress in corals (Hallock, 2005).

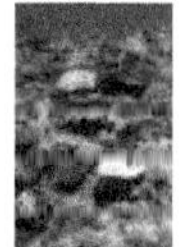

8.4.3.2 Incorporate Connectivity in Planning MPA Networks

Connectivity is the natural linkage between marine habitats (Crowder *et al.*, 2000; Stewart,

[39] See also **Sadovy**, Y., 2006: Protecting the spawning and nursery habitats of fish: the use of MPAs to safeguard critical life-history stages for marine life. *MPA News, International News and Analysis on Marine Protected Areas*, **8(2)**, 1-3.

[40] **Salm**, R.V. and S.L. Coles, 2001: *Coral Bleaching and Marine Protected Areas. Proceedings of the Workshop on Mitigating Coral Bleaching Impact Through MPA Design, Bishop Museum, Honolulu, HI, 29-31 May 2001.* Asia Pacific Coastal Marine Program Report #0102, The Nature Conservancy, Honolulu, HI, 118 pp.

Noyce, and Possingham, 2003; Roberts *et al.*, 2003b), which occurs through advection by ocean currents and includes larval dispersal and movements of adults and juveniles. Connectivity is an important part of ensuring larval exchange and the replenishment of populations in areas damaged by natural or human-related agents. Salm *et al.* (2006) recommend that patterns of connectivity be identified among source and sink reefs to inform reef selection in the design of MPA networks and enhance recovery following disturbance events. This principle applies to other marine systems, such as mangroves, as well. For example, healthy mangroves could be selected up-current from areas that may succumb to sea level rise, and areas could be selected that would be suitable habitat for mangroves in the future following sea level rise. These areas of healthy mangroves could provide secure sources of propagules to replenish down-current mangroves following a disturbance event.

A suspected benefit of MPAs is the dispersal of larvae to areas surrounding MPAs, but there are few data that can be used to estimate the exchange of larvae among local populations (Palumbi, 2004). Understanding larval dispersal and transport are critical to determining connectivity, and thus the design of MPAs. The size of an individual MPA should be based on the movement of adults of species of interest (Hastings and Botsford, 2003; Botsford, Micheli, and Hastings, 2003) and be large enough to contain the different habitats used and daily movements. The distance between adjacent MPAs should take into account the potential dispersal distances of larvae of fish, invertebrates, and other species of interest.[41]

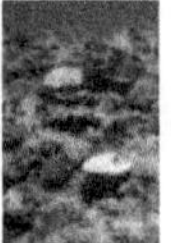

One approach in MPA design has been to establish the size of MPAs based on the spatial scale of movements of adults of heavily fished species, and to space MPAs based on scales of larval dispersal (Palumbi, 2004). However, guidelines for the minimum size of MPAs and no-take reserves, and spacing between adjacent MPAs, vary dramatically depending on the goals for the MPAs (Hastings and Botsford, 2003). Friedlander *et al.* (2003) suggested that no-take zones should measure ca. 10 km^2

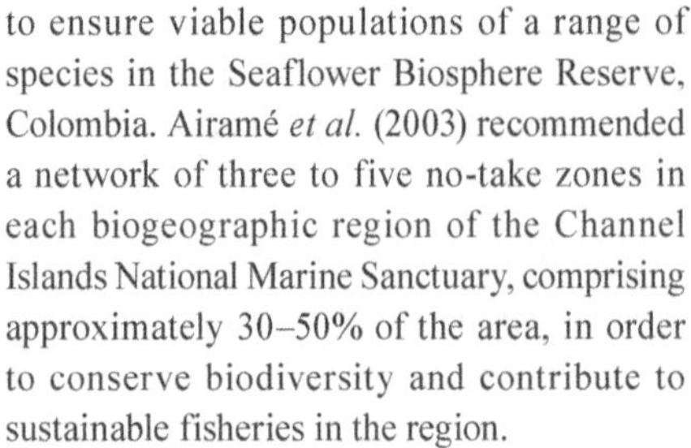

to ensure viable populations of a range of species in the Seaflower Biosphere Reserve, Colombia. Airamé *et al.* (2003) recommended a network of three to five no-take zones in each biogeographic region of the Channel Islands National Marine Sanctuary, comprising approximately 30–50% of the area, in order to conserve biodiversity and contribute to sustainable fisheries in the region.

Recent studies confirm that larval dispersal is more localized than previously thought, and short-lived species may require regular recruitment from oceanographically connected sites (Cowen, Paris, and Srinivasan, 2006; Steneck, 2006). Palumbi (2003) concluded that marine reserves tens of km apart may exchange larvae in a single generation. Shanks, Grantham, and Carr (2003) similarly concluded that marine reserves spaced 20 km apart would allow larvae to be carried to adjacent reserves. The Science Advisory Team to California's Marine Life Protection Act Initiative recommended spacing high protection MPAs, such as marine reserves, within 50–100 km in order to accommodate larval dispersal distances of a wide range of species of interest. Halpern *et al.* (2006) corroborated these estimates using an uncertainty-modeling approach.

No-take zones measuring a minimum of 20 km in diameter will accommodate short-distance dispersers in addition to including a significant proportion of local benthic fish species, thus generating fisheries benefits (Shanks, Grantham, and Carr, 2003; Fernandes *et al.*, 2005; Mora *et al.*, 2006). While this recommendation is likely to protect the majority of small benthic fish and benthic invertebrates, it is unlikely to protect large pelagic fish and large migratory species (Roberts *et al.*, 2003b; Palumbi, 2004). Recommendations to protect highly migratory and pelagic species include designing MPAs to protect predictable breeding and foraging habits, ensuring these have dynamic boundaries and extensive buffers, and establishing dynamic MPAs that are defined by the extent and location of large-scale oceanographic features, such as oceanic fronts, where changes in types and abundances of marine organisms often occur (Hyrenbach, Forney, and Dayton, 2000).

A system-wide approach should be taken that addresses patterns of connectivity among

[41] **California Department of Fish and Game**, 2007: *California Marine Life Protection Act: Master Plan for MPAs*. California Department of Fish and Game.

ecosystems such as mangroves, coral reefs, and seagrass beds (Mumby *et al.*, 2004). For example, mangroves and seagrass beds in the Caribbean enhance the biomass of coral reef fish communities because they provide essential nursery habitat. Coral reefs can protect mangroves and seagress beds by buffering the impacts of wave erosion, while mangroves can protect reefs and seagrass beds from siltation. Thus, connectivity among functionally linked habitats helps maintain ecosystem function and resilience (Ogden and Gladfelter, 1983; Roberts, 1996; Nagelkerken *et al.*, 2000). Entire ecological units (*e.g.*, coral reefs with their associated mangroves and seagrasses) should be included in MPA design where possible. If entire biological units cannot be included, then larger areas should be chosen over smaller areas to accommodate local-scale recruitment.

Although maintaining connectivity within and between MPAs may help maintain marine biodiversity, ecosystem function, and resilience, many challenges exist. For example, the same currents and pathways that allow for larval recruitment following a disturbance event can expose an ecosystem to invasive species, pathogens, parasites, or pollutants, which can undermine the resilience of a system (McClanahan, Polunin, and Done, 2002). Numerous challenges also exist in estimating larval dispersal patterns. Although there have been detailed studies addressing dispersal *potential* of marine species based on their larval biology (*e.g.*, Shanks, Grantham, and Carr, 2003; Kinlan and Gaines, 2003), little is known about where in the oceans larvae go and how far they travel. A single network design is unlikely to satisfy the potential dispersal ranges for all species; Roberts *et al.* (2003b) recommended an approach using various sizes and spacing of MPAs in a network to accommodate the diversity of dispersal ranges. Larval duration in the plankton varies from minutes to years, and the more time that propagules spend in the water column, the farther they tend to be dispersed (Shanks, Grantham, and Carr, 2003; Steneck, 2006). Evidence from hydrodynamic models and genetic structure data indicates that, in addition to large variation of larval dispersal distances among species, the average scale of dispersal can vary widely—even within a given species—at different locations in space and time (*e.g.*, Cowen *et al.*, 2003; Sotka *et al.*, 2004; Engie and Klinger, 2007). Some information suggests long-distance dispersal is common, but other emerging information suggests that larval dispersal may be limited (Jones *et al.*, 1999; Swearer *et al.*, 1999; Warner, Swearer, and Caselle, 2000; Thorrold *et al.*, 2001; Palumbi, 2003; Paris and Cowen, 2004; Jones, Planes, and Thorrold, 2005). Additional research will be required to better understand where and how far larvae travel in various marine ecosystems.

8.4.3.3 Replicate Multiple Habitat Types in MPA Networks

Recognizing that the science underlying our understanding of resilience is developing and that climate change will not affect marine habitats and species equally everywhere, an element of risk spreading must be built into MPA design. To help avoid the loss of a particular habitat type, managers can protect multiple examples of all habitats (Hockey and Branch, 1994; Roberts *et al.*, 2001; Friedlander *et al.*, 2003; Roberts *et al.*, 2003b; Salm, Done, and McLeod, 2006; Wells, 2006).[21] For example, marine habitat types include coral reefs with varying degrees of exposure to wave energy (*e.g.*, offshore, mid-shelf, and inshore reefs), seagrass beds dominated by various seagrass species and in different environments, and a range of mangrove communities (riverine, basin, and fringe forests in areas of varying salinity, tidal fluctuation, and sea level) (Salm, Done, and McLeod, 2006). Reflecting the current federal goal of protecting at least 30% of lifetime stock spawning potential (Ault, Bohnsack, and Meester, 1998; National Marine Fisheries

Service, 2003), it has been recommended that more than 30% of appropriate habitats should be included in no-take zones (Bohnsack *et al.*, 2002). In 2004, the Great Barrier Reef Marine Park Authority increased the area of no-take zones from less than 5% to approximately 33% of the area of the Marine Park, ensuring that at least 20% of each bioregion (area of every region of biodiversity) was zoned as no-take (Fernandes *et al.*, 2005; Day *et al.*, 2002).

For both terrestrial and marine systems, species diversity often increases with habitat diversity, and species richness increases with habitat complexity; the greater the variety of habitats protected, the greater the biodiversity conserved (Friedlander *et al.*, 2003; Carr *et al.*, 2003). High species diversity may increase ecosystem resilience by ensuring sufficient redundancy to maintain ecological processes and protect against environmental disturbance (McNaughton, 1977; McClanahan, Polunin, and Done, 2002). This is particularly true in the context of additive or synergistic stressors. Maximizing habitat heterogeneity is critical for maintaining ecological health; thus MPAs should include large areas and depth gradients (Hansen, Biringer, and Hoffman, 2003; Roberts *et al.*, 2003a).[38] By protecting a representative range of habitat types and communities, MPAs have a higher potential to protect a region's biodiversity, biological connections between habitats, and ecological functions (Day *et al.*, 2002).

Replication of habitat types in multiple areas provides a further way to spread risks associated with climate change. If a habitat type is destroyed in one area, similar habitat in another area may provide larvae for recovery. While the number of replicates will be determined by a balance of desired representation and practical concerns such as funding and enforcement capacity (Airamé *et al.*, 2003), generally at least three to five replicates are recommended to effectively protect a particular habitat or community type (Airamé *et al.*, 2003; Roberts *et al.*, 2003b; Fernandes *et al.*, 2005). Wherever possible, multiple examples of each habitat type should be included in MPA networks or larger management frameworks such as multiple-use MPAs or areas under rigorous integrated management regimes (Salm, Done, and McLeod, 2006). This approach has the added advantage of protecting essential habitat for a wide variety of commercially valuable fish and macroinvertebrates.

While a risk-spreading approach to address the uncertainty of impacts of climate change makes practical sense, there are challenges to adequate representation. Managers must have classification maps (or local knowledge) of marine habitat types and communities to determine which representative examples should be included in MPA design. Replication of habitat types may not always be feasible due to limited monitoring and enforcement resources, conflicting needs of resource users, and existence of certain habitat types within an MPA.

8.4.4 Integrate Climate Change into MPA Planning, Management, and Evaluation

A number of tools exist to help managers address climate impacts and build resilience into MPA design and management. Ecological changes that are common in marine reserves worldwide and guidelines for marine reserve design are summarized in an educational booklet for policymakers, managers, and educators, entitled "The Science of Marine Reserves."[42] The Reef Resilience toolkit[43] provides marine resource managers with strategies to address coral bleaching and conserve reef fish spawning aggregations, helping to build resilience into coral reef conservation programs. "A Reef Manager's Guide to Coral Bleaching" provides information on the causes and consequences of coral bleaching and management strategies to help local and regional reef managers reduce this threat to coral reef ecosystems (Marshall and Shuttenberg, 2006). The application of some of these strategies is discussed in a recent report by the U.S. Environmental Protection Agency, which applies resilience theory in a case study for the reefs of American Samoa and proposes climate adaptation strategies that can be leveraged with existing local management plans,

[42] **Partnership for Interdisciplinary Studies of Coastal Oceans**, 2005: The science of marine reserves. Partnership for Interdisciplinary Studies of Coastal Oceans Website, http://www.piscoweb.org/outreach/pubs/reserves, accessed on 5-23-2007.

[43] **The Nature Conservancy and Partners**, 2004: *R² - Reef Resilience: Building Resilience into Coral Reef Conservation; Additional Tools for Managers.* Volume 2.0. CD ROM Toolkit, The Nature Conservancy, http://www.reefresilience.org/.

processes, and mandates (U.S. Environmental Protection Agency, 2007).

In contrast, with regard to the impacts on marine organisms of reductions in ocean pH due to CO_2 emissions (Caldeira and Wickett, 2003), management strategies have not yet been developed. Adding chemicals to counter acidification is not a viable option, as it would likely be only partly effective and, if so, only at a very local scale (The Royal Society, 2005). Therefore, further research is needed on impacts of high concentrations of CO_2 in the oceans, possible acclimatization or evolution of organisms in response to changes in ocean chemistry, and how management might respond (The Royal Society, 2005).

Determining management effectiveness is important for gauging the success of an MPA or network, and also can inform adaptive management strategies to address shortcomings in a particular MPA or network. To help managers improve the management of MPAs, the IUCN World Commission on Protected Areas and the World Wide Fund for Nature developed an MPA management effectiveness guidebook. This guidebook, "How is Your MPA Doing? A Guidebook of Natural and Social Indicators for Evaluating Marine Protected Area Management Effectiveness," helps managers and other decision-makers assess management effectiveness through the selection and use of biophysical, socioeconomic, and governance indicators.[44] The goal of the guidebook is to enhance the capability for adaptive management in MPAs. The "Framework for Measuring Success" (Parks and Salafsky, 2001) also provides a suite of tools to analyze community response to an MPA, and replicable methodologies to assess both social and ecological criteria.

National marine sanctuaries are preparing a series of Condition Reports for each site, which provide a summary of resources, pressures on those resources, current condition and trends, and management responses to the pressures.[36] This information is intended to be used in reviews of management plans and to help sanctuary staff identify monitoring, characterization, and research priorities to address gaps, day-to-day information needs, and new threats.

Managers in the United States can benefit from the example set by the Great Barrier Reef Marine Park Authority (GBRMPA), which is implementing a Climate Change Response Program[45] designed to: (1) understand climate change implications for the Great Barrier Reef; (2) share knowledge about climate change impacts and response options; (3) encourage and support reductions in greenhouse gas emissions; (4) maximize resilience of the Great Barrier Reef ecosystem; and (5) encourage and support Great Barrier Reef communities and industries to adapt to climate change. To further several of these objectives, GBRMPA has published a thorough assessment of vulnerabilities to climate change (Johnson and Marshall, 2007). This approach is a model for MPAs to consider worldwide.

8.4.4.1 MPA Monitoring and Research

MPAs must be effectively monitored to ensure the success of MPA design and management. If MPA design and management are not successful, then adaptations need to be made to meet the challenges posed by anthropogenic and natural stresses. As the number of pristine areas is decreasing rapidly, establishing baseline data for marine habitats is urgent and essential. Once baseline data are established, managers should monitor to determine the effects of climate change on local resources and populations. Retrospective testing of resistance to climate change impacts is difficult, so rapid response strategies should be in place to assess ecological effects of extreme events as they occur. For coral reefs, coral bleaching patterns either disappear with time or become confounded with other causes of mortality, such as predation by the crown-of-thorns starfish, disease, or multiple other stressors (Salm, Done, and McLeod, 2006). Therefore, response strategies must be implemented immediately following a

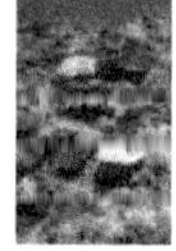

[44] **Pomeroy**, R.S., J.E. Parks, and L.M. Watson, 2004: *How Is Your MPA Doing? A Guidebook of Natural and Social Indicators for Evaluating Marine Protected Area Management Effectiveness.* http://effectivempa.noaa.gov/guidebook/guidebook.html, International Union for Conservation of Nature and Natural Resources, The World Conservation Union, Gland, Switzerland.

[45] **Great Barrier Reef Marine Park Authority**, 2007: Management responses. Great Barrier Reef Marine Park Authority Website, http://www.gbrmpa.gov.au/corp_site/key_issues/climate_change/management_responses, accessed on 12-24-2007.

mass bleaching event or other climate-related event to determine bleaching impacts. For coral reefs, bleaching and mortality responses of corals to heat stress, the recovery rates of coral communities, and the physiological response of certain corals to bleaching should be monitored. After the degree of damage from a mass bleaching or other climate-related event has been evaluated, MPA managers can consider whether active restoration may be an option for supporting natural recovery (Marshall and Schuttenberg, 2006). For coral reefs, restoration efforts may include transplanting coral colonies, introducing large numbers of coral larvae, and increasing densities of herbivores such as the sea urchin *Diadema antillarum* (in the Caribbean) or herbivorous reef fishes.

Monitoring also can be an effective way to engage community members and raise awareness of impacts of climate change on marine systems. For example, the Reef Check program enables community volunteers to collect coral reef monitoring data to supplement other monitoring data from researchers and government agencies. Programs that engage coral reef users (such as local fishermen and tourism operators) in monitoring can help raise awareness of impacts on marine systems and can help support the need to manage for local threats. The Nature Conservancy is managing the Florida Reef Resilience Program to develop strategies to improve the condition of Florida's coral reefs and support human dimensions investigations.[46] The program includes annual surveys of coral bleaching effects at reefs along the Florida Keys and the southeast Florida coast, using trained divers from agencies, universities, and non-governmental organizations.

[46] **The Nature Conservancy**, 2007: Florida Keys reef resilience program. The Nature Conservancy Website, http://www.nature.org/wherewework/northamerica/states/florida/preserves/art17499.html, accessed on 7-27-2007.

Changes in ocean chemistry (CO_2 and O_2 levels and salinity), hydrography (sea level, currents, vertical mixing, storms, and waves), and temperature should be monitored over long time scales to determine climate changes and possible climate trends. A location that is well isolated from local-scale anthropogenic effects and has a history of relevant investigations, such as Palmyra Atoll, is well-suited for this. Such an analysis could help determine the efficacy of MPA management in the context of climate change that is relatively independent of other anthropogenic effects, similar to the situation in the Northwestern Hawaiian Islands (see Case Study Summary 8.3).

NOAA's Coral Reef Watch program[47] provides products that can warn managers of potential impending bleaching events. In addition, Coral Reef Watch is developing bleaching forecasts that will provide outlooks of bleaching potential months in advance. These tools can help managers prepare for bleaching events so that when the event occurs, managers can have the necessary capacity in place to respond. In addition to a number of guides to help managers understand resilience and incorporate the concept in management actions, global information databases exist that consolidate climate change impacts on marine systems such as coral reefs. Reefbase[48] is a global information system and is the database of the Global Coral Reef Monitoring Network and the International Coral Reef Action Network. Coral bleaching reports, maps, photographs, and publications are freely available on the website, and bleaching reports can be submitted for inclusion in the database. Reefbase provides an essential mechanism for collecting bleaching data from around the world, thus helping researchers and managers to identify potential patterns in reef vulnerability.

[47] http://coralreefwatch.noaa.gov/

[48] www.reefbase.org

8.4.4.2 Social Resilience, Stakeholder Participation, and Education and Outreach

In addition to identifying and building ecological resilience into MPA design and management, it is equally important for managers to address social resilience (*i.e.*, social, economic, and political factors that influence MPAs and networks). Social resilience is the "ability of groups or communities to cope with external stresses and disturbances as a result of social, political, and environmental change" (Adger, 2000). MPAs that reinforce social resilience can provide communities with the opportunity to strengthen social relations and political stability, and diversify economic options (Corrigan, 2006). A variety of management actions have been identified to reinforce social resilience (Corrigan, 2006) including: (1) provide opportunities for shared leadership roles within government and management systems (Adger *et al.*, 2005; Cinner *et al.*, 2005; McClanahan *et al.*, 2006); (2) integrate MPAs and networks into broader coastal management initiatives to increase public awareness and support of management goals (U.S. Environmental Protection Agency, 2007; Marshall and Shuttenberg, 2006); (3) encourage local economic diversification so that communities are able to deal with environmental, economic, and social changes (Adger *et al.*, 2005; Marschke and Berkes, 2006); (4) encourage stakeholder participation and incorporate their ecological knowledge in a multi-governance system (Tompkins and Adger, 2004; Granek and Brown, 2005; Lebel *et al.*, 2006); and (5) make culturally appropriate conflict resolution mechanisms accessible to local communities (Christie, 2004; Marschke and Berkes, 2006).

Some MPA managers may feel that engaging in supporting human adaptive capacity to climate change impacts is beyond the scope of their work. However, it is important to recognize that resource use patterns will change in response to changing environmental conditions. For example, recent studies suggest that when fishers are meaningfully engaged in decision-making processes for management of natural resources, their confidence and social resilience to changes in resource access can be increased (Marshall, forthcoming). Furthermore, as management is adapted to address changing conditions, engagement with stakeholders during this process will help MPA managers build the alliances, knowledge, and influence needed to implement adaptive approaches (Schuttenberg and Marshall, 2007). For example, national marine sanctuaries have Sanctuary Advisory Councils composed of a wide range of stakeholder representatives, who provide advice to sanctuary managers and help develop sanctuary management plans.[49] Education and outreach programs can help inform the public about effects of climate change on marine ecosystems and the pressing need to ameliorate existing stressors in coastal waters. Such programs should be strengthened in national marine sanctuaries and all agencies that manage MPAs.

8.5 CONCLUSIONS

8.5.1 Management Considerations

Adaptive management of MPAs in the context of climate change includes the concept that intact marine ecosystems are more resistant and resilient to change than are degraded systems (Harley *et al.*, 2006). Marine reserves develop less-disrupted community structure as populations of heavily fished species recover and abundance patterns and size structures return to states reflecting lower fishing mortality. Implementing networks of MPAs, including large areas of the ocean, will help "spread the risk" posed by climate change by protecting multiple representatives of habitats and communities within ecosystems (Soto, 2001; Palumbi, 2003; Halpern, 2003; Halpern and Warner, 2003; Roberts *et al.*, 2003b; Palumbi, 2004; Kaufman *et al.*, 2004; Salm, Done, and McLeod, 2006).

The most effective configuration of MPAs may be a network of highly protected areas and other types of zones nested within a broader management framework (Botsford, 2005; Hilborn, Micheli, and De Leo, 2006; Crowder *et al.*, 2006; Almany *et al.*, 2007; Young *et al.*, 2007). As part of this configuration, areas that are ecologically and physically significant and connected by currents should be identified and included as a way of enhancing resilience in

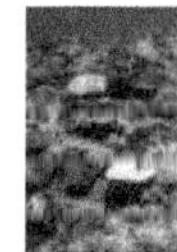

[49] **National Marine Sanctuary Program**, 2-6-2007: National Marine Sanctuaries advisory council's information. NOAA Website, http://sanctuaries.noaa.gov/management/ac/welcome.html, accessed on 7-27-2007.

the context of climate change. Critical areas to consider include nursery grounds, spawning grounds, areas of high species diversity, areas that contain a variety of habitat types in close proximity, and potential climate refugia. At the site level, managers can build resilience to climate change by protecting marine habitats from direct anthropogenic threats such as pollution, sedimentation, destructive fishing, and overfishing; ecosystem-based management, rather than single-species or other less-holistic approaches, will become increasingly important in the context of climate change. The healthier the ecosystem, the greater the potential will be for resistance to—and recovery from—climate-related disturbances.

In designing networks, managers should consider information on areas that may be refugia from climate change impacts, as well as information on connectivity (current patterns that support larval replenishment and recovery) among sites that vary in their sensitivities to climate change. Protection of seascapes creates areas sufficiently large to resist basic changes to entire ecosystems (Kaufman *et al.*, 2004). Large reserves may benefit individual species by enabling entire adult phases of life cycles to be completed without fishing mortality, with concomitant increases in reproductive output (Sobel and Dahlgren, 2004) and quality (Berkeley, Chapman, and Sogard, 2004).

A key issue for MPA managers concerns achieving the goals and objectives of a local-scale management plan in the context of larger-scale stressors from atmospheric, terrestrial, and marine sources (Jameson, Tupper, and Ridley, 2002). Another issue concerns maintaining a focus on immediate, deleterious effects of overexploitation, coastal pollution, and nonindigenous species as climate change impacts increase in magnitude or frequency over time (Paine, 1993). A conclusion of this report is that this focus is in fact an important element of adaptation to climate change. Within sites, managers can increase resilience to climate change by managing other anthropogenic stressors that also degrade ecosystems, such as overfishing and overexploitation; excessive inputs of nutrients, sediments, and pollutants; and habitat damage and destruction. Efforts by MPA managers to enhance resilience and resistance of marine communities may at least "buy some time" against threats of climate change by slowing the rate of decline caused by other, more manageable stressors (Hansen, Biringer, and Hoffman, 2003; Hoffman, 2003; Marshall and Schuttenberg, 2006).

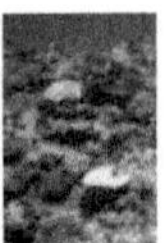

Resilience is also affected by trophic linkages, which are key characteristics maintaining ecosystem integrity. An approach that has been identified to maintain resilience is the management of functional groups, specifically herbivores. In some cases, the species that are necessary for recovery after a phase shift may be different from the species that had previously maintained the original state (*e.g.*, Bellwood, Hughes, and Hoey, 2006). This highlights the need to provide broad protection of species to maintain resilience and the need for further research on key species and ecological processes. However, abundant herbivores may not prevent shifts in algal-coral dominance in coral reef ecosystems (Ledlie *et al.*, 2007), and management for reduced levels of grazing may be necessary in plant-dominated systems such as kelp forests.

The challenges of climate change require creative solutions and collaboration among a variety of stakeholders to generate the necessary finances and support to respond to climate change stress. Global, regional, and local partnerships across a range of sectors such as agriculture, tourism, water resource management, conservation, and infrastructure development can help alleviate the financial burdens of responding to climate change in MPAs. Finally, effective implementation of the above strategies in support of ecological resilience will only be possible in the presence of human social resilience.

8.5.2 Research Priorities

The scientific knowledge required to reach general conclusions related to the impact of multiple stressors at community and ecosystem levels is for the most part absent for marine systems, and this gap impedes the ability of MPA managers to take management actions that have predictable outcomes. Existing levels of uncertainty will only increase as impacts of climate change strengthen. Within marine communities, temperature changes may result in new species assemblages and biological interactions that affect ecological processes such as productivity, nutrient fluxes, energy flow, and trophic webs. How such outcomes affect trophic links and other biological processes within communities is not clear, and is a high-priority area of research.

The extent of larval recruitment from local and longer-distance sources has been and must remain an active area of modeling and empirical investigations. Additional research will be required to better understand where and how far larvae travel in various marine ecosystems, to improve our understanding of where to implement MPAs and MPA networks.

The ability of corals to adapt or acclimatize to increasing seawater temperature is largely unknown (Berkelmans and van Oppen, 2006). Further, corals are sensitive to light and ultraviolet radiation, and thermal stress exacerbates this sensitivity (Hoegh-Guldberg *et al.*, 2007). The roles of temperature, light, holobiont characteristics and history, and other factors in coral bleaching are research topics of paramount importance.

Because of the greater solubility of CO_2 in cooler waters and at depth, reefs at the latitudinal margins of coral reef development (*e.g.*, Florida Keys and Hawaiian Islands) and deep-water coral formations may show the most rapid and dramatic response to changing pH. Further research is needed on impacts of high concentrations of CO_2 in the oceans, possible acclimatization or evolution of organisms in response to changes in ocean chemistry, and how management might respond (The Royal Society, 2005).

While there is no clear indication that ocean circulation patterns have changed recently (Bindoff *et al.*, 2007), future modifications could have large effects within and among ecosystems. Changing circulation would impact ecosystem and community connectivity in terms of nutrient fluxes, larval recruitment, spread of pollution, and other factors. Further modeling efforts may elucidate implications of potential changes in ocean circulation to MPA management.

Because pollution is usually more local in scope, it historically could be managed within individual MPAs; however, the addition of climate change stressors such as increased oceanic temperature, decreased pH, and greater fluctuations in salinity present greater challenges. Research in coral genomics may provide diagnostic tools for identifying stressors in coral reefs and other marine communities (*e.g.*, Edge *et al.*, 2005). MPA managers could benefit greatly from such tools, both in terms of distinguishing sources of stress and as potential "early-warning" signs for some factors such as thermal stress.

Research on marine ecosystems and climate change impacts continues to be a high-priority need, particularly in the context of using management actions as experiments in an adaptive-management framework. Although there is considerable research on physical impacts of climate change in marine systems (IPCC, 2007a), research on biological effects and ecological consequences is not as well developed.

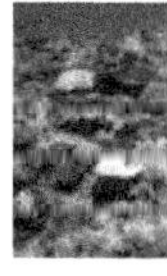

Synthesis and Conclusions

Lead Author: Peter Kareiva, The Nature Conservancy

Contributing Authors: Carolyn Enquist, The Nature Conservancy; Ayana Johnson, University of California, San Diego; Susan Herrod Julius, U.S. Environmental Protection Agency; Joshua Lawler, Oregon State University; Brian Petersen, University of California, Santa Cruz; Louis Pitelka, University of Maryland; Rebecca Shaw, The Nature Conservancy; Jordan M. West, U.S. Environmental Protection Agency

9.1 SUMMARY

The Nation's public lands and waters traditionally have been managed using frameworks and objectives that were established under an implicit assumption of stable climate and the potential of achieving specific desirable conditions. Climate change implies that past experience may not apply and that the assumption of a stable climate is in some regions untenable. Previous chapters in this report examine a selected group of management systems (National Forests, National Parks, National Wildlife Refuges, Wild and Scenic Rivers, National Estuaries, and Marine Protected Areas) and assess how these management systems can adapt to climate change. Using these chapters and their case studies, as well as more general scientific literature concerning adaptive management and climate change, this chapter presents a synthesis of suggested principles and management approaches for federal management agencies as well as other resource managers.

KEY FINDINGS

A useful starting point for adaptation is to analyze management goals, assess impacts, and characterize uncertainty. To inform adaptation decisions, the first step is to clarify the management goals that have been established for the system being studied. This information may then be used to define the boundaries of the impact assessment, including geographic scope, focal species, and other parameters. Within these boundaries, components of the assessment may then include developing conceptual models, assessing available ecological data and establishing current baseline information on system functioning, assessing available climate data, selecting impacts models, conducting scenario and sensitivity analyses that depict alternative futures, and characterizing uncertainty. Information from impact assessments helps determine whether existing monitoring programs need to be adjusted, or new ones established, to track changes in variables that represent triggers for threshold changes in ecosystems or that reflect overall resilience. Such monitoring programs can inform the location and timing of needed adaptation actions as well as the effectiveness of such actions once they are implemented. However, because of the high degree of uncertainty about the magnitude and temporal/spatial scale of climate change impacts, managers may find it difficult to translate results from impact assessments into practical management actions. The solution is not to view a scenario result as a "prediction" that supports planning for "most likely" outcomes. Rather, it is to select a range of future scenarios that capture the breadth of plausible outcomes and develop robust adaptation responses that address this full range.

A variety of adaptation approaches can be used to apply existing and new practices to promote resilience to climate change. Resilience may be defined as the amount of change or disturbance that an ecosystem can absorb without undergoing a fundamental shift to a different set of processes and structures. Many adaptation approaches suggested below are already being used to address a variety of other environmental stressors; however, their application may need to be adjusted to ensure their effectiveness for climate adaptation. These approaches include (1) protecting key ecosystem features that form the underpinnings of a system; (2) reducing anthropogenic stresses that erode resilience; (3) increasing representation of different genotypes, species, and communities under protection; (4) increasing the number of replicate units of each ecosystem type or population under protection; (5) restoring ecosystems that have been compromised or lost; (6) identifying and using areas that are "refuges" from climate change; and (7) relocating organisms to appropriate habitats as conditions change.

Reducing anthropogenic stresses is an approach for which there is considerable scientific confidence in its ability to promote resilience for virtually any situation. The effectiveness of the other approaches—including *protecting key ecosystem features, representation, replication, restoration, identifying refuges,* and especially *relocation*—is much more uncertain and will depend on a clear understanding of how the ecosystem in question functions, the extent and type of climate change that will occur, and the resulting ecosystem impacts. One method to implement adaptation approaches under such conditions of uncertainty is adaptive management. Adaptive management is a process that promotes flexible decision making, such that adjustments are made in decisions as outcomes from management actions and other events are better understood. This method requires careful monitoring of management results to advance scientific understanding and to help adjust policies or operations as part of an iterative learning process.

Barriers to implementing existing and new adaptation practices may be used as opportunities for strategic thinking. Providing information on adaptation approaches and specific strategies may not be enough to assist managers in addressing climate change impacts. Actual or perceived barriers may inhibit or prevent implementation of some types of adaptation. Identifying and understanding those barriers could facilitate critical adjustments to increase successful implementation and adaptive capacity of organizations. Four main types of barriers that affect implementation are (1) interpretation of legislative goals, (2) restrictive management procedures, (3) limitations on human and financial capital, and (4) gaps in information. Identifying a potential barrier, such as gaps in information or expertise necessary for implementing adaptation strategies, provides the basis for finding a solution, such as linking with other managers to coordinate training and research activities or sharing data and monitoring strategies to test scientific hypotheses. The challenge of turning barriers into opportunities may vary in the amount and degree of effort required, the levels of management necessary to engage, and the length of time needed. For example, re-evaluating management capabilities in light of existing authorities and legislation to expand their breadth may require more time, effort, and involvement of high level decision makers compared with altering the timing of management activities to take advantage of seasonal changes. Nevertheless, it should be possible to undertake strategic thinking and reshape priorities to convert barriers into opportunities to successfully implement adaptation.

Beyond the adaptation options reviewed in this report, key activities to ensure the Nation's capability to adapt include applying triage, determining appropriate scales of response, and reassessing management goals. Our capability to respond appropriately to climate change impacts will depend on (1) developing systematic approaches for triage (*i.e.*, a form of prioritizing adaptation actions), (2) determining the appropriate geographic and temporal scales of response to climate change, and (3) assessing whether current management goals will continue to be relevant in the future, or whether they need to be adjusted. Triage involves maximizing the effectiveness of existing resources by re-evaluating current goals and management targets in light of observed and projected ecological changes. The goal is to determine those management actions that are worthwhile to continue and those that may need to be abandoned.

To assess the appropriate scales of response, consideration of observed and projected ecological changes are again needed. In the event that impacts are broader than single management units or occur at predictable periods through time, the spatial, temporal, and biological scope of management plans may need to be systematically broadened and integrated to increase the capacity to adapt beyond that of any given unit.

Over time, some ecosystems may undergo state changes such that managing for resilience will no longer be feasible. In these cases, adapting to climate change would require more than simply changing management practices—it could require changing management goals. In other words, when climate change has such strong impacts that original management goals are untenable, the prudent course may be to alter the goals. At such a point, it will be necessary to manage for and embrace change. Climate change requires new patterns of thinking and greater agility in management planning and activities in order to respond to the inherent uncertainty of the challenge.

9.2 INTRODUCTION

Today's natural resource planning and management practices were developed under relatively stable climatic conditions in the last century, and under a theoretical notion that ecological systems tend toward a natural equilibrium state for which one could manage. Most natural resource planning, management, and monitoring methodologies that are in place today are still based on the assumption that climate, species distributions, and ecological processes will remain stable, save for the direct impacts of management actions and historical interannual variability. Indeed, many government entities identify a "reference condition" based on historical ranges of variability as a guide to future desired conditions (Dixon, 2003).

Although mainstream management practices typically follow these traditional assumptions, in recent years resource managers have recognized that climatic influences on ecosystems in the future will be increasingly complex and often outside the range of historical variability and, accordingly, more sophisticated management plans are needed to ensure that goals can continue to be met. By transforming management and goal-setting approaches from a static, equilibrium view of the natural world to a highly dynamic, uncertain, and variable framework, major advances in managing for change can be made, and thus adaptation is possible.

As resource managers become aware of climate change and the challenges it poses, a major limitation is lack of guidance on what steps to take, especially guidance that is commensurate with agency cultures and the practical experiences that managers have accumulated from years of dealing with other stresses such as droughts, fires, and pest and pathogen outbreaks. Thus, it is the intent in this chapter to synthesize the lessons learned from across the previous chapters together with recent theoretical work concerning adaptive management and resource management under uncertainty, and discuss how managers can (1) assess the impacts of climate change on their systems and goals (Section 9.3), (2) identify best practice approaches for adaptation (Section 9.4), and (3) evaluate barriers and opportunities associated with implementation (Section 9.5). When it comes to management, the institutional mandates and objectives determine the management constraints and in turn the response to changing climate. As a result, this discussion and synthesis are framed around the institutions that manage lands and waters, as opposed to the ecosystems

themselves. It may be the case that certain management goals are unattainable in the future and no adaptation options exist. In that case the adaption that takes place would be an alteration of institutional objectives. The final sections of this chapter address these circumstances and conclude with observations about how to advance our capability to adapt (Sections 9.6 and 9.7), including suggested approaches for making fundamental shifts in how ecosystems are managed to anticipate potential future ecosystem states. These discussions build on the other chapters of this report, and have benefited from helpful comments received during the public and and expert review periods.

9.3 ASSESSING IMPACTS TO SUPPORT ADAPTATION

9.3.1 Mental Models for Making Adaptation Decisions

Within the context of natural resource management, an impact assessment is a means of evaluating the sensitivity of a natural system to climate change. Sensitivity is defined by the IPCC (2001) as "the degree to which a system is affected, either adversely or beneficially, by climate-related stimuli." An impact assessment is part of a larger process to understand the risks posed by climate change, including those social and economic factors that may contribute to or ameliorate potential impacts, in order to decide where and when to adapt. In the climate change community, this process is well established (see Fig. 9.1a). It begins with an assessment of impacts, followed by an evaluation of an entity's capacity to respond (adaptive capacity). The information on impacts is then combined with information on adaptive capacity to determine a system's overall vulnerability. This information becomes the basis for selecting adaptation options to implement. The resource managers' mental model for this larger decision making process (see Fig. 9.1b) contains similar elements to the climate community's model, but addresses them in a different sequence of evaluation to planning. The managers' process begins with estimating potential impacts, reviewing all possible management options, evaluating the human capacity to respond, and finally deciding on specific management responses. The resource management community implicitly combines the information on potential impacts with knowledge of their capacity to respond during their planning processes. Since the primary audience for this report is the resource management community, the remainder of this discussion will follow their conceptual approach to decision making.

The following sub-sections lay out in greater detail some of the key issues and elements of an impact assessment, which must necessarily

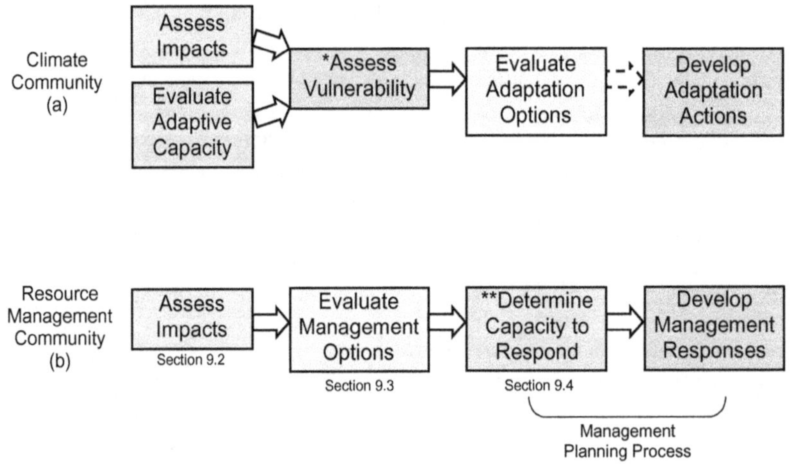

* Vulnerability is the sum of projected impacts and adaptive capacity; this step is done by managers when they evaluate the projected impacts and their capacity to respond during their planning process

** Assessing the capacity to respond in the management community is equivalent to assessing adaptive capacity in the climate community

Figure 9.1. Two conceptual models for describing different processes used by (a) the climate community and (b) the resource management community to support adaptation decision making. Colors are used to represent similar elements of the different processes.

begin with a clear articulation of the goals and objectives of the assessment and the decisions that will be informed. This specification largely determines the technical approach to be taken in an assessment, including its scope and scale, the focal ecosystem components and processes to be studied, the types of tools most appropriate to use, and the baseline data and monitoring needed. The final subsection discusses ways in which uncertainty inherent in assessments of climate change impacts may be explicitly addressed.

9.3.2 Elements of an Impact Assessment

Impact assessments combine (1) our understanding of the current state of the system and its processes and functions with (2) drivers of environmental change in order to (3) project potential responses to future changes in those drivers. Knowledge of the current state of the system, including its critical thresholds and coping ranges, provides the fundamental basis for understanding the implications of changes in future conditions. A coping range is the breadth of conditions under which a system continues to persist without significant, observable consequences, taking into account the system's natural resilience (Yohe and Tol, 2002). Change is not necessarily "bad," and the fact that a system responds by shifting to a new equilibrium or state may not necessarily be a negative outcome. Regardless of the change, it will behoove managers to adjust to or take advantage of the anticipated change. Several examples of approaches to conducting impact assessments are provided below along with a discussion of the types of tools needed and key issues related to conducting impact assessments.

9.3.2.1 A Guiding Framework for Impact Assessments

The aim of a framework to assess impacts is to provide a logical and consistent approach for eliciting the information needs of a decision maker, for conducting an assessment as efficiently as possible, and for producing credible and useful results. While impact assessments are routinely done to examine the ecological effects of various environmental stressors, the need to incorporate changes in climate variables adds significantly to the spatial and temporal scales of the assessment, and hence its complexity. One example framework, developed by Johnson and Weaver (2008) for natural resource managers, is responsive to these and other concerns that have been raised by those who work with climate data to conduct impact assessments. This framework is described in Box 9.1.

BOX 9.1. An example framework for incorporating climate change information into impact assessments.

Step 1 – Define decision context: Clarify management goals and endpoints of concern, as well as risk preferences and tradeoffs, time horizons for monitoring and management, and planning processes related to established endpoints.

Step 2 – Develop conceptual model: Develop a conceptual model linking the spatial and temporal scales of interaction between and among drivers and endpoints to determine the most important dependencies, sensitivities, and uncertainties in the system.

Step 3 – Assess available climate data: Determine whether available information about climate change is adequate for achieving the specified goals and endpoints. Data sources that may be used include historical weather observations, palaeoclimate data, and data from climate model experiments.

Step 4 – Downscale climate data: Where necessary, develop finer resolution datasets from coarser scale data, e.g., using statistical relationships ("statistical" downscaling) or computer models ("dynamical" downscaling), to drive impacts models. For guidance on downscaling techniques, see IPCC-TGICA (2007).

Step 5 – Select impact assessment models: Review and select impact assessment models that capture the processes and causal pathways represented in the conceptual model.

Step 6 – Conduct sensitivity analyses and scenario planning: Conduct analyses to evaluate the basic sensitivities in the system. Specify a number of climate scenarios that are consistent with associated global-scale scenarios, physically plausible, and sufficiently detailed to support an assessment of the specified endpoints of concern. Use these scenarios to learn the potential ranges of the system's response to changes in the climate drivers.

Step 7 – Manage risks through adaptation: Evaluate the information generated to determine potential management responses, recognizing that the consequences of decisions are generally not known and hence decisions are made to reduce the net negative effects of risk.

A number of other frameworks have been developed as well. For example, within the international conservation arena, a successful framework for managers was developed by The Nature Conservancy (Parrish, Braun, and Unnasch, 2003). The steps include (1) identifying the management goal, management targets, and threats (including climate change); (2) selecting measurable indicators; (3) determining the limits of acceptable variation in the indicators; and (4) assessing the current status of the system with respect to meeting management targets, as well as with respect to the indicators. An additional step would be to analyze data on indicators to decide whether a change in management is required. The steps were further refined by the Conservation Measures Partnership,[1] which includes the African Wildlife Foundation, Conservation International, The Nature Conservancy, the Wildlife Conservation Society, and the World Wide Fund for Nature/World Wildlife Fund. By melding these steps with an assessment of the costs of any management response (including "no response" as one option), it should be possible to offer practical guidance.

9.3.2.2 Tools to Assess Impacts

The example frameworks described in the previous section reference two key types of tools: models that represent the climate system as a driver of ecological change and models that embody the physical world to trace the effect of climate drivers through relevant pathways to impacts on management endpoints of concern. There are numerous tools that begin to help managers anticipate and manage for climate change (see Appendix), although characterization of uncertainty could be improved in these tools, along with "user friendliness" and the ability to frame management endpoints in a manner that more closely meshes with the needs of decision makers. Fortunately, tool development for impact analysis is one of the most active areas of climate research, and greatly improved tools can be expected within the next few years.

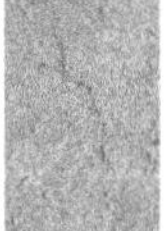

Climate Models

Across all types of federal lands, the most widely recognized need for information is the need for climate projections at useable scales—scales much finer than those associated with most general circulation model (GCM) projections (Chapter 6, Wild and Scenic Rivers). In particular, the resolution of current climate-change projections from GCMs is on the order of degrees of latitude and longitude (200–500 km^2). Projections from regional climate models are finer in resolution (*e.g.*, 10 km^2), but are not available for most regions. All climate projections can be downscaled using methods that take local topography and local climate patterns into account (Wilby *et al.*, 1998). Although relatively coarse climate projections may be useful for anticipating general trends, the effects of local topography, large water bodies, and specific ecological systems can make coarse predictions highly inaccurate. To be more useful to managers, projections will need to be downscaled using methods that account for local climate patterns. In addition, climate-change projections will need to be summarized in a way that takes their inherent uncertainty into account. That uncertainty arises from the basic model structure, the model parameters, and the path of global emissions into the future. Useful future projections will provide summaries that take this uncertainty into account and inform managers where the projections are more and less certain and, specifically, how confident we can be in a given level of change. Several different approaches exist for capturing the range of projected future climates (see comparison of approaches in Dettinger, 2005). It also will be important to work with climate modelers to ensure that they provide the biologically relevant output variables from the model results.

There are various methods of downscaling GCM data, including dynamical downscaling using regional climate models, statistical downscaling, and the change factor approach (a type of statistical downscaling). Dynamical downscaling uses physically based regional climate models that originate from numerical weather prediction and generate results at a scale of 50 km, although some generate results at 10km and finer scales (Georgi, Hewitson, and Christensen, 2001; Christensen *et al.*, 2007). As their name implies, they

[1] **Conservation Measures Partnership**, 2007: Open Standards for the Practice of Conservation, Version 2.0, http://conservationmeasures.org/CMP/Site_Docs/CMP_Open_Standards_Version_2.0.pdf, accessed on 4-11-2008.

are typically run for a region of the globe, using GCM outputs as boundary conditions. Statistical downscaling uses various methods to estimate a relationship between large-scale climate variables ("predictors") and finer-scale regional or local variables ("predictands"). This relationship is derived from an observed period of climate and then applied to the output from GCMs for future projections. This method is also used for temporal downscaling to project daily or hourly variables, typically for hydrologic analyses (Wilby *et al.*, 2004). Due to the complexity of determining a significant relationship between the "predictors" and "predictands," most studies that use statistical downscaling only use the results from one GCM (*e.g.*, Shongwe, Landman, and Mason, 2006; Spak *et al.*, 2007; Benestad, Hanssen-Bauer, and Fairland, 2007). The change factor approach to downscaling involves subtracting the modeled future climate from the control run at the native coarse resolution of the GCM. These modeled climate "anomalies" are then interpolated to create a seamless surface of modeled change at a finer resolution. These interpolated data are then added to the current climate to provide an estimate of future climate. Researchers use the change factor approach when a rapid assessment of multiple GCMs and emissions scenarios is required (*e.g.*, Mitchell *et al.*, 2004; Wilby *et al.*, 2004; Scholze *et al.*, 2006; Malcolm *et al.*, 2006).

It is becoming increasingly possible to examine multiple GCMs and look for more robust results. As this approach becomes widespread, the consequences of choosing one particular GCM will become less important. Moreover, all GCMs are undergoing refinement in models and parameter estimates. At this point, the key to applying any climate modeling technique is understanding the sensitivity of results to model selection before results are used to conduct impact assessments.

Impact Models to Assess Endpoints of Concern

Climate change impacts may be defined by two factors, (1) the types and magnitude of climate changes that are likely to affect the target in a given location, and (2) the sensitivity of a given conservation target to climate change. Assessing the types and magnitude of climate changes that a population or system is likely to experience will require climate-change projections as well as projected changes in climate-driven processes such as fire, hydrology, vegetation, and sea level rise (Chapter 4, National Parks; Chapter 5, National Wildlife Refuges). For example, managing forests in a changing climate will require data on projected potential changes to vegetation, as well as detailed data on the current condition of vegetation (Chapter 3, National Forests).

As another example, to support managing coastlines, a detailed sea level rise assessment was undertaken by the USGS for the lower 48 states, and specifically for coastal national parks.[2] More accurate projections of coastal inundation and saltwater intrusion, such as those based on LIDAR conducted for the Blackwater National Wildlife Refuge, will require more detailed elevation data and targeted hydrological modeling (Chapter 5, National Wildlife Refuges). One report that provides information on ongoing mapping efforts by federal and non-federal researchers related to the implications of sea level rise is Synthesis and Assessment Product 4.1 (in press), produced by the U.S. Climate Change Science Program. Various data layers are overlaid to develop new results, focusing on a contiguous portion of the U.S. coastal zone (New York to North Carolina).

Sensitivity of target organisms to climate change depends on several aspects of the biology of a species or the ecological composition and functioning of a system. For example, species that are physiologically sensitive to changes in temperature or moisture; species that occupy climate-sensitive habitats such as shallow wetlands, perennial streams, and alpine areas; and species with limited dispersal abilities will all be more sensitive to climate change (Root and Schneider, 2002). Populations with slow growth rates and populations at a species range boundary are also likely to be more sensitive to climate change (Pianka, 1970; Lovejoy and Hannah, 2005). Species, communities,

2 **U.S. Geological Survey**, 2007: Coastal vulnerability assessment of National Park units to sea-level rise. U.S. Geological Survey Website, http://woodshole.er.usgs.gov/project-pages/nps-cvi/, accessed on 6-11-2007.

or ecosystems that are highly dependant on specific climate-driven processes—such as fire regimes, sea level rise, and hydrology—will also be highly sensitive to climate change.

Projected shifts in individual species distributions are generally based on relatively coarse-scale data (*e.g.*, Pearson *et al.*, 2002; Thuiller *et al.*, 2005). Regional projections of species range shifts will require more detailed species distribution data. Some of these data already exist (*e.g.*, through the state Natural Heritage programs), but they need to be organized, catalogued and standardized. Even when built with finer-scale data, these species-distribution models have their limitations (Botkin *et al.*, 2007). They should not be seen as providing accurate projections of the future ranges of individual species, but instead should be viewed as assessments of the likely responses of plants and animals in general. They can be useful for identifying areas that are likely to experience more or less change in flora or fauna in a changing climate. In addition, as with the climate projections, all projections of climate-change impacts will need to include estimates of the inherent uncertainty and variability associated with the particular model that is used (*e.g.*, Araújo and New, 2007). For example, recent analyses of range shift models indicate that some models perform better than others. A model-averaging approach (*e.g.*, random forest models) was compared with five other modeling approaches and was found to have the greatest potential for accurately predicting range shifts in response to climate change (Lawler *et al.*, 2006).

An important consideration for impact analyses is to provide information on endpoints that are relevant to managers (*e.g.*, loss of valued species such as salmon) rather than those that might come naturally to ecologists (*e.g.*, changes in species composition or species richness). An exemplary impact analysis in this regard was a study of climate change impacts in California funded by the Union of Concerned Scientists.[3] The UCS study used a statistically downscaled version of two GCMs to consider future emissions conditions for the state. It produced compelling climate-related outputs. Projections of impacts, in the absence of aggressive emissions regulations, included heat waves that could cause two to three times more heat-related deaths by mid century than occur today in urban centers such as Los Angeles, a shorter ski season, declines in milk production by up to 20 percent by the end of the century for the dairy industry, and bad-tasting wine from the Napa Valley. Because the impacts chosen were relevant to management concerns, the study was covered extensively by national and California newspapers, radio stations, and TV stations (Tallis and Kareiva, 2006).

There are many new ecological models that would help managers address climate change, but the most important modeling tools will be those that integrate diverse information for decision making and prioritize areas for different management activities. Planners and managers need the capability to evaluate the vulnerability of each site to climate change and the social and economic costs of addressing those vulnerabilities. One could provide this help with models that allow the exploration of alternative future climate-change scenarios and different funding limitations that could be used for priority-setting and triage decisions. Comprehensive, dynamic, priority-setting tools have been developed for other management activities, such as watershed restoration (Lamy *et al.*, 2002). Developing a dynamic tool for priority-setting will be critical for effectively allocating limited resources.

9.3.2.3 Establishing Baseline Information

Collecting Information on Past and Current Condition

To estimate current and potential future impacts, a literature review of expected climate impacts may be conducted to provide a screening process that identifies "what trends to worry about." The next step beyond a literature review is a more focused elicitation of the ecological properties or components needed to reach management goals for lands and waters. For each of these properties or components, it will be important to determine the key to maintaining them (see Table 9.1 for examples). If the literature review reveals that any of the general climate trends may influence the ecological attributes or processes critical to meeting management goals, then the

[3] **Union of Concerned Scientists**, 2006: Union of Concerned Scientists homepage. Website, http://www.ucsusa.org/assets/documents/global_warming/Our-Changing-Climate-final.pdf, accessed on 6-11-2007.

Table 9.1. Examples of potential climate change-related effects on key ecosystem attributes upon which management goals depend.

Federal Lands	Ecosystem Attributes Critical to Management Goals	Potential Climate-Related Changes That Could Influence Management Goals
National forests	• Fire tolerance • Insect tolerance • Tolerance to invasives	• Altered fire regimes • Vegetation changes • Changes in species dominance
National wildlife refuges	• Persistence of threatened and endangered species • Wetland water replenishment • Coastal wetland habitat	• Threatened and endangered species decline or loss • Altered hydrology • Sea level rise
Marine protected areas	• Structural "foundation" species (e.g., corals, kelp) • Biodiversity • Water quality	• Increased ocean temperatures and decreased pH • Increased bleaching and disease • Altered precipitation and runoff
National estuaries	• Sediment filtration • Elevation and slope • Community composition	• Altered stream flow • Sea level rise • Salt water intrusion/species shifts
Wild and scenic rivers	• Anadromous fish habitat • Water quality • "Natural" flow	• Increased water temperatures • Changes in runoff • Altered stream flow
National parks	• Fire tolerance • Snow pack • Community composition	• Vegetation shifts • Changes in snow pack amount • Temperature-related species shifts

next steps are to identify baselines, establish monitoring programs, and consider specific management tools and models. For example, suppose the management goal is to maintain a particular vegetation type, such as classical Mediterranean vegetation. Mediterranean vegetation is restricted to the following five conditions (Aschmann, 1973):

- at least 65% of the annual precipitation occurs in the winter half of the year (November–April in the northern hemisphere and May–September in the southern hemisphere);
- annual precipitation is greater than 275 mm;
- annual precipitation is less than 900 mm;
- the coldest month of the year is below 15°C; and
- the annual hours below 0°C account for less than 3% of the total.

If the general literature review indicates that climate trends have a reasonable likelihood of influencing any of these defining features of Mediterranean plant communities, there will be a need for deeper analysis. Sensitivity to current or past climate variability may be a good indicator of potential future sensitivity. In the event that these analyses indicate that it will be very unlikely that the region will be able to sustain Mediterranean plant communities in the future, it may be necessary to cease management at particular sites and to consider protecting or managing other areas where these communities could persist. Triage decisions like this will be very difficult, and should be based not only on future predictions but also on the outcome of targeted monitoring.

Once the important ecological attributes or processes are identified, a manager needs to have a clear idea of the baseline set of conditions for the system. Ecologists, especially marine ecologists, have drawn attention to the fact that the world has changed so much that it can be hard to determine an accurate historical baseline for any system (Pauly, 1995). The reason that an understanding of a system's long history can be so valuable is that the historical record may include information about how systems respond to extreme stresses and perturbations. When dealing with sensitive, endangered, or stressed systems, experimental perturbation is not feasible. Where available, paleoecological records should be used to examine past ranges of natural environmental variability and past organismal responses to climate change (Willis and Birks, 2006). Although in an experimental sense "uncontrolled," there is no lack of both

historic and recent examples of perturbations (of various magnitudes) and recoveries through which to examine resilience.

Historic baselines have the potential to offer insights into how to manage for climate change. For example, while the authority to acquire land interests and water rights exists under the Wild and Scenic Rivers Act, lack of baseline data on flow regimes makes it difficult to determine how, when, and where to use this authority (Chapter 6, Wild and Scenic Rivers). Other examples of baseline data important for making management decisions and understanding potential effects of climate change include species composition and distribution of trees in forests; rates of freshwater discharge into estuaries; river flooding regimes; forest fire regimes; magnitude and timing of anadromous fish runs; and home ranges, migration patterns, and reproductive dynamics of sensitive organisms.

However, baselines also have the potential to be misleading. For example, in Chapter 3 (National Forests), it is noted that historic baselines are useful only if climate is incorporated into those past baselines and the relationship of vegetation to climate is explored. If a baseline is held up as a goal, and the baseline depends on historic climates that will never again be seen in a region, then the baseline could be misleading. Adjusting baselines to accommodate changing conditions is an approach that would require caution to avoid unnecessarily compromising ecosystem integrity for the future and losing valuable historical knowledge.

Monitoring to Inform Management Decisions

Monitoring is needed to support a manager's ability to detect changes in baseline conditions as well as to facilitate timely adaptation actions. Monitoring also provides a means to gauge whether management actions are effective. Some monitoring may be designed to detect general ecological trends in poorly understood systems. However, most monitoring programs should be designed with specific hypotheses in mind and trigger points that will initiate a policy or management re-evaluation (Gregory, Ohlson, and Arvai, 2006). For instance, using a combination of baseline and historical data, a monitoring program could be set up with pre-defined thresholds for a species' abundance or growth rate, or a river's flow rate, which, once exceeded, would cause a re-examination of management approaches and management objectives.

A second important feature of any monitoring program is the decision of what to monitor. Ideally several attributes should be monitored, and those that are selected should be chosen to represent the system in a tractable way and to give clear information about possible management options (Gregory and Failing, 2002). Otherwise there is a risk of collecting volumes of data but not really using it to alter management. Sometimes managers seek one aggregate indicator—the risk in this is that the indicator is harder to interpret because so many different processes could alter it.

Some systems will require site-specific monitoring programs, whereas others will be able to take advantage of more general monitoring programs (see Table 9.2 for examples of potential monitoring targets). For example, the analysis of National Forests (Chapter 3, National Forests) highlights the need for monitoring both native plant species and non-native and invasive species. In addition, the severity and frequency of forest fires are clearly linked to climate (Bessie and Johnson, 1995; Fried, Torn, and Mills, 2004; Westerling *et al.*, 2006). Thus, managing for changing fire regimes will require assessing fire risk by detecting changes in fuel loads and weather patterns. Detecting climate-driven changes in insect outbreaks and disease prevalence will require monitoring the occurrence and prevalence of key insects, pathogens, and disease vectors (Logan, Regniere, and Powell, 2003). Detecting early changes in forests will also require monitoring changes in hydrology and phenology, and in tree establishment, growth, and mortality. Some key monitoring efforts are already in place. For example, the Forest Service conducts an extensive inventory through its Forest Inventory and Analysis program, and the collaborative National Phenology Network collects data on the timing of ecological events across the country to inform climate change research.[4]

[4] **University of Wisconsin-Milwaukee**, 2007: National phenological network. University of Wisconsin-Milwaukee Website, http://www.uwm.edu/Dept/Geography/npn/, accessed on 6-11-2007.

Table 9.2. Examples of hypothesis-driven monitoring for adaptive management in a changing climate.

Chapter	Monitoring Target	Hypothesis (Why Monitored)	Management Implications (How Used)
Forests (Chapter 3)	Invasive species	Climate change will alter species distributions, creating new invasive species (Lovejoy and Hannah, 2005).	• Inform proactive actions to remove and block invasions
Parks (Chapter 4)/ National Wildlife Refuges (Chapter 5)	Species composition	Species are shifting ranges in response to climate change (Parmesan, 1996).	• Manage for species lost from one park or refuge at a different site • Inform translocation efforts
Wild and Scenic Rivers (Chapter 6)	River flow	Increased temperatures will decrease snow pack and increase evaporation, changing the timing and amount of flows (Poff, Brinson, and Day, Jr., 2002).	• Manage flows • Increase connectivity
National Estuaries (Chapter 7)	Ecosystem functioning and species composition	As sea level rises, marshes will be lost and uplands will be converted to marshes (Moore *et al.*, 2003).	• Facilitate upland conversion, species translocation
Marine Protected Areas (Chapter 8)	Water quality	Changes in temperature and runoff will affect acidity, oxygen levels, turbidity, and pollutant concentrations (Behrenfeld *et al.*, 2006; Guinotte *et al.*, 2006; Portner and Knust, 2007).	• Address pollution sources • Inform coastal watershed policies

In the National Wildlife Refuge System, monitoring might include targets associated with sea level rise, hydrology, and the dynamics of sensitive species populations. Monitoring of marine protected areas should address coral bleaching and disease, as well as the composition of plankton, seagrass, and microbial communities. In the national estuaries, the most effective monitoring will be of salinity, sea level, stream flow, sediment loads, disease prevalence, and invasive species. Wild and scenic rivers should be monitored for changes in flow regimes and shifts in species composition. Finally, national parks, which encompass a diversity of ecosystem types, should be monitored for any number of the biotic and abiotic factors listed for the other federal lands.

Although developing directed, intensive monitoring programs may seem daunting, there are several opportunities to build on existing and developing efforts. In addition to the Forest Service's Forest Inventory and Analysis program and the National Phenology Network mentioned above, other opportunities include the National Science Foundation's National Ecological Observation Network and the Park Service's Vital Signs program (*e.g.*, Mau-Crimmins *et al.*, 2005). Some federal lands have detailed species inventories (*e.g.*, the national parks are developing extensive species inventories for the Natural Resource Challenge) or detailed stream flow measurements. Despite the importance of monitoring, it is critical to recognize that monitoring is only one step in the management process and that monitoring alone will not address the affects of climate change on federal lands.

9.3.3 Uncertainty and How to Incorporate It into Assessments

The high degree of uncertainty inherent in assessments of climate change impacts can make it difficult for a manager to translate results from those assessments into practical management action. However, uncertainty is not the same thing as ignorance or lack of information—it simply means that there is more than one outcome possible as a result of climate change. Fortunately, there are approaches for dealing with uncertainty that allow progress.

9.3.3.1 Examples of Sources of Uncertainty

To project future climate change, climate modelers have applied seven "families" of greenhouse gas emissions scenarios that encompass a range of energy futures to a suite of 23 GCMs (IPCC, 2007), all differing in their climatic projections. Based on a doubling of CO_2, global mean temperatures are projected to increase from 1.4–5.8°C (2.5–10.5°F) with considerable discrepancies in the distribution of the temperature and precipitation change. These direct outputs are typically not very useful to managers because they lack the resolution at local and regional scales where environmental impacts relevant for natural resource management can be evaluated. However, as mentioned above, GCM model outputs derived at the very coarse grid scales of 2.5° x 3.25° (roughly 200–500 km^2, depending on latitude) can be downscaled (Melillo *et al.*, 1995; Pan *et al.*, 2001; Leung *et al.*, 2003; Salathé, Jr., 2003; Wood *et al.*, 2004; IPCC, 2007). But when GCM output data are downscaled, uncertainties are amplified. In Region 6 of the Forest Service, the regional office recommended that the National Forest not model climatic change as a part of a management plan revision process after science reviewers acknowledged the high degree of uncertainty associated with the application of climate change models at the forest level (Chapter 3, National Forests). In the Northwest, management of rivers in the face of climate change is complicated by the fact that the uncertainty is so great that 67% of the modeled futures predict a decrease in runoff, while 33% predict an increase. Thus the uncertainty can be about the direction of change as well as the magnitude of change (Chapter 6, Wild and Scenic Rivers).

Changes in temperature, precipitation, and CO_2 will drive changes in species interactions, species distributions and ranges, community assemblages, ecological processes, and, therefore, ecosystem services. To understand the implications of these changes on species and/or vegetation distribution, models have been designed to assess the responses of biomes to climate change—but this of course introduces more uncertainty, and therefore management risk, into the final analysis. For terrestrial research, dynamic global vegetation models (DGVM) and Species Distributions Models (SDM) have been developed to help predict biological and species impacts. These models have weaknesses that make managers reluctant to use them. For example DGCM vegetation models, which should be useful to forest managers, are limited by the fact that they do not simulate actual vegetation (only potential natural vegetation), or the full suite of species migration patterns and dispersal capabilities, or the integration of the impacts of other global changes such as land use change (fragmentation and human barriers to dispersal) and invasive species (Field, 1999). Where vegetation cover is more natural and the impacts of other global changes are not prominent, the model simulations are likely to have a higher probability of providing useful information of future change. For regions where there is low percentage of natural cover, where fragmentation is great, and large areas are under some form of management, the models will provide limited insight into future vegetation distribution. It is unclear how climate change will interact with these other global and local changes, as well as unanticipated evolutionary changes and tolerance responses, and the models do not address this.

9.3.3.2 Using Scenarios as a Means of Managing Under Uncertainty

It is not possible to *predict* the changes that will occur, but managers can get an indication of the *range* of changes possible. By working with a range of possible changes rather than a single projection, managers can focus on developing the most appropriate responses based on that range rather than on a "most likely" outcome. To develop a set of scenarios—*e.g.*, internally consistent views of reasonably plausible futures in which decisions may be explored (adapted from Porter, 1985; Schwartz, 1996)—quantitative or qualitative visions of the future are developed or described. These scenarios explore current assumptions and serve to expand viewpoints of the future. In the climate change impacts area, approaches for developing scenarios may range from using a number of different realizations from climate models representing a range of emissions growths, to analog scenarios, to informal synthetic scenario exercises that, for example, perturbate temperature and precipitation changes by percentage increments (*e.g.*, -5% change from baseline conditions, 0, +5%, +10%).

Model-based scenarios explore plausible future conditions through direct representations of complex patterns of change. These scenarios have the advantage of helping to further our understanding of potential system responses to a range of changes in drivers. When using spatially downscaled climate models and a large number of emissions scenarios and climate model combinations (as many as 30 or more), a subset of "highly likely" climate expectations may be identifiable for a subset of regions and ecosystems. More typically, results among models will disagree for many places, precluding any unambiguous conclusions. Where there is a high level of agreement, statements may be made such as, "for 80% of the different model runs, peak daily summer temperatures are expected to rise by at least x degrees." When downscaled and multiple runs are available (see the Appendix for possible sources), managers can use them to explore the consequences of different management options. For instance, Battin *et al.* (2007) were able to identify specific places where habitat restoration was likely to be effective in the face of climate change if the goal was recovery of salmon populations, and in specific places where restoration efforts would be fruitless given anticipated climate change.

Analog scenarios use historical data and previously observed sensitivity to weather and climate variability. When developing analog scenarios, if historical data are incomplete or non-existent for one location, observations from a different region may used. Synthetic scenarios specify changes in particular variables and apply those changes to an observed time series. For example, an historic time series of annual mean precipitation for the northeastern United States would be increased by 2% to create a synthetic scenario, but no other characteristics of precipitation would change. Developing a synthetic scenario might start by simply stating that in the future, it is possible that summers will be hotter and drier. That scenario would be used to alter the sets of historic time series, and decision makers would explore how management might respond.

Along with developing multiple scenarios using the methods described above, it may be helpful to do sensitivity analyses to discover a system's response to a range of possible changes in drivers. In such analyses, the key attributes of the system are examined to see how they respond to systematic changes in the climate drivers. This approach may allow managers to identify thresholds beyond which key management goals become unattainable.

All of these scenario-building approaches and sensitivity analyses provide the foundation for "if/then" planning, or scenario planning. One of the most practical ways of dealing with uncertainty is scenario planning—that is, making plans for more than one potential future. If one were planning an outdoor event (picnic, wedding, family reunion), it is likely that an alternate plan would be prepared in case of rain. Scenario planning has become a scientific version of this common sense approach. It is appropriate and prudent when there are large uncertainties that cannot be reduced in the near future, as is the case with climate change. The key to scenario planning is limiting the scenarios to a set of possibilities, typically anywhere from two to five. If sensitivity analyses are performed, those results can be used to select the most relevant scenarios that both address managers' needs and represent the widest possible, but still plausible, futures. The strategy is to then design a variety of management strategies that are robust across the whole range of scenarios and associated impacts. Ideally scenarios represent clusters of future projections that fit together as one bundled storyline that is easy to communicate to managers (*e.g.*, warmer and wetter, warmer and drier, negligible change). When used deftly, scenario planning can alleviate decision-makers' and managers' frustration at facing so much uncertainty and allow them to proactively manage risks. For detailed guidance on using scenario data for climate impact assessments, see IPCC-TGICA (2007).

9.4 BEST PRACTICES FOR ADAPTATION

Another element essential to the process of adaptation decision making is to know the possible management options (*e.g.*, adaptation options) available to address the breadth of projected impacts, and how those options may function to lessen the impacts. As defined in this report, the goal of adaptation is to reduce the risk of adverse environmental outcomes through activities that *increase the resilience* of ecological systems to climate change (Scheffer

et al., 2001; Turner, II *et al.*, 2003; Tompkins and Adger, 2004). Here, resilience refers to the amount of change or disturbance that a system can absorb *before it undergoes a fundamental shift* to a different set of processes and structures (Holling, 1973; Gunderson, 2000; Bennett, Cumming, and Peterson, 2005). Therefore, all of the adaptation approaches reviewed below involve strategies for supporting the ability of ecosystems to persist at local or regional scales.

The suites of characteristics that distinguish different ecosystems and regions determine the potential for successful adaptation to support resilience. This section begins with a description of resilience theory, including examples of some types of biological and physical factors that may confer resilience to climate change. This is followed by a review of seven major adaptation approaches gleaned from across the chapters of this report, a discussion of the confidence levels associated with these approaches, and an examination of adaptive management as an effective means of implementing adaptation strategies.

9.4.1 Resilience

Management of ecosystems for any objective will be made easier if the systems are resilient to change—whether it is climate change or any other disturbance. Resilience is the ability of a system to return to its initial state and function in spite of some major perturbation. For example, a highly resilient coral reef might bleach but would be able to recover rapidly. Similarly, a resilient forest ecosystem would quickly re-establish plant cover following a major forest fire, with negligible loss of soils or fertility. An important contributing factor to overall resilience is *resistance*, which is the ability of an organism or a system to remain unimpacted by major disturbance or stress. "Unimpacted," in this sense, means that the species or system can continue to provide the desired ecosystem services. Resistance is derived from intrinsic biological characteristics at the level of species or genetic varieties. Resistance contributes to resilience since ecosystems that contain resistant individuals or communities will exhibit faster overall recovery (through recruitment and regrowth) after a disturbance. It is certainly possible that if systems are not resilient, the change that results could produce some benefits. However, from the perspective of a resource manager responsible for managing the ecosystems in question, a lack of resilience would mean that it would be difficult to establish clear objectives for that system and a consistent plan for achieving those objectives.

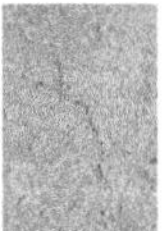

The science and theory of resilience may soon be sufficiently advanced to be able to confidently predict what confers resilience upon a system; the scientific literature is rapidly developing in this area and provides plausible hypotheses and likely resilience factors. Perhaps more importantly, common sense indicates that healthier ecosystems will generally be more resilient to disturbances. Activities that promote overall ecosystem health, whether they are restorative (*e.g.*, planting trees, captive breeding, and reintroduction) or protective (*e.g.*, restrictive of destructive uses) will tend to build resilience.

On the broadest level, working from the assumption that more intact and pristine ecosystems are more resilient to disturbances such as climate change, there are a number of ways to manage for resilience. The appropriate approach depends largely on the current state of the area being protected and the available resources with which to execute that protection. Options include (1) protecting intact systems (*e.g.*, Papahānaumokuākea Marine National Monument), (2) restoring systems to more pristine states (*e.g.*, restoring marshes and wetlands), and (3) preventing further degradation (*e.g.*, control of invasive species).

Beyond simply managing for pristine systems, which can be hard to identify, a quantifiable objective is to manage for biodiversity and key structural components or features. An important challenge associated with resilience is what might be called a "timescale mismatch." Resilience can be destroyed quickly, but often is "derived from things that can be restored only slowly, such as reservoirs of soil nutrients, heterogeneity of ecosystems on a landscape, or a variety of genotypes and species" (Folke *et al.*, 2002). This implies that while taking the necessary steps to prevent extinctions, management should worry most about species that have long generation times and low reproductive potential.

Our understanding of specific resilience factors for particular systems is sparse, making managing for resilience currently more an art than a science. Fortunately, two general concepts provide a simple framework for thinking about and managing for resilience. One is to ensure that ecosystems have all the components they need in order to recover from disturbances. This may be termed the biodiversity concept. The other is to support the species composing the structural foundation of the ecosystem, such as corals or large trees as habitat. This may be termed the structural concept. Although resource managers may not explicitly use these terms, examples of both concepts may be found in their decision-making.

Biodiversity Concept

Much academic research on managing for resilience invokes the precautionary principle. In this context, the precautionary principle calls for ensuring that ecosystems have all the biotic building blocks (functional groups, species, genes) that they need for recovery. These building blocks can also be thought of as *ecological memory*: the "network of species, their dynamic interactions among each other and with the environment, and the combination of structures that make reorganization after disturbance possible" (Bengtsson *et al.*, 2003).

A recent meta-analysis of ocean ecosystem services provides support for the biodiversity approach with its conclusion that in general, rates of resource collapse increased—and recovery rates decreased—exponentially with declining diversity. In contrast, with restoration of biodiversity, productivity increased fourfold and variability decreased by 21% on average (Worm *et al.*, 2006). Several other studies have concluded that diversity at numerous levels—*i.e.*, of functional groups, of species in functional groups, and within species and populations—appears to be critical for resilience and for the provision of ecosystem services (Chapin *et al.*, 1997; Luck, Daily, and Ehrlich, 2003; Folke *et al.*, 2004). National parks, national wildlife refuges, and marine protected areas all manage for maintaining as many native species as possible, and in so doing promote diversity as a resilience factor. The call for ecosystem-based management in the chapter on national estuaries represents a move toward a multi-species focus that could also enhance resilience. Although the detailed dynamics of the connection between biodiversity and resilience are not yet understood, evidence previously cited indicates that it is both practical and sensible as a precautionary act to protect biodiversity as a means of promoting resilience.

Biodiversity exists at multiple levels: genetic, species, function, and ecosystem. Table 9.3 briefly provides definitions and examples of management options for each of these four levels of biodiversity. It is worth noting that national parks, national wildlife refuges, and marine protected areas are all aimed at

Table 9.3. Levels of biodiversity and associated management options.

Levels of Biodiversity	Definition	Management Activities That Support Diversity
Genetic diversity	Allelic diversity and the presence/absence of rare alleles (foundation for all higher level diversity)	• Gene banks • Transplantation: re-introduction of lost genes (*e.g.*, transplanting and/or releasing hatchery-reared larvae/juveniles) • Protected areas and corridors
Species diversity	Quantity of species in a given area	• *Ex situ* conservation measures such as captive breeding programs • ESA listings • Protected areas
Functional diversity	Full representation of species within functional groups.	• Special protections for imperiled species within functional groups (*e.g.*, herbivorous fishes) • Protected areas
Ecosystem/landscape diversity	All important habitats represented as well as appropriately large scale of metapopulations	• Large protected areas • Networks of protected areas

supporting diversity to the extent that any "reserve" or "protected area" is. Wild and scenic rivers, national estuaries, and national forests have not traditionally had diversity as a core management goal. It is noteworthy, however, that the 2004–2008 USDA Forest Service Strategic plan does describe the Forest Service mission in terms of sustaining "diversity" (Chapter 3, National Forests).

Structural Concept

Organisms that provide ecosystem structure include trees in forests, corals on coral reefs, kelp in kelp forests, and grasses on prairies. These structure-providing groups represent the successional climax of their respective ecosystems—a climax that often takes a long time to reach. Logically, managers are concerned with loss of these species (whether due to disease, overharvesting, pollution, or natural disturbances) because of consequent cascading effects.

One approach to managing for resilience is to evaluate options in terms of what they mean for the recovery rate of fundamental structural aspects of an ecosystem. For example, the fishing technique of bottom trawling and the forestry technique of clear-cutting destroy biological structure, thus hindering recovery because the ecosystem is so degraded that either succession has to start from a more barren state or the community may even shift into an entirely new stable state. Thus, management plans should protect these structural species whose life histories dictate that if they are damaged, recovery time will increase.

It is important to note that while structural species are often representative of the ecosystem state most desirable to humans in terms of production of ecosystem services, they are still only representative of one of several states that are natural for that system. The expectation that these structural organisms will always dominate is unreasonable. In temperate forests, stand-replacing fires can be critical to resetting ecosystem dynamics; in kelp forests, kelp is periodically decimated by storms. Thus maintaining structural species does not mean management for permanence—it simply means managing for processes that will keep structural species in the system, albeit perhaps in a shifting mosaic of dominant trees in a forest, for example.

9.4.2 Adaptation Approaches

Managers' past experiences with unpredictable and extreme events such as hurricanes, floods, pest and disease outbreaks, invasions, and forest fires have already led to some existing approaches that can be used to adapt to climate change. Ecological studies combined with managers' expertise reveal several common themes for managing natural systems for resilience in the face of disturbance. A clear exposition of these themes is the starting point for developing best practices aimed at climate adaptation.

The seven approaches discussed below—(1) protection of key ecosystem features, (2) reduction of anthropogenic stresses, (3) representation, (4) replication, (5) restoration, (6) refugia, and (7) relocation—involve techniques that manipulate or take advantage of ecosystem properties to enhance their resilience to climatic changes. All of these adaptation approaches ultimately contribute to resilience as defined above, whether at the scale of individual protected area units, or at the scale of regional/national systems. While different chapters vary in their perspectives and terminologies regarding adaptation, the seven categories presented are inclusive of the range of adaptation options found throughout this report.

9.4.2.1 Protect Key Ecosystem Features

Within ecosystems, there may be particular structural characteristics (*e.g.*, three-dimensional complexity, growth patterns), organisms (*e.g.*, functional groups, native species), or areas (*e.g.*, buffer zones, migration corridors) that are particularly important for promoting the resilience of the overall system. Such key ecosystem features could be important focal points for special management protections or actions. For example, managers of national forests may proactively promote stand resilience to diseases and fires by using silviculture techniques such as widely spaced thinnings or shelterwood cuttings (Chapter 3, National Forests). Another example would be to aggressively prevent or reverse the establishment of invasive non-native species that threaten native species or impede current ecosystem function (Chapter 4, National Parks). Preserving the structural complexity of vegetation in tidal marshes, seagrass meadows, and mangroves may render estuaries more resilient (Chapter 7, National Estuaries). Finally, establishing and protecting corridors of connectivity that enable migrations can enhance resilience across landscapes in national wildlife refuges (Chapter 5, National Wildlife Refuges). Box 9.2 draws additional examples of this adaptation approach from across the chapters of this report.

BOX 9.2. Examples of adaptation actions that focus on protection of key ecosystem features as a means of supporting resilience.

Adaptation Approach: Protect Key Ecosystem Features

National Forests

- Facilitate natural (evolutionary) adaptation through management practices (e.g., prescribed fire and other silvicultural treatments) that shorten regeneration times and promote interspecific competition.
- Promote connected landscapes to facilitate species movements and gene flow, sustain key ecosystem processes (e.g., pollination and dispersal), and protect critical habitats for threatened and endangered species.

National Parks

- Remove barriers to upstream migration in rivers and streams.
- Reduce fragmentation and maintain or restore species migration corridors to facilitate natural flow of genes, species and populations.
- Use wildland fire, mechanical thinning, or prescribed burns where it is documented to reduce risk of anomalously severe fires.
- Minimize alteration of natural disturbance regimes, for example through protection of natural flow regimes in rivers or removal of infrastructure that prohibits the allowance of wildland fire.
- Aggressively prevent establishment of invasive non-native species or diseases where they are documented to threaten native species or current ecosystem function.

National Wildlife Refuges

- Manage risk of catastrophic fires through prescribed burns.
- Reduce or eliminate stressors on conservation target species.
- Improve the matrix surrounding the refuge by partnering with adjacent owners to improve/build new habitats.
- Install levees and other engineering works to alter water flows to benefit refuge species.
- Remove dispersal barriers and establish dispersal bridges for species.
- Use conservation easements around the refuge to allow species dispersal and maintain ecosystem function.
- Facilitate migration through the establishment and maintenance of wildlife corridors.

Wild & Scenic Rivers

- Maintain the natural flow regime through managing dam flow releases upstream of the wild and scenic river (through option agreements with willing partners) to protect flora and fauna in drier downstream river reaches, or to prevent losses from extreme flooding.
- Use drought-tolerant plant varieties to help protect riparian buffers.
- Create wetlands or off-channel storage basins to reduce erosion during high flow periods.
- Actively remove invasive species that threaten key native species.

National Estuaries

- Help protect tidal marshes from erosion with oyster breakwaters and rock sills and thus preserve their water filtration and fisheries enhancement functions.
- Preserve and restore the structural complexity and biodiversity of vegetation in tidal marshes, seagrass meadows, and mangroves.
- Adjust protections of important biogeochemical zones and critical habitats as the locations of these areas change with climate.
- Connect landscapes with corridors to enable migrations to sustain wildlife biodiversity across the landscape.
- Develop practical approaches to apply the principle of rolling easements to prevent engineered barriers from blocking landward retreat of coastal marshes and other shoreline habitats as sea level rises.

Marine Protected Areas

- Identify ecological connections among ecosystems and use them to inform the design of MPAs and management decisions such as protecting resistant areas to ensure sources of recruitment for recovery of populations in damaged areas.
- Manage functional species groups necessary to maintaining the health of reefs and other ecosystems.
- Design MPAs with dynamic boundaries and buffers to protect breeding and foraging habits of highly migratory and pelagic species.
- Monitor ecosystems and have rapid-response strategies prepared to assess ecological effects of extreme events as they occur.
- Identify and protect ecologically significant ("critical") areas such as nursery grounds, spawning grounds, and areas of high species diversity.

9.4.2.2 Reduce Anthropogenic Stresses

Managing for resilience often implies minimizing anthropogenic stressors (*e.g.*, pollution, overfishing, development) that hinder the ability of species or ecosystems to withstand a stressful climatic event. For example, one way of enhancing resilience in wildlife refuges is to reduce other stresses on native vegetation such as erosion or altered hydrology caused by human activities (Chapter 5, National Wildlife Refuges). Marine protected area managers may focus on human stressors such as fishing and inputs of nutrients, sediments, and pollutants both inside the protected area and outside the protected area on adjacent land and waters (Chapter 8, Marine Protected Areas). The resilience of rivers could be enhanced by strategically shifting access points or moving existing trails for wildlife or river enthusiasts, in order to protect important riparian zones (Chapter 6, Wild and Scenic Rivers). Box 9.3 presents additional examples of this adaptation approach drawn from across the chapters of this report.

9.4.2.3 Representation

Representation is based on the idea that biological systems come in a variety of forms. Species include locally adapted populations as opposed to one monotypic taxon, and major

BOX 9.3. Examples of adaptation actions that focus on reduction of anthropogenic stresses as a means of supporting resilience.

Adaptation Approach: Reduce Anthropogenic Stresses

National Forests

- Reduce the impact of current anthropogenic stressors such as fragmentation (e.g., by creating larger management units and migration corridors) and uncharacteristically severe wildfires and insect outbreaks (e.g., by reducing stand densities and abating fuels).
- Identify and take early proactive action against non-native invasive species (e.g., by using early detection and rapid response approaches).

National Parks

- Remove structures that harden the coastlines, impede natural regeneration of sediments, and prevent natural inland migration of sand and vegetation after disturbances.
- Reduce or eliminate water pollution by working with watershed coalitions to reduce non-point sources and with local, state and federal agencies to reduce atmospheric deposition.
- Manage Park Service and visitor use practices to prevent people from inadvertently contributing to climate change.

National Wildlife Refuges

- Reduce human water withdrawals to restore natural hydrologic regimes.

Wild & Scenic Rivers

- Purchase or lease water rights to enhance flow management options.
- Manage water storage and withdrawals to smooth the supply of available water throughout the year.
- Develop more effective stormwater infrastructure to reduce future occurrences of severe erosion.
- Consider shifting access points or moving existing trails for wildlife or river enthusiasts.

National Estuaries

- Conduct integrated management of nutrient sources and wetland treatment of nutrients to limit hypoxia and eutrophication.
- Manage water resources to ensure sustainable use in the face of changing recharge rates and saltwater infiltration.
- Prohibit bulkheads and other engineered structures on estuarine shores to preserve or delay the loss of important shallow-water habitats by permitting their inland migration as sea levels rise.

Marine Protected Areas

- Manage human stressors such as overfishing and excessive inputs of nutrients, sediments, and pollutants within MPAs.
- Improve water quality by raising awareness of adverse effects of land-based activities on marine environments, implementing integrated coastal and watershed management, and developing options for advanced wastewater treatment.

BOX 9.4. Examples of adaptation actions that focus on representation as a means of supporting resilience.

Adaptation Approach: Representation

National Forests
- Modify genetic diversity guidelines to increase the range of species, maintain high effective population sizes, and favor genotypes known for broad tolerance ranges.
- Where ecosystems will very likely become more water limited, manage for drought- and heat-tolerant species and populations, and where climate trends are less certain, manage for a variety of species and genotypes with a range of tolerances to low soil moisture and higher temperatures.

National Parks
- Allow the establishment of species that are non-native locally, but which maintain native biodiversity or enhance ecosystem function in the overall region.
- Actively plant or introduce desired species after disturbances or in anticipation of the loss of some species.

National Wildlife Refuges
- Strategically expand the boundaries of NWRs to increase ecological, genetic, geographical, behavioral and morphological variation in species.
- Facilitate the growth of plant species more adapted to future climate conditions.

Wild & Scenic Rivers
- Increase genetic diversity through plantings or by stocking fish.
- Increase physical habitat heterogeneity in channels to support diverse biotic assemblages.

National Estuaries
- Maintain high genetic diversity through strategies such as the establishment of reserves specifically for this purpose.
- Maintain landscape complexity of salt marsh landscapes, especially preserving marsh edge environments.

Marine Protected Areas
- Maximize habitat heterogeneity within MPAs and consider protecting larger areas to preserve biodiversity, biological connections among habitats, and ecological functions.
- Include entire ecological units (e.g., coral reefs with their associated mangroves and seagrasses) in MPA design to maintain ecosystem function and resilience.
- Ensure that the full breadth of habitat types is protected (e.g., fringing reef, fore reef, back reef, patch reef).

habitat types or community types include variations on a theme with different species compositions, as opposed to one invariant community. The idea behind representation as a strategy for resilience is simply that a portfolio of several slightly different forms of a species or an ecosystem increases the likelihood that, among those variants, there will be one or more that are suited to the new climate. A management plan for a large ecosystem that includes representation of all possible combinations of physical environments and biological communities increases the chances that, regardless of the climatic change that occurs, somewhere in the system there will be areas that survive and provide a source for recovery. Employing this approach with wildlife refuges may be particularly important for migrating birds because they use a diverse array of habitats at different stages of their life cycles and along their migration routes, and all of these habitats will be affected by climate change (Chapter 5, National Wildlife Refuges). At the level of species, it may be possible to increase genetic diversity in river systems through plantings or via stocking fish (Chapter 6, Wild and Scenic Rivers), or maintain complexity of salt marsh landscapes by preserving marsh edge environments (Chapter 7, National Estuaries). Box 9.4 presents additional examples of this adaptation approach drawn from across the chapters of this report.

9.4.2.4 Replication

Replication is simply managing for the continued survival of more than one example of each ecosystem or species, even if the replicated examples are identical. When one recognizes that climate change stress includes unpredictable extreme events and storms, then replication represents a strategy of having

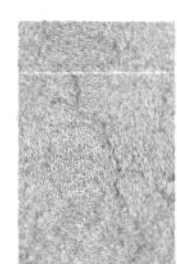

multiple bets in a game of chance. With marine protected areas, replication is explicitly used as a way to spread risk: if one area is negatively affected by a disturbance, then species, genotypes, and habitats in another area provide both insurance against extinction and a larval supply that may facilitate recovery of affected areas (Chapter 8, Marine Protected Areas). The analogy for forests would be spreading risks by increasing ecosystem redundancy and buffers in both natural environments and plantations (Chapter 3, National Forests). It is prudent to use replication in all systems. In practice, most replication strategies also serve as representation strategies (since no two populations or ecosystems can ever be truly identical), and conversely most representation strategies provide some form of replication. Box 9.5 provides examples of this adaptation approach drawn from across the chapters of this report.

BOX 9.5. Examples of adaptation actions that focus on replication as a means of supporting resilience.

Adaptation Approach: Replication

National Forests
- Spread risks by increasing ecosystem redundancy and buffers in both natural environments and plantations.

National Parks
- Practice bet-hedging by replicating populations and gene pools of desired species.

National Wildlife Refuges
- Provide redundant refuge types to reduce risk to trust species.

Wild & Scenic Rivers
- Establish special protection for multiple headwater reaches that support keystone processes or sensitive species.

National Estuaries
- When restoring oyster reefs, replicate reefs along a depth gradient to allow fish and crustaceans to survive when depth-dependant environmental degradation occurs.
- Support migrating shorebirds by ensuring protection of replicated estuaries along the flyway.

Marine Protected Areas
- Replicate habitat types in multiple areas to spread risks associated with climate change.

9.4.2.5 Restoration

In many cases natural intact ecosystems confer resilience to extreme events such as floods and storms. One strategy for adapting to climate change thus entails restoring intact ecosystems. For example the restoration of wetlands and natural floodplains will often confer resilience to floods. Restoration of particular species complexes may also be key to managing for resilience—a good example of this would be fire-adapted vegetation in forests that are expected to see more fires as a result of hotter and drier summers (Chapter 3, National Forests). At Blackwater National Wildlife Refuge, the USFWS is planning to restore wetlands that may otherwise be inundated by 2100 (Chapter 5, National Wildlife Refuges). In the case of estuaries, restoring the vegetational layering and structure of tidal marshes, seagrass meadows, and mangroves can stabilize estuary function (Chapter 7, National Estuaries). Box 9.6 provides additional examples of this adaptation approach drawn from across the chapters of this report.

9.4.2.6 Refugia and Relocation

The term *refugia* refers to physical environments that are less affected by climate change than other areas (*e.g.*, due to local currents, geographic location, etc.) and are thus a "refuge" from climate change for organisms. *Relocation* refers to human-facilitated transplantation of organisms from one location to another in order to bypass a barrier (*e.g.*, an urban area). Refugia and relocation, while major concepts, are actually subsets of one or more of the approaches listed above. For example, if refugia can be identified locally, they can be considered sites for long-term retention of species (*e.g.*, for representation and to maintain resilience) in forests (Chapter 3, National Forests). Or, in national wildlife refuges, it may be possible to use restoration techniques to reforest riparian boundaries with native species to create shaded thermal refugia for fish species (Chapter 5, National Wildlife Refuges). In the case of relocation, an example would be transport of fish populations in the Southwest that become stranded as water levels drop to river reaches with appropriate flows (*e.g.*, to preserve system-wide resilience and species representation) (Chapter 6, Wild and Scenic Rivers). Similarly, transplantation of organisms

BOX 9.6. Examples of adaptation actions that focus on restoration as a means of supporting resilience.

Adaptation Approach: Restoration

National Forests

- Use the paleological record and historical ecological studies to revise and update restoration goals so that selected species will be tolerant of anticipated climate.
- Where appropriate after large-scale disturbances, reset succession and manage for asynchrony at the landscape scale by promoting diverse age classes and species mixes, a variety of successional stages, and spatially complex and heterogeneous vegetation structure.

National Parks

- Restore vegetation where it confers biophysical protection to increase resilience, including riparian areas that shade streams and coastal wetland vegetation that buffers shorelines.
- Minimize soil loss after fire or vegetation dieback using native vegetation and debris.

National Wildlife Refuges

- Restore and increase habitat availability and reduce stressors in order to capture the full geographical, geophysical, and ecological ranges of species on as many refuges as possible.

Wild & Scenic Rivers

- Conduct river restoration projects to stabilize eroding banks, repair in-stream habitat, or promote fish passages from areas with high temperatures and less precipitation.
- Restore the natural capacity of rivers to buffer climate-change impacts (e.g., through land acquisition around rivers, levee setbacks to free the floodplain of infrastructure, riparian buffer repairs).

National Estuaries

- Restore important native species and remove invasive non-natives to improve marsh characteristics that promote propagation and production of fish and wildlife.
- Direct estuarine habitat restoration projects to places where the restored ecosystem has room to retreat as sea level rises.

Marine Protected Areas

- Following extreme events, consider whether actions should be taken to enhance natural recovery processes through active restoration.
- Consider mangrove restoration for potential benefits including shoreline protection, expansion of nursery habitat, and release of tannins and other dissolved organic compounds that may reduce photo-oxidative stress in corals.

among national parks could preserve system-wide representation of species that would not otherwise be able to overcome barriers to dispersal (Chapter 4, National Parks). Boxes 9.7 and 9.8 provide additional examples of these adaptation approaches drawn from across the chapters of this report.

9.4.3 Confidence

Due to uncertainties associated with climate change projections as well as uncertainties in species and ecosystem responses, there is also uncertainty as to how effective the different adaptation approaches listed above will be at supporting resilience. It is therefore essential to assess the level of confidence associated with each adaptation approach. For this report, the levels of confidence for each adaptation approach are based on the expert judgment of the authors, using a conceptual methodology developed by the IPCC (2007).

Confidence levels are presented for each of the seven adaptation approaches for each management system (Table 9.4). The goal of these adaptation approaches is to support the resilience of ecosystems to persist *in their current form* (*i.e.*, without major shifts to entirely redefined systems) under changing climatic conditions. Thus it is important to note at this point that promoting resilience may be a management strategy that is useful only on shorter time scales of a few decades rather than centuries, because as climate change continues, various thresholds of resilience will eventually be exceeded. Therefore, each of the authors'

BOX 9.7. Examples of adaptation actions that focus on the use of refugia as a means of supporting resilience.

Adaptation Approach: Refugia

National Forests
- Use the paleological record and historical ecological studies to identify environments buffered against climate change, which would be good candidates for long-term conservation.

National Parks
- Create or protect refugia for valued aquatic species at risk to the effects of early snowmelt on river flow.

National Wildlife Refuges
- Reforest riparian boundaries with native species to create shaded thermal refugia for fish species in rivers and streams.
- Identify climate change refugia and acquire necessary land.

Wild & Scenic Rivers
- Plant riparian vegetation to provide fish and other organisms with refugia.
- Acquire additional river reaches for the wild and scenic river where they contain naturally occurring refugia from climate change stressors.
- Create side-channels and adjacent wetlands to provide refugia for species during droughts and floods.

National Estuaries
- Restore oyster reefs along a depth gradient to provide shallow water refugia for mobile species such as fish and crustaceans to retreat to in response to climate-induced deep water hypoxia/anoxia.

Marine Protected Areas
- Identify and protect areas observed to be resistant to climate change effects or to recover quickly from climate-induced disturbances.
- Establish dynamic MPAs defined by large-scale oceanographic features such as oceanic fronts where changes in types and abundances of organisms often occur.

BOX 9.8. Examples of adaptation actions that focus on relocation as a means of supporting resilience.

Adaptation Approach: Relocation

National Forests
- Establish or strengthen long-term seed banks to create the option of re-establishing extirpated populations in new/more appropriate locations.

National Parks
- Assist in species migrations.

National Wildlife Refuges
- Facilitate long-distance transport of threatened endemic species.
- Facilitate interim propagation and sheltering or feeding of mistimed migrants, holding them until suitable habitat becomes available.

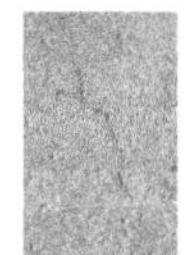

Wild & Scenic Rivers
- Establish programs to move isolated populations of species of interest that become stranded when water levels drop.

National Estuaries – none

Marine Protected Areas – none

confidence estimates are based solely on how effectively—in the near term—the adaptation approach will be at achieving positive ecological outcomes with respect to increased resilience to climate change. Through time, as ecosystem thresholds are exceeded, these approaches will cease to be effective, at which point major shifts in ecosystem processes, structures and components will be unavoidable. This eventuality is discussed in a later section (9.6.3, *Manage for Change*), where adaptation strategies associated with planning for major shifts are presented. In addition to limiting their confidence assessments to the near term, the authors also excluded from consideration any non-ecological factor (such as confidence in the ability to put particular approaches into practice) and only evaluated those adaptation approaches for which they had adaptation strategies discussed in their chapter.

9.4.3.1 Approach to Estimating Levels of Confidence

The authors considered two separate but related elements of confidence (IPCC, 2007). The first element is the amount of evidence that is available to assess the effectiveness of a given adaptation approach to support resilience. The second is the level of agreement or consensus in the expert community regarding the different lines of evidence. From each chapter, specific adaptation options were grouped according to the seven categories of "adaptation approaches" described in the previous section (see Boxes 9.2–9.8). The authors then developed confidence estimates for each adaptation approach based on consideration of the specific adaptation options and the following questions.

High/low amount of evidence. Is this adaptation approach well-studied and understood, or instead is it mostly experimental or theoretical and not well-studied? Does your experience in the field, your analyses of data, and your understanding of the literature and performance of specific adaptation options under this type of adaptation approach indicate that there is a high or low amount of information on the effectiveness of this approach?

High/low amount of agreement. Do the studies, reports, and your experience in the field, analyzing data, or implementing the types

Table 9.4. Confidence levels associated with seven different adaptation approaches, examined across six management system types. Estimates reflect the expert opinions of the authors and are based on the literature and personal experience.

Confidence Estimates for SAP 4.4 Adaptation Approaches

Agreement ↑		
	LH Low evidence High agreement	HH High evidence High agreement
	LL Low evidence Low agreement	HL High evidence Low agreement
	Evidence →	

	Protecting key ecosystem features	Reducing anthropogenic stresses	Representation	Replication	Restoration	Refugia	Relocation
National Forests	HL	HH	LL	LL	HL	HL	HL
National Parks	HH	HH	LL	LL	HH	LL	LL
National Wildlife Refuges	HH	HH	HH	HH	LL	HL	LL
Wild and Scenic Rivers	LH	HH	HH	HH	LL	HH	LL
National Estuaries	LL	HH	LH	LH	HH	LL	N/A
Marine Protected Areas	LH	LH	LH	LH	LL	LH	N/A

of adaptation strategies that comprise this approach reflect a high degree of agreement on the effectiveness of this approach, or does it lead to competing interpretations?

Because of the qualitative nature of this confidence exercise, the author teams provided explanations of the basis for each of their estimates under each adaptation approach (see Annex B, Confidence Estimates). The evidence they considered in making their judgments included peer-reviewed and gray literature (journal articles, reports, working papers, management plans, workshop reports, other management literature, other gray literature), data and observations, model results, and the authors' own experience, including their experiences in the field, their analyses of data, and their knowledge of the performance of specific adaptation options under each type of adaptation approach.

Confidence estimates are presented in Table 9.4 by management system type for each of the seven adaptation approaches. Such confidence estimates should be a key consideration when deciding which adaptation approaches to implement for a given system.

9.4.3.2 Findings

To take action today using the best available information, *reducing anthropogenic stresses* is currently the adaptation approach that ranks highest in confidence, in terms of both evidence and agreement across all six management systems. This may be due partly to the fact that managers have been dealing with anthropogenic stresses for a long time, so there are a lot of data and good agreement among the experts that this approach is effective in increasing resilience to any kind of stress, including climate change.

Protecting key ecosystem features, representation, replication, restoration, and *refugia* all received variable confidence rankings across the management system chapters. This could be due to a number of factors related to both evidence and agreement. One explanation could be differences in the amount and nature of research and other information available on an approach depending on the management system. For example, one management system may have a great deal of evidence for the effectiveness of an approach at the species level, but little evidence that it would be effective in enhancing resilience at the ecosystem level; in contrast, another management system may have more evidence at the ecosystem as well as species level. Also, regardless of the amount of evidence, different groups can arrive at different interpretations of what constitutes agreement based on management goals, institutional perspectives, and experiences with particular ecosystem types. Even though the variability in confidence in these approaches suggests that caution is warranted, many of the individual adaptation options under these approaches may still be effective. In these cases, a more detailed assessment of confidence is needed for each specific adaptation option and ecosystem in which it would be applied.

Relocation stands out as being the weakest in terms of confidence *at the current time*, based on available information. There appears to be little information (evidence) about *relocation* or its implications for ecosystem resilience, and thus there is little agreement among experts that it is a robust approach. Future research may change this ranking (as well as the rankings for other approaches) at any time.

9.4.3.3 Improving Confidence Estimates

Management planning to select and prioritize adaptation approaches will always involve some assessment of confidence, whether implicitly or explicitly. Explicit estimations of confidence, while difficult, afford managers a better understanding of the nature, implications, and risks of different adaptation approaches. The confidence exercise in this report is a first attempt at evaluating a series of seven conceptual approaches to adaptation that each represents an aggregation of various adaptation options. The next level of refinement for confidence assessments may involve evaluating confidence in individual adaptation options within each approach. This will be especially important in those cases where levels of confidence in an approach are highly variable across management systems or across ecosystems.

There are a number of challenges associated with improving confidence estimates for adaptation. One challenge is removing the inherent subjectivity of judgments about evidence and agreement. This could be addressed by more clearly defining terminology (*e.g.*, evidence and agreement) and developing more systematic rules (*e.g.*, weighting criteria for different sources of evidence). The goal of such improvements would be to move from a qualitative to a more quantitative method of expressing confidence, thereby facilitating more effective use of scientific information for adaptation planning. Finally, any confidence exercise would benefit from the largest number of participants as possible to improve the robustness of the results.

9.4.4 Adaptive Management

Once adaptation approaches have been selected after taking into account confidence levels, adaptive management is likely to be an effective method for implementing those approaches. It emphasizes managing based on observation and continuous learning and provides a means for effectively addressing varying degrees of uncertainty in our knowledge of current and future climate change impacts. Adaptive management is typically divided into two types: passive and active (Arvai *et al.*, 2006; Gregory, Ohlson, and Arvai, 2006). Passive adaptive management refers to using historical data to develop hypotheses about the best management action, followed by action and monitoring. Often models are used to guide the decisions and the monitoring can improve the models. Active adaptive management refers to actually conducting a management experiment, ideally with several different management actions implemented at once as a means of testing competing hypotheses. Examples include flood release experiments in the Grand Canyon (Chapter 4, National Parks) and at the Glen Canyon dam (National Research Council, 1999). Releasing water from a dam allows for the application of highly regulated experimental treatments and assessments of effects. For more information on adaptive management, see the Technical Guide[5] released in the spring of 2007 by the Department of Interior. It provides a robust analytical framework that is based on the experience, in-depth consultation, and best practices of scientists and natural resource managers.

[5] **Williams**, B. K., R. C. Szaro, and C. D. Shapiro. 2007. Adaptive Management: The U.S. Department of the Interior Technical Guide. Adaptive Management Working Group, U.S. Department of the Interior, Washington, DC.

Adaptive management to address climate change is an iterative process that involves the consideration of potential climate impacts, the design of management actions and experiments that take those impacts into account, monitoring of climate-sensitive species and processes to measure management effectiveness, and the redesign and implementation of improved (or new) management actions (Fig. 9.2). To maximize the implementation of climate-sensitive adaptive management within federal systems, managers can focus on (1) previously established strategies that were designed for other management issues but have strong potential for application toward climate change impacts, and (2) new strategies that are not yet in place but appear to be feasible and within reasonable reach of current management structures. In other words, at a minimum, managers need to vigorously pursue changes that are relatively easily accomplished under existing programs and management cultures.

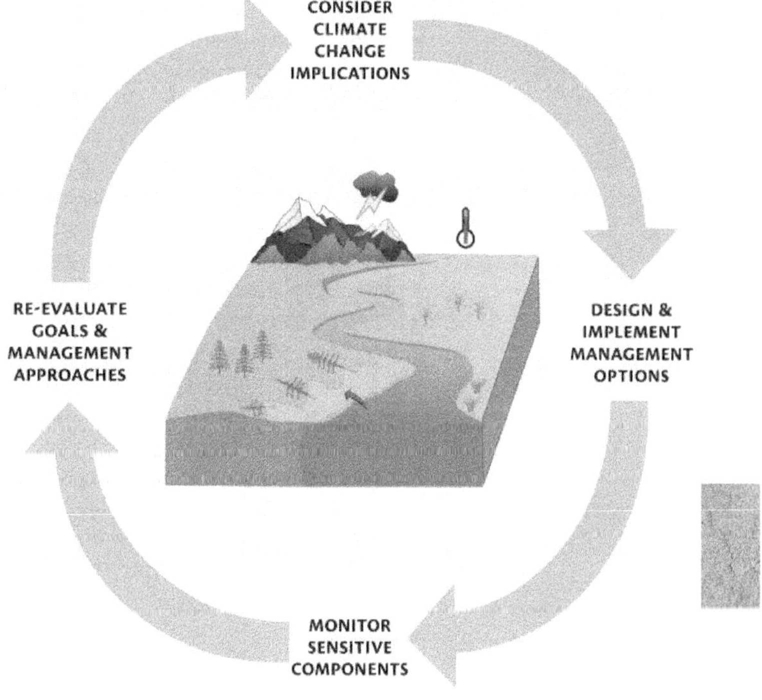

Figure 9.2. The process of adaptive management.

Recent examinations of the difficulty of actually using adaptive management have emphasized that the temporal and spatial scale, dimension of uncertainty, risks, and institutional support can create major difficulties with applying adaptive management. When one considers adaptive management (whether active or passive) in response to climate change, every one of these potential difficulties is at play (Arvai *et al.*, 2006; Gregory, Ohlson, and Arvai, 2006). The critical challenge will be stating explicit scientific hypotheses, establishing monitoring programs with predefined triggers that initiate a re-examination of management approaches, and a flexible policy or institutional framework (Gregory, Ohlson, and Arvai, 2006). These challenges do not mean adaptive management is impossible—only that attention to hypotheses, monitoring, periodic re-evaluations, and flexibility are necessary.

Even in the absence of an ability to experimentally manipulate systems, rapid, climate-induced ecological changes provide excellent opportunities to observe the effects of climate change in relatively short time frames. Managers and scientists can design studies to take advantage of increased climatic variability and climate trends to inform management. Some examples of such studies could include observing: which riparian plant species are best adapted to extreme variations in flow regime and flooding, how increased variability in climatic conditions affects population dynamics of target insect pests or focal wildlife species, and the effects of marine reserve size on recruitment and survival of key species. In order to make this approach effective, specific hypotheses should be proposed about which life history traits will predispose species to (biologically) adapt to climate change (Kelly and Adger, 2000). Otherwise the data collection will be less focused and efficient. Using climate-driven changes as treatments *per se* will be much less exact and less predictable than controlled experiments, so taking advantage of such situations for adaptive management studies will require increased flexibility, foresight, and creativity on the part of managers and scientists.

Another key element of adaptive management is monitoring of sensitive species and processes in order to measure the effectiveness of experimental management actions. In the case of adaptive management for climate change, this step is critical, not only for measuring the degree to which management actions result in positive outcomes on the ground, but also for supporting a better scientific understanding of how to characterize and measure ecological resilience. Most resource agencies already have monitoring programs and sets of indicators. As long as management goals are not changed (see Section 9.6.1), then these existing monitoring programs should reflect the outcomes of management actions on the ground. If management goals are altered because climate change is perceived to be so severe that historical goals are untenable, then entirely new indicators and monitoring programs may need to be designed. Whatever the case, monitoring is fundamental to supporting the reevaluation and refinement of management strategies as part of the adaptive process.

The same monitoring can also foster an improved understanding of how best to characterize and quantify resilience. For some systems, the ecology of climate stress (*e.g.*, coral bleaching) has been studied for decades, and resilience theory continues to develop rapidly. For other ecosystems, the impacts of climate change are less well understood, and understanding resilience is more difficult. In any event, while there may be some good conceptual models that describe resilience characteristics for species and ecosystems, there is generally a paucity of empirical data to confirm and resolve the relative importance of these characteristics. Such information is needed for the next generation of techniques and tools for quantification and prediction of resilience across species and ecosystems. If monitoring programs are designed with explicit hypotheses about resilience, they will be more likely to yield useful information.

The idea of "adaptive management" has been widely advocated among natural resource managers for decades and has been ascribed to many management decisions. However, due largely to the challenges cited above, it is not as widely or rigorously applied as it could be. Yet the prospect of uncertain, widespread, and severe climatic changes may galvanize managers to embrace adaptive management as an essential strategy. Climate change creates new situations of added complexity for which an adaptive management approach may be the only way to take management action today while allowing for increased understanding and refinement tomorrow.

9.5 BARRIERS AND OPPORTUNITIES FOR ADAPTATION

Although there may be many adaptation strategies that could be implemented, a very real consideration for managers is whether all of the possibilities are feasible. Factors limiting or enhancing managers' ability to implement options may be technical, economic, social, or political. As noted previously in this chapter, the climate community refers to such opportunities and constraints (or barriers) as adaptive capacity. It may be helpful to understand the types of barriers to implementation that exist in order to assess the feasibility of specific adaptation options, and even more so to identify corresponding ways in which barriers may be overcome. The barriers and opportunities discussed below are based on the expert opinions of the authors of this report and feedback from the expert workshops and are associated with implementation of adaptation options today, assuming no significant changes in institutional frameworks and authorities.

A useful way of thinking about both barriers and opportunities is in terms of the following four categories: (1) legislation and regulations, (2) management policies and procedures, (3) human and financial capital, and (4) information and science (see Tables 9.5–9.8). All of the federal land and water management systems reviewed in the preceding chapters are mandated by law to preserve and protect the nation's natural resources. Specific management goals vary across systems, however, due to the unique mission statements articulated in their founding legislation, or organic acts. Organic acts are fundamental pieces of legislation that either signify the organization of an agency or provide a charter for a network of public lands, such as the National Park Service Organic Act that established the National Park System. Accordingly, goals are manifested through management principles that could interpret those goals in ways that may inhibit or enhance the capability to adapt.

No matter how management goals are approached, achievement of goals may be difficult even without climate change. For example, in the case of the National Forest System, managers are asked to provide high-quality recreational opportunities and to develop means of meeting the nation's energy needs through biofuel production while reducing the risk of wildfire and invasive species and protecting both watersheds and biodiversity. Successful management requires not only significant resources (*e.g.*, staff capacity and access to information), but also the ability of managers to apply resources strategically and effectively (*e.g.*, for monitoring and management experiments) (Spittlehouse and Stewart, 2003).

Resources are managed carefully across federal agencies to deal with a growing human population that puts new and expanding pressures on managers' ability to meet management goals. Examples of these existing pressures include economic development near management unit boundaries (Chapter 5, National Wildlife Refuges), air pollution (Chapter 4, National Parks), increased wildfire-related costs and risks (Chapter 3, National Forests), habitat degradation and destruction (Chapter 8, Marine Protected Areas), pollutant loading (Chapter 7, National Estuaries), and excessive water withdrawals (Chapter 6, Wild and Scenic Rivers). The added threat of climate change may exceed the capacity of the federal management systems to protect the species and ecological systems that each is mandated to protect. However, as many of the previous chapters point out, this threat also represents an opportunity to undertake strategic thinking, reshape priorities, and use carefully considered actions to initiate the development of management adaptations to more effectively protect resources.

Adaptation responses to climate change are meant to reduce the risk of failing to achieve management goals. A better understanding of the barriers and opportunities that affect implementation of adaptation strategies could facilitate the identification of critical adjustments within the constraints of management structures and policies, and subsequently could foster increased adaptive capacity within and across federal management systems as those

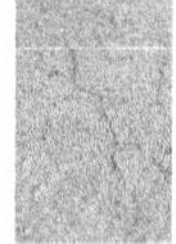

constraints are addressed in the longer term (see Section 9.6).

9.5.1 Legislation and Regulation

9.5.1.1 Perceived Barriers

In general, existing agency experience and law, taken together, provide the flexibility needed to adapt to climate change. However, an individual organic act or other enabling legislation, or its interpretation may sometimes be perceived as a barrier to adaptation. While original organic acts represented progressive policy and management frameworks at the time were written, many reflect a past era (Table 9.5). For example, the first unit of the National Wildlife Refuge System, Pelican Island, was designated in 1903 to protect waterfowl from being over-hunted when that was the greatest threat. At that time, the U.S. population was half of what it is now, and the interstate highway system was decades away from establishment (Chapter 5, National Wildlife Refuges). In addition, ambiguous language in enabling legislation poses challenges to addressing issues related to climate change, such as determining what "impaired" means (Chapter 4, National Parks). It also has been recognized that specific environmental policies such as the Endangered Species Act, National Environmental Policy Act, and the National Forest Management Act are highly static, making dynamic planning difficult and potentially impeding adaptive responses.[6] Even recently implemented legislation and management plans have not directly addressed climate change (Chapter 7, National Estuaries). In general, while community-focused approaches are more flexible, many existing laws force a species-specific approach to management (Chapter 3, National Forests), limiting agency action to address issues related to climate change.

Furthermore, organic acts and pursuant enabling legislation may limit the capacity to effectively manage some resources. For example, the chief legal limitation on intensive management to adapt to climate change for the National Wildlife Refuge System is the limited jurisdiction of many refuges over their water (Chapter 5, National Wildlife Refuges). Both the timing of water flows as well as the quantity of water flowing through refuges are often subject to state permitting and control by other federal agencies. Similarly, legal frameworks such as the Colorado River Compact establish water rights, compacts, and property rights that all serve to constrain the ability to use adaptive strategies to address climate change (Chapter 6, Wild and Scenic Rivers).

Protected areas have political rather than ecological boundaries as an artifact of legislation. These boundaries may pose a barrier to effectively addressing climate change. Climate change will likely lead to shifts in species and habitat distribution (Chapter 3, National Forests; Chapter 4, National Parks; Chapter 7, National Estuaries; Chapter 8, Marine Protected Areas), potentially moving them outside the bounds of federal jurisdiction or introducing new species that cause changes

[6] **Levings**, W., 2003: [illegible] Unpublished report on file at the Tahoe National Forest, pp.1-6.

Table 9.5. Examples of legislation and regulation as barriers to and opportunities for adaptation.

LEGISLATION AND REGULATION		
Perceived Barrier	**Opportunity**	**Examples**
Legislation and agency policies may be highly static, inhibit dynamic planning, impede flexible adaptive responses and force a fine-filter approach to management.	Re-evaluate capabilities of, or authorities under, existing legislation to determine how climate change can be addressed within the legislative boundaries.	• Use state wildlife action plans to manage lands adjacent to national wildlife refuges to enable climate-induced species emigration. • Re-evaluate specific ecosystem- and species-related legislation to use all capabilities within the legislation to address climate change. • Incorporate climate change impacts into priority setting for designation of new wild and scenic rivers (see Chapter 6 section 6.4.4).

in animal communities, such as changing predation and competition (Chapter 5, National Wildlife Refuges). Agencies often do not have the capacity or authority to address issues outside their jurisdiction, which could hamper efforts to adapt to climate change. This could affect smaller holdings more acutely than others (Chapter 5, National Wildlife Refuges).

Despite historical interpretations and organizational and geographic boundaries, existing legislation does not prohibit adaptation. Yet uncertainty surrounding application of certain management techniques can lead to costly and time-consuming challenges from particular stakeholders or the public (Chapter 3, National Forests). Fuel treatments and other adaptive projects that have ground-disturbing elements, such as salvage harvest after disturbance and use of herbicides before revegetation, have been strongly opposed by the public.[7] While using adaptation approaches in management poses the risk of spurring costly litigation from stakeholders, every chapter in this volume concludes that inaction with regard to climate change may prove more damaging and costly than acting with insufficient knowledge of the outcomes.

9.5.1.2 Opportunities

Federal land and water managers can use existing legislative tools in opportunistic ways (Table 9.5). Managers can strategically apply existing legislation or regulations at the national or state level by applying traditional features or levers in non-traditional ways. For example, while still operating within the legislative framework, features of existing legislation can be effectively used to coordinate management outside of jurisdictional boundaries. Generally, the USFWS has ample proprietary authority to engage in transplantation-relocation, habitat engineering (including irrigation-hydrologic management), and captive breeding to support conservation (Chapter 5, National Wildlife Refuges). These activities are especially applicable to managing shifts in species distributions and in potentially preventing species extirpations likely to result from climate change. Portions of existing legislation could also be used to influence dam operations at the state level as a means of providing adaptive flow controls under future climate changes (*e.g.*, using the Clean Water Act to prevent low flows in vulnerable stream reaches, adjusting thermal properties of flows). As these examples suggest, managers can influence change within the legislative framework to address climate change impacts.

9.5.2 Management Policies and Procedures

9.5.2.1 Perceived Barriers

Some management systems have a history of static policies that are counter to the dynamic management actions called for today (Table 9.6) and do not recognize climatic change as a significant problem or stressor. These agency policies do not allow for sufficient flexibility under uncertainty and change. Without flexibility, existing management goals and priorities—though potentially unrealistic given climate change—may have to be pursued without adjustments. Yet, with limited resources and staff time, priorities need to be established and adaptation efforts focused to make best use of limited resources. There are several specific hindrances to such management changes that are worth mentioning in detail.

First, addressing climate change will require flexible and long-term planning horizons. Existing issues on public lands, coupled with insufficient resources (described below), force many agencies and managers to operate under crisis conditions, focusing on short-term and narrow objectives (Chapter 4, National Parks). Agencies often put priority on maintaining, retaining, and restoring historic conditions. These imperatives can lead to static as opposed to dynamic management (Chapter 3, National Forests) and may not be possible to achieve as a result of climate change. Additionally, place-based management paradigms may direct management at inappropriate spatial and temporal scales for climate change. Managing on a landscape scale, as opposed to smaller-scale piecemeal planning, would enable greater adaptability to climate-related changes (Chapter 3, National Forests).

A number of factors may limit the usefulness of management plans. The extent to which plans are followed and updated is highly variable across management systems. Further, plans may

[7] **Levings**, W., 2003: [illegible] Unpublished report on file at the Tahoe National Forest, pp.1-6.

not always adequately address evolving issues or directly identify actions necessary to address climate change (Chapter 3, National Forests; Chapter 8, Marine Protected Areas). If a plan is not updated regularly, or a planning horizon is too short-sighted in view of climate change, a plan's management goals may become outdated or inappropriate. To date, few management plans address or incorporate climate change directly. Fortunately, many agencies recognize the need for management plans to identify the risks posed by climate change and to have the ability to adapt in response (Chapter 6, Wild and Scenic Rivers). Some proactive steps to address climate change will likely cost very little and could be included in policy and management plans (Chapter 7, National Estuaries). These include documenting baseline conditions to aid in identifying future changes and threats, identifying protection options, and developing techniques and methods to help predict climate related changes at various scales (Chapter 3, National Forests; Chapter 6, Wild and Scenic Rivers).

Last, even if the plan for a particular management system addresses climate change appropriately, many federal lands and waters are affected by neighboring lands for which they have limited or no control (Chapter 4, National Parks). National wildlife refuges and wild and scenic rivers are subject to water regulation by other agencies or entities. This fragmented jurisdiction means that collaboration among agencies is required so that they are all working toward common goals using common management approaches. Although such collaboration does occur, formal co-management remains the exception, not the rule. Despite this lack of collaboration, there is widespread recognition that managing surrounding lands and waters is important to meeting management objectives (Chapter 5, National Wildlife Refuges; Chapter 8, Marine Protected Areas), which may lead to more effective management across borders in the future.

9.5.2.2 Opportunities

Each management system mandates the development of a management plan. Incorporating climate change adaptation could be made a part of all planning exercises, both at the level of individual units and collaboratively with other management units. This might encourage more units in the same broad geographical areas to look for opportunities to coordinate and collaborate on the development of regional management plans (Table 9.6). A natural next step would then be to prioritize actions within the management plan. Different approaches may be used at different scales to decide on management activities across the public lands network or at specific sites. If planning and prioritizing occurs across a network of sites, then not only does this approach facilitate sharing of information between units, but this broader landscape approach also lends itself well to climate change planning. This has already occurred in the National Forest System, where the Olympic, Mt. Baker, and Gifford Pinchot National Forests have combined resources to produce coordinated plans. The Olympic National Forest's approach to its strategic planning process is also exemplary of an entity already possessing the capacity to incorporate climate change through its specific guidance on prioritization.

In some cases, existing management plans may already set the stage for climate adaptation. A good example is the Forest Service's adoption of an early detection/rapid response strategy for invasive species. This same type of thinking could easily be translated to an early detection/rapid response management approach to climate impacts. Even destructive extreme climate events can be viewed as management opportunities by providing valuable post-disturbance data. For example, reforestation techniques following a fire or windfall event can be better honed and implemented with such data (*e.g.*, use of genotypes that are better adjusted to the new or unfolding regional climate, use of nursery stock tolerant to low soil moisture and high temperature, or use of a variety of genotypes in the nursery stocks) (see Chapter 3, National Forests).

Management plans that are allowed to incorporate climate change adaptation strategies but that have not yet done so provide a blank canvas of opportunity. In the near term, state wildlife action plans are an example of this type of leveraging opportunity. Another example is the Forest Service's involvement with the Puget Sound Coalition and the National Estuary Program's involvement in Coastal Habitat Protection Plans for fish, an ecosystem-based fisheries management approach at the state

Table 9.6. Examples of management policies and procedures as barriers to and opportunities for adaptation.

MANAGEMENT POLICIES AND PROCEDURES		
Perceived Barrier	**Opportunity**	**Examples**
Seasonal management activities may be affected by changes in timing and duration of seasons.	Review timing of management activities and take advantage of seasonal changes that provide more opportunities to implement beneficial adaptation actions.	• Take advantage of shorter winter seasons (longer prescribed fire seasons) to do fuel treatments on more national forest acres (see the Tahoe National Forest Case Study, Annex A1.1).
Agency policies do not recognize climatic change as a significant problem or stressor.	Take advantage of flexibility in the planning guidelines and processes to develop management actions that address climate change impacts.	• Where guidelines are flexible for meeting strategic planning goals (e.g., maintain biodiversity), re-prioritize management actions to address effect of climate change on achievement of goals (see the Olympic National Forest Case Study, Annex A1.2).
Political boundaries do not necessarily align with ecological processes; some resources cross boundaries; checkerboard ownership pattern with lands alternating between public and private ownership at odds with landscape-scale management (see Chapter 3 section 3.4.5).	Identify management authorities/agencies with similar goals and adjacent lands; share information and create coalitions and partnerships that extend beyond political boundaries to coordinate management; acquire property for system expansion.	• Develop management plans that encompass multiple forest units such as the Pacific Northwest Forest Plan that includes Olympic National Forest-Mt. Baker-Gifford Pinchot National Forest (see the Olympic National Forest Case Study, Annex A1.2). • Implement active management at broader landscape scales through existing multi-agency management processes such as (1) the Herger-Feinstein Quincy Library Group Pilot and the FPA Adaptive Management project on Tahoe National Forest (see the Tahoe National Forest Case Study, Annex A1.1), (2) the Greater Yellowstone Coordinating Committee, and the Southern Appalachian Man and the Biosphere Program with relationships across jurisdictional boundaries (see Chapter 4 section 4.4.3), (3) The Delaware River, managed cooperatively as a partnership river (see the Upper Delaware River Case Study, Annex A4.3). • Coordinate dam management at the landscape level for species that cross political boundaries by using dam operations prospectively as thermal controls under future climate changes (see Chapter 6 section 6.4.4.2). • Coordinate habitat and thermal needs for fish species with entities that control the timing and amount of up-stream water releases (see Chapter 6 section 6.4.4.2).

level. Stakeholder processes, described above as a barrier, might be an opportunity to move forward with new management approaches if public education campaigns precede the stakeholder involvement. The issue of climate change has received sufficient attention that many people in the public have begun to demand actions by the agencies to address it.

As suggested by the many themes identified by the federal land and water management systems, the key to successful adaptation is to turn barriers into opportunities. This should be possible with increased availability of practical information, corresponding flexibility in management goals, and strong leadership. At the very least, managers (and corresponding management plans) may need to recognize climate change and its synergistic effects as an overarching threat to their resources.

9.5.3 Human and Financial Capital

9.5.3.1 Perceived Barriers

Level of funding and staff capacity (or regular staff turnover) may pose significant barriers to adaptation to climate change (Table 9.7). Agencies may also lack adaptive capacity due to the reward systems in place. Currently, in some agencies a reward system exists that focuses primarily on achieving narrowly prescribed targets, and funding is directed at achieving these specific activities. This system

provides few incentives for creative project development and implementation, instead creating a culture that prioritizes projects with easily attainable goals.

Budgets may also curtail adaptation efforts. Managers may lack sufficient resources to deal with routine needs. Managers may have even fewer resources available to address unexpected events, which will likely increase as a result of climate change. In addition, staff capacity may not be sufficient to address climate change. While climate change stands to increase the scope of management by increasing both the area of land requiring active management and the planning burden per unit area (because of adaptive management techniques), agencies such as the USFWS face decreasing personnel in some regions. Additionally, minimal institutional capacity exists to capture experience and expand learning (Chapter 4, National Parks). As a result, many agency personnel do not have adequate training, expertise, or understanding to effectively address emerging issues (Chapter 3, National Forests). All of these factors work to constrain the ability of managers to alter or supplement practices that would enable adaptation to climate change.

9.5.3.2 Opportunities

Agency employees play important roles as crafters and ultimate implementers of management plans and strategies. In fact, with respect to whether the implementation of adaptation strategies is successful or unsuccessful, the management of people can be as—or more—important than managing the natural resource. A lack of risk-taking coupled with the uncertainty surrounding climate change could lead to a situation where managers opt for the no-action approach (*e.g.*, Hall and Fagre, 2003). On the other hand, climate change could cause the opposite response if managers perceive that risks must be taken because of the uncertainties surrounding climate change. Implementation of human resource policies that minimize risk for action and protect people when mistakes are made will be critical to enabling managers to make difficult choices under climate change (Table 9.7). A "safe-to-fail" policy would be exemplary of this approach (Chapter 4, National Parks). A safe-to-fail policy or action is one in which the system can recover without irreversible damage to either natural or human resources (*e.g.* careers and livelihoods). Because the uncertainties associated with projections of climate change are substantial, expected outcomes or targets of agency policies and actions may be equally likely to be correct or incorrect. Although managers aim to implement a "correct" action, it must be expected that when the behavior of drivers and system responses is uncertain, failures are likely to occur when attempting to manage for impacts of climate change (Chapter 4, National Parks).

Tackling the challenge of managing natural resources in the face of climate change may require that staff members not only feel valued but also empowered by their institutions. Scores of federal land management employees began their careers as passionate stewards of the nation's natural resources. With the threat of climate change further compounding management challenges, it is important that this passion be reinvigorated and fully cultivated. Existing employees could be effectively trained (or specialist positions designated) for tackling climate change issues within the context of their current job descriptions and management frameworks (Chapter 3, National Forests). For example, the National Park Service has recently implemented a program to educate park staff on climate change issues, in addition to offering training for presenting this information to park visitors in 11 national parks. Called the "Climate Friendly Parks" program, it includes guidelines for inventorying a park's greenhouse gas emissions, park-specific suggestions to reduce greenhouse gas emissions, and help for setting realistic emissions reduction goals. Additionally, the Park Service's Pacific West Regional Office has been proactive in educating western park managers on issues related to climate change as well as promoting messages to communicate to the public and actions to address the challenge of climate change (Chapter 4, National Parks). Such "no regrets" activities offer a cost-effective mechanism for empowering existing employees with both knowledge and public outreach skills.

Table 9.7. Examples of human and financial capital as barriers to and opportunities for adaptation.

HUMAN AND FINANCIAL CAPITAL		
Perceived Barrier	**Opportunity**	**Examples**
Lack of incentive to take risks, develop creative projects; reward system focuses on achieving narrowly prescribed targets; funds allocated to achieve targets encourage routine, easily accomplished activities.	Shift from a culture of punishing failure to one that values creative thinking and supports incremental learning and gradual achievement of management goals.	• Develop incentives that reward risk taking and innovative thinking. • Build into performance expectations of a gradient between success and failure. • Set up a systematic method for (1) learning from mistakes and successes, and (2) eliciting the experience and empirical data of front line managers, resource management personnel, and scientific staff. (Drawn from Chapter 4 section 4.4.2).
Little to no climate expertise within many management units at the regional and local level; disconnect between science and management that impedes access to information.	Use newly created positions or staff openings as opportunities to add climate change expertise; train resource managers and other personnel in climate change science.	• Use incremental changes in staff to "reinvent and redefine" organizations' institutional ability to better respond to climate change impacts (see the Tahoe National Forest Case Study, Annex A1.1). • Develop expertise through incorporation into existing Forest Service training programs like the silvicultural certification program, regional integrated resource training workshops, and regional training sessions for resource staffs (see Chapter 3 section 3.5). • Develop managers' guides, climate primers, management toolkits, a Web clearinghouse, and video presentations (see Chapter 3 section 3.5).
National and regional budget policies/processes constrain the potential for altering or supplementing current management practices to enable adaptation to climate change (see Chapter 3 section 3.5; general decline in staff resources and capacity (see Chapter 3 section 3.4.5).	Look for creative ways to augment the workforce and stretch budgets to institute adaptation practices (e.g., individuals or parties with mutual interests in learning about or addressing climate change that may be engaged at no additional cost).	• Augment budget and workforce through volunteers from the public or other sources such as institutions with compatible educational requirements, neighborhood groups, environmental associations, etc., such as the Reef Check Program that help collect coral reef monitoring data (see Chapter 8 sections 8.3.3, 8.4.4.1 and 8.4.4.2). • Identify organizations or private citizens that benefit from adaptation actions to share implementation costs in order to avoid more costly impacts/damages. • Use emerging carbon markets to promote (re-) development of regional biomass and biofuels industries, providing economic incentives for active adaptive management, funds from these industries could be used to promote thinning and fuel-reduction projects (see Tahoe National Forest Case Study, Annex A1.1).

9.5.4 Information and Science

9.5.4.1 Perceived Barriers

Adaptation is predicated upon research and scientific information. Addressing emerging issues that arise as a result of climate change will require new research and information to use in developing strategic management plans. Critical gaps in scientific information, such as understanding of ecosystem function and structure, coupled with the high degree of uncertainty surrounding potential impacts of climate change, hinder the potential for effective implementation of adaptation (Table 9.8; Chapter 8, Marine Protected Areas). A lack of climate-related data from monitoring precludes managers from assessing the extent to which climate has affected their systems. Staff and budget limitations may not only constrain the ability to monitor but may also preclude managers from analyzing data from the monitoring programs that do receive support. Without adequate monitoring, it remains difficult to move forward confidently with appropriate adaptation efforts (Chapter 6, Wild and Scenic Rivers).

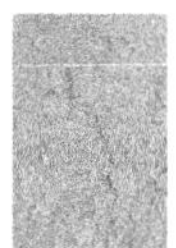

Even if managers had sufficient information, decision-making would still prove problematic. Managers often lack sufficient tools to help guide them in selecting appropriate management

Table 9.8. Examples of information and science as barriers to and opportunities for adaptation.

INFORMATION AND SCIENCE		
Perceived Barrier	**Opportunity**	**Examples**
Often no inventory or baseline information on condition exists, and nothing is in place to detect climate change impacts.	Identify existing monitoring programs for management; develop a suite of climate change indicators and incorporate them into existing programs.	• Use monitoring programs such as the NPS vital signs for the Inventory and Monitoring Program, Global Fiducial Program, LTER networks, and NEON to monitor for climate change impacts and effectiveness of adaptation options (see Chapter 4 section 4.4.3).
Historic conditions may no longer sufficiently inform future planning (e.g., "100-year" flood events may occur more often and dams need to be constructed accordingly).	Evaluate policies that use historic conditions and determine how to better reflect accurate baselines in the face of climate change; modify design assumptions to account for changing climate conditions.	• Change emphasis from maintenance of "minimum flows" to the more sophisticated and scientifically based "natural flow paradigm," as is happening in some places (see Chapter 6 section 6.3.4.2).
Lack of decision support tools and models, uncertainty in climate change science, and critical gaps in scientific information that limit assessment of risks and efficacy and sustainability of actions.	Identify and use all available tools/mechanisms currently in place to deal with existing problems to apply to climate-change related impacts.	• Use early detection/rapid response approaches (such as that used to manage invasive species) to respond quickly to the impacts of extreme events (e.g., disturbances, floods, windstorms) with an eye towards adaptation (see Chapter 3 section 3.3.3). • Diversify existing portfolio of management approaches to address high levels of uncertainty. • Hedge bets and optimize practices in situations where system dynamics and responses are fairly certain. • Use adaptive management in situations with greater uncertainty (see Chapter 4 section 4.4.3).
Occurrence of extreme climate events outside historical experience.	Use disturbed landscapes as templates for "management experiments" that provide data to improve adaptive management of natural resources.	• After fire, reforest with genotypes of species that are better adjusted to the new or unfolding regional climate with nursery stock tolerant to low soil moisture and high temperature, or with a variety of genotypes in the nursery stock (see Chapter 3 section 3.4.1.2).
Stakeholders/public may have insufficient information to properly evaluate adaptation actions, and thus may oppose/prevent implementation of adaptive projects (e.g., such as those that have ground-disturbing elements like salvaging harvests after disturbance and using herbicides before revegetating). Appeals and litigation from external public often results in the default of no action (See Chapter 3 section 3.4.5).	Inform public and promote consensus-building on tough decisions; invite input from a broad range of sources to generate buy-in across stakeholder interests.	• Conduct public outreach activities with information on climate impacts and adaptation options—including demonstration projects with concrete results—through workshops, scoping meetings, face-to-face dialog, and informal disposition processes to raise public awareness and buy-in for specific management actions (e.g., like Tahoe NF, Annex A1.1 and Partnership for the Sounds (the Estuarium) and North Carolina Aquariums, Annex A5.1). • Use state and local stakeholders to develop management plans to gain support and participation in implementation and oversight of planning activities, as the National Estuary CCMPs do (see Chapter 7 section 7.2.2), the Coastal Habitat Protection Plans do for fisheries management (see Chapter 7 section 7.5), and some National Forests do (Chapter 3 section 3.5).

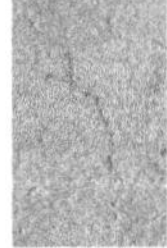

approaches that address climate change. The complexity of climate models poses a barrier to adequately understanding future scenarios and how to react to them, and gaps in tools and resource availability limit the ability of managers to prioritize actions to address climate change (Chapter 3, National Forests). Of particular importance is the need to establish tools to help identify tradeoffs in different management decisions and understand how those tradeoffs would affect particular variables of interest (*e.g.*, air quality levels from prescribed fires versus high-intensity natural fires).

Another gap exists between stakeholder information and expertise compared with that held by resource managers and scientists. Stakeholders often do not have full information, sufficient expertise, or a long-term perspective that allow them to evaluate the relative merit of adaptation options. Therefore, they may act to inhibit or even block the use of adaptation in management planning. Strong local preferences can contradict broader agency goals and drive non-optimal decision-making, all of which act to limit or preclude acceptance of proactive management (Chapter 3, National Forests).

9.5.4.2 Opportunities

Although barriers exist, effective collaboration and linkages among managers and resource scientists are possible (Table 9.8). Scientists can support management by targeting their research to provide managers with information relevant to major management challenges, which would enable managers to make better-informed decisions as new resource issues emerge. Resource scientists have monitoring data and research results that are often underused or ignored. Monitoring efforts that have specific objectives and are conducted with information use in mind would make the data more useful for managers. The need for monitoring efforts may provide impetus for a more unified approach across agencies or management regions. This would serve to not only provide more comprehensive information but would also serve to minimize costs associated with monitoring efforts.

A unified effort is also needed to invest resources and training into the promotion of agile approaches to adaptation management across all federal resource agencies and land or water managers. This would include producing general guidance in terms of the likely impacts of concern, and the implications of these impacts for ecosystem services and management. It would also mean expending efforts to develop "climate science translators" who are capable of translating the projections of climate models to managers and planners who are not trained in the highly specialized field of GCMs. These translators would be scientists adept at responding to climate change who help design adaptive responses. They would also function as outreach staff who would explain to the public what climate change might mean to long-standing recreational opportunities or management goals.

Many federal lands and waters provide excellent opportunities for educating the public about climate change. The national parks and wildlife refuges already put extensive resources into education and outreach for environmental, ecological, and cultural subjects. There are several ways in which the agencies can inform the public about climate change and climate-change impacts. The first of these uses traditional communication venues such as information kiosks and signs, documentaries, and brochures. Interactive video displays are well suited to demonstrating the potential effects of climate change. Such displays could demonstrate the effects of different climate-change scenarios on specific places or systems, making use, for example, of photos or video documenting coral bleaching and retreating glaciers, or modeling studies projecting changes in specific lands or waters (Kerr, 2004; 2005).

The second major way that agencies can inform the public is to provide examples of sustainable practices that reduce greenhouse gas emissions. The National Park Service's Climate Friendly Parks program is a good example of such an outreach effort. The program involves a baseline inventory of park emissions using Environmental Protection Agency models and then uses that inventory to develop methods for reducing emissions, including coordinating transportation, implementing energy-saving technology, and reducing solid waste. Similar programs could easily be developed for other agencies.

9.6 ADVANCING THE NATION'S CAPABILITY TO ADAPT

Until now, we have discussed specific details and concepts for managers to consider relating to adapting to climate change. When all of these details and case studies are pulled together it is the opinion of the authors of this report that the following fundamental strategic foci will aid in achieving adaptation to climate change: (1) have a rational approach for establishing priorities and triage; (2) make sure the management is done at appropriate scales, and not necessarily simply the scales of convenience or tradition; (3) manage expecting change; and (4) increase collaboration among agencies in research and management activities.

In order to understand how these conclusions were reached, one needs only to appreciate that for virtually every category of federal land and water management, one is likely to find situations that exist in which currently available adaptation strategies will not enable a manager to meet specific goals, especially where those goals are related to keeping ecosystems unchanged or species where they are. The expert opinion of the report authors is that these circumstances may require fundamental shifts in how ecosystems are managed. Such shifts may entail reformulating goals, managing cooperatively across landscapes, and looking forward to potential future ecosystem states and facilitating movement toward those preferred states. These sorts of fundamental shifts in management at local-to-regional scales may only be possible with coincident changes in organizations at the national level that empower managers to make the necessary shifts. Thus, fundamental shifts in national-level policies may also be needed.

Even with actions taken to limit greenhouse gas emissions in the future, such shifts in management and policies may be necessary since concentrations resident in the atmosphere are significant enough to require planning for adaptation actions today (Myers, 1979). Ecosystem responses to the consequences of increasing concentrations are likely to be unusually fast, large, and non-linear in character. More areas are becoming vulnerable to climate change because of anthropogenic constraints compounding natural barriers to biological adaptations.

The types of changes that may be needed at the national level include modification of priorities across systems and species and use of new rules for triage; enabling management to occur at larger scales and for projected ecological changes; and expansion of interagency collaboration and access to expertise in climate change science and adaptation, data, and tools. Although many agencies have embraced subsets of these needed changes, there are no examples of the full suite of these changes being implemented as a best practices approach.

9.6.1 Re-Evaluate Priorities and Consider Triage

Climate change not only requires consideration of how to adapt management approaches, it also requires reconsideration of management objectives. In a world with unlimited resources and staff time, climate adaptation would simply be a matter of management innovation, monitoring, and more accessible and useable science. In reality, priorities may need to be re-examined and re-established to focus adaptation efforts appropriately and make the best use of limited resources. At the regional scale, one example of the type of change that may be needed is in selected estuaries where freshwater runoff is expected to increase and salt water is expected to penetrate further upstream. Given this scenario, combined with the goal of protecting anadromous fishes, models could be used to project shifts in critical propagation habitats and management efforts could be refocused to those sites (Chapter 7, National Estuaries). In Rocky Mountain National Park, because warmer winters are expected to result in greatly increased elk populations, a plan to reduce elk populations to appropriate numbers is being prepared with the goal of population control (Annex A2).

In the situations above, the goals are still attainable with some modifications. However, in general, resource managers could face significant constraints on their authority to re-prioritize and make decisions about which goals to modify and how to accomplish those modifications. National-level policies may have to be re-examined with thought toward how to accommodate and even enable such changes in management at the regional level. This re-examination of policies at the national level is another form of priority-setting. Similar to regional-level prioritization, prioritization at the national level would require information at larger scales about the distribution of natural resources and conservation targets, the vulnerability of those targets to climate change, and costs of different management actions in different systems. Prioritization schemes may weight these three factors in different ways, depending on goals and needs. Knowing where resources and conservation targets are is relatively straightforward, although even baseline information on species distributions is often lacking (Chapter 5, National Wildlife Refuges; Chapter 6, Wild and Scenic Rivers). Prioritization schemes that weight rare species or systems heavily would likely target lands with more threatened and endangered species and unique ecosystems.

Because climate-driven changes in some ecological systems are likely to be extreme, priority-setting may, in some instances, involve triage (Metzger, Leemans, and Schröter, 2005). Some goals may have to be abandoned and new goals established if climate change effects are severe enough. Even with substantial focused and creative management efforts, some systems may not be able to maintain the ecological properties and services that they provide in today's climate. In other systems, the cost of adaptation may far outweigh the ecological, social, or economic returns it would provide. In such cases, resources may be better invested in other systems. One simple example of triage would be the decision to abandon habitat management efforts for a population of an endangered species on land at the "trailing" edge of its shifting range. If the refuge or park that currently provides habitat for the species will be unsuitable for the species in the next 50 years, it might be best to actively manage for habitat elsewhere and, depending on the species and the circumstances, investigate the potential for relocation. Such decisions will have to be made with extreme care. In addition to evaluating projected trends in climate and habitat suitability, it will be necessary to monitor the species or habitats in question to determine whether the projected trends are being realized. All of the changes in management approaches discussed throughout the rest of this section would likely require fundamental changes in policy and engagement in triage at the national level.

9.6.2 Manage at Appropriate Scales

Experience gained from natural resource management programs and other activities may offer insights into the application of integrated ecosystem management under changing climatic conditions. Integrated ecosystems management seeks to optimize the positive ecological and socioeconomic benefits of activities aimed at maintaining ecosystem services under a multitude of existing stressors. One lesson learned from this approach is that it may be necessary to define the management scale beyond the boundaries of a single habitat type, conservation area, or political or administrative unit to encompass an entire ecosystem or region. Currently, management plans for forests, rivers, marine protected areas, estuaries, national parks, and wildlife refuges are often developed for discrete geographics with specific attributes (species, ecosystems, commodities), without recognition that they may be nested within other systems. For example, marine protected areas are often within national estuaries; wild and scenic rivers are often within national parks. With few exceptions (see Section 9.5.2), plans are not developed with the ability to fully consider the matrix in which they are embedded and the extent to which those attributes may vary over time in response to drivers external to the management system. Climate change adaptation opportunities may be missed if land and water resources are thought of as distinct, static, or out of context of a regional and even continental arena. A better approach would be to systematically broaden and integrate management plans, where possible. Although a single national park or national forest may have limited capacity for adaptation, the entire system of parks and forests and refuges in a region may have the capacity for adaptation. When spatial scales of consideration are larger, federal agencies often have mutually reinforcing goals that may result in the enhancement of their ability to manage cooperatively across landscapes (Leeworthy and Wiley, 2003).

9.6.3 Manage for Change

Agencies have established best practices based on many years of past experience. Unfortunately, dramatic climate change may change the rules of the game, rendering yesterday's best practices tomorrow's bad practices. Experienced managers have begun to realize that they can anticipate changes in conditions, especially conditions that might alter the impacts of grazing, fire, logging, harvesting, park visitation, and so forth. Such anticipatory thinking will be critical, as climate change will likely exceed ecosystem thresholds over time such that strategies to increase ecosystem resilience will no longer be effective. At this point, major shifts in ecosystem processes, structures, and components will be unavoidable, and adaptation will require planning for management of major ecosystem shifts.

For example, some existing management plans identify a desired state (based on structural, ecosystem service, or ecosystem process attributes of the past) and then prescribe practices to achieve that state. While there is clarity and accountability in such fixed management objectives, these objectives may be unrealistic in light of dramatic environmental change. A desirable alternative management approach may be to "manage for change." For example, when revegetation and silviculture are used for post-disturbance rehabilitation, species properly suited to the expected future climate could be used. In Tahoe National Forest, white fir could be favored over red fir, pines could be preferentially harvested at high elevations over fir, and species could be shifted upslope within expanded seed transfer guides (Chapter 3, National Forests). It is also possible that, after accounting for change, restoration may cease to be an appropriate undertaking. Again, in Tahoe National Forest, warming waters may render selected river reaches no longer suitable for salmon, so restoration of those reaches may not be a realistic management activity (Chapter 3, National Forests). The same applies to meadows in Tahoe National Forest, where restoration efforts may be abandoned due to possible succession to non-meadow conditions. Management will not be able to prevent change, so it may also be important to manage the public's expectations. For example, the goal of the Park Service is to maintain a park exactly as it always has been, composed of the same tree species (Chapter 4, National Parks),

BOX 9.9. Adaptation options for managing in the context of major climatic and ecological changes.

Adaptation Options for Managing for Change

- Assist transitions, population adjustments, and range shifts through manipulation of species mixes, altered genotype selections, modified age structures, and novel silivicultural techniques.
- Rather than focusing only on historic distributions, spread species over a range of environments according to modeled future conditions.
- Proactively manage early successional stages that follow widespread climate-related mortality by promoting diverse age classes, species mixes, stand diversities, genetic diversity, etc., at landscape scales.
- Identify areas that supported species in the past under similar conditions to those projected for the future and consider these sites for establishment of "neo-native" plantations or restoration sites.
- Favor the natural regeneration of species better adapted to projected future conditions.
- Realign management targets to recognize significantly disrupted conditions, rather than continuing to manage for restoration to a "reference" condition that is no longer realistic given climate change.
- Manage the public's expectations as to what ecological states will be possible (or impossible) given the discrepancy between historical climate conditions and current/future climate conditions.
- Develop guidelines for scenarios under which restoration projects or rebuilding of human structures should occur after climate disturbances.

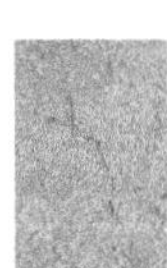

and the public may not recognize the potential impossibility of this goal. Some additional examples of adaptation options for managing for change are presented in Box 9.9.

Scenario-based planning can be a useful approach in efforts to manage for change. As discussed in Section 9.3.3.2, this is a qualitative process that involves exploration of a broad set of scenarios, which are plausible—yet very uncertain—stories or narratives about what might happen in the future. Protected-area managers, along with subject matter experts, can engage in scenario planning related to climate change and resources of interest and put into place plans for both high-probability and low-probability, high-risk events. Development of realistic plans may require a philosophical shift concerning when restoration is an appropriate post-disturbance response. It is impractical to attempt to keep ecosystem boundaries static. Estuaries display this poignantly. After a flood, there is often intense pressure to restore to the pre-flooding state (Chapter 7, National Estuaries). To ensure sound management responses, guidelines for the scenarios under which restoration and rebuilding should occur could be established in advance of disturbances. In this sense, disturbances could become opportunities for managing toward a distribution of human population and infrastructure that is more realistic given changing climate.

9.6.4 Expand Interagency Collaboration, Integration, and Lesson-Sharing

The scale of the challenge posed by climate disruption and the uncertainty surrounding future changes demand coordinated, collaborative responses that go far beyond traditional "agency-by-agency" responses to stressors and threats. Every chapter in this volume has noted the need for a structured, interagency effort and for partnerships and collaboration in everything from research to management and land acquisition. Scientists and mangers across agencies and management systems would benefit from greater sharing of data, models, and experiences. It may be necessary to develop formal structures and policies that foster extensive interagency cooperation.

One example of how to enhance the incorporation of climate information into management could be to designate climate experts to advise agency scientists and managers on climate change related issues. They could advise agency scientists and managers both at the national and at the site level, providing guidance, translating climate-impact projections, and coordinating interagency collaborations.

In the area of climate change science, one interagency program established specifically to address climate change research is the U.S. Climate Change Science Program (CCSP). The goals of this program are to develop scientific knowledge of the climate system; the causes of changes in this system; and the effects of such changes on ecosystems, society, and the economy; and also to determine how best to apply that knowledge to decision-making. Climate change research conducted across 13 U.S. government departments and agencies is coordinated through the CCSP. The CCSP could be expanded to include management research and coordination to bridge the gap between resource management needs and scientific research priorities. This may enhance the goal of the CCSP to apply existing knowledge to decision-making.

There are also other examples of existing collaborations across agencies that could be used as models. Several examples of interagency initiatives established to address universal threats to resources include the National Invasive Species Council, the Joint Fire Science Program, and National Interagency Fire Center. The analogy for climate change adaptation would be a group that would coordinate management activities, interpret research findings, inform on priority-setting, and disseminate data and tools.

Any collaborative interagency effort would benefit from coordinating regional and national databases with scientific and monitoring data to increase the capacity to make informed decisions related to climate-induced changes. Pooling resources would allow for more effective data generation and sharing. Coordination could be done through easily accessible databases that can access and readily provide comprehensive information and serve to better inform managers and decision-makers in their efforts to adapt to climate change. Information on climate-change

projections and climate-change-related research could also be included. Ideally, this would be a web-based clearinghouse with maps, a literature database, and pertinent models (*e.g.* sea level projection models such as the Sea Level Affecting Marshes Model [SLAMM] and hydrology models such as those developed and used by the USGS[8] and EPA.[9] All maps, data, models, and papers could be easily downloaded and updated frequently as new information becomes available.

Collaborations through national councils or interagency efforts may gain the greatest momentum and credibility when they address on-the-ground management challenges. There are several nascent collaborative networks that may provide models for success, such as the Greater Yellowstone Coalition and some collaborative research and management coalitions built around marine protected areas and wild and scenic rivers. These sorts of networks are critical to illustrating how to overcome the challenges posed by lack of funding, and how to create critical ecological and sociological connectivity. With strong leadership, a systematic national network of such coalitions could lead to increased adaptive capacity across agencies and may set precedents for coordinating approaches among regional, state, and local-level management agencies.

9.7 CONCLUSIONS

Information on climate trends and climate impacts has increased dramatically within the last few years. The public, business leaders, and political leaders now widely recognize the risks of climate change and are beginning to take action. While a great deal of discussion has focused on emissions reductions and policies to limit climate change, many may not realize that—no matter which policy path is taken—some substantial climate change, uncertainty, and risk are inevitable. Moreover, the climate change that is already occurring will be here for years to come. Adaptation to climate change will therefore be necessary. Although there are constraints and limits to adaptation, some adaptation measures can go a long way toward reducing the loss of ecosystem services and limiting the economic or social burden of climate disruption. However, if the management cultures and planning approaches of agencies continue with a business-as-usual approach, it is likely that ecosystem services will suffer major degradation. It is the opinion of this report's authors and expert stakeholders that we may be seeing a tipping point in terms of the need to plan and take appropriate action on climate adaptation.

These experts believe that the current mindset toward management of natural resources and ecosystems may have to change. The spatial scale and ecological scope of climate change may necessitate that we broaden our thinking to view the natural resources of the United States as one large interlocking and interacting system, including state, federal, and private lands, with resilience emerging from coordinated stewardship of all of the parts. To achieve this, institutions may have to collaborate and cooperate more. Under conditions of uncertain climatic changes combined with uncertain ecosystem responses, agile management may have to become the rule rather than the exception. While energy corporations, insurance firms, and coastal developers are beginning to adapt to climate change, it is essential that federal agencies responsible for managing the nation's land and water resources also develop management agility and deftness in dealing with climate disruptions. Maladaptation—adaptation that does not succeed in reducing vulnerability but increases it instead—must be avoided. Finally, to adapt to climate change, managers need to know in advance where the greatest vulnerabilities lie. In response to vulnerability analyses, agencies and the public can work together to bolster the resilience of those ecosystems and ecosystem services that are both valuable and capable of remaining viable into the future.

It is crucial to emphasize that adaptation is not simply a matter of managers figuring out what to do, and then setting about to change their practices. All management is conducted within a broader context of socioeconomic incentives

[8] **U.S. Geological Survey**, 1-4-2007: USGS water resources National Research Program (NRP) models. USGS Website, http://water.usgs.gov/nrp/models.html, accessed on 6-12-2007.

[9] **U.S. Environmental Protection Agency**, 4-27-2007: Better assessment science integrating point & nonpoint sources. U.S.Environmental Protection Agency Website, http://www.epa.gov/waterscience/basins, accessed on 6-12-2007.

and institutional behaviors. This means it is essential to make sure that polices that seem external to the federal land and water resource management agencies do not undermine adaptation to climate change. One of the best examples of this danger is private, federal, and state insurance for coastal properties that are at risk of repeated storm damage or flooding. As long as insurance and mortgages are available for coastal building, coasts will be developed with seawalls and other hardened structures that ultimately interfere with beach replenishment, rollback of marshes, and natural floodplains. At first glance one would not think that mortgages and insurance had anything to do with the adaptation of national estuaries to climate change, but in fact these economic incentives and constraints largely dictate the pattern of coastal development.

Federal lands and waters do not function in isolation from human systems or from private land or water uses. For this reason, mechanisms for reducing conflict among private property uses and federal lands and waters are essential. For example, the National Park Service is working cooperatively with landowners bordering the Rio Grande in Texas to establish binding agreements that offer them technical assistance with measures to alleviate potentially adverse impacts on the river resulting from their land-use activities. In addition, landowners may voluntarily donate or sell lands or interests in lands (*i.e.*, easements) as part of a cooperative agreement. In the absence of agreements with private landowners, withdrawals from rivers and loss of riparian vegetation could foreclose opportunities for adaptation, potentially exacerbating the impacts of climate change.

One adaptive response is large protected areas and replicated protected areas, but they are often associated with taking areas of land or ocean away from productive activities such as ranching, farming, or fishing. However, protected areas have multiple beneficial effects on the economy that are also important to consider. For example, in the Florida Keys it has been shown that total annual spending by recreating visitors to the Florida Keys was $1.2 billion between June 2000 and May 2001 (IPCC, 2007).

Society can adapt to climate change through technological solutions and infrastructure, through behavioral choices (altered food and recreational choices), through land management practices, and through planning responses (Johnson and Weaver, 2008). Although federal resource management agencies will tend to adapt by altering management policies, the effectiveness of those policies will be constrained by or enhanced by all of the other societal responses. In general, the federal government's authority over national parks, national forests, and other public resources is most likely to remain effective if management is aligned with the public's well-being and perception of well-being. Experienced resource managers recognize this and regularly invest in public education. This means that education and communication regarding managing for adaptation needs just as much attention as does the science of adaptation.

Repeatedly, in response to crises and national challenges, the nation's executive and congressional leadership have mandated new collaboration among agencies, extended existing authorities, and encouraged innovation. The report authors and expert stakeholders conclude that this is exactly what is needed to adapt to climate change. The security of land and water resources and critical ecosystem services requires a national initiative and leadership. Greater agility will be required than has ever before been demanded from major land or water managers. The public has become accustomed to stakeholder involvement in major resource use decisions. This involvement cannot be sacrificed, but decision-making processes could be streamlined so that management approaches do not stand still while climate change proceeds rapidly. The specific recommendations for adaptation that emerge from studies of national forests, national parks, national wildlife refuges, wild and scenic rivers, national estuaries, and marine protected areas will not take root unless there is leadership at the highest level to address climate adaptation.

APPENDIX

Resources for Assessing Climate Vulnerability and Impacts

NCAR's MAGICC and SCENGEN
http://www.cgd.ucar.edu/cas/wigley/magicc/index.html
Coupled, user-friendly interactive software suites that allow users to investigate future climate change and its uncertainties at both the global-mean and regional levels.

WALTER
http://java.arid.arizona.edu/ahp/
Fire-Climate-Society (FCS-1) is an online, spatially explicit strategic wildfire planning model with an embedded multi-criteria decision process that facilitates the construction of user-designed risk assessment maps under alternative climate scenarios and varying perspectives of fire probability and values at risk.

North American Regional Climate Change Assessment Program
http://www.narccap.ucar.edu/

Regional Hydro-Ecologic Simulation Tool
http://geography.sdsu.edu/Research/Projects/RHESSYS

U.S. Climate Division Dataset Mapping Tool
http://www.cdc.noaa.gov/USclimate/USclimdivs.html
http://www.cdc.noaa.gov/cgi-bin/PublicData/getpage.pl
This tool can generate regional maps.

ISPE/Weiss/Overpeck climate change projections for West (based on IPCC)
http://www.geo.arizona.edu/dgesl/research/regional/projected_US_climate_change/projected_US_climate_change.htm

High Plains Regional Climate Center
http://www.hprcc.unl.edu/

Intergovernmental Panel On Climate Change
http://www.ipcc.ch/
Climate change reports, graphics, summaries.

The Hadley Centre
http://www.metoffice.gov.uk/research/hadleycentre/index.html
Coarse scale global temperature, soil moisture, sea level, and sea-ice volume and area projections.

National Center for Atmospheric Research (NCAR)
http://www.ucar.edu/research/climate/
Coarse resolution climate-change projections, regional climate model.

Pew Center on Global Climate Change
http://www.pewclimate.org/what_s_being_done/
Background on climate change, policy implications.

NOAA Earth System Research Lab (Climate Analysis Branch)
http://www.cdc.noaa.gov/
Current climate data and near-term forecasts.

The Climate Institute
http://www.climate.org/climate_main.shtml
Basic background information on climate change.

U.S. Global Change Research Information Office
http://www.gcrio.org/
Reports and information about climate change.

Real Climate
http://www.realclimate.org/
In-depth discussions with scientists about many different aspects of climate change.

EPA Sea Level Rise
http://yosemite.epa.gov/oar/globalwarming.nsf/content/ResourceCenterPublicationsSeaLevelRiseIndex.html
Reports and impact projections.

CLIMAS, Climate Assessment for the Southwest
http://www.ispe.arizona.edu/climas/
A source for climate change related research, short-term forecasts and climate reconstructions for the southwestern United States.

Climate Impacts Group, University of Washington
http://www.cses.washington.edu/cig/
Climate-change research and projections for the Pacific Northwest.

GLOSSARY

adaptation
Adaptation to climate change refers to adjustment in natural or human systems in response to actual or expected climatic stimuli or their effects, which moderates harm or exploits beneficial opportunities.

adaptive capacity
The ability of a system to adjust to climate change (including climate variability and extremes) to moderate potential damages, to take advantage of opportunities, or to cope with the consequences.

adaptive governance
Institutional and political frameworks designed to adapt to changing relationships between society and ecosystems in ways that sustain ecosystem services; expands the focus from adaptive management of ecosystems to address the broader social contexts that enable ecosystem-based management.

adaptive management
A process that promotes flexible decision making that can be adjusted in the face of uncertainties as outcomes from management actions and other events become better understood. Careful monitoring of these outcomes both advances scientific understanding and helps adjust policies or operations as part of an iterative learning process. It also recognizes the importance of natural variability in contributing to ecological resilience and productivity.

anthropogenic stress
(1) Stressors resulting from or produced by human beings (see "stressor" definition below); (2) Any human activity that causes an ecosystem response that is considered negative.

anticipatory adaptation
Adaptation that takes place before impacts of climate change are observed. Also referred to as proactive adaptation.

biodiversity
(1) The variability among living organisms from all sources including, inter alia, terrestrial, marine and other aquatic ecosystems and the ecological complexes of which they are part; this includes diversity within species, between species and of ecosystems. (2) The diversity of genes, populations, species, communities, and ecosystems, which underlies all ecosystem processes and determines the environment on which organisms, including people, depend.

catastrophic event
(1) A sudden natural or man-made disturbance that causes widespread destruction. (2) In the context of climate change, a suddenly occurring event having wide distribution and large impacts on human and/or natural systems (*e.g.*, mass extinctions, rapid sea level rise, or shifts in atmospheric or oceanic circulation patterns over less than a decade). Such events have occurred in the past due to natural causes.

climate change
Climate change refers to any change in climate over time, whether due to natural variability or as a result of human activity. This usage differs from that in the United Nations Framework Convention on Climate Change, which defines "climate change" as: "a change of climate which is attributed directly or indirectly to human activity that alters the composition of the global atmosphere and which is in addition to natural climate variability observed over comparable time periods."

climate change scenario
A plausible and often simplified representation of the future climate, based on an internally consistent set of climatological relationships, typically constructed for explicit use as input to impact models. A "climate change scenario" is the difference between a climate scenario and the current climate.

climate variability
Climate variability refers to variations in the mean state and other statistics (such as standard deviations, statistics of extremes, etc.) of the climate on all temporal and spatial scales beyond that of individual weather events. Variability may be due to natural internal processes within the climate system (internal variability), or to variations in natural or anthropogenic external forcing (external variability).

confidence (in an adaptation approach)
Degree of belief that an event will occur given observations, modeling results, and current knowledge. In this report, confidence is based on the expert opinion of the authors and is composed of two elements: (1) the amount of evidence available to support the determination that the effectiveness of a given adaptation approach is well-studied and understood and (2) the level of agreement or

consensus within the scientific community about the different lines of evidence on the effectiveness of that adaptation approach.

disturbance regime
Frequency, intensity, and types of disturbances, such as fires, insect or pest outbreaks, floods, and droughts.

ecoregions
Areas of general similarity in ecosystems and in the type, quality, and quantity of environmental resources.

ecosystem
A system of interacting living organisms together with their physical environment.

ecosystem management, or ecosystem-based management
There are many definitions for this term, and different agencies interpret the term in slightly different ways. Three definitions follow; the first is frequently cited. (1) Management that integrates scientific knowledge of ecological relationships within a complex sociopolitical and values framework toward the general goal of protecting native ecosystem integrity over the long term. (2) Any land-management system that seeks to protect viable populations of all native species, perpetuate natural disturbance regimes on the regional scale, adopt a planning timeline of centuries, and allow human use at levels that do not result in long-term ecological degradation. (3) The application of ecological and social information, options, and constraints to achieve desired social benefits within a defined geographic area over a specified period.

ecosystem services
The conditions and processes through which natural ecosystems, and the species that make them up, sustain and fulfill human life.

extreme weather event
An event that is rare within its statistical reference distribution at a particular place. Definitions of "rare" vary, but an extreme weather event would normally be as rare as or rarer than the 10th or 90th percentile. By definition, the characteristics of what is called "extreme weather" may vary from place to place. Extreme weather events may typically include floods and droughts.

global change
Changes in the global environment (including alterations in climate, land productivity, oceans or other water resources, atmospheric chemistry, and ecological systems) that may alter the capacity of the Earth to sustain life.

human social resilience
The capacity to absorb shocks while maintaining function.

impacts (climate change)
The effects of climate change on natural and human systems. Depending on the consideration of adaptation, one can distinguish between potential impacts and residual impacts:

Potential impacts: All impacts that may occur given a projected change in climate, without considering adaptation.

Residual impacts: The impacts of climate change that would occur after adaptation.

invasive species
Non-native species whose introduction does or is likely to cause economic or environmental harm or harm to human health.

likelihood
The likelihood of an occurrence, an outcome or a result, where this can be estimated probabilistically.

maladaptation
Any changes in natural or human systems that inadvertently increase vulnerability to climatic stimuli; an adaptation that does not succeed in reducing vulnerability but increases it instead.

management plan
In general, a document that provides guidance regarding all activities on federally managed lands. However, the meaning for National Forests is quite distinct. Specifically, the National Forest Management Act (NFMA (16 U.S.C. 1660(6)) requires the Forest Service to manage the National Forest System lands according to land and resource management plans that provide for multiple-uses and sustained-yield in accordance with MUSYA (16 U.S.C. 1604(e) and (g)(1)), in particular include coordination of outdoor recreation, range, timber, watershed, wildlife and fish, and wilderness and determine forest management systems, harvesting levels, and procedures in the light of all of the uses set forth in the Multiple-Use Sustained Yield Act of 1960, and the availability of lands and their suitability for resource management.

mitigation
An anthropogenic intervention to reduce the anthropogenic forcing of the climate system; strategies to reduce greenhouse gas sources and emissions and enhance greenhouse gas sinks.

native species
With respect to a particular ecosystem, a species that, other than as a result of an introduction, historically occurred or currently occurs in that ecosystem.

non-native species
Also referred to as "alien," "exotic," and ""introduced" species. This term refers to any species (including its seeds, eggs, spores, or other biological material capable of propagating that species) that is not native to a particular geographic region. Non-native species may, or may not be, invasive.

organic acts
Organic acts are fundamental pieces of legislation that signify the organization of an agency and/or provide a charter for a network of public lands. The first "organic act" was the Organic Administration Act of 1897, which outlined the primary purposes of national forests as (1) securing favorable conditions of water flows, and (2) furnishing a continuous supply of timber for the use and necessities of the citizens of the United States.

phenology
The timing of behavior cued by environmental information.

reactive adaptation
Adaptation that takes place after impacts of climate change have been observed.

realignment
Considered in the context of restoration, realignment refers to an adjustment in management or planning goals to account for substantially altered reference conditions and new ecosystem dynamics. The rationale for this adaptation approach is that historical (pre-disturbance) baselines may be inappropriate in the face of a changing climate.

refugia
Physical environments that are less affected by climate change than other areas (*e.g.*, due to local currents, geographic location, etc.) and are thus a "refuge" from climate change for organisms.

relocation
Human-facilitated transplantation of organisms from one location to another in order to bypass a barrier (*e.g.*, an urban area). Also referred to as "assisted migration."

replication
Multiple replicates of a habitat type (*e.g.*, multiple fore reef areas throughout the reef system) or population are protected as a "bet hedging" strategy against loss of the habitat type due to a localized disaster.

representation
Includes both (1) ensuring that the full breadth of habitat types is protected (*e.g.*, fringing reef, fore reef, back reef, patch reef) and (2) ensuring that full breadth of species diversity is included within sites; both concepts relate to maximizing overall biodiversity of the larger system.

resilience
The amount of change or disturbance that can be absorbed by a system before the system is redefined by a different set of processes and structures (*i.e.*, the ecosystem recovers from the disturbance without a major phase shift).

resistance
Ecological resistance is the ability of an organism, population, community, or ecosystem to withstand perturbations without significant loss of structure or function. From a management perspective, resistance includes both (1) the concept of taking advantage of/boosting the inherent (biological) degree to which species are able to resist change and (2) manipulation of the physical environment to counteract/resist physical/biological change.

restoration
Manipulation of the physical and biological environment in order to restore a desired ecological state or set of ecological processes.

sensitivity
Sensitivity is the degree to which a system is affected, either adversely or beneficially, by climate-related stimuli. The effect may be direct (*e.g.*, a change in crop yield in response to a change in the mean, range, or variability of temperature) or indirect (*e.g.*, damages caused by an increase in the frequency of coastal flooding due to sea-level rise).

stressor
Any physical, chemical, or biological entity that can induce an adverse response.

surprises
(1) Sudden, unexpected changes in the environment (biotic or abiotic) that may have disproportionately large ecological consequences. (2) In the context of climate change, unexpected events resulting from climate change (such as a shift in ocean circulation) that may have both positive and negative consequences. (3) In the context of social-ecological systems, a qualitative disagreement between ecosystem behavior and *a priori* expectations—an environmental cognitive dissonance.

trust species
All species where the federal government has primary jurisdiction including federally endangered or threatened species, migratory birds, anadromous fish, and certain marine mammals.

unimpaired
Refers to language in the NPS Organic Act that describes the purpose for which National Parks were established: "... to conserve the scenery and the natural and historic objects and the wild life therein and to provide for the enjoyment of the same in such manner and by such means as will leave

them unimpaired for the enjoyment of future generations." "Unimpaired" generally means "not damaged or diminished in any respect."

vulnerability
The degree to which a system is susceptible to, or unable to cope with, adverse effects of climate change, including climate variability and extremes. Vulnerability is a function of the character, magnitude, and rate of climate variation to which a system is exposed, its sensitivity, and its adaptive capacity.

wilderness management
(1) Management activities that aim to preserve the wilderness character of designated wilderness areas, which are "...area[s] where the earth and its community of life are untrammeled by man, where man himself is a visitor who does not remain." (2) The planning for and management of wilderness resources.

ACRONYMS AND INITIALISMS

ADCP Acoustic Doppler Current Profilers
ANILCA Alaska National Interest Lands Conservation Act
AOGCM Atmosphere-Ocean Coupled General Circulation Model
APES Albemarle-Pamlico Estuarine System
APHIS Animal and Plant Health Inspection Service
APNEP Abemarle-Pamlico National Estuarine Program
AQRV Air Quality Related Values
ATBA Area to Be Avoided
ATBI All Taxa-Biodiversity Inventory
ATV All-Terrain vehicle
AVHRR Advanced Very High Resolution Radiometer
BLM Bureau of Land Management
$CaCO_3$ Calcium Carbonate
CCMP Comprehensive Conservation and Management Plan
CCP Comprehensive Conservation Plan
CCSP Climate Change Science Program
CDFG California Department of Fish and Game
CERP Comprehensive Everglades Restoration Plan
CHPP Coastal Habitat Protection Plan
CINMS Channel Islands National Marine Sanctuary
CO_2 Carbon Dioxide
CoRIS Coral Reef Information System
CRED Coral Reef Ecosystem Division
CREIOS Coral Reef Ecosystem Integrated Observing System
CREWS Coral Reef Early Warning System
CRMP Comprehensive River Management Plan
CRP Conservation Reserve Program
CTD casts Water Conductivity-Temperature-Depth profiles
CWA Clean Water Act
CWMTF Clean Water Management Trust Fund
DDT Dichloro-diphenyl-trichloroethane
DEFRA United Kingdom Department for Environment Food and Rural Affairs
DGVM Dynamic Global Vegetation Model
DO Dissolved Oxygen
DRBC Delaware River Basin Commission
EBM Ecosystem-Based Management
EDRR Early Detection and Rapid Response
EEP Ecosystem Enhancement Program
EMA Existing Management Area

EMS	Environmental Management System
ENSO	El Niño/Southern Oscillation
EPA	Environmental Protection Agency
ERA	Estuary Restoration Act
ESA	Endangered Species Act
EU	European Union
FEMA	Federal Emergency Management Agency
FHP	U.S. Forest Service Forest Health Protection Program
FKNMS	Florida Keys National Marine Sanctuary
FKNMS Act	Florida Keys National Marine Sanctuary and Protection Act
FMP	Fishery Management Plan
FONSI	Finding of No Significant Importance
FPA	Forest Plan Amendment
FPR	Forest Plan Revision
GBR	Great Barrier Reef
GBRMPA	Great Barrier Reef Marine Park Authority
GBRNP	Great Barrier Reef National Park
GCM	General Circulation Model
GDP	Gross Domestic Product
GIS	Geographic Information Systems
GtC	Gigaton Carbon
HINWR	Hawaiian Islands National Wildlife Refuge
ICRAN	International Coral Reef Action Network
IOOS	Integrated Ocean Observing System
IPCC	Intergovernmental Panel on Climate Change
IUCN	International Union for the Conservation of Nature/World Conservation Union
LAPS	Land Acquisition Priority System
LIDAR	Light Detection and Ranging
LMP	Land and Resource Management Plan
LTER	Long-Term Ecological Research
MHI	Main Hawaiian Islands
MMA	Marine Managed Area
MPA	Marine Protected Area
MSA	Magnuson-Stevens Fishery Conservation Management Reauthorization Act
MSX	Multinucleate Sphere X, a parasite affecting oysters
NAO/NHM	North Atlantic Oscillation/Northern Hemisphere Annular Mode
NAWQA	National Water Quality Assessment
NEON	National Ecological Observatory Network
NEP	National Estuary Program
NEPA	National Environmental Policy Act
NF	National Forest
NFMA	National Forest Management Act
NFS	National Forest System
NGO	Non-Governmental Organization
NMSA	National Marine Sanctuaries Act
NMSP	National Marine Sanctuary Program
NOAA	National Oceanic and Atmospheric Administration
NOx	Nitrogen Oxides
NPDES	National Pollutant Discharge Elimination System
NPS	National Park Service
NRE	Neuse River Estuary
NRI	National Rivers Inventory
NWFP	Northwest Forest Plan
NWHI	Northwestern Hawaiian Islands
NWRS	National Wildlife Refuge System
NWRSIA	National Wildlife Refuge System Improvement Act
OHV	Off-Highway Vehicle
ONF	Olympic National Forest
ONFP	Olympic National Forest Plan
ONP	Olympic National Park
ORION	Ocean Research Interactive Observatory Networks
PCB	Polychlorinated biphenyl
PDO	Pacific Decadal Oscillation
PMNM	Papahānaumokuākea Marine National Monument
PPR	Prairie Pothole Region
PRE	Pamlico River Estuary
RMNP	Rocky Mountain National Park
RPA	Resource Planning Act (1974)
SAC	Sanctuary Advisory Council
SAMAB	Southern Appalachian Man and the Biosphere
SAP 4.4	Synthesis and Assessment Product 4.4.
SAV	Submerged Aquatic Vegetation
SDM	Species Distribution Model
SFA	Sustainable Fisheries Act
SJRWMD	St. Johns River Water Management District
SLAMM	Sea Level Affecting Marshes Model
SPA	Sanctuary Protection Area
SRES	Special Report on Emissions Scenarios
SST	Summer Sea Surface Temperature
SVP	Surface Velocity Program
SW	Southwest
TMDL	Total Maximum Daily Load
TNF	Tahoe National Forest

U.S. EEZ	U.S. Exclusive Economic Zone
UNESCO	United Nations Educational Scientific and Cultural Organization
UNF	Uwharrie National Forest
USACE	U.S. Army Corps of Engineers
USDA	U.S. Department of Agriculture
USFWS	U.S. Fish and Wildlife Service
USGS	U.S. Geological Survey
UW-CIG	University of Washington's Climate Impacts Group
VMS	Vessel Monitoring System
WCA	Watershed Condition Assessment
WMA	Wildlife Management Area
WQPP	Water Quality Protection Program
WSR	Wild and Scenic Rivers
WUI	Wildland Urban Interface
ZIMM	Zonal Inundation and Marsh Model

In order to ensure that the proposed structure and content of each chapter was assessed for technical rigor and feasibility from a management perspective, workshops for a limited set of expert stakeholders were held during the report's earliest development stages. Stakeholders from the management and adaptation research communities were selected from across federal and state governments, territories, non-governmental organizations, and academia to participate in a series of workshops to advise the authors of the report on its content. At each of the six workshops (one for each "management system" chapter), no more than 20 stakeholder participants gathered to have chapter lead and contributing authors present draft information on their chapters and case studies. Stakeholders were able to provide feedback, and authors incorporated the expert input into their revisions.

Name	Affiliation
National Forests	
Paul Arndt	United States Department of Agriculture (USDA) Forest Service, Region 8
Chris Bernabo*	National Council on Science for the Environment (NCSE)
Michael Case	World Wildlife Fund (WWF) Global Climate Change Programme
Bob Davis*	United States Forest Service (USFS), Region 3
Steve Eubanks	USFS Tahoe National Forest
Lee Frelich*	The University of Minnesota Center for Hardwood Ecology
Greg Kujawa	USDA Forest Service
Jeremy Littell*	University of Washington, Climate Impacts Group
Douglas W. MacCleery*	USDA Forest Service
Duane Nelson	USFS Regional Forest Revegetation, Region 5
Kathy A. O'Halloran*	Olympic National Forest
Frank Roth	USDA Forest Service, Region 4
Lindsey Rustad*	USDA Forest Service, Northern Research Station
Hugh Safford*	USFS, Region 5
Charles Sams	USDA Forest Service, Region 9
Allen Solomon*	USDA Forest Service, Washington Office
Jeff Sorkin*	USDA Forest Service
Peter Stine	Sierra Nevada Research Center
John Townsley*	Okanogan and Wenatchee National Forests
Mary Vasse	National Forest Foundation
Bonnie Wyatt	USDA Forest Service
Christina Zarrella	National Commission on Science for Sustainable Forestry
National Parks	
Stan Austin*	Rocky Mountain National Park
Jane Belnap	United States Geological Survey (USGS)
Gillian Bowser*	Texas A&M University
Gregg Bruff	Pictured Rocks National Lakeshore
Hannah Campbell**	National Oceanic and Atmospheric Administration (NOAA), Office of Global Programs, Regional Integrated Sciences and Assessments Program (RISA)
John Dennis**	National Park Service (NPS) Headquarters
Dan Fagre	USGS Northern Rocky Mountain Science Center

Name	Affiliation
Steve Fancy	NPS
David Graber*	Sequoia and Kings Canyon National Parks
John Gross*	NPS Vital Signs Program
Jon Jarvis	NPS
Beth Johnson	NPS
Kathy Jope	NPS
Sharon Kliwinski**	NPS Water Resources Division, Washington Liaison
Bob Krumenaker*	Apostle Islands National Lakeshore
Lloyde Loope	USGS
Abby Miller*	The Coalition of NPS Retirees
Jim Nations	National Parks and Conservation Association
Shawn Norton*	NPS Headquarters
David Parsons	USFS
David Peterson	USDA Forest Service
Mike Soukup*	NPS Headquarters
Lee Tarnay*	Yosemite National Park
Julie Thomas*	NPS
Kathy Tonnessen	University of Montana
Leigh Welling*	Crown of the Continent Research Learning Center
Mark Wenzler*	National Parks Conservation Association (NPCA)
Aaron Worstell	NPS, Air Resources Division
National Wildlife Refuges	
Dan Ashe*	United States Fish and Wildlife Service (USFWS), Refuges and Wildlife
Don Barry	Wilderness Society
Dawn Browne*	Ducks Unlimited
Tom Franklin*	Izaak Walton League
Patrick Gonzalez*	The Nature Conservancy
Lara Hansen	WWF Climate Change Program
Evan Hirsche	National Wildlife Refuge Association (NWRA)
Matt Hogan	Association of Fish and Wildlife Agencies
Doug Inkley*	National Wildlife Federation (NWF)
Danielle G. Jerry*	USFWS (Alaska)
Kurt Johnson*	USFWS
John Kostyack	NWF
James Kurth*	USFWS
Tom Lovejoy	The Heinz Center
Noah Matson*	Defenders of Wildlife
Sean McMahon*	NWF
Claudia Nierenberg	The Heinz Center
Maribeth Oakes*	Wilderness Society
Amber Pairis	Association of Fish and Wildlife Agencies
Camille Parmesan	The University of Texas at Austin
Caryn Rea	ConocoPhillips (Alaska)
Terry Rich	USFWS
John Schoen	Alaska Audubon

Name	Affiliation
Mike Slimak**	United States Environmental Protection Agency (USEPA), National Center for Environmental Assessment (NCEA)
Lisa Sorenson	Boston University
Kim Titus	Alaska Department of Fish and Game
Alan Wentz	Ducks Unlimited
John Wiens	The Nature Conservancy
Michael Woodbridge*	NWRA
Wild and Scenic Rivers	
Daniel M. Ashe*	USFWS, Refuges and Wildlife
Tom Beard	Far West Texas Water Planning Group
Donita Cotter*	National Wildlife Refuge System, Div. of Natural Resources
Jackie Diedrich*	USFS, Region 6
Karen Dunlap	USFS, Region 9, Ottawa National Forest
Andrew Fahlund*	American Rivers, Conservation
Dave Forney*	NPS, Upper Delaware Scenic and Recreational River (SRR)
Dan Haas*	USFWS, Hanford Reach National Monument
Kristy Hajny*	Niobrara, National Scenic River (NSR)
Joan Harn	NPS
Steve Harris	Rio Grande Restoration
John Haubert	NPS, Washington, D.C.
Peter Henn	Land Manager of Wekiva River Buffer Conservation Area for St Johns River Water Management District
Mike Higgins*	USFWS, National Wildlife Refuge System
Phil Horning	USFS, Tahoe National Forest
Quinn McKew*	American Rivers, Wild Rivers Program
Teri McMillan	Alaska Wild and Scenic Rivers (WSR) Program
Jerry Mosier	Klamath, California, multiple agencies
Tim O'Halloran	Yolo County Flood Control and Water Conservation District
David Purkey*	Stockholm Environment Institute-US Center
Jason Robertson*	U.S. Department of the Interior, Bureau of Land Management
Cassie Thomas*	NPS
Richard (Omar) Warner*	Kinni Consulting
National Estuaries	
Mark Alderson*	Sarasota Bay Project
Carol Auer*	NOAA/National Ocean Service
Rich Batiuk*	USEPA Region 3 – Chesapeake Bay Program
Suzanne Bricker	NOAA/National Ocean Service
Dean E. Carpenter*	Albemarle-Pamlico National Estuary Program (NEP)
Derb Carter*	Southern Environmental Law Center, Chapel Hill
James E. Cloern	USGS
Pamela Emerson	City of Seattle
Holly Greening*	Tampa Bay Estuary Program
Michael J. Kennish*	Rutgers University
Wim Kimmerer	Romberg Tiburon Center for Environmental Studies
Karen L. McKee*	USGS National Wetlands Research Center

Name	Affiliation
Doug Rader*	Environmental Defense, Raleigh Regional Office
Curtis J. Richardson*	Duke University Wetland Center, Nicholas School of the Environment and Earth Sciences
Stan Riggs*	East Carolina University, Greenville
Mary Ruckelshaus	NOAA National Marine Fisheries Service (NMFS), Seattle
Mark Schexnayder	Louisiana State University (LSU) Ag Center/Sea Grant
Ron Shultz*	Puget Sound Water Quality Action Team
Jan Smith*	Massachusetts Bays NEP
Katrina Smith Korfmacher*	University of Rochester
Kerry St. Pe	Barataria-Terrebonne NEP
Marine Protected Areas	
Peter Auster	National Undersea Research Center
Maria Brown*	Gulf of the Farallones National Marine Sanctuary (NMS)
Deborah Cramer*	Stellwagen Bank NMS Advisory Council
Andrew DeVogelaere	Monterey Bay NMS
Barbara Emley	Gulf of the Farallones NMS Advisory Council
Daniel Gleason*	Georgia Southern University
Lynne Hale*	The Nature Conservancy
Lara Hansen*	WWF
Sean Hastings	Channel Islands NMS
Terrie Klinger*	University of Washington (UW) School of Marine Affairs
Irina Kogan*	Gulf of the Farallones NMS
David Loomis*	University of Massachusetts
Steve Palumbi	Stanford University
Linda Paul*	Hawaii Audubon Society
Bruce Popham*	Florida Keys NMS Advisory Council
Steve Roady	Earthjustice; Duke University Nicholas School of the Environment and Earth Sciences
Teresa Scott*	Washington Department of Fish & Wildlife
Jack Sobel*	The Ocean Conservancy
Steve Tucker*	Cape Cod Commission
Charles M. Wahle, Ph.D.	NOAA National Marine Protected Areas Center
Lauren Wenzel*	NOAA National Marine Protected Areas Center
Bob Wilson*	Gulf of the Farallones NMS Advisory Council

* Indicates invitees who participated in the workshops
** Indicates participants in the workshops who were not on the original invite list

CHAPTER 2 REFERENCES

Bennett, E.M., G.S. Cumming, and G.D. Peterson, 2005: A systems model approach to determining resilience surrogates for case studies. *Ecosystems*, **8**, 945-957.

Bradshaw, W.E. and C.M. Holzapfel, 2006: Climate change: evolutionary response to rapid climate change. *Science*, **312(5779)**, 1477 1478.

Caldeira, K. and M.E. Wickett, 2005: Ocean model predictions of chemistry changes from carbon dioxide emissions to the atmosphere and ocean. *Journal of Geophysical Research*, **110**, 1-12.

Diffenbaugh, N.S., 2005: Atmosphere-land cover feedbacks alter the response of surface temperature to CO_2 forcing in the western United States. *Climate Dynamics*, **24(2)**, 237-251.

Diffenbaugh, N.S., J.S. Pal, R.J. Trapp, and F. Giorgi, 2005. Fine-scale processes regulate the response of extreme events to global climate change. *Proceedings of the National Academy of Sciences of the United States of America*, **102(44)**, 15774-15778.

Doney, S.C., 2006: The dangers of ocean acidification. *Scientific American*, **294(3)**, 58-65.

Donnelly, J.P. and J.D. Woodruff, 2007: Intense hurricane activity over the past 5,000 years controlled by El Niño and the West African Monsoon. *Nature*, **447**, 465-468.

Franks, S.J., S. Sim, and A.E. Weis, 2007: Rapid evolution of flowering time by an annual plant in response to a climate fluctuation. *Proceedings of the National Academy of Sciences of the United States of America*, **104**, 1278-1282.

Groisman, P.Y., R.W. Knight, D.R. Easterling, T.R. Karl, G.C. Hegerl, and V.N. Razuvaev, 2005: Trends in intense precipitation in the climate record. *Journal of Climate*, **18(9)**, 1326-1350.

Groisman, P.Y., R.W. Knight, and T.R. Karl, 2001: Heavy precipitation and high streamflow in the contiguous United States: trends in the twentieth century. *Bulletin of the American Meteorological Society*, **82(2)**, 219-246.

Groom, M.J., G.K. Meffe, and C.R. Carroll, 2006: *Principles of Conservation Biology*. [Groom, M.J., G.K. Meffe, and C.R. Carroll (eds.)]. Sinauer Press, Sunderland, MA, pp. 1-701.

Gunderson, L.H., 2000: Ecological resilience-in theory and application. *Annual Review of Ecology and Systematics*, **31**, 425-439.

Holling, C.S., 1973: Resilience and stability of ecological systems. *Annual Review of Ecology and Systematics*, **4**, 1-23.

IPCC, 2001: *Climate Change 2001: Impacts, Adaptation, and Vulnerability. Contribution of Working Group II to the Third Assessment Report of the Intergovernmental Panel on Climate Change.* [McCarthy, J.J., O.F. Canziani, N.A. Leary, D.J. Dokken, and K.S. White (eds.)]. Cambridge University Press, Cambridge, UK.

IPCC, 2007a: *Climate Change 2007: Impacts, Adaptation and Vulnerability. Contribution of Working Group II to the Fourth Assessment Report of the Intergovernmental Panel on Climate Change.* Cambridge University Press, Cambridge.

IPCC, 2007b: *Climate Change 2007: the Physical Science Basis. Contribution of Working Group I to the Fourth Assessment Report of the Intergovernmental Panel on Climate Change.* [Solomon, S., D. Quin, M. Manning, Z. Chen, M. Marquis, K.B. Averyt, M. Tignor, and H.L. Miller (eds.)]. Cambridge University Press, Cambridge, United Kingdom and New York, NY, USA, pp. 1-996.

IPCC, 2007c: Summary for policymakers, In: *Climate Change 2007: the Physical Science Basis. Contribution of Working Group I to the Fourth Assessment Report of the Intergovernmental Panel on Climate Change*, [Solomon, S., D. Qin, M. Manning, Z. Chen, M. Marquis, K.B. Averyt, M. Tignor, and H.L. Miller (eds.)]. Cambridge University Press, Cambridge, United Kingdom and New York, NY, USA.

Karl, T.R. and R.W. Knight, 1998: Secular trends of precipitation amount, frequency, and intensity in the United States. *Bulletin of the American Meteorological Society*, **79(2)**, 231-241.

Kleypas, J.A., R.A. Feely, V.J. Fabry, C. Langdon, C.L. Sabine, and L.L. Robbins, 2006: *Impacts of Ocean Acidification on Coral Reefs and Other Marine Calcifiers: a Guide for Future Research*. Workshop Report, National Science Foundation, National Oceanic and Atmospheric Administration, and the U.S. Geological Survey.

Knowles, N., M.D. Dettinger, and D.R. Cayan, 2006: Trends in snowfall versus rainfall in the Western United States. *Journal of Climate*, **19(18)**, 4545-4559.

Landsea, C.W., 2007: Counting Atlantic Tropical Cyclones Back to 1900. *Eos, Transactions American Geophysical Union*, **88(18)**, 197-202.

McCarty, J.P., 2001: Ecological consequences of recent climate change. *Conservation Biology*, **15(2)**, 320-331.

Mote, P.W., A.F. Hamlet, M.P. Clark, and D.P. Lettenmaier, 2005: Declining mountain snowpack in Western North America. *Bulletin of the American Meteorological Society*, **86(1)**, 39-49.

National Research Council, 2007: *Analysis of Global Change Assessments: Lessons Learned.* Committee on Analysis of Global Change Assessments, National Research Council, National Academies Press, Washington, D.C..

Parmesan, C. and H. Galbraith, 2004: *Observed Impacts of Global Climate Change in the US.* Pew Center on Global Climate Change.

Parmesan, C., 2006: Ecological and evolutionary responses to recent climate change. *Annual Review of Ecology, Evolution and Systematics*, **37**, 637-669.

Parmesan, C. and G. Yohe, 2003: A globally coherent fingerprint of climate change impacts across natural systems. *Nature*, **421**, 37-42.

Root, T.L., D.P. MacMynowski, M.D. Mastrandrea, and S.H. Schneider, 2005: Human-modified temperatures induce species changes: joint attribution. *Proceedings of the National Academy of Sciences of the United States of America*, **102(21)**, 7465-7469.

Scheffer, M., S. Carpenter, J.A. Foley, C. Folke, and B.H. Walker, 2001: Catastrophic shifts in ecosystems. *Nature*, **413**, 591-596.

Smith, T.M. and R.W. Reynolds, 2005: A global merged land-air-sea surface temperature reconstruction based on historical observations (1880-1997). *Journal of Climate*, **18(12)**, 2021-2036.

Stewart, I.T., D.R. Cayan, and M.D. Dettinger, 2004: Changes in snowmelt runoff timing in Western North America under a 'business as usual' climate change scenario. *Climatic Change*, **62**, 217-232.

Tebaldi, C., K. Hayhoe, J. Arblaster, and G. Meehl, 2006: Going to the extremes: an intercomparison of model-simulated historical and future changes in extreme events. *Climatic Change*, **79(3-4)**, 185-211.

The Royal Society, 2005: *Ocean Acidification Due to Increasing Atmospheric Carbon Dioxide.* Policy document 12/05, Royal Society.

Tompkins, E.L. and N.W. Adger, 2004: Does adaptive management of natural resources enhance resilience to climate change? *Ecology and Society*, **19(2)**.

Turner, B.L., II, R.E. Kasperson, P.A. Matsone, J.J. McCarthy, R.W. Corell, L. Christensene, N. Eckley, J.X. Kasperson, A. Luerse, M.L. Martello, C. Polsky, A. Pulsipher, and A. Schiller, 2003: A framework for vulnerability analysis in sustainability science. *Proceedings of the National Academy of Sciences of the United States of America Early Edition*, **100(14)**.

U.S. Environmental Protection Agency, 2000: *Stressor Identification Guidance Document.* EPA-822-B-00-025, U.S. Environmental Protection Agency, Office of Water and Office of Research and Development, Washington, DC, pp.1-208.

U.S. Environmental Protection Agency, 2007: *Proposed Indicators for the U.S. EPA's Report on the Environment (External Peer Review).* U.S. Environmental Protection Agency.

Vecchi, G.A. and B.J. Soden, 2007: Increased tropical Atlantic wind shear in model projections of global warming. *Geophysical Research Letters*, **34**.

Williams, K., K.C. Ewel, R.P. Stumpf, F.E. Putz, and T.W. Workman, 1999: Sea-level rise and coastal forest retreat on the west coast of Florida, USA. *Ecology*, **80(6)**, 2045-2063.

CHAPTER 3 REFERENCES

Aber, J.D., C.L. Goodale, S.V. Ollinger, M.L. Smith, A.H. Magill, M.E. Martin, R.A. Hallett, and J.L. Stoddard, 2001: Is nitrogen deposition altering the nitrogen status of northeastern forests? *BioScience*, **53(4)**, 375-389.

Adams, M.B., L. Loughry, and L. Plaugher, 2004: *Experimental Forests and Ranges of the USDA Forest Service.* General Technical Report NE-321, United States Department of Agriculture Forest Service, Northeastern Research Station.

Adger, W.N., 1999: Social vulnerability to climate change and extremes in coastal Vietnam. *World Development*, **27(2)**, 249-269.

Adger, W.N., 2003: Social aspects of adaptive capacity, In: *Climate Change, Adaptive Capacity and Development*, Imperial College Press, London, UK, pp. 29-49.

Adger, W.N., 2006: Vulnerability. *Global Environmental Change*, **16(3)**, 268-281.

Adger, W.N. and P.M. Kelly, 2001: *Living With Environmental Change: Social Vulnerability and Resilience in Vietnam*. Routledge.

Agee, J.K., 1996: *Fire Ecology of Pacific Northwest Forests*. Island Press, Washington, D.C..

Agee, J.K., 1998: The landscape ecology of western forest fire regimes. *Northwest Science*, **72**, 24-34.

Ainsworth, E.A. and S.P. Long, 2005: What have we learned from 15 years of free-air CO_2 enrichment (FACE)? A meta-analytic review of the responses of photosynthesis, canopy properties and plant production to rising CO_2. *New Phytologist*, **165**, 351-372.

Alberto, A.M.P., L.H. Ziska, C.R. Cervancia, and P.A. Manalo, 1996: The influence of increasing carbon dioxide and temperature on competitive interactions between aC 3 crop, rice (*Oryza sativa*) and aC 4 weed (*Echinochloa glabrescens*). *Australian Journal of Plant Physiology*, **23(6)**, 795-802.

Alig, R.J., J.D. Kline, and M. Lichtenstein, 2004: Urbanization on the US landscape: looking ahead in the 21st century. *Landscape and Urban Planning*, **69(2-3)**, 219-234.

Allegheny National Forest, 2006: *Ecological Context*. USDA Forest Service.

Allen, C.D. and D.D. Breshears, 1998: Drought-induced shift of a forest-woodland ecotone: Rapid landscape response to climate variation. *Proceedings of the National Academy of Sciences of the United States of America*, **95(25)**, 14839-14842.

Amman, G.D., 1973: Population changes of the Mountain Pine Beetle in relation to elevation. *Environmental Entomology*, **2(541)**, 547.

Anderson, J.L., R. W. Hilborn, R. T. Lackey, and D. Ludwig, 2003: Watershed restoration - adaptive decision making in the face of uncertainty, In: *Strategies for Restoring River Ecosystems: Sources of Variability and Uncertainty in Natural and Managed Systems*, American Fisheries Society, Bethesda, MD, pp. 202-232.

Angert, A., S. Biraud, C. Bonfils, C.C. Henning, W. Buermann, J. Pinzon, C.J. Tucker, and I. Fung, 2005: Drier summers cancel out the CO_2 uptake enhancement induced by warmer springs. *Proceedings of the National Academy of Sciences of the United States of America*, **102(31)**, 10823-10827.

Arnell, N.W., 1999: Climate change and global water resources. *Global Environmental Change, Part A: Human and Policy Dimensions*, **9**, S31-S49.

Aspen Global Change Institute, 2006: *Climate Change and Aspen: an Assessment of Impacts and Potential Responses*. Aspen Global Change Institute, Aspen, CO, pp. 1-178.

Auclair, A.N.D., P.D. Eglinton, and S.L. Minnemeyer, 1997: Principal forest dieback episodes in northern hardwoods: development of numeric indices of area extent and severity. *Water, Air and Soil Pollution*, **93**, 175-198.

Auclair, A.N.D., J.T. Lill, and C. Revenga, 1996: The role of climate variability and global warming in the dieback of northern hardwoods. *Water, Air and Soil Pollution*, **91**, 163-169.

Bachelet, D., J. Lenihan, R. Drapek, and R. Neilson, in press: VEMAP vs VINCERA: A DGVM sensitivity to differences in climate scenarios. *Global and Planetary Change*.

Bachelet, D., J. Lenihan, R. Neilson, R. Drapek, and T. Kittel, 2005: Simulating the response of natural ecosystems and their fire regimes to climatic variability in Alaska. *Canadian Journal of Forest Research*, **35(9)**, 2244-2257.

Bachelet, D., R.P. Neilson, T. Hickler, R.J. Drapek, J.M. Lenihan, M.T. Sykes, B. Smith, S. Sitch, and K. Thonicke, 2003: Simulating past and future dynamics of natural ecosystems in the United States. *Global Biogeochemical Cycles*, **17(2)**, 1045.

Bachelet, D., R.P. Neilson, J.M. Lenihan, and R.J. Drapek, 2001: Climate change effects on vegetation distribution and carbon budget in the United States. *Ecosystems*, **4**, 164-185.

Balvanera, P., A.B. Pfisterer, N. Buchmann, J.S. He, T. Nakashizuka, D. Raffaelli, and B. Schmid, 2006: Quantifying the evidence for biodiversity effects on ecosystem functioning and services. *Ecology Letters*, **9(10)**, 1146-1156.

Barden, L.S., 1987: Invasion of *Microstegium vimineum* (poaceae), an exotic, annual, shade-tolerant, C4 grass, into a North Carolina floodplain. *American Midland Naturalist*, **118(1)**, 40-45.

Barnett, T.P., J.C. Adam, and D.P. Lettenmaier, 2005: Potential impacts of a warming climate on water availability in snow-dominated regions. *Nature*, **438(7066)**, 303-309.

Baron, J.S., M.D. Hartman, L.E. Band, and R.B. Lamers, 2000: Sensitivity of a high-elevation rocky mountain watershed altered climate and CO_2. *Water Resources Research*, **36(1)**, 89-99.

Bayley, P.B., 1995: Understanding large river: floodplain ecosystems. *BioScience*, **45(3)**, 153-158.

Bazzaz, F.A., 1990: The response of natural ecosystems to the rising global CO_2 levels. *Annual Review of Ecology and Systematics*, **21**, 167-196.

Bennett, E.M., S.R. Carpenter, G.D. Peterson, G.S. Cumming, M. Zurek, and P. Pingali, 2003: Why global scenarios need ecology. *Frontiers in Ecology and the Environment*, **1**, 322-329.

Benning, T.L., D. LaPointe, C.T. Atkinson, and P.M. Vitousek, 2002: Interactions of climate change with biological invasions and land use in the Hawaiian Islands: modeling the fate of endemic birds using a geographic information system. *Proceedings of the National Academy of Sciences of the United States of America*, pp. 14246-14249.

Berg, E.E., J.D. Henry, C.L. Fastie, A.D. De Volder, and S.M. Matsuoka, 2006: Spruce beetle outbreaks on the Kenai Peninsula, Alaska, and Kluane National Park and Reserve, Yukon Territory: Relationship to summer temperatures and regional differences in disturbance regimes. *Forest Ecology and Management*, **227(3)**, 219-232.

Birch, T.W., 1996: *Private Forest-Land Owners of the United States, 1994*. Resour. Bull. NE-134 U.S. Department of Agriculture, Forest Service, Northeastern Forest Experiment Station, Radnor, PA, pp.183 p.

Birdsey, R., R. Alig, and D. Adams, 2000: Mitigation activities in the forest sector to reduce emissions and enhance sinks of greenhouse gases, In: *The Impact of Climate Change on America's Forests, Publication Number RMRS-GTR59*, [Joyce, L.A. and R. Birdsey (eds.)]. Rocky Mountain Research Station, Fort Collins, CO, pp. 112-131.

Blaikie, P., T. Cannon, I. Davis, and B. Wisner, 1994: *At Risk: Natural Hazards, People's Vulnerability and Disasters*. London: Routledge.

Blaikie, P.M., H.C. Brookfield, and B.J. Allen, 1987: *Land Degradation and Society*. Methuen.

Boisvenue, C. and S.W. Running, 2006: Impacts of climate change on natural forest productivity - evidence since the middle of the 20th century. *Global Change Biology*, **12(5)**, 862-882.

Bormann, B., R. Haynes, and J.R. Martin, 2007: Adaptive management of forest ecosystems: did some rubber hit the road? *BioScience*, **57(2)**, 186-191.

Boucher, T.V. and B.R. Mead, 2006: Vegetation change and forest regeneration on the Kenai Peninsula, Alaska following a spruce beetle outbreak, 1987-2000. *Forest Ecology and Management*, **227(3)**, 233-246.

Breshears, D.D., N.S. Cobb, P.M. Rich, K.P. Price, C.D. Allen, R.G. Balice, W.H. Romme, J.H. Kastens, M.L. Floyd, and J. Belnap, 2005: Regional vegetation die-off in response to global-change-type drought. *Proceedings of the National Academy of Sciences of the United States of America*, **102(42)**, 15144-15148.

Briceno-Elizondo, E., J. Garcia-Gonzalo, H. Peltola, and S. Kellomaki, 2006: Carbon stocks and timber yield in two boreal forest ecosystems under current and changing climatic conditions subjected to varying management regimes. *Environmental Science & Policy*, **9(3)**, 237-252.

Brooks, M.L., C.M. D'Antonio, D.M. Richardson, J.B. Grace, J.E. Keeley, J.M. DiTomaso, R.J. Hobbs, M. Pellant, and D. Pyke, 2004: Effects of invasive alien plants on fire regimes. *BioScience*, **54**, 677-688.

Brown, T.J., B.L. Hall, and A.L. Westerling, 2004: The impact of twenty-first century climate change on wildland fire danger in the Western United States: an applications perspective. *Climatic Change*, **62(1-3)**, 365-388.

Bruederle, L.P. and F.W. Stearns, 1985: Ice storm damage to a southern Wisconsin mesic forest. *Bulletin of the Torrey Botanical Club*, **112(2)**, 167-175.

Bugmann, H., B. Zierl, and S. Schumacher, 2005: Projecting the impacts of climate change on mountain forests and landscapes, In: *Global Change and Mountain Regions: an Overview of Current Knowledge*, [Huber, U.M., H. Bugmann, and M.A. Reasoner (eds.)]. Springer, Berlin, pp. 477-488.

Burdon, J.J. and T. Elmqvist, 1996: Selective sieves in the epidemiology of Melampsora lini. *Plant pathology*, **45(5)**, 933-943.

Burton, I., 1996: The growth of adaptation capacity: practice and policy, In: *Adapting to Climate Change: an International Perspective*, [Smith, J.B., N. Bhatti, G. Menzhulin, R. Benioff, M. Budyko, M. Campos, B. Jallow, and F. Rijsberman (eds.)]. Springer, New York, pp. 55-67.

Cameron, P., G. Jelinek, A.M. Kelly, L. Murray, and J. Heyworth, 2000: Triage. Textbook of adult emergency medicine. *Emergency Medicine*, **12**, 155-156.

Camp, A.E., P.F. Hessburg, R.L. Everett, and C.D. Oliver, 1995: *Spatial Changes in Forest Landscape Patterns Resulting From Altered Disturbance Regimes on the Eastern Slope of the Washington Cascades*. General Technical Report INT-GTR-320, US Department of Agriculture, Forest Service, Intermountain Research Station, Ogden, Utah, pp.1-283.

Carroll, A.L., S. W. Taylor, J. Régnière, and L. Safranyik, 2004: Effects of climate change on range expansion by the mountain pine beetle in British Columbia, In: *Mountain Pine Beetle Symposium: Challenges and Solutions*, [Shore, T.L., J.E. Brooks, and J.E. Stone (eds.)]. 30-31 October 2003, Information Report BC-X-399. Natural Resources Canada, Canadian Forest Service, Pacific Forestry Centre, Victoria, British Columbia, pp. 223-232, 298.

Cash, D. and J. Borck, 2006: Countering the 'loading dock' approach to linking science and decision making: a comparative analysis of ENSO forecasting systems. *Science, Technology, and Human Values*, **31(4)**, 465-494.

Cash, D.W., 2001: In order to aid in diffusing useful and practical information: Agricultural extension and boundary organizations. *Science, Technology, & Human Values*, **26(4)**, 431-453.

Cash, D.W., W.C. Clark, F. Alcock, N.M. Dickson, N. Eckley, D.H. Guston, J. Jaeger, and R.B. Mitchell, 2003: Knowledge systems for sustainable development. *Proceedings of the National Academy of Sciences of the United States of America*, **100(14)**, 8086-8091.

Cathcart, J. and M. Delaney, 2006: *Carbon Accounting: Determining Offsets From Forest Products*. Cloughesy, M. (Ed.), Forests, Carbon, and Climate Change: A synthesis of science findings, Oregon Forest Resources Institute, pp.157-176.

Certini, G., 2005: Effects of fire on properties of forest soils: a review. *Oecologia*, **143(1)**, 1-10.

Chen, C.Y., R.S. Stemberger, N.C. Kamman, B.M. Mayes, and C.L. Folt, 2005: Patterns of Hg bioaccumulation and transfer in aquatic food webs across multi-lake studies in the Northeast US. *Ecotoxicology*, **14(1)**, 135-147.

Chornesky, E.A., A.M. Bartuska, G.H. Aplet, K.O. Britton, J. Cummings-carlson, F.W. Davis, J. Eskow, D.R. Gordon, K.W. Gottschalk, and R.A. Haack, 2005: Science priorities for reducing the threat of invasive species to sustainable forestry. *BioScience*, **55(4)**, 335-348.

Christen, D. and G.R. Matlack, 2006: Essays: the role of roadsides in plant invasions: a demographic approach. *Conservation Biology*, **20(2)**, 385-391.

Clark, M.E., K.A. Rose, D.A. Levine, and W.W. Hargrove, 2001: Predicting climate change effects on Appalachian trout: combining GIS and individual-based modeling. *Ecological Applications*, **11(1)**, 161-178.

Collingham, Y.C. and B. Huntley, 2000: Impacts of habitat fragmentation and patch size upon migration rates. *Ecological Applications*, **10(1)**, 131-144.

Colorado Department of Natural Resources, 2005: *Report on the Health of Colorado's Forests*. Colorado Department of Natural Resources, Division of Forestry, Denver,Colorado, pp.1-32.

Coquard, J., P.B. Duffy, K.E. Taylor, and J.P. Iorio, 2004: Present and future surface climate in the western USA as simulated by 15 global climate models. *Climate Dynamics*, **23**, 455-472.

Cordell, H.K., C. Betz, J.M. Bowker, D.B.K. English, S.H. Mou, J.C. Bergstrom, J.R. Teasley, M.A. Tarrant, and J. Loomis, 1999: *Outdoor Recreation in American Life: a National Assessment of Demand and Supply Trends*. Sagamore Publishing, Champaign, IL, pp. xii-449.

Covington, W.W., R.L. Everett, R. Steele, L.L. Irwin, and T.A. Daer, 1994: Historical and anticipated changes in forest ecosystems of the Inland West of the United States. *Journal of Sustainable Forestry*, **2**, 13-63.

Currie, D.J., 2001: Projected effects of climate change on patterns of vertebrate and tree species richness in the conterminous United States. *Ecosystems*, **4(3)**, 216-225.

D'Antonio, C.M., 2000: Fire, plant invasions, and global changes, In: *Invasive Species in a Changing World*, [Mooney, H.A. and R.J. Hobbs (eds.)].

D'Antonio, C.M., J.T. Tunison, and R. Loh, 2000: Variation in the impact of exotic grass on native plant composition in relation to fire across an elevation gradient in Hawaii. *Austral Ecology*, **25**, 507-522.

D'Antonio, C.M. and P.M. Vitousek, 1992: Biological invasions by exotic grasses, the grass/fire cycle, and global change. *Annual Review of Ecology and Systematics*, **23**, 63-87.

Dale, V.H., L.A. Joyce, S. McNulty, R.P. Neilson, M.P. Ayres, M.D. Flannigan, P.J. Hanson, L.C. Irland, A.E. Lugo, and C.J. Peterson, 2001: Climate change and forest disturbances. *BioScience*.

Davis, M.B., 1989: Lags in vegetation response to greenhouse warming. *Climatic Change*, **15(1)**, 75-82.

Davis, M.B. and R.G. Shaw, 2001: Range shifts and adaptive responses to quaternary climate change: paleoclimate. *Science*, **292(5517)**, 673-679.

De Steven, D., J. Kline, and P.E. Matthiae, 1991: Long-term changes in a Wisconsin Fagus-Acer forest in relation to glaze storm disturbance. *Journal of Vegetation Science*, 201-208.

Decruyenaere, J.G. and J.S. Holt, 2005: Ramet demography of a clonal invader, *Arundo donax* (poaceae), in southern California. *Plant and Soil*, **(277)**, 41-52.

Deluca, T.H. and A. Sala, 2006: Frequent fire alters nitrogen transformations in ponderosa pine stands of the inland northwest. *Ecology*, **87(10)**, 2511-2522.

Dettinger, M.D., D.R. Cayan, M.K. Meyer, and A.E. Jeton, 2004: Simulated hydrologic responses to climate variations and change in the Merced, Carson, and American River basins, Sierra Nevada, California, 1900-2099. *Climatic Change*, **62(1/3)**, 283-317.

Díaz, S., J. Fargione, F.S. Chapin, and D. Tilman, 2006: Biodiversity loss threatens human well-being. *PLoS Biology*, **4(8)**, e277.

Dixon, G.E., 2003: *Essential FVS: A User's Guide to the Forest Vegetation Simulator*. U.S. Department of Agriculture, Forest Service, Forest Management Service Center, Fort Collins, CO, pp. 193.

Drake, B.G., M.A. Gonzalez-Meler, and S.P. Long, 1997: More efficient plants: A consequence of rising atmospheric CO_2? *Annual Review Plant Physiology Plant Molecular Biology*, **48**, 609-639.

Driscoll, C.T., Y.J.I. Han, C.Y. Chen, D.C. Evers, K.F. Lambert, T.M. Holsen, N.C. Kamman, and R.K. Munson, 2007: Mercury contamination in forest and freshwater ecosystems in the Northeastern United States. *BioScience*, **57**, 17-28.

Duffy, P.A., J.E. Walsh, J.M. Graham, D.H. Mann, and T.S. Rupp, 2005: Impacts of large-scale atmospheric-ocean variability on Alaskan fire season severity. *Ecological Applications*, **15(4)**, 1317-1330.

Dukes, J.S., 2000: Will the increasing atmospheric CO_2 concentration affect the success of invasive species. *Invasive Species in a Changing World.* Island Press, Washington.

Dukes, J.S. and H.A. Mooney, 1999: Does global change increase the success of biological invaders? *Trends in Ecology and Evolution*, **14(4)**, 135-139.

Dupouey, J.L., E. Dambrine, J.D. Laffite, and C. Moares, 2002: Irreversible impact of past land use on forest soils and biodiversity. *Ecology*, **83(11)**, 2978-2984.

Earle, T.C. and G.T. Cvetkovich, 1995: *Social Trust: Toward a Cosmopolitan Society*. Praeger/Greenwood.

Eaton, J.G. and R.M. Scheller, 1996: Effects of climate warming on fish thermal habitat in streams of the United States. *Limnology and Oceanography*, **41(5)**, 1109-1115.

Ebersole, J.L., W.J. Liss, and C.A. Frissell, 2001: Relationship between stream temperature, thermal refugia and rainbow trout *Oncorhynchus mykiss* abundance in arid-land streams in the northwestern United States. *Ecology of Freshwater Fish*, **10(1)**, 1-10.

Englin, J., P.C. Boxall, K. Chakraborty, and D.O. Watson, 1996: Valuing the impacts of forest fires on backcountry forest recreation. *Forest Science*, **42(6)**, 450-455.

Erickson, J.E., J.P. Megonigal, G. Peresta, and B.G. Drake, 2007: Salinity and sea level mediate elevated CO_2 effects on C_3-C_4 plant interactions and tissue nitrogen in a Chesapeake Bay tidal wetland. *Global Change Biology*, **13**, 202-215.

Ewers, R.M. and R.K. Didham, 2006: Confounding factors in the detection of species responses to habitat fragmentation. *Biological Reviews of the Cambridge Philosophical Society*, **81(1)**, 117-142.

Felzer, B., D. Kicklighter, J. Melillo, C. Wang, Q. Zhuang, and R. Prinn, 2004: Effects of ozone on net primary production and carbon sequestration in the conterminous United States using a biogeochemistry model. *Tellus Series B-Chemical and Physical Meteorology*, **56(3)**, 230-248.

Fenn, M.E., J.S. Baron, E.B. Allen, H.M. Rueth, K.R. Nydick, L. Geiser, W.D. Bowman, J.O. Sickman, T. Meixner, and D.W. Johnson, 2003: Ecological effects of nitrogen deposition in the western United States. *BioScience*, **53(4)**, 404-420.

Ferrell, G.T., 1996: *The Influence of Insect Pests and Pathogens on Sierra Forests.* Sierra Nevada Ecosystem Project: final report to Congress II, University of California, Centers for Water and Wildland Resources, Davis, CA, pp.1177-1192.

Fiore, A.M., D.J. Jacob, I. Bey, R.M. Yantosca, B.D. Field, A.C. Fusco, and J.G. Wilkinson, 2002: Background ozone over the United States in summer- Origin, trend, and contribution to pollution episodes. *Journal of Geophysical Research*, **107(D15)**.

Flannigan, M.D., B.J. Stocks, and B.M. Wotton, 2000: Climate change and forest fires. *Science of the Total Environment*, **262(3)**, 221-229.

Fleming, R.A., J.N. Candau, and R.S. McAlpine, 2002: Landscape-scale analysis of interactions between insect defoliation and forest fire in Central Canada. *Climatic Change*, **55(1)**, 251-272.

Foley, J.A., R. DeFries, G.P. Asner, C. Barford, G. Bonan, S.R. Carpenter, F.S. Chapin, M.T. Coe, G.C. Daily, H.K. Gibbs, J.H. Helkowski, T. Holloway, E.A. Howard, C.J. Kucharik, C. Monfreda, J.A. Patz, I.C. Prentice, N. Ramankutty, and P.K. Snyder, 2005: Global consequences of land use. *Science*, **309(5734)**, 570-574.

Foster, D., F. Swanson, J. Aber, I. Burke, N. Brokaw, D. Tilman, and A. Knapp, 2003: The importance of land-use legacies to ecology and conservation. *BioScience*, **53(1)**, 77-88.

Fox, D., 2007: Back to the no-analog future? *Science*, **316**, 823-825.

Frelich, L., C. Hale, S. Scheu, A. Holdsworth, L. Heneghan, P. Bohlen, and P. Reich, 2006: Earthworm invasion into previously earthworm-free temperate and boreal forests. *Biological Invasions*, **8(6)**, 1235-1245.

Frelich, L.E., 2002: *Forest Dynamics and Disturbance Regimes: Studies From Temperate Evergreen-Deciduous Forests.* Cambridge University Press, New York, NY.

Fu, Q., C.M. Johanson, J.M. Wallace, and T. Reichler, 2006: Enhanced mid-latitude tropospheric warming in satellite measurements. *Science*, **312(5777)**, 1179.

GAO, 2004: *Challenges to Agency Decision and Opportunities for BLM to Standardize Data Collection.* Oil and gas development GAO-05-124, U.S. Government Accountability Office, Washington, DC.

Gates, D.M., 1993: *Climate Change and Its Biological Consequences.* Sinauer Associates, Inc., Massachusetts.

Gottschalk, K.W., 1995: *Using Silviculture to Improve Health in Northeastern Conifer and Eastern Hardwood Forests.* Forest Health through Silviculture General Technical Report RM-267, USDA Forest Service, Ft. Collins, CO, pp.219-226.

Graham, R.T., 2003: *Hayman Fire Case Study: Summary.* General Technical Report RMRS-GTR-115.

Graham, R.T., A.E. Harvey, T.B. Jain, and J.R. Tonn, 1999: *The Effects of Thinning and Similar Stand Treatments on Fire Behavior in Western Forests.* General Technical Report PNW-GTR-463, USDA Forest Service.

Groisman, P.Y., R.W. Knight, D.R. Easterling, T.R. Karl, G.C. Hegerl, and V.N. Razuvaev, 2005: Trends in intense precipitation in the climate record. *Journal of Climate*, **18(9)**, 1326-1350.

Guarin, A. and A.H. Taylor, 2005: Drought triggered tree mortality in mixed conifer forests in Yosemite National Park, California, USA. *Forest Ecology and Management*, **218(1)**, 229-244.

Gutierrez, R.J., M. Cody, S. Courtney, and A.B. Franklin, 2007: The invasion of barred owls and its potential effect on the spotted owl: a conservation conundrum. *Biological Invasions*, **9(2)**, 181-196.

Guyette, R.P. and C.F. Rabeni, 1995: Climate response among growth increments of fish and trees. *Oecologia*, **104(3)**, 272-279.

Hale, C.M., L.E. Frelich, P.B. Reich, and J. Pastor, 2005: Effects of European earthworm invasion on soil characteristics in northern hardwood forests of Minnesota, USA. *Ecosystems*, **8(8)**, 911-927.

Halpin, P.N., 1997: Global climate change and natural-area protection: management responses and research directions. *Ecological Applications*, **7(3)**, 828-843.

Hance, T., J. van Baaren, P. Vernon, and G. Boivin, 2007: Impact of extreme temperatures on parasitoids in a climate change perspective. *Annual Review of Entomology*, **52**, 107-126.

Handmer, J.W., S. Dovers, and T.E. Downing, 1999: Societal vulnerability to climate change and variability. *Mitigation and Adaptation Strategies for Global Change*, **4(3)**, 267-281.

Hansen, A.J. and R. DeFries, 2007: Ecological mechanisms linking protected areas to surrounding lands. *Ecological Applications*, **17**, 974-988.

Hansen, A.J., R.R. Neilson, V.H. Dale, C.H. Flather, L.R. Iverson, D.J. Currie, S. Shafer, R. Cook, and P.J. Bartlein, 2001: Global change in forests: responses of species, communities, and biomes. *BioScience*, **51(9)**, 765-779.

Hanson, P.J. and J.F. Weltzin, 2000: Drought disturbance from climate change: response of United States forests. *Science of the Total Environment*, **262(3)**, 205-220.

Hanson, P.J., S.D. Wullschleger, R.J. Norby, T.J. Tschaplinski, and C.A. Gunderson, 2005: Importance of changing CO_2, temperature, precipitation, and ozone on carbon and water cycles of an upland-oak forest: incorporating experimental results into model simulations. *Global Change Biology*, **11(9)**, 1402-1423.

Hardy, C.C., K.M. Schmidt, J.M. Menakis, and N.R. Samson, 2001a: Spatial data for national fire planning and fuel management. *International Journal of Wildland Fire*, **10(4)**, 353-372.

Hardy, J.P., P.M. Groffman, R.D. Fitzhugh, K.S. Henry, A.T. Welman, J.D. Demers, T.J. Fahey, C.T. Driscoll, G.L. Tierney, and S. Nolan, 2001b: Snow depth manipulation and its influence on soil frost and water dynamics in a northern hardwood forest. *Biogeochemistry*, **56(2)**, 151-174.

Harmon, M.E. and B. Marks, 2002: Effects of silvicultural treatments on carbon stores in forest stands. *Canadian Journal of Forest Research*, **32**, 863-877.

Harris, J.A., R.J. Hobbs, E. Higgs, and J. Aronson, 2006: Ecological restoration and global climate change. *Restoration Ecology*, **14(2)**, 170-176.

Harvell, C.D., C.E. Mitchell, J.R. Ward, S. Altizer, A.P. Dobson, R.S. Ostfeld, and M.D. Samuel, 2002: Ecology - climate warming and disease risks for terrestrial and marine biota. *Science*, **296(5576)**, 2158-2162.

Hauer, F.R., J.A. Stanford, and M.S. Lorang, 2007: Pattern and process in northern rocky mountain headwaters: ecological linkages in the headwaters of the crown of the continent. *Journal of the American Water Resources Association*, **43(1)**, 104-117.

Hawbaker, T.J., V.C. Radeloff, M.K. Clayton, R.B. Hammer, and C.E. Gonzalez-Abraham, 2006: Road development, housing growth, and landscape fragmentation in northern Wisconsin: 1937-1999. *Ecological Applications*, **16(3)**, 1222-1237.

Hayhoe, K., D. Cayan, C.B. Field, P.C. Frumhoff, E.P. Maurer, N.L. Miller, S.C. Moser, S.H. Schneider, K.N. Cahill, E.E. Cleland, L. Dale, R. Drapek, R.M. Hanemann, L.S. Kalkstein, J. Lenihan, C.K. Lunch, R.P. Neilson, S.C. Sheridan, and J.H. Verville, 2004: Emissions pathways, climate change, and impacts on California. *Proceedings of the National Academy of Sciences of the United States of America*, **101**, 34.

Haynes, R.W., D.M. Adams, R.J. Alig, P.J. Ince, J.R. Mills, and X. Zhou, 2007: *The 2005 RPA Timber Assessment Update*. General Technical Report PNW-GTR-699, USDA Forest Service, Pacific Northwest Research Station, Portland, OR.

Haynes, R.W., B.T. Bormann, D.C. Lee, and J.R. Martin, 2006: *Northwest Forest Plan—the First 10 Years (1994-2003): Synthesis of Monitoring and Research Results*. General Technical Report PNW-GTR-651, U.S. Department of Agriculture Forest Service, Pacific Northwest Research Station, Portland, OR, pp.1-292.

Hendrix, P.F. and P.J. Bohlen, 2002: Exotic earthworm invasions in North America: ecological and policy implications. *BioScience*, **52(9)**, 801-811.

Hesseln, H., J.B. Loomis, A. González-Cabán, and S. Alexander, 2003: Wildfire effects on hiking and biking demand in New Mexico: a travel cost study. *Journal of Environmental Management*, **69**, 359-369.

Hicke, J.A., G.P. Asner, J.T. Randerson, C. Tucker, S. Los, R. Birdsey, J.C. Jenkins, and C. Field, 2002: Trends in North American net primary productivity derived from satellite observations, 1982-1998. *Global Biogeochemical Cycles*, **16(2)**, 1019.

Hobbs, R.J., S. Arico, J. Aronson, J.S. Baron, P. Bridgewater, V.A. Cramer, P.R. Epstein, J.J. Ewel, C.A. Klink, A.E. Lugo, D. Norton, D. Ojima, D.M. Richardson, E.W. Sanderson, F. Valladares, M. Vilà, R. Zamora, and M. Zobel, 2006: Novel ecosystems: theoretical and management aspects of the new ecological world order. *Global Ecology and Biogeography*, **15**, 1-7.

Hobbs, R.J. and H.A. Mooney, 1991: Effects of rainfall variability and gopher disturbance on serpentine annual grassland dynamics. *Ecology*, **72(1)**, 59-68.

Holling, C.S., 2001: Understanding the complexity of economic, ecological, and social systems. *Ecosystems*, **4**, 390-405.

Houlton, B.Z., C.T. Driscoll, T.J. Fahey, G.E. Likens, P.M. Groffman, E.S. Bernhardt, and D.C. Buso, 2003: Nitrogen dynamics in ice storm-damaged forest ecosystems: implications for nitrogen limitation theory. *Ecosystems*, **6(5)**, 431-443.

Huntington, T.G., 2003: Climate warming could reduce runoff significantly in New England, USA. *Agricultural and Forest Meteorology*, **117(3)**, 193-201.

Inamdar, S.P., N. O'Leary, M.J. Mitchell, and J.T. Riley, 2006: The impact of storm events on solute exports from a glaciated forested watershed in western New York, USA. *Hydrological Processes*, **20(16)**, 3423-3439.

Inouye, D.W., B. Barr, K.B. Armitage, and B.D. Inouye, 2000: Climate change is affecting altitudinal migrants and hibernating species. *Proceedings of the National Academy of Sciences of the United States of America*, **97(4)**, 1630-1633.

IPCC, 2001a: *Climate Change 2001: Impacts, Adaptation, and Vulnerability. Contribution of Working Group II to the Third Assessment Report of the Intergovernmental Panel on Climate Change*. [McCarthy, J.J., O.F. Canziani, N.A. Leary, D.J. Dokken, and K.S. White (eds.)]. Cambridge University Press, Cambridge, UK.

IPCC, 2001b: Summary for policymakers, In: *Climate Change 2001: the Science Basis. Contribution of Working Group I to the Third Assessment Report of the Intergovernmental Panel on Climate Change*, [Houghton, J.T., Y. Ding, D.J. Griggs, M. Noguer, P.J. van der Linden, and D. Xiaosu (eds.)]. Cambridge University Press, Cambridge and New York.

IPCC, 2007: Summary for policymakers, In: *Climate Change 2007: the Physical Science Basis. Contribution of Working Group I to the Fourth Assessment Report of the Intergovernmental Panel on Climate Change*, [Solomon, S., D. Qin, M. Manning, Z. Chen, M. Marquis, K.B. Averyt, M. Tignor, and H.L. Miller (eds.)]. Cambridge University Press, Cambridge, United Kingdom and New York, NY, USA.

Irland, L.C., 2000: Ice storms and forest impacts. *Science of the Total Environment*, **262(3)**, 231-242.

Irland, L.C., D. Adams, R. Alig, C.J. Betz, C.C. Chen, M. Hutchins, B.A. McCarl, K. Skog, and B.L. Sohngen, 2001: Assessing socioeconomic impacts of climate change on US forests, wood-product markets, and forest recreation. *BioScience*, **51(9)**, 753-764.

Iverson, L.R. and A.M. Prasad, 2001: Potential changes in tree species richness and forest community types following climate change. *Ecosystems*, **4(3)**, 186-199.

Johnson, D.W. and S.E. Lindberg, 1992: *Atmospheric Deposition and Forest Nutrient Cycling*. Springer.

Johnstone, J.F., F.S. Chapin, III, J. Foote, S. Kemmett, K. Price, and L. Viereck, 2004: Decadal observations of tree regeneration following fire in boreal forests. *Canadian Journal of Forest Research*, **34(2)**, 267-273.

Joyce, L.A., M.A. Fosberg, and J. Comandor, 1990: *Climate Change and America's Forests*. General Technical Report RM-187, USDA Forest Service, Rocky Mountain Forest and Range Experiment Station, Fort Collins, CO.

Joyce, L.A., 1995: *Productivity of America's Forests and Climate Change*. General Technical Report RM-271, US Dept. of Agriculture, Forest Service, Rocky Mountain Forest and Range Experiment Station, Fort Collins, CO.

Joyce, L.A., 2007: The impacts of climate change on forestry, In: *Resource and Market Projections for Forest Policy Development: Twenty Five Years of Experience With the U.S. RPA Timber Assessment*, [Adams, D.M. and R.W. Haynes (eds.)]. Springer.

Joyce, L.A., J. Aber, S. McNulty, D. H. Vale, A. Hansen, L. C. Irland, R. P. Neilson, and K. Skog, 2001: Potential consequences of climate variability and change for the forests of the United States, In: *Climate Change Impacts on the United States: the Potential Consequences of Climate Variability and Change, National Assessment Synthesis Team Report for the US Global Change Research Program*, Cambridge University Press, Cambridge, UK, pp. 489-522.

Joyce, L.A. and R. Birdsey, 2000: *The Impact of Climate Change on America's Forests*. RMRS-GTR-59, USDA Forest Service, Rocky Mountain Research Station, Fort Collins, CO.

Joyce, L.A. and M. Nungesser, 2000: *Ecosystem Productivity and the Impact of Climate Change*. The impact of climate change on America's forests General Technical Report RMRS-GTR-59, US Department of Agriculture, Forest Service, Rocky Mountain Research Station, Fort Collins, CO, pp.45-68.

Kareiva, P.M., J.G. Kingsolver, and R.B. Huey, 1993: *Biotic Interactions and Global Change*. Sinauer Associates Inc., Sunderland, MA.

Karl, T.R. and R.W. Knight, 1998: Secular trends of precipitation amount, frequency, and intensity in the United States. *Bulletin of the American Meteorological Society*, **79(2)**, 231-241.

Karnosky, D.F., K.S. Pregitzer, D.R. Zak, M.E. Kubiske, G.R. Hendrey, D. Weinstein, M. Nosal, and K.E. Percy, 2005: Scaling ozone responses of forest trees to the ecosystem level in a changing climate. *Plant, Cell and Environment*, **28(8)**, 965-981.

Karnosky, D.F., D.R. Zak, and K.S. Pregitzer, 2003: Tropospheric O_3 moderates responses of temperate hardwood forests to elevated CO_2: a synthesis of molecular to ecosystem results from the Aspen FACE project. *Functional Ecology*, **17(3)**, 289-304.

Kasperson, R.E. and J.X. Kasperson, 2001: *Climate Change, Vulnerability and Social Justice*. Stockholm Environment Institute, Stockholm.

Keane, R.E., K.C. Ryan, T.T. Veblen, C.D. Allen, J. Logan, and B. Hawkes, 2002: *Cascading Effects of Fire Exclusion in Rocky Mountain Ecosystems: a Literature Review*. General Technical Report RMRS-GTR-91, US Department of Agriculture Forest Service, Rocky Mountain Research Station, Fort Collins, CO, pp.1-24.

Keleher, C.J. and F.J. Rahel, 1996: Thermal limits to salmonid distributions in the Rocky Mountain region and potential habitat loss due to global warming: a geographic information system (GIS) approach. *Transactions of the American Fisheries Society*, **125(1)**, 1-13.

Kelkar, V.M., B.W. Geils, D.R. Becker, S.T. Overby, and D.G. Neary, 2006: How to recover more value from small pine trees: essential oils and resins. *Biomass and Bioenergy*, **30**, 316-320.

Kelly, E.G., E.D. Forsman, and R.G. Anthony, 2003: Are barred owls displacing spotted owls? *Condor*, **105(1)**, 45-53.

Kim, J., T.-K. Kim, R.W. Arritt, and N.L. Miller, 2002: Impacts of increased atmospheric CO_2 on the hydroclimate of the western United States. *Journal of Climate*, **15(14)**, 1926-1942.

King, J.S., M.E. Kubiske, K.S. Pregitzer, G.R. Hendrey, E.P. McDonald, C.P. Giardina, V.S. Quinn, and D.F. Karnosky, 2005: Tropospheric O_3 compromises net primary production in young stands of trembling aspen, paper birch and sugar maple in response to elevated atmospheric CO_2. *New Phytologist*, **168(3)**, 623-636.

Klein, R.J.T., S. Hug, F. Denton, T. E. Downing, R. G. Richels, J. B. Robinson, and F. L. Toth, 2007: Inter-relationships between adaptation and mitigation, In: *Climate Change 2007: Impacts, Adaptation and Vulnerability. Contribution of Working Group II to the Fourth Assessment Report of the Intergovernmental Panel on Climate Change*, [Parry, M.L., O.F. Caziani, J.P. Palutikof, P.J. van der Linden, and C.E. Hanson (eds.)]. Cambridge University Press, Cambridge, UK, pp. 745-777.

Kling, G., K. Hayhoe, L.B. Johnson, J.J. Magnuson, S. Polasky, S.K. Robinson, B.J. Shuter, M.M. Wander, D.J. Wuebbles, and D.R. Zak, 2003: *Confronting Climate Change in the Great Lakes Region: Impacts on Our Communities and Ecosystems*. Union of Concerned Scientists and The Ecological Society of America.

Krankina, O.N. and M. E. Harmon, 2006: Forest management strategies for carbon storage, In: *Forests, Carbon, and Climate Change: A Synthesis of Science Findings*, [Cloughesy, M. (ed.)]. Oregon Forest Resources Institute, pp. 77-92.

Kulakowski, D., T.T. Veblen, and P. Bebi, 2003: Effects of fire and spruce beetle outbreak legacies on the disturbance regime of a subalpine forest in Colorado. *Journal of Biogeography*, **30(9)**, 1445-1456.

Lacey, J.R., C.B. Marlow, and J.R. Lane, 1989: Influence of spotted knapweed (*Centaurea-maculosa*) on surface runoff and sediment yield. *Weed Technology*, **3(4)**, 627-631.

Lafon, C.W., 2006: Forest disturbance by ice storms in Quercus forests of the southern Appalachian Mountains, USA. *Ecoscience*, **13(1)**, 30-43.

Lafon, C.W., J.A. Hoss, and H.D. Grissino-Mayer, 2005: The contemporary fire regime of the central Appalachian Mountains and its relation to climate. *Physical Geography*, **26(2)**, 126-146.

Ledig, F.T. and J.H. Kitzmiller, 1992: Genetic strategies for reforestation in the face of global climate change. *Forest Ecology and Management*, **50(1)**, 153-169.

Lemmen, D.S. and F.J. Warren, 2004: *Climate Change: a Canadian Perspective*. Natural Resources Canada, Ottowa.

Lenihan, J.M., D. Bachelet, R. Drapek, and R.P. Neilson, 2006: *The Response of Vegetation, Distribution, Ecosystem Productivity, and Fire in California to Future Climate Scenarios Simulated by the MC1 Dynamic Vegetation Model*. Climate action team report to the Governor and Legislators, available from http://www.energy.ca.gov/2005publications/CEC-500-2005-191/CEC-500-2005-191-SF.PDF.

Lenihan, J.M., D. Bachelet, R.P. Neilson, and R. Drapek, in press: Simulated response of conterminous United States ecosystems to climate change at different levels of fire suppression, CO_2, and growth response to CO_2. *Global and Planetary Change*.

Lindner, M., P. Lasch, and M. Erhard, 2000: Alternative forest management strategies under climatic change- prospects for gap model applications in risk analyses. *Silva Fennica*, **34(2)**, 101-111.

Lippincott, C.L., 2000: Effects of *Imperata cylindrica* (L.) Beauv. (Cogongrass) invasion on fire regime in Florida sandhill (USA). *Natural Areas Journal*, **20(2)**, 140-149.

Littell, J.S., 2006: Climate impacts to forest ecosystem processes: Douglas-fir growth in northwestern U.S. mountain landscapes and area burned by wildfire in western U.S. ecoprovinces. *PhD Dissertation, University of Washington, Seattle*.

Littell, J.S. and D.L. Peterson, 2005: A method for estimating vulnerability of Douglas-fir growth to climate change in the northwestern US. *Forestry Chronicle*, **81(3)**, 369-374.

Logan, J.A. and B.J. Bentz, 1999: Model analysis of Mountain Pine Beetle (*Coleoptera: Scolytidae*) seasonality. *Environmental Entomology*, **28(6)**, 924-934.

Logan, J.A. and J.A. Powell, 2001: Ghost forests, global warming, and the mountain pine beetle (*Coleoptera: Scolytidae*). *American Entomologist*, **47(3)**, 160-172.

Logan, J.A. and J. A. Powell, 2005: Ecological consequences of climate change altered forest insect disturbance regimes, In: *Climate Change in Western North America: Evidence and Environmental Effects*, [Wagner, F.H. (ed.)]. Allen Press, Lawrence, KS.

Logan, J.A., J. Regniere, and J.A. Powell, 2003: Assessing the impacts of global warming on forest pest dynamics. *Frontiers in Ecology and the Environment*, **1(3)**, 130-137.

Mack, R.N., 1981: Invasion of bromus tectorum L. into western North America: an ecological chronicle. *Agro-Ecosystems*, **7(2)**, 145-165.

Malcolm, J.R., C. Liu, R.P. Neilson, L. Hansen, and L. Hannah, 2006: Global warming and extinctions of endemic species from biodiversity hotspots. *Conservation Biology*, **20(2)**, 538-548.

Malhi, Y., P. Meir, and S. Brown, 2002: Forests, carbon and global climate. *Philos Transact Ser A Math Phys Eng Sci*, **360(1797)**, 1567-1591.

Manion, P.D., 1991: *Tree Disease Concepts*. Prentice Hall, Englewood Cliffs, NJ.

Marchetti, M.P. and P.B. Moyle, 2001: Effects of flow regime on fish assemblages in a regulated California stream. *Ecological Applications*, **11(2)**, 530-539.

Markham, A., 1996: Potential impacts of climate change on ecosystems: a review of implications for policymakers and conservation biologists. *Climate Research*, **6**, 171-191.

Mastrandrea, M.D. and S.H. Schneider, 2001: Integrated assessment of abrupt climatic changes. *Climate Policy*, **1(4)**, 433-449.

McDonald, K.A. and J.H. Brown, 1992: Using montane mammals to model extinctions due to global change. *Conservation Biology*, **6(3)**, 409-415.

McKenzie, D.H., Z. Gedalof, D.L. Peterson, and P. Mote, 2004: Climatic change, wildfire, and conservation. *Conservation Biology*, **18(4)**, 890-902.

McKenzie, D.H., A.E. Hessl, and D.L. Peterson, 2001: Recent growth of conifer species of western North America: assessing spatial patterns of radial growth trends. *Canadian Journal of Forest Research*, **31(3)**, 526-538.

McKenzie, D.H., D.L. Peterson, and J. Littell, in press: Global warming and stress complexes in forests of western North America. *Forest fires and air pollution issues*. [Bytnerowicz, A., M. Arbaugh, C. Anderson, and A. Riebau (ed.)]. Elsevier Science Ltd.

McKinney, M.L. and J.L. Lockwood, 1999: Biotic homogenization: a few winners replacing many losers in the next mass extinction. *Trends in Ecology and Evolution*, **14(11)**, 450-453.

McLachlan, J.S., J.J. Hellmann, and M.W. Schwartz, 2007: A framework for debate of assisted migration in an era of climate change. *Conservation Biology*, **21(2)**, 297-302.

McLaughlin, S.B., M. Nosal, S.D. Wullschleger, and G. Sun, 2007a: Interactive effects of ozone and climate on tree growth and water use in a southern Appalachian forest in the USA. *New Phytologist*, **174(1)**, 109-124.

McLaughlin, S.B., S.D. Wullschleger, G. Sun, and M. Nosal, 2007b: Interactive effects of ozone and climate on water use, soil moisture content and streamflow in a southern Appalachian forest in the USA. *New Phytologist*, **174(1)**, 125-136.

McNulty, S.G., 2002: Hurricane impacts on US forest carbon sequestration. *Environmental Pollution*, **116(Supplement 1)**, S17-S24.

McNulty, S.G., J.A. Moore Myers, T.J. Sullivan, and H. Li, in press: Estimates of critical acid loads and exceedances for forest soils across the conterminous United States. *Environmental Pollution*.

Meehl, G.A., C. Tebaldi, and D. Nychka, 2004: Changes in frost days in simulations of twentyfirst century climate. *Climate Dynamics*, **23(5)**, 495-511.

Melack, J.M., J. Dozier, C.R. Goldman, D. Greenland, A.M. Milner, and R.J. Naiman, 1997: Effects of climate change on inland waters of the Pacific Coastal Mountains and Western Great Basin of North America. *Hydrological Processes*, **11(8)**, 971-992.

Melillo, J., A.D. McGuire, D.W. Kicklighter, B. Moore, III, C.J. Vorosmarty, and A.L. Schloss, 1993: Global climate change and terrestrial net primary production. *Nature*, **363(6426)**, 234-240.

Michener, W.K. and R.A. Haeuber, 1998: Flooding: natural and managed disturbances. *BioScience*, **48(9)**, 677-680.

Milchunas, D.G. and W.K. Lauenroth, 1995: Inertia in plant community structure - state changes after cessation of nutrient-enrichment stress. *Ecological Applications*, **5(2)**, 452-458.

Millar, C.I., 1989: Allozyme variation of bishop pine associated with pygmy-forest soils in northern California. *Canadian Journal of Forest Research*, **19(7)**, 870-879.

Millar, C.I., 1998: Reconsidering the conservation of monterey pine. *Fremontia*, **26(3)**, 12-16.

Millar, C.I., 1999: Evolution and biogeography of Pinus radiata, with a proposed revision of its Quaternary history. *New Zealand Journal of Forestry Science*, **29(3)**, 335-365.

Millar, C.I. and L. B. Brubaker, 2006: Climate change and paleoecology: New contexts for restoration ecology, Island Press.

Millar, C.I., J.C. King, R.D. Westfall, H.A. Alden, and D.L. Delany, 2006: Late Holocene forest dynamics, volcanism, and climate change at Whitewing Mountain and San Joaquin Ridge, Mono County, Sierra Nevada, CA, USA. *Quaternary research*, **66**, 273-287.

Millar, C.I., R.D. Westfall, and D.L. Delany, in press: Mortality and growth suppression in high elevation limber pine (*Pinus flexilis*) forests in response to multi-year droughts and 20th century warming. *Canadian Journal of Forest Research*.

Millar, C.I. and W.B. Woolfenden, 1999: The role of climate change in interpreting historical variability. *Ecological Applications*, **9(4)**, 1207-1216.

Millennium Ecosystem Assessment, 2005: *Ecosystems and Human Well-Being: General Synthesis*. Island Press, Washington, DC.

Miller, J.H., 2003: *Nonnative Invasive Plants of Southern Forests: a Field Guide for Identification and Control*. General Technical Report SRS-62, U.S. Department of Agriculture, Forest Service, South Research Station, Asheville, NC, pp.1-93.

Miller, P.R., 1992: Mixed conifer forests of the San Bernardino Mountains, California, In: *The Response of Western Forests to Air Pollution*, [Olson, R.K., D. Binkley, and M. Bohm (eds.)]. Springer-Verlag, New York, pp. 461-497.

Milly, P.C.D., K.A. Dunne, and A.V. Vecchia, 2005: Global pattern of trends in streamflow and water availability in a changing climate. *Nature*, **438(7066)**, 347-350.

Milne, B.T., V.K. Gupta, and C. Restrepo, 2002: A scale invariant coupling of plants, water, energy, and terrain. *Ecoscience*, **9(2)**, 191-199.

Milne, B.T., A.R. Johnson, T.H. Keitt, C.A. Hatfield, J. David, and P.T. Hraber, 1996: Detection of critical densities associated with pinon-juniper woodland ecotones. *Ecology*, **77(3)**, 805-821.

Minnich, R.A., 2001: An integrated model of two fire regimes. *Conservation Biology*, **15(6)**, 1549-1553.

Mitchell, J.E., 2000: *Rangeland Resource Trends in the United States: a Technical Document Supporting the 2000 USDA Forest Service RPA Assessment*. General Technical Report RMRS-GTR-68, U.S. Department of Agriculture, Forest Service, Rocky Mountain Research Station, Fort Collins, CO, pp.1-84.

Mohseni, O., H.G. Stefan, and J.G. Eaton, 2003: Global warming and potential changes in fish habitat in U.S. streams. *Climatic Change*, **59(3)**, 389-409.

Mooney, H.A. and E.E. Cleland, 2001: The evolutionary impact of invasive species. *Proceedings of the National Academy of Sciences of the United States of America*, **98(10)**, 5446-5451.

Mooney, H.A. and R.J. Hobbs, 2000: *Invasive Species in a Changing World*. Island Press, Washington, DC.

Moore, A., B. Nelson, B. Pence, C. Hydock, J. Hickey, P. Thurston, S. Reitz, B. White, and R. Kandare, 2006: *Allegheny National Forest: Ecological Context*.

Moritz, C., 2002: Symposium on biodiversity, systematics, and conservation strategies to protect biological diversity and the evolutionary processes that sustain it. *Systematic Biology*, **51(2)**, 238-254.

Moritz, M.A., 2003: Spatiotemporal analysis of controls on shrubland fire regimes: Age dependency and fire hazard. *Ecology*, **84(2)**, 351-361.

Moser, S.C., 2005: Impact assessments and policy responses to sea-level rise in three US states: An exploration of human-dimension uncertainties. *Global Environmental Change, Part A: Human and Policy Dimensions*, **15(4)**, 353-369.

Moser, S.C., R.E. Kasperson, G. Yohe, and J. Agyeman, in press: Adaptation to climate change in the northeast United States: opportunities, processes, constraints. *Mitigation and Adaptation Strategies for Global Change*.

Mote, P.W., A.F. Hamlet, M.P. Clark, and D.P. Lettenmaier, 2005: Declining mountain snowpack in Western North America. *Bulletin of the American Meteorological Society*, **86(1)**, 39-49.

Multihazard Mitigation Council, 2006: *Natural Hazard Mitigation Saves: an Independent Study to Assess the Future Savings From Mitigation Activities*. Report to FEMA, National Institute for Building Sciences, Boston, MA.

Murphy, J.D., D.W. Johnson, W.W. Miller, R.F. Walker, E.F. Carrol, and R.R. Blank, 2006: Wildfire effects on soil nutrients and leaching in a Tahoe Basin watershed. *Journal of Environmental Quality*, **35**, 479-489.

National Assessment Synthesis Team, US Global Change Research Program, 2001: *Climate Change Impacts on the United States (Foundation Report)*. Cambridge University Press, Cambridge, United Kingdom.

National Drought Policy Commission, 2000: *Preparing for Drought in the 21st Century*. U.S. Department of Agriculture.

National Research Council, 1987: *The Mono Basin Ecosystem: Effects of Changing Lake Level*. Mono Basin Ecosystem Study Committee, Board on Environmental Studies and Toxicology and Commission on Physical Sciences, Mathematics and Resources. National Academy Press.

Nearing, M.A., 2001: Potential changes in rainfall erosivity in the U.S. with climate change during the 21st century. *Journal of Soil and Water Conservation*, **56(3)**, 229-232.

Neary, D.G., C.C. Klopatek, L.F. DeBano, and P.F. Ffolliott, 1999: Fire effects on belowground sustainability: a review and synthesis. *Forest Ecology and Management*, **122(1)**, 51-71.

Neary, D.G., K.C. Ryan, and L.F. DeBano, 2005: *Wildland Fire in Ecosystems: Effects of Fire on Soils and Water*. General Technical Report RMRS-GTR-42-volume 4, US Department of Agriculture Forest Service, Rocky Mountain Research Station, Ogden, UT, pp.1-250.

Neff, J.C., J.W. Harden, and G. Gleixner, 2005: Fire effects on soil organic matter content, composition, and nutrients in boreal interior Alaska. *Canadian Journal of Forest Research*, **35(9)**, 2178-2187.

Neilson, R.P., J.M. Lenihan, D. Bachelet, and R.J. Drapek, 2005a: Climate change implications for sagebrush ecosystem. In: *Transactions of the 70th North American Wildlife and Natural Resources Conference*, Wildlife Management Institute.

Neilson, R.P., L.F. Pitelka, A.M. Solomon, R. Nathan, G.F. Midgley, J. Fragoso, H. Lischke, and K. Thompson, 2005b: Forecasting regional to global plant migration in response to climate change. *BioScience*, **55(9)**, 749-760.

Neilson, R.P. and L.H. Wullstein, 1983: Biogeography of 2 southwest american oaks in relation to atmospheric dynamics. *Journal of Biogeography*, **10(4)**, 275-297.

Nilsson, S. and W. Schopfhauser, 1995: The carbon-sequestration potential of a global afforestation program. *Climatic Change*, **30(3)**, 267-293.

Noon, B.R. and J.A. Blakesley, 2006: Conservation of the northern spotted owl under the northwest forest plan. *Conservation Biology*, **20(2)**, 288-296.

Norby, R.J., L. A. Joyce, and S. D. Wullschleger, 2005: Modern and future forests in a changing atmosphere, In: *A History of Atmospheric CO_2 and Its Effects on Plants, Animals, and Ecosystems*, [Ehleringer, J.R., T.E. Cerling, and M.D. Dearing (eds.)]. Springer Science and Business Media, Inc., New York, pp. 394-414.

Noss, R.F., 1990: Indicators for monitoring biodiversity - a hierarchical approach. *Conservation Biology*, **4(4)**, 355-364.

Noss, R.F., 2001: Beyond Kyoto: Forest management in a time of rapid climate change. *Conservation Biology*, **15(3)**, 578-590.

Novacek, M.J. and E.E. Cleland, 2001: The current biodiversity extinction event: scenarios for mitigation and recovery. *Proceedings of the National Academy of Sciences of the United States of America*, **98(10)**, 5466-5470.

Nowacki, G.J. and M.G. Kramer, 1998: *The Effects of Wind Disturbance on Temperate Rain Forest Structure and Dynamics of Southeast Alaska*. Conservation and resource assessments for the Tongass land management plan revision. General Technical Report PNW-GTR-421, US Department of Agriculture Forest Service, Pacific Northwest Research Station, Portland, OR, pp.1-25.

Nowak, R.S., D.S. Ellsworth, and S.D. Smith, 2004: Functional responses of plants to elevated atmospheric CO_2 - do photosynthetic and productivity data from FACE experiments support early predictions? *New Phytologist*, **162(2)**, 253-280.

Olden, J.D., 2006: Biotic homogenization: a new research agenda for conservation biogeography. *Journal of Biogeography*, **33(12)**, 2027-2039.

Oliver, C. and B.C. Larson, 1996: *Forest Stand Dynamics*. John Wiley and Sons, New York, pp. 1-521.

Oswalt, C.M. and S.N. Oswalt, 2007: Winter litter disturbance facilitates the spread of the nonnative invasive grass *Microstegium vimineum* (Trin.) A. Camus. *Forest Ecology and Management*, **249**, 199-203.

Ottawa National Forest, 2006: Affected environment and environmental consequences, Chapter 3, In: *Draft Environmental Impact Assessment*, USDA Forest Service, available at http://www.fs.fed.us/r9/ottawa/forest_management/forest_plan/revision/fp_final/volume_1_final_eis/final_eis_chapter_3.pdf.

Papadopol, C.S., 2000: Impacts of climate warming on forests in Ontario: options for adaptation and mitigation. *Forestry Chronicle*, **76(1)**, 139-149.

Papaik, M.J. and C.D. Canham, 2006: Species resistance and community response to wind disturbance regimes in northern temperate forests. *Journal of Ecology*, **94(5)**, 1011-1026.

Parker, W.C., S.J. Colombo, M.L. Cherry, M.D. Flannigan, S. Greifenhagen, R.S. McAlpine, C. Papadopol, and T. Scarr, 2000: Third millennium forestry: what climate change might mean to forests and forest management in Ontario. *Forestry Chronicle*, **76(3)**, 445-463.

Parmesan, C., 2006: Ecological and evolutionary responses to recent climate change. *Annual Review of Ecology, Evolution and Systematics*, **37**, 637-669.

Paulson, A., 2007: A speck of a species - felling pines across West. *The Christian Science Monitor*, **February 22, 2007**.

Peterson, D.L., M.J. Arbaugh, and L.J. Robinson, 1991: Growth trends of ozone-stressed ponderosa pine (*Pinus ponderosa*) in the Sierra Nevada of California, USA. *The Holocene*, **1(50)**, 61.

Peterson, D.L., M.C. Johnson, D.H. McKenzie, J.K. Agee, T.B. Jain, and E.D. Reinhardt, 2005: *Forest Structure and Fire Hazard in Dry Forests of the Western United States*. General Technical Report GTR-PNW-628, USDA Forest Service.

Peterson, G.D., G.S. Cumming, and S.R. Carpenter, 2003. Scenario planning: A tool for conservation in an uncertain world. *Conservation Biology*, **17(2)**, 358-366.

Peterson, S.A., J. Van Sickle, A.T. Herlihy, and R.M. Hughes, 2007: Mercury concentration in fish from streams and rivers throughout the western united states. *Environmental Science & Technology*, **41(1)**, 58-65.

Pielke, R.A., Sr., J.O. Adegoke, T.N. Chase, C.H. Marshall, T. Matsui, and D. Niyogi, 2006: A new paradigm for assessing the role of agriculture in the climate system and in climate change. *Agriculture and Forest Meteorology*, **142(2-4)**, 234-254.

Pimentel, D., L. Lach, R. Zuniga, and D. Morrison, 2000: Environmental and economic costs of nonindigenous species in the United States. *BioScience*, **50(1)**, 53-64.

Pounds, A.J., M.R. Bustamante, L.A. Coloma, J.A. Consuegra, M.P.L. Fogden, P.N. Foster, E. La Marca, K.L. Masters, A. Merino-Viteri, R. Puschendorf, S.R. Ron, G.A. Sanchez-Azofeifa, C.J. Still, and B.E. Young, 2006: Widespread amphibian extinctions from epidemic disease driven by global warming. *Nature*, **439(7073)**, 161-167.

Preston, B.L., 2006: Risk-based reanalysis of the effects of climate change on US cold-water habitat. *Climatic Change*, **76(1-2)**, 91-119.

Price, J.T. and T.L. Root, 2005: *Potential Impacts of Climate Change on Neotropical Migrants: Management Implications*. General Technical Report PSW-GTR-191, USDA Forest Service.

Price, M.F. and G.R. Neville, 2003: Designing strategies to increase the resilience of alpine montane systems to climate change. *Restoration Ecology*, **14(2)**, 170-176.

Pyne, S.J., P.L. Andrews, and R.D. Laven, 1996: *Introduction to Wildland Fire*. John Wiley & Sons, New York.

Radeloff, V.C., R.B. Hammer, S.I. Stewart, J.S. Fried, S.S. Holcomb, and J.F. McKeefry, 2005: The wildland-urban interface in the United States. *Ecological Applications*, **15(3)**, 799-805.

Rahel, F.J., 2000: Homogenization of fish faunas across the United States. *Science*, **288(5467)**, 854-856.

Rahel, F.J., C.J. Keleher, and J.L. Anderson, 1996: Potential habitat loss and population fragmentation for cold water fish in the North Platte River drainage of the Rocky Mountains: response to climate warming. *Limnology and Oceanography*, **41(5)**, 1116-1123.

Reed, D.H. and R. Frankham, 2003: Correlation between fitness and genetic diversity. *Conservation Biology*, **17**, 230-237.

Regier, H.A. and J.D. Meisner, 1990: Anticipated effects of climate change on freshwater fishes and their habitat. *Fisheries*, **15(6)**, 10-15.

Rehfeldt, G.E., N.L. Crookston, M.V. Warwell, and J.S. Evans, 2006: Empirical analyses of plant-climate relationships for the western United States. *International Journal of Plant Science*, **167(6)**, 1123-1150.

Reusch, T.B.H., A. Ehlers, A. Hammerli, and B. Worm, 2005: Ecosystem recovery after climatic extremes enhanced by genotypic diversity. *Proceedings of the National Academy of Sciences of the United States of America*, **102**, 2826-2831.

Rhoads, A.G., S.P. Hamburg, T.J. Fahey, T.G. Siccama, E.N. Hane, J. Battles, C. Cogbill, J. Randall, and G. Wilson, 2002: Effects of an intense ice storm on the structure of a northern hardwood forest. *Canadian Journal of Forest Research*, **32(10)**, 1763-1775.

Rice, K.J. and N.C. Emery, 2003: Managing microevolution: restoration in the face of global change. *Frontiers in Ecology and the Environment*, **(1)**, 469-478.

Richards, K.R., R.N. Sampson, and S. Brown, 2006: *Agricultural & Forestlands: U.S. Carbon Policy Strategies*. Pew Center on Global Climate Change.

Riitters, K.H. and J.W. Coulston, 2005: Hot spots of perforated forest in the eastern United States. *Environmental Management*, **35(4)**, 483-492.

Riitters, K.H. and J.D. Wickham, 2003: How far to the nearest road? *Frontiers in Ecology and the Environment*, **1(3)**, 125-129.

Root, T.L., J.T. Price, K.R. Hall, S.H. Schneider, C. Rosenzweig, and J.A. Pounds, 2003: Fingerprints of global warming on wild animals and plants. *Nature*, **421**, 57-60.

Ross, D.W., G.E. Daterman, J.L. Boughton, and T.M. Quigley, 2001: *Forest Health Restoration in South-Central Alaska: a Problem Analysis*. General Technical Report PNW-GTR-523, USDA Forest Service, Pacific Northwest Research Station, Portland, OR.

Rouault, G., J.N. Candau, F. Lieutier, L.M. Nageleisen, J.C. Martin, and N. Warzee, 2006: Effects of drought and heat on forest insect populations in relation to the 2003 drought in Western Europe. *Annals of Forest Science*, **63(6)**, 613-624.

Rowley, W.D., 1985: *US Forest Service Grazing and Rangelands*. Texas A & M University Press, College Station, Texas, 270.

Rueth, H.M., J. S. Baron, and L. A. Joyce, 2002: Natural resource extraction: past, present, and future, In: *Rocky Mountain Futures: an Ecological Perspective*, [Baron, J.S. (ed.)]. Island Press, Washington, DC, pp. 85-112.

Running, S.W., 2006: Is global warming causing more, larger wildfires? *Science*, **313(5789)**, 927-928.

Sampson, N. and L. DeCoster, 2000: Forest fragmentation: implications for sustainable private forests. *Journal of Forestry*, **98(3)**, 4-8.

Sampson, R.N., R. J. Scholes, C. Cerri, L. Erda, D. O. Hall, M. Handa, P. Hill, M. Howden, H. Janzen, J. Kimble, R. Lal, G. Marland, K. Minami, K. Paustian, P. Read, P. A. Sanchez, C. Scoppa, B. Solberg, M. A. Trossero, S. Trumbore, O. Van Cleemput, A. Whitmore, and D. Xu, 2000: Additional human-induced activities - article 3.4, [Watson, R.T., I.R. Noble, B. Bolin, N.H. Ravindranath, D.J. Verardo, and D.J. Dokken (eds.)]. Cambridge University Press, pp. 181-281.

Sasek, T.W. and B.R. Strain, 1990: Implications of atmospheric CO_2 enrichment and climatic change for the geographical distribution of two introduced vines in the U.S.A. *Climatic Change*, **16**, 31-51.

Sax, D.F. and S.D. Gaines, 2003: Species diversity: from global decreases to local increases. *Trends in Ecology and Evolution*, **18(11)**, 561-566.

Schmidt, K.M., J.P. Menakis, C.C. Hardy, W.J. Hann, and D.L. Bunnell, 2002: *Development of Coarse-Scale Spatial Data for Wildland Fire and Fuel Management*. U.S. Department of Agriculture Forest Service, Rocky Mountain Research Station, USDA Forest Service.

Schneider, S.H., S. Semenov, A. Patwardhan, I. Burton, C. H. D. Magadza, M. Oppenheimer, A. B. Pittock, A. Rahman, J. B. Smith, A. Suarez, and F. Yamin, 2007: Assessing key vulnerabilities and the risk from climate change, In: *Climate Change 2007: Impacts, Adaptation and Vulnerability. Contribution of Working Group II to the Fourth Assessment Report of the Intergovernmental Panel on Climate Change*, [Parry, M.L., O.F. Canziani, J.P. Palutikof, P.J. van der Linden, and C.E. Hanson (eds.)]. Cambridge University Press, Cambridge, UK, pp. 779-810.

Schneider, S.H. and T.L. Root, 2002: *Wildlife Responses to Climate Change: North American Case Studies*. Island Press, Washington, D.C., pp. 437.

Schoennagel, T., T.T. Veblen, and W.H. Romme, 2004: The interaction of fire, fuels, and climate across rocky mountain forests. *BioScience*, **54(7)**, 661-676.

Schoettle, A.W. and R.A. Sniezko, in press: Proactive management options for high elevation white pines threatened by an exotic pathogen. *Journal of Forest Research*.

Scholze, M., W. Knorr, N.W. Arnell, and I.C. Prentice, 2006: A climate-change risk analysis for world ecosystems. *Proceedings of the National Academy of Sciences of the United States of America*, **103(35)**, 13116-13120.

Seager, R., M. Ting, I. Held, Y. Kushnir, J. Lu, G. Vecchi, H. Huang, N. Harnik, A. Leetmaa, N.L. Lau, J. Velez, and N. Naik, 2007: Model projections of an imminent transition to a more arid climate in southwestern North America. *Science*, **316(1181)**, 1184.

Seymour, R.S., A.S. White, and P.G. deMaynadier, 2002: Natural disturbance regimes in northeastern North America-evaluating silvicultural systems using natural scales and frequencies. *Forest Ecology and Management*, **155(1-3)**, 357-367.

Simberloff, D., 2000: Global climate change and introduced species in United States forests. *Science of the Total Environment*, **262(3)**, 253-261.

Sitch, S., P.M. Cox, W.J. Collins, and C. Huntingford, 2007: Indirect radiative forcing of climate change through ozone effects on the land-carbon sink. *Nature*, **448**, 791-794.

Skinner, W.R., A. Shabbar, and M.D. Flanningan, 2006: Large forest fires in Canada and the relationship to global sea surface temperatures. *Journal of Geophysical Research-Atmospheres*, **111(D14)**, D14106.

Slovic, P., 1993: Perceived risk, trust, and democracy. *Risk Analysis*, **13(6)**, 675-682.

Smit, B., I. Burton, R.J.T. Klein, and J. Wandel, 2000: An anatomy of adaptation to climate change and variability. *Climatic Change*, **45(1)**, 223-251.

Smit, B. and O. Pilifosova, 2003: From adaptation to adaptive capacity and vulnerability reduction, In: *Climate Change, Adaptive Capacity and Development*, [Smith, J.B., R.J.T. Klein, and S. Hug (eds.)]. Imperial College Press, London, pp. 9-28.

Smit, B., O. Pilifosova, I. Burton, B. Challenger, S. Huq, R.J.T. Klein, G. Yohe, N. Adger, T. Downing, and E. Harvey, 2001: Adaptation to climate change in the context of sustainable development and equity. *Climate Change 2001: Impacts, Adaptation, and Vulnerability, Contribution of Working Group II to the Third Assessment Report of the Intergovernmental Panel on Climate Change*. [McCarthy, J.J., O.F. Canziani, N.A. Leary, D.J. Dokken, and K.S. White (eds.)]. Cambridge University Press, Cambridge, UK. 877-912.

Smit, B. and J. Wandel, 2006: Adaptation, adaptive capacity and vulnerability. *Global Environmental Change*, **16**, 282-292.

Smith, S.D., T.E. Huxman, S.F. Zitzer, T.N. Charlet, D.C. Housman, J.S. Coleman, L.K. Fenstermaker, J.R. Seemann, and R.S. Nowak, 2000: Elevated CO_2 increases productivity and invasive species success in an arid ecosystem. *Nature*, **408**, 79-82.

Sohngen, B. and R. Mendelsohn, 1998: Valuing the impact of large-scale ecological change in a market: the effect of climate change on U.S. timber. *American Economic Review*, **88**, 686-710.

Spencer, C.N., K.O. Gabel, and F.R. Hauer, 2003: Wildfire effects on stream food webs and nutrient dynamics in Glacier National Park, USA. *Forest Ecology and Management*, **178(1-2)**, 141-153.

Spittlehouse, D.L. and R.B. Stewart, 2003: Adaptation to climate change in forest management. *BC Journal of Ecosystems and Management*, **4(1)**.

Stein, B.A., S.R. Flack, N.B. Benton, and N. Conservancy, 1996: *America's Least Wanted: Alien Species Invasions of US Ecosystems*. The Nature Conservancy, Arlington, VA.

Stenseth, N.C., N.I. Samia, H. Viljugrein, K.L. Kausrud, M. Begon, S. Davis, H. Leirs, V.M. Dubyanskiy, J. Esper, V.S. Ageyev, N.L. Klassovskiy, S.B. Pole, and K.S. Chan, 2006: Plague dynamics are driven by climate variation. *Proceedings of the National Academy of Sciences of the United States of America*, **103(35)**, 13110-13115.

Stephenson, N.L., 1998: Actual evapotranspiration and deficit: biologically meaningful correlates of vegetation distribution across spatial scales. *Journal of Biogeography*, **25(5)**, 855-870.

Stewart, I.T., D.R. Cayan, and M.D. Dettinger, 2004: Changes in snowmelt runoff timing in Western North America under a 'business as usual' climate change scenario. *Climatic Change*, **62**, 217-232.

Stewart, S.I., V.C. Radeloff, and R.B. Hammer, 2006: *The Wildland-Urban Interface in the United States*. The public and wildland fire management: social science findings for managers Gen. Tech. Rep. NRS-1, U.S. Department of Agriculture, Forest Service, Northern Research Station, Newtown Square, PA: pp.197-202.

Stine, S., 1996: *Climate, (1650-1850)*. Version II, University of California, Centers for Water and Wildland Resources, Davis, pp.25-30.

Suffling, R. and D. Scott, 2002: Assessment of climate change effects on Canada's National Park system. *Environmental monitoring and assessment*, **74(2)**, 117-139.

Sun, G., S.G. McNulty, J. Lu, D.M. Amatya, Y. Liang, and R.K. Kolka, 2005: Regional annual water yield from forest lands and its response to potential deforestation across the southeastern United States. *Journal of Hydrology*, **308(1)**, 258-268.

Sutherst, R.W., 2004: Global change and human vulnerability to vector-borne diseases. *Clinical Microbiology Reviews*, **17(1)**.

Suttle, K.B., M.A. Thompsen, and M.E. Power, 2007: Species interactions reverse grassland responses to changing climates. *Science*, **315**, 640-642.

Swanson, F., S.L. Johnson, S.V. Gregory, and S.A. Acker, 1998: Flood disturbance in a forested mountain landscape. *BioScience*, **48(9)**, 681-689.

Swetnam, T.W. and J.L. Betancourt, 1998: Mesoscale disturbance and ecological response to decadal climatic variability in the American southwest. *Journal of Climate*, **11(12)**, 3128-3147.

Taylor, A.H. and R.M. Beaty, 2005: Climatic influences on fire regimes in the northern Sierra Nevada Mountains, Lake Tahoe Basin, Nevada, USA. *Journal of Biogeography*, **32(3)**, 425-438.

Tebaldi, C., K. Hayhoe, J.M. Arblaster, and G.A. Meehl, 2006: Going to the extremes. *Climatic Change*, **79(3-4)**, 185-211.

Thomas, C.D., A. Cameron, R.E. Green, M. Bakkenes, L.J. Beaumont, Y.C. Collingham, B.F.N. Erasmus, M.F. de Siqueira, A. Grainger, L. Hannah, L. Hughes, B. Huntley, A.S. Van Jaarsveld, G.F. Midgley, L. Miles, M.A. Ortega-Huerta, A.T. Peterson, O.L. Phillips, and S.E. Williams, 2004: Extinction risk from climate change. *Nature*, **427(6970)**, 145-148.

Tol, R.S., 2002: Estimates of the damage costs of climate change, Part II: dynamic estimates. *Environmental & Resource Economics*, **21(2)**, 135-160.

Toth, F.L., 1999: *Fair Weather? Equity Concerns in Climate Change*. Earthscan, London.

Truscott, A.M., C. Soulsby, S.C.F. Palmer, L. Newell, and P.E. Hulme, 2006: The dispersal characteristics of the invasive plant *Mimulus guttatus* and the ecological significance of increased occurrence of high-flow events. *Journal of Ecology*, **94**, 1080-1091.

Turner, M.G., W.L. Baker, C.J. Peterson, and R.K. Peet, 1998: Factors influencing succession: lessons from large, infrequent natural disturbances. *Ecosystems*, **1**, 511-523.

U.S. Environmental Protection Agency, 2000: *Stressor Identification Guidance Document*. EPA-822-B-00-025, U.S. Environmental Protection Agency, Office of Water & Office of Research and Development, Cincinnati, OH.

Ungerer, M.J., M.P. Ayres, and M.J. Lombardero, 1999: Climate and the northern distribution limits of Dendroctonus frontalis Zimmermann (*Coleoptera: Scolytidae*). *Journal of Biogeography*, **26(6)**, 1133-1145.

USDA, USDI, and DOE, 2006: *Scientific Inventory of Onshore Federal Lands' Oil and Gas Resources and the Extent and Nature of Restrictions or Impediments to Their Development.* pp.1-344.

USDA Forest Service, 1993: *Report of the Forest Service - Fiscal Year 1992.* Washington, DC.

USDA Forest Service, 2000: *2000 RPA Assessment of Forest and Rangelands.* FS 687, USDA Forest Service, Washington, DC.

USDA Forest Service, 2003: *An Analysis of the Timber Situation in the United States: 1952 to 2050.* General Technical Report PNW-GTR-560, Pacific Northwest Research Station, Portland, OR.

USDA Forest Service, 2004: *National Strategy and Implementation Plan for Invasive Species Management.* FS-805.

USDA Forest Service, in press: 2005 RPA assessment of forest and rangelands. *USDA Forest Service.*

USDA Forest Service, 2007a: *Conservation Education Strategic Plan to Advance Environmental Literacy 2007-2012.* FS-879, United States Department of Agriculture Forest Service.

USDA Forest Service, 2007b: *USDA Forest Service Strategic Plan FY 2007-2012.* FS-880, United States Department of Agriculture Forest Service.

USDA Forest Service Health Protection, 2005: *Forest Insect and Disease Conditions 2004.*

Van Mantgem, P.J., N.L. Stephenson, M.B. Keifer, and J. Keeley, 2004: Effects of an introduced pathogen and fire exclusion on the demography of sugar pine. *Ecological Applications*, **14(5)**, 1590-1602.

Vaux, H.J., P.D. Gardner, and J. Thomas, 1984: *Methods for Assessing the Impact of Fire on Forest Recreation.* General Technical Report PSW-79, US Deptment of Agriculture, Forest Service, Pacific Southwest Forest and Range Experiment Station, Berkeley, CA, pp.1-13.

Veblen, T.T., K.S. Hadley, E.M. Nel, T. Kitzberger, M. Reid, and R. Villalba, 1994: Disturbance regime and disturbance interactions in a Rocky Mountain subalpine forest. *Journal of Ecology*, **82(1)**, 125-135.

Veblen, T.T., K.S. Hadley, and M.S. Reid, 1991: Disturbance and stand development of a Colorado subalpine forest. *Journal of Biogeography*, **18(6)**, 707-716.

Vogel, C., S.C. Moser, R.E. Kasperson, and G. Dabelko, in press: Linking vulnerability, adaptation and resilience science to practice: players, pathways and partnerships. *Global Environmental Change.*

Volney, W.J.A. and R.A. Fleming, 2000: Climate change and impacts of boreal forest insects. *Agriculture Ecosystems & Environment*, **82(1-3)**, 283-294.

Von Hagen, B. and M. Burnett, 2006: Emerging markets for carbon stored by northwest forests, In: *Forests, Carbon, and Climate Change: A Synthesis of Science Findings*, [Cloughesy, M. (ed.)]. Oregon Forest Resources Institute, pp. 131-156.

Von Holle, B. and D. Simberloff, 2005: Ecological resistance to biological invasion overwhelmed by propagule pressure. *Ecology*, **86(12)**, 3212-3218.

Wagle, R.F. and J.H. Kitchen, Jr., 1972: Influence of fire on soil nutrients in a ponderosa pine type. *Ecology*, **53(1)**, 118-125.

Walker, B., S. Carpenter, J. Anderies, N. Abel, G.S. Cumming, M. Janssen, L. Lebel, J. Norberg, G.D. Peterson, and R. Pritchard, 2002: Resilience management in social-ecological systems: a working hypothesis for a participatory approach. *Conservation Ecology*, **6(1)**, 14.

Walker, J.C.G. and J.F. Kasting, 1992: Effects of fuel and forest conservation on future levels of atmospheric carbon-dioxide. *Global and Planetary Change*, **97(3)**, 151-189.

Watterson, N.A. and J.A. Jones, 2006: Flood and debris flow interactions with roads promote the invasion of exotic plants along steep mountain streams, western Oregon. *Geomorphology*, **78**, 107-123.

Watts, M.J. and H.G. Bohle, 1993: The space of vulnerability: the causal structure of hunger and famine. *Progress in Human Geography*, **17(1)**, 43-67.

Webb, T., III, 1986: Is vegetation in equilibrium with climate? How to interpret Late-Quaternary pollen data. *Plant Ecology*, **67(2)**, 75-91.

Weltzin, J.F., R.T. Belote, and N.J. Sanders, 2003: Biological invaders in a greenhouse world: will elevated CO_2 fuel plant invasions? *Frontiers in Ecology and the Environment*, **1(3)**, 146-153.

Werner, R.A., E.H. Holsten, S.M. Matsuoka, and R.E. Burnside, 2006: Spruce beetles and forest ecosystems in south-central Alaska: A review of 30 years of research. *Forest Ecology and Management*, **227(3)**, 195-206.

Westbrooks, R.G., 1998: *Invasive Plants: Changing the Landscape of America: Fact Book*. Federal Interagency Committee for the Management of Noxious and Exotic Weeds, Washington, DC.

Westerling, A.L. and B. Bryant, 2005: *Climate Change and Wildfire in and Around California: Fire Modeling and Loss Modeling*.

Westerling, A.L., A. Gershunov, T.J. Brown, D.R. Cayan, and M.D. Dettinger, 2003: Climate and wildfire in the Western United States. *Bulletin of the American Meteorological Society*, **84(5)**, 595-604.

Westerling, A.L., H.G. Hidalgo, D.R. Cayan, and T.W. Swetnam, 2006: Warming and earlier spring increase western U.S. forest wildfire activity. *Science*, **313(5789)**, 940-943.

Western Governors' Association, 2006: *A Collaborative Approach for Reducing Wildland Fire Risks to Communities and the Environment*. 10-year strategy implementation plan, available from http://www.westgov.org/wga/publicat/TYIP.pdf.

Wheaton, E., 2001: *Changing Fire Risk in a Changing Climate: a Literature Review and Assessment*. Saskatchewan Research Council SRC Publication No. 11341-2E01, Saskatchewan Research Council.

Whiles, M.R. and J.E. Garvey, 2004: *Freshwater Resources in the Hoosier-Shawnee Ecological Assessment Area*. General Technical Report NC-244, North Central Forest Experiment Station, USDA Forest Service, St. Paul, MN, pp.1-267.

Whitlock, C., S.L. Shafer, and J. Marlon, 2003: The role of climate and vegetation change in shaping past and future fire regimes in the northwestern U.S. and the implications for ecosystem management. *Forest Ecology Management*, **178(1-2)**, 5-21.

Wilbanks, T.J. and R.W. Kates, 1999: Global change in local places: how scale matters. *Climatic Change*, **43(3)**, 601-628.

Wilcove, D.S. and L.Y. Chen, 1998: Management costs for endangered species. *Conservation Biology*, **12(6)**, 1405-1407.

Williams, D.G. and Z. Baruch, 2000: African grass invasion in the Americas: ecosystem consequences and the role of ecophysiology. *Biological Invasions*, **2(2)**, 123-140.

Williams, D.W. and A.M. Liebhold, 2002: Climate change and the outbreak ranges of two North American bark beetles. *Agricultural and Forest Entomology*, **4(2)**, 87-99.

Williams, J.W., S.T. Jackson, and J.E. Kutzbach, 2007: Projected distributions of novel and disappearing climates by 2100 AD. *Proceedings of the National Academy of Sciences of the United States of America*, **104(14)**, 5738-5742.

Willis, K.J. and H.J.B. Birks, 2006: What is natural? The need for a long-term perspective in biodiversity conservation. *Science*, **314(5803)**, 1261.

Wilmking, M., G.P. Juday, V.A. Barber, and H.S.J. Zald, 2004: Recent climate warming forces contrasting growth responses of white spruce at treeline in Alaska through temperature thresholds. *Global Change Biology*, **10(10)**, 1724-1736.

Wilson, J., 2006: Using wood products to reduce global warming, In: *Forests, Carbon, and Climate: a Synthesis of Science Findings*, [Cloughesy, M. (ed.)]. Oregon Forest Resources Institute, pp. 114-129.

Wisner, B., P. Blaikie, T. Cannon, and A. Davis, 2004: *At Risk: Natural Hazards, People's Vulnerability and Disasters*. Routedge, London.

Woodworth, B.L., C.T. Atkinson, D.A. LaPointe, P.J. Hart, C.S. Spiegel, E.J. Tweed, C. Henneman, J. LeBrun, T. Denette, R. DeMots, K.L. Kozar, D. Triglia, D. Lease, A. Gregor, T. Smith, and D. Duffy, 2005: Host population persistence in the face of introduced vector-borne diseases: Hawaii amakihi and avian malaria. *Proceedings of the National Academy of Sciences of the United States of America*, **102(5)**, 1531-1536.

Worrall, J.J., T.D. Lee, and T.C. Harrington, 2005: Forest dynamics and agents that initiate and expand canopy gaps in Picea-Abies forests of Crawford Notch, New Hampshire, USA. *Journal of Ecology*, **93(1)**, 178-190.

Yarie, J., L. Viereck, K. van Cleve, and P. Adams, 1998: Flooding and ecosystem dynamics along the Tanana river. *BioScience*, **48(9)**, 690-695.

Ying, C.C. and A.D. Yanchuk, 2006: The development of British Columbia's tree seed transfer guidelines: purpose, concept, methodology, and implementation. *Forest Ecology and Management*, **227**, 1-13.

Yohe, G., N. Andronova, and M. Schlesinger, 2004: To hedge or not against an uncertain climate future? *Science*, **305(5695)**, 416-417.

Yohe, G.W. and R.S.J. Tol, 2002: Indicators for social and economic coping capacity--moving toward a working definition of adaptive capacity. *Global Environmental Change*, **12**, 25-40.

Yorks, T.E. and K.B. Adams, 2005: Ice storm impact and management implications for jack pine and pitch pine stands in New York, USA. *Forestry Chronicle*, **81.4**, 502-515.

Ziska, L.H., 2003: Evaluation of the growth response of six invasive species to past, present and future atmospheric carbon dioxide. *Journal of Experimental Botany*, **54(381)**, 395-404.

Ziska, L.H., S. Faulkner, and J. Lydon, 2004: Changes in biomass and root: shoot ratio of field-grown Canada thistle (*Cirsium arvense*), a noxious, invasive weed, with elevated CO2: implications for control with glyphosate. *Weed Science*, **52(4)**, 584-588.

Ziska, L.H., J.B. Reeves, and B. Blank, 2005: The impact of recent increases in atmospheric CO_2 on biomass production and vegetative retention of Cheatgrass (*Bromus tectorum*): implications for fire disturbance. *Global Change Biology*, **11(8)**, 1325-1332.

CHAPTER 4 REFERENCES

Albright, H.M. and M.A. Schenck, 1999: *Creating the National Park Service*. Norman Publishing.

Allen, C.D., 2007: Interactions across spatial scales among forest dieback, fire, and erosion in northern New Mexico landscapes. *Ecosystems*, **10(5)**, 797-808.

Allen, C.D. and D.D. Breshears, 1998: Drought-induced shift of a forest-woodland ecotone: rapid landscape response to climate variation. *Proceedings of the National Academy of Sciences of the United States of America*, **95(25)**, 14839-14842.

Allen, C.D., M. Savage, D.A. Falk, K.F. Suckling, T. Schulke, T.W. Swetnam, P.B. Stacey, P. Morgan, M. Hoffman, and J.T. Klingel, 2002: Ecological restoration of southwestern ponderosa pine ecosystems: a broad perspective. *Ecological Applications*, **12(5)**, 1418-1433.

Baron, J.S., 2004: Research in National Parks. *Ecological Applications*, **14(1)**, 3-4.

Baron, J.S., H.M. Rueth, A.M. Wolfe, K.R. Nydick, E.J. Allstott, J.T. Minear, and B. Moraska, 2000: Ecosystem responses to nitrogen deposition in the Colorado Front Range. *Ecosystems*, **3(4)**, 352-368.

Barron, M.G., S.E. Duvall, and K.J. Barron, 2004: Retrospective and current risks of mercury to panthers in the Florida Everglades. *Ecotoxicology*, **13(3)**, 223-229.

Beckage, B., L.J. Gross, and W.J. Platt, 2006: Modelling responses of pine savannas to climate change and large-scale disturbance. *Applied Vegetation Science*, **9(1)**, 75-82.

Beever, E.A., P.F. Brussard, and J. Berger, 2003: Patterns of apparent extirpation among isolated populations of pikas(Ochotona princeps) in the Great Basin. *Journal of Mammalogy*, **84(1)**, 37-54.

Belnap, J., 2003: The world at your feet: desert biological soil crusts. *Frontiers in Ecology and the Environment*, **1(4)**, 181-189.

Bowman, W.D., J.R. Gartner, K. Holland, and M. Wiedermann, 2006: Nitrogen critical loads for alpine vegetation and terrestrial ecosystem response: are we there yet? *Ecological Applications*, **16(3)**, 1183-1193.

Brooks, M.L., 1999: Habitat invasibility and dominance by alien annual plants in the western Mojave Desert. *Biological Invasions*, **1(4)**, 325-337.

Brooks, M.L., C.M. D'Antonio, D.M. Richardson, J.B. Grace, J.E. Keeley, J.M. DiTomaso, R.J. Hobbs, M. Pellant, and D. Pyke, 2004: Effects of invasive alien plants on fire regimes. *BioScience*, **54(7)**, 677-688.

Brown, D.G., K.M. Johnson, T.R. Loveland, and D.M. Theobald, 2005: Rural land-use trends in the conterminous United States, 1950-2000. *Ecological Applications*, **15(6)**, 1851-1863.

Bulger, A.J., B.J. Cosby, and J.R. Webb, 2000: Current, reconstructed past, and projected future status of brook trout (*Salvelinus fontinalis*) streams in Virginia. *Canadian Journal of Fisheries and Aquatic Sciences*, **57(7)**, 1515-1523.

Burkett, V., D. Wilcox, R. Stottlemyer, W. Barrow, D. Fagre, J. Baron, J. Price, J.L. Nielsen, C.D. Allen, D.L. Peterson, G. Ruggerone, and T. Doyle, 2005: Nonlinear dynamics in ecosystem response to climatic change: case studies and policy implications. *Ecological Complexity*, **2(4)**, 357-394.

Carpenter, S.R., 2002: Ecological futures: building an ecology of the long now. *Ecology*, **83(8)**, 2069-2083.

Chase, A., 1987: *Playing God in Yellowstone: the Destruction of America's First National Park*. Harcourt Brace, Orlando, FL.

Dale, V.H., L.A. Joyce, S. McNulty, R.P. Neilson, M.P. Ayres, M.D. Flannigan, P.J. Hanson, L.C. Irland, A.E. Lugo, and C.J. Peterson, 2001: Climate change and forest disturbances. *BioScience*.

Davis, S.M., D.L. Childers, J.J. Lorenz, H.R. Wanless, and T.E. Hopkins, 2005: A conceptual model of ecological interactions in the mangrove estuaries of the Florida Everglades. *Wetlands*, **25(4)**, 832-842.

Driscoll, C.T., G.B. Lawrence, A.J. Bulger, T.J. Butler, C.S. Cronan, C. Eagar, K.F. Lambert, G.E. Likens, J.L. Stoddard, and K.C. Weathers, 2001: Acidic deposition in the Northeastern United States: Sources and inputs, ecosystem effects, and management strategies. *BioScience*, **51(3)**, 180-198.

Edwards, M.E., L.B. Brubaker, A.V. Lozhkin, and P.M. Anderson, 2005: Structurally novel biomes: a response to past warming in Beringia. *Ecology*, **86(7)**, 1696-1703.

Fagre, D.B., D.L. Peterson, and A.E. Hessl, 2003: Taking the pulse of mountains: ecosystem responses to climatic variability. *Climatic Change*, **59(1-2)**, 263-282.

Fahrig, L., 2003: Effects of habitat fragmentation on biodiversity. *Annual Review of Ecology, Evolution and Systematics*, **34**, 487-515.

Finney, M.A., C.W. McHugh, and I.C. Grenfell, 2004: Stand- and landscape-level effects of prescribed burning on two Arizona wildfires. *Canadian Journal of Forest Research*, **35(7)**, 1714-1722.

Ford, D. and P. Williams, 1989: *Karst Geomorphology and Hydrology*. Chapman and Hall, Winchester, MA, pp. 1-320.

Gleick, P.H., 2006: *The World's Water 2006-2007: the Biennial Report on Freshwater Resources*. Island Press, Washington, DC.

Granshaw, F.D. and A.G. Fountain, 2006: Glacier change (1958-1998) in the North Cascades National Park Complex, Washington, USA. *Journal of Glaciology*, **52(177)**, 251-256.

Grayson, D.K., 2005: A brief history of Great Basin pikas. *Journal of Biogeography*, **32(12)**, 2103-2111.

Gunderson, L.H., 2000: Ecological resilience-in theory and application. *Annual Review of Ecology and Systematics*, **31**, 425-439.

Gunderson, L.H., 2001: Managing surprising ecosystems in southern Florida. *Ecological Economics*, **37(3)**, 371-378.

Gunderson, L.H. and C.S. Holling, 2002: *Panarchy: Understanding Transformations in Systems of Humans and Nature*. Island Press, Washington, DC.

Hall, M.H.P. and D.B. Fagre, 2003: Modeled climate-induced glacier change in Glacier National Park, 1850-2100. *BioScience*, **53(2)**, 131-140.

Hessl, A.E. and W.L. Baker, 1997: Spruce-fir growth form changes in the forest-tundra ecotone of Rocky Mountain National Park, Colorado, USA. *Ecography*, **20(4)**, 356-367.

Hobbs, R.J., S. Arico, J. Aronson, J.S. Baron, P. Bridgewater, V.A. Cramer, P.R. Epstein, J.J. Ewel, C.A. Klink, A.E. Lugo, D. Norton, D. Ojima, D.M. Richardson, E.W. Sanderson, F. Valladares, M. Vilà, R. Zamora, and M. Zobel, 2006: Novel ecosystems: theoretical and management aspects of the new ecological world order. *Global Ecology and Biogeography*, **15**, 1-7.

Hodgkins, G.A. and R.W. Dudley, 2006: Changes in the timing of winter-spring streamflows in eastern North America, 1913-2002. *Geophysical Research Letters*, **33(6)**.

Holling, C.S., 1973: Resilience and stability of ecological systems. *Annual Review of Ecology and Systematics*, **(4)**, 1-23.

Holling, C.S., 1978: *Adaptive Environmental Assessment and Management*. Blackburn Press, Caldwell, NJ.

Huntington, T.G., G.A. Hodgkins, B.D. Keim, and R.W. Dudley, 2004: Changes in the proportion of precipitation occurring as snow in New England (1949-2000). *Journal of Climate*, **17(13)**, 2626-2636.

IPCC, 2007: Summary for policymakers, In: *Climate Change 2007: the Physical Science Basis. Contribution of Working Group I to the Fourth Assessment Report of the Intergovernmental Panel on Climate Change*, [Solomon, S., D. Qin, M. Manning, Z. Chen, M. Marquis, K.B. Averyt, M. Tignor, and H.L. Miller (eds.)]. Cambridge University Press, Cambridge, United Kingdom and New York, NY, USA.

Keane, R.E., K.C. Ryan, T.T. Veblen, C.D. Allen, J. Logan, and B. Hawkes, 2002: *Cascading Effects of Fire Exclusion in the Rocky Mountain Ecosystems: a Literature Review*. General Technical Report RMRSGTR-91, U.S. Department of Agriculture, Forest Service, Rocky Mountain Research Station, Fort Collins, CO, pp.1-24.

Knowles, G., 2003: Aquatic life in the Sonoran Desert. *Endangered Species Bulletin*, **28(3)**, 22-23.

Knowles, N., M.D. Dettinger, and D.R. Cayan, 2006: Trends in snowfall versus rainfall in the Western United States. *Journal of Climate*, **19(18)**, 4545-4559.

Kohut, R.J., 2007: *Ozone Risk Assessments for Vital Signs Monitoring Networks, Appalachian National Scenic Trail, and Natchez Trace National Scenic Trail*. Natural Resource Report NPS/NRPC/ARD/NRTR-2007/001, US Department of Interior, National Park Service, Fort Collins, Colorado.

Kokko, H. and A. López-Sepulcre, 2006: From individual dispersal to species ranges: perspectives for a changing world. *Science*, **313(5788)**, 789-791.

Langner, J., R. Bergström, and V. Foltescu, 2005: Impact of climate change on surface ozone and deposition of sulphur and nitrogen in Europe. *Atmospheric Environment*, **39(6)**, 1129-1141.

Lee, K.N., 1993: *Compass and Gyroscope: Integrating Science and Politics for the Environment*. Island Press, Washington, DC.

Leopold, A.S., 1963: *Wildlife Management in the National Parks*. Report submitted by Advisory Board on Wildlife Management, appointed by Secretary of the Interior Udall, pp.1-23.

Lettenmaier, D.P., A.W. Wood, R.N. Palmer, E.F. Wood, and E.Z. Stakhiv, 1999: Water resources implications of global warming: a U.S. regional perspective. *Climatic Change*, **43(3)**, 537-579.

MacAvoy, S.E. and A.J. Bulger, 1995: Survival of brook trout (*Salvelinus fontinalis*) embryos and fry in streams of different acid sensitivity in Shenandoah National Park, USA. *Water, Air, & Soil Pollution*, **85(2)**, 445-450.

McKenzie, D., Z. Gedalof, D.L. Peterson, and P. Mote, 2004. Climatic change, wildfire, and conservation. *Conservation Biology*, **18(4)**, 890-902.

McKenzie, D.H., S. O'Neill, N.K. Larkin, and R.A. Norheim, 2006: How will climatic change affect air quality in parks and wilderness? In: *Proceedings of the 2005 George Wright Society Annual Meeting* [Harmon, D. (ed.)].

Mote, P.W., 2006: Climate-driven variability and trends in mountain snowpack in western North America. *Journal of Climate*, **19(23)**, 6209-6220.

Muir, J., 1911: *My First Summer in the Sierra*. Houghton Mifflin Company, New York.

Murdoch, P.S., J.S. Baron, and T.L. Miller, 2000: Potential effects of climate change on surface-water quality in North America. *Journal of the American Water Resources Association*, **36(2)**, 347-366.

National Invasive Species Council, 2001: *Meeting the Invasive Species Challenge: National Invasive Species Management Plan*. pp.1-80.

National Park Service, 1998: *Natural Resource Year in Review, 1997*. Publication D-1247, Department of the Interior.

National Park Service, 1999: *Natural Resource Challenge: The National Park Service's Action Plan for Preserving Natural Resources*. Natural Resource Information Division, National Park Service, Fort Collins, CO.

National Park Service, 2004a: *Final Yosemite Fire Management Plan: Environmental Impact Statement*.

National Park Service, 2004b: *Funding the Natural Resource Challenge*. Report to Congress Fiscal Year 2004, National Park Service, US Department of Interior., Washington, D.C..

National Park Service, 2006: *Management Policies 2006*. U.S. Department of the Interior, National Park Service.

National Research Council, 1999: *Our Common Journey: a Transition Toward Sustainability*. National Academy Press, Washington, DC.

National Research Council, 2003: *Adaptive Monitoring and Assessment for the Comprehensive Everglades Restoration Plan*. National Academy Press, pp.1-122.

Neufville, R., 2003: Real options: dealing with uncertainty in systems planning and design. *Integrated Assessment*, **4(1)**, 26-34.

Paine, R.T., M.J. Tegner, and E.A. Johnson, 1998: Compounded perturbations yield ecological surprises. *Ecosystems*, **1(6)**, 535-545.

Parmesan, C., 2006: Ecological and evolutionary responses to recent climate change. *Annual Review of Ecology, Evolution and Systematics*, **37**, 637-669.

Parsons, D.J., 2004: Supporting basic ecological research in U. S. National Parks: challenges and opportunities. *Ecological Applications*, **14(1)**, 5-13.

Pauly, D., 1995: Anecdotes and the shifting baseline syndrome of fisheries. *Trends in Ecology and Evolution*, **10(10)**, 430-430.

Peterson, G.D., G.S. Cumming, and S.R. Carpenter, 2003: Scenario planning: a tool for conservation in an uncertain world. *Conservation Biology*, **17(2)**, 358-366.

Pitcaithley, D.T., 2001: Philosophical underpinnings of the National Park idea. *Ranger*.

Poff, N.L.R., J.D. Olden, D.M. Merritt, and D.M. Pepin, 2007: Homogenization of regional river dynamics by dams and global biodiversity implications. *Proceedings of the National Academy of Sciences of the United States of America*, **104(14)**, 5732-5737.

Raskin, P.D., 2005: Global scenarios: background review for the Millennium Ecosystem Assessment. *Ecosystems*, **8(2)**, 133-142.

Regan, H.M., M. Colyvan, and M.A. Burgman, 2002: A taxonomy and treatment of uncertainty for ecology and conservation biology. *Ecological Applications*, **12(2)**, 618-628.

Ripple, W.J. and R.L. Beschta, 2005: Linking wolves and plants: Aldo Leopold on trophic cascades. *BioScience*, **55(7)**, 613-621.

Rittel, H.W.J. and M.M. Webber, 1973: Dilemmas in a general theory of planning. *Policy Sciences*, **4(2)**, 155-169.

Rodgers, J.A., Jr. and H.T. Smith, 1995: Set-back distances to protect nesting bird colonies from human disturbance in Florida. *Conservation Biology*, **9(1)**, 89-99.

Romme, W.H. and D.G. Despain, 1989: Historical perspective on the Yellowstone firs of 1988. *BioScience*, **39(10)**, 695-699.

Saxon, E., B. Baker, W. Hargrove, F. Hoffman, and C. Zganjar, 2005: Mapping environments at risk under different global climate change scenarios. *Ecology Letters*, **8(1)**, 53-60.

Sellars, R.W., 1999: *Preserving Nature in the National Parks: a History*. Yale University Press.

Shaw, J.D., B.E. Steed, and L.T. DeBlander, 2005: Forest inventory and analysis (FIA) annual inventory answers the question: what is happening to pinyon-juniper woodlands? *Journal of Forestry*, **103(6)**, 280-285.

Singer, J.F., C.V. Bleich, and A.M. Gudorf, 2000: Restoration of bighorn sheep metapopulations in and near western National Parks. *Restoration Ecology*, **8(4)**, 14-24.

Smith, S.M., D.E. Gawlik, K. Rutchey, G.E. Crozier, and S. Gray, 2003: Assessing drought-related ecological risk in the Florida Everglades. *Journal of Environmental Management*, **68(4)**, 355-366.

Stanford, J.A. and B. K. Ellis, 2002: Natural and cultural influences on ecosystem processes in the Flathead River Basin (Montana, British Columbia), In: *Rocky Mountain Futures: an Ecological Perspective*, Island Press, Covelo, CA, pp. 269-284.

Stewart, I.T., D.R. Cayan, and M.D. Dettinger, 2005: Changes toward earlier streamflow timing across western North America. *Journal of Climate*, **18(8)**, 1136-1155.

Stohlgren, T.J., G.W. Chong, L.D. Schell, K.A. Rimar, Y. Otsuki, M. Lee, M.A. Kalkhan, and C.A. Villa, 2002: Assessing vulnerability to invasion by nonnative plant species at multiple spatial scales. *Environmental Management*, **29(4)**, 566-577.

Sydoriak, C.A., C.D. Allen, and B.F. Jacobs, 2000: Would ecological landscape restoration make the Bandelier Wilderness more or less of a wilderness? *Proceedings: Wilderness Science in a Time of Change Conference-Volume 5: Wilderness Ecosystems, Threats, and Management*, **Proceedings RMRS-P-15-VOL-5**, 209-215.

Tomback, D.F. and K. C. Kendall, 2002: Rocky road in the Rockies: challenges to biodiversity, In: *Rocky Mountain Futures, an Ecological Perspective*, [Baron, J. (ed.)]. Island Press, Washington, DC, pp. 153-180.

U.S. Geological Survey, 2005: *The State of the Colorado River Ecosystem in Grand Canyon*. USGS Circular 1282, U.S. Department of the Interior, U.S. Geological Survey, pp.1-220.

Unger, S., 1999: *The Restoration of an Ecosystem*. Everglades National Park.

Walters, C., 1986: *Adaptive Management of Renewable Resources*. McGraw Hill, New York.

Walters, C., J. Korman, L.E. Stevens, and B. Gold, 2000: Ecosystem modeling for evaluation of adaptive management policies in the Grand Canyon. *Conservation Ecology*, **4(2)**1 (online).

Walters, C.J. and C.S. Holling, 1990: Large-scale management experiments and learning by doing. *Ecology*, **71(6)**, 2060-2068.

Weiss, J.L. and J.T. Overpeck, 2005: Is the Sonoran Desert losing its cool? *Global Change Biology*, **11(12)**, 2065-2077.

Westerling, A.L., H.G. Hidalgo, D.R. Cayan, and T.W. Swetnam, 2006: Warming and earlier spring increase western U.S. forest wildfire activity. *Science*, **313(5789)**, 940-943.

Williams, J.W., S.T. Jackson, and J.E. Kutzbach, 2007: Projected distributions of novel and disappearing climates by 2100 AD. *Proceedings of the National Academy of Sciences of the United States of America*, **104(14)**, 5738-5742.

Willis, K.J. and H.J.B. Birks, 2006: What is natural? The need for a long-term perspective in biodiversity conservation. *Science*, **314(5803)**, 1261.

Winks, R.W., 1997: The National Park Service Act of 1916: a contradictory mandate? *Denver University Law Review*, **74(3)**, 575-623.

CHAPTER 5 REFERENCES

Allan, J.D., M. A. Palmer, and N. L. Poff, 2005: Climate change and freshwater ecosystems, In: *Climate Change and Biodiversity*, [Lovejoy, T.E. and L. Hannah (eds.)]. Yale University Press, New Haven.

Alward, R.D., J.K. Detling, and D.G. Milchunas, 1999: Grassland vegetation changes and nocturnal global warming. *Science*, **283(5399)**, 229-231.

Anderson, D.R., K.P. Burnham, J.D. Nichols, and M.J. Conroy, 1987: The need for experiments to understand population dynamics of American black ducks. *Wildlife Society Bulletin*, **15(2)**, 282-284.

Araújo, M.B. and M. New, 2007: Ensemble forecasting of species distributions. *Trends in Ecology and Evolution*, **22**, 42-47.

Araújo, M.B. and C. Rahbek, 2006: How does climate change affect biodiversity? *Science*, **313(5792)**, 1396-1397.

Bachelet, D., R.P. Neilson, T. Hickler, R.J. Drapek, J.M. Lenihan, M.T. Sykes, B. Smith, S. Sitch, and K. Thonicke, 2003: Simulating past and future dynamics of natural ecosystems in the United States. *Global Biogeochemical Cycles*, **17(2)**, 1045-1066.

Bachelet, D., R.P. Neilson, J.M. Lenihan, and R.J. Drapek, 2001: Climate change effects on vegetation distribution and carbon budget in the United States. *Ecosystems*, **4**, 164-185.

Barnett, T.P., J.C. Adam, and D.P. Lettenmaier, 2005: Potential impacts of a warming climate on water availability in snow-dominated regions. *Nature*, **438(7066)**, 303-309.

Batt, B.D.J., M. G. Anderson, C. D. Anderson, and F. D. Caswell, 1989: The use of prairie potholes by North American ducks, In: *Northern Prairie Wetlands*, Iowa State University Press, Ames, IA, pp. 204-227.

Berteaux, D., D. Reale, A.G. McAdam, and S. Boutin, 2004: Keeping pace with fast climate change: can arctic life count on evolution? *Integrative and Comparative Biology*, **44(2)**, 140-151.

Bildstein, K.L., 1998: Long-term counts of migrating raptors: a role for volunteers in wildlife research. *Journal of Wildlife Management*, **62(2)**, 435-445.

Blades, E., 2007: The National Wildlife Refuge System: providing a conservation advantage to threatened and endangered species in the United States. *Thesis*.

Both, C., S. Bouwhuis, C.M. Lessells, and M.E. Visser, 2006: Climate change and population declines in a long-distance migratory bird. *Nature*, **441(7089)**, 81-83.

Botkin, D.B., 1990: *Discordant Harmonies: a New Ecology for the Twenty-First Century*. Oxford University Press, New York.

Breiman, L., 2001: Random forests. *Machine Learning*, **45(1)**, 5-32.

Brooks, M.L., C.M. D'Antonio, D.M. Richardson, J.B. Grace, J.E. Keeley, J.M. DiTomaso, R.J. Hobbs, M. Pellant, and D. Pyke, 2004: Effects of invasive alien plants on fire regimes. *BioScience*, **54**, 677-688.

Bunn, D., A. Mummert, M. Hoshovsky, K. Gilardi, and S. Shanks, 2007: *California Wildlife: Conservation Challenges*. California's Wildlife Action Plan. California Department of Fish and Game, Sacramento, CA.

Burkett, V. and J. Kusler, 2000: Climate change: potential impacts and interactions in wetlands of the United States. *Journal of American Water Resources Association*, **36(2)**, 313-320.

Caudill, J. and E. Henderson, 2003: *Banking on Nature 2002: the Economic Benefits to Local Communities of National Wildlife Refuge Visitation*. U.S. Fish and Wildlife Service, Division of Economics, Washington, DC.

Chao, P., 1999: Great Lakes water resources: climate change impact analysis with transient GCM scenarios. *Journal of American Water Resources Association*, **35**, 1499-1507.

Cinq-Mars, J. and A.W. Diamond, 1991: The effects of global climate change on fish and wildlife resources. *Transactions of the North American Wildlife and Natural Resources Conference*, **NAWTA 6**, 171-176.

Curtin, C.G., 1993: The evolution of the U.S. National Wildlife Refuge System and the doctrine of compatibility. *Conservation Biology*, **7(1)**, 29-38.

Czech, B., 2005: The capacity of the National Wildlife Refuge System to conserve threatened and endangered animal species in the United States. *Conservation Biology*, **19(4)**, 1246-1253.

D'Antonio, C.M. and P.M. Vitousek, 1992: Biological invasions by exotic grasses, the grass/fire cycle, and global change. *Annual Review of Ecology and Systematics*, **23**, 63-87.

Daly, C., D. Bachelet, J.M. Lenihan, R.P. Neilson, W. Parton, and D. Ojima, 2000: Dynamic simulation of tree-grass interactions for global change studies. *Ecological Applications*, **10(2)**, 449-469.

Daniels, R.C., T.W. White, and K.K. Chapman, 1993: Sea-level rise: destruction of threatened and endangered species habitat in South Carolina. *Environmental Management*, **17(3)**, 373-385.

Davis, M.B., R.G. Shaw, and J.R. Etterson, 2005: Evolutionary responses to changing climate. *Ecology*, **86(7)**, 1704-1714.

Davison, R.P., A. Falcucci,.L. Maiorano, and J. M. Scott, 2006: The National Wildlife Refuge System, In: *The Endangered Species Act at Thirty*, [Goble, D.D., J.M. Scott, and F.W. Davis (eds.)]. Island Press, Washington, Covelo, London, pp. 90-100.

Dettinger, M.D., 2005: From climate-change spaghetti to climate-change distributions for 21st century California. *San Francisco Estuary and Watershed Science*, **3(1)**, Article 4.

Donahue, D.L., 1999: *Western Range Revisited: Removing Livestock From Public Lands to Conserve Native Biodiversity*. University of Oklahoma Press, pp. 1-388.

Dukes, J.S. and H.A. Mooney, 1999: Does global change increase the success of biological invaders? *Trends in Ecology and Evolution*, **14(4)**, 135-139.

Elith, J., C.H. Graham, R.P. Anderson, M. Dudík, S. Ferrier, A. Guisan, R.J. Hijmans, F. Huettmann, J.R. Leathwick, A. Lehmann, J. Li, L.G. Lohmann, B.A. Loiselle, G. Manion, C. Moritz, M. Nakamura, Y. Nakazawa, J.M. Overton, A.T. Peterson, S.J. Phillips, K.S. Richardson, R. Scachetti-Pereira, R.E. Schapire, J. Soberón, S. Williams, M.S. Wisz, and N.E. Zimmermann, 2006: Novel methods improve prediction of species' distributions from occurrence data. *Ecography*, **29(2)**, 129-151.

Erwin, R.M., G.M. Sanders, and D.J. Prosser, 2004: Changes in lagoonal marsh morphology at selected northeastern Atlantic coast sites of signi cance to migratory waterbirds. *Wetlands*, **24(4)**, 891-903.

Fischman, R.L., 2003: *The National Wildlife Refuges: Coordinating a Conservation System Through Law*. Island Press, Washington, Covelo, and London.

Fischman, R.L., 2004: The meanings of biological integrity, diversity, and environmental health. *Natural Resources Journal*, **44**, 989-1026.

Fischman, R.L., 2005: The signi cance of national wildlife refuges in the development of U.S. conservation policy. *The Journal of Land Use and Environmental Law*, **21**, 1-22.

Flannigan, M.D., K.A. Logan, B.D. Amiro, W.R. Skinner, and B.J. Stocks, 2005: Future area burned in Canada. *Climatic Change*, **72(1)**, 1-16.

Ford, S.E., 1996: Range extension by the oyster parasite Perkinsus marinus into the northeastern United States: response to climate change? *Journal of Shellfish Research*, **15**, 45-56.

Frederick, K.D. and P.H. Gleick, 1999: *Water and Global Climate Change: Potential Impacts on U.S. Water Resources*. Pew Center on Global Climate Change, Arlington, VA, pp.1-55.

Gabrielson, I.N., 1943: *Wildlife Refuges*. The Macmillan Company, New York.

Galbraith, H., R. Jones, R. Park, J. Clough, S. Herrod-Julius, B. Harrington, and G. Page, 2002: Global climate change and sea level rise: potential losses of intertidal habitat for shorebirds. *Waterbirds*, **25(2)**, 173-183.

Garrott, R.A., P.J. White, and C.A.V. White, 1993: Overabundance: an Issue for conservation biologists? *Conservation Biology*, **7(4)**, 946-949.

Gergely, K., J.M. Scott, and D. Goble, 2000: A new direction for the U.S. National Wildlife Refuges: the National Wildlife Refuge System Improvement Act of 1997. *Natural Areas Journal*, **20(2)**, 107-118.

Giorgi, F., 1990: Simulation of regional climate using a limited area model nested in a general circulation model. *Journal of Climate*, **3(9)**, 941-963.

Gonzalez, P., R.P. Neilson, and R.J. Drapek, 2005: Climate change vegetation shifts across global ecoregions. *Ecological Society of America Annual Meeting Abstracts*, **90**, 228.

Grumbine, R.E., 1990: Viable populations, reserve size, and federal lands management: a critique. *Conservation Biology*, **4(2)**, 127-134.

Guinotte, J.M., J. Orr, S. Cairns, A. Freiwald, L. Morgan, and R. George, 2006: Will human-induced changes in seawater chemistry alter the distribution of deep-sea scleractinian corals? *Frontiers in Ecology and the Environment*, **4(3)**, 141-146.

Hampe, A. and R.J. Petit, 2005: Conserving biodiversity under climate change: the rear edge matters. *Ecology Letters*, **8(5)**, 461-467.

Hannah, L., G.F. Midgley, G.O. Hughes, and B. Bomhard, 2005: The view from the Cape: extinction risk, protected areas, and climate change. *BioScience*, **55(3)**.

Hannah, L., G.F. Midgley, T. Lovejoy, W.J. Bond, M. Bush, J.C. Lovett, D. Scott, and F.I. Woodward, 2002: Conservation of biodiversity in a changing climate. *Conservation Biology*, **16(1)**, 264-268.

Harris, L.D., 1984: *The Fragmented Forest: Island Biogeography Theory and the Preservation of Biotic Diversity*. University of Chicago Press, Chicago, IL.

Harvell, C.D., K. Kim, J.M. Burkholder, R.R. Colwell, P.R. Epstein, D.J. Grimes, E.E. Hofmann, E.K. Lipp, A. Osterhaus, and R.M. Overstreet, 1999: Emerging marine diseases--climate links and anthropogenic factors. *Science*, **285**, 1505-1510.

Harvell, C.D., C.E. Mitchell, J.R. Ward, S. Altizer, A.P. Dobson, R.S. Ostfeld, and M.D. Samuel, 2002: Climate warming and disease risks for terrestrial and marine biota. *Science*, **296(5576)**, 2158-2162.

Harvell, D., K. Kim, C. Quirolo, J. Weir, and G. Smith, 2001: Coral bleaching and disease: contributors to 1998 mass mortality in Briareum asbestinum (*Octocorallia, Gorgonacea*). *Hydrobiologia*, **460(1)**, 97-104.

Hayhoe, K., D. Cayan, C.B. Field, P.C. Frumhoff, E.P. Maurer, N.L. Miller, S.C. Moser, S.H. Schneider, K.N. Cahill, E.E. Cleland, L. Dale, R. Drapek, R.M. Hanemann, L.S. Kalkstein, J. Lenihan, C.K. Lunch, R.P. Neilson, S.C. Sheridan, and J.H. Verville, 2004: Emissions pathways, climate change, and impacts on California. *Proceedings of the National Academy of Sciences of the United States of America*, **101(34)**, 12422-12427.

Hayhoe, K., C.P. Wake, T.G. Huntington, L. Luo, M.D. Schwartz, J. Sheffield, E. Wood, B. Anderson, J. Bradbury, A. DeGaetano, T.J. Troy, and D. Wolfe, 2007: Past and future changes in climate and hydrological indicators in the US Northeast. *Climate Dynamics*, **28(4)**, 381-407.

Hersteinsson, P. and D.W. Macdonald, 1992: Interspecific competition and the geographical distribution of red and arctic foxes Vulpes vulpes and Alopex lagopus. *Oikos*, **64(3)**, 505-515.

Hoekstra, J.M., T.M. Boucher, T.H. Ricketts, and C. Roberts, 2005: Confronting a biome crisis: global disparities of habitat loss and protection. *Ecology Letters*, **8(1)**, 23-29.

Holling, C.S., 1973: Resilience and stability of ecological systems. *Annual Review of Ecology and Systematics*, **4**, 1-23.

Holling, C.S., 1978: *Adaptive Environmental Assessment and Management*. Blackburn Press, Caldwell, NJ.

Houghton, J.T., Y. Ding, D.J. Griggs, M. Noguer, P.J. van der Linden, X. Dai, K. Maskell, and C.A. Johnson, 2001: *Climate Change 2001: the Scientific Basis*. Cambridge University Press, Cambridge.

Hulme, P.E., 2005: Adapting to climate change: is there scope for ecological management in the face of a global threat? *Journal of Applied Ecology*, **42(5)**, 784-794.

Huntley, B., Y.C. Collingham, R.E. Green, G.M. Hilton, C. Rahbek, and S.G. Willis, 2006: Potential impacts of climatic change upon geographical distributions of birds. *Ibis*, **148**, 8-28.

Hurd, B., N. Leary, R. Jones, and J. Smith, 1999: Relative regional vulnerability of water resources to climate change. *Journal of the American Planning Association*, **35(6)**, 1399-1409.

Inkley, D.B., M.G. Anderson, A.R. Blaustein, V.R. Burkett, B. Felzer, B. Griffith, J. Price, and T.L. Root, 2004. *Global Climate Change and Wildlife in North America*. The Wildlife Society, Bethesda, MD.

IPCC, 2000: *Special Report on Emissions Scenarios*. [Nakicenovic, N. and R. Swart (eds.)]. Cambridge University Press, Cambridge, UK, pp. 1-570.

IPCC, 2001: *Climate Change 2001: Impacts, Adaptation, and Vulnerability. Contribution of Working Group II to the Third Assessment Report of the Intergovernmental Panel on Climate Change*. [McCarthy, J.J., O.F. Canziani, N.A. Leary, D.J. Dokken, and K.S. White (eds.)]. Cambridge University Press, Cambridge, UK.

IPCC, 2007a: *Climate Change 2007: The Physical Science Basis. Contribution of Working Group I to the Fourth Assessment Report of the Intergovernmental Panel on Climate Change*. [Solomon, S., D. Qin, M. Manning, Z. Chen, M. Marquis, K.B. Averyt, M. Tignor, and H.L. Miller (eds.)]. Cambridge University Press, Cambridge, United Kingdom and New York, NY, USA, pp. 1-996.

IPCC, 2007b: Summary for policymakers, In: *Climate Change 2007: the Physical Science Basis. Contribution of Working Group I to the Fourth Assessment Report of the Intergovernmental Panel on Climate Change*, [Solomon, S., D. Qin, M. Manning, Z. Chen, M. Marquis, K.B. Averyt, M. Tignor, and H.L. Miller (eds.)]. Cambridge University Press, Cambridge, United Kingdom and New York, NY, USA.

Iverson, L.R., M.W. Schwartz, and A.M. Prasad, 2004: How fast and far might tree species migrate in the eastern United States due to climate change? *Global Ecology and Biogeography*, **13(3)**, 209-219.

Iverson, L.R. and A.M. Prasad, 1998. Predicting abundance of 80 tree species following climate change in the eastern United States. *Ecological Monographs*, **68(4)**, 465-485.

Jackson, J.B.C., M.X. Kirby, W.H. Berger, K.A. Bjorndal, L.W. Botsford, B.J. Bourque, R.H. Bradbury, R. Cooke, J. Erlandson, J.A. Estes, T.P. Hughes, S. Kidwell, C.B. Lange, H.S. Lenihan, J.M. Pandolfi, C.H. Peterson, R.S. Steneck, M.J. Tegner, and R.R. Warner, 2001: Historical overfishing and the recent collapse of coastal ecosystems. *Science*, **293**, 629-638.

Johnson, F.A., W.L. Kendall, and J.A. Dubovsky, 2002: Conditions and limitations on learning in the adaptive management of mallard harvests. *Wildlife Society Bulletin*, **30**, 176-185.

Johnson, W.C., B.V. Millett, T. Gilmanov, R.A. Voldseth, G.R. Guntenspergen, and D.E. Naugle, 2005: Vulnerability of northern prairie wetlands to climate change. *BioScience*, **55(10)**, 863-872.

Juanes, F., S. Gephard, and K.F. Beland, 2004: Long-term changes in migration timing of adult Atlantic salmon (*Salmo salar*) at the southern edge of the species distribution. *Canadian Journal of Fisheries and Aquatic Sciences*, **61(12)**, 2392-2400.

Jump, A.S. and J. Peñuelas, 2005: Running to stand still: adaptation and the response of plants to rapid climate change. *Ecology Letters*, **8(9)**, 1010-1020.

Kareiva, P.M., J. G. Kingsolver, and R. B. Huey, 1993: Introduction, In: *Biotic Interactions and Global Change*, Sinauer Associates Inc., Sunderland, MA, pp. 1-6.

Knox, J.C., 1993: Large increases in flood magnitude in response to modest changes in climate. *Nature*, **361(6411)**, 430-432.

Kutz, S.J., E.P. Hoberg, L. Polley, and E.J. Jenkins, 2005: Global warming is changing the dynamics of Arctic host-parasite systems. *Proceedings of the Royal Society of London, Series B: Biological Sciences*, **272(1581)**, 2571-2576.

LaPointe, D., T. L. Benning, and C. T. Atkinson, 2005: Avian malaria, climate change, and native birds of Hawaii, In: *Climate Change and Biodiversity*, [Lovejoy, T.E. and L. Hannah (eds.)]. Yale University Press, New Haven, pp. 317-321.

Larsen, C.F., R.J. Motyka, J.T. Freymueller, and K. Echelmeyer, 2004a: Rapid uplift of southern Alaska caused by recent ice loss. *Geophysical Journal International*, **158(3)**, 1118-1133.

Larsen, C.F., R.J. Motyka, J.T. Freymueller, K.A. Echelmeyer, and E.R. Ivins, 2005: Rapid viscoelastic uplift in southeast Alaska caused by post-Little Ice Age glacial retreat. *Earth and Planetary Science Letters*, **237(3)**, 548-560.

Larsen, C.I., G. Clark, G. Guntenspergen, D.R. Cahoon, V. Caruso, C. Huppo, and T. Yanosky, 2004b: *The Blackwater NWR Inundation Model. Rising Sea Level on a Low-Lying Coast: Land Use Planning for Wetlands.* U.S. Geological Survey, Reston, VA.

Larson, D.L., 1995: Effects of climate on numbers of northern prairie wetlands. *Climatic Change*, **30(2)**, 169-180.

Lawler, J.J., D. White, R.P. Neilson, and A.R. Blaustein, 2006: Predicting climate-induced range shifts: model differences and model reliability. *Global Change Biology*, **12**, 1568-1584.

Lemke, P., J. Ren, R. B. Alley, I. Allison, J. Carrasco, G. M. Flato, Y. Fujii, G. Kaser, P. Mote, R. H. Thomas, and T. Zhang, 2007: Observations: changes in snow, ice and frozen ground, In: *Climate Change 2007: the Physical Science Basis. Contribution of Working Group I to the Fourth Assessment Report of the Intergovernmental Panel on Climate Change*, [Solomon, S., D. Quin, M. Manning, Z. Chen, M. Marquis, K.B. Averyt, M. Tignor, and H.L. Miller (eds.)]. Cambridge University Press, Cambridge, UK.

Lenihan, J.M., R. Drapek, D. Bachelet, and R.P. Neilson, 2003: Climate change effects on vegetation distribution, carbon, and fire in California. *Ecological Applications*, **13(6)**, 1667-1681.

Logan, J.A., J. Regniere, and J.A. Powell, 2003: Assessing the impacts of global warming on forest pest dynamics. *Frontiers in Ecology and the Environment*, **1(3)**, 130-137.

Lovejoy, T.E. and L. Hannah, 2006: *Climate Change and Biodiversity*. [Lovejoy, T.E. and L. Hannah (eds.)]. Yale University Press, New Haven, CT.

MacPherson, A.H., 1964: A northward range extension of the red fox in the eastern Canadian arctic. *Journal of Mammalogy*, **45(1)**, 138-140.

Magnuson, J.J., K.E. Webster, R.A. Assel, C.J. Bowser, P.J. Dillon, J.G. Eaton, H.E. Evans, E.J. Fee, R.I. Hall, L.R. Mortsch, D.W. Schindler, and F.H. Quinn, 1997: Potential effects of climate changes on aquatic systems: Laurentian Great Lakes and Precambrian shield region. *Hydrological Processes*, **11**, 825-871.

Marsh, P. and N.N. Neumann, 2001: Processes controlling the rapid drainage of two ice-rich permafrost-dammed lakes in NW Canada. *Hydrological Processes*, **15(18)**, 3433-3446.

Matthews, W.J. and E.G. Zimmerman, 1990: Potential effects of global warming on native fishes of the southern Great Plains and the Southwest. *Fisheries*, **15**, 26-32.

McLachlan, J.S., J.J. Hellmann, and M.W. Schwartz, 2007: A framework for debate of assisted migration in an era of climate change. *Conservation Biology*, **21(2)**, 297-302.

Meehl, G.A., W.M. Washington, W.D. Collins, J.M. Arblaster, A. Hu, L.E. Buja, W.G. Strand, and H. Teng, 2005: How much more global warming and sea level rise? *Science*, **307(5716)**, 1769-1772.

Michener, W.K., E.R. Blood, K.L. Bildstein, M.M. Brinson, and L.R. Gardner, 1997: Climate change, hurricanes and tropical storms, and rising sea level in coastal wetlands. *Ecological Applications*, **7(3)**, 770-801.

Millennium Ecosystem Assessment, 2006: *Ecosystems and Human Well-Being: Current State and Trends*. Island Press, Washington, DC.

Milly, P.C.D., K.A. Dunne, and A.V. Vecchia, 2005: Global pattern of trends in streamflow and water availability in a changing climate. *Nature*, **438(7066)**, 347-350.

Moore, M.V., M.L. Pace, J.R. Mather, P.S. Murdoch, R.W. Howarth, C.L. Folt, C.Y. Chen, H.F. Hemond, P.A. Flebbe, and C.T. Driscoll, 1997: Potential effects of climate change on freshwater ecosystems of the New England/Mid-Atlantic Region. *Hydrological Processes*, **11**, 925-947.

Mote, P.W., E.A. Parson, A.F. Hamlet, W.S. Keeton, D. Lettenmaier, N. Mantua, E.L. Miles, D.W. Peterson, D.L. Peterson, R. Slaughter, and A.K. Snover, 2003: Preparing for climatic change: the water, salmon, and forests of the Pacific Northwest. *Climatic Change*, **61(1)**, 45-88.

National Ecological Assessment Team, 2006: *Strategic Habitat Conservation Initiative Final Report*. U.S. Geological Survey and U.S. Fish and Wildlife Service.

National Research Council, 2007: *Endangered and Threatened Fishes in the Klamath River Basin: Causes of Decline and Strategies for Recovery*. National Research Council, Washington, DC.

Neilson, R.P., I. C. Prentice, B. Smith, T. G. F. Kittel, and D. Viner, 1998: Simulated changes in vegetation distribution under global warming, In: *The Regional Impacts of Climate Change: an Assessment of Vulnerability*, Intergovernmental Panel on Climate Change. Cambridge University Press, Cambridge, UK, pp. 439-456.

Nichols, J.D., F.A. Johnson, and B.K. Williams, 1995: Managing North American waterfowl in the face of uncertainty. *Annual Review of Ecology and Systematics*, **26**, 177-199.

Noss, R.F., 1987: Protecting natural areas in fragmented landscapes. *Natural Areas Journal*, **7(1)**, 2-13.

Oakley, K.L., L.P. Thomas, and S.G. Fancy, 2003: Guidelines for long-term monitoring protocols. *Wildlife Society Bulletin*, **31(4)**, 1000-1003.

Omernik, J.M., 1987: Ecoregions of the conterminous United States. *Annals of the Association of American Geographers*, **77(1)**, 118-125.

Pamperin, N.J., E.H. Follmann, and B. Petersen, 2006: Interspecific killing of an arctic fox by a red fox at Prudhoe Bay, Alaska. *Arctic*, **59(4)**, 361-364.

Park, R.A., M. S. Treehan, P. W. Mausel, and R. C. Howe, 1989: The effects of sea level rise on US coastal wetlands, In: *Potential Effects of Global Climate Change on the United States*, U.S. Environmental Protection Agency, Washington, DC.

Parmesan, C., 2006: Ecological and evolutionary responses to recent climate change. *Annual Review of Ecology, Evolution and Systematics*, **37**, 637-669.

Parmesan, C. and G. Yohe, 2003: A globally coherent fingerprint of climate change impacts across natural systems. *Nature*, **421**, 37-42.

Payette, S., M.J. Fortin, and I. Gamache, 2001: The subarctic forest tundra: the structure of a biome in a changing climate. *BioScience*, **51(9)**, 709-718.

Pearson, R.G. and T.P. Dawson, 2003: Predicting the impacts of climate change on the distribution of species: are climate envelope models useful? *Global Ecology and Biogeography*, **12**, 361-371.

Pearson, R.G., T.P. Dawson, P.M. Berry, and P.A. Harrison, 2002: SPECIES: A spatial evaluation of climate impact on the envelope of species. *Ecological Modelling*, **154(3)**, 289-300.

Peters, R.L. and T. E. Lovejoy, 1994: Global warming and biological diversity, [Lovejoy, T.E. and R.L. Peters (eds.)]. Yale University Press, New Haven, CT.

Peterson, A.T., L.G. Ball, and K.C. Cohoon, 2002: Predicting distributions of tropical birds. *Ibis*, **144**, e27-e32.

Peterson, A.T., H. Tian, E. Martinez-Meyer, J. Soberon, V. Sanchez-Cordero, and B. Huntley, 2005: Modeling distributional shifts of individual species and biomes, In: *Climate Change and Biodiversity*, [Lovejoy, T.E. and L. Hannah (eds.)]. Yale University Press, New Haven, pp. 211-228.

Peterson, A.T. and D.A. Vieglais, 2001: Predicting species invasions using ecological niche modeling: new approaches from bioinformatics attack a pressing problem. *BioScience*, **51(5)**, 363-371.

Pidgorna, A.B., 2007: Representation, redundancy, and resilience: waterfowl and the National Wildlife Refuge System. *Dissertation*.

Poff, N.L., M.M. Brinson, and J.W. Day, Jr., 2002: *Aquatic Ecosystems & Global Climate Change: Potential Impacts on Inland Freshwater and Coastal Wetland Ecosystems in the United States*. Pew Center on Global Climate Change, Arlington, VA, pp.1-56.

Poiani, K.A. and W.C. Johnson, 1991: Global warming and prairie wetlands: potential consequences for waterfowl habitat. *BioScience*, **41(9)**, 611-618.

Pounds, A.J., M.R. Bustamante, L.A. Coloma, J.A. Consuegra, M.P.L. Fogden, P.N. Foster, E. La Marca, K.L. Masters, A. Merino-Viteri, R. Puschendorf, S.R. Ron, G.A. Sanchez-Azofeifa, C.J. Still, and B.E. Young, 2006: Widespread amphibian extinctions from epidemic disease driven by global warming. *Nature*, **439(7073)**, 161-167.

Price, J. and P. Glick, 2002: *The Bird Watcher's Guide to Global Warming*. National Wildlife Federation and the American Bird Conservancy, Reston, Virginia.

Randerson, J.T., H. Liu, M.G. Flanner, S.D. Chambers, Y. Jin, P.G. Hess, G. Pfister, M.C. Mack, K.K. Treseder, L.R. Welp, F.S. Chapin, III, J.W. Harden, M.L. Goulden, E. Lyons, J.C. Neff, E.A.G. Schuur, and C.S. Zender, 2006: The impact of boreal forest fire on climate warming. *Science*, **314(5802)**, 1130-1132.

Root, T.L., J.T. Price, K.R. Hall, S.H. Schneider, C. Rosenzweig, and J.A. Pounds, 2003: Fingerprints of global warming on wild animals and plants. *Nature*, **421**, 57-60.

Ross, M.S., J.J. O'Brien, and L.d.S.L. Sternberg, 1994: Sea-level rise and the reduction in pine forests in the Florida Keys. *Ecological Applications*, **4(1)**, 144-156.

Rouse, W.R., M.S.V. Douglas, R.E. Hecky, A.E. Hershey, G.W. Kling, L. Lesack, P. Marsh, M. McDonald, B.J. Nicholson, N.T. Roulet, and J.P. Smol, 1997: Effects of climate change on the freshwaters of Arctic and subarctic North America. *Hydrological Processes*, **11**, 873-902.

Roy, S.B., P.F. Ricci, K.V. Summers, C.F. Chung, and R.A. Goldstein, 2005: Evaluation of the sustainability of water withdrawals in the United States, 1995 to 2025. *Journal of the American Water Resources Association*, **41(5)**, 1091-1108.

Rueda, L.M., K.J. Patel, R.C. Axtell, and R.E. Stinner, 1990: Temperature-dependent development and survival rates of *Culex quinquefasciatus* and *Aedes aegypti* (Diptera: Culicidae). *Journal of Medical Entomology*, **27(5)**, 892-898.

Rupp, T.S., F.S. Chapin, and A.M. Starfield, 2000: Response of subarctic vegetation to transient climatic change on the Seward Peninsula in north-west Alaska. *Global Change Biology*, **6(5)**, 541-555.

Russell, F.L., D.B. Zippin, and N.L. Fowler, 2001: Effects of white-tailed deer (*Odocoileus virginianus*) on plants, plant populations and communities: a review. *American Midland Naturalist*, **146(1)**, 1-26.

Salafsky, N., R. Margoluis, and K.H. Redford, 2001: *Adaptive Management: a Tool for Conservation Practitioners*. Biodiversity Support Program, Washington, DC.

Satchell, M., 2003: Troubled waters. *National Wildlife*, **41(2)**, 35-41.

Sauer, J.R., G.W. Pendleton, and B.G. Peterjohn, 1996: Evaluating causes of population change in North American insectivorous songbirds. *Conservation Biology*, **10(2)**, 465-478.

Schindler, D.W., 1998: A dim future for boreal waters and landscapes. *BioScience*, **48(3)**, 157-164.

Schoennagel, T., T.T. Veblen, and W.H. Romme, 2004: The interaction of fire, fuels, and climate across rocky mountain forests. *BioScience*, **54(7)**, 661-676.

Scholze, M., W. Knorr, N.W. Arnell, and I.C. Prentice, 2006: A climate-change risk analysis for world ecosystems. *Proceedings of the National Academy of Sciences of the United States of America*, **103**, 13116-13120.

Scott, J.M., F. Davis, B. Csuti, R. Noss, B. Butterfield, C. Groves, H. Anderson, S. Caicco, F. Derchia, T.C. Edwards Jr, J. Ulliman, Jr., and R.G. Wright, 1993: GAP analysis: a geographical approach to protection of biological diversity. *Wildlife monographs*, **123**, 1-41.

Scott, J.M., P.J. Heglund, M.L. Morrison, J.B. Haufler, M.G. Raphael, W.A. Wall, and F.B. Samson, 2002: *Predicting Species Occurrences: Issues of Accuracy and Scale*. Island Press, Washington, pp. 1-868.

Scott, J.M., T. Loveland, K. Gergely, J. Strittholt, and N. Staus, 2004: National Wildlife Refuge System: ecological context and integrity. *Natural Resources Journal*, **44(4)**, 1041-1066.

Serreze, M.C., J.E. Walsh, F.S. Chapin III, T. Osterkamp, M. Dyurgerov, V. Romanovsky, W.C. Oechel, J. Morison, T. Zhang, and R.G. Barry, 2000: Observational evidence of recent change in the northern high-latitude environment. *Climatic Change*, **46(1-2)**, 159-207.

Shafer, S.L., P.J. Bartlein, and R.S. Thompson, 2001: Potential changes in the distribution of western North America tree and shrub taxa under future climate scenarios. *Ecosystems*, **4**, 200-215.

Shaffer, M.L. and B. A. Stein, 2000: Safeguarding our precious heritage, In: *Precious Heritage: the Status of Biodiversity in the United States*, [Stein, B.A., L.S. Kutner, and J.S. Adams (eds.)]. Oxford University Press, New York, pp. 301-321.

Sitch, S., B. Smith, I.C. Prentice, A. Arneth, A. Bondeau, W. Cramer, J.O. Kaplan, S. Levis, W. Lucht, M.T. Sykes, K. Thonicke, and S. Venevsky, 2003: Evaluation of ecosystem dynamics, plant geography and terrestrial carbon cycling in the LPJ dynamic global vegetation model. *Global Change Biology*, **9**, 161-185.

Small, C., V. Gornitz, and J.E. Cohen, 2000: Coastal hazards and the global distribution of human population. *Environmental Geosciences*, **7(1)**, 3-12.

Sorenson, L.G., R. Goldberg, T.L. Root, and M.G. Anderson, 1998: Potential effects of global warming on waterfowl populations breeding in the Northern Great Plains. *Climatic Change*, **40(2)**, 343-369.

Soulé, M.E., 1987: *Viable Populations*. Cambridge University Press, New York, NY.

Striegl, R.G., G.R. Aiken, M.M. Dornblaser, P.A. Raymond, and K.P. Wickland, 2005: A decrease in discharge-normalized DOC export by the Yukon River during summer through autumn. *Geophysical Research Letters*, **32(21)**, L21413.

Sutherst, R., 2000: Climate change and invasive species: a conceptual framework, In: *Invasive Species in a Changing World*, [Mooney, H.A. and R.J. Hobbs (eds.)]. Island Press, Washington, DC, pp. 211-240.

Tabb, D.C. and A.C. Jones, 1962: Effect of Hurricane Donna on the aquatic fauna of North Florida Bay. *Transactions of the American Fisheries Society*, **91(4)**, 375-378.

Thieler, E.R. and E.S. Hammar-Klose, 1999: *National Assessment of Coastal Vulnerability to Future Sea-Level Rise: Preliminary Results for the U.S. Atlantic Coast*. U.S. Geological Survey Open-File Report 99-593, U.S. Geological Survey, Woods Hole, MA.

Thieler, E.R. and E.S. Hammar-Klose, 2000a: *National Assessment of Coastal Vulnerability to Future Sea-Level Rise: Preliminary Results for the U.S. Gulf of Mexico Coast*. U.S. Geological Survey Open-File Report 00-179, U.S. Geological Survey, Woods Hole, MA.

Thieler, E.R. and E.S. Hammar-Klose, 2000b: *National Assessment of Coastal Vulnerability to Future Sea-Level Rise: Preliminary Results for the U.S. Pacific Coast*. U.S. Geological Survey Open-File Report 00-178, U.S. Geological Survey, Woods Hole, MA.

Thomas, C.D., A. Cameron, R.E. Green, M. Bakkenes, L.J. Beaumont, Y.C. Collingham, B.F.N. Erasmus, M.F. de Siqueira, A. Grainger, L. Hannah, L. Hughes, B. Huntley, A.S. Van Jaarsveld, G.F. Midgley, L. Miles, M.A. Ortega-Huerta, A.T. Peterson, O.L. Phillips, and S.E. Williams, 2004a: Extinction risk from climate change. *Nature*, **427(6970)**, 145-148.

Thomas, C.D., A. Cameron, R.E. Green, M. Bakkenes, L.J. Beaumont, Y.C. Collingham, B.F.N. Erasmus, M.F. de Siqueira, A. Grainger, L. Hannah, L. Hughes, B.R.I.A. Huntley, A.S. Van Jaarsveld, G.F. Midgley, L. Miles, M.A. Ortega-Huerta, A.T. Peterson, O.L. Phillips, and S.E. Williams, 2004b: Extinction risk from climate change. *Nature*, **427**, 145-148.

Thuiller, W., S. Lavorel, and M.B. Araujo, 2005: Niche properties and geographical extent as predictors of species sensitivity to climate change. *Global Ecology and Biogeography*, **14(4)**, 347-357.

Titus, J.G. and C. Richman, 2001: Maps of lands vulnerable to sea level rise: modeled elevations along the U.S. Atlantic and Gulf Coasts. *Climate Research*, **18**, 205-228.

Tompkins, E.L. and N.W. Adger, 2004: Does adaptive management of natural resources enhance resilience to climate change? *Ecology and Society*, **19(2)**.

Turner, B.L., II, R.E. Kasperson, P.A. Matsone, J.J. McCarthy, R.W. Corell, L. Christensene, N. Eckley, J.X. Kasperson, A. Luerse, M.L. Martello, C. Polsky, A. Pulsipher, and A. Schiller, 2003: A framework for vulnerability analysis in sustainability science. *Proceedings of the National Academy of Sciences of the United States of America*, **100(14)**, 8074-8079.

U.S. Climate Change Science Program, 2007: *Synthesis and Assessment Product 4.1: Coastal Elevation and Sensitivity to Sea Level Rise*. A report by the U.S. Climate Change Science Program and the Subcommittee on Global Change Research, U.S. Environmental Protection Agency.

U.S. Fish and Wildlife Service, 1989: Endangered species in the wake of hurricane Hugo. *Endangered Species Technical Bulletin*, **XIV(9-10)**, 3-7.

U.S. Fish and Wildlife Service, 1996. *Land Acquisition Planning*. 341 FW2.

U.S. Fish and Wildlife Service, 1999: *Fulfilling the Promise: the National Wildlife Refuge System*. The National Wildlife Refuge System, U.S. Fish and Wildlife Service, Department of Interior, Washington, DC.

U.S. Fish and Wildlife Service and Canadian Wildlife Service, 1986: *North American Waterfowl Management Plan*. US Department of the Interior, Environment Canada.

Urban, F.E., J.E. Cole, and J.T. Overpeck, 2000: Influence of mean climate change on climate variability from a 155-year tropical Pacific coral record. *Nature*, **407(6807)**, 989-993.

van Riper, C., III and D.J. Mattson, 2005: *The Colorado Plateau: Biophysical, Socioeconomic, and Cultural Research*. University of Arizona Press.

Walters, C., 1986: *Adaptive Management of Renewable Resources*. McGraw Hill, New York.

Walther, G.R., E. Post, P. Convey, A. Menzel, C. Parmesan, T.J.C. Beebee, J.M. Fromentin, O. Hoegh-Guldberg, and F. Bairlein, 2002: Ecological responses to recent climate change. *Nature*, **416**, 389-395.

Watson, R.T., M.C. Zinyowera, and R.H. Moss, 1996: *Climate Change 1995 - Impacts, Adaptations and Mitigation of Climate Change: Scientific-Technical Analyses*. Contribution of Working Group II to the Second Assessment Report of the Intergovernmental Panel on Climate Change, Cambridge University Press, Cambridge, MA.

Westbrooks, R.G., 2001: Potential impacts of global climate changes on the establishment and spread on invasive species. *Transactions of the North American Wildlife and Natural Resources Conference*, **66**, 344-370.

Westerling, A.L., H.G. Hidalgo, D.R. Cayan, and T.W. Swetnam, 2006: Warming and earlier spring increase western U.S. forest wildfire activity. *Science*, **313(5789)**, 940-943.

Wilby, R.L., T.M.L. Wigley, D. Conway, P.D. Jones, B.C. Hewitson, J. Main, and D.S. Wilks, 1998: Statistical downscaling of general circulation model output: a comparison of methods. *Water Resources Research*, **34(11)**, 2995-3008.

Williams, B.K., R.C. Szaro, and C.D. Shapiro, 2007: *Adaptive Management: The U.S. Department of the Interior Technical Guide*. Adaptive Management Working Group, U.S. Department of the Interior, Washington, DC.

Winter, T.C., 2000: The vulnerability of wetlands to climate change: a hydrologic landscape perspective. *Journal of the American Water Resources Association*, **36(2)**, 305-311.

Zimov, S.A., E.A.G. Schuur, and F.S. Chapin, III, 2006: Climate change: permafrost and the global carbon budget. *Science*, **312(5780)**, 1612-1613.

CHAPTER 6 REFERENCES

Ahearn, D.S., R.W. Sheibley, and R.A. Dahlgren, 2005: Effects of river regulation on water quality in the lower Mokelumne River, California. *River Research and Applications*, **21(6)**, 651-670.

Alcamo, J., P. Doell, T. Henrichs, F. Kaspar, B. Lehner, T. Roesch, and S. Siebert, 2003: Global estimates of water withdrawals and availability under current and future "business-as-usual" conditions. *Hydrological Sciences Journal*, **48(3)**, 339-348.

Alcamo, J., M. Flörke, and M. Märker, 2007: Future long-term changes in global water resources driven by socio-economic and climatic changes. *Hydrological Sciences Journal*, **52(2)**, 247-275.

Allan, J.D., 1995: *Stream Ecology: Structure and Function of Running Waters*. Kluwer Academic Pub.

Allan, J.D., 2004: Landscapes and riverscapes: the influence of land use on stream ecosystems. *Annual Review of Ecology, Evolution and Systematics*, **35**, 257-284.

Allan, J.D. and A.S. Flecker, 1993: Biodiversity conservation in running waters. *BioScience*, **43(1)**, 32-43.

Allan, J.D., M. A. Palmer, and N. L. Poff, 2005: Freshwater ecology, In: *Climate Change and Biodiversity*, [Lovejoy, T.E. and L. Hannah (eds.)]. Yale University Press, New Haven.

Arnold, C.L., Jr. and C.J. Gibbons, 1986: Impervious surface coverage: the emergence of a key environmental indicator. *Journal of the American Planning Association*, **62(2)**, 243-258.

Bachelet, D., R.P. Neilson, J.M. Lenihan, and R.J. Drapek, 2001: Climate change effects on vegetation distribution and carbon budget in the United States. *Ecosystems*, **4**, 164-185.

Baron, J.S., H.M. Rueth, A.M. Wolfe, K.R. Nydick, E.J. Allstott, J.T. Minear, and B. Moraska, 2000: Ecosystem responses to nitrogen deposition in the Colorado Front Range. *Ecosystems*, **3(4)**, 352-368.

Baron, J.S., N.L. Poff, P.L. Angermeier, C.N. Dahm, P.H. Gleick, N.G. Hairston, Jr., R.B. Jackson, C.A. Johnston, B.D. Richter, and A.D. Steinman, 2002: Meeting ecological and societal needs for freshwater. *Ecological Applications*, **12**, 1247-1260.

Beitinger, T.L., W.A. Bennett, and R.W. McCauley, 2000: Temperature tolerances of North American freshwater fishes exposed to dynamic changes in temperature. *Environmental Biology of Fishes*, **58**, 237-275.

Bergkamp, G., B. Orlando, and I. Burton, 2003: *Change: Adaptation of Water Management to Climate Change*. International Union for Conservation of Nature and Natural Resources, Gland, Switzerland and Cambridge, UK.

Bernhardt, E.S., E.B. Sudduth, M.A. Palmer, J.D. Allan, J.L. Meyer, G. Alexander, J. Follastad-Shah, B. Hassett, R. Jenkinson, R. Lave, J. Rumps, and L. Pagano, 2007: Restoring rivers one reach at a time: results from a survey of US river restoration practitioners. *Restoration Ecology*, **15(3)**, 482-493.

Bledsoe, B.P. and C.C. Watson, 2001: Effects of urbanization on channel instability. *Journal of American Water Resources Association*, **37**, 255-270.

Brown, B.L., 2003: Spatial heterogeneity reduces temporal variability in stream insect communities. *Ecology Letters*, **6(4)**, 316-325.

Bunn, S.E. and A.H. Arthington, 2002: Basic principles and ecological consequences of altered flow regimes for aquatic biodiversity. *Environmental Management*, **30(4)**, 492-507.

Clarkson, R.W. and M.R. Childs, 2000: Temperature effects of hypolimnial-release dams on early life stages of Colorado River basin big-river fishes. *Copeia*, **2000(2)**, 402-412.

Collier, M., R.H. Webb, and J.C. Schmidt, 1996: *Dams and Rivers: a Primer on the Downstream Effects of Dams*. U.S. Geological Survey, Denver, CO.

Dunne, T. and L.B. Leopold, 1978: *Water in Environmental Planning*. W.H. Freeman and Co., San Francisco, pp. 1-818.

Dynesius, M., R. Jansson, M.E. Johansson, and C. Nilsson, 2004: Intercontinental similarities in riparian-plant diversity and sensitivity to river regulation. *Ecological Applications*, **14(1)**, 173-191.

Eaton, J.G. and R.M. Scheller, 1996: Effects of climate warming on fish thermal habitat in streams of the United States. *Limnology and Oceanography*, **41**, 1109-1115.

Ewert, S.E.D., 2001: Evolution of an environmentalist: Senator Frank Church and the Hells Canyon controversy. *Montana*, **51(Spring)**, 36-51.

Fitzhugh, T.W. and B.D. Richter, 2004: Quenching urban thirst: growing cities and their impacts on freshwater ecosystems. *BioScience*, **54(8)**, 741-754.

Hassol, S.J., 2004: *Impacts of a Warming Arctic*. Cambridge University Press, Cambridge, UK.

Hilborn, R., T.P. Quinn, D.E. Schindler, and D.E. Rogers, 2003: Biocomplexity and fisheries sustainability. *Proceedings of the National Academy of Sciences of the United States of America*, **100(11)**, 6564-6568.

Hill, W.R., 1996: Effects of light, In: *Algal Ecology: Freshwater Benthic Ecosystems*, [Stevenson, R.J., M.L. Bothwell, and R.L. Lowe (eds.)]. Academic Press, San Diego, California, pp. 121-149.

Hodgkins, G.A. and R.W. Dudley, 2006: Changes in the timing of winter-spring streamflows in eastern North America, 1913-2002. *Geophysical Research Letters*, **33(6)**.

IPCC, 2007a: Summary for policymakers, In: *Climate Change 2007: the Physical Science Basis. Contribution of Working Group I to the Fourth Assessment Report of the Intergovernmental Panel on Climate Change*, [Solomon, S., D. Qin, M. Manning, Z. Chen, M. Marquis, K.B. Averyt, M. Tignor, and H.L. Miller (eds.)]. Cambridge University Press, Cambridge, United Kingdom and New York, NY, USA.

IPCC, 2007b: Summary for policymakers, In: *Climate Change 2007: Impacts, Adaptation and Vulnerability. Contribution of Working Group II to the Fourth Assessment Report of the Intergovernmental Panel on Climate Change*, [Parry, M.L., O.F. Canziani, J.P. Palutikof, P.J. van der Linden, and C.E. Hanson (eds.)]. Cambridge University Press, Cambridge, UK, pp. 7-22.

Knighton, D., 1998: *Fluvial Forms and Processes: a New Perspective*. Oxford University Press, New York.

Lehmkuhl, D.M., 1974: Thermal regime alterations and vital environmental physiological signals in aquatic systems, In: *Thermal Ecology*, [Gibbons, J.W. and R.R. Sharitz (eds.)]. AEC Symposium Series, CONF-730505, pp. 216-222.

Leopold, A., 1978: *A Sand County Almanac: With Essays on Conservation of Round River*. Ballantine Books, New York.

Lettenmaier, D.P., E.F. Wood, and J.R. Wallis, 1994: Hydro-climatological trends in the continental United States, 1948-88. *Journal of Climate*, **7(4)**, 586-607.

Ligon, F.K., W.E. Dietrich, and W.J. Trush, 1995: Downstream ecological effects of dams: a geomorphic perspective. *BioScience*, **45(3)**, 183-192.

Loomis, J., J. Koteen, and B. Hurd, 2003: Economic and institutional strategies for adapting to water resource effects of climate change, In: *Water and Climate in the Western United States*, [Lewis, W.M., Jr. (ed.)]. University Press of Colorado, Boulder.

Lowe, R.L. and Y. Pan, 1996: Benthic algal communities as biological monitors, In: *Algal Ecology: Freshwater Benthic Ecosystems*, [Stevenson, R.J., M.L. Bothwell, and R.L. Lowe (eds.)]. Academic Press, San Diego, California, pp. 705-739.

Lytle, D.A. and N.L. Poff, 2004: Adaptation to natural flow regimes. *Trends in Ecology and Evolution*, **19(2)**, 94-100.

Magilligan, F.J., K.H. Nislow, and B.E. Graber, 2003: Scale-independent assessment of discharge reduction and riparian disconnectivity following flow regulation by dams. *Geology*, **31(7)**, 569-572.

McCabe, G.J. and M.P. Clark, 2005: Trends and variability in snowmelt runoff in the western United States. *Journal of Hydrometeorology*, **6(4)**, 476-482.

McCully, P., 1996: *Silenced Rivers: the Ecology and Politics of Large Dams*. Zed Books, London, UK.

McMahon, G. and T.F. Cuffney, 2000: Quantifying urban intensity in drainage basins for assessing stream ecological conditions. *Journal of the American Water Resources Association*, **36(6)**, 1247-1262.

Milly, P.C.D., K.A. Dunne, and A.V. Vecchia, 2005: Global pattern of trends in streamflow and water availability in a changing climate. *Nature*, **438(7066)**, 347-350.

Moyle, P.B. and J.F. Mount, 2007: Homogenous rivers, homogenous faunas. *Proceedings of the National Academy of Sciences of the United States of America*, **104(14)**, 5711-5712.

Murdoch, P.S., J.S. Baron, and T.L. Miller, 2000: Potential effects of climate change on surface-water quality in North America. *Journal of the American Water Resources Association*, **36(2)**, 347-366.

Naiman, R.J., H. Décamps, and M.E. McClain, 2005: *Riparia: Ecology, Conservation, and Management of Streamside Communities*. Elsevier Academic Press, San Diego, pp. 1-430.

National Park Service, 2004: *Rio Grande Wild and Scenic River: Final General Management Plan / Environmental Impact Statement.*

National Research Council, 1993: *Soil and Water Quality: an Agenda for Agriculture*. National Academy Press, Washington, DC.

National Research Council, 2004: *Developing a Research and Restoration Plan for Arctic-Yukon-Kuskokwim (Western Alaska) Salmon*. Washington, DC.

Nelson, K.C. and M.A. Palmer, 2007: Stream temperature surges under urbanization and climate change: data, models, and responses. *Journal of the American Water Resources Association*, **43(2)**, 440-452.

Nowak, D.J. and J.T. Walton, 2005: Projected urban growth (2000-2050) and its estimated impact on the US forest resource. *Journal of Forestry*, **103(8)**, 383-389.

Palmer, M., A.P. Covich, B.J. Finlay, J. Gibert, K.D. Hyde, R.K. Johnson, T. Kairesalo, S. Lake, C.R. Lovell, R.J. Naiman, C. Ricci, F. Sabater, and D. Strayer, 1997: Biodiversity and ecosystem processes in freshwater sediments. *Ambio*, **26(8)**, 571-577.

Palmer, M.A., E.S. Bernhardt, J.D. Allan, P.S. Lake, G. Alexander, S. Brooks, J. Carr, S. Clayton, C.N. Dahm, and J.F. Shah, 2005: Standards for ecologically successful river restoration. *Journal of Applied Ecology*, **42(2)**, 208-217.

Palmer, M.A., C.A. Reidy, C. Nilsson, M. Flörke, J. Alcamo, P.S. Lake, and N. Bond, 2008: Climate change and the world's river basins: anticipating management options. *Frontiers in Ecology and the Environment*, DOI: 10.1890/060148.

Paul, M.J. and J.L. Meyer, 2001: Streams in the urban landscape. *Annual Review of Ecology and Systematics*, **32**, 333-365.

Pizzuto, J.E., G. E. Moglen, M. A. Palmer, and K. C. Nelson, 2008: Two model scenarios illustrating the effects of land use and climate change on gravel riverbeds of suburban Maryland, U.S.A., In: *Gravel River Beds VI: From Process Understanding to the Restoration of Mountain Rivers*, [Rinaldi, M., H. Habersack, and H. Piegay (eds.)]. Elsevier.

Poff, J.L., P. L. Angermeier, S. D. Cooper, P. S. Lake, K. D. Fausch, K. O. Winemiller, L. A. K. Mertes, and M. W. Oswood, 2006: Global change and stream fish diversity, In: *Future Scenarios of Global Biodiversity*, [Chapin III, F.S., O.E. Sala, and E. Huber-Sanwald (eds.)]. Oxford University Press, Oxford.

Poff, N.L., 2002: Ecological response to and management of increased flooding caused by climate change. *Philosophical Transactions: Mathematical, Physical & Engineering Sciences*, **360(1796)**, 1497-1510.

Poff, N.L. and J.V. Ward, 1990: Physical habitat template of lotic systems: recovery in the context of historical pattern of spatiotemporal heterogeneity. *Environmental Management*, **14(5)**, 629-646.

Poff, N.L.R., J.D. Olden, D.M. Merritt, and D.M. Pepin, 2007: Homogenization of regional river dynamics by dams and global biodiversity implications. *Proceedings of the National Academy of Sciences of the United States of America*, **104(14)**, 5732-5737.

Poff, N.L., J.D. Allan, M.B. Bain, J.R. Karr, K.L. Prestegaard, B.D. Richter, R. Sparks, and J. Stromberg, 1997: The natural flow regime: a new paradigm for riverine conservation and restoration. *BioScience*, **47**, 769-784.

Poff, N.L., M.M. Brinson, and J.W. Day, Jr., 2002: *Aquatic Ecosystems & Global Climate Change:Potential Impacts on Inland Freshwater and Coastal Wetland Ecosystems in the United States.* Pew Center on Global Climate Change, pp.1-56.

Postel, S. and B. Richter, 2003: *Rivers for Life: Managing Water for People and Nature*. Island Press, Washington, DC.

Postel, S.L., 2007: Aquatic ecosystem protection and drinking water utilities. *Journal American Water Works Association*, **99(2)**, 52-63.

Richter, B.D., J.V. Baumgartner, R. Wigington, and D.P. Braun, 1997: How much water does a river need? *Freshwater Biology*, **37(1)**, 231-249.

Richter, B.D., R. Mathews, D.L. Harrison, and R. Wigington, 2003: Ecologically sustainable water management: managing river flows for ecological integrity. *Ecological Applications*, **13(1)**, 206-224.

Rier, S.T., N.C. Tuchman, R.G. Wetzel, and J.A. Teeri, 2002: Elevated CO_2-induced changes in the chemistry of quaking aspen (*Populus tremuloides* Michaux) leaf litter: subsequent mass loss and microbial response in a stream ecosystem. *Journal of The North American Benthological Society*, **21**, 16-27.

Sala, O.E., F.S. Chapin III, J.J. Armesto, E. Berlow, J. Bloomfield, R. Dirzo, E. Huber-Sanwald, L.F. Huenneke, R.B. Jackson, A. Kinzig, R. Leemans, D.M. Lodge, H.A. Mooney, M. Oesterheld, N.L. Poff, M.T. Sykes, B.H. Walker, M. Walker, and D.H. Wall, 2000: Global biodiversity scenarios for the year 2100. *Science*, **287**, 1770-1774.

Shafroth, P.B., J.C. Stromberg, and D.T. Patten, 2002: Riparian vegetation response to altered disturbance and stress regimes. *Ecological Applications*, **12(1)**, 107-123.

Silk, N. and K. Ciruna, 2005: *A Practitioner's Guide to Freshwater Biodiversity Conservation*. Island Press, Washington, DC.

Smith, L.C., Y. Sheng, G.M. MacDonald, and L.D. Hinzman, 2005: Disappearing Arctic lakes. *Science*, **308(5727)**, 1429.

Stewart, I.T., D.R. Cayan, and M.D. Dettinger, 2004: Changes in snowmelt runoff timing in Western North America under a 'business as usual' climate change scenario. *Climatic Change*, **62**, 217-232.

Stewart, I.T., D.R. Cayan, and M.D. Dettinger, 2005: Changes toward earlier streamflow timing across western North America. *Journal of Climate*, **18(8)**, 1136-1155.

Thuiller, W., 2004: Patterns and uncertainties of species' range shifts under climate change. *Global Change Biology*, **10(12)**, 2020-2027.

Tockner, K. and J.A. Stanford, 2002: Riverine flood plains: present state and future trends. *Environmental Conservation*, **29(3)**, 308-330.

Todd, C.R., T. Ryan, S.J. Nicol, and A.R. Bearlin, 2005: The impact of cold water releases on the critical period of post-spawning survival and its implications for Murray cod (*Maccullochella peelii peelii*): a case study of the Mitta Mitta river, south-eastern Australia. *River Research and Applications*, **21**, 1035-1052.

Tuchman, N.C., K.A. Wahtera, R.G. Wetzel, and J.A. Teeri, 2003: Elevated atmospheric CO_2 alters leaf litter quality for stream ecosystems: an in situ leaf decomposition study. *Hydrobiologia*, **495(1)**, 203-211.

Tuchman, N.C., R.G. Wetzel, S.T. Rier, K.A. Wahtera, and J.A. Teeri, 2002: Elevated atmospheric CO_2 lowers leaf litter nutritional quality for stream ecosystem food webs. *Global Change Biology*, **8(2)**, 145-153.

U.S. Environmental Protection Agency, 2000: *National Water Quality Inventory*. EPA-841-R-02-001, U.S. Environmental Protection Agency, Washington, DC.

Van der Kraak, G. and N. W. Pankhurst, 1997: Temperature effects on the reproductive performance of fish, In: *Global Warming: Implications for Freshwater and Marine Fish*, [Wood, C.M. and D.G. McDonald (eds.)]. Cambridge University Press, Cambridge.

Vannote, R.L. and B.W. Sweeney, 1980: Geographic analysis of thermal equilibria: a conceptual model for evaluating the effect of natural and modified thermal regimes on aquatic insect communities. *The American Naturalist*, **115(5)**, 667-695.

Vörösmarty, C.J., M. Meybeck, B. Fekete, K. Sharma, P. Green, and J.P.M. Syvitski, 2003: Anthropogenic sediment retention: major global impact from registered river impoundments. *Global and Planetary Change*, **39(1-2)**, 169-190.

Vörösmarty, C.J., P. Green, J. Salisbury, and R.B. Lammers, 2000: Global water resources: vulnerability from climate change and population growth. *Science*, **289**, 284-288.

Walsh, C.J., T.D. Fletcher, and A.R. Ladson, 2005: Stream restoration in urban catchments through redesigning stormwater systems: looking to the catchment to save the stream. *Journal of The North American Benthological Society*, **24(3)**, 690-705.

Walsh, C.J., A.H. Roy, J.W. Feminella, P.D. Cottingham, P.M. Groffman, and R.P. Morgan, 2005: The urban stream syndrome: current knowledge and the search for a cure. *Journal of The North American Benthological Society*, **24**, 706-723.

Ward, J.V., 1992: *Aquatic Insect Ecology: Biology and Habitat*. John Wiley & Sons, New York, pp. 1-438.

Webster, J.R. and E.F. Benfield, 1986: Vascular plant breakdown in freshwater ecosystems. *Annual Review of Ecology and Systematics*, **17**, 567-594.

Williams, G.P. and M.G. Wolman, 1984: *Downstream Effects of Dams on Alluvial Rivers*. Professional Paper 1286, U.S. Geological Survey, Washington, DC.

Willis, C.M. and G.B. Griggs, 2003: Reductions in fluvial sediment discharge by coastal dams in California and implications for beach sustainability. *Journal of Geology*, **111(2)**, 167-182.

Wohl, E., M. A. Palmer, and J. M. Kondolf, 2008: The U.S. experience. Chapter 10, In: *River Futures*, [Brierley, G.J. and K.A. Fryirs (eds.)]. Island Press.

Wolman, M.G., 1967: A cycle of sedimentation and erosion in urban river channels. *Geografiska Annaler Series A: Physical Geography*, **49(2-4)**, 385-395.

World Commission on Dams, 2000: *Dams and Development: a New Framework for Decisions-Making*. Earthscan Publications, London.

Xenopoulos, M.A., D.M. Lodge, J. Alcamo, M. Marker, K. Schulze, and D.P. Van Vuuren, 2005: Scenarios of freshwater fish extinctions from climate change and water withdrawal. *Global Change Biology*, **11(10)**, 1557-1564.

CHAPTER 7 REFERENCES

Alber, M. and J.E. Sheldon, 1999: Use of a date-specific method to examine variability in the flushing times of Georgia estuaries. *Estuarine Coastal and Shelf Science*, **49(4)**, 469-482.

Alber, M., 2002: A conceptual model of estuarine freshwater inflow management. *Estuaries*, **25(6)**, 1246-1261.

Anderson, T.H. and G.T. Taylor, 2001: Nutrient pulses, plankton blooms, and seasonal hypoxia in western Long Island Sound. *Estuaries*, **24(2)**, 228-243.

Arora, V.K., F.H.S. Chiew, and R.B. Grayson, 2000: The use of river runoff to test CSIRO 9 land surface scheme in the Amazon and Mississippi River Basins. *International Journal of Climatology*, **20(10)**, 1077-1096.

Baird, D., R.R. Christian, C.H. Peterson, and G.A. Johnson, 2004: Consequences of hypoxia on estuarine ecosystem function: energy diversion from consumers to microbes. *Ecological Applications*, **14(3)**, 805-822.

Barry, J.P., C.H. Baxter, R.D. Sagarin, and S.E. Gilman, 1995: Climate-related, long-term faunal changes in a California Rocky intertidal community. *Science*, **267(5198)**, 672-675.

Bearden, D.M., 2001: *National Estuary Program: a Collaborative Approach to Protecting Coastal Water Quality*. CRS Report for Congress #97-644, Congressional Research Service.

Beaugrand, G., P.C. Reid, F. Ibanez, J.A. Lindley, and M. Edwards, 2002: Reorganization of North Atlantic Marine Copepod biodiversity and climate. *Science*, **296**, 1692-1694.

Boesch, D.F., M.N. Josselyn, A.J. Mehta, J.T. Morris, W.K. Nuttle, C.A. Simenstad, and D.J.P. Swift, 1994: *Scientific Assessment of Coastal Wetland Loss, Restoration and Management in Louisiana*. Coastal Education and Research Foundation, pp. 1-103.

Boicourt, W.C., 1992: Influences of circulation processes on dissolved oxygen in the Chesapeake Bay, In: *Oxygen Dynamics in Chesapeake Bay: a Synthesis of Research*, [Smith, D.E., M. Leffler, and G. Mackiernan (eds.)]. University of Maryland Sea Grant College Publications, College Park, Maryland, pp. 1-234.

Borsuk, M.E., D. Higdon, C.A. Stow, and K.H. Reckhow, 2001: A Bayesian hierarchical model to predict benthic oxygen demand from organic matter loading in estuaries and coastal zones. *Ecological Modelling*, **143(3)**, 165-181.

Boynton, W.R., J.H. Garber, R. Summers, and W.M. Kemp, 1995: Inputs, transformations, and transport of nitrogen and phosphorus in Chesapeake Bay and selected tributaries. *Estuaries*, **18(1)**, 285-314.

Bozek, C.M. and D.M. Burdick, 2005: Impacts of seawalls on saltmarsh plant communities in the Great Bay Estuary, New Hampshire USA. *Wetlands Ecology and Management*, **13(5)**, 553-568.

Breitburg, D., S. Seitzinger, and J. Sanders, 1999: *The Effects of Multiple Stressors on Freshwater and Marine Ecosystems*. American Society of Limnology and Oceanography.

Breitburg, D.L., C. A. Baxter, R. W. Hatfield, R. W. Howarth, C. G. Jones, G. M. Lovett, and C. Wigand, 1998: Understanding effects of multiple stressors: ideas and challenges, In: *Successes, Limitations, and Frontiers in Ecosystem Science*, [Pace, M.L. and P.M. Groffman (eds.)]. Springer, New York, pp. 416-431.

Breitburg, D.L., L.D. Coen, M.W. Luckenbach, R. Mann, M. Posey, and J.A. Wesson, 2000: Oyster reef restoration: convergence of harvest and conservation strategies. *Journal of Shellfish Research*, **19(1)**, 371-377.

Breitburg, D.L. and G. F. Riedel, 2005: Multiple stressors in marine systems, In: *Marine Conservation Biology: the Science of Maintaining the Sea's Biodiversity*, [Norse, E. and L.B. Crowder (eds.)]. Marine Conservation Biology Institute.

Breitburg, D.L., J.G. Sanders, C.C. Gilmour, C.A. Hatfield, R.W. Osman, G.F. Riedel, S.P. Seitzinger, and K.G. Sellner, 1999: Variability in responses to nutrients and trace elements, and transmission of stressor effects through an estuarine food web. *Limnology and Oceanography*, **44(3)**, 837-863.

Bricker, S.B., C.G. Clement, D.E. Pirhalla, S.P. Orlando, and D.R.G. Farrow, 1999: *National Estuarine Eutrophication Assessment: Effects of Nutrient Enrichment in the Nation's Estuaries*. National Centers for Coastal Ocean Science, National Oceanic and Atmospheric Administration, Silver Spring, MD, pp. 1-71.

Brinson, M.M., 1993: *A Hydrogeomorphic Classification for Wetlands*. Technical Report WRP-DE-4, US Army Engineer Waterways Experiment Station, Available from National Technical Information Service, Vicksburg, Mississippi.

Brinson, M.M. and R.R. Christian, 1999: Stability and response of *Juncus roemerianus* patches in a salt marsh. *Wetlands*, **19(1)**, 65-70.

Brinson, M.M., R.R. Christian, and L.K. Blum, 1995: Multiple states in the sea-level induced transition from terrestrial forest to estuary. *Estuaries*, **18(4)**, 648-659.

Brown, J.J., 1993: The State and Indian nations' water resource planning. *Occasional Paper*, **19**.

Brown, J.J., 1994: Treaty rights: twenty years after the Boldt decision. *Wicazo Sa Review*, **10(2)**, 1-16.

Bruno, J.F., K.E. Boyer, J.E. Duffy, S.C. Lee, and J.S. Kertesz, 2005: Effects of macroalgal species identity and richness on primary production in benthic marine communities. *Ecology Letters*, **8(11)**, 1165-1174.

Bruno, J.F., C.W. Kennedy, T.A. Rand, and M.B. Grant, 2004: Landscape-scale patterns of biological invasions in shoreline plant communities. *Oikos*, **107(3)**, 531-540.

Burkett, V., D. Wilcox, R. Stottlemyer, W. Barrow, D. Fagre, J. Baron, J. Price, J.L. Nielsen, C.D. Allen, D.L. Peterson, G. Ruggerone, and T. Doyle, 2005: Nonlinear dynamics in ecosystem response to climatic change: case studies and policy implications. *Ecological Complexity*, **2(4)**, 357-394.

Buzzelli, C.P., R.A. Luettich Jr, S.P. Powers, C.H. Peterson, J.E. McNinch, J.L. Pinckney, and H.W. Paerl, 2002: Estimating the spatial extent of bottom-water hypoxia and habitat degradation in a shallow estuary. *Marine Ecology Progress Series*, **230**, 103-112.

Caldeira, K. and M.E. Wickett, 2003: Anthropogenic carbon and ocean pH. *Nature*, **425(6956)**, 365-365.

Callaway, J.C., J.A. Nyman, and R.D. DeLaune, 1996: Sediment accretion in coastal wetlands: a review and a simulation model of processes. *Current Topics in Wetland Biogeochemistry*, **2**, 2-23.

Carpenter, S., B. Walker, J.M. Anderies, and N. Abel, 2001: From metaphor to neasurement: resilience of what to what? *Ecosystems*, **4(8)**, 765-781.

Carpenter, S.R. and O. Kinne, 2003: *Regime Shifts in Lake Ecosystems: Pattern and Variation*. International Ecology Institute, Luhe, Germany.

Carpenter, S.R., D. Ludwig, and W.A. Brock, 1999: Management of eutrophication for lakes subject to potentially irreversible change. *Ecological Applications*, **9(3)**, 751-771.

Carson, R., 1962: *Silent Spring*. Houghton Mifflin.

Chmura, G.L., S.C. Anisfeld, D.R. Cahoon, and J.C. Lynch, 2003: Global carbon sequestration in tidal, saline wetland soils. *Global Biogeochemical Cycles*, **17(4)**.

Christensen, N.L., A.M. Bartuska, J.H. Brown, S. Carpenter, C. D'Antonio, R. Francis, J.F. Franklin, J.A. MacMahon, R.F. Noss, D.J. Parsons, C.H. Peterson, M.G. Turner, and R.G. Woodmansee, 1996: The report of the Ecological Society of America Committee on the scientific basis for ecosystem management. *Ecological Applications*, **6(3)**, 665-691.

Christian, R.R., L. Stasavich, C. Thomas, and M. M. Brinson, 2000: Reference is a moving target in sea-level controlled wetlands, In: *Concepts and Controversies in Tidal Marsh Ecology*, [Weinstein, M.P. and D.A. Kreeger (eds.)]. Kluwer Press, The Netherlands, pp. 805-825.

Church, J.A., 2001: How fast are sea levels rising? *Science*, **294**, 802-803.

Cloern, J.E., A.E. Alpine, B.E. Cole, R.L.J. Wong, J.F. Arthur, and M.D. Ball, 1983: River discharge controls phytoplankton dynamics in the Northern San Francisco Bay Estuary. *Estuarine Coastal and Shelf Science*, **16(4)**.

Coastal Protection and Restoration Authority of Louisiana, 2007: *Louisiana's Comprehensive Master Plan for a Sustainable Coast*. Coastal Protection and Restoration Authority of Louisiana.

Coen, L.D., M. W. Luckenbach, and D. L. Breitburg, 1999: The role of oyster reefs as essential fish habitat: a review of current knowledge and some new perspectives, [Benaka, L.R. (ed.)]. Bethesda, Maryland, pp. 438-454.

Committee on Mitigating Shore Erosion along Sheltered Coasts, National Research Council, 2006: *Mitigating Shore Erosion Along Sheltered Coasts*. pp.1-188.

Committee on Mitigating Wetland Losses, National Research Council, 2001: *Compensating for Wetland Losses Under the Clean Water Act*. National Academies Press, Washington, DC.

Conley, D.J., S. Markager, J. Andersen, T. Ellermann, and L.M. Svendsen, 2002: Coastal eutrophication and the Danish national aquatic monitoring and assessment program. *Estuaries*, **25(4)**, 848-861.

Cooper, S.R. and G.S. Brush, 1993: A 2,500-year history of anoxia and eutrophication in Chesapeake Bay. *Estuaries*, **16**, 617-626.

Cooper, S.R., S.K. McGlothlin, M. Madritch, and D.L. Jones, 2004: Paleoecological evidence of human impacts on the Neuse and Pamlico Estuaries of North Carolina, USA. *Estuaries*, **27(4)**, 617-633.

Copeland, B.J., 1966: Effects of decreased river flow on estuarine ecology. *Journal Water Pollution Control Federation*, **38**, 1831-1839.

Costello, J.H., B.K. Sullivan, and D.J. Gifford, 2006: A physical-biological interaction underlying variable phenological responses to climate change by coastal zooplankton. *Journal of Plankton Research*, **28(11)**, 1099-1105.

Cropper, C.R., 2005: The study of endocrine-disrupting compounds: past approaches and new directions. *Integrative and Comparative Biology*, **45**, 194-2000.

Curtis, P.S., L.M. Balduman, B.G. Drake, and D.F. Whigham, 1990: Elevated Atmospheric CO_2 Effects on Belowground Processes in C3 and C4 Estuarine Marsh Communities. *Ecology*, **71(5)**, 2001-2006.

Dacey, J.W.H., B.G. Drake, and M.J. Klug, 1994: Stimulation of methane emission by carbon dioxide enrichment of marsh vegetation. *Nature*, **370(6484)**, 47-49.

Dakora, F. and B.G. Drake, 2000: Elevated CO_2 stimulates associative N_2 fixation in a C_3 plant of the Chesapeake Bay wetland. *Plant, Cell and Environment*, **23(9)**, 943-953.

Davis, M.B., 1983: Holocene vegetational history of the eastern United States, [Wright, H.E., Jr. (ed.)]. University of Minnesota Press, Minneapolis, MN, pp. 166-181.

Day, J.W., Jr., D.F. Boesch, E.J. Clairain, G.P. Kemp, S.B. Laska, W.J. Mitsch, K. Orth, H. Mashriqui, D.J. Reed, L. Shabman, C.A. Simenstad, B.J. Streever, R.R. Twilley, C.C. Watson, J.T. Wells, and D.F. Whigham, 2007: Restoration of the Mississippi Delta: lessons from Hurricanes Katrina and Rita. *Science*, **315(5819)**, 1679-1684.

Day, J.W., Jr., C.A.S. Hall, W.M. Kemp, and A. Yanez-Arancibia, 1989: *Estuarine Ecology*. Wiley and Sons, New York, NY.

Dayton, P.K., S. Thrush, and F.C. Coleman, 2002: *Ecological Effects of Fishing in Marine Ecosystems of the United States*. Pew Oceans Commission.

Dennison, W.C., R.J. Orth, K.A. Moore, J.C. Stevenson, V. Carter, S. Kollar, P.W. Bergstrom, and R.A. Batiuk, 1993: Assessing water quality with submersed aquatic vegetation. *BioScience*, **43(2)**, 86-94.

Diaz, R.J. and R. Rosenberg, 1995: Marine benthic hypoxia: a review of its ecological effects and the behavioural responses of benthic macrofauna. *Oceanography and Marine Biology Annual Review*, **33**, 245-303.

Dobson, A. and J. Foufopoulos, 2001: Emerging infectious pathogens of wildlife. *Philosophical Transactions: Biological Sciences*, **356(1411)**, 1001-1012.

Donnelly, J.P. and M.D. Bertness, 2001: Rapid shoreward encroachment of salt marsh cordgrass in response to accelerated sea-level rise. *Proceedings of the National Academy of Sciences of the United States of America*, **98(25)**, 14218-14223.

Dowdeswell, J.A., 2006: Atmospheric science: the Greenland ice sheet and global sea-level rise. *Science*, **311(5763)**, 963-964.

Drake, B.G., L. Hughes, E. A. Johnson, B. A. Seibel, M. A. Cochrane, V. J. Fabry, D. Rasse, and L. Hannah, 2005: Synergistic Effects, In: *Climate Change and Biodiversity*, [Lovejoy, T.E. and L. Hannah (eds.)]. Yale University Press, New Haven, pp. 296-316.

Drake, B.G., M.S. Muehe, G. Peresta, M.A. Gonzalez-Meler, and R. Matamala, 1995: Acclimation of photosynthesis, respiration and ecosystem carbon flux of a wetland on Chesapeake Bay, Maryland to elevated atmospheric CO_2 concentration. *Plant and Soil*, **187(2)**, 111-118.

Duarte, C.M., 1991: Seagrass depth limits. *Aquatic Botany*, **40(4)**, 363-377.

Duffy, J.E., 2002: Biodiversity and ecosystem function: the consumer connection. *Oikos*, **99(2)**, 201-219.

Emanuel, K., 2005: Increasing destructiveness of tropical cyclones over the past 30 years. *Nature*, **436(7051)**, 686-688.

Fagherazzi, S., L. Carniello, L. D'Alpaos, and A. Defina, 2006: Critical bifurcation of shallow microtidal landforms in tidal flats and salt marshes. *Proceedings of the National Academy of Sciences of the United States of America*, **103(22)**, 8337-8341.

Federal Emergency Management Agency, 1991: *Projected Impact of Relative Sea Level Rise on the National Flood Insurance Program*. Washington, DC.

Feely, R.A., C.L. Sabine, K. Lee, W. Berelson, J. Kleypas, V.J. Fabry, and F.J. Millero, 2004: Impact of anthropogenic CO_2 on the $CaCO_3$ system in the oceans. *Science*, **305(5682)**, 362-366.

Foley, J.A., R. DeFries, G.P. Asner, C.C. Barford, G.B. Bonan, S.R. Carpenter, F.S. Chapin III, M.T. Coe, G.C. Daily, H.K. Gibbs, J.H. Helkowski, T. Holloway, E.A. Howard, C.J. Kucharik, C. Monfreda, J. Patz, I.C. Prentice, N. Ramankutty, and P.K. Snyder, 2005: Global consequences of land use. *Science*, **309(5734)**, 570-574.

Folke, C., S. Carpenter, B. Walker, M. Scheffer, T. Elmqvist, L.H. Gunderson, and C.S. Holling, 2004: Regime shifts, resilience, and biodiversity in ecosystem management. *Annual Review of Ecology and Systematics*, **35**, 557-581.

Folke, C. and N. Kautsky, 1989: The role of ecosystems for a sustainable development of aquaculture. *Ambio*, **18(4)**, 234-243.

Fonseca, M.S., B.E. Julius, and W.J. Kenworthy, 2000: Integrating biology and economics in seagrass restoration: How much is enough and why? *Ecological Engineering*, **15(3)**, 227-237.

Ford, S.E., 1996: Range extension by the oyster parasite Perkinsus marinus into the northeastern United States: response to climate change? *Journal of Shellfish Research*, **15**, 45-56.

Galbraith, H., R. Jones, R. Park, J. Clough, S. Herrod-Julius, B. Harrington, and G. Page, 2002: Global climate change and sea level rise: potential losses of intertidal habitat for shorebirds. *Waterbirds*, **25(2)**, 173-183.

Gedney, N., P.M. Cox, R.A. Betts, O. Boucher, C. Huntingford, and P.A. Stott, 2006: Detection of a direct carbon dioxide effect in continental river runoff records. *Nature*, **439(7078)**, 835-838.

Goldburg, R. and T. Triplett, 1997: *Murky Waters: Environmental Effects of Aquaculture in the U.S.* Environmental Defense Fund.

González, J.L. and T.E. Törnqvist, 2006: Coastal Louisiana in crisis: subsidence or sea level rise? *Eos, Transactions American Geophysical Union*, **87(45)**, 493-498.

Greenberg, R., J.E. Maldonado, S. Droege, and M.V. McDonald, 2006: A global perspective on the evolution and conservation of their terrestrial vertebrates. *BioScience*, **56(8)**, 675-685.

Griffin, D.A. and P.H. LeBlond, 1990: Estuary/ocean exchange controlled by spring-neap tidal mixing. *Estuarine Coastal and Shelf Science*, **30(3)**, 275-297.

Groffman, P.M., J.S. Baron, T. Blett, A.J. Gold, I. Goodman, L.H. Gunderson, B.M. Levinson, M.A. Palmer, H.W. Paerl, G.D. Peterson, N.L. Poff, D.W. Rejesk, J. Reynolds, M.G. Turner, K.C. Weathers, and J. Wiens, 2006: Ecological thresholds: the key to successful environmental management or an important concept with no practical application? *Ecosystems*, **9(1)**, 1-13.

Grumbine, R.E., 1994: What is ecosystem management? *Conservation Biology*, **8(1)**, 27-38.

Guenette, S., T. Lauck, and C. Clark, 1998: Marine reserves: from Beverton and Holt to the present. *Reviews in Fish Biology and Fisheries*, **8(3)**, 251-272.

Gunderson, L.H., C. S. Holling, L. Pritchard, and G. D. Peterson, 2002: A summary and a synthesis of resilience in large scale systems, In: *Resilience and Behavior of Large-Scale Systems*, Island Press, Washington, DC, pp. 3-20.

Hagy, J.D., W.R. Boynton, C.W. Keefe, and K.V. Wood, 2004: Hypoxia in Chesapeake Bay, 1950-2001: long-term change in relation to nutrient loading and river flow. *Estuaries*, **27(4)**, 634-658.

Hakalahti, T., A. Karvonen, and E.T. Valtonen, 2006: Climate warming and disease risks in temperate regions Argulus coregoni and Diplostomum spathaceum as case studies. *Journal of Helminthology*, **80(2)**, 93-98.

Hall, S.R., A.J. Tessier, M.A. Duffy, M. Huebner, and C.E. Cbceres, 2006: Warmer does not have to mean sicker: temperature and predators can jointly drive timing of epidemics. *Ecology*, **87(7)**, 1684-1695.

Halpern, B.S., 2003: The impact of marine reserves: Do reserves work and does reserve size matter? *Ecological Applications*, **13(1)**, S117-S137.

Harley, C.D.G. and R. Hughes, 2006: Reviews and synthesis: the impacts of climate change in coastal marine systems. *Ecology Letters*, **9(2)**, 228-241.

Harris, L.D. and W. P. Cropper Jr., 1992: Between the devil and the deep blue sea: implications of climate change for Florida's fauna, In: *Global Warming and Biological Diversity*, [Peters, R.L. and T.E. Lovejoy (eds.)]. Yale University Press, New Haven, CT, pp. 309-324.

Harvell, C.D., C.E. Mitchell, J.R. Ward, S. Altizer, A.P. Dobson, R.S. Ostfeld, and M.D. Samuel, 2002: Climate warming and disease risks for terrestrial and marine biota. *Science*, **296(5576)**, 2158-2162.

Harvell, D., R. Aronson, N. Baron, J. Connell, A. Dobson, S. Ellner, L. Gerber, K. Kim, A. Kuris, and H. McCallum, 2004: The rising tide of ocean diseases: unsolved problems and research priorities. *Frontiers in Ecology and the Environment*, **2(7)**, 375-382.

Hauxwell, J., J. Cebrian, C. Furlong, and I. Valiela, 2001: Macroalgal canopies contribute to eelgrass (*Zostera marina*) decline in temperate estuarine ecosystems. *Ecology*, **82(4)**, 1007-1022.

Hayden, B.P., M.C.F.V. Santos, G. Shao, and R.C. Kochel, 1995: Geomorphological controls on coastal vegetation at the Virginia Coast Reserve. *Geomorphology*, **13**, 283-300.

Health Ecological and Economic Dimensions of Global Change Program, 1998: *Marine Ecosystems: Emerging Diseases As Indicators of Change. Health of the Oceans From Labrador to Venezuela.* Year of the ocean special report The Center for Conservation Medicine and CHGE Harvard Medical School, Boston, MA, pp.1-85.

Holling, C.S., 1972: Resilience and stability of ecological systems. *Research Report*, **4**, 1-23.

Howarth, R.W., J.R. Fruci, and D. Sherman, 1991: Inputs of sediment and carbon to an estuarine ecosystem: influence of land use. *Ecological Applications*, **1(1)**, 27-39.

Howarth, R.W., D.P. Swaney, T.J. Butler, and R. Marino, 2000: Climatic control on eutrophication of the Hudson River estuary. *Ecosystems*, **3(2)**, 210-215.

Hughes, A.R. and J.J. Stachowicz, 2004: Genetic diversity enhances the resistance of a seagrass ecosystem to disturbance. In: *Proceedings of the National Academy of Sciences of the United States of America* 2004.

IPCC, 2001: *Climate Change 2001: Impacts, Adaptation, and Vulnerability. Contribution of Working Group II to the Third Assessment Report of the Intergovernmental Panel on Climate Change.* [McCarthy, J.J., O.F. Canziani, N.A. Leary, D.J. Dokken, and K.S. White (eds.)]. Cambridge University Press, Cambridge, UK.

IPCC, 2007: Summary for policymakers, In: *Climate Change 2007: the Physical Science Basis. Contribution of Working Group I to the Fourth Assessment Report of the Intergovernmental Panel on Climate Change*, [Solomon, S., D. Qin, M. Manning, Z. Chen, M. Marquis, K.B. Averyt, M. Tignor, and H.L. Miller (eds.)]. Cambridge University Press, Cambridge, United Kingdom and New York, NY, USA.

Jackson, J.B.C., M.X. Kirby, W.H. Berger, K.A. Bjorndal, L.W. Botsford, B.J. Bourque, R.H. Bradbury, R. Cooke, J. Erlandson, and J.A. Estes, 2001: Historical overfishing and the recent collapse of coastal ecosystems. *Science*, **293(5530)**, 629-638.

Jones, C.G., J.H. Lawton, and M. Shachak, 1994: Organisms as ecosystem engineers. *Oikos*, **69(3)**, 373-386.

Kates, R.W., C.E. Colten, S. Laska, and S.P. Leatherman, 2006: Reconstruction of New Orleans after Hurricane Katrina: a research perspective. *Proceedings of the National Academy of Sciences of the United States of America*, **103(40)**, 14653-14660.

Kemp, W.M., P.A. Sampou, J. Garber, J. Tuttle, and W.R. Boynton, 1992: Seasonal depletion of oxygen from bottom waters of Chesapeake Bay: roles of benthic and planktonic respiration and physical exchange processes. *Marine Ecology Progress Series*, **85(1)**.

Kennedy, V.S., 1996: The ecological role of the Eastern oyster, *Crassostrea virginica*, with remarks on disease. *Journal of Shellfish Research*, **15**, 177-183.

Kennedy, V.S., R.R. Twilley, J.A. Kleypas, J.H. Cowan, Jr., and S.R. Hare, 2002: *Coastal and Marine Ecosystems & Global Climate Change: Potential Effects on U.S. Resources*. Pew Center on Global Climate Change, pp.1-64.

Kennish, M.J., 1999: *Estuary Restoration and Maintenance: the National Estuary Program*. CRC Press Inc.

Kleypas, J.A., R.A. Feely, V.J. Fabry, C. Langdon, C.L. Sabine, and L.L. Robbins, 2006: *Impacts of Ocean Acidification on Coral Reefs and Other Marine Calcifiers: a Guide for Future Research*. Workshop Report, National Science Foundation, National Oceanic and Atmospheric Administration, and the U.S. Geological Survey.

Kuenzler, E.J., P.J. Mulholland, L.A. Ruley, and R.P. Sniffen, 1977: *Water Quality in North Carolina Coastal Plain Streams and Effects of Channelization*. 127, Water Resources Research Institute of the University of North Carolina, Raleigh, NC.

Lafferty, K.D. and L.R. Gerber, 2002: Good medicine for conservation biology: the intersection of epidemiology and conservation theory. *Conservation Biology*, **16(3)**, 593-604.

Lafferty, K.D., J.W. Porter, and S.E. Ford, 2004: Are diseases increasing in the ocean? *Annual Review of Ecology, Evolution and Systematics*, **35**, 31-54.

Lee, K.N., 1993: *Compass and Gyroscope: Integrating Science and Politics for the Environment*. Island Press, Washington, DC.

Lenihan, H.S. and C.H. Peterson, 1998: How habitat degradation through fishery disturbance enhances impacts of hypoxia on oyster reefs. *Ecological Applications*, **8(1)**, 128-140.

Lenihan, H.S., C.H. Peterson, J.E. Byers, J.H. Grabowski, G.W. Thayer, and D.R. Colby, 2001: Cascading of habitat degradation: oyster reefs invaded by refugee fishes escaping stress. *Ecological Applications*, **11(3)**, 764-782.

Leonard, L., T. Clayton, and O. Pilkey, 1990: An analysis of replenished beach design parameters on U. S. East Coast barrier islands. *Journal of Coastal Research*, **6(1)**, 15-36.

Leung, L.Y.R. and Y. Qian, 2003: Changes in seasonal and extreme hydrologic conditions of the Georgia Basin/Puget Sound in an ensemble regional climate simulation for the mid-century. *Canadian Water Resources Journal*, **28(4)**, 605-631.

Levin, L.A., D.F. Boesch, A. Covich, C. Dahm, C. Erseus, K.C. Ewel, R.T. Kneib, A. Moldenke, M.A. Palmer, and P. Snelgrove, 2001: The function of marine critical transition zones and the importance of sediment biodiversity. *Ecosystems*, **4(5)**, 430-451.

Li, M., A. Gargett, and K. Denman, 2000: What determines seasonal and interannual variability of phytoplankton and zooplankton in strongly estuarine systems? *Estuarine Coastal and Shelf Science*, **50(4)**, 467-488.

Lotze, H.K., H.S. Lenihan, B.J. Bourque, R.H. Bradbury, R.G. Cooke, M.C. Kay, S.M. Kidwell, M.X. Kirby, C.H. Peterson, and J.B.C. Jackson, 2006: Depletion, degradation, and recovery potential of estuaries and coastal seas. *Science*, **312(5781)**, 1806-1809.

Lyman, J.M., J.K. Willis, and G.C. Johnson, 2006: Recent cooling of the upper ocean. *Geophysical Research Letters*, **33**, L18604.

Mallin, M.A., H.W. Paerl, J. Rudek, and P.W. Bates, 1993: Regulation of estuarine primary production by watershed rainfall and river flow. *Marine Ecology Progress Series*, **93(1/2)**, 1999-203.

Marks, D., J. Kimball, D. Tingey, and T. Link, 1998: The sensitivity of snowmelt processes to climate conditions and forest cover during rain-on-snow: a case study of the 1996 Pacific Northwest flood. *Hydrological Processes*, **12(10)**, 1569-1587.

Marsh, A.S., D.P. Rasse, B.G. Drake, and J.P. Megonigal, 2005: Effect of elevated CO_2 on carbon pools and fluxes in a brackish marsh. *Estuaries*, **28**, 695-704.

Meehl, G.A., W.M. Washington, W.D. Collins, J.M. Arblaster, A. Hu, L.E. Buja, W.G. Strand, and H. Teng, 2005: How much more global warming and sea level rise? *Science*, **307(5716)**, 1769-1772.

Meyer, D.L., E.C. Townsend, and G.W. Thayer, 1997: Stabilization and erosion control value of oyster cultch for intertidal marsh. *Restoration Ecology*, **5(1)**, 93-99.

Meyer, J.N. and R.T. Di Giulio, 2003: Heritable adaptation and fitness costs in killifish (*Fundulus heteroclitus*) inhabiting a polluted estuary. *Ecological Applications*, **13(2)**, 490-503.

Micheli, F., B.S. Halpern, L.W. Botsford, and R.R. Warner, 2004: Trajectories and correlates of community change in no-take marine reserves. *Ecological Applications*, **14(6)**, 1709-1723.

Micheli, F. and C.H. Peterson, 1999: Estuarine vegetated habitats as corridors for predator movements. *Conservation Biology*, **13(4)**, 869-881.

Mileti, D.S., 1999: *Disasters by Design: a Reassessment of Natural Hazards in the United States*. Joseph Henry Press.

Millennium Ecosystem Assessment, 2005: *Ecosystems and Human Well-Being: Wetlands and Water*. World Resources Institute, Washington, DC.

Mitsch, W.J. and J.W. Day Jr, 2006: Restoration of wetlands in the Mississippi-Ohio-Missouri (MOM) River Basin: experience and needed research. *Ecological Engineering*, **26**, 55-69.

Mitsch, W.J. and J.G. Gosselink, 2000: *Wetlands*. John Wiley, New York.

Mooney, H.A. and R.J. Hobbs, 2000: *Invasive Species in a Changing World*. Island Press, Washington, DC.

Moorhead, K.K. and M.M. Brinson, 1995: Response of wetlands to rising sea level in the lower coastal plain of North Carolina. *Ecological Applications*, **5(1)**, 261-271.

Morris, J.T., P.V. Sundareshwar, C.T. Nietch, B. Kjerfve, and D.R. Cahoon, 2002: Responses of coastal wetlands to rising sea level. *Ecology*, **83(10)**, 2869-2877.

Mote, P.W., 2006: Climate-driven variability and trends in mountain snowpack in western North America. *Journal of Climate*, **19(23)**, 6209-6220.

Mote, P.W., E.A. Parson, A.F. Hamlet, W.S. Keeton, D. Lettenmaier, N. Mantua, E.L. Miles, D.W. Peterson, D.L. Peterson, R. Slaughter, and A.K. Snover, 2003: Preparing for climatic change: the water, salmon, and forests of the Pacific Northwest. *Climatic Change*, **61(1)**, 45-88.

Mullins, P.H. and T.C. Marks, 1987: Flowering phenology and seed production of Spartina anglica. *Journal of Ecology*, **25(4)**, 1037-1048.

Mydlarz, L.D., L.E. Jones, and C.D. Harvell, 2006: Innate immunity, environmental drivers, and disease ecology of marine and freshwater invertebrates. *Annual Review of Ecology, Evolution and Systematics*, **37**, 251-288.

Myers, R.A., J.K. Baum, T.D. Shapherd, S.P. Powers, and C.H. Peterson, 2007: Cascading effects of the loss of apex predatory sharks from a coastal ocean. *Science*, **315(5820)**, 1846-1850.

Naeem, S., 2002: Ecosystem consequences of biodiversity loss: The evolution of a paradigm. *Ecology*, **83(6)**, 1537-1552.

National Assessment Synthesis Team, 2000: *Climate Change Impacts on the United States: the Potential Consequences of Climate Variability and Change*. U.S. Global Change Research Program, Washington, DC.

National Coastal Assessment Group, 2000: *Coastal: The Potential Consequences of Climate Variability and Change*. pp.1-181.

National Marine Fisheries Service, 2006: *Recovery Plan for the Hawaiian Monk Seal*. National Marine Fisheries Service, Silver Spring, MD, pp.1-148.

National Research Council, 2004: *Non-Native Oysters in the Chesapeake Bay*. Committee on Non-native Oysters in the Chesapeake Bay, National Research Council, National Academies Press, Washington, DC.

Newell, R.I.E., J.C. Cornwell, and M.S. Owens, 2002: Influence of simulated bivalve biodeposition and microphytobenthos on sediment nitrogen dynamics: a laboratory study. *Limnology and Oceanography*, **47(5)**, 1367-1379.

Newell, R.I.E. and J. A. Ott, 1999: Macrobenthic communities and eutrophication, In: *Ecosystems at the Land-Sea Margin: Drainage Basin to Coastal Sea*, [Malone, T.C., A. Malej, L. Harding, N. Smodlaka, and R. Turner (eds.)]. American Geophysical Union, Washington, DC, pp. 265-293.

Nixon, S.W., 1995: Coastal marine eutrophication: a definition, social causes, and future concerns. *Ophelia*, **41**, 199-219.

Officer, C.B., R.B. Biggs, J.L. Taft, L.E. Cronin, M.A. Tyler, and W.R. Boynton, 1984: Chesapeake Bay anoxia: origin, development, and significance. *Science*, **223(4631)**, 22-26.

Orr, J.C., V.J. Fabry, O. Aumont, L. Bopp, S.C. Doney, R.A. Feely, A. Gnanadesikan, N. Gruber, A. Ishida, and F. Joos, 2005: Anthropogenic ocean acidification over the twenty-first century and its impact on calcifying organisms. *Nature*, **437(7059)**, 681-686.

Orth, R.J., T.J.B. Carruthers, W.C. Dennison, C.M. Duarte, J.W. Fourqurean, K.L. Keck, Jr., A.R. Hughes, G.A. Kendrick, W.H. Kenworthy, S. Olyarnik, F.T. Short, M. Waycott, and S.L. Williams, 2006: A global crisis for seagrass ecosystems. *BioScience*, **56(12)**, 987-996.

Otto-Bliesner, B.L., S.J. Marshall, J.T. Overpeck, G.H. Miller, and A. Hu, 2006: Simulating arctic climate warmth and icefield retreat in the last interglaciation. *Science*, **311(5768)**, 1751-1753.

Overpeck, J.T., B.L. Otto-Bliesner, G.H. Miller, D.R. Muhs, R.B. Alley, and J.T. Kiehl, 2006: Paleoclimatic evidence for future ice-sheet instability and rapid sea-level rise. *Science*, **311(5768)**, 1747-1750.

Paerl, H.W. and J.D. Bales, 2001: Ecosystem impacts of three sequential hurricanes (Dennis, Floyd, and Irene) on the United States' largest lagoonal estuary, Pamlico Sound, NC. *Proceedings of the National Academy of Sciences of the United States of America*, **98(10)**, 5655-5660.

Paerl, H.W., J.L. Pinckney, J.M. Fear, and B.L. Peierls, 1998: Ecosystem responses to internal and watershed organic matter loading: consequences for hypoxia in the eutrophying Neuse River Estuary, North Carolina, USA. *Marine Ecology Progress Series*, **166**, 17-25.

Park, R.A., M.S. Treehan, P.W. Mausel, and R.C. Howe, 1989: *The Effects of Sea Level Rise on US Coastal Wetlands*. EPA-230-05-89-052, Office of Policy, Planning, and Evaluation, US Environmental Protection Agency, Washington, DC.

Parker Jr., R.O. and R.L. Dixon, 1998: Changes in a North Carolina reef fish community after 15 years of intense fishing- global warming implications. *Transactions of the American Fisheries Society*, **127(6)**, 908-920.

Parmesan, C. and H. Galbraith, 2004: *Observed Impacts of Global Climate Change in the US*. Pew Center on Global Climate Change, Arlington, VA.

Parmesan, C. and G. Yohe, 2003: A globally coherent fingerprint of climate change impacts across natural systems. *Nature*, **421**, 37-42.

Parmesan, C., 2006: Ecological and evolutionary responses to recent climate change. *Annual Review of Ecology, Evolution and Systematics*, **37**, 637-669.

Parson, E.A., P. W. Mote, A. Hamlet, N. Mantua, A. Snover, W. Keeton, E. Miles, D. Canning, and K. G. Ideker, 2001: Potential consequences of climate variability and change for the Pacific Northwest, In: *The Potential Consequences of Climate Variability and Change: Foundation Report*, Report by the National Assessment Synthesis Team for the US Global Change Research Program, Cambridge University Press, Cambridge, UK, pp. 247-281.

Pauly, D., V. Christensen, J. Dalsgaard, R. Froese, and F. Torres, Jr., 1998: Fishing down marine food webs. *Science*, **279(5352)**, 860-863.

Peterson, C.H. and R. Black, 1988: Density-dependent mortality caused by physical stress interacting with biotic history. *American Naturalist*, **131(2)**, 257-270.

Peterson, C.H. and J. A. Estes, 2001: Conservation and management of marine communities, [Bertness, M.D., S.D. Gaines, and M.E. Hay (eds.)]. pp. 469-508.

Peterson, C.H., S.D. Rice, J.W. Short, D. Esler, J.L. Bodkin, B.E. Ballachey, and D.B. Irons, 2003: Long-term ecosystem response to the Exxon Valdez oil spill. *Science*, **302(5653)**, 2082-2086.

Peterson, D., D. Cayan, J. DiLeo, M. Noble, and M. Dettinger, 1995: The role of climate in estuarine variability. *American Scientist*, **83(1)**, 58-67.

Peterson, G.W. and R.E. Turner, 1994: The value of salt marsh edge vs interior as a habitat for fish and decapod crustaceans in a Louisiana tidal marsh. *Estuaries*, **17(1)**, 235-262.

Petraitis, P.S. and S.R. Dudgeon, 2004: Detection of alternative stable states in marine communities. *Journal of Experimental Marine Biology and Ecology*, **300(1)**, 343-371.

Pew Center on Global Climate Change, 2003: *Innovative Policy Solutions to Global Climate Change: the U.S. Domestic Response to Climate Change. Key Elements of a Prospective Program*. pp.1-8.

Pielke, R., G. Prins, S. Rayner, and D. Sarewitz, 2007: Climate change 2007: lifting the taboo on adaptation. *Nature*, **445**, 597-598.

Pikitch, E.K., C. Santora, E.A. Babcock, A. Bakun, R. Bonfil, D.O. Conover, P. Dayton, P. Doukakis, D. Fluharty, and B. Heneman, 2004: Ecosystem-based fishery management. *Science*, **305(5682)**, 346-347.

Pilkey, O.H. and H.L. Wright III, 1988: Seawalls versus beaches. *Journal of Coastal Research*, **4**, 41-64.

Poff, N.L., J.D. Allan, M.B. Bain, J.R. Karr, K.L. Prestegaard, B.D. Richter, R. Sparks, and J. Stromberg, 1997: The natural flow regime: a new paradigm for riverine conservation and restoration. *BioScience*, **47**, 769-784.

Postel, S., 1992: *The Last Oasis-Facing Water Scarcity*. Norton & Co, New York.

Poulin, R., 2005: Global warming and temperature-mediated increases in cercarial emergence in trematode parasites. *Parasitology*, **132(01)**, 143-151.

Poulin, R. and K.N. Mouritsen, 2006: Climate change, parasitism and the structure of intertidal ecosystems. *Journal of Helminthology*, **80(2)**, 183-191.

Pritchard, D.W., 1967: *What Is an Estuary: Physical Viewpoint*. Publication Number 83, American Association for the Advancement of Science, Washington, DC, pp.3-5.

Purcell, J.E., F.P. Cresswell, D.G. Cargo, and V.S. Kennedy, 1991: Differential ingestion and digestion of bivalve larvae by the scyphozoan Chrysaora quinquecirrha and the ctenophore Mnemiopsis leidyi. *Biological Bulletin, Marine Biological Laboratory, Woods Hole*, **180(1)**, 103-111.

Rabalais, N.N., R.E. Turner, and D. Scavia, 2002: Beyond science into policy: Gulf of Mexico hypoxia and the Mississippi River. *BioScience*, **52(2)**, 129-142.

Rahmstorf, S., 2007: A semi-empirical approach to projecting future sea-level rise. *Science*, **315(5810)**, 368-370.

Ramus, J., L.A. Eby, C.M. McClellan, and L.B. Crowder, 2003: Phytoplankton forcing by a record freshwater discharge event into a large lagoonal estuary. *Estuaries*, **26(5)**, 1344-1352.

Raven, J., 2005: *Ocean Acidification Due to Increasing Atmospheric Carbon Dioxide*. The Royal Society, London.

Reed, D.J., 1995: The response of coastal marshes to sea-level rise: survival or submergence? *Earth Surface Processes and Landforms*, **20(1)**, 39-48.

Reed, D.J., 2002: Sea-level rise and coastal marsh sustainability: geological and ecological factors in the Mississippi Delta plain. *Geomorphology*, **48(1)**, 233-243.

Remane, A. and C. Schlieper, 1971: *Biology of Brackish Water*. Wiley Interscience Division, John Wiley & Sons, New York, NY.

Riggs, S.R., 2002: *The Soundfront Series: Shoreline Erosion in North Carolina Estuaries*. UNC-SG-01-11.

Riggs, S.R. and D.V. Ames, 2003: *Drowning the North Carolina Coast: Sea-Level Rise and Estuarine Dynamics*. UNC-SG-03-04, NC Sea Grant College Program, Raleigh, NC, pp.1-152.

Rignot, E. and P. Kanagaratnam, 2006: Changes in the velocity structure of the Greenland ice sheet. *Science*, **311(5763)**, 986-990.

Rinaldi, S. and M. Scheffer, 2000: Geometric analysis of ecological models with slow and fast processes. *Ecosystems*, **3(6)**, 507-521.

Ritchie, K., 2006: Regulation of microbial populations by coral surface mucus and mucus-associated bacteria. *Marine Ecology Progress Series*, **322**, 1-14.

Roberts, C.M., S. Andelman, G. Branch, R.H. Bustamante, J.C. Castilla, J. Dugan, B.S. Halpern, K.D. Lafferty, H. Leslie, and J. Lubchenco, 2003: Ecological criteria for evaluating candidate sites for marine reserves. *Ecological Applications*, **13(1)**, S199-S214.

Robins, J.B., I.A. Halliday, J. Staunton-Smith, D.G. Mayer, and M.J. Sellin, 2005: Freshwater-flow requirements of estuarine fisheries in tropical Australia: a review of the state of knowledge and application of a suggested approach. *Marine & Freshwater Research*, **56(3)**, 343-360.

Root, T.L., J. Price, K.R. Hall, S.H. Schnelder, C. Rosenzweig, and A.J. Pounds, 2003: Fingerprints of global warming on wild animals and plants. *Nature*, **421**, 57-60.

Rothschild, B.J., J.S. Ault, P. Goulletquer, and M. Heral, 1994: Decline of the Chesapeake Bay oyster population: a century of habitat destruction and overfishing. *Marine ecology progress series. Oldendorf*, **111(1)**, 29-39.

Roy, B.A., S. Guesewell, and J. Harte, 2004: Response of plant pathogens and herbivores to a warming experiment. *Ecology*, **85(9)**, 2570-2581.

Rozas, L.P., T.J. Minello, I. Munuera-Femandez, B. Fry, and B. Wissel, 2005: Macrofaunal distributions and habitat change following winter-spring releases of freshwater into the Breton Sound estuary, Louisiana(USA). *Estuarine Coastal and Shelf Science*, **65(1-2)**, 319-336.

Ruiz, G.M., J.T. Carlton, E.D. Grosholz, and A.H. Hines, 1997: Global invasions of marine and estuarine habitats by non-indigenous species: Mechanisms, extent, and consequences. *American Zoologist*, **37(6)**, 621-632.

Salathé, E.P., 2006: Influences of a shift in North Pacific storm tracks on western North American precipitation under global warming. *Geophysical Research Letters*, **33(19)**.

Sanchez-Arcilla, A. and J.A. Jimenez, 1997: Physical impacts of climatic change on deltaic coastal systems (I): an approach. *Climatic Change*, **35(1)**, 71-93.

Sarmiento, J.L., R. Slater, R. Barber, L. Bopp, S.C. Doney, A.C. Hirst, J. Kleypas, R. Matear, U. Mikolajewicz, and P. Monfray, 2004: Response of ocean ecosystems to climate warming. *Global Biogeochemical Cycles*, **18(3)**.

Scavia, D., J.C. Field, D.F. Boesch, R.W. Buddemeier, V. Burkett, D.R. Cayan, M. Fogarty, M.A. Harwell, R.W. Howarth, C. Mason, D.J. Reed, T.C. Royer, A.H. Sallenger, and J.G. Titus, 2002: Climate change impacts on U.S. coastal and marine ecosystems. *Estuaries*, **25(2)**, 149-164.

Scavia, D., E.L.A. Kelly, and J.D. Hagy, 2006: A simple model for forecasting the effects of nitrogen loads on chesapeake bay hypoxia. *Estuaries and coasts*, **29(4)**, 674-684.

Scheffer, M. and S.R. Carpenter, 2003: Catastrophic regime shifts in ecosystems: linking theory to observation. *Trends in Ecology and Evolution*, **18(12)**, 648-656.

Scheffer, M., S. Carpenter, J.A. Foley, C. Folke, and B.H. Walker, 2001: Catastrophic shifts in ecosystems. *Nature*, **413**, 591-596.

Schwimmer, R.A. and J.E. Pizzuto, 2000: A model for the evolution of marsh shorelines. *Journal of Sedimentary Research Section A: Sedimentary Petrology and Processes*, **70(5)**, 1026-1035.

Seitz, R.D., R.N. Lipcius, N.H. Olmstead, M.S. Seebo, and D.M. Lambert, 2006: Influence of shallow-water habitats and shoreline development on abundance, biomass, and diversity of benthic prey and predators in Chesapeake Bay. *Marine Ecology Progress Series*, **326**, 11-27.

Sheldon, J.E. and M. Alber, 2002: A comparison of residence time calculations using simple compartment models of the Altamaha River Estuary, Georgia. *Estuaries*, **25(6)**, 1304-1317.

Short, F.T. and D.M. Burdick, 1996: Quantifying eelgrass habitat loss in relation to housing development and nitrogen loading in Waquoit Bay, Massachusetts. *Estuaries*, **19(3)**, 730-739.

Short, F.T. and S. Wyllie-Echeverria, 1996: Natural and human induced disturbance of seagrass. *Environmental Conservation*, **23**, 17-27.

Simenstad, C.A., K. L. Fresh, and E. O. Salo, 1982. The role of Puget Sound and Washington coastal estuaries in the life history of Pacific salmon: an unappreciated function, In: *Estuarine Comparisons*, [Kennedy, V.S. (ed.)]. Academic Press, New York, NY, pp. 343-364.

Simenstad, C.A., R. M. Thom, D. A. Levy, and D. L. Bottom, 2000: Landscape structure and scale constraints on restoring estuarine wetlands for Pacific coast juvenile fishes, In: *Concepts and Controversies in Tidal Marsh Ecology*, [Weinstein, M.P. and D.A. Kreeger (eds.)]. Kluwer Academic Publishing, Dordrecht, pp. 597-630.

Sims, D.W., V.J. Wearmouth, M.J. Genner, A.J. Southward, and S.J. Hawkins, 2004: Low-temperature-driven early spawning migration of a temperate marine fish. *Journal of Animal Ecology*, **73(2)**, 333-341.

Sklar, F.H. and J.A. Browder, 1998: Coastal environmental impacts brought about by alterations to freshwater flow in the Gulf of Mexico. *Environmental Management*, **22(4)**, 547-562.

Snover, A.K., P.W. Mote, L. Whitley Binder, A.F. Hamlet, and N.J. Mantua, 2005: *Uncertain Future: Change and Its Effects on Puget Sound. A Report for the Puget Sound Action Team by the Climate Impacts Group*. Center for Science in the Earth System, Joint Institute for the Study of the Atmosphere and Oceans, University of Washington, Seattle.

Solan, M., B.J. Cardinale, A.L. Downing, K.A.M. Engelhardt, J.L. Ruesink, and D.S. Srivastava, 2004: Extinction and ecosystem function in the marine benthos. *Science*, **306(5699)**, 1177-1180.

Southward, A.J., S.J. Hawkins, and M.T. Burrows, 1995: Seventy years' observations of changes in distribution and abundance of zooplankton and intertidal organisms in the western English Channel in relation to rising sea temperature. *Journal of Thermal Biology*, **20(1)**, 127-155.

Southward, A.J., O. Langmead, N.J. Hardman-Mountford, J. Aiken, G.T. Boalch, P.R. Dando, M.J. Genner, I. Joint, M.A. Kendall, N.C. Halliday, R.P. Harris, R. Leaper, N. Mieszkowska, R.D. Pingree, A.J. Richardson, D.W. Sims, T. Smith, A.W. Walne, and S.J. Hawkins, 2004: Long-term oceanographic and ecological research in the western English Channel. *Advances in Marine Biology*, **47**, 1-105.

Stachowicz, J.J., R.B. Whitlatch, and R.W. Osman, 1999: Species diversity and invasion resistance in a marine ecosystem. *Science*, **286(5444)**, 1577-1579.

Stephan, C.D., R.L. Peuser, and M.S. Fonseca, 2001: *Evaluating Fishing Gear Impacts to Submreged Aquatic Vegetation and Determining Mitigation Strategies*. ASMFC Habitat Management Series No. 5, Atlantic States Marine Fisheries Commission, Washington, DC.

Syvitski, J.P.M., C.J. Voeroesmarty, A.J. Kettner, and P. Green, 2005: Impact of humans on the flux of terrestrial sediment to the global coastal ocean. *Science*, **308(5720)**, 376-380.

Tait, J.F. and G.B. Griggs, 1990: Beach response to the presence of a seawall; comparison of field observations. *Shore and Beach*, **58(2)**, 11-28.

Tartig, E.K., F. Mushacke, D. Fallon, and A. Kolker, 2000: *A Wetlands Climate Change Impact Assessment for the Metropolitan East Coast Region*. Center for International Earth Science Information Network.

Tenore, K.R., 1970: The macrobenthos of the Pamlico River estuary, North Carolina. *Ecological Monographs*, **42**, 51-69.

Tilman, D. and J.A. Downing, 1994: Biodiversity and stability in grasslands. *Nature*, **367(6461)**, 363-365.

Titus, J.G., 1989: *Sea Level Rise*. EPA 230-05-89-052, U.S. Environmental Protection Agency, Washington, DC.

Titus, J.G., 2000: Does the U.S. government realize that the sea is rising? How to restructure federal programs so that wetlands can survive. *Golden Gate University Law Review*, **30(4)**, 717-778.

Titus, J.G., 2004: *Maps That Depict the Business-As-Usual Response to Sea Level Rise in the Decentralized United States of America*. paper presented at the OECD Global Forum on Sustainable Development: Development and Climate Change ENV/EPOC/GF/SD/RD(2004)9/FINAL, Paris.

Titus, J.G., 1991: Greenhouse effect and coastal wetland policy: how americans could abandon an area the size of Massachusetts at minimum cost. *Environmental Management*, **15(1)**, 39-58.

Titus, J.G., 1998: Rising seas, coastal erosion, and the takings clause: how to save wetlands and beaches without hurting property owners. *Maryland Law Review*, **57(4)**, 1279-1399.

Titus, J.G., R. Park, S.P. Leatherman, J.R. Weggel, P.W. Mausel, S. Brown, G. Gaunt, M. Trehan, and G. Yohe, 1991: Greenhouse effect and sea level rise: the cost of holding back the sea. *Coastal Management*, **19**, 171-204.

Titus, J.G. and C. Richman, 2001: Maps of lands vulnerable to sea level rise: modeled elevations along the U.S. Atlantic and Gulf Coasts. *Climate Research*, **18**, 205-228.

Turgeon, J., R. Stoks, R.A. Thum, J.M. Brown, and M.A. McPeek, 2005: Simultaneous Quaternary radiations of three damselfly clades across the holarctic. *American Naturalist*, **165(4)**, E78-E107.

Turner, R.E., J.J. Baustian, E.M. Swenson, and J.S. Spicer, 2006: Wetland sedimentation from Hurricanes Katrina and Rita. *Science*, **314(5798)**, 449-452.

Turner, R.E., W.W. Schroeder, and W.J. Wiseman, 1987: Role of stratification in the deoxygenation of Mobile Bay and adjacent shelf bottom waters. *Estuaries*, **10(1)**, 13-19.

U.S. Army Corps of Engineers, in press: Louisiana coastal protection and restoration. *To be submitted to Congress.*

U.S. Climate Change Science Program, in press: Synthesis and assessment product 4.1: Coastal elevation and sensitivity to sea level rise. *A report by the U. S. Climate Change Science Program and the Subcommittee on Global Change Research, U. S. Environmental Protection Agency*. [Titus, J.G. (ed.)].

U.S. Commission on Ocean Policy, 2004: *An Ocean Blueprint for the 21st Century*. 9, pp.15-18.

U.S. Environmental Protection Agency, 1989: *The Potential Effects of Global Climate Change on the United States: Report to Congress*. EPA-230-05-89-052, Office of Policy, Planning, and Evaluation, US Environmental Protection Agency.

U.S. Environmental Protection Agency, 2000: *Stressor Identification Guidance Document*. EPA-822-B-00-025, U.S. Environmental Protection Agency, Office of Water and Office of Research and Development, Cincinnati, OH, pp.1-208.

Vinebrooke, R.D., K.L. Cottingham, M.S.J. Norberg, S.I. Dodson, S.C. Maberly, and U. Sommer, 2004: Impacts of multiple stressors on biodiversity and ecosystem functioning: the role of species co-tolerance. *Oikos*, **104(3)**, 451-457.

Walters, C.J., 1986: *Adaptive Management of Renewable Resources*. McMillan, New York, New York.

Walther, G.R., E. Post, P. Convey, A. Menzel, C. Parmesan, T.J.C. Beebee, J.M. Fromentin, O. Hoegh-Guldberg, and F. Bairlein, 2002: Ecological responses to recent climate change. *Nature*, **416**, 389-395.

Ward, J.R. and K.D. Lafferty, 2004: The elusive baseline of marine disease: are diseases in ocean ecosystems increasing? *PLoS Biology*, **2(4)**, 542-547.

Whitfield, A.K., 2005: Fishes and freshwater in southern African estuaries - a review. *Aquatic Living Resources*, **18(3)**, 275-289.

Whitfield, A.K., 1994: Abundance of larval and 0+ juvenile marine fishes in the lower reaches of 3 Southern African estuaries with differing fresh-water inputs. *Marine Ecology Progress Series*, **105(3)**, 257-267.

Williams, S.L. and K. L. Heck Jr., 2001: Seagrass community ecology, [Bertness, M.D., S.D. Gaines, and M.E. Hay (eds.)]. Sinauer Associates, Inc, MA, USA, pp. 317-337.

Woerner, L.S. and C.T. Hackney, 1997: Distribution of *Juncus roemerianus* in North Carolina marshes: the importance of physical and abiotic variables. *Wetlands*, **17(2)**, 284-291.

Wolfe, D.A., 1986: *Estuarine Variability*. Academic Press, New York, NY.

Wolock, D.M. and G.J. McCabe, 1999: Explaining spatial variability in mean annual runoff in the conterminous United States. *Climate Research*, **11**, 149-159.

Worm, B., E.B. Barbier, N. Beaumont, J.E. Duffy, C. Folke, B.S. Halpern, J.B.C. Jackson, H.K. Lotze, F. Micheli, S.R. Palumbi, E. Sala, K.A. Selkoe, J.J. Stachowicz, and R. Watson, 2006: Impacts of biodiversity loss on ocean ecosystem services. *Science*, **314(5800)**, 787-790.

Yohe, G., J. Neumann, P. Marshall, and H. Ameden, 1996: The economic cost of greenhouse-induced sea-level rise for developed property in the United States. *Climatic Change*, **32(4)**, 387-410.

Zedler, J.B., 1993: Canopy architecture of natural and planted cordgrass marshes: Selecting habitat evaluation criteria. *Ecological Applications*, **3(1)**, 123-138.

Zimmerman, R.J., T. J. Minello, and L. P. Rozas, 2000: Salt marsh linkages to productivity of penaeid shrimps and blue crabs in the northern Gulf of Mexico, In: *Concepts and Controversies in Tidal Marsh Ecology*, [Weinstein, M.P. and D.A. Kreeger (eds.)]. pp. 293-314.

CHAPTER 8 REFERENCES

Adger, W.N., 2000: Social and ecological resilience: are they related? *Progress in Human Geography*, **24(3)**, 347-364.

Adger, W.N., T.P. Hughes, C. Folke, S.R. Carpenter, and J. Rockstroem, 2005: Social-ecological resilience to coastal disasters. *Science*, **309(5737)**, 1036-1039.

Agardy, T., P. Bridgewater, M.P. Crosby, J. Day, P.K. Dayton, R. Kenchington, D. Laffoley, P. McConney, P.A. Murray, J.E. Parks, and L. Peau, 2003: Dangerous targets? Unresolved issues and ideological clashes around marine protected areas. *Aquatic Conservation: Marine and Freshwater Ecosystems*, **13(4)**, 353-367.

Agardy, T.S., 1997: *Marine Protected Areas and Ocean Conservation*. R.G. Landes Company and Academic Press, Austin, TX, pp. 1-244.

Airamé, S., J.E. Dugan, K.D. Lafferty, H. Leslie, D.A. McArdle, and R.R. Warner, 2003: Applying ecological criteria to marine reserve design: a case study from the California Channel Islands. *Ecological Applications*, **13(1)**, S170-S184.

Albritton, D.L. and L. G. M. Filho, 2001: Technical summary, In: *Climate Change 2001. the Scientific Basis. Contribution of Working Group I to the Third Assessment Report of the Intergovernmental Panel on Climate Change*, [Houghton, J.T., Y. Ding, D.J. Griggs, M. Noguer, P.J. van der Linden, X. Dai, K. Maskell, and C.A. Johnson (eds.)]. Cambridge University Press, Cambridge, United Kingdom and New York, NY, USA.

Allison, G.W., S.D. Gaines, J. Lubchenco, and H.P. Possingham, 2003: Ensuring persistence of marine reserves: catastrophes require adopting an insurance factor. *Ecological Applications*, **13(1)**, S8-S24.

Allison, G.W., J. Lubchenco, and M.H. Carr, 1998: Marine reserves are necessary but not sufficient for marine conservation. *Ecological Applications*, **8 Supplement-Ecosystem Management for Sustainable Marine Fisheries(1)**, S79-S92.

Almany, G.R., M.L. Berumen, S.R. Therrold, S. Planes, and G.P. Jones, 2007: Local replenishment of coral reef fish populations in a marine reserve. *Science*, **316(5825)**, 742-744.

Andrew, N.L., Y. Agatsuma, E. Ballesteros, A.G. Bazhin, E.P. Creaser, D.K.A. Barnes, L.W. Botsford, A. Bradbury, A. Campbell, and J.D. Dixon, 2002: Status and management of world sea urchin fisheries. *Oceanography and Marine Biology: an Annual review*, **40**, 343-425.

Aronson, R.B. and W.F. Precht, 2006: Conservation, precaution, and Caribbean reefs. *Coral Reefs*, **25**, 441-450.

Ault, J.S., J.A. Bohnsack, and G.A. Meester, 1998: A retrospective (1979-1996) multispecies assessment of coral reef fish stocks in the Florida Keys. *Fishery Bulletin*, **96(3)**, 395-414.

Bakun, A., 1990: Global climate change and intensification of coastal ocean upwelling. *Science*, **247**, 198-201.

Baldwin, A., M. Egnotovich, M. Ford, and W. Platt, 2001: Regeneration in fringe mangrove forests damaged by Hurricane Andrew. *Plant Ecology*, **157(2)**, 151-164.

Baltz, D.M. and P.B. Moyle, 1993: Invasion resistance to introduced species by a native assemblage of California stream fishes. *Ecological Applications*, **3**, 246-255.

Barr, B.W., 2004: A seamless network of ocean parks and marine sanctuaries: The National Park Service/National Marine Sanctuary partnership. *The George Wright Forum*, **21**, 42-48.

Barry, J.P., C.H. Baxter, R.D. Sagarin, and S.E. Gilman, 1995: Climate-related, long-term faunal changes in a California rocky intertidal community. *Science*, **267(5198)**, 672-675.

Beger, M., G.P. Jones, and P.L. Munday, 2003: Conservation of coral reef biodiversity: a comparison of reserve selection procedures for corals and fishes. *Biological Conservation*, **111(1)**, 53-62.

Behrens, M.D. and K.D. Lafferty, 2004: Effects of marine reserves and urchin disease on southern Californian rocky reef communities. *Marine Ecology Progress Series*, **279**, 129-139.

Bell, R.E., M. Studinger, C.A. Shuman, M.A. Fahnestock, and I. Joughin, 2007: Large subglacial lakes in East Antarctica at the onset of fast-flowing ice streams. *Nature*, **445**, 904-907.

Bellwood, D.R. and T.P. Hughes, 2001: Regional-scale assembly rules and biodiversity of coral reefs. *Science*, **292(5521)**, 1532-1534.

Bellwood, D.R., T.P. Hughes, C. Folke, and M. Nystroem, 2004: Confronting the coral reef crisis. *Nature*, **429(6994)**, 827-833.

Bellwood, D.R., T.P. Hughes, and A.S. Hoey, 2006: Sleeping functional group drives coral-reef recovery. *Current Biology*, **16(24)**, 2434-2439.

Berkeley, S.A., C. Chapman, and S.M. Sogard, 2004: Maternal age as a determinant of larval growth and survival in a marine fish, *Sebastes melanops*. *Ecology*, **85(5)**, 1258-1264.

Berkelmans, R. and M.J.H. van Oppen, 2006: The role of zooxanthellae in the thermal tolerance of corals: a "nugget of hope" for coral reefs in an era of climate change. *Proceedings of the Royal Society B: Biological Sciences*, **273(1599)**, 2305-2312.

Bertness, M.D., S.D. Gaines, and M.E. Hay, 2001: *Marine Community Ecology*. Sinauer Associates, Sunderland, MA, pp. 1-550.

Bindoff, N.L., J. Willebrand, V. Artale, A. Cazenave, J. Gregory, S. Gulev, K. Hanawa, C. Le Quere, S. Levitus, Y. Nojiri, C. K. Shum, L. D. Talley, and A. Unnikrishnan, 2007: Observations: Oceanic Climate Change and Sea Level, In: *Climate Change 2007: the Physical Science Basis. Contribution of Working Group I to the Fourth Assessment Report of the Intergovernmental Panel on Climate Change*, [Solomon, S., D. Quin, M. Manning, Z. Chen, M. Marquis, K.B. Averyt, M. Tignor, and H.L. Miller (eds.)]. Cambridge University Press, Cambridge, United Kingdom and New York, NY, USA, pp. 385-432.

Black, K., P. Moran, D. Burrage, and G. De'ath, 1995: Association of low-frequency currents and crown-of-thorns starfish outbreaks. *Marine Ecology Progress Series*, **125(1)**, 185-194.

Black, K.P., P.J. Moran, and L.S. Hammond, 1991: Numerical models show coral reefs can be self-seeding. *Marine Ecology Progress Series*, **69**, 55-65.

Boessenkool, K.P., I.R. Hall, H. Elderfield, and I. Yashayaev, 2007: North Atlantic climate and deep-ocean flow speed changes during the last 230 years. *Geophysical Research Letters*, **34(L13614)**, 1-6.

Bohnsack, J.A., B. Causey, M.P. Crosby, R.B. Griffis, M.A. Hixon, T.F. Hourigan, K.H. Koltes, J.E. Maragos, A. Simons, and J.T. Tilmant, 2002: A rationale for minimum 20–30% no-take protection. In: *Proceedings of the Ninth International Coral Reef Symposium* 23, October 2000, pp. 615-619.

Botsford, L.W., 2005: Potential contributions of marine reserves to sustainable fisheries: recent modeling results. *Bulletin of Marine Science*, **76(2)**, 245-260.

Botsford, L.W., F. Micheli, and A. Hastings, 2003: Principles for the design of marine reserves. *Ecological Applications*, **13(1)**, S25-S31.

Boyett, H.V., D.G. Bourne, and B.L. Willis, 2007: Elevated temperature and light enhance progression and spread of black band disease on staghorn corals of the Great Barrier. *Marine Biology*, **151(5)**.

Breitburg, D.L. and G. F. Riedel, 2005: Multiple stressors in marine systems, In: *Marine Conservation Biology: the Science of Maintaining the Sea's Biodiversity*, [Norse, E. and L.B. Crowder (eds.)]. Island Press, Washington, DC, pp. 167-182.

Buddemeier, R.W., J.A. Kleypas, and R. Aronson, 2004: *Coral Reefs and Global Climate Change: Potential Contributions of Climate Change to Stresses on Coral Reef Ecosystems*. Pew Center on Global Climate Change.

Burton, G.A., Jr. and R. Pitt, 2001: *Stormwater Effects Handbook: a Toolbox for Watershed Mangers, Scientists and Engineers*. Lewis Publishers, Boca Raton, FL, pp.1-911.

Cabanes, C., A. Cazenave, and C. Le Provost, 2001: Sea level rise during past 40 years determined from satellite and in situ observations. *Science*, **294(5543)**, 840-842.

Caldeira, K. and M.E. Wickett, 2003: Anthropogenic carbon and ocean pH. *Nature*, **425(6956)**, 365-365.

Caldeira, K. and M.E. Wickett, 2005: Ocean model predictions of chemistry changes from carbon dioxide emissions to the atmosphere and ocean. *Journal of Geophysical Research*, **110(C09S04)**.

Carlton, J.T., 1996: Biological invasions and cryptogenic species. *Ecology*, **77(6)**, 1653-1655.

Carlton, J.T., 2000: Global change and biological invasions in the oceans, In: *Invasive Species in a Changing World*, [Mooney, H.A. and R.J. Hobbs (eds.)]. Island Press, Washington, DC, pp. 31-53.

Carpenter, S.R., N.F. Caraco, D.L. Correll, R.W. Howarth, A.N. Sharpley, and V.H. Smith, 1998: Nonpoint pollution of surface waters with phosphorus and nitrogen. *Ecological Applications*, **8(3)**, 559-568.

Carr, M.H., J.E. Neigel, J.A. Estes, S. Andelman, R.R. Warner, and J.L. Largier, 2003: Comparing marine and terrestrial ecosystems: Implications for the design of coastal marine reserves. *Ecological Applications*, **13(1)**, S90-S107.

Chen, J.L., C.R. Wilson, and B.D. Tapley, 2006: Satellite gravity measurements confirm accelerated melting of Greenland ice sheet. *Science*, **313**, 1958-1960.

Cho, L., 2005: Marine protected areas: a tool for integrated coastal management in Belize. *Ocean & Coastal Management*, **48(11)**, 932-947.

Christie, P., 2004: Marine protected areas as biological successes and social failures in Southeast Asia. *American Fisheries Society Symposium*, **42**, 155-164.

Cinner, J., M.J. Marnane, T.R. McClanahan, and G.R. Almany, 2005: Periodic closures as adaptive coral reef management in the Indo-Pacific. *Ecology and Society*, **11(1)**, 31.

Coe, J.M. and D. Rogers, 1997: *Marine Debris: Sources, Impacts, and Solutions*. Springer-Verlag, New York, NY, pp. 1-432.

Condrey, R. and D. Fuller, 1992: The US Gulf shrimp fishery, In: *Climate Variability, Climate Change, and Fisheries*, [Glantz, M.F. (ed.)]. Cambridge University Press, Cambridge, UK, pp. 89-119.

Corrigan, C., 2006: *The Marine Learning Partnership: Effective Design and Management of Tropical Marine Protected Area Networks Through Cross-Institutional Learning*. Year End Report, The Nature Conservancy.

Costanza, R., R. d'Arge, R. de Groot, S. Farber, M. Grasso, B. Hannon, K. Limburg, S. Naeem, R.V. O'Neill, J. Paruelo, R.G. Raskin, P. Sutton, and M. van den Belt, 1997. The value of the world's ecosystem services and natural capital. *Nature*, **387(6630)**, 253-260.

Cowen, R.K., C.B. Paris, D.B. Olson, and J.L. Fortuna, 2003: The role of long distance dispersal versus local retention in replenishing marine populations. *Gulf and Caribbean Research Supplement*, **14**, 129-137.

Cowen, R.K., C.B. Paris, and A. Srinivasan, 2006: Scaling of connectivity in marine populations. *Science*, **311(5760)**, 522-527.

Cowie-Haskell, B.D. and J.M. Delaney, 2003: Integrating science into the design of the Tortugas Ecological Reserve. *Marine Technology Society Journal*, **37(1)**, 68-79.

Crossett, K.M., T.J. Culliton, P.C. Wiley, and T.R. Goodspeed, 2004: *Population Trends Along the Coastal United States: 1980-2008*. National Oceanographic and Atmospheric Administration, Washington, DC, pp.1-54.

Crowder, L.B., S.J. Lyman, W.F. Figueira, and J. Priddy, 2000: Source-sink population dynamics and the problem of siting marine reserves. *Bulletin of Marine Science*, **66(3)**, 799-820.

Crowder, L.B., G. Osherenko, O.R. Young, S. Airamé, E.A. Norse, N. Baron, J.C. Day, F. Douvere, C.N. Ehler, B.S. Halpern, S.J. Langdon, K.L. McLeod, J.C. Ogden, R.E. Peach, A.A. Rosenberg, and J.A. Wilson, 2006: Resolving mismatches in U.S. ocean governance. *Science*, **313(5787)**, 617-618.

Curry, R., B. Dickson, and I. Yashayaev, 2003: A change in the freshwater balance in the Atlantic Ocean over the past four decades. *Nature*, **426**, 826-829.

Curry, R. and C. Mauritzen, 2005: Dilution of the northern atlantic current in recent decades. *Science*, **308(5729)**, 1772-1774.

Davis, G.E., 2004: Maintaining unimpaired ocean resources and experiences: a National Park Service ocean stewardship strategy. *The George Wright Forum*, **21**, 22-41.

Davis, G.E., L.L. Loope, C.T. Roman, G. Smith, and J.T. Tilmant, 1994: *Assessment of Hurricane Andrew Impacts on Natural and Archeological Resources of Big Cypress National Preserve, Biscayne National Park, and Everglades National Park*. National Park Service, pp.1-158.

Day, J., L. Fernandes, A. Lewis, G. De'ath, S. Slegers, B. Barnett, B. Kerrigan, D. Breen, J. Innes, J. Oliver, T. Ward, and D. Lowe, 2002: The representative areas program for protecting biodiversity in the Great Barrier Reef World Heritage Area. In: *Proceedings of the Ninth International Coral Reef Symposium* 23, October 2000, pp. 687-696.

Dayton, P.K., S. Thrush, and F.C. Coleman, 2002: *Ecological Effects of Fishing in Marine Ecosystems of the United States*. Pew Oceans Commission, Arlington, VA, pp. 1-45.

Delaney, J.M., 2003: Community capacity building in the designation of the Tortugas Ecological Reserve. *Gulf and Caribbean Research*, **12(2)**, 163-169.

Denny, M.W., 1988: *Biology and the Mechanics of the Wave-Swept Environment*. Princeton University Press, Princeton, NJ, pp. 1-218.

Dickson, B., I. Yashayaev, J. Meincke, B. Turrell, S. Dye, and J. Holfort, 2002: Rapid freshening of the deep North Atlantic Ocean over the past four decades. *Nature*, **(416)**, 832-837.

Done, T. and R. Jones, 2006: Tropical coastal ecosystems and climate change prediction: global and local risks, In: *Coral Reefs and Climate Change: Science and Management*, [Phinney, J.T., O. Hoegh-Guldberg, J. Kleypas, W. Skirving, and A. Strong (eds.)]. American Geophysical Union, Washington, DC.

Done, T.J. and R.E. Reichelt, 1998: Integrated coastal zone and fisheries ecosystem management: generic goals and performance indices. *Ecological Applications*, **8(1)**, S110-S118.

Donner, S.D., T.R. Knutson, and M. Oppenheimer, 2007: Model-based assessment of the role of human-induced climate change in the 2005 Caribbean coral bleaching event.

Proceedings of the National Academy of Sciences of the United States of America, **104(13)**, 5483-5488.

Donner, S.D., W.J. Skirving, C.M. Little, M. Oppenheimer, and O. Hoegh-Guldberg, 2005: Global assessment of coral bleaching and required rates of adaptation under climate change. *Global Change Biology*, **11(12)**, 2251.

Douglas, A.E., 2003: Coral bleaching--how and why? *Marine Pollution Bulletin*, **46(4)**, 385-392.

Dulvy, N.K., Y. Sadovy, and J.D. Reynolds, 2003: Extinction vulnerability in marine populations. *Fish and Fisheries*, **4(1)**, 25-64.

Edge, S.E., M.B. Morgan, D.F. Gleason, and T.W. Snell, 2005: Development of a coral cDNA array to examine gene expression profiles in Montastraea faveolata exposed to environmental stress. *Marine Pollution Bulletin*, **51**, 507-523.

Eng, C.T., J.N. Paw, and F.Y. Guarin, 1989: Environmental impact of aquaculture and the effects of pollution on coastal aquaculture development in southeast Asia. *Marine Pollution Bulletin*, **20(7)**, 335-343.

Engel, J. and R. Kvitek, 1998: Effects of otter trawling on a benthic Community in Monterey Bay National Marine Sanctuary. *Conservation Biology*, **12(6)**, 1204-1214.

Engie, K. and T. Klinger, 2007: Modeling passive dispersal through a large estuarine system to evaluate marine reserve network connections. *Estuaries and coasts*, **30(2)**, 201-213.

Estes, J.A., 2005: Carnivory and trophic connectivity in kelp forests, In: *Large Carnivores and the Conservation of Biodiversity*, [Ray, J.C., K.H. Redford, R.S. Steneck, and J. Berger (eds.)]. Island Press, Washington, DC, pp. 61-81.

Fabricius, K.E. and G. De'ath, 2004: Identifying ecological change and its causes: a case study on coral reefs. *Ecological Applications*, **14(5)**, 1448-1465.

Feely, R.A., C.L. Sabine, K. Lee, W. Berelson, J. Kleypas, V.J. Fabry, and F.J. Millero, 2004: Impact of anthropogenic CO_2 on the $CaCO_3$ system in the oceans. *Science*, **305(5682)**, 362-366.

Fernandes, L., J. Day, A. Lewis, S. Slegers, B. Kerrigan, D. Breen, D. Cameron, B. Jago, J. Hall, D. Lowe, J. Tanzer, V. Chadwick, L. Thompson, K. Gorman, M. Simmons, B. Barnett, K. Sampson, G. De'ath, B. Mapstone, H. Marsh, H. Possingham, I. Ball, T. Ward, K. Dobbs, J. Aumend, D. Slater, and K. Stapleton, 2005: Establishing representative no-take areas in the Great Barrier Reef: large-scale implementation of theory on Marine Protected Areas. *Conservation Biology*, **19(6)**, 1733-1744.

Fields, P.A., J.B. Graham, R.H. Rosenblatt, and G.N. Somero, 1993: Effects of expected global climate change on marine faunas. *Trends in Ecology and Evolution*, **8**, 361-367.

Fitt, W.K., B.E. Brown, M.E. Warner, and R.P. Dunne, 2001: Coral bleaching: interpretation of thermal tolerance limits and thermal thresholds in tropical corals. *Coral Reefs*, **20(1)**, 51-65.

Fosså, J.H., P.B. Mortensen, and D.M. Furevik, 2002: The deep-water coral *Lophelia pertusa* in Norwegian waters: distribution and fishery impacts. *Hydrobiologia*, **471**, 1-12.

Frank, K.T., B. Petrie, J.S. Choi, and W.C. Leggett, 2005: Trophic cascades in a formerly cod-dominated ecosystem. *Science*, **308(5728)**, 1621-1623.

Friedlander, A., J.S. Nowlis, J.A. Sanchez, R. Appeldoorn, P. Usseglio, C. Mccormick, S. Bejarano, and A. Mitchell-Chui, 2003: Designing effective Marine Protected Areas in Seaflower biosphere reserve, Colombia, based on biological and sociological information. *Conservation Biology*, **17(6)**, 1769-1784.

Galbraith, H., R. Jones, R. Park, J. Clough, S. Herrod-Julius, B. Harrington, and G. Page, 2002: Global climate change and sea level rise: potential losses of intertidal habitat for shorebirds. *Waterbirds*, **25(2)**, 173-183.

Gardner, T.A., I.M. Cote, J.A. Gill, A. Grant, and A.R. Watkinson, 2003: Long-term region-wide declines in Caribbean corals. *Science*, **301(5635)**, 958-960.

Gardner, T.A., I.M. Cote, J.A. Gill, A. Grant, and A.R. Watkinson, 2005: Hurricanes and Caribbean coral reefs: impacts, recovery patterns, and role in long-term decline. *Ecology*, **86(1)**, 174-184.

Gattuso, J.P., M. Frankignoulle, and R. Wollast, 1998: Carbon and carbonate metabolism in coastal aquatic ecosystems. *Annual Review of Ecology and Systematics*, **29**, 405-434.

Gerber, L.R. and S.S. Heppell, 2004: The use of demographic sensitivity analysis in marine species conservation planning. *Biological Conservation*, **120(1)**, 121-128.

Gleason, D.F., P.J. Edmunds, and R.D. Gates, 2006: Ultraviolet radiation effects on the behavior and recruitment of larvae from the reef coral *Porites astreoides*. *Marine Biology*, **148(3)**, 503-512.

Gleason, D.F. and G.M. Wellington, 1993: Ultraviolet radiation and coral bleaching. *Nature*, **365(836)**, 838.

Glenn, R.P. and T.L. Pugh, 2006: Epizootic shell disease in American lobster (*Homarus americanus*) in Massachusetts coastal waters: interactions of temperature, maturity, and intermolt duration. *Journal of Crustacean Biology*, **26(4)**, 639-645.

Glynn, P.W., 1991: Coral reef bleaching in the 1980s and possible connections with global warming. *Trends in Ecology and Evolution*, **6(6)**, 175-179.

Glynn, P.W., 1993: Coral reef bleaching: ecological perspectives. *Coral Reefs*, **12(1)**, 1-17.

Goldberg, J. and C. Wilkinson, 2004: Global threats to coral reefs: coral bleaching, global climate change, disease, predator plagues, and invasive species, In: *Status of Coral Reefs of the World: 2004*, [Wilkinson, C. (ed.)]. Australian Institute of Marine Science, Townsville, Queensland, pp. 67-92.

Graham, N.A.J., R.D. Evans, and G.R. Russ, 2003: The effects of marine reserve protection on the trophic relationships of reef fishes on the Great Barrier Reef. *Environmental Conservation*, **30(2)**, 200-208.

Granek, E.F. and M.A. Brown, 2005: Co-management approach to marine conservation in Moheli, Comoros Islands. *Conservation Biology*, **19(6)**, 1724-1732.

Greene, C.H. and A.J. Pershing, 2007: Climate drives sea change. *Science*, **315(5815)**, 1084-1085.

Grigg, R.W., S.J. Dollar, A. Huppert, B.D. Causey, K. Andrews, W.L. Kruczynski, W.F. Precht, S.L. Miller, R.B. Aronson, and J.F. Bruno, 2005: Reassessing U.S. coral reefs. *Science*, **308(5729)**, 1740-1742.

Grober-Dunsmore, R., L. Wooninck, and C. Wahle, in press: Vertical zoning in marine protected areas: ecological considerations for balancing pelagic fishing with conservation of benthic communities. *Fisheries*.

Gucinski, H., R.T. Lackey, and B.C. Spence, 1990: Global climate change: policy implications for fisheries. *Fisheries*, **15(6)**, 33-38.

Gunderson, L.H., 2000: Resilience in theory and practice. *Annual Review of Ecology and Systematics*, **31**, 425-439.

Hallock, P., 2005: Global change and modern coral reefs: new opportunities to understand shallow-water carbonate depositional processes. *Sedimentary Geology*, **175(1-4)**, 19-33.

Halpern, B.S., 2003: The impact of marine reserves: do reserves work and does reserve size matter? *Ecological Applications*, **13(1)**, S117-S137.

Halpern, B.S. and K. Cottenie, 2007: Little evidence for climate effects on local-scale structure and dynamics of California kelp forest communities. *Global Change Biology*, **13(1)**, 236-251.

Halpern, B.S., H.M. Regan, H.P. Possingham, and M.A. McCarthy, 2006: Accounting for uncertainty in marine reserve design. *Ecological Letters*, **9(1)**, 2-11.

Halpern, B.S. and R.R. Warner, 2003: Matching marine reserve design to reserve objectives. *Proceedings of the Royal Society of London, Series B: Biological Sciences*, **270(1527)**, 1871-1878.

Hansen, L., J.L. Biringer, and J.R. Hoffman, 2003: *Buying Time: a User's Manual for Building Resistance and Resilience to Climate Change in Natural Systems*. [Hansen, L.J., J.L. Biringer, and J.R. Hoffman (eds.)]. World Wildlife Foundation, Washington, DC, pp. 1-244.

Harley, C.D.G., R. Hughes, K.M. Hultgren, B.G. Miner, C.J.B. Sorte, C.S. Thornber, L.F. Rodriguez, L. Tomanek, and S.L. Williams, 2006: The impacts of climate change in coastal marine systems. *Ecology Letters*, **9(2)**, 228-241.

Harvell, C.D., K. Kim, J.M. Burkholder, R.R. Colwell, P.R. Epstein, D.J. Grimes, E.E. Hofmann, E.K. Lipp, A. Osterhaus, and R.M. Overstreet, 1999: Emerging marine diseases--climate links and anthropogenic factors. *Science*, **285**, 1505-1510.

Harvell, C.D., C.E. Mitchell, J.R. Ward, S. Altizer, A.P. Dobson, R.S. Ostfeld, and M.D. Samuel, 2002: Climate warming and disease risks for terrestrial and marine biota. *Science*, **296(5576)**, 2158-2162.

Harvell, D., K. Kim, C. Quirolo, J. Weir, and G. Smith, 2001: Coral bleaching and disease: contributors to 1998 mass mortality in Briareum asbestinum (*Octocorallia, Gorgonacea*). *Hydrobiologia*, **460(1)**, 97-104.

Hastings, A. and L.W. Botsford, 2003: Comparing designs of marine reserves for fisheries and for biodiversity. *Ecological Applications*, **13(1)**, S65-S70.

Hatcher, B.G. and A.W.D. Larkum, 1983: An experimental analysis of factors controlling the standing crop of the epilithic algal community on a coral reef. *Journal of Experimental Marine Biology and Ecology*, **69(1)**, 61-84.

Helmuth, B., 2002: How do we measure the environment? Linking intertidal thermal physiology and ecology through biophysics. *Integrative and Comparative Biology*, **42(4)**, 837-845.

Hiddink, J.G., S. Jennings, and M.J. Kaiser, 2006: Indicators of the ecological impact of bottom-trawl disturbance on seabed communities. *Ecosystems*, **9(7)**, 1190-1199.

Hilborn, R., F. Micheli, and G.A. De Leo, 2006: Integrating marine protected areas with catch regulation. *Canadian Journal of Fisheries and Aquatic Sciences*, **63(3)**, 642-649.

Hixon, M.A. and B.N. Tissot, 2007: Comparison of trawled vs untrawled mud seafloor assemblages of fishes and macroinvertebrates at Coquille Bank, Oregon. *Journal of Experimental Marine Biology and Ecology*, **344**, 23-24.

Hockey, P.A.R. and G.M. Branch, 1994: Conserving marine biodiversity on the African coast: Implications of a terrestrial perspective. *Aquatic Conservation: Marine and Freshwater Ecosystems*, **4(4)**, 345-362.

Hoegh-Guldberg, O., 1999: Climate change, coral bleaching and the future of the world's coral reefs. *Marine & Freshwater Research*, **50(8)**, 839-866.

Hoegh-Guldberg, O., K. Anthony, R. Berkelmans, S. Dove, K. Fabricius, J. Lough, P. A. Marshall, M. J. H. van Oppen, A. Negri, and B. Willis, 2007: Vulnerability of reef-building corals on the Great Barrier Reef to Climate Change, In: *Climate Change and the Great Barrier Reef*, [Johnson, J.E. and P.A. Marshall (eds.)]. Great Barrier Reef Marine Park Authority & Australian Greenhouse Office.

Hoffman, J., 2003: Designing reserves to sustain temperate marine ecosystems in the face of global climate change, In: *Buying Time: a User's Manual for Building Resistance and Resilience to Climate Change in Natural Systems*, [Hansen, L.J., J.L. Biringer, and J.R. Hoffman (eds.)]. WWF Climate Change Program, Washington, DC, pp. 123-155.

Holling, C.S., 1995: What barriers? What bridges?, In: *Barriers and Bridges to the Renewal of Ecosystems and Institutions*, [Gunderson, L.H., C.S. Holling, and S.S. Light (eds.)]. Columbia University Press, New York, NY, pp. 3-34.

Hughes, T.P., A.H. Baird, D.R. Bellwood, M. Card, S.R. Connolly, C. Folke, R. Grosberg, O. Hoegh-Guldberg, J.B.C. Jackson, J. Kleypas, J.M. Lough, P. Marshall, M. Nystrom, S.R. Palumbi, J.M. Pandolfi, B. Rosen, and J. Roughgarden, 2003: Climate change, human impacts, and the resilience of coral reefs. *Science*, **301(5635)**, 929-933.

Hughes, T.P., D.R. Bellwood, C. Folke, R.S. Steneck, and J. Wilson, 2005: New paradigms for supporting the resilience of marine ecosystems. *Trends in Ecology and Evolution*, **20(7)**, 380-386.

Husebø, Å., L. Nøttestad, J.H. Fosså, D.M. Furevik, and S.B. Jørgensen, 2002: Distribution and abundance of fish in deep-sea coral habitats. *Hydrobiologia*, **471(1)**, 91-99.

Hyrenbach, K.D., K.A. Forney, and P.K. Dayton, 2000: Marine protected areas and ocean basin management. *Aquatic Conservation: Marine and Freshwater Ecosystems*, **10(6)**, 437-458.

IPCC, 2001: *Climate Change 2001: Impacts, Adaptation, and Vulnerability. Contribution of Working Group II to the Third Assessment Report of the Intergovernmental Panel on Climate Change.* [McCarthy, J.J., O.F. Canziani, N.A. Leary, D.J. Dokken, and K.S. White (eds.)]. Cambridge University Press, Cambridge, UK.

IPCC, 2007a: *Climate Change 2007: the Physical Science Basis. Contribution of Working Group I to the Fourth Assessment Report of the Intergovernmental Panel on Climate Change.* [Solomon, S., D. Quin, M. Manning, Z. Chen, M. Marquis, K.B. Averyt, M. Tignor, and H.L. Miller (eds.)]. Cambridge University Press, Cambridge, United Kingdom and New York, NY, USA, pp. 1-996.

IPCC, 2007b: Summary for policymakers, In: *Climate Change 2007: Impacts, Adaptation and Vulnerability. Contribution of Working Group II to the Fourth Assessment Report of the Intergovernmental Panel on Climate Change*, [Parry, M.L., O.F. Canziani, J.P. Palutikof, P.J. van der Linden, and C.E. Hanson (eds.)]. Cambridge University Press, Cambridge, UK, pp. 7-22.

IPCC, 2007c: Summary for policymakers, In: *Climate Change 2007: the Physical Science Basis. Contribution of Working Group I to the Fourth Assessment Report of the Intergovernmental Panel on Climate Change*, [Solomon, S., D. Qin, M. Manning, Z. Chen, M. Marquis, K.B. Averyt, M. Tignor, and H.L. Miller (eds.)]. Cambridge University Press, Cambridge, United Kingdom and New York, NY, USA, pp. 1-21.

Iwama, G.K., 1991: Interactions between aquaculture and the environment. *Critical Reviews in Environmental Control*, **21(2)**, 177-216.

Jackson, J.B.C., M.X. Kirby, W.H. Berger, K.A. Bjorndal, L.W. Botsford, B.J. Bourque, R.H. Bradbury, R. Cooke, J. Erlandson, J.A. Estes, T.P. Hughes, S. Kidwell, C.B. Lange, H.S. Lenihan, J.M. Pandolfi, C.H. Peterson, R.S. Steneck, M.J. Tegner, and R.R. Warner, 2001: Historical overfishing and the recent collapse of coastal ecosystems. *Science*, **293**, 629-638.

Jameson, S.C., M.H. Tupper, and J.M. Ridley, 2002: The three screen doors: can marine" protected" areas be effective? *Marine Pollution Bulletin*, **44(11)**, 1177-1183.

Johnson, J. and P. Marshall, 2007: *Climate Change and the Great Barrier Reef: a Vulnerability Assessment.* Great Barrier Reef Marine Park Authority.

Jones, G.P., M.J. Milicich, M.J. Emslie, and C. Lunow, 1999: Self-recruitment in a coral reef fish population. *Nature*, **402(6763)**, 802-804.

Jones, G.P., S. Planes, and S.R. Thorrold, 2005: Coral reef fish larvae settle close to home. *Current Biology*, **15(14)**, 1314-1318.

Jonsson, L.G., P.G. Nilsson, F. Floruta, and T. Lundaelv, 2004: Distributional patterns of macro- and megafauna associated with a reef of the cold-water coral *Lophelia pertusa* on the Swedish west coast. *Marine Ecology Progress Series*, **284**, 163-171.

Kaiser, M.J., 2005: Are marine protected areas a red herring or fisheries panacea? *Canadian Journal of Fisheries and Aquatic Sciences*, **62(5)**, 1194-1199.

Kaufman, L., J. B. C. Jackson, E. Sala, P. Chisolm, E. D. Gomez, C. Peterson, R. V. Salm, and G. Llewellyn, 2004: Restoring and maintaining marine ecosystem function, In: *Defying Ocean's End*, [Glover, L.K. and S.A. Earle (eds.)]. Island Press, Washington, DC, pp. 165-181.

Kelleher, G., C. Bleakley, and S. Wells, 1995: A global system of Marine Protected Areas. *Vols I to IV*.

Keller, B.D. and B.D. Causey, 2005: Linkages between the Florida Keys National Marine Sanctuary and the South Florida Ecosystem Restoration Initiative. *Ocean & Coastal Management*, **48(11-12)**, 869-900.

Kennedy, V.S., R.R. Twilley, J.A. Kleypas, J.H. Cowan, Jr., and S.R. Hare, 2002: *Coastal and Marine Ecosystems & Global Climate Change: Potential Effects on U.S. Resources*. Prepared for the Pew Center on Global Climate Change, Pew Center on Global Climate Change, Arlington, VA.

Khamer, M., D. Bouya, and C. Ronneau, 2000: Metallic and organic pollutants associated with urban wastewater in the waters and sediments of a Moroccan river. *Water Quality Research Journal of Canada*, **35**, 147-161.

Kim, K. and C.D. Harvell, 2004: The rise and fall of a six-year coral-fungal epizootic. *American Naturalist*, **164**, S52-S63.

Kimball, M.E., J.M. Miller, P.E. Whitfield, and J.A. Hare, 2004: Thermal tolerance and potential distribution of invasive lionfish (*Pterois volitans/miles* complex) on the east coast of the United States. *Marine Ecology Progress Series*, **283**, 269-278.

Kinlan, B.P. and S.D. Gaines, 2003: Propagule dispersal in marine and terrestrial environments: a community perspective. *Ecology*, **84(8)**, 2007-2020.

Kleypas, J.A., R.W. Buddemeier, D. Archer, J.P. Gattuso, C. Langdon, and B.N. Opdyke, 1999: Geochemical consequences of increased atmospheric carbon dioxide on coral reefs. *Science*, **284(5411)**, 118-120.

Kleypas, J.A. and C. Langdon, 2006: Coral reefs and changing seawater chemistry, In: *Coral Reefs and Climate Change: Science and Management*, [Phinney, J.T., O. Hoegh-Guldberg, J. Kleypas, W.J. Skirving, and A. Strong (eds.)]. American Geophysical Union, Washington, DC, pp. 73-110.

Koslow, J.A., K. Gowlett-Holmes, J.K. Lowry, T. O'Hara, G.C.B. Poore, and A. Williams, 2001: Seamount benthic macrofauna off southern Tasmania: community structure and impacts of trawling. *Marine Ecology Progress Series*, **213**, 111-125.

Krieger, K.J. and B.L. Wing, 2002: Megafauna associations with deepwater corals (*Primnoa spp.*) in the Gulf of Alaska. *Hydrobiologia*, **471(1-3)**, 83-90.

Law, R. and K. Stokes, 2005: Evolutionary impacts of fishing on target populations, In: *Marine Conservation Biology: the Science of Maintaining the Sea's Biodiversity*, [Norse, E. and L.B. Crowder (eds.)]. Island Press, Washington, DC, pp. 232-246.

Lebel, L., J.M. Anderies, B. Campbell, C. Folke, S. Hatfield-Dodds, T.P. Hughes, and J. Wilson, 2006: Governance and the capacity to manage resilience in regional social-ecological systems. *Ecology and Society*, **11(1)**, 19.

Lecchini, D., S. Planes, and R. Galzin, 2005: Experimental assessment of sensory modalities of coral-reef fish larvae in the recognition of their settlement habitat. *Behavioral Ecology and Sociobiology*, **58(1)**, 18-26.

Lecchini, D., J. Shima, B. Banaigs, and R. Galzin, 2005: Larval sensory abilities and mechanisms of habitat selection of a coral reef fish during settlement. *Oecologia*, **143(2)**, 326-334.

Ledlie, M.H., N.A.J. Graham, J.C. Bythell, S.K. Wilson, S. Jennings, N.V.C. Polunin, and J.W. Harden, 2007: Phase shifts and the role of herbivory in the resilience of coral reefs. *Coral Reefs*, **26**, 641-653.

Lee, S.Y., 1995: Mangrove outwelling: a review. *Hydrobiologia*, **295(1-3)**, 203-212.

Lee, T.N., C. Rooth, E. Williams, M. McGowan, and A.F. Szmant, 1992: Influence of Florida current, gyres and wind-driven circulation on transport of larvae and recruitment in the Florida Keys coral reefs. *Continental Shelf Research*, **12(7/8)**, 971-1002.

Leis, J.M., B.M. Carson-Ewart, and J. Webley, 2002: Settlement behaviour of coral-reef fish larvae at subsurface artificial-reef moorings. *Marine & Freshwater Research*, **53(2)**, 319-327.

Leis, J.M. and M. I. McCormick, 2002: The biology, behavior, and ecology of the pelagic, larval stage of coral reef fishes, Academic Press, San Diego, CA, pp. 171-199.

Lesser, M.P., J.C. Bythell, R.D. Gates, R.W. Johnstone, and O. Hoegh-Guldberg, 2007: Are infectious diseases really killing corals? Alternative interpretations of the experimental and ecological data. *Journal of Experimental Marine Biology and Ecology*, **346(1-2)**, 36-44.

Levin, L.A., 2006: Recent progress in understanding larval dispersal: new directions and digressions. *Integrative and Comparative Biology*, **46(3)**, 282-297.

Lirman, D. and P. Fong, 2007: Is proximity to land-based sources of coral stressors an appropriate measure of risk to coral reefs? An example from the Florida Reef tract. *Marine Pollution Bulletin*, **54**, 779-791.

Lotze, H.K., H.S. Lenihan, B.J. Bourque, R.H. Bradbury, R.G. Cooke, M.C. Kay, S.M. Kidwell, M.X. Kirby, C.H. Peterson, and J.B.C. Jackson, 2006: Depletion, degradation, and recovery potential of estuaries and coastal seas. *Science*, **312(5781)**, 1806-1809.

Lovelace, J.K. and B.F. MacPherson, 1998: *Effects of Hurricane Andrew on Wetlands in Southern Florida and Louisiana*. National Water Summary on Wetland Resources USGS Water Supply Paper #2425.

Lugo-Fernandez, A., K.J.P. Deslarzes, J.M. Price, G.S. Boland, and M.V. Morin, 2001: Inferring probable dispersal of Flower Garden Banks Coral Larvae (Gulf of Mexico) using observed and simulated drifter trajectories. *Continental Shelf Research*, **21(1)**, 47-67.

Mann, K.H. and J.R.N. Lazier, 2006: *Dynamics of Marine Ecosystems*. Blackwell Publishing, Malden, MA, pp. 1-496.

Mantua, N.J., S.R. Hare, Y. Zhang, J.M. Wallace, and R.C. Francis, 1997: A pacific interdecadal climate oscillation with impacts on salmon production. *Bulletin of the American Meteorological Society*, **78(6)**, 1069-1079.

Marchetti, M.P., P.B. Moyle, and R. Levine, 2004: Invasive species profiling? Exploring the characteristics of non-native fishes across invasion stages in California. *Freshwater Biology*, **49(5)**, 646-661.

Marschke, M.J. and F. Berkes, 2006: Exploring strategies that build livelihood resilience: a case from Cambodia. *Ecology and Society*, **11(1)**, 42.

Marshall, N., in press: Can policy perception influence social resilience to policy change? *Fisheries Research*.

Marshall, P. and H. Schuttenberg, 2006: A Reef Manager's Guide to Coral Bleaching. Great Barrier Reef Marine Park Authority, http://www.coris.noaa.gov/activities/reef_managers_guide/, pp.1-178.

Marshall, P. and H. Schuttenberg, 2006: Adapting coral reef management in the face of climate change, In: *Coral Reefs and Climate Change: Science and Management*, [Phinney, J.T., O. Hoegh-Guldberg, J. Kleypas, W.J. Skirving, and A. Strong (eds.)]. American Geophysical Union, Washington, DC, pp. 223-241.

McCarty, J.P., 2001: Ecological consequences of recent climate change. *Conservation Biology*, **15(2)**, 320-331.

McClanahan, T.R., M.J. Marnane, J.E. Cinner, and W.E. Kiene, 2006: A comparison of marine protected areas and alternative approaches to coral-reef management. *Current Biology*, **16(14)**, 1408-1413.

McClanahan, T.R., N. V. C. Polunin, and T. J. Done, 2002: Resilience of coral reefs, In: *Resilience and Behaviour of Large-Scale Ecosystems*, [Gunderson, L.H. and L. Pritchard, Jr. (eds.)]. Island Press, Washington, DC, pp. 111-163.

McCoy, E.D., H.R. Mushinsky, D. Johnson, and W.E. Meshaka Jr, 1996: Mangrove damage caused by Hurricane Andrew on the southwestern coast of Florida. *Bulletin of Marine Science*, **59(1)**, 1-8.

McGowan, J.A., D.R. Cayan, L.M. Dorman, and A. Butler, 1998: Climate-ocean variability and ecosystem response in the Northeast Pacific. *Science*, **281(5374)**, 210-217.

McGregor, H.V., M. Dima, H.W. Fischer, and S. Mulitza, 2007: Rapid 20th-century increase in coastal upwelling off northwest Africa. *Science*, **315(5812)**, 637.

McLeod, E. and R.V. Salm, 2006: *Managing Mangroves for Resilience to Climate Change*. The World Conservation Union, Gland, Switzerland, pp.1-66.

McNaughton, S.J., 1977: Diversity and stability of ecological communities: a comment on the role of empiricism in ecology. *The American Naturalist*, **111(979)**, 515-525.

McNeil, B.I., R.J. Matear, and D.J. Barnes, 2004: Coral reef calcification and climate change: the effect of ocean warming. *Geophysical Research Letters*, **31(22)**, L22309.

McPhaden, M.J. and D. Zhang, 2002: Slowdown of the meridional overturning circulation in the upper Pacific Ocean. *Nature*, **415(6872)**, 603-608.

Michaelidis, B., B.C. Ouzounis, A. Paleras, and H.O. Portner, 2005: Effects of long-term moderate hypercapnia on acid-base balance and growth rate in marine mussels *Mytilus galloprovinciallis*. *Marine Ecology Progress Series*, **293**, 109-118.

Micheli, F., B.S. Halpern, L.W. Botsford, and R.R. Warner, 2004: Trajectories and correlates of community change in no-take marine reserves. *Ecological Applications*, **14(6)**, 1709-1723.

Millennium Ecosystem Assessment, 2005: *Ecosystems and Human Well-Being: Current State and Trends. Findings of the Condition and Trends Working Group*. Island Press, Washington, DC.

Mills, L.S., M.E. Soulé, and D.F. Doak, 1993: The keystone-species concept in ecology and conservation. *BioScience*, **43(4)**, 219-224.

Moore, M.V., M.L. Pace, J.R. Mather, P.S. Murdoch, R.W. Howarth, C.L. Folt, C.Y. Chen, H.F. Hemond, P.A. Flebbe, and C.T. Driscoll, 1997: Potential effects of climate change on freshwater ecosystems of the New England/Mid-Atlantic Region. *Hydrological Processes*, **11**, 925-947.

Mora, C., S. Andréfouët, M.J. Costello, C. Kranenburg, A. Rollo, J. Veron, K.J. Gaston, and R.A. Myers, 2006: Coral Reefs and the Global Network of Marine Protected Areas. *Science*, **312**, 1750-1751.

Mosquera, I., I.M. Cote, S. Jennings, and J.D. Reynolds, 2000: Conservation benefits of marine reserves for fish populations. *Animal Conservation*, **3(4)**, 321-332.

Moyle, P.B., 1986: Fish introductions into North America: patterns and ecological impact, In: *Ecology of Biological Invasions of North America and Hawaii*, [Mooney, H.A. and J.A. Drake (eds.)]. Springer, NY, pp. 27-43.

Mullineaux, L.S. and C.A. Butman, 1991: Initial contact, exploration and attachment of barnacle (*Balanus amphitrite*) cyprids settling in flow. *Marine Biology*, **110(1)**, 93-103.

Mumby, P.J., C.P. Dahlgren, A.R. Harborne, C.V. Kappel, F. Micheli, D.R. Brumbaugh, K.E. Holmes, J.M. Mendes, K. Broad, and J.N. Sanchirico, 2006: Fishing, trophic cascades, and the process of grazing on coral reefs. *Science*, **311(5757)**, 98-101.

Mumby, P.J., A.J. Edwards, J.E. Arias-Gonzalez, K.C. Lindeman, P.G. Blackwell, A. Gall, M.I. Gorczynska, A.R. Harborne, C.L. Pescod, H. Renken, C.C.C. Wabnitz, and G. Llewellyn, 2004: Mangroves enhance the biomass of coral reef fish communities in the Caribbean. *Nature*, **427(6974)**, 533-536.

Mumby, P.J., A.R. Harborne, J. Williams, C.V. Kappel, D.R. Brumbaugh, F. Micheli, K.E. Holmes, C.P. Dahlgren, C.B. Paris, and P.G. Blackwell, 2007: Trophic cascade facilitates coral recruitment in a marine reserve. *Proceedings of the National Academy of Sciences of the United States of America*, **104(20)**, 8362-8367.

Munn, C.B., 2006: Viruses as pathogens of marine organismsùfrom bacteria to whales. *Journal of the Marine Biological Association of the UK*, **86(3)**, 453-467.

Murawski, S.A., 1993. Climate change and marine fish distributions: forecasting from historical analogy. *Transactions of the American Fisheries Society*, **122(5)**, 647-658.

Mydlarz, L.D., L.E. Jones, and C.D. Harvell, 2006: Innate immunity, environmental drivers, and disease ecology of marine and freshwater invertebrates. *Annual Review of Ecology, Evolution and Systematics*, **37**, 251-288.

Nagelkerken, I., 2007: Are non-estuarine mangroves connected to coral reefs through fish migration? *Bulletin of Marine Science*, **80(3)**, 595-607.

Nagelkerken, I., M. Dorenbosch, W. Verberk, E.C. de la Moriniere, and G. van der Velde, 2000: Importance of shallow-water biotopes of a Caribbean bay for juvenile coral reef fishes: patterns in biotope association, community structure and spatial distribution. *Marine Ecology Progress Series*, **202**, 175-192.

National Marine Fisheries Service, 2003: *Annual Report to Congress on the Status of U.S. Fisheries - 2002*. U.S. Department of Commerce, National Oceanic and Atmospheric Administration, National Marine Fisheries Service, Silver Spring, MD, pp.1-156.

National Marine Fisheries Service, 2005: *2005 Report on the Status of U.S. Marine Fish Stocks*. National Marine Fisheries Service, Silver Spring, MD, pp.1-20.

National Research Council, 1999: *Sustaining Marine Fisheries*. National Academy Press, Washington, DC, pp.1-164.

National Research Council, 2001: *Marine Protected Areas: Tools for Sustaining Ocean Ecosystems*. National Academy Press, Washington, DC, pp.1-272.

National Safety Council, 1998: *Coastal Challenges: a Guide to Coastal and Marine Issues*. Environmental Health Center, Washington, DC, pp.1-365.

Naylor, R.L., R.J. Goldburg, J. Primavera, N. Kautsky, M.C.M. Beveridge, J. Clay, C. Folke, J. Lubchenco, H. Mooney, and M. Troell, 2000: Effects of aquaculture on world fish supplies. *Nature*, **405(6790)**, 1017-1024.

Nicholls, R.J. and S.P. Leatherman, 1996: Adapting to sea-level rise: relative sea-level trends to 2100 for the United States. *Coastal Management*, **24(4)**, 301-324.

Norse, E.A., 1993: Global marine biological diversity: a strategy for building conservation into decision making, [Norse, E.A. (ed.)]. Island Press, Washington, DC, pp. 1-383.

Nyström, M., C. Folke, and F. Moberg, 2000: Coral reef disturbance and resilience in a human-dominated environment. *Trends in Ecology and Evolution*, **15(10)**, 413-420.

O'Connor, M.I., J.F. Bruno, S.D. Gaines, B.S. Halpern, S.E. Lester, B.P. Kinlan, and J.M. Weiss, 2007: Temperature control of larval dispersal and the implications for marine ecology, evolution, and conservation. *Proceedings of the National Academy of Sciences of the United States of America*, **104**, 1266-1271.

Obura, D., B. D. Causey, and J. Church, 2006: Management response to a bleaching event, In: *Coral Reefs and Climate Change: Science and Management*, [Phinney, J.T., O. Hoegh-Guldberg, J. Kleypas, W.J. Skirving, and A. Strong (eds.)]. American Geophysical Union, Washington, DC, pp. 181-206.

Ogden, J.C. and E.H. Gladfelter, 1983: *Coral Reefs, Seagrass Beds and Mangroves: Their Interaction in the Coastal Zones of the Caribbean*. UNESCO Reports in Marine Science 23, pp.1-133.

Ogden, J.C. and R. Wicklund, 1988: *Mass Bleaching of Coral Reefs in the Caribbean: a Research Strategy*. Research Report 88-2, National Oceanic and Atmospheric Administration, Oceanic and Atmospheric Research, Office of Undersea Research, pp.1-51.

Orr, J.C., V.J. Fabry, O. Aumont, L. Bopp, S.C. Doney, R.A. Feely, A. Gnanadesikan, N. Gruber, A. Ishida, F. Joos, R.M. Key, K. Lindsay, E. Maier-Reimer, R. Matear, P. Monfray, A. Mouchet, R.G. Najjar, G.K. Plattner, K.B. Rodgers, C.L. Sabine, J.L. Sarmiento, R. Schlitzer, R.D. Slater, I.J. Totterdell, M.F. Weirig, Y. Yamanaka, and A. Yool, 2005: Anthropogenic ocean acidification over the twenty-first century and its impact on calcifying organisms. *Nature*, **437**, 681-686.

Orth, R.J., T.J.B. Carruthers, W.C. Dennison, C.M. Duarte, J.W. Fourqurean, K.L.Jr. Heck, A.R. Hughes, G.A. Kendrick, W.J. Kenworthy, S. Olyarnik, F.T. Short, M. Waycott, and S.L. Williams, 2006: A global crisis for seagrass ecosystems. *BioScience*, **56(12)**, 987-996.

Paine, R.T., 1993: A salty and salutary perspective on global change, In: *Biotic Interactions and Global Change*, [Kareiva, P.M., J.G. Kingsolver, and R.B. Huey (eds.)]. Sinauer Associates, Inc., Sunderland, Massachusetts, pp. 347-355.

Paine, R.T., M.J. Tegner, and E.A. Johnson, 1998: Compounded perturbations yield ecological surprises. *Ecosystems*, **1(6)**, 535-545.

Palumbi, S.R., 2001: The ecology of marine protected areas, In: *Marine Community Ecology*, [Bertness, M.D., S.D. Gaines, and M.E. Hay (eds.)]. Sinauer Associates, Inc., Sunderland, MA, pp. 509-530.

Palumbi, S.R., 2002: *Marine Reserves: a Tool for Ecosystem Management and Conservation*. Pew Oceans Commission, Arlington, VA, pp.1-45.

Palumbi, S.R., 2003: Population genetics, demographic connectivity, and the design of marine reserves. *Ecological Applications*, **13(1)**, S146-S158.

Palumbi, S.R., 2004: Marine reserves and ocean neighborhoods: the spatial scale of marine populations and their management. *Annual Review of Environment and Resources*, **29**, 31-68.

Palumbi, S.R., G. Grabowsky, T. Duda, L. Geyer, and N. Tachino, 1997: Speciation and population genetic structure in tropical Pacific sea urchins. *Evolution*, **51(5)**, 1506-1517.

Pandolfi, J.M., R.H. Bradbury, E. Sala, T.P. Hughes, K.A. Bjorndal, R.G. Cooke, D. McArdle, L. McClenachan, M.J.H. Newman, G. Paredes, R.R. Warner, and J.B.C. Jackson, 2003: Global trajectories of the long-term decline of coral reef ecosystems. *Science*, **301(5635)**, 955-958.

Pandolfi, J.M., J.B.C. Jackson, N. Baron, R.H. Bradbury, H.M. Guzman, T.P. Hughes, C.V. Kappel, F. Micheli, J.C. Ogden, H.P. Possingham, and E. Sala, 2005: Are U. S. coral reefs on the slippery slope to slime? *Science*, **307(5716)**, 1725-1726.

Pane, E.F. and J.P. Barry, 2007: Extracellular acid-base regulation during short-term hypercapnia is effective in a shallow-water crab, but ineffective in a deep-sea crab. *Marine Ecology Progress Series*, **334**, 1-9.

Paris, C.B. and R.K. Cowen, 2004: Direct evidence of a biophysical retention mechanism for coral reef fish larvae. *Limnology and Oceanography*, **49(6)**, 1964-1979.

Parks, J. and N. Salafsky, 2001: *Fish for the Future? A Collaborative Test of Locally-Managed Marine Areas As a Biodiversity Conservation and Fisheries Management Tool in the Indo-Pacific Region: Report on the Initiation of a Learning Portfolio*. World Resources Institute, Washington, DC, pp.1-82.

Pauly, D., J. Alder, E. Bennett, V. Christensen, P. Tyedmers, and R. Watson, 2003: The future for fisheries. *Science*, **302(5649)**, 1359-1361.

Pauly, D., V. Christensen, J. Dalsgaard, R. Froese, and F. Torres, Jr., 1998: Fishing down marine food webs. *Science*, **279(5352)**, 860-863.

Pauly, D., V. Christensen, S. Guqnette, T.J. Pitcher, U.R. Sumaila, C.J. Walters, R. Watson, and D. Zeller, 2002: Towards sustainability in world fisheries. *Nature*, **418**, 689-695.

Pechenik, J.A., 1999: On the advantages and disadvantages of larval stages in benthic marine invertebrate life cycles. *Marine Ecology Progress Series*, **177**, 269-297.

Perry, A.L., P.J. Low, J.R. Ellis, and J.D. Reynolds, 2005: Climate change and distribution shifts in marine fishes. *Science*, **308(5730)**, 1912-1915.

Peterson, B.J., J. McClelland, R. Curry, R.M. Holmes, J.E. Walsh, and K. Aagaard, 2006: Trajectory shifts in the Arctic and subarctic freshwater cycle. *Science*, **(313)**, 1061-1066.

Pew Oceans Commission, 2003: *America's Living Oceans: Charting a Course for Sea Change - a Report to the Nation*. Pew Oceans Commission, Arlington, VA, pp.1-144.

Philip, S. and G.J. Van Oldenborgh, 2006: Shifts in ENSO coupling processes under global warming. *Geophysical Research Letters*, **33(11)**, L11704.

Phinney, J.T., O. Hoegh-Guldberg, J. Kleypas, W. Skirving, and A. Strong, 2006: *Coral Reefs and Climate Change: Science and Management*. American Geophysical Union, Washington, DC, pp. 1-244.

Porter, J.W., P. Dustan, W.C. Jaap, K.L. Patterson, V. Kosmynin, O.W. Meier, M.E. Patterson, and M. Parsons, 2001: Patterns of spread of coral disease in the Florida Keys. *Hydrobiologia*, **460(1-3)**, 1-24.

Porter, J.W., V. Kosmynin, K. L. Patterson, K. G. Porter, W. C. Jaap, J. L. Wheaton, K. Hackett, M. Lybolt, C. P. Tsokos, G. Yanev, G. M. Marcinek, J. Dotten, D. Eaken, M. Patterson, O. W. Meier, M. Brill, and P. Dustan, 2002: Detection of coral reef change by the Florida Keys Coral Reef Monitoring Project, In: *The Everglades, Florida Bay, and Coral Reefs of the Florida Keys: an Ecosystem Sourcebook*, [Porter, J.W. and K.G. Porter (eds.)]. CRC Press, Boca Raton, FL, pp. 749-769.

Precht, W.F. and R.B. Aronson, 2004: Climate flickers and range shifts of reef corals. *Frontiers in Ecology and the Environment*, **2(6)**, 307-314.

Rabalais, N.N., R.E. Turner, and W.J. Wiseman Jr, 2002: Gulf of Mexico hypoxia, aka "the dead zone". *Annual Review of Ecology and Systematics*, **33**, 235-263.

Raimondi, P.T. and A.N.C. Morse, 2000: The consequences of complex larval behavior in a coral. *Ecology*, **81(11)**, 3193-3211.

Reaser, J.K., R. Pomerance, and P.O. Thomas, 2000: Coral bleaching and global climate change: scientific findings and policy recommendations. *Conservation Biology*, **14(5)**, 1500-1511.

Reed, J.K., 2002: Deep-water *Oculina* coral reefs of Florida: biology, impacts, and management. *Hydrobiologia*, **471(1)**, 43-55.

Rignot, E. and P. Kanagaratnam, 2006: Changes in the velocity structure of the Greenland icesheet. *Science*, **311(986)**, 990.

Roberts, C.M., 1996: Settlement and beyond: population regulation and community structure of reef fishes, [Polunin, N.V.C. (ed.)]. Chapman and Hall Ltd, London, England, UK and New York, New York, USA, pp. 85-112.

Roberts, C.M., 1997a: Connectivity and management of Caribbean coral reefs. *Science*, **278(5342)**, 1454-1457.

Roberts, C.M., 1997b: Ecological advice for the global fisheries crisis. *Trends in Ecology and Evolution*, **12(1)**, 35-38.

Roberts, C.M., 2005: Marine protected areas and biodiversity conservation, In: *Marine Conservation Biology: the Science of Maintaining the Sea's Biodiversity*, [Norse, E. and L.B. Crowder (eds.)]. Island Press, Washington, DC, pp. 265-279.

Roberts, C.M., S. Andelman, G. Branch, R.H. Bustamante, J.C. Castilla, J. Dugan, B.S. Halpern, K.D. Lafferty, H. Leslie, J. Lubchenco, D. MacArdle, H.P. Possingham, M. Ruckelshaus, and R.R. Warner, 2003a: Ecological criteria for evaluating candidate sites for marine reserves. *Ecological Applications*, **13(1)**, S199-S214.

Roberts, C.M., G. Branch, R.H. Bustamante, J.C. Castilla, J. Dugan, B.S. Halpern, K.D. Lafferty, H. Leslie, J. Lubchenco, D. McArdle, M. Ruckelshaus, and R.R. Warner, 2003b: Application of ecological criteria in selecting marine reserves and developing reserve networks. *Ecological Applications*, **13**, S215-S228.

Roberts, C.M., B. Halpern, S.R. Palumbi, and R.R. Warner, 2001: Designing marine reserve networks: why small, isolated protected areas are not enough. *Conservation Biology in Practice*, **2(3)**, 11-17.

Roberts, J.M., A.J. Wheeler, and A. Freiwald, 2006: Reefs of the deep: the biology and geology of cold-water coral ecosystems. *Science*, **312(5773)**, 543-547.

Roberts, S. and M. Hirshfield, 2004: Deep-sea corals: out of sight, but no longer out of mind. *Frontiers in Ecology and the Environment*, **2(3)**, 123-130.

Robertson, D.R., J.H. Choat, J.M. Posada, J. Pitt, and J.L. Ackerman, 2005: Ocean surgeonfish *Acanthurus bahianus*. II. Fishing effects on longevity, size and abundance? *Marine Ecology Progress Series*, **295**, 245-256.

Roessig, J.M., C.M. Woodley, J.J. Cech, and L.J. Hansen, 2004: Effects of global climate change on marine and estuarine fishes and fisheries. *Reviews in Fish Biology and Fisheries*, **14(2)**, 251-275.

Rogers, A.D., 1999: The biology of *Lophelia pertusa* (Linnaeus 1758) and other deep-water reef-forming corals and impacts from human activities. *International Review of Hydrobiology*, **84(4)**, 315-406.

Ruiz, G.M., P.W. Fofonoff, J.T. Carlton, M.J. Wonham, and A.H. Hines, 2000: Invasion of coastal marine communities in North America: apparent patterns, processes, and biases. *Annual Review of Ecology and Systematics*, **31**, 481-531.

Sale, P.F., R.K. Cowen, B.S. Danilowicz, G.P. Jones, J.P. Kritzer, K.C. Lindeman, S. Planes, N.V.C. Polunin, G.R. Russ, Y.J. Sadovy, and R.S. Steneck, 2005: Critical science gaps impede use of no-take fishery reserves. *Trends in Ecology and Evolution*, **20(2)**, 74-80.

Salm, R., J. Clark, and E. Siirila, 2000: *Marine and Coastal Protected Areas: a Guide for Planners and Managers*. Report Number 3, International Union for Conservation of Nature and Natural Resources, Washington, DC, pp.1-387.

Salm, R.V., T. Done, and E. McLeod, 2006: Marine protected area planning in a changing climate, In: *Coral Reefs and Climate Change: Science and Management,* [Phinney, J.T., O. Hoegh-Guldberg, J. Kleypas, W. Skirving, and A. Strong (eds.)]. American Geophysical Union, Washington, DC, pp. 207-221.

Sammarco, P.W., 1980: Diadema and its relationship to coral spat mortality: grazing, competition, and biological disturbance. *Journal of Experimental Marine Biology and Ecology*, **45**, 245-272.

Sanford, E., 1999: Regulation of keystone predation by small changes in ocean temperature. *Science*, **283(5410)**, 2095-2097.

Scavia, D., J.C. Field, D.F. Boesch, R.W. Buddemeier, V. Burkett, D.R. Cayan, M. Fogarty, M.A. Harwell, R.W. Howarth, C. Mason, D.J. Reed, T.C. Royer, A.H. Sallenger, and J.G. Titus, 2002: Climate change impacts on U.S. coastal and marine ecosystems. *Estuaries*, **25(2)**, 149-164.

Scheltema, R.S., 1986: On dispersal and planktonic larvae of marine invertebrates: an ecletic overview and summary of problems. *Bulletin of Marine Science*, **39**, 290-322.

Schuttenberg, H.Z. and P. Marshall, 2007: *Managing for Mass Coral Bleaching: Strategies for Supporting Socio-Ecological Resilience*. Status of Caribbean coral reefs after bleaching and hurricanes in 2005 Reef and Rainforest Research Centre, Townsville.

Shanks, A.L., B.A. Grantham, and M.H. Carr, 2003: Propagule dispersal distance and the size and spacing of marine reserves. *Ecological Applications*, **13(1)**, S159-S169.

Shepherd, A. and D. Wingham, 2007: Recent sea-level contributions of the Antarctic and Greenland ice sheets. *Science*, **315**, 1529-1532.

Shirayama, Y. and H. Thornton, 2005: Effect of increased atmospheric CO_2 on shallow water marine benthos. *Journal of Geophysical Research*, **110(C9)**.

Smith, S.V. and R.W. Buddemeier, 1992: Global change and coral reef ecosystems. *Annual Review of Ecology and Systematics*, **23**, 89-118.

Snyder, D.B. and G.H. Burgess, 2007: The Indo-Pacific red lionfish, *Pterois volitans* (Pisces: Scorpaenidae), new to Bahamian ichthyofauna. *Coral Reefs*, **26(1)**, 175.

Snyder, M.A., L.C. Sloan, N.S. Diffenbaugh, and J.L. Bell, 2003: Future climate change and upwelling in the California current. *Geophysical Research Letters*, **30(15)**.

Sobel, J.A. and C. Dahlgren, 2004: *Marine Reserves: a Guide to Science, Design, and Use*. Island Press, Washington, DC, pp. 1-383.

Sotka, E.E., J.P. Wares, J.A. Barth, R.K. Grosberg, and S.R. Palumbi, 2004: Strong genetic clines and geographical variation in gene flow in the rocky intertidal barnacle *Balanus glandula*. *Molecular Ecology*, **13(8)**, 2143-2156.

Soto, C.G., 2001: The potential impacts of global climate change on marine protected areas. *Reviews in Fish Biology and Fisheries*, **11(3)**, 181-195.

Sousa, W.P., 1984: The role of disturbance in natural communities. *Annual Review of Ecology and Systematics*, **15**, 353-391.

Stachowicz, J.J., J.R. Terwin, R.B. Whitlatch, and R.W. Osman, 2002: Linking climate change and biological invasions: ocean warming facilitates nonindigenous species invasions. *Proceedings of the National Academy of Sciences of the United States of America*, **99(24)**, 15497-15500.

Steneck, R.S., 2006: Staying connected in a turbulent world. *Science*, **311(5760)**, 480-481.

Steneck, R.S. and J. T. Carlton, 2001: Human alterations of marine communities: students beware!, In: *Marine Community Ecology*, [Bertness, M.D., S.D. Gaines, and M.E. Hay (eds.)]. Sinauer Associates, Inc., Sunderland, MA, pp. 445-468.

Steneck, R.S. and M.N. Dethier, 1994: A functional group approach to the structure of algal-dominated communities. *Oikos*, **69(3)**, 476-498.

Steneck, R.S., M.H. Graham, B.J. Bourque, D. Corbett, J.M. Erlandson, J.A. Estes, and M.J. Tegner, 2002: Kelp forest ecosystems: biodiversity, stability, resilience and future. *Environmental Conservation*, **29(4)**, 436-459.

Steneck, R.S. and E. Sala, 2005: Large marine carnivores: trophic cascades and top-down controls in coastal ecosystems past and present, In: *Large Carnivores and the Conservation of Biodiversity*, [Ray, J.C., K.H. Redford, R.S. Steneck, and J. Berger (eds.)]. Island Press, Washington, DC, pp. 110-137.

Steneck, R.S., J. Vavrinec, and A.V. Leland, 2004: Accelerating trophic level dysfunction in kelp forest ecosystems of the western North Atlantic. *Ecosystems*, **7(4)**, 323-331.

Steneck, R.S. and C.J. Wilson, 2001: Large-scale and long-term, spatial and temporal patterns in demography and landings of the American lobster, Homarus americanus. *Journal of Marine and Freshwater Research*, **52**, 1303-1319.

Stewart, R.R., T. Noyce, and H.P. Possingham, 2003: Opportunity cost of ad hoc marine reserve design decisions: an example from South Australia. *Marine Ecology Progress Series*, **253**, 25-38.

Stillman, J.H., 2003: Acclimation capacity underlies susceptibility to climate change. *Science*, **301(5629)**, 65.

Stobutzki, I.C. and D.R. Bellwood, 1997: Sustained swimming abilities of the late pelagic stages of coral reef fishes. *Marine Ecology Progress Series*, **149(1)**, 35-41.

Stocker, T.F. and O. Marchal, 2000: Abrupt climate change in the computer: is it real? In: *Proceedings of the National Academy of Sciences of the United States of America* , pp. 1362-1365.

Swearer, S.E., J.E. Caselle, D.W. Lea, and R.R. Warner, 1999: Larval retention and recruitment in an island population of a coral-reef fish. *Nature*, **402(6763)**, 799-802.

Taylor, M.S. and M.E. Hellberg, 2003: Genetic evidence for local retention of pelagic larvae in a Caribbean reef fish. *Science*, **299(5603)**, 107-109.

The Group of Experts on Scientific Aspects of Marine Environmental Protection, 2001: *Protecting the Oceans From Land-Based Activities.* Land-based sources and activities affecting the quality and uses of the marine, coastal and associated freshwater environment United Nations Environment Program, Nairobi.

The Royal Society, 2005: *Ocean Acidification Due to Increasing Atmospheric Carbon Dioxide*. The Royal Society, London, -60.

The State of Queensland and Commonwealth of Australia, 2003: *Reef Water Quality Protection Plan; for Catchments Adjacent to the Great Barrier Reef World Heritage Area.* Queensland Department of Premier and Cabinet, Brisbane.

Thorrold, S.R., C. Latkoczy, P.K. Swart, and C.M. Jones, 2001: Natal homing in a marine fish metapopulation. *Science*, **291(5502)**, 297-299.

Thrush, S.F. and P.K. Dayton, 2002: Disturbance to marine benthic habitats by trawling and dredging: implications for Marine Biodiversity. *Annual Review of Ecology and Systematics*, **33**, 449-473.

Tilmant, J.T., R.W. Curry, R. Jones, A. Szmant, J.C. Zieman, M. Flora, M.B. Robblee, D. Smith, R.W. Snow, and H. Wanless, 1994: Hurricane Andrew's effects on marine resources: the small underwater impact contrasts sharply with the destruction in mangrove and upland-forest communities. *BioScience*, **44(4)**, 230-237.

Tolimieri, N., A. Jeffs, and J.C. Montgomery, 2000: Ambient sound as a cue for navigation by the pelagic larvae of reef fishes. *Marine Ecology Progress Series*, **207**, 219-224.

Tomanek, L. and G.N. Somero, 1999: Evolutionary and acclimation-induced variation in the heat-shock responses of congeneric marine snails (genus *Tegula*) from different thermal habitats: implications for limits of thermotolerance and biogeography. *Journal of Experimental Biology*, **202**, 2925-2936.

Tompkins, E.L. and W.N. Adger, 2004: Does adaptive management of natural resources enhance resilience to climate change? *Ecology and Society*, **9(2)**, 10.

Turgeon, D.D., R.G. Asch, B.D. Causey, R.E. Dodge, W. Jaap, K. Banks, J. Delaney, B.D. Keller, R. Speiler, C.A. Matos, J.R. Garcia, E. Diaz, D. Catanzaro, C.S. Rogers, Z. Hillis-Starr, R. Nemeth, M. Taylor, G.P. Schmahl, M.W. Miller, D.A. Gulko, J.E. Maragos, A.M. Friedlander, C.L. Hunter, R.S. Brainard, P. Craig, R.H. Richmond, G. Davis, J. Starmer, M. Trianni, P. Houk, C.E. Birkeland, A. Edwards, Y. Golbuu, J. Gutierrez, N. Idechong, G. Paulay, A. Tafileichig, and N. Vander Velde, 2002: *The State of Coral Reef Ecosystems of the United States and Pacific Freely Associated States: 2002.* National Oceanic and Atmospheric Administration/National Ocean Service/National Centers for Coastal Ocean Science, Silver Spring, MD, pp.1-265.

U.S. Census Bureau, 2001: County and city data book: 2000. **(13th Edition)**, 1-895.

U.S. Climate Change Science Program and Subcommittee on Global Change Research, 2003: *Vision for the Program and Highlights of the Scientific Strategic Plan.* U.S. Climate Change Science Program, Washington, D.C..

U.S. Commission on Ocean Policy, 2004: *An Ocean Blueprint for the 21st Century. Final Report.* U.S. Commission on Ocean Policy, Washington, D.C., pp.1-522.

U.S. Department of Commerce, 1996: *Final Management Plan/Environmental Impact Statement for the Florida Keys National Marine Sanctuary, Volume I.* National Oceanic and Atmospheric Administration, Silver Spring, MD, pp.1-319.

U.S. Environmental Protection Agency, 2007: *Climate Change and Interacting Stressors: Implications for Coral Reef Management in American Samoa.* EPA/600/R-07/069, Global Change Research Program, National Center for Environmental Assessment, Washignton, DC. Available from the National Technical Information Service, Springfield, VA, and online at http://www.epa.gov/ncea.

Wadell, J.E., 2005: *The State of Coral Reef Ecosystems of the United States and Pacific Freely Associated States: 2005.* NOAA Technical Memorandum NOS NCCOS 11, NOAA/NCCOS Center for Coastal Monitoring and Assessment's Biogeography Team, Silver Spring, MD, pp.1-522.

Wainwright, P.C., 1994: Functional morphology as a tool in ecological research, In: *Ecological Morphology: Integrative Organismal Biology*, [Wainwright, P.C. and S.M. Reilly (eds.)]. University of Chicago Press, Chicago, IL, pp. 42-59.

Walther, G.R., E. Post, P. Convey, A. Menzel, C. Parmesan, T.J.C. Beebee, J.M. Fromentin, O. Hoegh-Guldberg, and F. Bairlein, 2002: Ecological responses to recent climate change. *Nature*, **416**, 389-395.

Warner, R.R., S.E. Swearer, and J.E. Caselle, 2000: Larval accumulation and retention: Implications for the design of marine reserves and essential fish habitat. *Bulletin of Marine Science*, **66(3)**, 821-830.

Watling, L. and M. Risk, 2002: Special issue on biology of cold water corals: proceedings of the first international deep-sea coral symposium. *Hydrobiologia*, **471**.

Wells, S., 2006: *Establishing National and Regional Systems of MPAs – a Review of Progress With Lessons Learned.* Second Draft, UNEP World Conservation Monitoring Centre, UNEP Regional Seas Programme, ICRAN, IUCN/WCPA – Marine.

West, J.M. and R.V. Salm, 2003: Resistance and resilience to coral bleaching: implications for coral reef conservation and management. *Conservation Biology*, **17(4)**, 956-967.

Whelan, K.R.T., J. Miller, O. Sanchez, and M. Patterson, 2007: Impact of the 2005 coral bleaching event on *Porites porites* and *Colpophyllia natans* at Tektite Reef, US Virgin Islands. *Coral Reefs*, **26**, 689-693.

Whitfield, P.E., J.A. Hare, A.W. David, S.L. Harter, R.C. Mu±oz, and C.M. Addison, 2007: Abundance estimates of the Indo-Pacific lionfish *Pterois volitans/miles* complex in the Western North Atlantic. *Biological Invasions*, **9(1)**, 53-64.

Whitfield, P.E., W.J. Kenworthy, K.K. Hammerstrom, and M.S. Fonseca, 2002: The role of a hurricane in the expansion of disturbances initiated by motor vessels on seagrass banks. *Journal of Coastal Research*, **37**, 86-99.

Wilkinson, C., O. Linden, H. Cesar, G. Hodgson, J. Rubens, and A.E. Strong, 1999: Ecological and socioeconomic impacts of 1998 coral mortality in the Indian Ocean: an ENSO impact and a warning of future change? *Ambio*, **28(2)**, 188.

Wilkinson, C.R., 1998: *Status of Coral Reefs of the World: 1998.* Australian Institute of Marine Science, Townsville, Australia.

Wilkinson, C.R., 2000: *Status of Coral Reefs of the World: 2000.* Australian Institute of Marine Science, Townsville, Australia.

Wilkinson, C.R., 2002: *Status of Coral Reefs of the World: 2002.* Australian Institute of Marine Science, Townsville, Australia.

Wilkinson, C.R., 2004: *Status of Coral Reefs of the World: 2004.* Australian Institute of Marine Science, Townsville, Australia.

Williams, D.M.B., E. Wolanski, and J.C. Andrews, 1984: Transport mechanisms and the potential movement of planktonic larvae in the central region of the Great Barrier Reef. *Coral Reefs*, **3(4)**, 229-236.

Williams, E.H.Jr. and L. Bunkley-Williams, 1990: The world-wide coral reef bleaching cycle and related sources of coral mortality. *Atoll Research Bulletin*, **(355)**, 1-72.

Williams, E.H.Jr., C. Goenaga, and V. Vicente, 1987: Mass bleachings on Atlantic coral reefs. *Science*, **238**, 877-878.

Wooldridge, S., T. Done, R. Berkelmans, R. Jones, and P. Marshall, 2005: Precursors for resilience in coral communities in a warming climate: a belief network approach. *Marine Ecology Progress Series*, **295**, 157-169.

Wooninck, L. and C. Bertrand, 2004: Marine managed areas designated by NOAA fisheries: a characterization study and preliminary assessment. *American Fisheries Society Symposium*, **42**, 89-103.

World Resources Institute, 1996: *World Resources 1996-97: the Urban Environment.* United Nations Environment Programme, United Nations Development Programme, and the World Bank, pp.1-384.

Young, O.R., G. Osherenko, J. Ekstrom, L.B. Crowder, J. Ogden, J.A. Wilson, J.C. Day, F. Douvere, C.N. Ehler, K.L. McLeod, B.S. Halpern, and R. Peach, 2007: Solving the crisis in ocean governance: place-based management of marine ecosystems. *Environment: Science and Policy for Sustainable Development*, **49(4)**, 20-32.

Zervas, C., 2001: *Sea Level Variations of the United States, 1854-1999*. Technical Report NOS CO-OPS 36, US Department of Commerce, National Oceanic and Atmospheric Administration, National Ocean Service, Silver Spring, MD.

CHAPTER 9 REFERENCES

Araújo, M.B. and M. New, 2007: Ensemble forecasting of species distributions. *Trends in Ecology and Evolution*, **22**, 42-47.

Arvai, J., G. Bridge, N. Dolsak, R. Franzese, T. Koontz, A. Luginbuhl, P. Robbins, K. Richards, K.S. Korfmacher, B. Sohngen, J. Tansey, and A. Thompson, 2006: Adaptive management of the global climate problem: gridging the gap between climate research and climate policy. *Climatic Change*, **78**, 217-225.

Aschmann, H., 1973: Distribution and peculiarity of Mediterranean ecosystems, In: *Mediterranean Type Ecosystems: Origin and Structure*, [Castri, F.D. and H. Mooney (eds.)]. Springer-Verlag, New York, NY, pp. 11-19.

Battin, J., M.W. Wiley, M.H. Ruckelshaus, R.N. Palmer, E. Korb, K.K. Bartz, and H. Imaki, 2007: Projected impacts of climate change on salmon habitat restoration. *Proceedings of the National Academy of Sciences of the United States of America*, **104(16)**, 6720-6725.

Behrenfeld, M.J., R.T. O'Malley, D.A. Siegel, C.R. McClain, J.L. Sarmiento, G.C. Feldman, A.J. Milligan, P.G. Falkowski, R.M. Letelier, and E.S. Boss, 2006: Climate-driven trends in contemporary ocean productivity. *Nature*, **444(7120)**, 752-755.

Benestad, R.E., I. Hanssen-Bauer, and E.J. Fairland, 2007: An evaluation of statistical models for downscaling precipitation and their ability to capture long-term trends *International Journal of Climatology*, **27(5)**, 649-665.

Bengtsson, J., P. Angelstam, T. Elmqvist, U. Emanuelsson, C. Folke, M. Ihse, F. Moberg, and M. Nystroem, 2003: Reserves, resilience and dynamic landscapes. *Ambio*, **32(6)**, 389-396.

Bennett, E.M., G.S. Cumming, and G.D. Peterson, 2005: A systems model approach to determining resilience surrogates for case studies. *Ecosystems*, **8**, 945-957.

Bessie, W.C. and E.A. Johnson, 1995: The relative importance of fuels and weather on fire behavior in subalpine forests. *Ecology*, **76(3)**, 747-762.

Botkin, D.B., H. Saxe, M.B. Araújo, R. Betts, R.H.W. Bradshaw, T. Cedhagen, P. Chesson, T.P. Dawson, J.R. Etterson, D.P. Faith, S. Ferrier, A. Guisan, A.S. Hansen, D.W. Hilbert, C. Loehle, C. Margules, M. New, M.J. Sobel, and D.R.B. Stockwell, 2007: Forecasting the effects of global warming on biodiversity. *BioScience*, **57(3)**, 227-236.

Chapin, F.S., B.H. Walker, R.J. Hobbs, D.U. Hooper, J.H. Lawton, O.E. Sala, and D. Tilman, 1997: Biotic control over the functioning of ecosystems. *Science*, **277(5325)**, 500-504.

Christensen, J., B. C. Hewistson, A. Busuioc, A. Chen, X. Gao, I. Held, R. Jones, R. K. Kolli, W.-T. Kwon, R. Laprise, V. Magaña Rueda, L. Mearns, C. G. Menéndez, J. Räisänen, A. Rinke, A. Sarr, and P. Whetton, 2007: Regional climate projections, In: *Climate Change 2007: the Physical Science Basis. Contribution of Working Group I to Fourth Assessment Report of the Intergovernmental Panel on Climate Change*, [Solomon, S., D. Qin, M. Manning, Z. Chen, M. Marquis, K.B. Averyt, M. Tignor, and H.L. Miller (eds.)]. Cambridge University Press, Cambridge, United Kingdom and New York, NY, USA, pp. 848-940.

Dettinger, M.D., 2005: From climate-change spaghetti to climate-change distributions for 21st century California. *San Francisco Estuary and Watershed Science*, **3(1)**.

Dixon, G.E., 2003: *Essential FVS: A User's Guide to the Forest Vegetation Simulator*. U.S. Department of Agriculture, Forest Service, Forest Management Service Center, Fort Collins, CO, pp.193p.

Field, C.B., 1999: Diverse controls on carbon storage under elevated CO_2: toward a synthesis, In: *Carbon Dioxide and Environmental Stress*, [Luo, Y. (ed.)]. Academic Press, San Diego, California, pp. 373-391.

Folke, C., S. Carpenter, T. Elmqvist, L. Gunderson, C. Holling, and B. Walker, 2002: Resilience and sustainable development: building adaptive capacity in a world of transformations. *Ambio*, **31(5)**, 437-440.

Folke, C., S. Carpenter, B. Walker, M. Scheffer, T. Elmqvist, L.H. Gunderson, and C.S. Holling, 2004: Regime shifts, resilience, and biodiversity in ecosystem management. *Annual Review of Ecology and Systematics*, **35**, 557-581.

Fried, J.S., M.S. Torn, and E. Mills, 2004: The impact of climate change on wildfire severity: a regional forecast for Northern California. *Climatic Change*, **64(1)**, 169-191.

Georgi, F., B. Hewitson, and J. Christensen, 2001: Regional climate information - evaluation and predictions, In: *Climate Change 2001: the Scientific Basis. Contribution of Working Group I to Third Assessment Report of IPCC*, Cambridge University Press, Cambridge, UK, pp. 583-638.

Gregory, R. and L. Failing, 2002: Using decision analysis to encourage sound deliberation: water use planning in British Columbia, Canada. *Journal of Policy Analysis and Management*, **21**, 492-499.

Gregory, R., D. Ohlson, and J. Arvai, 2006: Deconstructing adaptive management: criteria for applications to environmental management. *Ecological Applications*, **16(6)**, 2411-2425.

Guinotte, J.M., J. Orr, S. Cairns, A. Freiwald, L. Morgan, and R. George, 2006: Will human-induced changes in seawater chemistry alter the distribution of deep-sea scleractinian corals? *Frontiers in Ecology and the Environment*, **4(3)**, 141-146.

Gunderson, L.H., 2000: Ecological resilience-in theory and application. *Annual Review of Ecology and Systematics*, **31**, 425-439.

Hall, M.H.P. and D.B. Fagre, 2003: Modeled climate-induced glacier change in Glacier National Park, 1850-2100. *BioScience*, **53(2)**, 131-140.

Holling, C.S., 1973: Resilience and stability of ecological systems. *Annual Review of Ecology and Systematics*, **4**, 1-23.

IPCC, 2001: *Climate Change 2001: the Scientific Basis. Contribution of Working Group I to the Third Assessment Report of the Intergovernmental Panel on Climate Change*. [Houghton, J.T., Y. Ding, D.J. Griggs, M. Noguer, P.J. van der Linden, X. Dai, K. Maskell, and C.A. Johnson (eds.)]. Cambridge University Press, Cambridge, United Kingdom and New York, NY, USA.

IPCC, 2007: Summary for policymakers, In: *Climate Change 2007: the Physical Science Basis. Contribution of Working Group I to the Fourth Assessment Report of the Intergovernmental Panel on Climate Change*, [Solomon, S., D. Qin, M. Manning, Z. Chen, M. Marquis, K.B. Averyt, M. Tignor, and H.L. Miller (eds.)]. University of Cambridge Press, Cambridge, United Kingdom and New York, NY, USA.

IPCC-TGICA, 2007: *General Guidelines on the Use of Scenario Data for Climate Impact and Adaptation Assessment*. Version 2, Prepared by T.R. Carter on behalf of the Intergovernmental Panel on Climate Change, Task Group on Data and Scenario Support for Impact and Climate Assessment, pp.1-66.

Johnson, T. and C. Weaver: A framework for assessing climate impacts on water and watershed systems. *Environmental Management*. Published online: October 2, 2008. DOI: 10.1007/s00267-008-9205-4.

Kelly, P.M. and W.N. Adger, 2000: Theory and practice in assessing vulnerability to climate change and facilitating adaptation. *Climatic Change*, **47(4)**, 325-352.

Kerr, R.A., 2004: Climate change: three degrees of consensus. *Science*, **305**, 932-934.

Kerr, R.A., 2005: How hot will the greenhouse world be? *Science*, **309(5731)**, 100.

Lamy, F., J. Bolte, M. Santelmann, and C. Smith, 2002: Development and evaluation of multiple-objective decision-making methods for watershed management planning. *Journal of the American Water Resources Association*, **38(2)**, 517-529.

Lawler, J.J., D. White, R.P. Neilson, and A.R. Blaustein, 2006: Predicting climate-induced range shifts: model differences and model reliability. *Global Change Biology*, **12**, 1568-1584.

Leeworthy, V.R. and P.C. Wiley, 2003: *Profiles and Economic Contribution: General Visitors to Monroe County, Florida 2000-2001*. National Oceanic and Atmospheric Administration, Silver Spring, MD, pp.1-24.

Leung, L.R., L.O. Mearns, F. Giorgi, and R.L. Wilby, 2003: Regional climate research: needs and opportunities. *Bulletin of the American Meteorological Society*, **84**, 89-95.

Logan, J.A., J. Regniere, and J.A. Powell, 2003: Assessing the impacts of global warming on forest pest dynamics. *Frontiers in Ecology and the Environment*, **1(3)**, 130-137.

Lovejoy, T.E. and L. Hannah, 2005: *Climate Change and Biodiversity*. Yale University Press, New Haven.

Luck, G.W., G.C. Daily, and P.R. Ehrlich, 2003: Population diversity and ecosystem services. *Trends in Ecology and Evolution*, **18(7)**, 331-336.

Malcolm, J.R., C. Liu, R.P. Neilson, L. Hansen, and L. Hannah, 2006: Global warming and extinctions of endemic species from biodiversity hotspots. *Conservation Biology*, **20(2)**, 538-548.

Mau-Crimmins, T., A. Hubbard, D. Angell, C. Filippone, and N. Kline, 2005: *Sonoran Desert Network: Vitals Signs Monitoring Plan*. National Park Service, Intermountain Region, Denver, CO.

Melillo, J.M., J. Borchers, J. Chaney, H. Fisher, S. Fox, A. Haxeltine, A. Janetos, D.W. Kicklighter, T.G.F. Kittel, A.D. McGuire, R. McKeown, R. Neilson, R. Nemani, D.S. Ojima, T. Painter, Y. Pan, W.J. Parton, L. Pierce, L. Pitelka, C. Prentice, B. Rizzo, N.A. Rosenbloom, S. Running, D.S. Schimel, S. Sitch, T. Smith, and I. Woodward, 1995: Vegetation/ecosystem modeling and analysis project: comparing biogeography and geochemistry models in a continental-scale study of terrestrial ecosystem responses to climate change and CO_2 doubling. *Global Biogeochemical Cycles*, **9(4)**, 407-437.

Metzger, M.J., R. Leemans, and D. Schroter, 2005: A multidisciplinary multi-scale framework for assessing vulnerability to global change. *International Journal of Applied Earth Observation and Geoinformation*, **7**, 253-267.

Mitchell, T.D., T.R. Carter, P.D. Jones, M. Hulme, and M. New, 2004: *A Comprehensive Set of High-Resolution Grids of Monthly Climate for Europe and the Globe: the Observed Record (1901-2000) and 16 Scenarios (2001-2100)*. Working Paper 55, Tyndall Centre for Climate Change Research.

Moore, J.L., A. Balmford, T. Brooks, N.D. Burgess, L.A. Hansen, C. Rahbek, and P.H. Williams, 2003: Performance of sub-Saharan vertebrates as indicator groups for identifying priority areas for conservation. *Conservation Biology*, **17(1)**, 207-218.

Myers, N., 1979: *The Sinking Arc*. Pergamon Press, New York, NY.

National Research Council, 1999: *Downstream: Adaptive Management of the Glen Canyon Dam and the Colorado River Ecosystem*. National Academies Press, Washington, DC.

Pan, Z., J.H. Christensen, R.W. Arritt, and W.J. Gutowski, 2001: Evaluation of uncertainties in regional climate change simulations. *Journal of Geophysical Research*, **106**, 17735-17751.

Parmesan, C., 1996: Climate and species' range. *Nature*, **382**, 765-766.

Parrish, J.D., D.P. Braun, and R.S. Unnasch, 2003: Are we conserving what we say we are? Measuring ecological integrity within protected areas. *BioScience*, **53(9)**, 851-860.

Pauly, D., 1995: Anecdotes and the shifting baseline syndrome of fisheries. *Trends in Ecology and Evolution*, **10(10)**, 430-430.

Pearson, R.G., T.P. Dawson, P.M. Berry, and P.A. Harrison, 2002: SPECIES: A spatial evaluation of climate impact on the envelope of species. *Ecological Modelling*, **154(3)**, 289-300.

Pianka, E.R., 1970: On r- and K-selection. *The American Naturalist*, **104(940)**, 592-597.

Poff, N.L., M.M. Brinson, and J.W. Day, Jr., 2002: *Aquatic Ecosystems & Global Climate Change: Potential Impacts on Inland Freshwater and Coastal Wetland Ecosystems in the United States*. Pew Center on Global Climate Change, pp.1-56.

Porter, M.E., 1985: *The Competitive Advantage*. Free Press, New York, NY.

Portner, H.O. and R. Knust, 2007: Climate change affects marine fishes through the oxygen limitation of thermal tolerance. *Science*, **315(5808)**, 95-97.

Root, T.L. and S. H. Schneider, 2002: Climate change: overview and implications for wildlife, In: *Wildlife Responses to Climate Change: North American Case Studies*, [Schneider, S.H. and T.L. Root (eds.)]. Island Press, Washington, DC, pp. 1-56.

Salathé, E.P., Jr., 2003: Comparison of various precipitation downscaling methods for the simulation of streamflow in a rainshadow river basin. *International Journal of Climatology*, **23(8)**, 887-901.

Scheffer, M., S. Carpenter, J.A. Foley, C. Folke, and B.H. Walker, 2001: Catastrophic shifts in ecosystems. *Nature*, **413**, 591-596.

Scholze, M., W. Knorr, N.W. Arnell, and I.C. Prentice, 2006: A climate-change risk analysis for world ecosystems. *Proceedings of the National Academy of Sciences of the United States of America*, **103(35)**, 13116-13120.

Schwartz, P., 1996: *Art of the Long View: Planning for the Future in an Uncertain World*. Currency Doubleday, New York, NY, pp. 1-258.

Shongwe, M.E., W.A. Landman, and S.J. Mason, 2006: Performance of recalibration systems for GCM forecasts for southern Africa. *International Journal of Climatology*, **26(12)**, 1567-1585.

Spak, S., T. Holloway, B. Lynn, and R. Goldberg, 2007: A comparison of statistical and dynamical downscaling for surface temperature in North America. *Journal of Geophysical Research*, **112**, 1029-1034.

Spittlehouse, D.L. and R.B. Stewart, 2003: Adaptation to climate change in forest management. *BC Journal of Ecosystems and Management*, **4(1)**, 7-17.

Tallis, H.M. and P. Kareiva, 2006: Shaping global environmental decisions using socio-ecological models. *Trends in Ecology and Evolution*, **21**, 562-568.

Thuiller, W., S. Lavorel, M.B. Araujo, M.T. Sykes, and I.C. Prentice, 2005: Climate change threats to plant diversity in Europe. *Proceedings of the National Academy of Sciences of the United States of America*, **102(23)**, 8245-8250.

Tompkins, E.L. and N.W. Adger, 2004: Does adaptive management of natural resources enhance resilience to climate change? *Ecology and Society*, **19(2)**.

Turner, B.L., II, R.E. Kasperson, P.A. Matsone, J.J. McCarthy, R.W. Corell, L. Christensene, N. Eckley, J.X. Kasperson, A. Luerse, M.L. Martello, C. Polsky, A. Pulsipher, and A. Schiller, 2003: A framework for vulnerability analysis in sustainability science. *Proceedings of the National Academy of Sciences of the United States of America Early Edition*, **100(14)**.

U.S. Climate Change Science Program, in press: Synthesis and assessment product 4.1: Coastal elevation and sensitivity to sea level rise. *A report by the U. S. Climate Change Science Program and the Subcommittee on Global Change Research, U. S. Environmental Protection Agency*. [Titus, J.G. (ed.)].

Westerling, A.L., H.G. Hidalgo, D.R. Cayan, and T.W. Swetnam, 2006: Warming and earlier spring increase western U.S. forest wildfire activity. *Science*, **313(5789)**, 940-943.

Wilby, R.L., S.P. Charles, E. Zorita, B. Timbal, P. Whetton, and L.O. Mearns, 2004: *Guidelines for Use of Climate Scenarios Developed From Statistical Downscaling Methods*. Intergovernmental Panel on Climate Change, Task Group on Data and Scenarios Support for Impact and Climate Assessment, pp.1-27.

Wilby, R.L., T.M.L. Wigley, D. Conway, P.D. Jones, B.C. Hewitson, J. Main, and D.S. Wilks, 1998: Statistical downscaling of general circulation model output: a comparison of methods. *Water Resources Research*, **34(11)**, 2995-3008.

Willis, K.J. and H.J.B. Birks, 2006: What is natural? The need for a long-term perspective in biodiversity conservation. *Science*, **314(5803)**, 1261.

Wood, A.W., L.R. Leung, V. Sridhar, and D.P. Lettenmaier, 2004: Hydrologic implications of dynamical and statistical approaches to downscaling climate model outputs. *Climatic Change*, **62(1)**, 189-216.

Worm, B., E.B. Barbier, N. Beaumont, J.E. Duffy, C. Folke, B.S. Halpern, J.B.C. Jackson, H.K. Lotze, F. Micheli, S.R. Palumbi, E. Sala, K.A. Selkoe, J.J. Stachowicz, and R. Watson, 2006: Impacts of biodiversity loss on ocean ecosystem services. *Science*, **314(5800)**, 787-790.

Yohe, G.W. and R.S.J. Tol, 2002: Indicators for social and economic coping capacity--moving toward a working definition of adaptive capacity. *Global Environmental Change*, **12**, 25-40.

ANNEX A REFERENCES

Aeby, G.S., J.C. Kenyon, J.E. Maragos, and D.C. Potts, 2003: First record of mass coral bleaching in the Northwestern Hawaiian Islands. *Coral Reefs*, **22**, 256-256.

Aeby, G.S., 2006: Baseline levels of coral disease in the Northwestern Hawaiian Islands. *Atoll Research Bulletin*, **543**, 471-488.

Anil, A.C., D. Desai, and L. Khandeparker, 2001: Larval development and metamorphosis in *Balanus amphitrite* Darwin (Cirripedia; Thoracica): significance of food concentration, temperature and nucleic acids. *Journal of Experimental Marine Biology and Ecology*, **263(2)**, 125-141.

Antonelis, G.A., J.D. Baker, T.C. Johanos, R.C. Braun, and A.L. Harting, 2006: Hawaiian monk seal: status and conservation issues. *Atoll Research Bulletin*, **543**, 75-101.

Apple, D.D., 1996: Changing social and legal forces affecting the management of national forests. *Women in Natural Resources*, **18**, 1-13.

Arzel, C., J. Elmberg, and M. Guillemain, 2006: Ecology of spring-migrating Anatidae: a review. *Journal of Ornithology*, **147(2)**, 167-184.

Ault, J.S., J.A. Bohnsack, and G.A. Meester, 1998: A retrospective (1979-1996) multispecies assessment of coral reef fish stocks in the Florida Keys. *Fishery Bulletin*, **96(3)**, 395-414.

Ault, J.S., S.G. Smith, J.A. Bohnsack, J. Luo, D.E. Harper, and D.B. McClellan, 2006: Building sustainable fisheries in Florida's coral reef ecosystem: positive signs in the Dry Tortugas. *Bulletin of Marine Science*, **78(3)**, 633-654.

Austin, J.E., A.D. Afton, M.G. Anderson, R.G. Clark, C.M. Custer, J.S. Lawrence, J.B. Pollard, and J.K. Ringelman, 2000: Declining scaup populations: issues, hypotheses, and research needs. *Wildlife Society Bulletin*, **28(1)**, 254-263.

Bachelet, D., R.P. Neilson, J.M. Lenihan, and R.J. Drapek, 2001: Climate change effects on vegetation distribution and carbon budget in the United States. *Ecosystems*, **4**, 164-185.

Bachelet, D., R.P. Neilson, T. Hickler, R.J. Drapek, J.M. Lenihan, M.T. Sykes, B. Smith, S. Sitch, and K. Thonicke, 2003: Simulating past and future dynamics of natural ecosystems in the United States. *Global Biogeochemical Cycles*, **17(2)**, 1045-1066.

Baker, J.D., C.L. Littnan, and D.W. Johnston, 2006: Potential effects of sea level rise on the terrestrial habitats of endangered and endemic megafauna in the Northwestern Hawaiian Islands. *Endangered Species Research*, **4**, 1-10.

Balazs, G.H. and M. Chaloupka, 2006: Recovery trend over 32 years at the Hawaiian green turtle rookery of French Frigate Shoals. *Atoll Research Bulletin*, **543**, 147-158.

Barber, V.A., G.P. Juday, and B.P. Finney, 2000: Reduced growth of Alaskan white spruce in the twentieth century from temperature-induced drought stress. *Nature*, **405(6787)**, 668-673.

Barnett, T.P., D.W. Pierce, and R. Schnur, 2001: Detection of anthropogenic climate change in the world's oceans. *Science*, **292**, 270-274.

Barry, J.P., C.H. Baxter, R.D. Sagarin, and S.E. Gilman, 1995: Climate-related, long-term faunal changes in a California rocky intertidal community. *Science*, **267(5198)**, 672-675.

Barth, J.A., B.A. Menge, J. Lubchenco, F. Chan, J.M. Bane, A.R. Kirincich, M.A. McManus, K.J. Nielsen, S.D. Pierce, and L. Washburn, 2007: Delayed upwelling alters nearshore coastal ocean ecosystems in the northern California current. *Proceedings of the National Academy of Sciences of the United States of America*, **104(10)**, 3719-3724.

Battin, J., M.W. Wiley, M.H. Ruckelshaus, R.N. Palmer, E. Korb, K.K. Bartz, and H. Imaki, 2007: Projected impacts of climate change on salmon habitat restoration. *Proceedings of the National Academy of Sciences of the United States of America*, **104(16)**, 6720-.

Bayne, B.L., R. J. Thompson, and J. Widdows, 1973: Some effects of temperature and food on the rate of oxygen consumption by *Mytilus edulus*. In: *Effects of Temperature on Ectothermic Organisms*, [Weiser, W. (ed.)]. Springer-Verlag, Berlin, pp. 181-193.

Beesley, D., 1996: Reconstructing the landscape: An environmental history, 1820-1960. *Sierra Nevada Ecosystem Project: Final report to Congress*, **2**, 3-24.

Beever, E.A., P.F. Brussard, and J. Berger, 2003: Patterns of apparent extirpation among isolated populations of pikas(Ochotona princeps) in the Great Basin. *Journal of Mammalogy*, **84(1)**, 37-54.

Behrens, M.D. and K.D. Lafferty, 2004: Effects of marine reserves and urchin disease on southern Californian rocky reef communities. *Marine Ecology Progress Series*, **279**, 129-139.

Bellwood, D.R., T.P. Hughes, C. Folke, and M. Nystroem, 2004: Confronting the coral reef crisis. *Nature*, **429(6994)**, 827-833.

Bitz, C.M. and D.S. Battisti, 1999: Interannual to decadal variability in climate and the glacier mass balance in Washington, Western Canada, and Alaska. *Journal of Climate*, **12(11)**, 3181-3196.

Bohnsack, J.A., D.E. Harper, and D.B. McClellan, 1994: Fisheries trends from Monroe County, Florida. *Bulletin of Marine Science*, **54(3)**, 982-1018.

Boland, R., B. Zgliczynski, J. Asher, A. Hall, K. Hogrefe, and M. Timmers, 2006: Dynamics of debris densities and removal at the northwestern Hawaiian Islands coral reefs. *Atoll Research Bulletin*, **543**, 461-470.

Bricker, S.B., C.G. Clement, D.E. Pirhalla, S.P. Orlando, and D.R.G. Farrow, 1999. *National Estuarine Eutrophication Assessment: Effects of Nutrient Enrichment in the Nation's Estuaries*. National Centers for Coastal Ocean Science, National Oceanic and Atmospheric Administration, Silver Spring, MD, pp. 1-71.

Brinson, M.M., 1991: *Ecology of a Nontidal Brackish Marsh in Coastal North Carolina*. [Brinson, M.M. (ed.)]. U. S. Fish and Wildlife Service, National Wetlands Research Center, Slidell, Louisiana.

Buddemeier, R.W., J.A. Kleypas, and R. Aronson, 2004: *Coral Reefs and Global Climate Change: Potential Contributions of Climate Change to Stresses on Coral Reef Ecosystems*. Pew Center on Global Climate Change.

Bureau of Land Management, 2000: *The Rio Grande Corridor Final Plan*. U.S. Department of Interior, pp.1-54.

Burkholder, J.M., E.J. Noga, C.H. Hobbs, and H.B. Glasgow Jr, 1992: New 'phantom' dinoflagellate is the causative agent of major estuarine fish kills. *Nature*, **358(6385)**, 407-410

Burns, T.P., 1985: Hard-coral distribution and cold-water disturbances in South Florida: variation with depth and location. *Coral Reefs*, **4**, 117-124.

Busenberg, G., 2004: Wildfire management in the United States: The evolution of a policy failure. *Review of Policy Research*, **21(2)**, 145-156.

Buzzelli, C.P., R.A. Luettich Jr, S.P. Powers, C.H. Peterson, J.E. McNinch, J.L. Pinckney, and H.W. Paerl, 2002: Estimating the spatial extent of bottom-water hypoxia and habitat degradation in a shallow estuary. *Marine Ecology Progress Series*, **230**, 103-112.

Cabanes, C., A. Cazenave, and C. Le Provost, 2001: Sea level rise during past 40 years determined from satellite and in situ observations. *Science*, **294(5543)**, 840-842.

Caldeira, K. and M.E. Wickett, 2003: Anthropogenic carbon and ocean pH. *Nature*, **425(6956)**, 365-365.

California Climate Action Team, 2005: *First Annual Report to the Governor and Legislators (Draft)*.

Cayan, D., P. Bromirski, K. Hayhoe, M. Tyree, M. Dettinger, and R. Flick, 2006a: *Projecting Future Sea Level*. CEC-500-2005-202-SF, White paper prepared for the California Climate Change Center.

Cayan, D., A.L. Luers, M. Hanemann, and G. Franco, 2006b: *Scenarios of Climate Change in California: an Overview*. Climate action team report to the Governor and Legislators. California Climate Change Center.

Cayan, D.R., M.D. Dettinger, H.F. Diaz, and N.E. Graham, 1998: Decadal variability of precipitation over Western North America. *Journal of Climate*, **11(12)**, 3148-3166.

Chin, A., P. M. Kyne, T. I. Walker, R. B. McAuley, J. D. Stevens, C. L. Dudgeon, and R. D. Pillans, 2007: Vulnerability of chondrichthyan fishes of the Great Barrier Reef to climate change, In: *Climate Change and the Great Barrier Reef*, [Johnson, J. and P. Marshall (eds.)]. Great Barrier Reef Marine Park Authority, Townsville.

Christian, R.R., L. Stasavich, C. Thomas, and M. M. Brinson, 2000: Reference is a moving target in sea-level controlled wetlands, In: *Concepts and Controversies in Tidal Marsh Ecology*, [Weinstein, M.P. and D.A. Kreeger (eds.)]. Kluwer Press, The Netherlands, pp. 805-825.

Climate Impacts Group, University of Washington, 2004: *Overview of Climate Change Impacts in the U.S. Pacific Northwest*. Climate Impacts Group, University of Washington, Seattle.

Clow, D.W., L. Schrott, R. Webb, D.H. Campbell, A. Torizzo, and M. Dornblaser, 2003: Ground water occurrence and contributions to streamflow in an alpine catchment, Colorado front range. *Ground Water*, **41(7)**, 937-950.

Coles, S.L. and Y.H. Fadlallah, 1991: Reef coral survival and mortality at low temperatures in the Arabian Gulf: new species-specific lower temperature limits. *Coral Reefs*, **9(4)**, 231-237.

Conference of the Upper Delaware Townships, 1986: *Final Management Plan: Upper Delaware Scenic and Recreational River*. pp.1-197.

Congdon, B.C., C. A. Erwin, D. R. Peck, G. B. Baker, M. C. Double, and P. O'Neill, 2007: Vulnerability of seabirds on the Great Barrier Reef to climate change, In: *Climate Change and the Great Barrier Reef*, [Johnson, J. and P. Marshall (eds.)]. Great Barrier Reef Marine Park Authority, Townsville.

Conley, D.J., S. Markager, J. Andersen, T. Ellermann, and L.M. Svendsen, 2002: Coastal eutrophication and the Danish national aquatic monitoring and assessment program. *Estuaries*, **25(4)**, 848-861.

Cook, G.D., R.J. Williams, L.B. Hutley, A.P. O'Grady, and A.C. Liedloff, 2002: Variation in vegetative water use in the savannas of the North Australian Tropical Transect. *Journal of Vegetation Science*, **13(3)**, 413-418.

Cooper, D.J., J. Dickens, N. Thompson Hobbs, L. Christensen, and L. Landrum, 2006: Hydrologic, geomorphic and climatic processes controlling willow establishment in a montane ecosystem. *Hydrological Processes*, **20(8)**, 1845-1864.

Cooper, S.R., S.K. McGlothlin, M. Madritch, and D.L. Jones, 2004: Paleoecological evidence of human impacts on the Neuse and Pamlico Estuaries of North Carolina, USA. *Estuaries*, **27(4)**, 617-633.

Copeland, B.J. and J.E. Hobbie, 1972: *Phosphorus and Eutrophication in the Pamlico River Estuary, N. C., 1966-1969- A SUMMARY*. 1972-65, University of North Carolina Water Resources Research Institute, Raleigh, North Carolina.

Cowie-Haskell, B.D. and J.M. Delaney, 2003: Integrating science into the design of the Tortugas Ecological Reserve. *Marine Technology Society Journal*, **37(1)**, 68-79.

Cubasch, U., G. A. Meehl, G. J. Boer, R. J. Stouffer, M. Dix, A. Noda, C. A. Senior, S. Raper, and K. S. Yap, 2001: Projections of future climate change, In: *Climate Change 2001: The Scientific Basis. Contribution of Working Group I to the Third Assessment Report of the Intergovernmental Panel on Climate Change*, [Houghton, J.T., Y. Ding, D.J. Griggs, M. Noguer, P.J. van der Linden, X. Dai, K. Maskell, and C.A. Johnson (eds.)]. Cambridge University Press, Cambridge, United Kingdom and New York, NY, USA, pp. 525-582.

Dameron, O.J., M. Parke, M.A. Albins, and R. Brainard, 2007: Marine debris accumulation in the Northwestern Hawaiian Islands: an examination of rates and processes. *Marine Pollution Bulletin*, **54(4)**, 423-433.

Davis, G.E., 1982: A century of natural change in coral distribution at the Dry Tortugas: a comparison of reef maps from 1881 and 1976. *Bulletin of Marine Science*, **32(2)**, 608-623.

Dayton, P.K., M.J. Tegner, P.B. Edwards, and K.L. Riser, 1999: Temporal and spatial scales of kelp demography: the role of oceanographic climate. *Ecological Monographs*, **69(2)**, 219-250.

Dayton, P.K., M.J. Tegner, P.E. Parnell, and P.B. Edwards, 1992: Temporal and spatial patterns of disturbance and recovery in a kelp forest community. *Ecological Monographs*, **62(3)**, 421-445.

Delaney, J.M., 2003: Community capacity building in the designation of the Tortugas Ecological Reserve. *Gulf and Caribbean Research*, **12(2)**, 163-169.

Delaware River Basin Commission, 2004: *Water Resource Plan for the Delaware River Basin*. Delaware River Basin Commission, pp.1-100.

DeMartini, E.E. and A.M. Friedlander, 2004: Spatial patterns of endemism in shallow-water reef fish populations of the Northwestern Hawaiian Islands. *Marine Ecology Progress Series*, **271**, 281-296.

DeMartini, E.E. and A.M. Friedlander, 2006: Predation, endemism, and related processes structuring shallow-water reef fish assemblages of the Northwestern Hawaiian Islands. *Atoll Research Bulletin*, **543**, 237-256.

DeMartini, E.E., A.M. Friedlander, and S.R. Holzwarth, 2005: Size at sex change in protogynous labroids, prey body size distributions, and apex predator densities at NW Hawaiian atolls. *Marine Ecology Progress Series*, **297**, 259-271.

Dettinger, M.D., D.R. Cayan, M.K. Meyer, and A.E. Jeton, 2004: Simulated hydrologic responses to climate variations and change in the Merced, Carson, and American River basins, Sierra Nevada, California, 1900-2099. *Climatic Change*, **62(1/3)**, 283-317.

Dollar, S.J. and R.W. Grigg, 2004: Anthropogenic and natural stresses on selected coral reefs in Hawaii: a multidecade synthesis of impact and recovery. *Pacific Science*, **58(2)**, 281-304.

Done, T., P. Whetton, R. Jones, R. Berkelmans, J. Lough, W. Skirving, and S. Wooldridge, 2003: *Global Climate Change and Coral Bleaching on the Great Barrier Reef*. State of Queensland Greenhouse Taskforce through the Department of Natural Resources and Mines.

Done, T.J., 1999: Coral community adaptability to environmental change at the scales of regions, reefs and reef zones. *American Zoologist*, **39(1)**, 66-79.

Donner, S.D., T.R. Knutson, and M. Oppenheimer, 2007: Model-based assessment of the role of human-induced climate change in the 2005 Caribbean coral bleaching event. *Proceedings of the National Academy of Sciences of the United States of America*, **104(13)**, 5483-5488.

Duane, T., 1996: *Sierra Nevada Ecosystem Project Final Report to Congress: Status of the Sierra Nevada*. Centers for Water and Wildland Resources, University of California.

Dye, D.G., 2002: Variability and trends in the annual snow-cover cycle in Northern Hemisphere land areas, 1972-2000. *Hydrological Processes*, **16(15)**, 3065-3077.

Edwards, M.S., 2004: Estimating scale-dependency in disturbance impacts: El Niños and giant kelp forests in the northeast Pacific. *Oecologia*, **138(3)**, 436-447.

Ettl, G.J. and D.L. Peterson, 1995: Growth response of subalpine fir (*Abies lasiocarpa*) to climate in the Olympic Mountains Washington, USA. *Global Change Biology*, **1(3)**, 213-230.

Euskirchen, S., A.D. McGuire, D.W. Kicklighter, Q. Zhuang, J.S. Clein, R.J. Dargaville, D.G. Dye, J.S. Kimball, K.C. McDonald, J.M. Melillo, V.E. Romanovsky, and N.V. Smith, 2006: Importance of recent shifts in soil thermal dynamics on growing season length, productivity, and carbon sequestration in terrestrial high-latitude ecosystems. *Global Change Biology*, **12**, 731-750.

Field, J.C., D. F. Boesch, D. Scavia, R. H. Buddemeier, V. R. Burkett, D. Cayan, M. Fogerty, M. A. Harwell, R. W. Howarth, C. Mason, L. J. Pietrafesa, D. J. Reed, T. C. Royer, A. H. Sallenger, M. Spranger, and J. G. Titus, 2001: Potential consequences of climate variability and change on coastal and marine resources, In: *Climate Change Impacts in the United States: Potential Consequences of Climate Change and Variability and Change*, Report for the U.S. Global Change Research Program, Cambridge University Press, Cambridge, UK.

Firing, J. and R.E. Brainard, 2006: Ten years of shipboard ADCP measurements along the Northwestern Hawaiian Islands. *Atoll Research Bulletin*, **543**, 331-368.

Florida Department of Environmental Protection, 2005: *Wekiva River Basin State Parks, Multi-Unit Management Plan*. pp.1-98.

Florida Keys National Marine Sanctuary, 2002: *Comprehensive Science Plan*. Available from http://floridakeys.noaa.gov/research_monitoring/fknms_science_plan.pdf.

Foster, M.S. and D.R. Schiel, 1985: *The Ecology of Giant Kelp Forests in California: a Community Profile*. Biological Report 85(7.2), U.S. Fish and Wildlife Service, Slidell, LA, pp.1-153.

Friedlander, A., G. S. Aeby, R. S. Brainard, A. Clark, E. DeMartini, S. Godwin, J. Kenyon, R. Kosaki, J. Maragos, and P. Vroom, 2005: The state of coral reef ecosystems of the northwestern Hawaiian islands, In: *The State of Coral Reef Ecosystems of the United States and Pacific Freely Associated States: 2005*, [Wadell, J.E. (ed.)]. NOAA/NCCOS Center for Coastal Monitoring and Assessment's Biogeography Team, Silver Spring, MD, pp. 270-311.

Friedlander, A.M. and E.E. DeMartini, 2002: Contrasts in density, size, and biomass of reef fishes between the northwestern and the main Hawaiian Islands: the effects of fishing down apex predators. *Marine Ecology Progress Series*, **230**, 253-264.

Galbraith, H., D. Yates, D.D. Purkey, A. Huber-Lee, J. Sieber, J. West, S. Herrod-Julius, and B. Joyce, in press: Climate warming, water storage, and chinook salmon in California's Sacramento Valley. *Climatic Change*.

Gardner, T.A., I.M. Cote, J.A. Gill, A. Grant, and A.R. Watkinson, 2003: Long-term region-wide declines in Caribbean corals. *Science*, **301(5635)**, 958-960.

Gavin, D.G., J.S. McLachlan, L.B. Brubaker, and K.A. Young, 2001: Postglacial history of subalpine forests, Olympic Peninsula, Washington, USA. *The Holocene*, **11(2)**, 177-188.

Giese, G.L., H.B. Wilder, and G.G. Parker, 1985: *Hydrology of Major Estuaries and Sounds of North Carolina*. USGS Water-Supply Paper 2221, USGS, pp.1-108.

Glynn, P.W., 1993: Coral reef bleaching: ecological perspectives. *Coral Reefs*, **12(1)**, 1-17.

Graham, M.H., 2004: Effects of local deforestation on the diversity and structure of southern California giant kelp forest food webs. *Ecosystems*, **7(4)**, 341-357.

Graham, M.H., P.K. Dayton, and J.M. Erlandson, 2003: Ice ages and ecological transitions on temperate coasts. *Trends in Ecology and Evolution*, **18(1)**, 33-40.

Great Barrier Reef Marine Park Authority, 2007: *Measuring the Economic and Financial Value of the Great Barrier Reef Marine Park 2005/06*. Access Economics.

Grigg, R.W., 1981: Acropora in Hawaii. Part 2: zoogeography. *Pacific Science*, **35**, 15-24.

Grigg, R.W., 1982: Darwin point: a threshold for atoll formation. *Coral Reefs*, **1(1)**, 29-34.

Grigg, R.W., 1988: Paleoceanography of coral reefs in the Hawaiian-Emperor chain. *Science*, **240(4860)**, 1737-1743.

Grigg, R.W., 1998: Holocene coral reef accretion in Hawaii: a function of wave exposure and sea level history. *Coral Reefs*, **17(3)**, 263-272.

Grigg, R.W., 2006: The history of marine research in the Northwestern Hawaiian Islands: lessons from the past and hopes for the future. *Atoll Research Bulletin*, **543**, 13-22.

Grigg, R.W., J. Polovina, A. Friedlander, and S. Rohman, 2007: Biology and paleoceanography of the coral reefs in the northwestern Hawaiian Islands, In: *Coral Reefs of the United States*, [Riegl, B. and R. Dodge (eds.)]. Springer-Vergal Publishing.

Grigg, R.W., J. Wells, and C. Wallace, 1981: Acropora in Hawaii, Part 1: history of the scientific record, systematics and ecology. *Pacific Science*, **35**, 1-13.

Guzmán, H.M. and J. Cortés, 2001: Changes in reef community structure after fifteen years of natural disturbances in the eastern pacific (Costa Rica). *Bulletin of Marine Science*, **69(1)**, 133-149.

Halpern, B.S., 2003: The impact of marine reserves: do reserves work and does reserve size matter? *Ecological Applications*, **13(1)**, S117-S137.

Halpern, B.S. and K. Cottenie, 2007: Little evidence for climate effects on local-scale structure and dynamics of California kelp forest communities. *Global Change Biology*, **13(1)**, 236-251.

Hamlet, A.F., P.W. Mote, M.P. Clark, and D.P. Lettenmaier, 2005: Effects of temperature and precipitation variability on snowpack trends in the western United States. *Journal of Climate*, **18(21)**, 4545-4561.

Hamlet, A.F., P.W. Mote, M.P. Clark, and D.P. Lettenmaier, 2007: Twentieth-century trends in runoff, evapotranspiration, and soil moisture in the western United States. *Journal of Climate*, **20(8)**, 1468-1486.

Harvell, C.D., K. Kim, J.M. Burkholder, R.R. Colwell, P.R. Epstein, D.J. Grimes, E.E. Hofmann, E.K. Lipp, A. Osterhaus, and R.M. Overstreet, 1999: Emerging marine diseases--climate links and anthropogenic factors. *Science*, **285**, 1505-1510.

Harvell, C.D., C.E. Mitchell, J.R. Ward, S. Altizer, A.P. Dobson, R.S. Ostfeld, and M.D. Samuel, 2002: Climate warming and disease risks for terrestrial and marine biota. *Science*, **296(5576)**, 2158-2162.

Hatton, T., P. Reece, P. Taylor, and K. McEwan, 1998: Does leaf water efficiency vary among eucalypts in water-limited environments? *Tree Physiology*, **18(8)**, 529-536.

Hawkings, J., 1996: Case study 1: Canada old crow flats, Yukon territory, In: *Wetlands, Biodiversity and the Ramsar Convention: the Role of the Convention on Wetlands in the Conservation and Wise Use of Biodiversity*, [Hails, A.J. (ed.)]. Ramsar Convention Bureau, Gland, Switzerland.

Hayhoe, K., D. Cayan, C.B. Field, P.C. Frumhoff, E.P. Maurer, N.L. Miller, S.C. Moser, S.H. Schneider, K.N. Cahill, E.E. Cleland, L. Dale, R. Drapek, R.M. Hanemann, L.S. Kalkstein, J. Lenihan, C.K. Lunch, R.P. Neilson, S.C. Sheridan, and J.H. Verville, 2004: Emissions pathways, climate change, and impacts on California. *Proceedings of the National Academy of Sciences of the United States of America*, **101**, 34-.

Haynes, D., 2001: *Great Barrier Reef Water Quality: Current Issues*. [Haynes, D. (ed.)]. Great Barrier Reef Marine Park Authority, Townsville, Australia.

Helmuth, B., B.R. Broitman, C.A. Blanchette, S. Gilman, P. Halpin, C.D.G. Harley, M.J. O'Donnell, G.E. Hofmann, B. Menge, and D. Strickland, 2006: Mosaic patterns of thermal stress in the rocky intertidal zone: implications for climate change. *Ecological Monographs*, **76(4)**, 461-479.

Herrlinger, T.J., 1981: *Range Extension of Kelletia Kelletii*. Veliger, pp. 1-78.

Heusser, C.J., 1974: Quaternary vegetation, climate, and glaciation of the Hoh River Valley, Washington. *Geological Society of America Bulletin*, **85(10)**, 1547-1560.

Hewatt, W.G., 1937: Ecological studies on selected marine intertidal communities of Monterey Bay, California. *American Midland Naturalist*, **18(2)**, 161-206.

Heyward, F., 1939: The relation of fire to stand composition of longleaf pine forests. *Ecology*, **20(2)**, 287-304.

Hinzman, L.D., N.D. Bettez, W.R. Bolton, F.S. Chapin, III, M.B. Dyurgerov, C.L. Fastie, B. Griffith, R.D. Hollister, A. Hope, H.P. Huntington, A.M. Jensen, G.J. Jia, T. Jorgenson, D.L. Kane, D.R. Klein, G. Kofinas, A.H. Lynch, A.H. Lloyd, A.D. McGuire, F.E. Nelson, M. Nolan, W.C. Oechel, T.E. Osterkamp, C.H. Racine, V.E. Romanovsky, R.S. Stone, D.A. Stow, M. Sturm, C.E. Tweedie, G.L. Vourlitis, M.D. Walker, P.J. Webber, J. Welker, K.S. Winker, and K. Yoshikawa, 2005: Evidence and implications of recent climate change in northern Alaska and other arctic regions. *Climatic Change*, **72(3)**, 251-298.

Hobbie, J.E., B. J. Copeland, and W. G. Harrison, 1975: Sources and fates of nutrients in the Pamlico River estuary, North Carolina, In: *Chemistry, Biology and the Estuarine System*, [Cronin, L.E. (ed.)]. Academic Press, New York, NY, pp. 287-302.

Hoegh-Guldberg, O., 1999: Climate change, coral bleaching and the future of the world's coral reefs. *Marine & Freshwater Research*, **50(8)**, 839-866.

Hoegh-Guldberg, O., 2004: Coral reefs and projections of future change, In: *Coral Health and Disease*, [Rosenberg, E. and Y. Loya (eds.)]. Springer, Berlin, Germany, pp. 463-484.

Hoegh-Guldberg, O., K. Anthony, R. Berkelmans, S. Dove, K. Fabricius, J. Lough, P. A. Marshall, M. J. H. van Oppen, A. Negri, and B. Willis, 2007: Vulnerability of reef-building corals on the Great Barrier Reef to Climate Change, In: *Climate Change and the Great Barrier Reef*, [Johnson, J.E. and P.A. Marshall (eds.)]. Great Barrier Reef Marine Park Authority & Australian Greenhouse Office.

Hoegh-Guldberg, O. and J.S. Pearse, 1995: Temperature, food availability, and the development of marine invertebrate larvae. *American Zoologist*, **35(4)**, 415-425.

Hoeke, R., R. Brainard, R. Moffitt, and M. Merrifield, 2006: The role of oceanographic conditions and reef morphology in the 2002 coral bleaching event in the Northwestern Hawaiian Islands. *Atoll Research Bulletin*, **543**, 489-503.

Hoffman, J., 2003: Designing reserves to sustain temperate marine ecosystems in the face of global climate change, In: *Buying Time: a User's Manual for Building Resistance and Resilience to Climate Change in Natural Systems*, [Hansen, L.J., J.L. Biringer, and J.R. Hoffman (eds.)]. WWF Climate Change Program, Washington, DC, pp. 123-155.

Hofmann, E.E., J.M. Klinck, S.E. Ford, and E.N. Powell, 1999: Disease dynamics: modeling of the effect of climate change on oyster disease. *National Shellfisheries Association*, **19(1)**, 329-.

Hogg, E.H., 2005: Impacts of drought on forest growth and regeneration following fire in southwestern Yukon, Canada. *Canadian Journal of Forest Research*, **35(9)**, 2141-2150.

Hogg, E.H. and P.Y. Bernier, 2005: Climate change impacts on drought-prone forests in western Canada. *Forestry Chronicle*, **81(5)**, 675-682.

Hogg, E.H., J.P. Brandt, and P. Hochtubajda, 2005: Factors affecting interannual variation in growth of western Canadian aspen forests during 1951-2000. *Canadian Journal of Forest Research*, **35(3)**, 610-622.

Holbrook, S.J., R.J. Schmitt, and J.S. Stephens, Jr., 1997: Changes in an assemblage of temperate reef fishes associated with a climate shift. *Ecological Applications*, **7(4)**, 1299-1310.

Holman, M.L. and D.L. Peterson, 2006: Spatial and temporal variability in forest growth in the Olympic Mountains, Washington: sensitivity to climatic variability. *Canadian Journal of Forest Research*, **36(1)**, 92-104.

Hughes, T.P., A.H. Baird, D.R. Bellwood, M. Card, S.R. Connolly, C. Folke, R. Grosberg, O. Hoegh-Guldberg, J.B.C. Jackson, J. Kleypas, J.M. Lough, P. Marshall, M. Nystrom, S.R. Palumbi, J.M. Pandolfi, B. Rosen, and J. Roughgarden, 2003: Climate change, human impacts, and the resilience of coral reefs. *Science*, **301(5635)**, 929-933.

Hutchins, L.W., 1947: The bases for temperature zonation in geographical distribution. *Ecological Monographs*, **17(3)**, 325-335.

Inouye, D.W., B. Barr, K.B. Armitage, and B.D. Inouye, 2000: Climate change is affecting altitudinal migrants and hibernating species. *Proceedings of the National Academy of Sciences of the United States of America*, **97(4)**, 1630-1633.

IPCC, 2001: *Climate Change 2001: the Scientific Basis. Contribution of Working Group I to the Third Assessment Report of the Intergovernmental Panel on Climate Change.* [Houghton, J.T., Y. Ding, D.J. Griggs, M. Noguer, P.J. van der Linden, X. Dai, K. Maskell, and C.A. Johnson (eds.)]. Cambridge University Press, Cambridge, United Kingdom and New York, NY, USA.

IPCC, 2007b: Summary for policymakers, In: *Climate Change 2007: Impacts, Adaptation and Vulnerability. Contribution of Working Group II to the Fourth Assessment Report of the Intergovernmental Panel on Climate Change*, [Parry, M.L., O.F. Canziani, J.P. Palutikof, P.J. van der Linden, and C.E. Hanson (eds.)]. Cambridge University Press, Cambridge, UK, pp. 7-22.

IPCC, 2007a: Summary for policymakers, In: *Climate Change 2007: the Physical Science Basis. Contribution of Working Group I to the Fourth Assessment Report of the Intergovernmental Panel on Climate Change*, [Solomon, S., D. Qin, M. Manning, Z. Chen, M. Marquis, K.B. Averyt, M. Tignor, and H.L. Miller (eds.)]. Cambridge University Press, Cambridge, United Kingdom and New York, NY, USA.

Jaap, W.C., 1979: Observation on zooxanthellae expulsion at Middle Sambo Reef, Florida Keys. *Bulletin of Marine Science*, **29**, 414-422.

Jaap, W.C., 1984: *The Ecology of the South Florida Coral Reefs: a Community Profile*. FWS OBS-82/08 and MMS 84-0038, U.S. Fish and Wildlife Service, Metaine, LA, pp.1-152.

Jaap, W.C. and P. Hallock, 1990: Coral reefs, In: *Ecosystems of Florida*, [Meyers, R.L. and J.J. Ewel (eds.)]. University of Central Florida Press, Orlando, Florida, pp. 574-616.

Jackson, J.B.C., M.X. Kirby, W.H. Berger, K.A. Bjorndal, L.W. Botsford, B.J. Bourque, R.H. Bradbury, R. Cooke, J. Erlandson, J.A. Estes, T.P. Hughes, S. Kidwell, C.B. Lange, H.S. Lenihan, J.M. Pandolfi, C.H. Peterson, R.S. Steneck, M.J. Tegner, and R.R. Warner, 2001: Historical overfishing and the recent collapse of coastal ecosystems. *Science*, **293**, 629-638.

Johnson, J. and P. Marshall, 2007: *Climate Change and the Great Barrier Reef: A Vulnerability Assessment*. Great Barrier Reef Marine Park Authority.

Jokiel, P.L., 1987: Ecology, biogeography and evolution of corals in Hawaii. *Trends in Ecology and Evolution*, **2(7)**, 179-182.

Jokiel, P.L. and E.K. Brown, 2004: Global warming, regional trends and inshore environmental conditions influence coral bleaching in Hawaii. *Global Change Biology*, **10(10)**, 1627-1641.

Jokiel, P.L., E.K. Brown, A. Friedlander, S.K. Rodgers, and W.R. Smith, 2004: Hawaii coral reef assessment and monitoring program: spatial patterns and temporal dynamics in reef coral communities. *Pacific Science*, **58(2)**, 159-174.

Jokiel, P.L. and S.L. Coles, 1990: Response of Hawaiian and other Indo-Pacific reef corals to elevated temperature. *Coral Reefs*, **8(4)**, 155-162.

Kattsov, V.M. and E. Källén, 2005: Future climate change: modeling and scenarios for the Arctic, In: *Arctic Climate Impact Assessment*, Cambridge University Press, Cambridge, UK, pp. 99-150.

Kay, E.A. and S.R. Palumbi, 1987: Endemism and evolution in Hawaiian marine invertebrates. *Trends in Ecology and Evolution*, **2**, 183-186.

Keller, B.D. and B.D. Causey, 2005: Linkages between the Florida Keys National Marine Sanctuary and the South Florida Ecosystem Restoration Initiative. *Ocean & Coastal Management*, **48(11-12)**, 869-900.

Keller, B.D. and S. Donahue, 2006: *2002-03 Florida Keys National Marine Sanctuary Science Report: an Ecosystem Report Card After Five Years of Marine Zoning*. Marine Sanctuaries Conservation Series NMSP-06-12, U.S. Department of Commerce, National Oceanic and Atmospheric Administration, National Marine Sanctuary Program, Silver Spring, MD, pp.1-358.

Kemp, W.M., P.A. Sampou, J. Garber, J. Tuttle, and W.R. Boynton, 1992: Seasonal depletion of oxygen from bottom waters of Chesapeake Bay: roles of benthic and planktonic respiration and physical exchange processes. *Marine Ecology Progress Series*, **85(1)**.

Kenyon, J. and R.E. Brainard, 2006: Second recorded episode of mass coral bleaching in the Northwestern Hawaiian Islands. *Atoll Research Bulletin*, **543**, 505-523.

Kenyon, J.C., P.S. Vroom, K.N. Page, M.J. Dunlap, C.B. Wilkinson, and G.S. Aeby, 2006: Community structure of hermatypic corals at French Frigate Shoals, Northwestern Hawaiian Islands: capacity for resistance and resilience to selective stressors. *Pacific Science*, **60(2)**, 153-175.

Kiessling, W., 2001: Paleoclimatic significance of Phanerozoic reefs. *Geology*, **29(8)**, 751-754.

Klein, E., E.E. Berg, and R. Dial, 2005: Wetland drying and succession across the Kenai Peninsula Lowlands, south-central Alaska. *Canadian Journal of Forest Research*, **35(8)**, 1931-1941.

Kleypas, J.A., 2006: Constraints on predicting coral reef response to climate change, In: *Geological Approaches to Coral Reef Ecology*, [Aronson, R. (ed.)]. Springer, Verlag, NY, pp. 386-424.

Kleypas, J.A., R.W. Buddemeier, and J.P. Gattuso, 2001: The future of coral reefs in an age of global change. *International Journal of Earth Sciences*, **90(2)**, 426-437.

Kleypas, J.A., J.W. McManus, and L.A.B. Mendez, 1999: Environmental limits to coral reef development: where do we draw the line? *Integrative and Comparative Biology*, **39(1)**, 146-159.

Knowles, N., M.D. Dettinger, and D.R. Cayan, 2006: Trends in snowfall versus rainfall in the Western United States. *Journal of Climate*, **19(18)**, 4545-4559.

Korfmacher, K.S., 1998: Invisible successes, visible failures: paradoxes of ecosystem management in the Albemarle-Pamlico estuarine study. *Coastal Management*, **26(3)**, 191-212.

Korfmacher, K.S., 2002: Science and ecosystem management in the Albemarle-Pamlico Estuarine study. *Ocean & Coastal Management*, **45**, 277-300.

Krapu, G.L., D.A. Brandt, and R.R. Cox, Jr., 2004: Less waste corn, more land in soybeans, and the switch to genetically modified crops: trends with important implications for wildlife management. *Wildlife Society Bulletin*, **32(1)**, 127-136.

Kuta, K.G. and L.L. Richardson, 1996: Abundance and distribution of black band disease on coral reefs in the northern Florida Keys. *Coral Reefs*, **15(4)**, 219-223.

Ladah, L., J. Zertuche-Gonzalez, and G. Hernandez-Carmona, 1999: Rapid recovery giant kelp (*Macrocystis pyrifera*, Phaeophyceae) recruitment near its southern limit in Baja California after mass disappearance during ENSO 1997-1998. *Journal of Phycology*, **35**, 1106-1112.

Lafleur, P.M., 1993: Potential water balance response to climatic warming: the case of a coastal wetland ecosystem of the James Bay lowland. *Wetlands*, **13(4)**, 270-276.

Lang, J.C., H.R. Lasker, E.H. Gladfelter, P. Hallock, W.C. Jaap, F.J. Losada, and R.G. Muller, 1992: Spatial and temporal variability during periods of "recovery" after mass bleaching on Western Atlantic coral reefs. *American Zoologist*, **32(6)**, 696-706.

Larson, D.L., 1995: Effects of climate on numbers of northern prairie wetlands. *Climatic Change*, **30(2)**, 169-180.

Lawrence, D.M. and A.G. Slater, 2005: A projection of severe near-surface permafrost degradation during the 21st century. *Geophysical Research Letters*, **32(L24401)**.

Lee, T.N., E. Williams, E. Johns, D. Wilson, and N. P. Smith, 2002. Transport processes linking south Florida coastal ecosystems, In: *The Everglades, Florida Bay, and Coral Reefs of the Florida Keys: an Ecosystem Sourcebook*, [Porter, J.W. and K.G. Porter (eds.)]. CRC Press, Boca Raton, FL, pp. 309-342.

Lemieux, C.J. and D.J. Scott, 2005: Climate change, biodiversity conservation and protected area planning in Canada. *The Canadian Geographer*, **49(4)**, 384-399.

Lenihan, J.M., D. Bachelet, R. Drapek, and R.P. Neilson, 2006: *The Response of Vegetation, Distribution, Ecosystem Productivity, and Fire in California to Future Climate Scenarios Simulated by the MC1 Dynamic Vegetation Model*. Climate action team report to the Governor and Legislators, available from http://www.energy.ca.gov/2005publications/CEC-500-2005-191/CEC-500-2005-191-SF.PDF.

Lessios, H.A., D.R. Robertson, and J.D. Cubit, 1984: Spread of *Diadema* mass mortality through the Caribbean. *Science*, **226(4672)**, 335-337.

Levitus, S., J.I. Antonov, T.P. Boyer, and C. Stephens, 2000: Warming of the world ocean. *Science*, **287**, 2225-2229.

Lighty, R.G., I.G. Macintyre, and R. Stuckenrath, 1978: Submerged early Holocene barrier reef south-east Florida shelf. *Nature*, **276(5683)**, 59-60.

Lins, H.F. and J.R. Slack, 1999: Streamflow trends in the United States. *Geophysical Research Letters*, **26(2)**, 227-230.

Littell, J.S., 2006: Climate impacts to forest ecosystem processes: douglas-fir growth in northwestern U.S. mountain landscapes and area burned by wildfire in western U.S. ecoprovinces. *PhD Dissertation, University of Washington, Seattle*.

Logan, J.A. and J.A. Powell, 2001: Ghost forests, global warming, and the mountain pine beetle (*Coleoptera: Scolytidae*). *American Entomologist*, **47(3)**, 160-172.

Logan, J.A., J. Regniere, and J.A. Powell, 2003: Assessing the impacts of global warming on forest pest dynamics. *Frontiers in Ecology and the Environment*, **1(3)**, 130-137.

Lotze, H.K., H.S. Lenihan, B.J. Bourque, R.H. Bradbury, R.G. Cooke, M.C. Kay, S.M. Kidwell, M.X. Kirby, C.H. Peterson, and J.B.C. Jackson, 2006: Depletion, degradation, and recovery potential of estuaries and coastal seas. *Science*, **312(5781)**, 1806-1809.

Lough, J., 2007: Climate and climate change scenarios for the Great Barrier Reef, In: *Climate Change and the Great Barrier Reef*, [Johnson, J. and P. Marshall (eds.)]. Great Barrier Reef Marine Park Authority, Townsville, Australia, pp. 15-50.

Luppi, T.A., E.D. Spivak, and C.C. Bas, 2003: The effects of temperature and salinity on larval development of *Armases rubripes* Rathbun, 1897 (Brachyura, Grapsoidea, Sesarmidae), and the southern limit of its geographical distribution. *Estuarine, Coastal and Shelf Science*, **58(3)**, 575-585.

Magnuson, J.J., D.M. Robertson, B.J. Benson, R.H. Wynne, D.M. Livingstone, T. Arai, R.A. Assel, R.G. Barry, V. Card, E. Kuusisto, N.G. Granin, T.D. Prowse, K.M. Stewart, and V.S. Vuglinski, 2000: Historical trends in lake and river ice cover in the Northern Hemisphere. *Science*, **289(5485)**, 1743-1746.

Mallin, M.A., J.M. Burkholder, L.B. Cahoon, and M.H. Posey, 2000: North and South Carolina coasts. *Marine Pollution Bulletin*, **41(1)**, 56-75.

Mantua, N.J., S.R. Hare, Y. Zhang, J.M. Wallace, and R.C. Francis, 1997: A pacific interdecadal climate oscillation with impacts on salmon production. *Bulletin of the American Meteorological Society*, **78(6)**, 1069-1079.

Maragos, J.E., D.C. Potts, G. Aeby, D. Gulko, J. Kenyon, D. Siciliano, and D. VanRavenswaay, 2004: 2000-2002 Rapid ecological assessment of corals (*Anthozoa*) on shallow reefs of the Northwestern Hawaiian Islands. Part 1: species and distribution. *Pacific Science*, **58(2)**, 211-230.

Marshall, P. and H. Schuttenberg, 2006: Adapting coral reef management in the face of climate change, In: *Coral Reefs and Climate Change: Science and Management*, [Phinney, J.T., O. Hoegh-Guldberg, J. Kleypas, W.J. Skirving, and A. Strong (eds.)]. American Geophysical Union, Washington, DC, pp. 223-241.

McBean, G.A., G. Alekseev, D. Chen, E. Forland, J. Fyfe, P. Y. Groisman, R. King, H. Melling, R. Vose, and P. H. Whitefield, 2005: Arctic climate - past and present, In: *Arctic Climate Impact Assessment*, [Corell, R.W. (ed.)]. Cambridge University Press, Cambridge, UK, pp. 21-60.

McCabe, G.J. and D.M. Wolock, 2002: Trends and temperature sensitivity of moisture conditions in the conterminous United States. *Climate Research*, **20(1)**, 19-29.

McDonald, K.C., J.S. Kimball, E. Njoku, R. Zimmermann, and M. Zhao, 2004: Variability in springtime thaw in the terrestrial high latitudes: monitoring a major control on the biospheric assimilation of atmospheric CO_2 with spaceborne microwave remote sensing. *Earth Interactions*, **8(20)**, 1-23.

McGowan, J.A., D.R. Cayan, L.M. Dorman, and A. Butler, 1998: Climate-ocean variability and ecosystem response in the Northeast Pacific. *Science*, **281(5374)**, 210-217.

McGuire, A.D., M. Apps, F. S. Chapin III, R. Dargaville, M. D. Flannigan, E. S. Kasischke, D. Kicklighter, J. Kimball, W. Kurz, D. J. McCrae, K. A. McDonald, J. Melillo, R. Myneni, B. J. Stocks, D. L. Verbyla, and Q. Zhuang, 2004: Land cover disturbances and feedbacks to the climate system in Canada and Alaska, In: *Land Change Science: Observing, Monitoring, and Understanding Trajectories of Change on the Earth's Surface*, [Gutman, G. and A.C. Janetos (eds.)]. Kluwer Academic Publisher, Netherlands, pp. 139-162.

McLachlan, J.S. and L.B. Brubaker, 1995: Local and regional vegetation change on the northeastern Olympic Peninsula during the Holocene. *Canadian Journal of Botany*, **73(10)**, 1618-1627.

Melillo, J., A.D. McGuire, D.W. Kicklighter, B. Moore, III, C.J. Vorosmarty, and A.L. Schloss, 1993: Global climate change and terrestrial net primary production. *Nature*, **363(6426)**, 234-240.

Mid-Atlantic Regional Assessment Team, 2000: *Preparing for a Changing Climate: Mid-Atlantic Overview*. U.S. Global Change Research Program, U.S. Environmental Protection Agency and Pennsylvania State University.

Millar, C.I., R.D. Westfall, D.L. Delany, J.C. King, and L.J. Graumlich, 2004: Response of subalpine conifers in the Sierra Nevada, California, USA, to 20th-century warming and decadal climate variability. *Arctic, Antarctic, and Alpine Research*, **36(2)**, 181-200.

Miller, J., R. Waara, E. Muller, and C. Rogers, 2006: Coral bleaching and disease combine to cause extensive mortality on reefs in US Virgin Islands. *Coral Reefs*, **25(3)**, 418-418.

Miller, S.L., M. Chiappone, D.W. Swanson, J.S. Ault, S.G. Smith, G.A. Meester, J. Luo, E.C. Franklin, J.A. Bohnsack, D.E. Harper, and D.B. McClellan, 2001: An extensive deep reef terrace on the Tortugas bank, Florida Keys National Marine Sanctuary. *Coral Reefs*, 299-300.

Moorhead, K.K. and M.M. Brinson, 1995: Response of wetlands to rising sea level in the lower coastal plain of North Carolina. *Ecological Applications*, **5(1)**, 261-271.

Mote, P.W., 2003: Trends in temperature and precipitation in the Pacific Northwest during the twentieth century. *Northwest Science*, **77(4)**, 271-282.

Mote, P.W., 2006: Climate-driven variability and trends in mountain snowpack in western North America. *Journal of Climate*, **19(23)**, 6209-6220.

Mote, P.W., A.F. Hamlet, M.P. Clark, and D.P. Lettenmaier, 2005: Declining mountain snowpack in Western North America. *Bulletin of the American Meteorological Society*, **86(1)**, 39-49.

Munday, P.L., G. P. Jones, M. Sheaves, A. J. Williams, and G. Goby, 2007: Vulnerability of fishes of the Great Barrier Reef to climate change, In: *Climate Change and the Great Barrier Reef*, [Johnson, J. and P. Marshall (eds.)]. Great Barrier Reef Marine Park Authority, Townsville.

Mundy, B.C., 2005: *Checklist of the Fishes of the Hawaiian Archipelago*. Bishop Museum Press, Honolulu, Hawaii.

Murray, S.N. and M.M. Littler, 1981: Biogeographical analysis of intertidal macrophyte floras of southern California. *Journal of Biogeography*, **8(5)**, 339-351.

Myers, R.A. and B. Worm, 2003: Rapid worldwide depletion of predatory fish communities. *Nature*, **423(6937)**, 280-283.

Myers, R.A. and B. Worm, 2005: Extinction, survival or recovery of large predatory fishes. *Philosophical Transactions of the Royal Society of London, Series B: Biological Sciences*, **360(1453)**, 13-20.

Najjar, R.G., H.A. Walker, P.J. Anderson, E.J. Barron, R.J. Bord, J.R. Gibson, V.S. Kennedy, C.G. Knight, J.P. Megonigal, and R.E. O'Connor, 2000: The potential impacts of climate change on the mid-Atlantic coastal region. *Climate Research*, **14**, 219-233.

Nakawatase, J.M. and D.L. Peterson, 2006: Spatial variability in forest growth- climate relationships in the Olympic Mountains, Washington. *Canadian Journal of Forest Research*, **36(1)**, 77-91.

National Assessment Synthesis Team, 2000: *Climate Change Impacts on the United States: the Potential Consequences of Climate Variability and Change*. U.S. Global Change Research Program, Washington, DC.

National Park Service, 1996: *Water Resources Management Plan - Big Bend National Park*. Department of Hydrology and Water Resources, Univ. of Arizona, Tucson, Big Bend National Park, Texas, and National Park Service - Water Resources Division, Fort Collins, CO, pp.1-163.

National Park Service, 2004: *Rio Grande Wild and Scenic River: Final General Management Plan / Environmental Impact Statement*.

Nearing, M.A., 2001: Potential changes in rainfall erosivity in the U.S. with climate change during the 21st century. *Journal of Soil and Water Conservation*, **56(3)**, 229-232.

Neilson, R.P. and R.J. Drapek, 1998: Potentially complex biosphere responses to transient global warming. *Global Change Biology*, **4(5)**, 505-521.

New Mexico Department of Game and Fish, 2006: *Comprehensive Wildlife Conservation Strategy for New Mexico*. New Mexico Department of Game and Fish, Santa Fe, New Mexico, pp.1-526.

O'Connor, M.I., J.F. Bruno, S.D. Gaines, B.S. Halpern, S.E. Lester, B.P. Kinlan, and J.M. Weiss, 2007: Temperature control of larval dispersal and the implications for marine ecology, evolution, and conservation. *Proceedings of the National Academy of Sciences of the United States of America*, **104**, 1266-1271.

Oechel, W.C., S.J. Hastings, R.C. Zulueta, G.L. Vourlitis, L. Hinzman, and D. Kane, 2000: Acclimation of ecosystem CO_2 exchange in the Alaskan Arctic in response to decadal climate warming. *Nature*, **406(6799)**, 978-981.

Paerl, H.W., R.L. Dennis, and D.R. Whitall, 2002: Atmospheric deposition of nitrogen: Implications for nutrient overenrichment of coastal waters. *Estuaries*, **25**, 677-693.

Paerl, H.W., L.M. Valdes, A.R. Joyner, B.L. Peierls, M.F. Piehler, S.R. Riggs, R.R. Christian, L.A. Eby, L.B. Crowder, J.S. Ramus, E.J. Clesceri, C.P. Buzzelli, and R.A. Luettich, Jr., 2006: Ecological response to hurricane events in the Pamlico Sound System, North Carolina, and implications for assessment and management in a regime of increased frequency. *Estuaries and coasts*, **29(6A)**, 1033-1045.

Pagano, T., P. Pasteris, M. Dettinger, D. Cayan, and K. Redmond, 2004: Water year 2004: western water managers feel the heat. *EOS Transactions*, **85(40)**, 385-392.

Pandolfi, J.M., J.B.C. Jackson, N. Baron, R.H. Bradbury, H.M. Guzman, T.P. Hughes, C.V. Kappel, F. Micheli, J.C. Ogden, H.P. Possingham, and E. Sala, 2005: Are U. S. coral reefs on the slippery slope to slime? *Science*, **307(5716)**, 1725-1726.

Parasiewicz, P., undated: Strategy for sustainable management of the Upper Delaware River basin.

Peierls, B.L., R.R. Christian, and H.W. Paerl, 2003: Water quality and phytoplankton as indicators of hurricane impacts on a large estuarine ecosystem. *Estuaries*, **26(5)**, 1329-1343.

Peterson, C.H. and M.J. Bishop, 2005: Assessing the environmental impacts of beach nourishment. *BioScience*, **55(10)**, 887-896.

Peterson, C.H., M.J. Bishop, G.A. Johnson, L.M. D'Anna, and L.M. Manning, 2006: Exploiting beach filling as an unaffordable experiment: benthic intertidal impacts propagating upwards to shorebirds. *Journal of Experimental Marine Biology and Ecology*, **338(2)**, 205-221.

Peterson, C.H. and J. A. Estes, 2001: Conservation and management of marine communities, [Bertness, M.D., S.D. Gaines, and M.E. Hay (eds.)]. pp. 469-508.

Peterson, D.W. and D.L. Peterson, 2001: Mountain hemlock growth responds to climatic variability at annual and decadal time scales. *Ecology*, **82(12)**, 3330-3345.

Petraitis, P.S., 1992: Effects of body size and water temperature on grazing rates of four intertidal gastropods. *Australian Journal of Ecology*, **17(4)**, 409-414.

Philippart, C.J.M., H.M. van Aken, J.J. Beukema, O.G. Bos, G.C. Cadee, and R. Dekker, 2003: Climate-related changes in recruitment of the bivalve *Macoma balthica*. *Limnology and Oceanography*, **48(6)**, 2171-2185.

Phillips, N.E., 2005: Growth of filter-feeding benthic invertebrates from a region with variable upwelling intensity. *Marine Ecology Progress Series*, **295**, 79-89.

Piehler, M.F., L.J. Twomey, N.S. Hall, and H.W. Paerl, 2004: Impacts of inorganic nutrient enrichment on the phytoplankton community structure and function in Pamlico Sound, NC USA. *Estuarine Coastal and Shelf Science*, **61(197)**, 207-.

Pimentel, D. and N. Kounang, 1998: Ecology of soil erosion in ecosystems. *Ecosystems*, **1(5)**, 416-426.

Podestá, G.P. and P.W. Glynn, 2001: The 1997-98 El Niño event in Panama and Galapagos: an update of thermal stress indices relative to coral bleaching. *Bulletin of Marine Science*, **69(1)**, 43-59.

Poff, N.L., M.M. Brinson, and J.W. Day, Jr., 2002: *Aquatic Ecosystems & Global Climate Change:Potential Impacts on Inland Freshwater and Coastal Wetland Ecosystems in the United States*. Pew Center on Global Climate Change, pp.1-56.

Polovina, J.J., E. Howell, D.R. Kobayashi, and M.P. Seki, 2001: The transition zone chlorophyll front, a dynamic global feature defining migration and forage habitat for marine resources. *Progress in Oceanography*, **49(1)**, 469-483.

Polovina, J.J., P. Kleiber, and D.R. Kobayashi, 1999: Application of TOPEX-POSEIDON satellite altimetry to simulate transport dynamics of larvae of spiny lobster, *Panulirus marginatus*, in the Northwestern Hawaiian Islands, 1993-1996. *Fishery Bulletin*, **97(1)**, 132-143.

Polovina, J.J., G.T. Mitchum, N.E. Graham, M.G. Craig, E.E. DeMartini, and E.N. Flint, 1994: Physical and biological consequences of a climate event in the central North Pacific. *Fisheries Oceanography*, **3(1)**, 15-21.

Polovina, J.P., G.T. Mitchem, and G.T. Evans, 1995: Decadal and basin-scale variation in mixed layer depth and the impact on biological production in the Central and North Pacific, 1960-1988. *Deep-sea Research*, **42**, 1701-1716.

Porter, J.W., J.F. Battey, and G.J. Smith, 1982: Perturbation and change in coral reef communities. *Proceedings of the National Academy of Sciences of the United States of America*, **79**, 1678-1681.

Porter, J.W. and O.W. Meier, 1992: Quantification of loss and change in Floridian reef coral populations. *Integrative and Comparative Biology*, **32(6)**, 625-.

Porter, J.W. and J.I. Tougas, 2001: Reef ecosystems: threats to their biodiversity. *Encyclopedia of Biodiversity*, **5**, 73-95.

Precht, W.F. and R.B. Aronson, 2004: Climate flickers and range shifts of reef corals. *Frontiers in Ecology and the Environment*, **2(6)**, 307-314.

Precht, W.F. and S. L. Miller, 2006: Ecological shifts along the Florida reef tract: the past as a key to the future, In: *Geological Approaches to Coral Reef Ecology*, [Aronson, R.B. (ed.)]. Springer, New York, NY, pp. 237-312.

Puglise, K.A. and R. Kelty, 2007: *NOAA Coral Reef Ecosystem Research Plan for Fiscal Years 2007 to 2011*. NOAA Technical Memorandum CRCP 1, NOAA Coral Reef Conservation Program, Silver Spring, MD, pp.1-128.

Randall, J.E., 1998: Zoogeography of shore fishes of Indo-Pacific region. *Zoological Studies*, **37(4)**, 227-268.

Randall, J.E., J.L. Earle, R.L. Pyle, J.D. Parrish, and T. Hayes, 1993: Annotated checklist of the fishes of Midway Atoll, Northwestern Hawaiian Islands. *Pacific Science*, **47**, 356-400.

Rauzon, M.J., 2001: *Isles of Refuge: Wildlife and History of the North-Western Hawaiian Islands*. University of Hawaii Press.

Reaser, J.K., R. Pomerance, and P.O. Thomas, 2000: Coral bleaching and global climate change: scientific findings and policy recommendations. *Conservation Biology*, **14(5)**, 1500-1511.

Richardson, R.B. and J.B. Loomis, 2004: Adaptive recreation planning and climate change: a contingent visitation approach. *Ecological Economics*, **50**, 83-99.

Riggs, S.R., 1996: Sediment evolution and habitat function of organic-rich muds within the Albemarle estuarine system, North Carolina. *Estuaries*, **19(2A)**, 169-185.

Riggs, S.R. and D.V. Ames, 2003: *Drowning the North Carolina Coast: Sea-Level Rise and Estuarine Dynamics*. UNC-SG-03-04, NC Sea Grant College Program, Raleigh, NC, pp.1-152.

Riordan, B., D. Verbyla, and A.D. McGuire, 2006: Shrinking ponds in subarctic Alaska based on 1950-2002 remotely sensed images. *Journal of Geophysical Research-Biogeosciences*, **111**, G04002-.

Robblee, M.B., T.R. Barber, P.R. Carlson Jr, M.J. Durako, J.W. Fourqurean, L.K. Muehlstein, D. Porter, L.A. Yarbro, R.T. Zieman, and J.C. Zieman, 1991: Mass mortality of the tropical seagrass *Thalassia testudinum* in Florida Bay (USA). *Marine Ecology Progress Series*, **71(3)**, 297-299.

Roberts, C.M., C.J. McClean, J.E.N. Veron, J.P. Hawkins, G.R. Allen, D.E. McAllister, C.G. Mittermeier, F.W. Schueler, M. Spalding, and F. Wells, 2002: Marine biodiversity hotspots and conservation priorities for tropical reefs. *Science*, **295(5558)**, 1280-1284.

Roberts, C.M., J. D. Reynolds, I. M. Cote, and J. P. Hawkins, 2006: Redesigning coral reef conservation, In: *Coral Reef Conservation*, Cambridge University Press, Cambridge, UK, pp. 515-537.

Roberts, H.H., L.J.Jr. Rouse, and N.D. Walker, 1983: Evolution of cold-water stress conditions in high-latitude reef systems: Florida Reef Tract and the Bahama Banks. *Caribbean Journal of Science*, **19(55)**, 60-.

Rogers, C.E. and J.P. McCarty, 2000: Climate change and ecosystems of the mid-Atlantic region. *Climate Research*, **14**, 235-244.

Rouse, W.R., 1998: A water balance model for a subarctic sedge fen and its application to climatic change. *Climatic Change*, **38(2)**, 207-234.

Running, S.W., J.B. Way, K.C. McDonald, J.S. Kimball, S. Frolking, A.R. Keyser, and R. Zimmerman, 1999: Radar remote sensing proposed for monitoring freeze-thaw transitions in boreal regions. *Eos Transactions, American Geophysical Union*, **80(19)**, 220-221.

Saavedra, F., D.W. Inouye, M.V. Price, and J. Harte, 2003: Changes in flowering and abundance of *Delphinium nuttallianum* (Ranunculaceae) in response to a subalpine climate warming experiment. *Global Change Biology*, **9(6)**, 885-894.

Saether, B.E., J. Tufto, S. Engen, K. Jerstad, O.W. Roestad, and J.E. Skaatan, 2000: Population dynamical consequences of climate change for a small temperate songbird. *Science*, **287(5454)**, 854-856.

Sagarin, R.D., J.P. Barry, S.E. Gilman, and C.H. Baxter, 1999: Climate-related change in an intertidal community over short and long time scales. *Ecological Monographs*, **69(4)**, 465-490.

Sala, E., 2006: Top predators provide insurance against climate change. *Trends in Ecology and Evolution*, **21**, 479-480.

Salathé, E.P., Jr., 2005: Downscaling simulations of future global climate with application to hydrologic modelling. *International Journal of Climatology*, **25(4)**, 419-436.

Sanford, E., 1999: Regulation of keystone predation by small changes in ocean temperature. *Science*, **283(5410)**, 2095-2097.

Sanford, E., 2002: The feeding, growth, and energetics of two rocky intertidal predators (*Pisaster ochraceus* and *Nucella canaliculata*) under water temperatures simulating episodic upwelling. *Journal of Experimental Marine Biology and Ecology*, **273(2)**, 199-218.

Schmidt, J.C., B. L. Everitt, and G. A. Richard, 2003: Hydrology and geomorphology of the Rio Grande and implications for river rehabilitation, In: *Aquatic Fauna of the Northern Chihuahuan Desert.Museum of Texas Tech University*, [Garrett, G.P. and N.L. Allan (eds.)]. Museum of Texas Tech University, Special Publications, Lubbock, TX, pp. 25-45.

Serreze, M.C., J.E. Walsh, F.S. Chapin III, T. Osterkamp, M. Dyurgerov, V. Romanovsky, W.C. Oechel, J. Morison, T. Zhang, and R.G. Barry, 2000: Observational evidence of recent change in the northern high-latitude environment. *Climatic Change*, **46(1-2)**, 159-207.

Shallenberger, R.J., 2006: History of management in the Northwestern Hawaiian Islands. *Atoll Research Bulletin*, **543**, 23-32.

Sheppard, C., 2006: Longer-term impacts of climate change on coral reefs, In: *Coral Reef Conservation*, [Côté, I.M. and J.D. Reynolds (eds.)]. Cambridge University Press, Cambridge, UK, pp. 264-290.

Shevock, J.R., 1996: *Status of Rare and Endemic Plants*. Sierra Nevada Ecosystem Project: final report to Congress, Vol. II, Assessments and scientific basis for management options University of California, Centers for Water and Wildland Resources, Davis, pp.691-707.

Shinn, E.A., 1989: What is really killing the corals. *Sea Frontiers*, **35**, 72-81.

Sierra Nevada Ecosystem Project Science Team, 1996: *Fire and Fuels*. Sierra Nevada Ecosystem Project, final report to Congress, Volume I, Assessment Summaries and Management Strategies Report No. 37, Chapter 4, Centers for Water and Wildland Resources, University of California, Davis, pp.61-71.

Skeat, H., 2003: *Sustainable Tourism in the Great Barrier Reef Marine Park*. 2003 Environment by numbers: selected articles on Australia's environment 4617, Australian Bureau of Statistics.

Smith, A., J. Monkivitch, P. Koloi, J. Hassall, and G. Hamilton, 2004: Environmental impact assessment in the Great Barrier Reef Marine Park. *The Environmental Engineer*, **5(4)**, 14-18.

Smith, J.B., S.E. Ragland, and G.J. Pitts, 1996: Process for evaluating anticipatory adaptation measures for climate change. *Water, Air, & Soil Pollution*, **92(1)**, 229-238.

Smith, N.V., S.S. Saatchi, and J.T. Randerson, 2004: Trends in high northern latitude soil freeze and thaw cycles from 1988 to 2002. *Journal of Geophysical Research*, **109**, D12101-.

Smith, S.G., D.W. Swanson, J.S. Ault, M. Chiappone, and S.L. Miller, forthcoming: Sampling survey design for multiple spatial scale coral reef assessments in the Florida Keys. *Coral Reefs*.

Smith, S.V. and R.W. Buddemeier, 1992: Global change and coral reef ecosystems. *Annual Review of Ecology and Systematics*, **23**, 89-118.

Smith, T.M. and R.W. Reynolds, 2004: Improved extended seconstruction of SST (1854-1997). *Journal of Climate*, **17(12)**, 2466-2477.

Soto, C.G., 2001: The potential impacts of global climate change on marine protected areas. *Reviews in Fish Biology and Fisheries*, **11(3)**, 181-195.

St. Johns River Water Management District, 2002: *Middle St. Johns River Basin Surface Water Improvement and Management Plan*. Palatka, Florida, pp.1-78.

St. Johns River Water Management District, 2006a: *Middle St. Johns River Basin Initiative: Fiscal Year 2007-2008*. Palatka, Florida, pp.1-22.

St. Johns River Water Management District, 2006b: *Water Supply Assessment and Water Supply Plan*. Palatka, Florida, pp.1-4.

Stanley, D.W., 1992: *Historical Trends: Water Quality and Fisheries, Albemarle-Pamlico Sounds, With Emphasis on the Pamlico River Estuary*. UNC-SG-92-04, University of North Carolina Sea Grant College Program Publication, Institute for Coastal and Marine Resources, East Carolina University, Greenville, NC.

Stanley, D.W. and S.W. Nixon, 1992: Stratification and bottom-water hypoxia in the Pamlico River Estuary. *Estuaries*, **15(3)**, 270-281.

Stanturf, J.A., D. D. Wade, T. A. Waldrop, D. K. Kennard, and G. L. Achtemeier, 2002: Background paper: fire in southern forest landscapes, In: *Southern Forest Resource Assessment, General Technical Report SRS-53*, [Wear, D.N. and J.G. Greis (eds.)]. U.S. Department of Agriculture, Forest Service, Southern Research Station, Asheville, NC, pp. 607-630.

Steel, J. and N. Carolina, 1991: *Albemarle-Pamlico Estuarine System: Technical Analysis of Status and Trends*. Albemarle-Pamlico Estuarine Study Report 91-01, Environmental Protection Agency National Estuary Program, Raleigh, NC.

Steneck, R.S., M.H. Graham, B.J. Bourque, D. Corbett, J.M. Erlandson, J.A. Estes, and M.J. Tegner, 2002: Kelp forest ecosystems: biodiversity, stability, resilience and future. *Environmental Conservation*, **29(4)**, 436-459.

Stewart, I.T., D.R. Cayan, and M.D. Dettinger, 2004: Changes in snowmelt runoff timing in Western North America under a 'business as usual' climate change scenario. *Climatic Change*, **62**, 217-232.

Stewart, I.T., D.R. Cayan, and M.D. Dettinger, 2005: Changes toward earlier streamflow timing across western North America. *Journal of Climate*, **18(8)**, 1136-1155.

Sun, G., S.G. McNulty, J. Lu, D.M. Amatya, Y. Liang, and R.K. Kolka, 2005: Regional annual water yield from forest lands and its response to potential deforestation across the southeastern United States. *Journal of Hydrology*, **308(1)**, 258-268.

Tahoe National Forest, 1990: *Tahoe National Forest Land and Resource Management Plan*. USDA Forest Service, Pacific Southwest Region.

The State of Queensland and Commonwealth of Australia, 2003: *Reef Water Quality Protection Plan; for Catchments Adjacent to the Great Barrier Reef World Heritage Area*. Queensland Department of Premier and Cabinet, Brisbane.

Titus, J.G., 2000: Does the U.S. government realize that the sea is rising? How to restructure federal programs so that wetlands can survive. *Golden Gate University Law Review*, **30(4)**, 717-778.

Toy, T.J., G.R. Foster, and K.G. Renard, 2002: *Soil Erosion: Processes, Predicition, Measurement, and Control*. John Wiley and Sons.

U.S. Climate Change Science Program, 2007: *Synthesis and Assessment Product 4.1: Coastal Elevation and Sensitivity to Sea Level Rise*. A report by the U.S. Climate Change Science Program and the Subcommittee on Global Change Research, U.S. Environmental Protection Agency.

U.S. Department of Commerce, 1996: *Final Management Plan/Environmental Impact Statement for the Florida Keys National Marine Sanctuary, Volume I*. National Oceanic and Atmospheric Administration, Silver Spring, MD, pp.1-319.

U.S. Department of Commerce, 2000: *Tortugas Ecological Reserve: Final Supplemental Environmental Impact Statement/Final Supplemental Management Plan*. National Oceanic and Atmospheric Administration, Silver Spring, MD, pp.1-315.

U.S. Department of Commerce, 2006: *Channel Islands National Marine Sanctuary Draft Management Plan / Draft Environmental Impact Statement*. National Oceanic and Atmospheric Administration, National Marine Sanctuary Program, Silver Spring, MD.

U.S. Fish and Wildlife Service, 2006: *Waterfowl Population Status 2006*. U.S. Department of the Interior, Washington, DC.

USDA Forest Service, 2000a: *National Fire Plan*.

USDA Forest Service, 2000b: *Water and the Forest Service*. FS-660, Washington, DC.

USDA Forest Service, 2003: *An Analysis of the Timber Situation in the United States: 1952 to 2050*. General Technical Report PNW-GTR-560, Pacific Northwest Research Station, Portland, OR.

USDA Forest Service, 2004: *Sierra Nevada Forest Plan Amendment (SNFPA)*. 2004-ROD, USDA Forest Service, Pacific Southwest Region.

Vargas-Ángel, B., J.D. Thomas, and S.M. Hoke, 2003: High-latitude *Acropora Cervicornis* thickets off Fort Lauderdale, Florida, USA. *Coral Reefs*, **22(4)**, 465-473.

Wadell, J.E., 2005: *The State of Coral Reef Ecosystems of the United States and Pacific Freely Associated States: 2005*. NOAA Technical Memorandum NOS NCCOS 11, NOAA/NCCOS Center for Coastal Monitoring and Assessment's Biogeography Team, Silver Spring, MD, pp.1-522.

Walker, N.D., H.H. Roberts, L.J. Rouse, and O.K. Huh, 1982: Thermal history of reef-associated environments during a record cold-air outbreak event. *Coral Reefs*, **1**, 83-87.

Walker, N.D., L.J. Rouse, and O.K. Huh, 1987: Response of subtropical shallow-water environments to cold-air outbreak events: satellite radiometry and heat flux modeling. *Continental Shelf Research*, **7**, 735-757.

Wang, G., N.T. Hobbs, K.M. Giesen, H. Galbraith, D.S. Ojima, and C.E. Braun, 2002a: Relationships between climate and population dynamics of white-tailed ptarmigan (*Lagopus leucurus*) in Rocky Mountain National Park, Colorado, USA. *Climate Research*, **23**, 81-87.

Wang, G., N.T. Hobbs, H. Galbraith, and K.M. Giesen, 2002b: Signatures of large-scale and local climates on the demography of white-tailed ptarmigans in Rocky Mountain National Park, Colorado, USA. *International Journal of Biometeorology*, **46**, 197-201.

Wang, G., N.T. Hobbs, F.J. Singer, D.S. Ojima, and B.C. Lubow, 2002c: Impacts of climate changes on elk population dynamics in Rocky Mountain National Park, Colorado, U.S.A. *Climate Change*, **54(1-2)**, 205-224.

Wear, D.N. and J.G. Greis, 2002: *The Southern Forest Resource Assessment: Summary Report: United States Forest Service*. General Technical Report SRS-54, Washington, DC, USA, -103.

Weiler, S., J. Loomis, R. Richardson, and S. Shwiff, 2002: Driving regional economic models with a statistical model: hypothesis testing for economic impact analysis. *Review of Regional Studies*, **32(1)**, 97-111.

West, J.M., P.A. Marshall, R.V. Salm, and H.Z. Schuttenberg, 2006: Coral bleaching: managing for resilience in a changing world, In: *Principles of Conservation Biology*, [Groom, M.J., G.K. Meffe, and C.R. Carroll (eds.)].

West, J.M. and R.V. Salm, 2003: Resistance and resilience to coral bleaching: implications for coral reef conservation and management. *Conservation Biology*, **17(4)**, 956-967.

Westerling, A.L., H.G. Hidalgo, D.R. Cayan, and T.W. Swetnam, 2006: Warming and earlier spring increase western U.S. forest wildfire activity. *Science*, **313(5789)**, 940-943.

Westmacott, S., K. Teleki, S. Wells, and J. West, 2000: *Management of Bleached and Severely Damaged Coral Reefs*. IUCN, The World Conservation Union, Washington, DC.

Whitney, G.G., 1994: *From Coastal Wilderness to Fruited Plain: a History of Environmental Change in Temperate North America, 1500 to the Present*. Cambridge University Press, Cambridge, pp. 1-451.

Wilkinson, C.R., 2004: *Status of Coral Reefs of the World: 2004*. Australian Institute of Marine Science, Townsville, Australia.

Williams, J.W., S.T. Jackson, and J.E. Kutzbach, 2007: Projected distributions of novel and disappearing climates by 2100 AD. *Proceedings of the National Academy of Sciences of the United States of America*, **104(14)**, 5738-5742.

Willis, B.L., C. A. Page, and E. A. Dinsdale, 2004: Coral disease on the Great Barrier Reef, In: *Coral Health and Disease*, [Rosenberg, E. and Y. Loya (eds.)]. Springer-Verlag, Berlin, Germany, pp. 69-104.

Woodhouse, B., 2005: An end to Mexico's Rio Grande deficit? *Southwest Hydrology*, **4(5)**, 19-.

Woodward, A., E.G. Schreiner, and D.G. Silsbee, 1995: Climate, geography, and tree establishment in subalpine meadows of the Olympic Mountains, Washington, USA. *Arctic and Alpine Research*, **27(3)**, 217-225.

Wooldridge, S., T. Done, R. Berkelmans, R. Jones, and P. Marshall, 2005: Precursors for resilience in coral communities in a warming climate: a belief network approach. *Marine Ecology Progress Series*, **295**, 157-169.

Yamada, S.B., B.R. Dumbauld, A. Kalin, C.E. Hunt, R. Figlar-Barnes, and A. Randall, 2005: Growth and persistence of a recent invader Carcinus maenas in estuaries of the northeastern Pacific. *Biological Invasions*, **7(2)**, 309-321.

Yoshikawa, K. and L.D. Hinzman, 2003: Shrinking thermokarst ponds and groundwater dynamics in discontinuous permafrost near council, Alaska. *Permafrost and Periglacial Processes*, **14(2)**, 151-160.

Zacherl, D., S.D. Gaines, and S.I. Lonhart, 2003: The limits to biogeographical distributions: insights from the northward range extension of the marine snail, *Kelletia kelletii* (Forbes, 1852). *Journal of Biogeography*, **30(6)**, 913-924.

Zervas, C., 2001: *Sea Level Variations of the United States, 1854-1999*. Technical Report NOS CO-OPS 36, US Dept. of Commerce, National Oceanic and Atmospheric Administration, National Ocean Service, -201.

Zolbrod, A.N. and D.L.U.S. Peterson, 1999: Response of high-elevation forests in the Olympic Mountains to climatic change. *Canadian Journal of Forest Restoration*, **29(12)**, 1966-1979.

ANNEX B REFERENCES

Adger, W.N., S. Agrawala, M. M. Q. Mirza, C. Conde, K. O'Brien, J. Pulhin, R. Pulwarty, B. Smit, and K. Takahashi, 2007: Assessment of adaptation practices, options, constraints and capacity, In: *Climate Change 2007: Impacts, Adaptation and Vulnerability. Contribution of Working Group II to the Fourth Assessment Report of the Intergovernmental Panel on Climate Change*, [Parry, M.L., O.F. Canziani, J.P. Palutikof, P.J. van der Linden, and C.E. Hanson (eds.)]. Cambridge University Press, Cambridge, UK, pp. 717-743.

Allen, C.D., M. Savage, D.A. Falk, K.F. Suckling, T. Schulke, T.W. Swetnam, P.B. Stacey, P. Morgan, M. Hoffman, and J.T. Klingel, 2002: Ecological restoration of southwestern ponderosa pine ecosystems: a broad perspective. *Ecological Applications*, **12(5)**, 1418-1433.

Allison, G.W., J. Lubchenco, and M.H. Carr, 1998: Marine reserves are necessary but not sufficient for marine conservation. *Ecological Applications*, **8 Supplement-Ecosystem Management for Sustainable Marine Fisheries(1)**, S79-S92.

Babcock, R.C., S. Kelly, N.T. Shears, J.W. Walker, and T.J. Willis, 1999: Changes in community structure in temperate marine reserves. *Marine Ecology Progress Series*, **189**, 125-134.

Baron, J.S., H.M. Rueth, A.M. Wolfe, K.R. Nydick, E.J. Allstott, J.T. Minear, and B. Moraska, 2000: Ecosystem responses to nitrogen deposition in the Colorado Front Range. *Ecosystems*, **3(4)**, 352-368.

Battin, J., M.W. Wiley, M.H. Ruckelshaus, R.N. Palmer, E. Korb, K.K. Bartz, and H. Imaki, 2007: Projected impacts of climate change on salmon habitat restoration. *Proceedings of the National Academy of Sciences of the United States of America*, **104(16)**, 6720.

Bengtsson, J., P. Angelstam, T. Elmqvist, U. Emanuelsson, C. Folke, M. Ihse, F. Moberg, and M. Nystroem, 2003: Reserves, resilience and dynamic landscapes. *Ambio*, **32(6)**, 389-396.

Bernhardt, E.S., M.A. Palmer, J.D. Allan, G. Alexander, K. Barnas, S. Brooks, J. Carr, S. Clayton, C. Dahm, J. Follstad-Shah, D. Galat, S. Gloss, P. Goodwin, D. Hart, B. Hassett, R. Jenkinson, S. Katz, G.M. Kondolf, P.S. Lake, R. Lave, J.L. Meyer, T.K. O'Donnell, L. Pagano, B. Powell, and E. Sudduth, 2005: Synthesizing U.S. river restoration efforts. *Science*, **308**, 636-637.

Boesch, D.F., J. C. Field, and D. Scavia, 2000: The potential consequences of climate variability and change on coastal areas and marine resources, In: *U.S. National Assessment of the Potential Consequences of Climate Variability and Change*, Report of the Coastal Areas and Marine Resources Sector Team, for the U.S. Global Change Research Program, Silver Spring, MD.

Bruno, J.F. and E.R. Selig, 2007: Regional decline of coral cover in the Indo-Pacific: timing, extent, and subregional comparisons. *PLoS ONE*, **2(8)**, 1-8.

Brussaard, L., P.C. de Ruiter, and G.G. Brown, 2007: Soil biodiversity for agricultural sustainability. *Agriculture, Ecosystems and Environment*, **121(3)**, 233-244.

Burke, L.M. and J. Maidens, 2004: *Reefs at Risk in the Caribbean*. World Resources Institute, Washington, DC.

Camp, A.E., P.F. Hessburg, R.L. Everett, and C.D. Oliver, 1995: *Spatial Changes in Forest Landscape Patterns Resulting From Altered Disturbance Regimes on the Eastern Slope of the Washington Cascades*. General Technical Report INT-GTR-320, US Department of Agriculture, Forest Service, Intermountain Research Station, Ogden, Utah, pp.1-283.

Chen, C.Y., R.S. Stemberger, N.C. Kamman, B.M. Mayes, and C.L. Folt, 2005: Patterns of Hg bioaccumulation and transfer in aquatic food webs across multi-lake studies in the Northeast US. *Ecotoxicology*, **14(1)**, 135-147.

Chornesky, E.A., A.M. Bartuska, G.H. Aplet, K.O. Britton, J. Cummings-carlson, F.W. Davis, J. Eskow, D.R. Gordon, K.W. Gottschalk, and R.A. Haack, 2005: Science priorities for reducing the threat of invasive species to sustainable forestry. *BioScience*, **55(4)**, 335-348.

Coles, S.L. and B.E. Brown, 2003: Coral bleaching-capacity for acclimatization and adaptation. *Advances in Marine Biology*, **46**, 183-223.

Collingham, Y.C. and B. Huntley, 2000: Impacts of habitat fragmentation and patch size upon migration rates. *Ecological Applications*, **10(1)**, 131-144.

Driscoll, C.T., Y.J.I. Han, C.Y. Chen, D.C. Evers, K.F. Lambert, T.M. Holsen, N.C. Kamman, and R.K. Munson, 2007: Mercury contamination in forest and freshwater ecosystems in the Northeastern United States. *BioScience*, **57**, 17-28.

Elmqvist, T., C. Folke, M. Nyström, G. Peterson, J. Bengtsson, B. Walker, and J. Norberg, 2003: Response diversity, ecosystem change, and resilience. *Frontiers in Ecology and the Environment*, **1(9)**, 488-494.

Felzer, B., D. Kicklighter, J. Melillo, C. Wang, Q. Zhuang, and R. Prinn, 2004: Effects of ozone on net primary production and carbon sequestration in the conterminous United States using a biogeochemistry model. *Tellus Series B-Chemical and Physical Meteorology*, **56(3)**, 230-248.

Fenn, M.E., J.S. Baron, E.B. Allen, H.M. Rueth, K.R. Nydick, L. Geiser, W.D. Bowman, J.O. Sickman, T. Meixner, and D.W. Johnson, 2003: Ecological effects of nitrogen deposition in the western United States. *BioScience*, **53(4)**, 404-420.

Fernandes, L., J. Day, A. Lewis, S. Slegers, B. Kerrigan, D. Breen, D. Cameron, B. Jago, J. Hall, D. Lowe, J. Tanzer, V. Chadwick, L. Thompson, K. Gorman, M. Simmons, B. Barnett, K. Sampson, G. De'ath, B. Mapstone, H. Marsh, H. Possingham, I. Ball, T. Ward, K. Dobbs, J. Aumend, D. Slater, and K. Stapleton, 2005: Establishing representative no-take areas in the Great Barrier Reef: large-scale implementation of theory on Marine Protected Areas. *Conservation Biology*, **19(6)**, 1733-1744.

Fischer, J., D.B. Lindenmayer, and A.D. Manning, 2006: Biodiversity, ecosystem function, and resilience: ten guiding principles for commodity production landscapes. *Frontiers in Ecology and the Environment*, **4(2)**, 80-86.

Gunderson, L., 2007: Ecology: a different route to recovery for coral reefs. *Current Biology*, **17(1)**, 27-28.

Gunderson, L.H., 2000: Ecological resilience-in theory and practice. *Annual Review of Ecology and Systematics*, **31**, 425-439.

Halpin, P.N., 1997: Global climate change and natural-area protection: management responses and research directions. *Ecological Applications*, **7(3)**, 828-843.

Harris, J.A., R.J. Hobbs, E. Higgs, and J. Aronson, 2006: Ecological restoration and global climate change. *Restoration Ecology*, **14(2)**, 170-176.

Hooper, D.U., F.S. Chapin, III, J.J. Ewel, A. Hector, P. Inchausti, S. Lavorel, J.H. Lawton, D.M. Lodge, M. Loreau, S. Naeem, B. Schmid, H. Setälä, A.J. Symstad, J. Vandermeer, and D.A. Wardle, 2005: Effects of biodiversity on ecosystem functioning: a consensus of current knowledge. *Ecological Monographs*, **75(1)**, 3-35.

Hughes, A.R. and J.J. Stachowicz, 2004: Genetic diversity enhances the resistance of a seagrass ecosystem to disturbance. In: *Proceedings of the National Academy of Sciences of the United States of America* 2004.

Hughes, T.P., M.J. Rodrigues, D.R. Bellwood, D. Ceccarelli, O. Hoegh-Guldberg, L. McCook, N. Moltschaniwskyj, M.S. Pratchett, R.S. Steneck, and B. Willis, 2007: Phase shifts, herbivory, and the resilience of coral reefs to climate change. *Current Biology*, **17(4)**, 360-365.

IPCC, 2007: *Climate Change 2007: Impacts, Adaptation and Vulnerability. Contribution of Working Group II to the Fourth Assessment Report of the Intergovernmental Panel on Climate Change*. [Parry, M.L., O.F. Canziani, J.P. Palutikof, P.J. van der Linden, and C.E. Hanson (eds.)]. Cambridge University Press, Cambridge, UK, pp. 1-976.

Jaap, W.C., J. H. Hudson, R. E. Dodge, D. Gilliam, and R. Shaul, 2006: Coral reef restoration with case studies from Florida, In: *Coral Reef Conservation*, [Côté, I.M. and J.D. Reynolds (eds.)]. Cambridge University Press, Cambridge, UK, pp. 478-514.

Jameson, S.C., M.H. Tupper, and J.M. Ridley, 2002: The three screen doors: can marine" protected" areas be effective? *Marine Pollution Bulletin*, **44(11)**, 1177-1183.

Joyce, L.A., J. Aber, S. McNulty, D. H. Vale, A. Hansen, L. C. Irland, R. P. Neilson, and K. Skog, 2001: Potential consequences of climate variability and change for the forests of the United States, In: *Climate Change Impacts on the United States: the Potential Consequences of Climate Variability and Change, National Assessment Synthesis Team Report for the US Global Change Research Program*, Cambridge University Press, Cambridge, UK, pp. 489-522.

Lacey, J.R., C.B. Marlow, and J.R. Lane, 1989: Influence of spotted knapweed (*Centaurea-maculosa*) on surface runoff and sediment yield. *Weed Technology*, **3(4)**, 627-631.

Landres, P.B., P. Morgan, and F.J. Swanson, 1999: Overview of the use of natural variability concepts in managing ecological systems. *Ecological Applications*, **9(4)**, 1179-1188.

Lenihan, H.S., C.H. Peterson, J.E. Byers, J.H. Grabowski, G.W. Thayer, and D.R. Colby, 2001: Cascading of habitat degradation: oyster reefs invaded by refugee fishes escaping stress. *Ecological Applications*, **11(3)**, 764-782.

Levey, D.J., B.M. Bolker, J.J. Tewksbury, S. Sargent, and N.M. Haddad, 2005: Effects of landscape corridors on seed dispersal by birds. *Science*, **309**, 146-148.

Lippincott, C.L., 2000: Effects of *Imperata cylindrica* (L.) Beauv. (Cogongrass) invasion on fire regime in Florida sandhill (USA). *Natural Areas Journal*, **20(2)**, 140-149.

Loreau, M., S. Naeem, P. Inchausti, J. Bengtsson, J.P. Grime, A. Hector, D.U. Hooper, M.A. Huston, D. Raffaelli, B. Schmid, D. Tilman, and D.A. Wardle, 2001: Biodiversity and ecosystem functioning: current knowledge and future challenges. *Science*, **294(5543)**, 804-808.

Markham, A., 1996: Potential impacts of climate change on ecosystems: a review of implications for policymakers and conservation biologists. *Climate Research*, **6**, 171-191.

Marshall, P. and J. Johnson, 2007: The Great Barrier Reef and climate change: vulnerability and management implications, Chapter 24, In: *Climate Change and the Great Barrier Reef: a Vulnerability Assessment*, Great Barrier Reef Marine Park Authority.

Marshall, P. and H. Schuttenberg, 2006: Adapting coral reef management in the face of climate change, In: *Coral Reefs and Climate Change: Science and Management*, [Phinney, J.T., O. Hoegh-Guldberg, J. Kleypas, W.J. Skirving, and A. Strong (eds.)]. American Geophysical Union, Washington, DC, pp. 223-241.

McClanahan, T.R., S. Mwaguni, and N.A. Muthiga, 2005: Management of the Kenyan coast. *Ocean and Coastal Management*, **48**, 901-931.

McClanahan, T.R., N. V. C. Polunin, and T. J. Done, 2002: Ecological states and the resilience of coral reefs, In: *Resilience and Behaviour of Large-Scale Ecosystems*, [Gunderson, L.H. and L. Pritchard, Jr. (eds.)]. Island Press, Washington, DC, pp. 111-163.

McCook, L.J., C. Folke, T. Hughes, M. Nyström, D. Obura, and R. Salm, 2007: Ecological resilience, climate change and the Great Barrier Reef, Chapter 4, In: *Climate Change and the Great Barrier Reef: a Vulnerability Assessment*, Great Barrier Reef Marine Park Authority.

McLachlan, J.S., J.J. Hellmann, and M.W. Schwartz, 2007: A framework for debate of assisted migration in an era of climate change. *Conservation Biology*, **21(2)**, 297-302.

McLaughlin, S.B., M. Nosal, S.D. Wullschleger, and G. Sun, 2007a: Interactive effects of ozone and climate on tree growth and water use in a southern Appalachian forest in the USA. *New Phytologist*, **174(1)**, 109-124.

McLaughlin, S.B., S.D. Wullschleger, G. Sun, and M. Nosal, 2007b: Interactive effects of ozone and climate on water use, soil moisture content and streamflow in a southern Appalachian forest in the USA. *New Phytologist*, **174(1)**, 125-136.

Millar, C.I., 1989: Allozyme variation of bishop pine associated with pygmy-forest soils in northern California. *Canadian Journal of Forest Research*, **19(7)**, 870-879.

Millar, C.I., J.C. King, R.D. Westfall, H.A. Alden, and D.L. Delany, 2006: Late Holocene forest dynamics, volcanism, and climate change at Whitewing Mountain and San Joaquin Ridge, Mono County, Sierra Nevada, CA, USA. *Quaternary research*, **66**, 273-287.

Millar, C.I., R.D. Westfall, and D.L. Delany, in press: Mortality and growth suppression in high elevation limber pine (*Pinus flexilis*) forests in response to multi-year droughts and 20th century warming. *Canadian Journal of Forest Research*.

Millar, C.I. and W.B. Woolfenden, 1999: The role of climate change in interpreting historical variability. *Ecological Applications*, **9(4)**, 1207-1216.

Mooney, H.A. and R.J. Hobbs, 2000: *Invasive Species in a Changing World*. Island Press, Washington, DC.

Mumby, P.J., A.R. Harborne, J. Williams, C.V. Kappel, D.R. Brumbaugh, F. Micheli, K.E. Holmes, C.P. Dahlgren, C.B. Paris, and P.G. Blackwell, 2007: Trophic cascade facilitates coral recruitment in a marine reserve. *Proceedings of the National Academy of Sciences of the United States of America*, **104(20)**, 8362-8367.

National Research Council, 2004: *Air Quality Management in the United States*. National Academies Press, National Academies of Science.

Noss, R.F., 2001: Beyond Kyoto: Forest management in a time of rapid climate change. *Conservation Biology*, **15(3)**, 578-590.

Novacek, M.J. and E.E. Cleland, 2001: The current biodiversity extinction event: scenarios for mitigation and recovery. *Proceedings of the National Academy of Sciences of the United States of America*, **98(10)**, 5466-5470.

Ottawa National Forest, 2006: Affected environment and environmental consequences, Chapter 3, In: *Draft Environmental Impact Assessment*, USDA Forest Service, available at http://www.fs.fed.us/r9/ottawa/forest_management/forest_plan/revision/fp_final/volume_1_final_eis/final_eis_chapter_3.pdf.

Palumbi, S.R., 2002: *Marine Reserves: a Tool for Ecosystem Management and Conservation*. Pew Oceans Commission, Arlington, VA, pp.1-45.

Parker, B.R. and D.W. Schindler, 2006: Cascading trophic interactions in an oligotrophic species-poor alpine lake. *Ecosystems*, **9(2)**, 157-166.

Peterson, S.A., J. Van Sickle, A.T. Herlihy, and R.M. Hughes, 2007: Mercury concentration in fish from streams and rivers throughout the western united states. *Environmental Science & Technology*, **41(1)**, 58-65.

Pimentel, D., L. Lach, R. Zuniga, and D. Morrison, 2000: Environmental and economic costs of nonindigenous species in the United States. *BioScience*, **50(1)**, 53-64.

Poff, N.L.R., J.D. Olden, D.M. Merritt, and D.M. Pepin, 2007: Homogenization of regional river dynamics by dams and global biodiversity implications. *Proceedings of the National Academy of Sciences of the United States of America*, **104(14)**, 5732-5737.

Precht, W.F. and R. B. Aronson, 2006: Death and the resurrection of Caribbean reefs: a palaeoocological perspective, In: *Coral Reef Conservation*, [Cote, I.M. and J. Reynolds (eds.)]. Cambridge University Press, Cambridge, pp. 40-77.

Precht, W.F. and S. L. Miller, 2006: Ecological shifts along the Florida reef tract: the past as a key to the future, In: *Geological Approaches to Coral Reef Ecology*, [Aronson, R.B. (ed.)]. Springer, New York, NY, pp. 237-312.

Rahel, F.J., 2000: Homogenization of fish faunas across the United States. *Science*, **288(5467)**, 854-856.

Rice, K.J. and N.C. Emery, 2003: Managing microevolution: restoration in the face of global change. *Frontiers in Ecology and the Environment*, **(1)**, 469-478.

Riegl, B. and W.E. Piller, 2003: Possible refugia for reefs in times of environmental stress. *International Journal of Earth Sciences*, **92(4)**, 520-531.

Roberts, C.M., J. D. Reynolds, I. M. Côté, and J. P. Hawkins, 2006: Redesigning coral reef conservation, In: *Coral Reef Conservation*, Cambridge University Press, Cambridge, UK, pp. 515-537.

Salm, R.V., T. Done, and E. McLeod, 2006: Marine protected area planning in a changing climate, In: *Coral Reefs and Climate Change: Science and Management*, [Phinney, J.T., O. Hoegh-Guldberg, J. Kleypas, W. Skirving, and A. Strong (eds.)]. American Geophysical Union, Washington, DC, pp. 207-221.

Scavia, D., J.C. Field, D.F. Boesch, R.W. Buddemeier, V. Burkett, D.R. Cayan, M. Fogarty, M.A. Harwell, R.W. Howarth, C. Mason, D.J. Reed, T.C. Royer, A.H. Sallenger, and J.G. Titus, 2002: Climate change impacts on U.S. coastal and marine ecosystems. *Estuaries*, **25(2)**, 149-164.

Schneider, S.H., S. Semenov, A. Patwardhan, I. Burton, C. H. D. Magadza, M. Oppenheimer, A. B. Pittock, A. Rahman, J. B. Smith, A. Suarez, and F. Yamin, 2007: Assessing key vulnerabilities and the risk from climate change, In: *Climate Change 2007: Impacts, Adaptation and Vulnerability. Contribution of Working Group II to the Fourth Assessment Report of the Intergovernmental Panel on Climate Change*, [Parry, M.L., O.F. Canziani, J.P. Palutikof, P.J. van der Linden, and C.E. Hanson (eds.)]. Cambridge University Press, Cambridge, UK, pp. 779-810.

Singer, J.F., C.V. Bleich, and A.M. Gudorf, 2000: Restoration of bighorn sheep metapopulations in and near western National Parks. *Restoration Ecology*, **8(4)**, 14-24.

Sobel, J.A. and C. Dahlgren, 2004: *Marine Reserves: a Guide to Science, Design, and Use*. Island Press, Washington, DC, pp. 1-383.

Spittlehouse, D.L. and R.B. Stewart, 2003: Adaptation to climate change in forest management. *BC Journal of Ecosystems and Management*, **4(1)**.

Stein, B.A., S.R. Flack, N.B. Benton, and N. Conservancy, 1996: *America's Least Wanted: Alien Species Invasions of US Ecosystems*. The Nature Conservancy, Arlington, VA.

Tilman, D., P.B. Reich, and J.M.H. Knops, 2006: Biodiversity and ecosystem stability in a decade-long grassland experiment. *Nature*, **441(7093)**, 629-632.

Tomback, D.F. and K. C. Kendall, 2002: Rocky road in the Rockies: challenges to biodiversity, In: *Rocky Mountain Futures, an Ecological Perspective*, [Baron, J. (ed.)]. Island Press, Washington, DC, pp. 153-180.

U.S. Geological Survey, 2005: *The State of the Colorado River Ecosystem in Grand Canyon*. USGS Circular 1282, U.S. Department of the Interior, U.S. Geological Survey, pp.1-220.

USDA Forest Service, 2004: *National Strategy and Implementation Plan for Invasive Species Management*. FS-805.

Von Holle, B. and D. Simberloff, 2005: Ecological resistance to biological invasion overwhelmed by propagule pressure. *Ecology*, **86(12)**, 3212-3218.

Walker, B., 1995: Conserving biological diversity through ecosystem resilience. *Conservation Biology*, **9(4)**, 747-752.

Walker, B., A. Kinzig, and J. Langridge, 1999: Plant attribute diversity, resilience, and ecosystem function: the nature and significance of dominant and minor species. *Ecosystems*, **2(2)**, 95-113.

Welch, D., 2005: What should protected areas managers do in the face of climate change? *The George Wright Forum*, **22(1)**, 75-93.

West, J.M. and R.V. Salm, 2003: Resistance and resilience to coral bleaching: implications for coral reef conservation and management. *Conservation Biology*, **17(4)**, 956-967.

Williams, D.G. and Z. Baruch, 2000: African grass invasion in the Americas: ecosystem consequences and the role of ecophysiology. *Biological Invasions*, **2(2)**, 123-140.

Williams, J.W., S.T. Jackson, and J.E. Kutzbach, 2007: Projected distributions of novel and disappearing climates by 2100 AD. *Proceedings of the National Academy of Sciences of the United States of America*, **104(14)**, 5738-5742.

Ziska, L.H., J.B. Reeves, and B. Blank, 2005: The impact of recent increases in atmospheric CO_2 on biomass production and vegetative retention of Cheatgrass (*Bromus tectorum*): implications for fire disturbance. *Global Change Biology*, **11(8)**, 1325-1332.

ACKNC

CHAPTER 3

Authors' Acknowledgements

We would like to thank the December 2006 Annapolis workshop participants for their thoughts on early drafts of the report, and the participants in the Tahoe Workshop and the Olympic case study for their participation in discussions on climate change and forest management. We also thank Douglas Powell, Stephen Solem, Nick Reyna, Ken Karkula, Al Abee, John Townsley, Allen Solomon, and Phil Mattson for their comments on the March draft. We would like to thank Sharon Friedman, Andy Kratz, and Claudia Regan for their comments on the March draft and the staff from Region 2 that participated in discussions on the draft report. We thank Tim Davis for several helpful comments on earlier drafts of this manuscript and Robert Norheim for the map in Figure A1.4 in Annex A and David P. Coulson for the map in Figure 3.3. Robin Stoddard helped us with access to photos of ONF. We would also like to thank the respondents to the public review and the members of the Peer Review panel.

Workshop Participants

- Chris Bernabo, National Council on Science for the Environment
- Bob Davis, U.S.D.A. Forest Service
- Lee Frelich, The University of Minnesota Center for Hardwood Ecology
- Jeremy Littell, University of Washington
- Douglas W. MacCleery, U.S.D.A. Forest Service
- Kathy A. O'Halloran, Olympic National Forest
- Lindsey Rustad, U.S.D.A. Forest Service
- Hugh Safford, U.S.D.A. Forest Service
- Allen Solomon, U.S.D.A. Forest Service
- Jeff Sorkin, U.S.D.A. Forest Service
- Chris Weaver, U.S. Environmental Protection Agency

CHAPTER 4

Authors' Acknowledgements

We wish to acknowledge the USGS Western Mountain Initiative, and advice, comments, and reviews from Abby Miller, Bob Krumenaker, David Graber, Vaughn Baker, Jeff Connor, and Ben Bobowski. Participants in the November 2006 workshop provided valuable comments and context.

Workshop Participants

- Stan Austin, Rocky Mountain National Park
- Gillian Bowser, National Park Service and Texas A&M University
- John Dennis, National Park Service
- David Graber, Sequoia and Kings Canyon National Parks
- John Gross, National Park Service Vital Signs Program
- Elizabeth Johnson, National Park Service Northeast Regional Office
- Sharon Klewinsky, National Park Service
- Bob Krumenaker, Apostle Islands National Lakeshore
- Abby Miller, The Coalition of National Park Service Retirees
- Shawn Norton, National Park Service
- Mike Soukup, National Park Service
- Lee Tarnay, Yosemite National Park
- Julie Thomas, National Park Service
- Leigh Welling, Crown of the Continent Research Learning Center
- Mark Wenzler, National Parks Conservation Association

CHAPTER 5

Authors' Acknowledgements

We extend our sincere thanks to Michael Higgins for writing the section on Water Quality and Quantity; David Rupp and Emmi Blades for use of their unpublished information; Jane Austin for reviewing earlier versions of this manuscript; Jennifer Roach for GIS assistance; Jenn Miller and Gina Wilson for citation assistance; and Mark Bertram, Larry Bright, Vernon Byrd, Danielle Jerry, Rex Reynolds, Ron Reynolds, and David Sharp for their invaluable comments and suggestions for the development of this report.

Workshop Participants

- Dawn Browne, Ducks Unlimited
- Tom Franklin, Izaak Walton League
- Doug Inkley, National Wildlife Federation
- Danielle G. Jerry, U.S. Fish and Wildlife Service
- Kurt Johnson, U.S. Fish and Wildlife Service
- James Kurth, U.S. Fish and Wildlife Service
- Noah Matson, Defenders of Wildlife
- Sean McMahon, National Wildlife Federation
- Maribeth Oakes, The Wilderness Society
- Michael Woodbridge, National Wildlife Refuge Association

Alaska Refuges Workshop Participants

- Mark Bertram, U.S. Fish and Wildlife Service
- Philip Martin, U.S. Fish and Wildlife Service
- Julian Fischer, U.S. Fish and Wildlife Service
- Vernon Byrd, U.S. Fish and Wildlife Service
- Keith Mueller, U.S. Fish and Wildlife Service
- David Douglas, U. S. Geological Survey
- Bill Hanson, U.S. Fish and Wildlife Service
- Cynthia Wentworth, U.S. Fish and Wildlife Service
- Patrick Walsh, U.S. Fish and Wildlife Service
- Cathy Rezabeck, U.S. Fish and Wildlife Service

CHAPTER 6

Authors' Acknowledgements

Cassie Thomas of the National Park Service provided assistance with the Alaska text box. Mary Brabham, Rob Mattson and Brian McGurk of the St. Johns River Water Management District; and Jaime Doubek-Racine of the National Park Service provided assistance with the Wekiva River Case Study. Jeff Bennett of the National Park Service, Gary Garrett of the Texas Parks and Wildlife Department, and Greg Gustina of the Bureau of Land Management provided input to the Rio Grande River Case Study. Don Hamilton and Dave Forney of the National Park Service provided assistance with the Delaware River Case Study.

Workshop Participants

- Daniel M. Ashe, U.S. Fish and Wildlife Service
- Donita Cotter, U.S. Fish and Wildlife Service
- Jackie Diedrich, U.S. Forest Service
- Andrew Fahlund, American Rivers
- Dave Forney, National Park Service
- Dan Haas, U.S. Fish and Wildlife Service
- Kristy Hajny, Niobrara National Scenic River
- Mike Huggins, U.S. Fish and Wildlife Service
- Quinn McKew, American Rivers
- David Purkey, Stockholm Environment Institute-U.S. Center
- Jason Robertson, Bureau of Land Management
- Cassie Thomas, National Park Service

CHAPTER 7

Workshop Participants

- Mark Alderson, Sarasota Bay Project
- Carol Auer, National Oceanic and Atmospheric Administration
- Rich Batiuk, U.S. Environmental Protection Agency
- Dean E. Carpenter, Albemarle-Pamlico National Estuary Program
- Derb Carter, Southern Environmental Law Center
- Holly Greening, Tampa Bay Estuary Program
- Michael J. Kennish, Rutgers University
- Karen L. McKee, U.S. Geological Survey National Wetlands Research Center
- Doug Rader, Environmental Defense
- Curtis J. Richardson, Duke University
- Stan Riggs, East Carolina University
- Ron Shultz, Puget Sound Water Quality Action Team
- Jan Smith, Massachusetts Bays National Estuary Program
- Katrina Smith Korfmacher, University of Rochester

CHAPTER 8

Authors' Acknowledgements

The case studies were prepared by Billy Causey and Steven Miller (Florida Keys National Marine Sanctuary), Johanna Johnson (Great Barrier Reef Marine Park), Alan Friedlander (Papahānaumokuākea Marine National Monument), and Satie Airamé (Channel Islands National Marine Sanctuary). Johanna Johnson would like to thank all the expert scientists who contributed to assessing the vulnerability of the Great Barrier Reef to climate change. Without their leadership and knowledge we would not have such an in-depth understanding of the implications of climate change for Great Barrier Reef species, habitats, key processes and the ecosystem, or have been able to develop the management strategies outlined in this case study. Elizabeth McLeod (The Nature Conservancy) drafted the section on adapting to climate change, and Christa Woodley (University of California at Davis) and Danny Gleason (Georgia Southern University) drafted the section on current status of management system. Rikki Grober-Dunsmore (National Oceanic and Atmospheric Administration, MPA Science Institute) prepared Table 8.2. We thank all the individuals who participated in the stakeholder workshop, 24–25 January 2007, and whose lively discussion provided information and comments that helped form the contents and conclusions of this chapter. We also thank the anonymous reviewers and the following people for comments on this chapter: R. Aronson, J. Brown, P. Bunje, D. Burden, A. DeVogelaere, E. Druffel, W. Fisher, H. Galbraith, P. Hallock Muller, J. Lang, J. Martinich, J. Ogden, W. Wiltse, and J. Yang. Finally, we are grateful to Susan Julius and Jordan West for their guidance, support, and suggestions for improving this chapter.

Workshop Participants

- Maria Brown, Gulf of the Farallones National Marine Sanctuary
- Deborah Cramer, Gloucester Maritime Heritage Center and Stellwagen Bank National Marine Sanctuary Advisory Council
- Daniel Gleason, Georgia Southern University and Gray's Reef National Marine Sanctuary Advisory Council
- Lynn Hale, The Nature Conservancy
- Lara Hansen, World Wildlife Fund
- Terrie Klinger, University of Washington and Olympic Coast National Marine Sanctuary Advisory Council
- Irina Kogan, Gulf of the Farallones National Marine Sanctuary
- David Loomis, University of Massachusetts
- Linda Paul, Hawaii Audubon Society
- Bruce Popham, Marathon Boat Yard and Florida Keys National Marine Sanctuary Advisory Council
- Teresa Scott, Washington Department of Fish and Wildlife and Olympic Coast National Marine Sanctuary Advisory Council
- Jack Sobel, The Ocean Conservancy
- Steve Tucker, Cape Cod Commission and Stellwagen Bank National Marine Sanctuary Advisory Council
- Lauren Wenzel, National Oceanic and Atmospheric Administration
- Bob Wilson, The Marine Mammal Center and Gulf of the Farallones National Marine Sanctuary Advisory Council

Case Studies

Editors:

Susan Herrod Julius, U.S. Environmental Protection Agency
Jordan M. West, U.S. Environmental Protection Agency

Authors:

National Forests Case Studies

Tahoe National Forest
Constance I. Millar, U.S.D.A. Forest Service
Linda A. Joyce, U.S.D.A. Forest Service
Geoffrey M. Blate, AAAS Fellow at U.S. Environmental Protection Agency

Olympic National Forest
David L. Peterson, U.S.D.A. Forest Service
Jeremy S. Littell, JISAO CSES Climate Impacts Group, University of Washington
Kathy A. O'Halloran, U.S.D.A. Forest Service

Uwharrie National Forest
Steven G. McNulty, U.S.D.A. Forest Service

National Parks Case Study

Rocky Mountain National Park
Jill S. Baron, U.S. Geological Survey and Colorado State University
Jill Oropeza, Colorado State University

National Wildlife Refuges Case Study

Alaska and the Central Flyway
Brad Griffith, U.S. Geological Survey and University of Alaska Fairbanks
A. David McGuire, U.S. Geological Survey and University of Alaska Fairbanks

Wild and Scenic Rivers Case Studies

Wekiva River
Rio Grande River
Upper Delaware River
Margaret A. Palmer, University of Maryland
Dennis Lettenmaier, University of Washington
N. LeRoy Poff, Colorado State University
Sandra Postel, Global Water Policy Project
Brian Richter, The Nature Conservancy
Richard Warner, Kinni Consulting

National Estuaries Case Study

The Albemarle-Pamlico Estuarine System
Robert R. Christian, East Carolina University
Charles H. Peterson, University of North Carolina
Michael F. Piehler, University of North Carolina
Richard T. Barber, Duke University
Kathryn L. Cottingham, Dartmouth College
Heike K. Lotze, Dalhousie University
Charles A. Simenstad, University of Washington
John W. Wilson, U.S. Environmental Protection Agency

Marine Protected Areas Case Studies

The Florida Keys National Marine Sanctuary
Billy Causey, National Oceanic and Atmospheric Administration
Steven L. Miller, University of North Carolina at Wilmington
Brian D. Keller, National Oceanic and Atmospheric Administration

The Great Barrier Reef Marine Park
Johanna Johnson, Great Barrier Reef Marine Park Authority

Papahānaumokuākea (Northwestern Hawaiian Islands) Marine National Monument
Alan Friedlander, National Oceanic and Atmospheric Administration

The Channel Islands National Marine Sanctuary
Satie Airame, University of California, Santa Barbara

A1. NATIONAL FORESTS CASE STUDIES

A1.1 Tahoe National Forest

A1.1.1 Setting and Context of Tahoe National Forest

Tahoe National Forest (TNF) is located in eastern California, where it straddles the northern Sierra Nevada (Fig. A1.1). The administrative boundary encompasses 475,722 ha (1,175,535 ac), of which one-third are privately owned forest industry lands arranged in alternate sections ("checkerboard") with TNF land. Elevations range from 365 m (1,200 ft) at the edge on the western slope to 2,788 m (9,148 ft) at the crest of the Sierra. The eastern slopes of TNF abut high-elevation (~1,525 m; 5,000 ft) arid steppes of the Great Basin. TNF experiences a Mediterranean-type climate with warm, dry summers alternating with cool, wet winters. The orientation of the Sierra Nevada paralleling the Pacific coast creates a steep west-east climatic gradient that contributes to strong orographic effects in temperature and a precipitation rainshadow. Near TNF's western boundary, average precipitation is low (125 cm; 50 in), highest at west-side mid-elevations (200 cm; 80 in), and lowest near the eastern boundary (50 cm; 20 in). Snow dominates winter precipitation in the upper elevations, providing critical water reserves for the long annual summer drought.

Floral and faunal diversity of TNF parallels the topographic and climatic gradients of the Sierra Nevada, with strong zonation along elevational bands. The long Mediterranean drought is a primary influence on the species that can grow and the natural disturbance regimes. Pine forests occupy low elevations on the western side. These grade upslope to a broad zone of economically and ecologically important mixed-conifer forests. Higher, at the elevation of the rain-snow zone, true-fir forests dominate; diverse subalpine forests are the highest-elevation tree communities. East of the crest, sparse eastside pine communities grade downslope to woodlands and shrublands of the Great Basin. Terrestrial and aquatic environments of TNF support critical habitat for a large number of plant and animal species, many of which have long been subjects of intense conservation concern. The TNF environments are used by 387 vertebrate species and more than 400 plant species (Tahoe National Forest, 1990; Shevock, 1996). Several keystone species at the Sierra rangewide scale depend on now-limited old-growth forest conditions or other rare habitats.

Cultural legacies have played significant roles in shaping present forest conditions and vulnerabilities in TNF. Timber, water, mining, and grazing, which started in the mid-1800s, remained intensive uses until the late 20th century. Low- to mid-elevation forests were denuded in the mid-1800s through early 1900s to provide wood for settlement (Beesley, 1996).

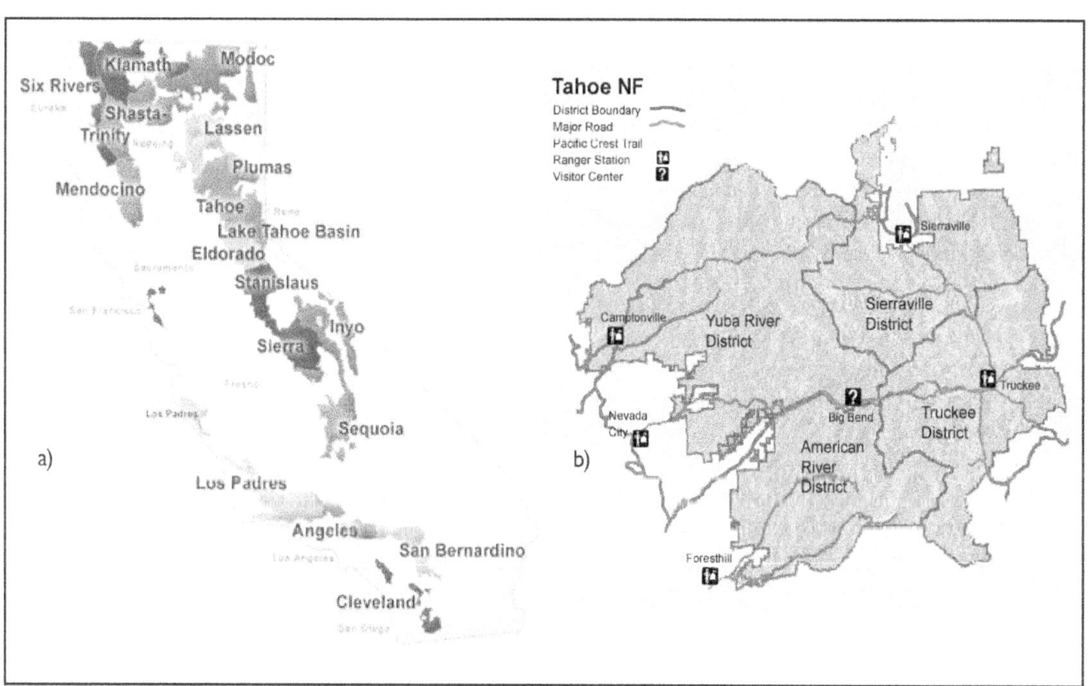

Figure A1.1. Map and location of the Tahoe National Forest, within California (a) and the Forest boundaries (b).[1]

[1] **USDA Forest Service**, 2007: Tahoe National Forest map. USDA Forest Service Website, http://www.fs.fed.us/r5/tahoe/maps_brochures/images/05_nov_01_tnf_map.jpg, accessed on 7-30-2007. And **USDA Forest Service**, 2007: National Forests in California. USDA Forest Service Website, http://www.fs.fed.us/r5/forests.html, accessed on 7-30-2007.

Subsequently the forests regrew, but although they continued to be extensively harvested until recently, decades of fire suppression contributed to extremely dense stands, even-age classes, and low structural diversity. These conditions led to extreme fire susceptibilities; large fire events have occurred in recent years, and fire vulnerability is the highest concern for management. Modern human use of TNF and adjacent lands has changed the way in which natural resources are managed. Population and development in the communities adjacent to the low elevations have exploded in the past decades, creating extensive wildland-urban interface issues (Duane, 1996). Changing demographies and consequent resource values of new residents have forced re-evaluation of TNF goals and practices, many of which limit the capacity of TNF to implement adaptive but manipulative practices in the face of changing climates. Recreation is now a primary use of TNF lands; timber management is minor. Fuels reduction is a key issue both for protection of TNF resources and of adjacent rural communities.

A1.1.2 Recent and Anticipated Regional Climate Changes and Impacts

The trend of temperature increase over the 20th century for California has paralleled the global pattern (IPCC, 2007a), although at greater magnitude (1.5–2°C; Millar *et al.*, 2004).[2] Precipitation has not shown strong directional changes, but has been variable at annual and interannual scales (Cayan *et al.*, 1998). Forest insect and disease, mortality, and fire events have become more severe in TNF, as throughout the West (Logan and Powell, 2001; Westerling *et al.*, 2006). Decreases in average snowpack up to 80% are documented throughout much of the West; snowpacks peak as much as 45 days earlier (Hamlet *et al.*, 2005; Mote *et al.*, 2005) and peak streamflow peaks up to three weeks earlier in spring (Stewart, Cayan, and Dettinger, 2005) than during the 1950s, based on an analysis of the last 50 years.

Many of the climate and ecological trends documented for the 20th century are projected to continue and exacerbate in the 21st century. Future climate scenarios and effects on water, forests, fires, insects, and disease for California are summarized in Hayhoe *et al.* (2004) and the California Climate Action Team reports (California Climate Action Team, 2005). All models project increased annual temperatures over California ranging from 2.3–5.8°C (4.1–10.4°F) (range of models to show model uncertainties). Model projections also indicate slight drying, especially in winter; interannual and interdecadal variability is projected to remain high in the next century. Snowpacks, however, are consistently projected to decline by as much as 97% at 1,000 m (3,280 ft.) elevation and 89% for all elevations. The combined effects of continued warming, declining snowpacks, and earlier stream runoff portend longer summer droughts for TNF, and increasing soil moisture deficits during the growing season. This would increase stress that an already long, dry Mediterranean summer imposes on vegetation and wildlife.

Coupling climate models with vegetation models yields major contractions and expansions in cover of dominant montane vegetation types by the late 21st century (Hayhoe *et al.*, 2004; Lenihan *et al.*, 2006). By 2070–2099, alpine and subalpine forest types are modeled to decline by up to 90%, shrublands by 75%, and mixed evergreen woodland by 50%. In contrast, mixed evergreen forest and grasslands are each projected to expand by 100%. The following conditions are expected to be exacerbated in TNF as a result of anticipated changes (Dettinger *et al.*, 2004; Hayhoe *et al.*, 2004; Cayan *et al.*, 2006b):

- Increased fuel build-up and risk of uncharacteristically severe and widespread forest fire.
- Longer fire seasons; year-round fires in some areas (winter fires have already occurred).
- Higher-elevation insect and disease and wildfire events (large fires already moving into true fir and subalpine forests, which is unprecedented).
- Increased interannual variability in precipitation, leading to fuels build up and causing additional forest stress. This situation promotes fire vulnerabilities and sensitivities.
- Increased water temperatures in rivers and lakes and lower water levels in late summer.
- Increased stress to forests during periodic multi-year droughts; heightened forest mortality.
- Decreased water quality as a result of increased watershed erosion and sediment flow.
- Increased likelihood of severe flood events.
- Loss of seed and other germplasm sources as a result of population extirpation events.

A1.1.3 Current TNF Natural-Resource Policy and Planning Context

In addition to national laws and regional management directives, management goals and direction for the lands and resources of TNF are specified by several overarching planning documents. These relate to different landscape scales and locations. The 1990 Tahoe National Forest Land and Resource Management Plan (LMP) (Tahoe National Forest, 1990) remains the comprehensive document for all resource management in TNF. The primary mission of TNF is to "serve as the public's steward of the land, and to manage the forest's resources for the benefit of all American people…[and]…to provide for the needs of both current and future generations" (Tahoe National Forest, 1990). Within this broad mission, specific goals, objectives, desired future

[2] See also, **Western Regional Climate Center**, 2005: Instrumental weather databases for western climate stations. Western Regional Climate Center Database, http://www.wrcc.dri.edu/, accessed on 4-27-2007.

conditions, and standards and guidelines are detailed for the following resource areas: recreation; interpretive services; visual management; cultural resources; wilderness; wildlife and fish; forage and wood resources; soil, water, and riparian areas; air quality; lands; minerals management; facilities; economic and environmental efficiency; security; human and community resources; and research.

Specific direction in the LMP has been amended by the Sierra Nevada Forest Plan Amendment (FPA; USDA Forest Service, 2004) and the Herger-Feinstein Quincy Library Group Forest Recovery Act.[3] The FPA is a multi-forest plan that specifies goals and direction for protecting old forests, wildlife habitats, watersheds, and communities on the 11 NFs of the Sierra Nevada and Modoc Plateau. Goals for old-growth forests focus on protection, enhancement, and maintenance of old forest ecosystems and their associated species through increasing density of large trees, increasing structural diversity of vegetation, and improving continuity of old forests at the landscape scale. A 2003 decision by the U.S. Fish and Wildlife Service to not list the California Spotted Owl as endangered was conditioned on the assumption that NFs (including TNF) would implement the direction of the FPA.

In regard to aquatic, riparian, and meadow habitat, the FPA goals and management direction are intended to improve the quantity, quality, and extent of highly degraded wetlands throughout the Sierra Nevada, and to improve habitat for aquatic and wetland-dependent wildlife species such as the willow flycatcher and the Yosemite toad.

Fire and fuels goals are among the most important in the FPA. In general, direction is given to provide a coordinated strategy for addressing the risk of catastrophic wildfire by reducing hazardous fuels while maintaining ecosystem functions and providing local economic benefits. The specific approaches to these goals are conditioned by the National Fire Plan of 2000 (USDA Forest Service, 2000a) and the Healthy Forests Restoration Act of 2003,[4] which emphasize strategic placement of fuel treatments across the landscape, removing only enough fuels to cause fires to burn at lower intensities and slower rates than in untreated areas, and are cost-efficient fuel treatments.

The FPA contained a Sierra-wide adaptive management and monitoring strategy. This strategy is being implemented as a pilot project on two NFs in the Sierra Nevada, one of which includes TNF. This seven-year pilot project, undertaken via a Memorandum of Understanding between the U.S. Forest Service, the U.S. Fish and Wildlife Service, and the University of California, applies scientifically rigorous design, treatment, and analysis approaches to fire and forest health, watershed health, and wildlife. Several watersheds of TNF are involved in each of the three issue areas of the FPA adaptive management project.

The Herger-Feinstein Quincy Library Group Forest Recovery Act of 1998 provides specific management goals and direction for a portion of TNF (the Sierraville Ranger District, 164,049 ac) and adjacent NFs. The Act derived from an agreement by a coalition of representatives of fisheries, timber, environmental, county government, citizen groups, and local communities that formed to develop a resource management program to promote ecologic and economic health for certain federal lands and communities in the northern Sierra Nevada. The Act launched a pilot project to test alternative strategies for managing sensitive species, a new fire and fuels strategy, and a new adaptive management strategy. The Herger-Feinstein Quincy Library Group Pilot is the resulting project with goals to test, assess, and demonstrate the effectiveness of fuelbreaks, group selection, individual tree selection, avoidance or protection of specified areas; and to implement a program for riparian restoration.

A1.1.4 TNF Management and Planning Approaches to Climate Change

Management practices identified by TNF staff as being relevant to climate issues are listed below, relative to the three categories of responses described in the National Forests chapter of this report: unplanned, reactive adaptation, or no adaptation measures planned or taken; management responses reacting to crisis conditions or targeting disturbance, extreme events; and proactive management anticipating climate changes.

A1.1.4.1 No Active Adaptation

Few if any of TNF's management policies or plans specifically mention or address climate or climate adaptation. Thus, while it would appear that "no adaptation" is the dominant paradigm at TNF, many practices are de-facto "climate-smart," where climatic trends or potential changes in climate are qualitatively or quantitatively incorporated into management consideration, as indicated in following sections.

A1.1.4.2 Management Responses Reacting to Changing Disturbance and Extreme Events

Most post-disturbance treatments planned by TNF were developed to meet goals of maintaining ecosystem health (*e.g.*, watershed protection, succession to forest after wildfire, fuel reduction after insect mortality) rather than catalyzing climate-adaptive conditions. Nonetheless, many of these best-forest-management practices are consistent with adaptive conditioning for climate contexts as well, as the example here suggests.

[3] Title 4, Section 401(j), P.L. 103-354

[4] H. R. 1904

Salvage and Planting Post-Fire

While in most cases the capacity cannot meet the need, TNF is able to respond adaptively on a small number of acres post-disturbance if the effort to develop NEPA documentation is adequate to defend against appeal and litigation.[5] In these circumstances, watershed protection measures are implemented and species-site needs are considered in decisions about what and where to plant, or what seed to use.

A1.1.4.3 Management Anticipating Climate Change

While TNF has not addressed climate directly through intentional proactive management, staff have been discussing climate change and climate implications for many years. This proactive thinking in itself has pre-conditioned TNF to taking climate into account in early management actions, and has started the discussion among staff regarding potential changes in strategic planning areas. Further, advances have been made in integrated planning processes that may be useful vehicles for incorporating climate-related treatments, thus pre-adapting TNF institutionally to move forward with proactive climate management. The following examples of actions and opportunities demonstrate how the TNF is moving forward with dynamic management.

Staff Support by Line Officers

The leadership team at TNF promotes broad science-based thinking and rewards adaptive and proactive behaviors. This practice clearly sets a stage where management responses to climate can be undertaken where possible, providing an incentive and the intellectual environment to do so.

Fireshed Assessment

The new Fireshed Assessment process is a major step toward integrated management of TNF lands. Effective implementation of this process already provides a vehicle for other dynamic and whole-landscape planning processes such as are needed for climate adaptation.

Fuel Reduction Projects

Strategies implemented by TNF as a result of FPA and Herger-Feinstein Quincy Library Group Pilot directions to reduce fuels and minimize chances of catastrophic fires are increasing the adaptability and resilience of TNF forests (Fig. A1.2). Strategically placed area treatments, a form of adaptive and dynamic approach to fuel management, are being tested on the adaptive management pilot of TNF.

Riparian Management Policies

New policies in the FPA for riparian and watershed management restrict road construction for timber management (*e.g.*, near or across perennial streams). Helicopters are used for logging in all situations where roads cannot be built. This allows more flexibility, adaptability, and reduces fragmentation and watershed erosion.

Figure A1.2. Thinned stands for fuel reduction and resilience management, part of the Herger-Feinstein Quincy Library Pilot Project. Photo courtesy of Tahoe National Forest.

Post-Event Recovery

While certain kinds of standardized post-fire restoration practices (*e.g.*, Burned Area Emergency Rehabilitation procedures) are not climate-proactive, a post-event recovery team at the Pacific Southwest regional level is investigating dynamic approaches to recovery post-major disturbance. These approaches might include planning for long-term changes on disturbed sites and taking advantage of new planting mixes, broadening gene pool mixes, planting in new spacing and designs, etc.

Revegetation and Silvicultural Choices

In stand improvement projects and revegetation efforts, choices are being considered to favor and/or plant different species and species mixes. For instance, where appropriate based on anticipated changes, white fir could be favored over red fir, pines would be preferentially harvested at high elevations over fir, and species would be shifted upslope within seed transfer guides.

Forest Plan Revision

The TNF LMP is due for revision. Climate considerations are being evaluated as the plan revision unfolds, including such options as flexible spotted owl (*Strix occidentalis occidentalis*) "Protected Activity Center" boundaries, species shifts in planting and thinning, and priority-setting for sensitive-species management.

Resisting Planned Projects That May Not Succeed Under Future Climate Conditions

Restoring salmon to TNF rivers is a goal in the current LMP (Fig. A1.3). With waters warming, however, future conditions of TNF rivers are not likely to provide suitable habitat for salmon. Thus, TNF is considering the option to not restore salmon. Meadow restoration is another

[5] **Levings**, W., 2003: *Economics of Delay*. Unpublished report on file at the Tahoe National Forest, pp.1-6.

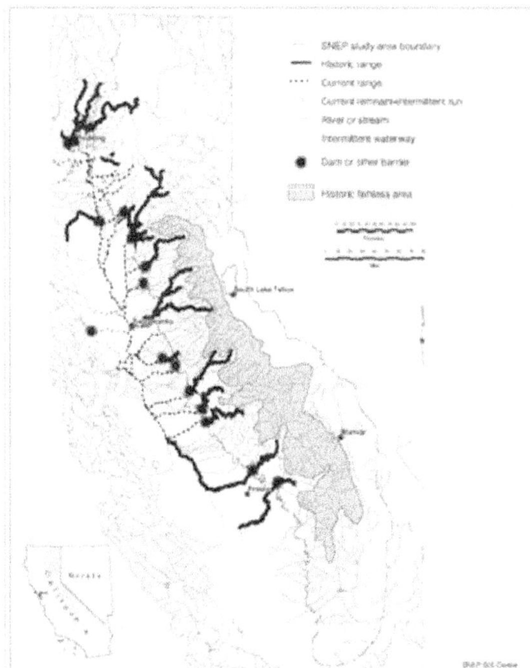

Figure A1.3. Former salmon habitat (rivers marked in bold black) of the Sierra Nevada. Tahoe National Forest (TNF) rivers are scheduled to have salmon restored to them in current national forest planning. Adaptive approaches suggest that future waters may be too warm on the TNF for salmon to survive, and thus restoration may be inappropriate to begin. Map adapted from (Sierra Nevada Ecosystem Project Science Team, 1996).

example: Rather than proceeding with plans for extensive and intensive meadow restoration, some areas are being considered for non-treatment due to possible succession of non-meadow conditions in these locations.

Resilience Management

All forms of proactive management that improve the resilience of natural resources are improving the adaptiveness of TNF by decreasing the number of situations where TNF must take crisis-reaction responses.

Dynamic Management

TNF staff is using opportunities available at present (*i.e.*, under current policy) to manage dynamically and experimentally. An example is cases in which plans treat critical species' range margins differently, favoring active management at advancing edges or optimal habitat rather than static or stressed margins.

Managing for Process

TNF staff is also using opportunities available at present to manage for process rather than structure or composition in proposed projects; for example, those involving succession after fires, where novel mixes of species and spacing may reflect likely natural dynamic processes of adaptation.

A1.1.5 Proactive Management Actions Anticipating Climate Change

A1.1.5.1 Examples of Potential Future Proactive Management Actions

The ideas listed below were identified by TNF staff as being examples of how management actions could be leveraged in the future to increase the TNF adaptive responses to climate change.

- Rapid assessments of current planning and policy. A science-based (*e.g.*, U.S. Forest Service research team) rapid assessment or "audit" of existing TNF planning documents (*e.g.*, the LMP and project plans) could focus on the level of climate adaptedness, pitfalls, and areas for improvement in current TNF plans and operations. Such an audit could focus on current management direction (written policy); current management practices (implementation); and priorities of species (*e.g.*, specific targeted species) and processes (fire, insects/disease). The audit would highlight concrete areas of the plans and projects that are ill-adapted as well as those that are proactive and already climate-proactive, and would recommend a set of specific areas where changes are needed and improvements could be made.
- Assessment/audit of the Sierra Nevada FPA. This would be a similar assessment to that above, but would be undertaken at the FPA scale. The FPA did not originally include climate, and the science consistency review highlighted this problem. A more comprehensive assessment of the FPA's strengths and weaknesses is needed, with a call for revision as appropriate.
- TNF as a pilot for the U.S. Forest Service Ecosystem Services program. Tapping into the ecosystem services market opportunities and acting as a pilot national forest within the ecosystem services goals and objectives may provide management flexibility needed for climate adaptation.
- Management unit size. Increase sizes of management units on the forest, so whole landscapes (watersheds, forest types) could be managed in a single resource plan; decrease administrative fragmentation. Whole ecosystem management, rather than piecemeal by small management unit or by single species or single issue, would favor adaptability to climate-related challenges.
- Watershed management; water storage. To increase groundwater storage capacities, treatments to improve infiltration could be implemented. For instance, in TNF, consider decreasing road densities and other activities (evaluate grazing) in order to change surfaces from impervious to permeable.
- Watershed management; salvage harvest. To decrease erosion and sediment loss following disturbance, there is widespread need in TNF to salvage-harvest affected trees and reforest soon after disturbance. This is the plan at present, but mostly cannot be implemented in adequate

time due to time required for NEPA processing and general public opposition.

- Event recovery. Post-disturbance mortality and shrub invasion must be dealt with swiftly to keep options open for forest regeneration on the site. The means are known; the capacity (money, legal defense) is needed.

A1.1.6 Barriers and Opportunities to Proactive Management for Climate Change at TNF

A1.1.6.1 Barriers

The situations listed below were identified by TNF staff as barriers that limit TNF's capacity to respond adaptively to climate change.

- Public opposition. Appeals and litigation of proposed active management projects directly restrict ability of TNF to implement adaptive practices.[5] There is a large public constituency that opposes active management of any kind. Thus, no matter the purpose, if adaptive management proposals involve on-the-ground disturbance, these publics attempt to prohibit their implementation. The likelihood of appeals and litigation means that a large proportion of staff time must necessarily be used to develop "appeal-proof" NEPA documents, rather than undertaking active management projects on the ground. This often results in a situation in which no-management action can be taken, regardless of the knowledge and intent to implement active and adaptive practices.
- Funding. Overall lack of funds means that adaptive projects, while identified and prioritized, cannot be implemented. General funding limitations are barriers throughout TNF operations. The annual federal budget process limits capacity to plan or implement long-term projects.
- Staff capacity. Loss of key staff areas (*e.g.*, silviculture) and general decline in resource staff and planning capacity translate to lower capacity to respond adaptively to needed changes.
- Scope of on-the-ground needs. As a result of legacy issues (fire-suppression, land-use history, etc.), as well as responses to changing climates (increasing densification of forests, increasing forest mortality), the area of land needing active management is rapidly escalating, and far exceeds staff capacity or available funds to treat it.
- Crisis reaction as routine planning approach. Inadequate TNF funding and staff capacity, combined with persistent legal opposition by external publics, force a continuous reactive approach to priority-setting. This results in crisis-management being the only approach to decision-making that is possible, as opposed to conducting or implementing long-term, skillful, or phased management plans.
- Checkerboard ownership pattern. The alternating sections of TNF and private land create barriers to planning or implementing landscape-scale management, which is needed for adaptive responses to climate challenges. Achieving mutually agreeable management goals regarding prescribed fire, road building, fire suppression, post-fire recovery, and many other landscape treatments is extremely difficult; thus, often no management can be done. This is especially challenging in the central part of TNF, where important corridors, riparian forests, and continuous wildlife habitat would be actively enhanced by management, but cannot be due to mixed ownership barriers.
- Existing environmental laws. Many current important environmental laws that regulate national forest actions such as the Endangered Species Act, the National Forest Management Act, and the National Environmental Policy Act are highly static, inhibit dynamic planning, and impede adaptive responses.[5] Further, these laws do not allow the option of not managing any specific situation—such choices may be necessary as triage-based adaptation in the future. Finally, while coarse-filter approaches are more adaptive, many existing laws force a fine-filter approach to management.
- Current agency management concepts and policies. Current agency-wide management paradigms limit capacity to plan in a proactive, forward-looking manner. For instance, the policies requiring use of historic-range-of-variability or other historic-reference approaches for goal-setting restrict dynamic, adaptive approaches to management. This problem was identified in vegetation management, dam construction ("100-year" flood references), and sensitive-species management (owls, salmon). Certain current regional policies and procedures limit adaptive responses. An example is the Burned Area Emergency Rehabilitation approach to post-fire rehabilitation. Burned Area Emergency Rehabilitation is a static and short-term set of practices that does not incorporate the capacity to respond flexibly and adaptively post-fire, such as taking actions to actively move the site in new ecological trajectories with different germplasm sources and different species mixes.
- Static management. Other current management paradigms that limit dynamic planning and managing include the focus on "maintaining," "retaining," and "restoring" conditions. The consequence of these imperatives in planning documents is to enforce static rather than dynamic management.
- Air quality standards. Regional regulatory standards for smoke and particulates are set low in order to optimize air quality. These levels, however, limit the capacity of TNF to conduct prescribed fires for adaptive fuel reduction or silvicultural stand treatment purposes.
- Community demographics and air quality/urban fuels. Changing demographics of foothill Sierran communities adjacent to TNF are moving toward less acceptance of smoke. Older and urban residents moving into the area in the past few years have little experience with fire and its effects, and have little understanding of or tolerance for smoke from prescribed fire treatments. Similarly, these

residents are not apt to subscribe to Fire-Safe Council home ownership/maintenance recommendations, thus putting their homes and landscaping at high risk from wildfire.

- Agency target and reward system. The current system at the national agency level for successful accomplishments (*i.e.*, the reward system) focuses on achieving narrowly prescribed targets ("building widgets"). Funds are allocated to achieving targets; thus simplistic, in-the-box thinking, and routine, easily accomplished activities are encouraged. There are few incentives for creative project development or implementation.
- Small landscape management units. Fragmentation and inflexibility result from partitioning TNF into small management units; small unit sizes also restrict the capacity for full understanding of ongoing dynamics and process. For instance, even the adaptive management pilot projects under the FPA are too small to be meaningful under the conditions anticipated in the future—at least 20,000 acres (8,093 ha) are needed.

A1.1.6.2 Barriers Opportunities

The activities listed below were identified by TNF staff as current or potential future opportunities to enhance managers' ability to proactively manage for climate change, some of which are currently employed at TNF.

- Year-round management opportunities. TNF is experiencing later winters (snow arriving later in the year), lower snowpacks, and earlier runoff. The TNF staff has taken advantage of these changes by continuing fuel treatments far beyond the season where historically these treatments could be done. At present, winter-prescribed fires are being undertaken, and conditions are ideal to do so. This enables treating more acres in adaptive practices than could be done if only summer were available for these management activities.
- Responses to public concerns through active dialog. TNF has effectively maintained a capacity to implement adaptive projects when in-depth, comprehensive analysis has been done on NEPA process. In addition, intensive education of the interested publics through workshops, scoping meetings, face-to-face dialog, and informal disposition processes have helped to develop support for plans (avoiding appeal), and thus these activities are enabling TNF's adaptive projects to be conducted.
- Responses to public concerns by demonstration. Specifically, TNF was able to gain public approval to cut larger diameter classes (needed for active management to achieve dynamic goals) than had been previously acceptable, through the use of 3-D computer simulations (visualizations), on-the-ground demonstration projects, "show-me" field trips, and other field-based educational efforts.
- Emerging carbon markets are likely to promote the (re-) development of regional biomass and biofuels industries. These industries will provide economic incentives for active adaptive management, in particular funds to support thinning and fuel-reduction projects.
- Planning flexibility in policy. The existence of the Herger-Feinstein Quincy Library Group Pilot and the FPA Adaptive Management project on TNF mean that there is more opportunity than in most other Sierra Nevada NFs to implement active management, especially at broader landscape scales.
- New staff areas defined. When capacity to add staff arises, new positions (climate-smart) may be added. Through incremental changes in staff, TNF may "re-invent and redefine" its institutional ability to better respond adaptively to novel challenges.
- Public education. There is an opportunity to further educate the local public about the scientific bases for climate change, the implications for the northern Sierra Nevada and TNF, and the need for active resource management.

A1.1.7 Increasing Adaptive Capacity to Respond to Climate Change

The ideas listed below were identified by TNF staff as being scientific, administrative, legal, or societal needs that would improve the capacity to respond adaptively to climate change challenges.

- New management strategies. Operationally appropriate and practical management strategies to address the many challenges and contexts implied by changing climates are needed.
- Scientifically supported practices for integrated management. Integration of resource management goals (*e.g.*, fuels, sensitive species, water, fire) rather than partitioning tasks into individual plans is already a barrier to effective ecosystem management. Changing climates are anticipated to increase the need for integration and integrated plans. Input from the science community on integrated knowledge, synthesis assessments, and toolboxes for integrated modeling, etc. will improve the capacity to respond adaptively.
- Projections and models. Modeled simulations of future climate, vegetation, species movements; rates of changes of all of these; and probabilities/uncertainties associated with the projections are needed.
- Case studies. Case studies of management planning and practices implemented as adaptive responses to climate are needed. Demonstration and template examples would allow ideas to disseminate quickly and be iteratively improved.
- Prioritization tools for managing a range of species and diverse ecosystems on TNF. Given the large number of species in the forest, it is impossible to manage all of them. Thus, new tools for adaptive decision-making are needed, as well as development of strategic processes to assist effective prioritizing of actions.
- Dynamic landscape and project planning. Scientific assistance is needed to help define targets and management goals that are appropriate in a changing climate context.

Additional work on probabilistic management units, ranges of conditions likely, continuingly variable habitat probabilities, and habitat suitability contour mapping would be useful. Management planning guidelines that allow rules to change adaptively as conditions change need to be developed.

- Scientific clearinghouse on climate information. In high demand is a reference/resource center, such as a website, with current and practical climate-related material. To be useful at the scale of individual forests such as TNF, the information needs to be locally relevant, simply written, and presented in one clear, consistent voice.
- Scientific support and assistance to individual and specific TNF proposed actions. A consistent, clear voice from science is needed to help build the most appropriate and adaptive plans and actions. There is also a need for clear scientific evidence that demonstrates both the appropriateness of proposed TNF actions and the problems that would result from no action. A website could include such information as brief and extended fact sheets, regional assessments, archives of relevant long-term data or links to other websites with climate-relevant data, model output and primers (climate-relevant ecological, economic, and planning models), training packages on climate change that can be delivered through workshops and online tutorials, and access to climate-based decision-support tools.
- Seed banks. Seed banks need to be stocked to capacity as buffer for fire, insects and disease, and other population extirpation events.

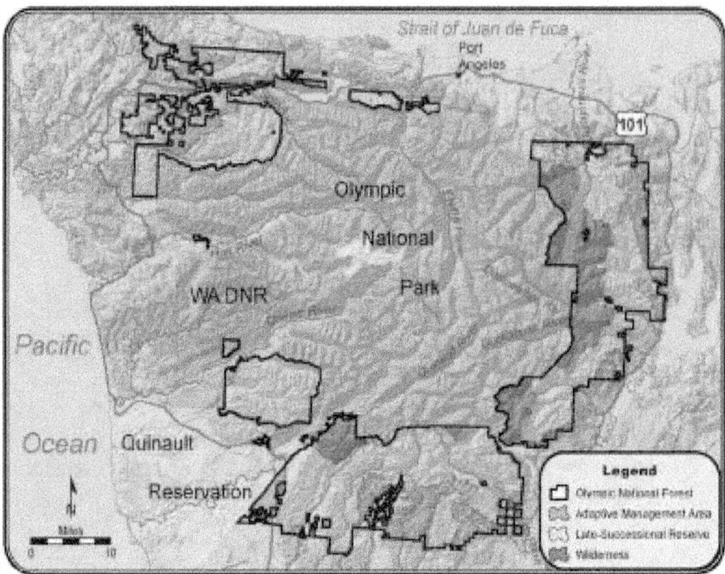

Figure A1.4. Olympic Peninsula land ownership and Northwest Forest Plan allocation map. Olympic National Forest contains lands (dark boundary) with different land use mandates and regulations. These include adaptive management areas, late-successional reserves, and Wilderness areas. Map courtesy of Robert Norheim, Climate Impacts Group, University of Washington.

A1.2 Olympic National Forest

A1.2.1 Setting and Context of the Olympic National Forest

A1.2.1.1 Biogeographic Description

The Olympic Peninsula, in western Washington State (Fig. A1.4), consists of a mountain range and foothills surrounded by the Pacific Ocean (west); the Strait of Juan de Fuca (north); Puget Sound (east); and low elevation, forested land (south). Its elevation profile extends from sea level to nearly 2,500 m (8,200 ft.) at Mount Olympus in the Olympic Mountains. The range creates a strong precipitation gradient, with historic precipitation averages of about 500 cm (197 in.) in the lowlands of the southwestern peninsula, 750 cm (295 in.) in the high mountains, and only 40 cm (16 in.) in the drier northeastern lowlands. The climate is mild temperate rainy, with a Mediterranean (dry) summer. Most of the precipitation falls in winter and at higher elevations; nearly all of it is snow that persists well into summer. The resulting biophysical landscape is a diverse array of seasonal climates and ecological conditions, including coastal estuaries and forests, mountain streams and lakes, temperate rainforests, alpine tundra, mixed conifer forests, and prairies.

The ecosystems on the peninsula are contained within a mosaic of federal, state, tribal, and private ownership. Olympic National Forest (ONF), comprising ~257,000 ha (~635,000 acres) (including five wilderness areas), surrounds Olympic National Park (ONP, ~364,000 ha (~899,000 acres)), the core of the peninsula. ONP is both a World Heritage Site and an International Biosphere Reserve. There are 12 Native American tribes on the peninsula. Approximately 3.5 million people live within four hours' travel of the ONF, and thus it is considered an urban forest because of its proximity to the cities of the greater Seattle area. Ecosystem services from ONF are notably diverse and include water supply to several municipal watersheds, nearly pristine air quality, abundant fish and wildlife (including several unique/endemic species of plants and animals, such as the Olympic marmot (*Marmota Olympus*) and the Roosevelt elk (*Cervus elaphus roosevelti*), as well as critical habitat for four threatened species of birds and anadromous fish), recreation, and timber following implementation of the Northwest Forest Plan amendment (NWFP) to the Olympic National Forest Plan. Hereafter, reference to the Olympic National Forest Plan (ONFP) refers to the 1990 Olympic National Forest Plan, as amended by the NWFP in 1994.

Managing ONF lands therefore requires consideration of complex geographical, climatological, ecological, and sociocultural issues. Climatic change is likely to influence the factors responsible for the Olympic Peninsula's diversity and biogeography, and numerous stakeholders and land management mandates will need to adapt to those changes to protect the natural and cultural resources on the Peninsula.

A1.2.2 Recent and Anticipated Climate Change and Impacts

The Pacific Northwest has warmed approximately 1°C (1.8°F) since 1920; most of this warming (0.9°C (1.6°F)) has been since 1950, and winter has warmed faster than summer (Mote, 2003). The trend in annual precipitation is less clear, though most sites show an increase between 1920 and 2000; decadal variability, rather than trends, best characterizes the region's 20th century precipitation (Mote, 2003). However, the winter temperature increase has caused the form of winter precipitation to change at mid- and low-elevation sites, and 30–60% declines in April 1 snow water equivalent have been observed in the Olympics and Cascade Range (Mote *et al.*, 2005). The timing of spring runoff was 10–30 days earlier in 2000 compared with 1948 (Stewart, Cayan, and Dettinger, 2004).

Proxy records indicate that climatic variability has affected ecological processes on the Olympic Peninsula for millennia (Heusser, 1974; Gavin *et al.*, 2001). For example, pollen spectra from subalpine lakes in the Olympics indicate common responses after the retreat of Pleistocene glaciers, divergent vegetation in the early Holocene, and convergent responses in the late Holocene (McLachlan and Brubaker, 1995). More recently, tree growth for many lower elevation species increased with water supply and decreased with high summer temperatures (Ettl and Peterson, 1995; Nakawatase and Peterson, 2006). A common lesson from both paleo and modern studies is that, for a given regional shift in climate, the ecological and climatic context of a particular site determines the degree and nature of the response (Holman and Peterson, 2006)—so much so that high versus low elevations and the wet versus the dry side of the Olympics may have very different responses to a uniform climatic change.

Hydrological resources also respond to climate. The timing, duration, and magnitude of stream runoff depend on the abundance of winter snowpack and winter-to-spring temperatures. The Olympic Mountains mirror regional patterns of decadal climatic variability and trends in climatic change. During the 20th century, snowpacks were smaller (especially at low elevations), temperatures were warmer (especially minimum temperatures), and precipitation varied significantly with the fluctuations of the Pacific Decadal Oscillation. Regional anadromous fish populations (Mantua *et al.*, 1997), tree growth (Peterson and Peterson, 2001), glacier mass balance (Bitz and Battisti, 1999), and forest fire activity (Littell, 2006; see also[6]) have responded to these changes.

Predictions of future climate for the Pacific Northwest are uncertain because of uncertainty about future fossil fuel emissions, global population, efficacy of mitigation, and the response and sensitivity of the climatic system. However, by comparing a range of scenarios and models for future events, climate modelers can estimate future climatic conditions. Regional climate models suggest an increase in mean temperature of 1.2–5.5°C (2.2–9.9°F), with a mean of 3.2°C (5.8°F) by 2090 (Salathé, Jr., 2005). Summer temperatures are projected to increase more than winter temperatures. Precipitation changes are less certain due to large natural variability, but slight increases in annual and winter precipitation are projected, while slight decreases in summer precipitation are possible (Salathé, Jr., 2005).

Projected changes in temperature and precipitation would lead to lower snowpacks at middle and lower elevations, shifts in timing of spring snowmelt and runoff, and increases in summer evapotranspiration (Mote *et al.*, 2005; Hamlet *et al.*, 2007). Runoff in winter (October to March) would increase, and summer runoff (April to September) would decrease (Hamlet *et al.*, 2007). For basins with vulnerable snowpack (*i.e.*, mid-elevations), streamflow would increase in winter and decrease in summer. Higher temperatures and lower summer flows would have serious consequences for anadromous and resident fish species (salmon, steelhead, bull trout). Floods may increase in frequency because the buffering effect of snowpacks would decrease and because the severity of storms is projected to increase (although less snow can decrease the maximum impacts of rain-on-snow events due to lower water storage in snow). Sea level rise would exacerbate flooding in coastal areas. Some effects, especially the timing of snowmelt and peak streamflow, are likely to vary substantially with topography

Increased summer temperature may lead to non-linear increases in evapotranspiration from vegetation and land surfaces (McCabe and Wolock, 2002). This, in turn, would decrease the growth (Littell, 2006; Nakawatase and Peterson, 2006), vigor, and fuel moisture in lower elevation (*e.g.*, Douglas-fir and western hemlock) forests while

[6] **Mote**, P.W., W.S. Keeton, and J.F. Franklin, 1999: Decadal variations in forest fire activity in the Pacific Northwest. In: *Proceedings of the 11th Conference on Applied Climatology*, American Meteorological Society, pp. 155-156.

increasing growth (Ettl and Peterson, 1995; Nakawatase and Peterson, 2006) and regeneration in high elevation (*e.g.*, subalpine fir and mountain hemlock) forests (Woodward, Schreiner, and Silsbee, 1995). Higher temperatures would also expand the range and decrease generation time of climatically limited forest insects such as the mountain pine beetle (Logan, Regniere, and Powell, 2003), as well as increase the area burned by fire in western Washington and Oregon (Littell, 2006).

The distribution and abundance of plant and animal species would change over time (Zolbrod and Peterson, 1999), given that paleoecological data show this has always been a result of climatic variability in the range projected for future warming. This change may be difficult to observe at small scales, and would be facilitated in many cases by large-scale disturbances such as fire or windstorms that remove much of the overstory and "clear the slate" for a new cohort of vegetation. The regeneration phase will be the key stage at which species will compete and establish in a warmer climate, thus determining the composition of future vegetative assemblages and habitat for animals.

Thus, ecosystem services in ONF are likely to be affected by climatic change. Water quality for threatened fish species may decline as temperatures increase and, potentially, as increasing storm intensity causes road failures. Water quantity may decline in summer when it is most needed, as streamflow timing shifts with temperature changes. Air quality will decline if drought frequencies or durations increase and cause increased area burned by fire. The influence of climate change on habitat for threatened species is less certain, but high elevation and currently rare species would be more vulnerable (*e.g.*, Olympic marmot, bull trout, whitebark pine).

A1.2.3 Current ONF Policy Environment, Planning Context and Management Goals

Current natural resources management in ONF is directed primarily from policy mandates and shaped by historical land use and forest fragmentation (Fig. A1.4). ONF is a "restoration forest" charged with managing large, contiguous areas of second-growth forest. Natural resource objectives include managing for native biodiversity and promoting the development of late-successional forests (*e.g.*, NWFP); restoring and protecting aquatic ecosystems from the impacts of an aging road infrastructure; and managing for individual threatened and endangered species as defined by the Endangered Species Act (ESA) or other policies related to the protection of other rare species.

Most ONF natural resources management activities are focused on restoring important habitats (*e.g.*, native prairies, old-growth forests, pristine waterways), rehabilitation or restoration of impacts related to unmaintained logging roads, invasive species control, and monitoring. Collaboration with other agencies occurs, and is a cornerstone of the NWFP. Without clear consensus on climate change, cross-boundary difficulties in solving problems may arise due to differing mandates, requirements, and strategies, but there is no evidence that this is currently a problem.

Planning guidelines for ONF are structured by mandates from the National Forest Management Act (NFMA) and the NWFP. The ONF land management plan (OLMP, to be revised in the future in coordination with other western Washington NFs) is influenced by the NWFP as well as regional Forest Service policy. Planning also is influenced by comments from the public served by ONF. Project planning is carried out at a site-specific level, so incorporating regional climatic change information into Environmental Assessment/Environmental Impact Statement documents can be difficult because assessment takes place at the site scale, while there is still substantial uncertainty surrounding climate change predictions—especially precipitation—at sub-regional scales.

Adaptation to climatic change is not yet addressed formally in the OLMP or included in planning for most management activities. Current management objectives are attempting to confer resilience by promoting landscape diversity and biodiversity and this is in keeping with adapting to climate change. To this end, tools available to ONF managers include restoration of aquatic systems (especially the minimization of the impacts of roads, bridges, and culverts); active management of terrestrial systems (through thinning and planting); and, increasingly, treatment of invasive species. Prescribed fire and wildland use fire are unlikely tools because of the low historical area burned, limitations of the Clean Air Act, and low funding levels. The range of strategies and information in using these tools varies across ONF land use designations. Late-successional reserves and wilderness have less leeway than adaptive management areas, because there are more explicit restrictions on land use and silvicultural treatment.

A1.2.4 Proactive Management Actions Anticipating Climate Change

ONF's policy and regulatory environment encompasses a great deal of responsibility, but little scientific information or specific guidance is available to guide adaptation to climatic change. The scope of possible adaptation, clear strategies for successful outcomes, and the tools available to managers are all limited. Under current funding restrictions, most tools would need to be adapted from management responses to current stresses (Table A1.1). Future impacts on ecological and socioeconomic sensitivities can result in potential tradeoffs or conflicts. For example, currently threatened species may become even more rare in the future (*e.g.*, bull [illegible] Case Study Outline Foci for the ONF: current ecosystem stresses, management goals, current management methods, and climate change impacts

Current Ecosystem Stresses	Management Goal(s)	Current Methods	Climate Impacts on Ecosystems and Management Practices
Historical timber harvest impacts on landscape	Promote species and landscape biodiversity	Silvicultural treatment to achieve a broad range of habitats for native species	Depends on how area and frequency of disturbances changes (windthrow, fire, endemic/exotic insect/pathogen outbreaks). Increases in the above, and their interactions, in ONF per se are understudied because they have not been large problems. All are climate mediated, and could become so, but unknown impact on management practices.
	Increase late seral habitat	Silvicultural treatments to increase rate of "old growth" structure development	
	Protect old-growth dependent species	Same as above	Currently, the main disturbance legacy on ONF is 20th century logging.
Aquatic ecosystem degradation	Restore aquatic ecosystems to conditions that support endangered species	Riparian restoration, culvert rehabilitation	Warming waters, changes in timing of seasonal snow/rain/runoff will increase need for restoration, but potentially limit its success rate as well.
Impacts of unmaintained, closed roads	Remove potential effects of unmaintained roads	Road restoration/ rehabilitation; occasionally removal	If intense storms, flooding, or rain-on-snow events increase in frequency, closed road failures will likely increase in frequency. Multiple failures on the same road limit response/access. This will require substantial investment in new management efforts.
Invasive exotic species	Limit spread of new invasives	Preventive educ./ strategies	If disturbances or recreational travel increase or if climate changes the competitive balance between natives and exotics , efficacy of current strategies uncertain
	Treat established invasive species	Treatment limited to hand pulling in most locations; herbicide where permitted.	
Endemic Insects	Currently none	Monitoring	Uncertain
Fire	Currently none	Suppression (rare)	Depends on interplay between climate-mediated fire and climate-mediated regeneration

trout, spotted owl, marbled murrelet, Olympic marmot) due to stress complexes, undermining the likelihood of successful protection. Another example is when short-term impacts must be weighed against long-term gains. Fish species may be vulnerable to failures of unmaintained, closed roads caused by increased precipitation/storminess, but road rehabilitation may produce temporary sedimentation and may invite invasive weeds. Ideally, triage situations could be avoided, but in the face of climate change and limited resources it may be necessary to prioritize management actions with the highest likelihood of success, at the expense of actions that divert resources and have less-certain outcomes.

Generally, success of adaptation strategies should be defined by their ability to reduce the vulnerability of resources to a changing climate while attaining current management goals. Strategies include prioritizing treatments with the greatest likelihood of being effective (resources are too limited to do otherwise) and recognizing that some treatments may cause short-term detrimental effects but have long-term benefits. For structures, using designs and engineering standards that match future conditions (*e.g.*, culvert size) will help minimize future crises. Specific strategies likely to be used in ONF terrestrial ecosystems are to increase landscape diversity, maintain biological diversity, and employ early detection/rapid response for invasive species.

Landscape diversity and resilience can be achieved by: (1) targeted thinning (increases diversity, can decrease vulnerability by increasing tree vigor, and can reduce vulnerability to disturbance); (2) avoiding a "one size fits all" toolkit, and using a variety of treatments even if new prescriptions are required; (3) creating openings large enough for elk habitat, but small enough to minimize invasive exotics; (4) considering preserves at many elevations, not just high-elevation wilderness; and (5) considering "blocking" ownerships (land trades) to reduce edges, maintain corridors, and consolidate habitat.

Biological diversity may be maintained by: (1) planting species in anticipation of climate change—using different geographical locations and nursery stock from outside current seed zones; (2) maintaining within-species diversity; and (3) providing corridors for wildlife. However, there must be credible rationale for decisions to use seed and seedlings other than local native plant species.

Early detection/rapid response focuses on solving small problems before they become large, unsolvable problems, and recognizes that proactive management is more effective than long delays in implementation. For example, the ONF strategic plan recognizes that invasive species often become established in small, treatable patches, and are best addressed at early stages of invasion. Although designed for other problems like invasives, it is also appropriate for climate change because it could allow managers to respond quickly to the impacts of extreme events (disturbances, floods, windstorms) with an eye toward adaptation.

Large-scale disturbance can cause sudden and major changes in ecosystems, but can be used as occasions to apply adaptation strategies. ONF is currently climatically buffered from chronic disturbance complexes already evident in drier forests, but age-class studies and paleoproxy evidence indicate that large-scale disturbances occurred in the past. For comparison, fire suppression and harvest practices in British Columbia played a role in the current pine beetle outbreak by homogenizing forest structure over very large areas. In ONF, the amount of young forest (as a result of 20th century harvest) is both a risk (hence ONF's "restoration" status) and an opportunity. Large disturbances that may occur in the future could be used to influence the future structure and function of forests. Carefully designed management experiments for adapting to climatic change could be implemented. There is a clear need to have concepts and plans in place in anticipation of large fire and wind events, so that maximum benefit can be realized.

Information and tools needed to assist adaptation are primarily a long-term, management-science partnership with decision-specific scientific information. ONF relayed a critical request of scientists: natural resource managers need a manager's guide with important scientific concepts and techniques. Critical gaps in scientific information hinder adaptation by limiting assessment of risks, efficacy, and sustainability of actions. Managers would also like assistance and consultation on interpreting climate and ecosystem model output so that the context and relevance of model predictions can be reconciled with managers' priorities for adaptation. Managers identified a need to determine effectiveness of prevention and control efforts for invasive species; monitoring is critical (and expensive). There is a strong need for data on genetic variability of key species, as well as recent results of hydrologic modeling relative to water supply, seasonal patterns, and temperature. In contrast, managers pointed out that ONF collects data on a large array of different topics, many of them important, but new data collection should be implemented only if it will be highly relevant, scientifically robust, and inform key decisions.

A1.2.5 Opportunities and Barriers to Proactive Management for Climate Change on the ONF

An important opportunity for adapting to climatic change at the regional scale is the coordinated development of forest plans among ONF, Mt. Baker-Snoqualmie National Forest, and Gifford Pinchot National Forest. The target date for beginning this forest planning effort is 2012. The effort would facilitate further cooperation and planning for adaptation in similar ecosystems subject to similar stressors. ONF has implemented a strategic plan that has similar capacity for guiding prioritization and can incorporate climatic change elements now, rather than waiting for the multi-forest plan effort. By explicitly addressing resilience to climatic change (and simultaneously developing any science needed to do so) in the OLMP, ONF can formalize the use of climate change information in management actions.

A second, related opportunity is to integrate climatic change into region-wide NWFP guidelines that amended Pacific Northwest forest plans. The legacy of the 20th century timber economy in the Pacific Northwest has created ecological problems, but also opportunities (Fig. A1.5). Landscapes predominately in early seral stages are more easily influenced by management actions, such as targeted thinning and planting, than are late seral forests, so there is an opportunity to anticipate climate change and prepare for its impacts with carefully considered management actions. By recognizing the likely future impacts of climatic change on forest ecosystems (such as shifts in disturbance regimes), the revised forest plans can become an evolving set of guidelines for forest managers. Specifically, will the NWFP network of late successional reserves remain resilient to climatic change and its influence on disturbance regimes? Are there specific management practices in adaptive management areas that would change given the likely impacts of climatic change?

Collaboration among multiple organizations is key to successful management. ONF staff believe that the "stage is set" for continued and future collaboration among organizations and agencies on the Olympic Peninsula. Climatic change and ecosystems do not recognize political boundaries, and significant adaptive leverage can be gained by cooperation. Initiatives by coalitions and partnerships can include climatic change (*e.g.*, the Puget Sound Partnership) and are conducive to an environment in which adaptation actions are well supported. In some cases, working with other agencies can improve the likelihood of success by increasing overall land base and resources for addressing problems.

Major barriers to adaptation are (1) limited resources, (2) policies that do not recognize climate change as a significant problem or stressor, and (3) the lack of a strong management-science partnership. National and regional budget policies and processes are significant barriers to adaptation, and represent a constraint on the potential for altering or supplementing current management practices to enable adaptation to climate change. Current emphasis on fire and fuel treatments in dry forest systems has greatly reduced resources for stand density management, pathogen management, etc. in forests that do not have as much fire on the ground but may, in the future, be equally vulnerable. Multiple agency collaboration can be difficult because of conflicting legislation, mandates, and cultures, but such collaboration is likely to be a hallmark of successful adaptation to climatic change. Certainly increased collaboration between scientists and managers could streamline the process of proposing testable scientific questions and applying knowledge to management decisions and actions.

Policies, laws, and regulations that are based on a more static view of the environment do not consider the flexibility required to adapt to changing conditions outside historical observations. The NFMA puts limitations on management actions, and NEPA delays implementation of actions. The ESA requires fine-scale management for many imperiled species, which may be unrealistic in a rapidly changing climate. Given the projected future rate of climate change and the resource limitations for land management agencies, it may be more sustainable and a more efficient use of funding to protect systems and landscape diversity than to plan for and protect many individual species. The NWFP partially embraces this strategy, but does not focus specifically on climate change. The Clean Water Act could become an important barrier in the future as stream temperatures increase; this may result

[illegible] Olympic National Forest is charged with mitigating the legacy of 20th century timber harvest. Landscape fragmentation and extensive road networks (upper left) are consequences of this legacy that influence strategies for adaptation to climate change. The old-growth forest dependent northern spotted owl (upper right) is one focus of the NWFP, which prescribes forest practices but does not address climatic change. Changes in the timing and intensity of runoff expected with climate change are likely to interact with this legacy to have negative impacts on unmaintained roads (lower left) that in turn will impact water quality for five threatened or endangered species of anadromous and resident fish. Photo Credits: All photos courtesy Olympic National Forest.

in unattainable standards that constrain management actions. NEPA, the ESA, the Clean Water Act, and the NWFP all focus on historical reference points in comparatively static environments, but climate change warrants looking to future impacts and the need for preparation.

Future barriers to adaptation may arise with the interaction of current policy restrictions and the potential need to adapt to climatically mediated changes in ecosystem processes. One example is the potential for using wildland fire for the benefit of forest ecosystems, which is not currently an authorized management tool on ONF. The benefits of wildland fire use (likely limited in ONF to natural ignitions within wilderness areas) would need to be weighed against the cost of authorization. Authorization to use this tool in the short term would require a Forest Plan amendment and associated NEPA process. A less costly but longer-horizon alternative is to include wildland fire use in the 2012 Forest Plan revision effort. Benefits would be limited to wildland fire use that could be approved within the confines of the ESA and other regulations. Olympic National Park recently completed a fire management plan that authorizes wildland fire use, but has restrictions related to ESA requirements. For ONF the role of wildland fire use in management would also be limited by the ESA and the adjacency of non-federal land concerns.

A1.2.6 Increasing the Adaptive Capacity to Respond to Climate Change

The ecosystem stressors ONF manages for currently (Table A1.1) are likely to be exacerbated by climatic change, but little work has focused on quantifying the direct linkages between the climate system and future ecosystem services on the Olympic Peninsula. Resilience to climate change is therefore only describable qualitatively. Past timber harvest has resulted in a very large area of lower-elevation forest consisting of second growth, in an ecosystem that was characterized by resilient old growth. This landscape homogenization has occurred in other forest types, and, at least in theory, results in less resilience to climate-mediated disturbances. However, such characterization is at the moment speculative. Aquatic ecosystems are probably less resilient, and measuring resilience there is similarly underdeveloped.

The primary conclusions of this case study are:

1. Climate change and its impacts are identifiable regionally, and adaptation to climate change is necessary to ensure the sustainability of ecosystem services.
2. ONF management priorities (Table A1.1) are consistent with adaptation to climatic change and promoting resilience to the impacts of climate change. However, available resources do not allow adaptation at sufficient scale. Moreover, scientific uncertainty remains about the best adaptation strategies and practices.
3. The current political and regulatory contexts limit adaptive capacity to current and future climatic changes by:
    a. failing to incorporate climatic change into policy, regulations, and guidelines;
    b. requiring lengthy planning processes for management actions, regardless of scope; and
    c. adopting priorities and guidelines that are not clear in intent and/or consistently applicable at national, regional, and forest levels.
4. These limitations can be overcome by:
    a. developing a manager's guide to climate impacts and adaptation;
    b. developing an ongoing science-management partnership focused on climate change;
    c. incorporating climatic change explicitly into national, regional, and forest-level policy;
    d. re-examining the appropriateness of laws, regulations, and policies on management actions in the context of adaptation to climatic change;
    e. creating clear, consistent priorities that provide guidance but allow for local/forest level strategies and management actions that increase resilience and reduce vulnerability to climatic change;
    f. allocating resources sufficient for adaptation; and
    g. increasing educational and outreach efforts to promote awareness of climate change impacts on ecosystem services.

ONF is at a crossroads. The effects of climatic change on forest ecosystems and natural resources are already detectable. Adapting to those changes and sustaining ecosystem services is an obvious and urgent priority, yet adaptive capacity is limited by the policy environment, current allocation of scarce resources, and lack of relevant scientific information on the effects of climate change and, more crucially, on the likely outcomes of adaptive strategies. Adaptive management is one potential strategy for learning how to predict, act on, and mitigate the impacts of climatic change on a forest ecosystem, but if there is no leeway for management actions or those actions must occur quickly, then adaptation options are limited in the current environment. ONF staff indicated that if they were managing for climate change, given what they know now and their current levels of funding and personnel, they would continue to emphasize management for biodiversity. It is possible, for example, that they might further increase their current emphasis on restoration and diversity. Another possible change, reminiscent of the earlier Forest Service priorities, would be to emphasize the role of forests as producers of hydrological commodities.

Key components of adaptation will be to (1) develop a vision of what is needed and remove as many barriers as possible; (2) increase collaboration among agencies, managers, and scientists at multiple scales; and (3) facilitate strategies (such as early detection/rapid response) that are proven to work. A functional forest ecosystem is most likely to persist if managers prioritize landscape diversity and biological diversity. Equally certain is that management actions should not, in aggregate, lead to the extirpation of rare species. Clear and consistent mandates, priorities, and policies are needed to support sustainability of ecosystem services in the face of a warmer climate and changing biophysical conditions.

We envision a future in which the policy, planning, and scientific aspects of ecosystem-based management co-evolve with changes in climate and ecosystems. This vision requires trust, collaboration, and education among policy makers, land managers, and scientists as well as the publics they serve. Climate will continue to change, effects on ecosystems will be complex, and land managers will struggle to adapt to those changes with limited resources. Collaboration with scientists is certain to produce information that relates directly to on-the-ground decision making. Less certain is how opportunities for adaptation will be realized while retaining public support for resource management actions. ONF has already transitioned from producing a few commodities to producing a broad array of ecosystem services, but the more ambitious vision of coevolution must progress rapidly in order for adaptation to keep pace with anticipated effects of climatic change.

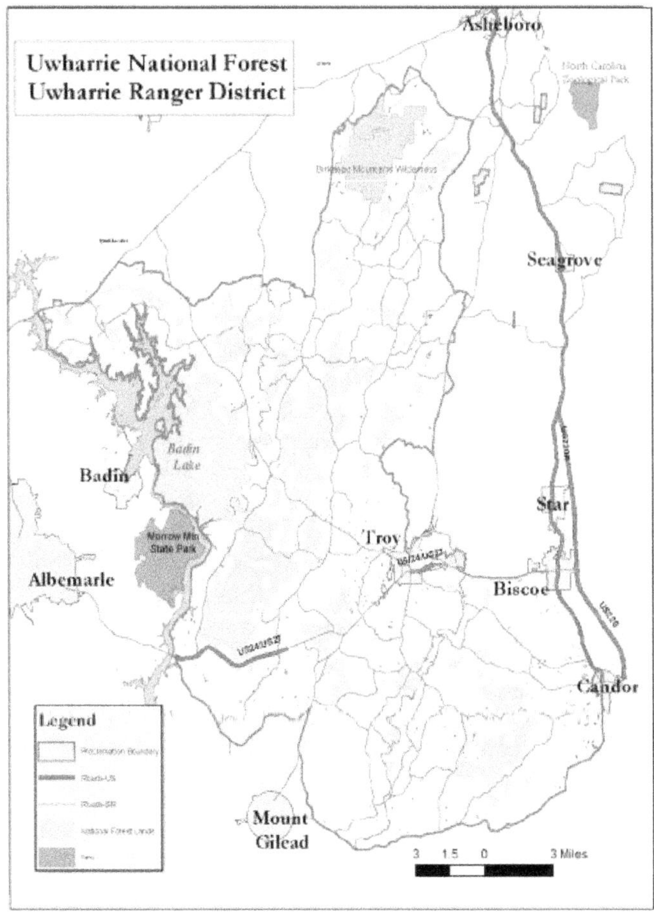

Map of the Uwharrie National Forest in North Carolina.[6]

A1.3.1 Setting and Context of the Uwharrie National Forest

The Uwharrie National Forest (originally called the Uwharrie Reservation) was first purchased by the federal government in 1931 during the Great Depression. In 1961, President John F. Kennedy proclaimed the federal lands in Montgomery, Randolph, and Davidson Counties (Fig. A1.6). The UNF is within a two-hour drive of North Carolina's largest population centers, including Winston-Salem, Greensboro, Charlotte, Raleigh, and Durham. The forest is fragmented into 61 separate parcels, which pose unique forest management challenges (Fig. A1.6). Therefore, much of UNF has been modified from a natural to a managed ecological condition. UNF has a rolling topography, with elevation ranging from 122 to 305 m above sea level. Although small by most national forest standards (20,383 ha), the UNF provides a variety of natural resources, including clean rivers and streams, diverse vegetation for scenery, wildlife habitat, and wood products. There is also a wide variety of recreational activities, and UNF is a natural setting for tourism and economic development.

The UNF is rich in history. It is named for the Uwharrie Mountains, some of the oldest in North America. According to geologists, the Uwharries were created from an ancient chain of volcanoes. The 1,000-foot hills of today were once 20,000-foot peaks.

[7] **USDA Forest Service**, 2007: Uwharrie National Forest Uwharrie Ranger District. University of North Carolina at Asheville National Forest Service Website, http://www.cs.unca.edu/nfsnc/uwharrie_plan/maps/uwharrie_map.pdf, accessed on 7-30-2007.

The UNF is located at the crossroads of both prehistoric and historic settlements. Their legacy is one of the greatest concentrations of archeological sites in the Southeast. Left undisturbed, these sites and artifacts give a record of our heritage. The first large gold discovery in the United States occurred around 1799 at the nearby Reed Gold Mine. In the early 1800s, gold was found in the Uwharries, with a later boom during the depression of the 1930s. Old mining sites still remain, and part-time prospectors still pan in the streams and find traces of gold dust.

Today, the UNF is dynamic and responsive to public needs. It continues to provide timber, wildlife, water, recreation opportunities, and a natural setting for tourism and economic development. Recreational use is growing, especially in the Badin Lake area and along the 20-mile Uwharrie National Recreation Trail. Badin Lake is one of the largest bodies of water included in the series of reservoirs within the Yadkin-PeeDee River drainage system. The entire watershed is known as the Uwharrie Lakes Region. Badin Lake is a popular setting for many different recreation activities, including camping, hiking, fishing, boating, and hunting. The area is rich game land for deer and wild turkey, and a home for bald eagles.

A1.3.2 Current Uwharrie NF Planning Context, Forest Plan Revision and Climate Change

The National Forest Management Act of 1976 requires that all NFs periodically revise their forest management plan.[8] Existing environmental and economic situations within the forest are examined. Then plans are revised to move the forest closer to a desired future condition. The current UNF forest management plan was originally developed in 1986, and UNF is now undergoing a Forest Plan Revision (FPR).

The revised forest plan focuses on three themes. Two of the themes—restoring the forest to a more natural ecological condition, and providing outstanding and environmentally friendly outdoor recreation opportunities—will likely be affected by a changing climate. The third theme of the FPR (*i.e.*, better managing heritage (historical and archeological) resources) will likely not be significantly affected by climate change. Thus, this case study examines potential impacts on the first two UNF FPR themes.

The revised forest plan will suggest management strategies that help reduce risks to the health and sustainability of UNF associated with projected impacts of a changing climate. Therefore, the UNF case study focuses on specific recommended modifications to the forest plan. This level of specificity was not possible with either the Tahoe or Olympic National Forest case studies because neither has recently undergone a forest plan revision that incorporates climate change impacts into forest management decision making.

A1.3.2.1 Revised Forest Plan Theme 1: Restoring the Forest to a More Natural Ecological Condition

Prior to the 1940s, fires were a regular occurrence in southern U.S. ecosystems (Whitney, 1994). The reoccurrence interval varied among vegetation types, with more frequent fires being less intense than less frequent fires (Wear and Greis, 2002). Upland oak (*Quercus* sp.) and hickory (*Carya* sp.) forests would burn at an interval of 7–20 years with flame heights of less than one m (3.3 ft.). These fires would kill thin-barked tree species such as red maple (*Acer rubrum*), sweetgum (*Liquidambar styraciflua*), and tulip poplar (*Liriodendron tulipifera*), while leaving the more fire-resistant oaks and hickories alive. Pine ecosystems had a shorter fire return interval of 3–5 years, with flame heights reaching 1–2 m (3.3–6.6 ft.), thus favoring fire- and drought-resistant longleaf (*Pinus palustris*) and shortleaf (*Pinus echinata*) pines more than loblolly pines. The fires also removed much of the mid-canopy vegetation and promoted light-demanding grasses and herbs.[9] Deciduous and coniferous tree species are equally represented in UNF. However, a higher percent of the conifers are in loblolly pine (*Pinus taeda*) plantations than would have historically occurred, because of the planting emphasis of this species over the past 40 years.[9]

Climate change is projected to increase the number and severity of wildfires across the southern United States in the coming years (Bachelet *et al.*, 2001). As part of its FPR, UNF plans to restore approximately 120 ha (296 acres) of loblolly pine plantation to more fire-resistant ecosystem types (*e.g.*, longleaf pine) each year.[9] This management shift will restore UNF to a more historically natural condition and reduce catastrophic wildfire risk associated with an increase in fuel loading (Stanturf *et al.*, 2002; Busenberg, 2004) and hotter climate (Bachelet *et al.*, 2001).

A1.3.2.2 Revised Forest Plan Theme 2: Provide Outstanding and Environmentally Friendly Outdoor Recreation Opportunities

Recreation opportunities provided by UNF are an important ecosystem service to the local and regional communities. The proximity to large population centers and diverse interest in outdoor activities make UNF a destination for many groups that use the trails and water bodies located within the forest. The continued quality of these trails, streams, and lakes are of very high importance to UNF's mission.

During the 20th century the frequency of extreme precipitation events has increased, and climate models

[8] 16 U.S.C. §1600-1614

[9] **Uwharrie National Forest**, 2007: *Proposed Uwharrie National Forest Land Management Plan*. Available from http://www.cs.unca.edu/nfsnc/uwharrie_plan/wo_review_draft_plan.pdf. USDA Forest Service, Asheville, NC.

suggest that rainfall intensity will continue to increase during the 21st century (Nearing, 2001). Soil erosion occurs when the surface soil is exposed to rainfall and surface runoff. Soil erosion is affected by many factors, including rainfall intensity, land cover, soil texture and structure (soil erodibility), and land topography (slope) (Toy, Foster, and Renard, 2002). Because soil erosion increases linearly with rainfall-runoff erosivity, it would be expected to increase over the next 50 years in the UNF region if no management measures are taken to control the current soil erosion problems. Soil erosion is limited to exposed (*i.e.*, without vegetative cover) soil surfaces (Pimentel and Kounang, 1998). Hiking, off-highway vehicles, and logging trails and forest harvest areas represent the major types of exposed soil surface in UNF.[9] Increased soil erosion would degrade both trail and water quality.

In response to current and projected increases in soil erosion potential, the UNF FPR proposes to repair authorized roads and trails, close unauthorized roads and trails, minimize new road construction, and reroute needed roads that increase soil erosion. In total, these measures should effectively reduce the potential impact of increased precipitation intensity on soil erosion in the UNF.

A1.3.3 Long-Term Natural Resource Services

In addition to the objectives outlined in the Uwharrie forest plan revision, forests in the United States provide valuable natural resources of clean water and wood products. While the demand for U.S. pulp and paper products has decreased in recent years, it is important to assess the long-term ability of the forests to supply wood resources if a future need should arise. The demand for clean, dependable water is increasing within the southern United States as population pressure on water resources increase. Therefore, climate change impacts on UNF water yield and timber supply were also assessed in the UNF Watershed Analysis Document of the FPR.

A1.3.3.1 Water Yield

Clean water is one of the most valuable commodities that our NFs provide. National forest lands are the largest single source of water in the United States and one of the original reasons that the NFS was established in 1891 (USDA Forest Service, 2000b). There is concern that climate change could reduce water yield from the Uwharrie. Currently, about 1,590 mm of precipitation falls in UNF every year, with close to 70% (or 1,100 mm) of it evapotranspiring back to the atmosphere. The other 30% (or 490 mm) leaves the forest as stream runoff and percolates downward becoming a part of the groundwater.[9] Climate change models suggest that precipitation may increase to 1,780 mm per year. Air temperature is also expected to increase, which will, in turn, increase forest evapotranspiration. In total, stream water flow is projected to decrease by approximately 10% by the middle of the 21st century if there is no change in forest management (Sun *et al.*, 2005).[10]

Forest water use increases with increased tree stocking density and leaf area (Hatton *et al.*, 1998; Cook *et al.*, 2002). The use of controlled fire and other forest management activities that will increase tree spacing and shift the forest toward more fire- and drought-tolerant tree species will also help to reduce forest water use (Heyward, 1939). Based on this line of research, most of the climate change-caused reductions in water yield can be compensated through this proposed change in forest management.

A1.3.3.2 Timber and Pulpwood Productivity

The southern United States has long been a major supplier of pulpwood and timber. But because an increasing amount of timber and pulpwood is being supplied to the United States by Canada, Europe, and countries in the Southern Hemisphere (USDA Forest Service, 2003), national forest managers have moved away from an emphasis on timber supply toward recreational opportunities and sustainable water (Apple, 1996).

Climate change will have variable impacts globally. Timber production in some countries, such as Canada, may benefit from warmer climate, while countries closer to the Equator may experience significant reductions in productivity (Melillo *et al.*, 1993). Although NFs are not currently major sources of wood products, this situation could change as timber production from other parts of the world shifts. Therefore, it is important to assess the impact of climate change on forest productivity in UNF. Forest productivity models suggest that although pine productivity may decrease, hardwood productivity is projected to increase and the net loss of total forest productivity would be small for the UNF over the next 40 years (National Assessment Synthesis Team, 2000). However, the analysis did not account for the potential for increased fire occurrence, which could significantly reduce overall forest volume and growth (Bachelet *et al.*, 2001). The proposed shift in forest tree types to more drought-tolerant and fire-resistant species should also help to assure that UNF remains a timber resource for future generations (Smith, Ragland, and Pitts, 1996).

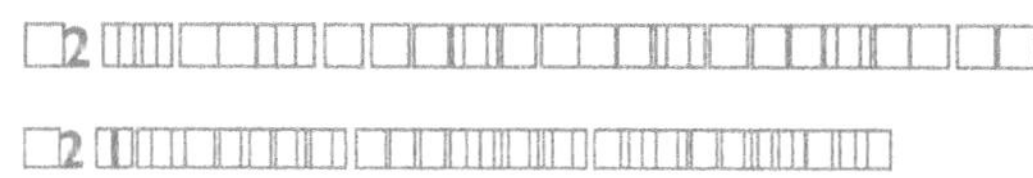

The climate is going to change continuously over at least the next 100 years. Ecosystems, species, and processes in each of the 270 natural resource parks will be affected by

[10] See also **Sun**, G., S.G. McNulty, E. Cohen, J.M. Myers, and D. Wear, 2005: Modeling the impacts of climate change, landuse change, and human population dynamics on water availability and demands in the Southeastern US. Paper number 052219. Proceedings of the 2005 ASAE Annual Meeting, St. Joseph, MI.

climate change over this time period. Therefore, it was not appropriate to select a case study based on its perceived current vulnerability to climate change. Some parks are beginning to face issues related to sea level rise; treasured species in others are at risk. Regardless of the apparent urgency in some parks, all will have to initiate adaptation actions in order to meet NPS mission and goals. Rocky Mountain National Park (RMNP), Colorado, was selected for a case study because it is a good example of the state at which most parks find themselves as they confront resource management in the face of climate change. Park managers know RMNP has some highly vulnerable and visible resources, including glaciers and alpine tundra communities, but there is high uncertainty regarding just how vulnerable they are, what specific changes might occur, how rapidly change might occur, or what to do. The following case study describes RMNP's first attempt to take stock of the Park with respect to climate change, and begin to think about management.

A2.1.1 Park Description and Management Goals

RMNP was established in 1915 and "is dedicated and set apart as a public park for the benefit and enjoyment of the people of the United States ...with regulations primarily aimed at the freest use of the said park and for the preservation of natural conditions and scenic beauties."[11] The Park is located in the Front Range of the southern Colorado Rocky Mountains, the first mountain range west of the Great Plains. RMNP's wide elevation gradient—from 8,000 to more than 14,000 feet—includes montane forests and grasslands, old-growth subalpine forests, and the largest expanse of alpine tundra in the lower 48 states. More than 150 lakes and 450 miles of streams form the headwaters of the Colorado River to the west, and the South Platte River to the east. Rich wetlands and riparian areas are regional hotspots of native biodiversity. Several small glaciers and rock glaciers persist in east-facing cirque basins along the Continental Divide. The snow that accumulates in these basins each winter provides water that supports downstream cities and agricultural activities in Colorado and neighboring states. RMNP is home to populations of migratory elk, mule deer, bighorn sheep, and charismatic predators such as golden eagles, cougars, and bobcats; many plant and animal species that live in the alpine, including white-tailed ptarmigan, pika, and yellow-bellied marmot; and several endangered species, including the boreal toad and the greenback cutthroat trout.

At slightly larger than 415 square miles, RMNP is not large compared with other western national parks (Yellowstone, by comparison, is more than eight times larger). RMNP is bordered on all four sides by national forests. The Roosevelt National Forest surrounds the Park on the north and east, the Routt National Forest is to the northwest, and the Arapahoe National Forest surrounds the southwest, southern, and eastern Park boundaries. Approximately half of the adjacent Forest Service land is in wilderness designation (Comanche Peak Wilderness, Neota Wilderness, Never Summer Wilderness, and Indian Peaks Wilderness), and 95% of RMNP is managed as if it was wilderness. A primary goal for RMNP, therefore, is to protect and manage the Park in its natural condition (see Box A2.1). Wilderness status has been proposed since 1974, and legislation is pending. RMNP is also designated a Clean Air Act Class I Area, meaning the superintendent has a responsibility to protect air-quality related values, including vegetation, visibility, water quality, wildlife, historic and prehistoric structures and objects, cultural landscapes, and most other elements of a park environment that are sensitive to air pollution. Several endangered species, such as the boreal toad and the greenback cutthroat trout, have management plans for enhancement and recovery. Other current management issues include fire, elk, and invasive exotic species. All told, there are more than 30 planning documents (Acts, Executive Orders, Plans, and Recommendations) that guide RMNP operations.

The towns of Estes Park and Grand Lake form gateway communities, and are connected by Trail Ridge Road which is open for traffic crossing the Continental Divide during the summer and fall months. Largely because of its spectacular vistas, the Park receives more than three million visitors each year, 25% of whom come from Colorado. Most visitor use is

[illegible]2 [illegible]Definition of Wilderness

A wilderness, in contrast with those areas where man and his own works dominate the landscape, is hereby recognized as an area where the earth and its community of life are untrammeled by man, where man himself is a visitor who does not remain. For the purposes of this chapter, an area of wilderness is further defined to mean an area of undeveloped Federal land retaining its primeval character and influence, without permanent improvements or human habitation, which is protected and managed so as to preserve its natural conditions and which (1) generally appears to have been affected primarily by the forces of nature, with the imprint of man's work substantially unnoticeable; (2) has outstanding opportunities for solitude or a primitive and unconfined type of recreation; (3) has at least five thousand acres of land or is of sufficient size as to make practicable its preservation and use in an unimpaired condition; and (4) may also contain ecological, geological, or other features of scientific, educational, scenic, or historical value.

[11] 16 U.S.C. § 191-198

in the summer, when hiking, camping, mountain climbing, viewing nature, and sightseeing are common. Fall visitation is also popular, when visitors arrive to view aspen leaves and watch and listen to elk go through their mating rituals.

A2.1.2 Observed Climate Change in the Western United States

Many climate change signals have been observed in the western United States, but not all of them in the southern Rocky Mountains or in RMNP. Strong trends in winter warming, increased proportions of winter precipitation falling as rain instead of snow, and earlier snowmelt are found throughout the western United States (Stewart, Cayan, and Dettinger, 2005; Knowles, Dettinger, and Cayan, 2006; Mote, 2006). All of these trends are more pronounced in the Pacific Northwest and the Sierra Nevada than they are in the Colorado Front Range of the southern Rocky Mountains. The less pronounced evidence for RMNP compared with the rest of western U.S. mountains should not be interpreted as a lack of climate change potential within the Park. The high (and thus cold) elevations and a shift over the past 40 years from a more even annual distribution of precipitation to more winter precipitation have contributed to Front Range mountain weather going against the trend seen across much of the rest of the West (Knowles, Dettinger, and Cayan, 2006).

Summer warming has been observed in RMNP, and while a ten year record is insufficient for an understanding of cause, July temperatures increased approximately 3°C, as measured at three high elevation sites from 1991-2001 (Clow *et al.*, 2003). RMNP, along with most of the rest of the western United States, experienced record-breaking extreme March temperatures and coincident early melting of winter snowpack in 2004. While not directly attributable to climate change, extreme heat events are consistent with climate change model projections that suggest increased rates of extreme events due to the warming atmosphere (Pagano *et al.*, 2004).

Figure A2.1. Photos of Arapahoe Glacier in 1898 and 2004.[12]

Figure A2.2. Photo pair of Rowe Glacier, with permissions, NSIDC and leachfam website.[13]

A2.1.3 Observed and Projected Effects of Climate Change in the Southern Rocky Mountains and Rocky Mountain National Park

A number of studies have indicated that climatic warming is being expressed in environmental change in the southern Rocky Mountains and in RMNP: mountain glacier retreat (evidence of climatic warming) is occurring adjacent to and within RMNP. Arapahoe Glacier, located 10 miles south of the Park on the Continental Divide, has thinned by more than 40 m since 1960 (Fig. A2.1). Photograph pairs of Rowe Glacier in RMNP also show the loss of ice mass over time (Fig. A2.2). Responses to climatic change are also showing up in ecological communities: a long-term study of the timing of marmot emergence from hibernation in central Colorado found marmots emerge on average 38 days earlier than they did in 1977 (Inouye *et al.*, 2000). This is triggered by warming spring temperatures. Similarly, the spring arrival of migratory robins to Crested Butte, Colorado, is two weeks earlier now than in 1977. This also signals biological changes in response to climate (Inouye *et al.*, 2000).

A number of species of plants and animals may be vulnerable to climate change. Dwarf larkspur (*Delphinium nuttalianum*)

---

[12] **NSIDC/WDC for Glaciology**, Boulder, Compiler, 2006: Online glacier photograph database. *National Snow and Ice Data Center/World Data Center for Glaciology*. Available at http://nsidc.org/data/g00472.html.

[13] **Lee**, W.T., 1916: Rowe Glacier photograph. In: Online glacier photograph database. National Snow and Ice Data Center/World Data Center for Glaciology.
Leach, A., 1994: Rowe Glacier photograph. Available from http://www.leachfam.com/securearea/album.php. Boulder, Colorado.

shows a strong positive correlation between snowpack and flower production (Saavedra *et al.*, 2003). Research findings suggest that reduced snowpacks that accompany global warming might reduce fitness of this flowering plant. Local weather, as opposed to regional patterns, exerts a strong influence on several species of birds found in the Park, including white-tailed ptarmigan, *Lagopus leucurus* (Wang *et al.*, 2002b). The median hatch rates of white-tailed ptarmigan in RMNP advanced significantly from 1975–1999 in response to warmer April and May temperatures. Population numbers have been declining along Trail Ridge Road, where they are routinely monitored (Wang *et al.*, 2002a), and where population growth rates were negatively correlated with warmer winter temperatures. The Wang *et al.* (2002b) study suggests that ptarmigan may likely be extinct in RMNP within another two or three decades. Dippers (*Cinclus mexicanus*) in RMNP may also be vulnerable, as has been shown by studies of the closely related white-fronted dipper (*Cinclus cinclus*) in Scandinavia (Saether *et al.*, 2000; Wang *et al.*, 2002b).

Some studies of animal responses to climate change in the Park reveal positive responses. Elk populations were projected to double under climate scenarios of warmer winters and possibly wetter summers, while model results for warmer winters with drier summers projected an increase in the elk population of 50% (Wang *et al.*, 2002c). Elk populations have been increasing within RMNP due to enhanced overwinter survival, and this may be another factor in the demise of white-tailed ptarmigan, as elk are now taking advantage of warmer springs to graze on high level tundra where they compete with ptarmigan for shrubby browse.

Greenback cutthroat trout, an endangered species, have been translocated into streams and lakes in RMNP as part of a recovery effort. Water temperatures in many of the translocation streams are colder than optimal for greenback cutthroat trout growth and reproduction. Of the ten streams where the fish were reintroduced by the Colorado Division of Wildlife, only three had temperatures within the range for successful growth and reproduction at the time of translocation. A modeling scenario that postulated warmer stream temperatures suggests that three additional streams will experience sufficient temperature increases to raise the probability of translocation success to >70%. In at least one of these streams, however, temperatures are projected to also warm enough to allow the establishment of whirling disease, caused by *Myxobolus cerebralis*, a parasite that is fatal to young trout.[14]

[14] **Cooney**, S., 2005: Modeling global warming scenarios in greenback cutthroat trout (Oncorhynchus clarki stomias) streams: implications for species recovery. M.S. thesis, Colorado State University, Fort Collins.

Other studies suggest that climate warming will diminish opportunities for willow establishment along riparian areas in RMNP (Cooper *et al.*, 2006), and the occurrence of longer and more severe fire seasons will increase throughout the western United States (Westerling *et al.*, 2006).

An analysis of recreation preferences under climate change scenarios projected a relatively small increase (10-15%) in visitation to RMNP for climate-related reasons under climate warming scenarios (Richardson and Loomis, 2004). An economic study of whether such an increased visitation would affect the economy and employment outlook for Estes Park similarly did not find climate change to be very important (Weiler *et al.*, 2002). A more important driver of economic change for the Town of Estes Park was projected increases in human population numbers within the State of Colorado (Weiler *et al.*, 2002).

A2.1.4 Adapting to Climate Change

RMNP is relatively rich in information about its ecosystems and natural resources, and has benefited from long-term research and monitoring projects and climate change assessments. Examples include research and monitoring, in Loch Vale Watershed[15], and the focused assessment of the effects of climate change on RMNP and its Gateway Community.[16] Even so, planning and resource management in the Park does not yet include considerations of climate change. A workshop in March 2007 provided the opportunity for Park managers and community members to begin thinking about the steps to take to increase preparedness for a climate that will be warmer and less predictable. Results of the workshop are summarized below.

In many ways, effective science-based management in RMNP has enhanced the ability of park natural resources to adapt to climate change. Most of the water rights have been purchased, dams and ditches have been removed, and many streams and lakes have been restored to free-flowing status since 1980. An exception is the Grand River Ditch. Park managers have also been proactive in removing or preventing invasive species such as leafy spurge, and invasive non-native species such as mountain goats; managing fire through controlled burns and thinning; reducing regional air pollution through partnerships with regulatory agencies; and preparing a plan to reduce elk populations to more sustainable numbers.

Despite these actions, RMNP managers are concerned over the potential for catastrophic wildfire, increasing insect

[15] **Natural Resource Ecology Laboratory**, 2007: Loch Vale Watershed research project. Colorado State University, www.nrel.colostate.edu/projects/lvws, accessed on 5-15-2007.

[16] **Natural Resource Ecology Laboratory**, 2002: Science to achieve results. Colorado State University, http://www.nrel.colostate.edu/projects/star/index.html, accessed on 4-6-2007.

infestations and outbreaks, and damage from large storm events with increasing climate change. A flooding event in the Grand River Ditch, while not necessarily caused by climate change, serves as an example of the potential effects from future storm-caused floods. The Grand Ditch diverts a significant percentage of annual Colorado River tributary streamflow into the east-flowing Poudre River. It was developed in 1894, and is privately owned and managed. A breach of the ditch during snowmelt in May 2003 caused significant erosion and damage to Kawuneechee Valley forests, wetlands, trails, bridges, and campsites.

Park managers are also concerned about the future of alpine tundra and species that live above treeline, but do not have much information about current alpine species populations and trends. Modest baseline data and monitoring programs are currently in place. Regional biogeographic models suggest that the treeline will rise and some alpine areas will diminish or disappear (Neilson and Drapek, 1998). Reduced tundra area, or its fragmentation by trees, could endanger many obligate tundra plants and animals. Species such as pika, white-tailed ptarmigan, and marmots are already known to be responsive to climate change (Inouye *et al.*, 2000; Wang *et al.*, 2002a; Beever, Brussard, and Berger, 2003).

RMNP managers have identified a strategy for increasing their ability to adapt to climate change built on their current activities, what they know, and what they do not know about upcoming challenges related to climate change. The strategy involves bringing teams of experts and regional resource managers together in a series of workshops to share information and help identify resources and processes that may be most susceptible to climate change. Support for high resolution models that project possible changes to species and processes can be used to establish scenarios of future ecological trajectories and end-states. Regularly held workshops with scientific experts offer the opportunity to develop planning scenarios, propose adaptive experiments and management opportunities, and keep abreast of the state of knowledge regarding climate change and its effects.

Managers also propose establishing a Rocky Mountain National Park Science Advisory Board. A Science Advisory Board could serve as a springboard for thinking strategically and enabling the Park to anticipate climate-related events. RMNP managers recognize the need to develop baselines for species or processes of highest concern (or of greatest indicator value) and plan to establish monitoring programs to track changes over time. The vital signs that have been identified for the Park need to be reviewed and possibly revised in order to capture effects that will occur with climate change.

Park managers identified a critical need to develop a series of learning activities and opportunities for all Park employees to increase their knowledge of climate change-related natural resource issues within RMNP. The Continental Divide Learning Center was recognized as an ideal venue for these activities. Managers have proposed that the Center be used as a hub for adaptive learning, articulating the value of natural resources better, and turning managers into consumers of science.

Finally, Park mangers have recognized the importance of building greater collaborations with regional partners in order to facilitate regional planning, especially for issues that cross Park boundaries. RMNP already has strong working relations with the Town of Estes Park, the Colorado Department of Public Health and Environment, the Colorado Division of Wildlife, the U.S. Fish and Wildlife Service, Larimer and Boulder Counties, and many local organizations and schools. Opportunities to work more closely with the Routt, Arapaho, and Roosevelt National Forest managers could be pursued with the objective of discussing shared management goals.

In summary, RMNP managers propose to continue current resource management activities to minimize damage from other threats, increase their knowledge of which species and ecosystems are subject to change from climate change, monitor rates of change for select species and processes, and work with experts to consider what management actions are appropriate to their protection. By developing working relations with neighboring and regional resource managers, the Park keeps its options open for allowing species to migrate in and out of the Park, considering assisted migrations, and promoting regional approaches toward fire management (Box A2.2).

A2.1.5 Needed: A New Approach Toward Resource Management

RMNP, like other national parks, often operates in reactive mode, with limited opportunity for long-term planning. Reactive management has a number of causes, only some of which are related to tight budgets and restrictive funding mechanisms. Partly because national parks are so visible to the public, there are public expectations and political pressures that trigger short-term management activities (tree thinning in lodgepole pine forest is one example of an activity that is visible to many, but of questionable value in reducing the risk of catastrophic fire). Natural resource issues are increasingly complex, and climate change adds greatly to this complexity.

RMNP managers have been proactive in addressing many of the resource issues faced by the Park. Yet they recognize there is still more to be done, particularly in human resource management. Complex issues require broad and flexible ways of thinking about them, and creative new tools for their management. Professional development programs for

[illegible] 2.2 Opportunities and Barriers for Rocky Mountain National Park in Adapting to Climate Change

[illegible]

- Cadre of highly trained natural resource professionals
- Extensive scientifically grounded knowledge of many natural resources and processes
- Continental Divide Learning Center serves as hub of learning and training
- Plan to establish a Science Advisory Board
- Climate Friendly Parks Program has enhanced climate change awareness
- Good working relations with city, county, state, and federal land and resource managers
- RMNP is surrounded on nearly all sides by protected national forest lands, including wilderness.
- Regionally, mountain and high valley lands to the north, west, and south of RMNP are mostly publicly owned and protected, or sparsely populated ranch and second home developments.
- RMNP is a headwater park and controls most of the water rights within its boundaries. As such, it has direct control over its aquatic ecosystems and water quality.

[illegible]

- Insufficient knowledge about individual species' status and trends
- Limited opportunity for long-term strategic planning
- Limited interagency coordination of management programs
- The large and growing urban, suburban, exurban Front Range urban corridor may hinder migration of species into or out of RMNP from the Great Plains and Foothills to the east.

current resource managers, rangers, and park managers could be strengthened so that all employees understand the natural resources that are under the protection of the NPS, the causes and consequences of threats to these resources, and the various management options that are available.

The skill sets for new National Park Service (NPS) employees should reflect broad systems training. University programs for natural resource management could shift from traditional training in fisheries, wildlife, or recreational management to providing more holistic ecosystems management training. Curricula at universities and colleges could also emphasize critical and strategic thinking that embraces science and scientific tools for managing adaptively, and recognizes the need for lifelong learning. Climate change can serve as the catalyst for this new way of managing national park resources. Indeed, if the natural resources entrusted to RMNP—and other parks—are to persist and thrive under future climates, the Park Service will need managers that see the whole as well as the parts, and act accordingly.

[illegible]

[illegible]

Warming trends in Alaska and the Arctic are more pronounced than in southerly regions of the United States, and the disproportionate rate of warming in Alaska is expected to continue throughout the coming century (IPCC, 2001) (see Fig. 5.3a in the National Wildlife Refuges chapter). Migratory birds are one of the major trust species groups of the National Wildlife Refuge System (NWRS), and birds that breed in Alaska traverse most of the system as they use portions of the Pacific, Central (see Fig. A3.1), Mississippi, and Atlantic Flyways during their annual cycle. Projected warming is expected to encompass much of the Central Flyway but is expected to be less pronounced in the remaining flyways (IPCC, 2001). Historical records show strong warming in the Dakotas and a tendency toward

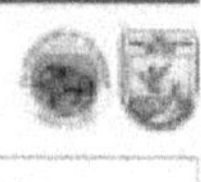

[illegible] Central Flyway Waterfowl Migration Corridor.[17]

17 U.S. Fish and Wildlife Service, 2007: Central flyway. U.S. Fish and Wildlife Service, Pacific Flyway Council Website, http://pacific-flyway.gov/Documents/Central_map.pdf, accessed on 6-2-0007.

cooling in the southern reaches of the flyway (see Fig. 5.3a in the National Wildlife Refuges chapter). Pervasive and dramatic habitat shifts (see Fig. 5.9 in the National Wildlife Refuges chapter) are projected in Alaska and especially throughout the Central Flyway by the end of the century.

Migration is an energetically costly and complex life history strategy (Arzel, Elmberg, and Guillemain, 2006). The heterogeneity in warming and additional stressors along migratory pathways along with their potential effects on productivity and population levels of migratory birds emphasize the importance of strong interconnections among units of the NWRS and the need for a national vision and a comprehensive management strategy to meet the challenge of climate change in the next century. The following case study examines warming and additional stressors, as well as management options in Alaska and the Central Flyway, which together produce 50–80% of the continent's ducks (Table A3.1).

A3.1.1 Current Environmental Conditions

A3.1.1.1 Changes in Climate and Growing Season Duration

Climate

In recent decades, warming has been very pronounced in Alaska, with most of the warming occurring in winter (December–February) and spring (March–May) (Serreze *et al.*, 2000; McBean *et al.*, 2005). In western and central Canada, the increases in air temperature have been somewhat less than those observed in Alaska (Serreze *et al.*, 2000). While precipitation has remained largely stable throughout Alaska and in Canada in recent decades, several lines of evidence indicate that Alaska and western Canada are experiencing increased drought stress due to increased summer water deficits (Barber, Juday, and Finney, 2000; Oechel *et al.*, 2000; Hogg and Bernier, 2005; Hogg, 2005; Hogg, Brandt, and Hochtubajda, 2005).

Growing Season Duration

The seasonal transition of northern ecosystems from a frozen to a thawed condition represents the closest analog to a biospheric "on-off switch" that exists in nature, dramatically affecting ecological, hydrologic, and meteorological processes (Running *et al.*, 1999). Several studies based on remote sensing indicate that growing seasons are changing in high-latitude regions (Dye, 2002; McDonald *et al.*, 2004; McGuire *et al.*, 2004; Smith, Saatchi, and Randerson, 2004; Euskirchen *et al.*, 2006). These studies identify earlier onset of thaw in northern North America, but the magnitude of change depends on the study. Putting together the trends in the onset of both thaw and freeze, Smith, Saatchi, and Randerson (2004) indicate that the trend for longer growing seasons in northern North America (3 days per decade) is primarily due to later freezing. However, other studies indicate that the lengthening growing season in North America is primarily due to earlier thaw (Dye, 2002; Euskirchen *et al.*, 2006). Consistent with earlier thaw of terrestrial ecosystems in northern North America, lake ice has also been observed to be melting earlier across much of the Northern Hemisphere in recent decades (Magnuson *et al.*, 2000). The study of Euskirchen *et al.* (2006) indicates that trends for earlier thaw are generally stronger in Alaska than in the Central Flyway of Canada and northern United States, but trends for later freeze are stronger in the Central Flyway of Canada and the northern United States than in Alaska.

A3.1.1.2 Changes in Agriculture

Agriculture and migratory waterfowl are intimately related because waterfowl make significant use of agricultural waste on staging and wintering areas. Much of the agricultural production in the United States is centered in the Central

Table A3.1. The annual cycle of migratory waterfowl that breed in Alaska may serve as an integrative focus for development of a national vision of climate effects and management adaptation options for the National Wildlife Refuge System. The complexity of potential interactions among locations, life history stages, climate mechanisms, non-climate stressors, and options for management adaptation for migratory waterfowl that breed in Alaska demonstrates that inter-regional assessment and timely communication will be essential to the development of a national vision.

Location	Life History	Climate Mechanisms	Non-Climate Stressors	Adaptation Options
Alaska	Production: • Breeding • Fledging	Early Thaw: • Resource access • Habitat area • Season length	• Minimal	• Assess System • Predict • Collaborate • Facilitate
Prairie Potholes (*Central Flyway*)	Staging: • Energy reserves	Late Freeze: • Habitat distribution • Migration timing • Harvest distribution	• Land use • Crop mix • Disturbance • Alternate Energy Sources	• Assess System • Predict • Partnerships • Secure Network
Southern United States	Wintering: • Survival • Nutrition	Sea Level: • Habitat access • Storms: • Frequency, Intensity	• Urbanization • Fragmentation • Pollution	• Partnerships • Education • Acquisition • Adaptive Mgmt.

Flyway. Dynamic markets, government subsidies, cleaner farming practices, and irrigation have changed the mix, area, and distribution of agricultural products during the past 50 years (Krapu, Brandt, and Cox, Jr., 2004). Genetically engineered crops and resultant changes in tillage practices and the use of pesticides and herbicides, as well as development of drought resistant crop varieties, will likely add heterogeneity to the dynamics of future crop production. While corn acreage has remained relatively stable during the past 50 years, waste corn available to waterfowl and other wildlife declined by one-quarter to one-half during the last two decades of the 20th century, primarily as a result of more efficient harvest (Krapu, Brandt, and Cox, Jr., 2004). While soybean acreage has increased by approximately 600% during the past 50 years, metabolizable energy and digestibility of soybeans is noticeably less than for corn, and waterfowl consume little, if any, soybeans (Krapu, Brandt, and Cox, Jr., 2004). These changes in availability of corn and soybeans suggest that nutrition of waterfowl on migratory staging areas may be compromised (Krapu, Brandt, and Cox, Jr., 2004). If a future emphasis on bio-fuels increases acreage in corn production, the potential negative effects of the recent increase in soybean production on waterfowl energetics may be ameliorated.

A3.1.1.3 Changes in Lake Area

Analyses of remotely sensed imagery indicate that there has been a significant loss of closed-basin water bodies (water bodies without an inlet or an outlet) over the past half century in many areas of Alaska (Riordan, Verbyla, and McGuire, 2006). Significant water body losses have occurred primarily in areas of discontinuous permafrost (Yoshikawa and Hinzman, 2003; Hinzman *et al.*, 2005; Riordan, Verbyla, and McGuire, 2006) and subarctic areas that are permafrost-free (Klein, Berg, and Dial, 2005). In an analysis of approximately 10,000 closed-basin ponds across eight study areas in Alaska with discontinuous permafrost, Riordan, Verbyla, and McGuire (2006) found that surface water area of the ponds decreased by 4–31% while the total number of closed-basin ponds surveyed within each study region decreased by 5–54% (Riordan, Verbyla, and McGuire, 2006). There was a significant increasing trend in annual mean surface air temperature and potential evapotranspiration since the 1950s for all the study regions, but there was no significant trend in annual precipitation during the same period. In contrast, it appears that lake area is not changing in regions of Alaska with continuous permafrost (Riordan, Verbyla, and McGuire, 2006). However, in adjacent Canada, significant water body losses have occurred in areas dominated by permafrost (Hawkings, 1996).[18]

[18] See also **Hawkings**, J. and E. Malta, 2000: Are northern wetlands drying up? A case study in the Old Crow Flats, Yukon. *51st AAAS Arctic Science Conference.*

Warming of permafrost may be causing a significant loss of lake area across the landscape because the loss of permafrost may allow surface waters to drain into groundwater (Yoshikawa and Hinzman, 2003; Hinzman *et al.*, 2005; Riordan, Verbyla, and McGuire, 2006). While permafrost generally restricts infiltration of surface water to the sub-surface groundwater, unfrozen zones called taliks may be found under lakes because of the ability of water to store and vertically transfer heat energy. As climate warming occurs, these talik regions can expand and provide lateral subsurface drainage to stream channels. This mechanism may be important in areas that have discontinuous permafrost such as the boreal forest region of Alaska. However, the reduction of open water bodies may also reflect increased evaporation under a warmer and effectively drier climate in Alaska, as the loss of open water has also been observed in permafrost-free areas (Klein, Berg, and Dial, 2005).

In the Prairie Pothole Region (PPR) of the Central Flyway, changes in climate accounted for 60% of the variation in the number of wet basins (Larson, 1995), with partially forested parklands being more sensitive to increasing temperature than treeless grasslands. When wet basins are limited, birds may overfly grasslands for parklands and then proceed even farther north to Alaska in particularly dry years in the pothole region. Small- and large-scale heterogeneity in lake drying may first cause a redistribution of birds and, if effects are pervasive enough, may ultimately cause changes in the productivity and abundance of birds. Fire and vegetation changes in the PPR and in Alaska may exacerbate these effects.

A3.1.2 Projections and Uncertainties of Future Climate Changes and Responses

A3.1.2.1 Projected Changes in Climate and Growing Season Duration

Climate

Projections of changes in climate during the 21st century for the region between 60° and 90° N indicate that air temperature may increase approximately 2°C (range ~1–4°C among models) and that precipitation may increase approximately 12% (range ~8–18% among models) (Kattsov and Källén, 2005). The increase in precipitation will be due largely to moisture transport from the south, as temperature-induced increases in evaporation put more moisture into the atmosphere. Across model projections, increases in temperature and precipitation are projected to be highest in winter and autumn. Across the region, there is much spatial variability in projected increases in temperature and precipitation, both within a model and among models. For any location, the scatter in projected temperature and precipitation changes among the models is larger than the mean temperature and precipitation change projected among the models (Kattsov and Källén, 2005).

In comparison with northern North America, climate model projections indicate that the Central Flyway of the United States will warm less with decreasing latitude (Cubasch *et al.*, 2001). Mid-continental regions such as the Central Flyway are generally projected to experience drying during the summer due to increased temperature and potential evapotranspiration that is not balanced by increases in precipitation (Cubasch *et al.*, 2001). Projections of changes in vegetation suggest that most of the Central Flyway (see Fig. A3.1 and Fig. 5.9d in the National Wildlife Refuges chapter) will experience a biome shift by the latter part of the 21st century (Bachelet *et al.*, 2003; Lemieux and Scott, 2005).

Growing Season Duration

One analysis suggests that projected climate change may increase growing season length in northern and temperate North America by 0.4–0.5 day per year during the 21st century (Euskirchen *et al.*, 2006), with stronger trends for more northern latitudes. This will be caused almost entirely by an earlier date of thaw in the spring, as the analysis indicated essentially no trend in the date of freeze. Analyses of this type need to be conducted across a broader range of climate scenarios to determine if this finding is robust. If so, then one inference is that lake ice would likely melt progressively earlier throughout northern and temperate North America during the 21st century.

A3.1.2.2 Changes in Lake Area

It is expected that the documented loss of surface water of closed-basin ponds in Alaska (Riordan, Verbyla, and McGuire, 2006) and adjacent Canada will continue if climate continues to warm in the 20th century. The ubiquitous loss of shallow permafrost (Lawrence and Slater, 2005) as well as the progressive loss of deep permafrost (Euskirchen *et al.*, 2006) are likely to enhance drainage by increasing the flow paths of lake water to ground water. Also, it is likely that enhanced evaporation will increase loss of water. While projections of climate change indicate that precipitation will increase, it is unlikely that increases in precipitation will compensate for water loss from lakes from increased evaporation. An analysis by Rouse (1998) estimated that if atmospheric CO_2 concentration doubles, an increase in precipitation of at least 20% would be needed to maintain the present-day water balance of a subarctic fen. Furthermore, Lafleur (1993) estimated that a summer temperature increase of 4°C would require an increase in summer precipitation of 25% to maintain present water balance. These changes in precipitation to maintain water balance are higher than the range of precipitation changes (8–18%) anticipated for the 60–90° N region in climate model projections (Kattsov and Källén, 2005).

A3.1.3 Non-Climate Stressors

In Alaska, climate is the primary driver of change in habitat value for breeding migrants through its effects on length of the ice-free season (U.S. Fish and Wildlife Service, 2006) and on lake drying (Riordan, Verbyla, and McGuire, 2006). Throughout the Central Flyway, projected major changes in vegetation are expected to occur by the end of the century (see Fig. 5.9d in the National Wildlife Refuges chapter) (Bachelet *et al.*, 2003; Lemieux and Scott, 2005). Additional stressors in the Central Flyway include competing land uses on staging areas outside the NWRS, changes in the distribution and mix of agricultural crops that may favor/disfavor foraging opportunities for migrants on migratory and winter ranges, and anthropogenic disturbance that may affect nutrient acquisition strategies for migrants in both spring and fall by restricting access to foraging areas. In southern regions of the Central Flyway, rising sea level and increasing urbanization may cause reductions in refuge area and increased insularity of remaining fragments. All stressors contribute to uncertainty in future distribution and abundance of birds. Climate dominates on Alaskan breeding grounds, and additional stressors complicate estimation of the net effects of climate on migrants and their use of staging and wintering areas in central and southern portions of the Central Flyway.

A3.1.4 Function of Alaska in the National Wildlife Refuge System

Alaska is a major breeding area for North American migratory waterfowl. Alaska and the adjacent Yukon Territory are particularly important breeding areas for American widgeon (~38% of total in 2006), green-winged teal (~31%), northern pintail (~31%) and greater and lesser scaup combined (~27%). Substantial proportions of the North American populations of western trumpeter swans, Brant geese, light geese (Snows) and greater sandhill cranes also breed in Alaska (U.S. Fish and Wildlife Service, 2006).

Alaska both contributes to NWRS waterfowl production and provides a vehicle to conceptually integrate most of the NWRS. Waterfowl that breed in Alaska make annual migrations throughout North America and are thus exposed to large-scale heterogeneity in potential climate warming effects. Migrants use the Pacific, Central, Mississippi, and to a lesser extent the Atlantic, Flyways on their annual spring and fall migrations. Their migration routes extend to wintering grounds as far south as Central and South America.

The spatial heterogeneity in warming, variable energetic demands among life history stages, and variable number and intensity of non-climate stressors along the migratory pathways creates substantial complexity within the NWRS. This complexity emphasizes that performance (*e.g.*, weight gain, survival, reproduction) of any species in any life history stage at any location within a region may be substantially affected by synergistic effects of climate and non-climate stressors elsewhere within the NWRS. A successful response to this complexity will require a national vision of the problems and solutions, and creative local action.

A3.1.4.1 Potential Effects of Climate Change on the Annual Cycle of Alaska Breeding Migrants

Abundance of waterfowl arriving on the breeding grounds is a function of survival and nutritional balance on the wintering grounds and on spring migration staging areas. Two types of breeding strategies are recognized. "Income" breeders obtain the energy for egg production primarily from the nesting area while "capital" breeders obtain energy for egg production primarily from wintering and spring staging areas. Regardless of whether species are income or capital breeders, food availability in the spring on breeding grounds in the Arctic is important to breeding success (Arzel, Elmberg, and Guillemain, 2006).

Breeding conditions for waterfowl in Alaska depend largely on the timing of spring ice melt (U.S. Fish and Wildlife Service, 2006). In the short term, earlier springs that result from warming likely advance green-up and ice melt, thus increasing access to open water and to new, highly digestible vegetation growth and to terrestrial and aquatic invertebrates. Such putative changes in open water and food resources in turn may influence the energetic balance and reproductive success of breeders and the performance of their offspring. Flexibility in arrival and breeding dates may allow some migrants to capitalize on earlier access to resources and increase the length of time available for re-nesting attempts and fledging of young. Some relatively late migrants, such as scaup (Austin *et al.*, 2000), may not be able to adapt to warming induced variable timing of open water and food resources, and thus may become decoupled from their primary resources at breeding.

In the long term, increased temperatures and greater length of the ice-free season on the breeding grounds may contribute to permafrost degradation and long-term reduction in the number and area of closed-basin ponds (Riordan, Verbyla, and McGuire, 2006), which may reduce habitat availability, particularly for diving ducks. Countering this potential reduction in habitat area may be changes in wetland chemistry and aquatic food resources. Reductions in water volume of remaining ponds may result in increased nutrient or contaminant concentrations, increases in phytoplankton, and a shift from an invertebrate community dominated by benthic amphipods to one dominated by zooplankton in the water column.[19] This has variable implications for foraging opportunities for waterfowl that make differential use of shallow and deep water for foraging. The net effects of lake drying on waterfowl populations in Alaska are not known at this time, but the heterogeneity in relatively local reductions and increases in lake area in relation to breeding waterfowl survey lines (see Fig. A3.2) may make it difficult to detect any effects that have occurred.

Departure of waterfowl from breeding grounds in the fall may be delayed by later freeze-up. The ability to prolong occupancy at northern latitudes may increase successful fledging and allow immature birds to begin fall migration in better body condition. Later freeze-up may allow immature birds, particularly large species such as swans, to delay their rate of travel southward and increase their opportunities for nutrient intake during migration. Changes in the timing of arrival at various southern staging areas may affect waterfowl's access to and availability of resources such as waste grain and may result in re-distribution of birds along the migration route as they attempt to optimize foraging opportunities. The primary effect of this later departure and reduced rate of southward migration may be observed in more northerly fall distributions of species and a northward shift in harvest locations as has already been observed for some species. Later freeze-up and warmer winters may allow species to "short-stop" their migrations and winter farther north. Observations by Central Flyway biologists indicate

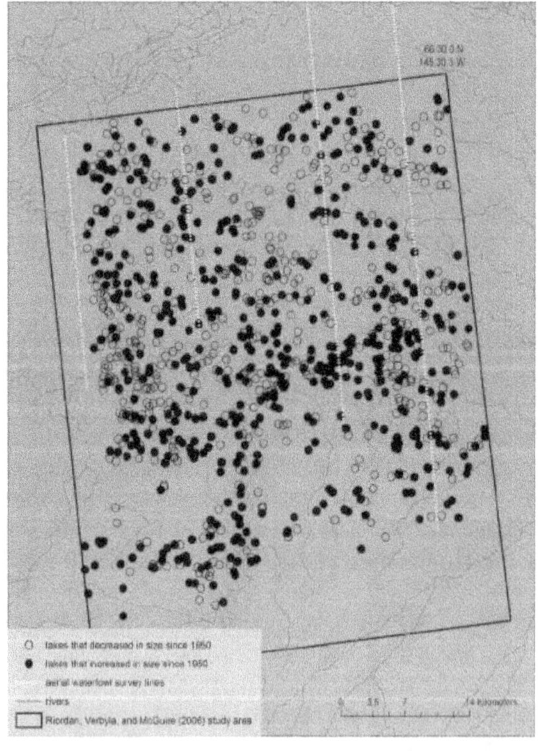

Figure A3. Heterogeneity in closed-basin lakes with increasing and decreasing surface area, 1950–2000, Yukon Flats NWR, Alaska. Net reduction in lake area was 18% with the area of 566 lakes decreasing, 364 lakes increasing, and 462 lakes remaining stable. Adapted from Riordan, Verbyla, and McGuire (2006).

[19] **Corcoran**, R.M., 2005: Lesser scaup nesting ecology in relation to water chemistry and macroinvertebrates on the Yukon Flats, Alaska. Masters Thesis. Department of Zoology and Physiology, University of Wyoming, Laramie, 1-83.

that 1) numbers of wintering white-fronted geese numbers have increased in Kansas in recent years, evidently as a result of diminished proclivity to travel further southward to Texas and Mexico for the winter; 2) portions of the tundra swan population now winter in Ontario rather than continuing southward; and 3) the winter distribution of Canada geese has shifted to more northern latitudes. The energetic and population implications of these putative northerly shifts in distribution in winter will ultimately be determined by the interaction of migratory costs, food availability, non-climate stressors such as anthropogenic disturbance and shifting agricultural practices, and harvest risk.

Earlier spring thaw may advance the timing of spring migration and increase the amount of time that some species, such as greater sandhill cranes, spend on their staging grounds in Nebraska. Increased foraging time during spring migration should benefit larger species, which tend to accumulate nutrients for breeding on the wintering grounds and on spring migration stopovers, more than smaller species, which tend to obtain nutrients necessary for breeding while on the breeding ground (Arzel, Elmberg, and Guillemain, 2006) although the explicit resolution of this concept needs to be quantified on a species-by-species basis. Warming-induced changes in the timing of forage availability on spring migration routes may cause redistribution of waterfowl or dietary shifts as they attempt to maximize the results of their strategic feeding prior to breeding. Increased understanding of the relative value of spring migration staging areas to reproductive success and annual population dynamics of different waterfowl species is a critical need in order to adapt management strategies to a changing climate.

A3.1.4.2 Implications for Migrants

Climate change adds temporal and spatial uncertainty to the problems associated with accessing resources necessary to meet energy requirements for migration and reproduction. Because birds are vagile, the primary near-term expected response to climate change is redistribution as birds seek to maintain energy balance.

Lengthened ice-free periods may result in earlier arrival on breeding grounds, delayed migration (*e.g.*, trumpeter swans and greater sandhill cranes), and wintering farther north (*e.g.*, white-fronted geese) among other phenomena. Warmer conditions that result in lake drying may result in birds over-flying normal breeding areas to areas farther north (*e.g.*, pintail ducks). Warmer temperatures may reduce water levels but increase nutrient levels in warmed lakes. Community composition of the invertebrate food base may change and life cycles of invertebrates may be shortened; amphipods may be disfavored and zooplankton favored with differential implications for birds with different feeding strategies. Changes in hydrologic periods may cause nest flooding or make nesting habitats that are normally isolated by floodwater accessible to predators. Either effect may alter nest and nesting hen survival.

The primary challenge to migratory waterfowl, and all other trust species for that matter, is that the spatial timing of resource availability may become decoupled from need. For example, late nesters such as lesser scaup may be hampered by pulsed resources that appear before nesting. Other species such as trumpeter swans may benefit from increased ice-free periods that enhance the potential to fledge young and provision them on southward migrations. Earlier and longer spring staging periods may benefit energetic status of migrating sandhill cranes. Harvest may shift northward as birds delay fall migrations.

Alaska and the Central Flyway (see Fig. A3.1) encompass substantial spatial variation in documented (see Fig. 5.3 in the National Wildlife Refuges chapter) and expected climate warming. This spatial variation in warming is superimposed on the variable demands of spatially distinct seasonal life history events (*e.g.*, nesting, staging, wintering) of migrants. Variance in success in any life history stage may affect waterfowl performance in subsequent stages at remote locations, as well as the long-term abundance and distribution of migrants. Performance of migrants at one location in one life history stage may be affected by climate in a different life history stage at a different location. The superimposition of spatially variable warming on spatially separated life history events creates substantial complexity in both documenting and developing an understanding of the potential effects of climate warming on major trust species of the NWRS. This unresolved complexity does offer a vehicle to focus on the interconnection of spatially separated units of the system and to foster a national and international vision of a management strategy for accommodating net climate warming effects on system trust species.

A3.1.5 Management Option Considerations

A3.1.5.1 Response Levels

Response to climate change challenges must occur at multiple integrated scales within the NWRS and among partner entities. Individual symptomatic challenges of climate change must be addressed at the refuge level, while NWRS planning is the most appropriate level for addressing systemic challenges to the system. Flyway Councils, if they can be encouraged to include a regular focus on climate change, may provide an essential mid-level integration mechanism. Regardless of the level of response, the immediate focus needs to be on what can be done.

A3.1.5.2 Necessary Management Tools

Foremost among necessary management tools are formal mechanisms to increase inter-agency communication and long-term national level planning. This could be accomplished through the establishment of an interagency public lands council or other entity that facilitates collaboration among federal land management agencies, NGOs, and private stakeholders. Institutional insularity of agencies and stakeholders at national and regional levels needs to be eliminated. The council should foster intra- and inter-agency climate change communication networks, because *ad hoc* communication within or among agencies is inadequate. Explicit outreach, partnerships and collaborations should be identified and target dates for their implementations drafted. In addition, the council should develop and implement national and regional coordination mechanisms and devise mechanisms for integrating potential climate effects into management decisions. The council needs to increase effective communication among wildlife, habitat, and climate specialists.

Within the NWRS there needs to be adequate support to insure the development of an increased capacity to rigorously model possible future conditions, and explicit recognition that spatial variation in climate has differential effects on life cycle stages of migrants; performance in one region may be affected by conditions outside a region. Enhanced ability to assist migratory trust species when "off-refuge" and enhanced ability to facilitate desirable range expansions within and across jurisdictions are needed.

Comprehensive Plans and Biological Reviews need to routinely address expected effects of climate change and identify potential mechanisms for adaptation to these challenges. The ability to effectively employ plans and reviews as focus mechanisms for potential climate change effects will be enhanced by institutionalization of climate change in job descriptions and increased training for refuge personnel.

A3.1.5.3 Barriers to Adaptation

The primary barriers to adaptation include the lack of a spatially explicit understanding of the heterogeneity and degree of uncertainty in effects of changing climate on seasonal habitats of trust species—breeding, staging and wintering—and their implications for populations. Currently there is concern about effects of climate change on trust species, but insufficient information on which to act. This lack of understanding hampers the development of an explicit national vision of potential net effects of climate change on migrants. In addition, the lack of a secure network of protected staging areas, similar to the established network of breeding and wintering areas, limits the ability of the NWRS to provide adequate security for migratory trust species in a changing climate. More efficient use of all types of resources will be needed to minimize these national-level barriers to adaptation of the NWRS to climate change.

A3.1.5.4 Opportunities for Adaptation

One of the greatest opportunities may lie in creating an institutional culture that rewards employees for being proactive catalysts for adaptation. This would require the acceptance of some degree of failure due to the uncertain nature of the magnitude and direction of climate change effects on habitats and populations. In addition, managers and their constituencies could be energized to mount successful adaptation to climate change by emphasizing the previous successful adaptations by the U.S. Fish and Wildlife Service (USFWS) to the first three management crises of market hunting, dust bowl habitat alteration, and threatened and endangered species management.

The capacity to provide more rigorous projections of possible future states will require the creative design of inventory and monitoring programs that enhance detection of climate change effects, particularly changing distributions of migratory trust species. Monitoring programs that establish baseline data regarding the synergy of climate change and other stressors (*e.g.,* contaminants, habitat fragmentation) will especially be needed. These monitoring programs will need to be coordinated with private, NGO and state and federal agency partners.

In stakeholder meetings, refuge biologists were emphatic that they needed more biological information in order to clearly define and to take preemptive management actions in anticipation of climate change. Thus, effective adaptation to climate change will require education, training and long-term research-management partnerships that are focused on adaptive responses to climate change. The following strategy is proposed for the activities of such a research-management partnership:

- Synthesize extant biological information relevant to biotic responses to climate change;
- Educate and train refuge mangers and other staff regarding climate change, its potential ecological effects, and the changes in management and planning that may be necessary;
- Evaluate possible management and policy responses to alternative climate change scenarios in multiple regional and national workshops;
- Conduct workshops involving managers, researchers and stakeholders to identify research questions relevant to managing species in the face of climate change;
- Conduct research on questions relevant to managing species in the face of climate change. This may require the development of tools that are useful for identifying the range of responses that are likely;
- Apply management actions in response to biotic responses that emerge as likely from such research; and
- Evaluate of the effectiveness of management actions and modification of management actions in the spirit of adaptive management.

Synthesis workshops should be held every few years to identify what has been learned and to redefine questions relevant to the management of species that depend on the NWRS.

There are a number of examples of recent climate-change-related challenges and potential and implemented adaptations in Alaska and the Central Flyway:

Potential adaptations:

- The development of a robust understanding of the relative contribution of various NWRS components to waterfowl performance in a warming climate is an immediate challenge. There is a clear research need to elucidate the relative contribution of staging and breeding areas to energetics and reproductive performance of waterfowl, and to clarify the interdependence of NWRS elements and their contributions to waterfowl demography. A flyway-scale perspective is necessary to understand the importance of migratory staging areas and to assess the relative importance of endogenous/exogenous energetics to reproduction and survival. These studies should address, in the explicit context of climate warming, strategic feeding by waterfowl, temporal shifts in diets, and the spatial and temporal implications of climate induced changes in the availability of various natural and agricultural foods (Arzel, Elmberg, and Guillemain, 2006).
- Providing adequate spatial and temporal distribution of migratory foraging opportunities is a chronic challenge to the NWRS. Spring staging areas are under-represented and this problem is likely to be exacerbated by a warming climate. It will be necessary to strengthen and clarify existing partnerships with private, NGO, and state and federal entities and to identify and develop new partnerships throughout the NWRS in order to provide a system of staging areas that are extensive and resilient enough to provide security for migratory trust species. Strategic system growth through fee-simple and conservation easement acquisition will be a necessary component of successful adaptation.

Implemented adaptations:

- Indigenous communities on the Aleutian Island chain (Alaska Maritime NWR) are concerned about the potential effects of increased shipping traffic in new routes that may become accessible in a more ice-free Arctic Ocean. Previous introductions of non-endemic species to islands have had severe negative effects on nesting Aleutian Canada geese. The ecosystem management mandate of the refuge facilitates a leadership role for the refuge that has been implemented through 1) development of monitoring partnerships that are designed to detect the appearance of invasive species and of contaminants, and 2) initiation of timely prevention/mitigation programs.
- Indigenous peoples that depend on Interior Alaska NWRs are concerned about the potential effects of climate-induced lake drying and changing snow conditions on their seasonal access to subsistence resources, and on the availability of waterfowl for subsistence harvest. The refuges have promoted enhanced capacity for projecting possible future conditions, and have educated users regarding observed and expected changes while clarifying conflicting information on the magnitude and extent of observed changes in lake number and area and in snow conditions.
- Warming-induced advances in the timing of ice-out can bias waterfowl population indices that are derived from traditional fixed-date surveys. The Office of Migratory Bird Management has developed quantitative models to project the arrival date of migrants based on weather and other records. This allows the office to dynamically adjust survey timing to match changing arrival dates and thereby reduce bias in population indices.

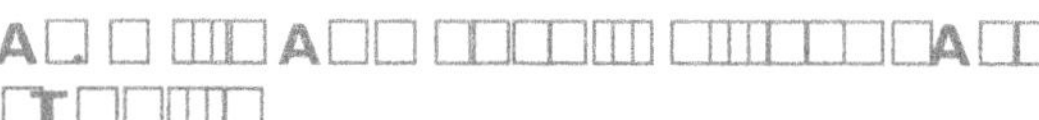

As emphasized throughout the Wild and Scenic Rivers (WSR) chapter, the effects of climate change on rivers will vary greatly throughout the United States depending on local geology, climate, land use, and a host of other factors. To illustrate the general "categories" of effects, we have selected three WSRs to highlight in the following case studies (Box A4.1). We selected these rivers because they span the range of some of the most obvious issues that managers will need to grapple with as they develop plans for protecting natural resources in the face of climate change. Rivers in the Southwest, such as the Rio Grande, will experience more severe droughts at a time when pressures for water extraction for growing populations are increasing. Rivers near coastal areas, such as the Wekiva, face potential impacts from sea level rise. A combination of groundwater withdrawals and sea level rise may lead to increases in salinity in the springs that feed this river. Rivers that are expected to experience both temperature increases and an increased frequency of flooding, such as the Upper Delaware, will need proactive management to prevent loss or damage to ecosystem services.

There are also key outstandingly remarkable values that the WSR program focuses on. One of those areas is anadromous fish. Box A4.2 provides an overview of potential climate change impacts to anadromous fish and offers management actions that may be taken to lessen those impacts.

The Wekiva River Basin, located north of Orlando, in east-central Florida, is a complex ecological system of streams, springs, seepage areas, lakes, sinkholes, wetland prairies, swamps, hardwood hammocks, pine flatwoods, and sand pine scrub communities. Several streams in the basin run crystal clear due to being spring-fed by the Floridan aquifer. Others are "blackwater" streams that receive most of their flow from precipitation, resulting in annual rainy season over-bank flows. (Fig. A4.1)

Box A.1. Climate Change, Multiple Stressors and WSRs

Examples are provided to illustrate categories of change and common complicating factors; however, a very large number of complicating stressors are expected around the United States and some factors may be present in all regions (e.g., invasive species). See the WSR Case Studies for literature citations.

Dominant Climate Change	Examples of Climate Change Impacts	Common Complicating Stressors	Example of Region	Case Study
More flooding	Flood mortality, channel erosion, poor water quality	Development in watershed	Northeast, Upper Midwest	Upper Delaware
Droughts, intense heat	Drought mortality, shrinking habitat, fragmentation	Over-extraction of water, invasive species	Southwest	Rio Grande
Little change in rainfall, moderately warmer	Impacts modest unless complicating stressors	Development in watershed	Northern Florida, Mississippi, parts of middle and western states	Wekiva River

In 2000, portions of the Wekiva River and its tributaries of Rock Springs Run, Wekiwa[20] Springs Run, and Black Water Creek were added to the National Wild and Scenic Rivers System. The designated segments total 66.9 km, including 50.5 km designated as Wild, 3.4 km as Scenic, and 13 km as Recreational. The National Park Service (NPS) has overall coordinating responsibility for the Wekiva River WSR, but there are no federal lands in the protected river corridor. Approximately 60%–70% of the 0.8-km-wide WSR corridor is in public ownership, primarily managed by the State of Florida Department of Environmental Protection and the St. Johns River Water Management District (SJRWMD). The long-term protection, preservation, and enhancement are provided through cooperation among the State of Florida, local political jurisdictions, landowners, and private organizations. The designated waterways that flow through publicly owned lands are managed by the agencies that have jurisdiction over the lands. SJRWMD has significant regulatory authority to manage surface and ground water resources throughout the Wekiva Basin.

One of the main tributaries to the Wekiva River is the Little Wekiva River. Running through the highly developed Orlando area, the Little Wekiva is the most heavily urbanized stream in the Wekiva River Basin, and consequently the most heavily affected. The Orlando metropolitan area has experienced rapid growth in the last two decades, and an estimated 1.3 million people now live within a 20-mile radius of the Wekiva River.

The sections of the Wekiva River and its tributaries that are designated as WSR are generally in superb ecological condition. The basin supports plant and animal species that are endangered, threatened, or of special concern, including the American Alligator, the Bald Eagle, the Wood Stork, the West Indian Manatee, and two invertebrates endemic to the Wekiva River, the Wekiwa hydrobe and the Wekiwa siltsnail. At the location of the U.S. Geological Survey's gauging station on the Wekiva River near Sanford, the drainage area of the basin is 489 square km. Elevations for the basin range from 1.5–53 m above sea level. The climate is subtropical, with an average annual temperature of around 22°C. Mean annual rainfall over the Wekiva basin is 132 cm, most of which occurs during the June–October rainy season.

The WSR management plan is being prepared with the leadership of the NPS. Based on information from the pre-legislation WSR study report,[21] and management plans for the state parks (Florida Department of Environmental Protection, 2005) and the SJRWMD (2006a), the priority management objectives for the WSR will likely include maintaining or improving: water quantity and quality in the springs, streams, and rivers; native aquatic and riparian ecosystems; viable populations of endangered and sensitive

[20] The term "Wekiwa" refers to the spring itself, from the Creek/Seminole "spring of water" or "bubbling water." "Wekiva" refers to the river, from the Creek/Seminole "flowing water."

[21] **National Park Service**, 1999: *Wekiva River, Rock Spring Run & Seminole Creek Wild and Scenic River Study*. U.S. Department of Interior, pp.1-49.

species; scenic values; and access and service for recreational users.

The Wekiva River was selected for a case study because it provides an example of a spring-fed WSR system, sub-tropical ecosystems, a coastal location with a history of tropical storms and hurricanes, and a system in a watershed dealing directly with large and expanding urban and suburban populations. In particular, the spring-fed systems combined with urban and suburban land uses require consideration of the relationship between groundwater and surface water and how they relate to management options in the context of climate change.

A4.1.1 Current Stressors and Management Methods Used to Address Them

The primary stressors of the Wekiva WSR are:

- water extraction for public, recreational and agricultural uses;
- land conversion to urban and suburban development;
- pollution, particularly nitrates, via groundwater pathways and surface water runoff; and
- invasive species.

The Wild and Scenic Wekiva River

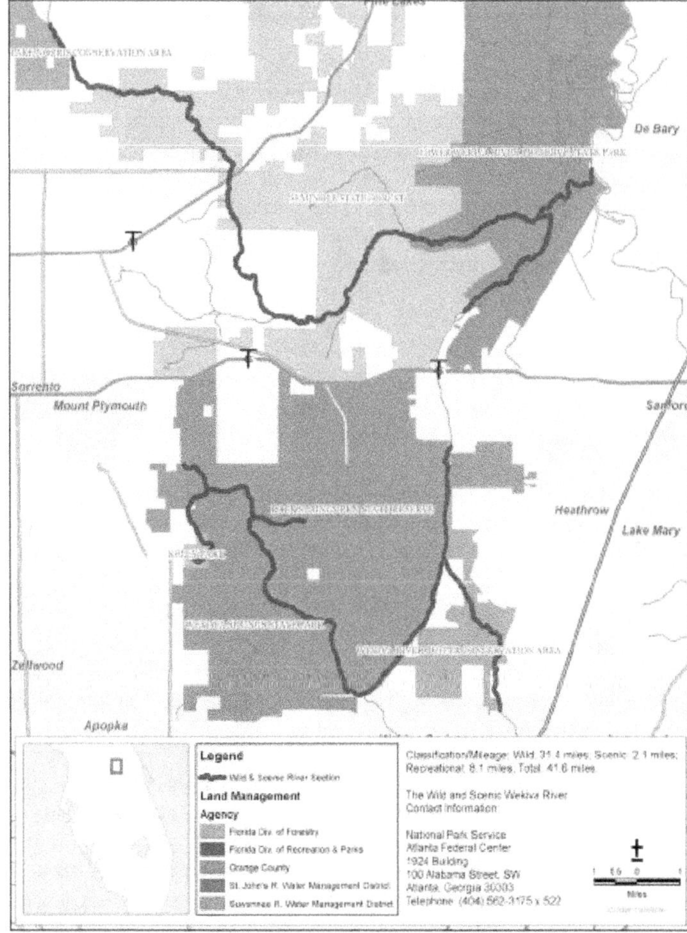

Figure A4.1. The Wild and Scenic portions of the Wekiva River. Data from USGS, National Atlas of the United States.[22]

The Floridan aquifer has a naturally high potentiometric surface (*i.e.*, the level that water will rise in an artesian well), which sustains the natural springs that are critical to the water regime of the Wekiva WSR. McGurk and Presley[23] cite numerous studies that show the long history of water extraction in East Central Florida and related these extractions to lowering of the potentiometric surface. Taking advantage of the high potentiometric surface, in the first half of the 20th century more than two thousands artesian (free-flowing) wells were drilled into the Upper Floridan aquifer, the water used to irrigate agriculture fields and the excess allowed to flow into the streams and rivers. Many of the artesian wells have since been plugged and otherwise regulated to reduce such squandering of the water resources.

Between 1970 and 1995, agricultural and recreational water use from the aquifer has increased nearly threefold to 958 million gallons per day (mgpd), with a significant part of the additional water supporting recreational uses (*i.e.*, golf courses). Over that same period, public (*e.g.*, city) use of water from the aquifer also increased threefold to 321 mgpd. Projections for the year 2020 are for water extraction for agricultural and recreational uses to barely increase, while extractions for public use will nearly double.[23] The St. Johns River, Southwest Florida, and South Florida Water Management Districts have jointly determined that the Floridan Aquifer will be at maximum sustainable yield by 2013, and by that date and into the future much of the water used by people will have to come from alternative sources.

Urban development prior to modern stormwater management controls is another stressor on aquatic systems in the Wekiva Basin. In particular, the Little Wekiva River exhibits extreme erosion and sedimentation caused by high flows and velocities during major storm events (St. Johns River

[22] **U.S. Geological Survey**, 2005: Federal land features of the United States - parkways and scenic rivers. *Federal Land Features of the United States*. http://www-atlas.usgs.gov/mld/fedlanl.html. Available from nationalatlas.gov.

[23] **McGurk**, B.E. and P.F. Presley, 2002: *Simulation of the Effects of Groundwater Withdrawals on the Floridan Aquifer System in East-Central Florida: Model Expansion and Revision*. St. Johns River Water Management District, pp.1-196.

A Migratory Fish

Many fish species are anadromous and adapted to cooler waters—living much of their lives in oceans, but migrating inland to spawn in colder reaches of freshwaters. Several species of salmon and sturgeon reproduce in the rivers of Alaska and the Pacific Northwest, while others, including Atlantic salmon, sturgeon, and striped bass, spawn in eastern seaboard rivers from the Rio Grande to the Canadian coast. Many of these species were also introduced to the Great Lakes, where they migrate up many of Michigan's WSRs. Such species played a significant role in the establishment of the Wild and Scenic Rivers Act and continue to be a primary focus in the management of WSRs. The life cycles of most of these species are determined largely by water temperatures and flows, driven by snowmelt or low water in the summer and fall.

Anadromous fish in the United States are exposed to several anthropogenic stressors that may be exacerbated by climate change. Dams impede or prevent fish migrations, including dams upstream of river stretches designated "wild and scenic." Water withdrawals and reservoir management have affected flow regimes, and water temperatures and pollutants—combined with increased sediment loads—have made many rivers uninhabitable for some migratory fish.

Climate change effects, including reduced streamflows, higher water temperatures, and altered frequencies and intensities of storms and droughts, will further degrade fish habitat (Climate Impacts Group, University of Washington, 2004). Battin et al. (2007) estimate a 20–40% decline in populations of Chinook salmon by 2050 due to higher water temperatures degrading thermal spawning habitat, and winter and early spring floods scouring riverbeds and destroying eggs. This may be a conservative estimate since the analysis did not address the effects that increased sea levels and ocean temperatures would have on Chinook during the oceanic phase of their life cycle, and the study focused on the run of Chinook salmon that spawns in late winter or spring and migrates to the sea by June. Yearlings that remain in freshwater throughout the summer months may be even more vulnerable.

Fish habitat restoration efforts are widespread throughout the United States. However, the models used to guide restoration efforts rarely include projected impacts of climate change. Nevertheless, Chinook salmon studies suggest that habitat restoration in lower elevation rivers (including reforesting narrow reaches to increase shade and decrease water temperatures) may reduce the adverse impacts of climate change (Battin et al., 2007). Galbraith et al.(forthcoming) also identify the potential importance of releases of cool water from existing dams for the preservation of thermal spawning and rearing habitat. Also, mitigating watershed-level anthropogenic stressors that could exacerbate climate change impacts (e.g., water withdrawals, pollutants) could be an effective adaptation option.

Ultimately, management of anadromous fish in WSR will need to reflect species and local circumstances. However, including climate change projections in habitat restoration plans, working to mitigate human-induced stressors, and implementing effective monitoring programs will likely be three of the most important actions managers can take to facilitate the adaptation of anadromous fish to climate change.

Water Management District, 2002). Approximately 479 drainage wells were completed in the Orlando area to control stormwater and control lake levels.[23] These drainage wells recharge the Floridan aquifer.

Declines in spring flows in the Wekiva River Basin are strongly correlated with urban development and ground water extraction (Florida Department of Environmental Protection, 2005). Projections based on current practices indicate that by 2020 water demand will surpass supply and recharge. By 2010, spring flows may decline to levels that will cause irreparable harm (Florida Department of Environmental Protection, 2005). In response to these projections, the SJRWMD has declared the central Florida region, which includes the Wekiva River Watershed, a "Priority Water Resource Caution Area" where measures are needed to protect ground water supplies and spring-dependent ecosystems. SJRWMD has developed "Minimum Flows and Levels" (a.k.a., instream flow criteria) for the Wekiva River and Blackwater Creek, and the district has identified minimum spring flows in selected major springs feeding the Wekiva and Rock Springs Run. These are an important regulatory tool to set limits on ground water withdrawals to prevent adverse reductions in spring flow.

The water management district recommends the following strategies for improving water management (St. Johns River Water Management District, 2006b):

- water conservation;
- use of reclaimed water; and
- water resource development, including:
    - artificial aquifer recharge
    - aquifer storage and recovery
    - avoidance of impacts through hydration
    - interconnectivity of water systems.

The SJRWMD, counties, and cities in the watershed are working on local water resources plans and an integrated basin-wide water plan that will guide water use and conservation land use changes for the coming decades.[24]

Water pollution is another significant stressor of the Wekiva WSR. The causes of water pollution are closely related to the water quantity issues discussed above. In particular, unusually high concentrations of nitrates emanating from the springs of the basin are stressing the native ecosystems in the spring runs. Nitrates promote algal blooms that deplete oxygen, shade-out native species, and may negatively affect invertebrate and fish habitat. Nitrates in spring water now may reflect more distant past inputs from agricultural operations and septic systems. The sources of the nitrogen in the springs are animal waste, sewage, and fertilizers (Florida Department of Environmental Protection, 2005), which readily leach to groundwater due to the karstic geology of the basin. Future spring discharges may reflect a newer type of input from reclaimed water application for both landscape irrigation and for direct recharge via rapid infiltration basins that have increased significantly within the past 10–15 years and continue to increase. The management solutions to reduce nitrate pollution include educating the public to use fewer chemicals and apply these with greater care, developing and applying agricultural best management practices, and increasing the use of central sewage treatment facilities in place of on-site systems such as septic tanks.

Recent data suggest that increases in dissolved chlorides in the springwaters may be related to sea level rise and groundwater withdrawals (Florida Department of Environmental Protection, 2005). To date, salinity changes in the Wekiva Basin springs are minor and the causes are unclear. Major increases in the salinity (increased chlorides) in the springwater would have significant impacts on the ecosystems of the WSR. Continued monitoring and further research are needed to determine the source of the chlorides (*e.g.*, recharge from polluted surface water or mixing with saltwater from below the Upper Floridan aquifer) and how to manage land and water to limit chlorides in the springflows.

Exotic plants are a major problem stressing ecosystems in the Wekiva WSR corridor. For example, wild taro (*Colocasia esculentum*) has infested Rock Springs Run and the lagoon area of Wekiwa Springs has hydrilla (*Hydrilla verticillata*), water hyacinth (*Eichhornia carssipes*), and water lettuce (*Pistia stratiotes*). The park managers use a combination of herbicides and manual labor to control invasive plant species (Florida Department of Environmental Protection, 2005).

Drought-related stress in upland areas has increased the vulnerability of trees to pest species, the Southern pine beetle (*Dendroctomus frontalis*) in particular. Infestations have prompted park managers to clear-cut infested stands and buffers to limit the spread of the beetles. Without these interventions, dead trees would contribute significant fuel, increasing the potential for destructive forest fires.

A4.1.2 Potential Effects of Climate Change on Ecosystems and Current Management Practices

For Central Florida, climate change models project average temperatures rising by perhaps 2.2–2.8°C and annual rainfall to total about the same as it does today.[25] However, the late summer and fall rainy season may see more frequent tropical storms and hurricanes, overwhelming the current stormwater management infrastructure and resulting in periodic surges of surface water with significant pollution and sedimentation loads. More runoff also means less recharge of the aquifer.

At other times of the year, droughts may be more frequent and of longer duration, leading to water shortages and increased withdrawals from the aquifer, which may reduce spring flows.

While there is only moderate confidence in projections of changes in patterns of precipitation, there is a high confidence that it will get warmer. Warmer temperatures over an extended period will change species composition in the WSR corridor. Some native species, particularly those with limited ranges, may no longer find suitable habitat, while invasive exotics, which often tolerate a broad range of conditions, would thrive. Current programs to control invasive species would face new challenges as some native species are lost and replaced by species that favor the warmer climate, particularly for terrestrial species. Where the cold spring waters can moderate water temperature in the streams and river, the current control programs for aquatic invasive species may still be successful in a moderately warmer climate. Warmer temperatures would also lead to increased evaporation and transpiration, which in turn may lead to more water used for irrigation; all of these factors

[24] **Florida Department of Community Affairs**, 2005: *Guidelines for Preparing Comprehensive Plan Amendments for the Wekiva Study Area Pursuant to the Wekiva Parkway and Protection Act*. pp.1-50.

[25] **University of Arizona**, Environmental Studies Laboratory, 2007: Climate change projections for the United States. University of Arizona, http://www.geo.arizona.edu/dgesl/, accessed on 5-17-2007.

combine to further reduce water available for ecosystems in the WSR. The warmer climate may also reduce or eliminate frost events that currently determine the range for some species in central Florida.

Climate change scenarios project sea level rising between 0.18–0.59 m by 2099 (IPCC, 2007b). There are two issues related to potential sea level rise relative to the Wekiva WSR: 1) how would changes in the tidal reach of the St. Johns River affect the Wekiva, and 2) how might the rising sea level affect the aquifer that supports the springflows? There are too few data available to answer these questions.

Finally, projected population increases in the Wekiva Basin and associated aquifer recharge area will add to the burden of managing for climate change impacts on water resources. Suburban expansion increases impermeable surfaces, thereby adding to polluted surface water runoff and reducing aquifer recharge. And groundwater will continue to be extracted for the public.

A4.1.3 Potential for Altering/Supplementing Current Management to Enable Adaptation to Climate Change

Future management adaptations for meeting ecosystem goals in the Wekiva WSR should include monitoring ecosystem health, including water quantity and quality; basin-wide modeling to protect future management needs; and implementation of management programs in advance of climatic changes. The water management district and other land management agencies have robust monitoring programs, though they may not be adequate to understand the complexity of applying reclaimed surface water in a the karst uplands. Current groundwater monitoring, which focuses on salinity, may need to be expanded to better understand how nitrates and other nutrients are transported to the springflows. Increasingly refined models are needed to understand how water and ecosystems in the Wekiva Basin respond to management.

In many ways, it appears that the SJRWMD and local government agencies are beginning to implement management programs that would be needed to maintain ecological processes in the Wekiva WSR in a climate change scenario. Aquifer management is widely recognized as among the most critical tools for ensuring public water supplies and ecological integrity of the Wekiva WSR. Most of the drinking water in and around the Wekiva Basin is extracted from the Floridan aquifer—the same water source for the springflows that are essential to ecosystems of the Wekiva WSR. The Floridan aquifer is a water reservoir that can be managed in ways analogous to a reservoir behind a dam. Like a dam, with each rain event, to the extent permitted by surface conditions, the aquifer is recharged; water otherwise runs into streams and rivers, effectively lost for most public uses and often negatively affecting riverine ecosystems. Different from a dam, aquifer recharge and replenishment operate in a delayed time frame. This characteristic makes reversal of any mitigation measures a slow process, and should be considered in adaptation planning for global climate changes. Recognizing these conditions, programs and plans are in place to minimize surface runoff and maximize groundwater recharge. Programs include, for example, minimizing impermeable surfaces (*e.g.*, roofs, driveways, and roads), and holding surface water in water gardens and artificial ponds.

Recharge water must be of sufficiently good quality in order to not adversely affect the WSR system. Current stormwater management programs, while quite good, are focused on capturing surface water runoff to prevent it from degrading water quality, but this then "re-routes" poor-quality water from a surface water load to a ground water load. The sandy soils and karst geology of the area may result in nitrate-loaded water recharged to the aquifer and then to the springs. There is a great deal to learn about the ultimate effects on groundwater quality of applying reclaimed water to land surface in the karstic uplands.

While the human population in the Wekiva Basin is expected to grow, climate change models suggest that annual rainfall will remain about the same over the next 100 years, presenting a challenge for meeting water demand. In response, programs in the basin are under development to conserve water (reduce water use per person) and to develop "new" water sources (hold and use more surface water). Similarly, programs are also being planned and implemented to reduce pollution, including educating the public and commercial users about what, when, and how to apply chemicals, including nitrate-based fertilizers.

Management adaptations to more intense rain events under climate change conditions would require more aggressive implementation of all these programs, to: maximize recharge of the aquifer during rain events, minimize withdrawals at all times and particularly during droughts, minimize pollution of surface water and groundwater, and monitor and prevent salt water intrusion in the surface water-groundwater-seawater balance system. Considering the importance of water to local residents and as a factor driving economic development, there is considerable political will to invest in water management technologies and programs in the Wekiva Basin. Through this century, current and emerging technologies will likely be adequate for meeting the water needs for human consumption and ecosystem services in the Wekiva Basin, if people are willing to make the investment in technologies and engineering and to allocate enough water to maintain ecosystems.

The Rio Grande, the second largest river in the American Southwest, rises in the snow-capped mountains of southern Colorado, flows south through the San Luis Valley, crosses into New Mexico and then flows south through Albuquerque and Las Cruces to El Paso, Texas, on the U.S.-Mexican border (see Figs. A4.2 and A4.3). A major tributary, the Rio Conchos, flows out of Mexico to join the Rio Grande below El Paso at Presidio and supplies most of the river's flow for the 1,254 miles of river corridor along the Texas-Mexico border. Since 1845, the Rio Grande has marked the boundary between Mexico and the United States from the twin border cities of Ciudad Juárez and El Paso to the Gulf of Mexico.

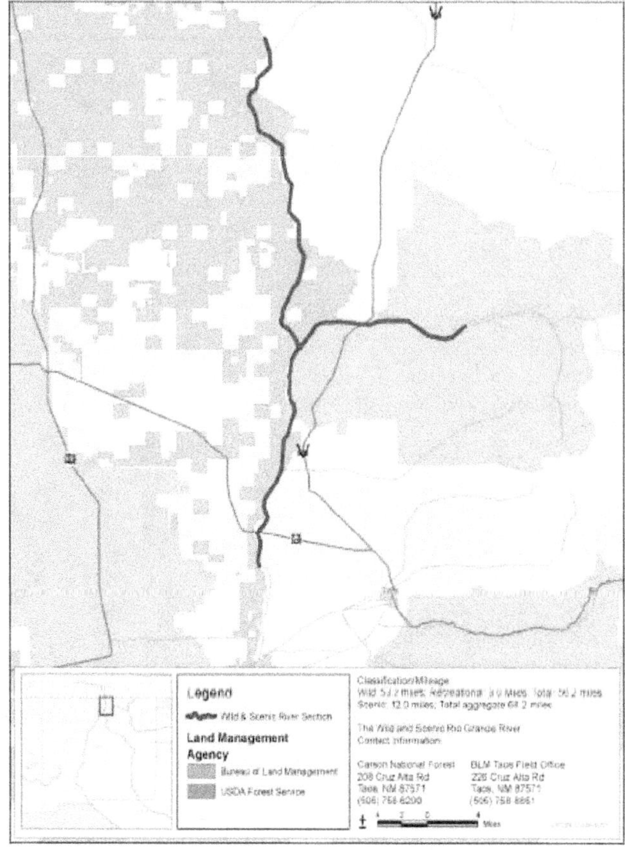

The Wild and Scenic portions of the Rio Grande WSR in New Mexico. Data from USGS, National Atlas of the United States.[20]

Three different segments of the Rio Grande that total 259.6 miles of stream have been designated as Wild, Scenic, and Recreational. Part of the 68.2-mile segment of the river south of the Colorado-New Mexico border was among the original eight river corridors designated as wild and scenic at the time of the system's creation in 1968. A total of 53.2 miles of this reach are designated as wild, passing through 800-foot chasms of the Rio Grande Gorge with limited development. This segment is administered by the Bureau of Land Management (BLM) and the U.S. Forest Service (USFS).[26] About 97% of the land in the New Mexico WSR management zones is owned and managed by the BLM or the USFS.

The longest segment of the Rio Grande WSR comprises 195.7 river miles in Texas (National Park Service, 2004) along the U.S.-Mexico border, with about half of this stretch classified as wild and half as scenic. This stretch, which was added to the system in 1978, is administered by the NPS at Big Bend National Park for the purpose of protecting the "outstanding remarkable" scenic, geologic, fish and wildlife, and recreational values (National Park Service, 2004). Land ownership is evenly divided between private and public (federal and state) owners on the United States side of the designated river segment.

In New Mexico, objectives for managing the WSR include (Bureau of Land Management, 2000):

- maintain water quality objectives designated by the New Mexico Environment Department;
- conserve or enhance riparian vegetation;
- preserve scenic qualities;
- provide for recreational access, including boating and fishing; and
- protect habitat for native species, particular federally listed species.

In Texas, the resource management goals for the wild and scenic river include (National Park Service, 2004):

- preserve the river in its natural, free-flowing character;
- conserve or restore wildlife, scenery, natural sights and sounds;
- achieve protection of cultural resources;
- prevent adverse impacts on natural and cultural resources;
- advocate for scientifically determined suitable instream flow levels to support fish and wildlife populations, riparian communities and recreational opportunities; and
- maintain or improve water quality to federal and state standards.

The Rio Grande WSR was selected for a case study because the distinct segments of the designated river provide examples of features typical of many rivers in the mountainous and arid Southwest. Attributes important

[26] **National Wild and Scenic Rivers System**, 2007: Homepage: National Wild and Scenic Rivers System. National Wild and Scenic Rivers System Website, http://www.rivers.gov, accessed on 5-30-2007.

to this paper include: significant federal and state ownership of the streamside in designated segments; an important influence of snowpack on river flow; complex water rights issues with a great deal of water being extracted upstream of the WSR; primary competition for water by agriculture; and an international component.

A4.2.1 Current Stressors and Management Methods Used to Address Them

The primary stressors of the Rio Grande WSR include (Bureau of Land Management, 2000; National Park Service, 2004; New Mexico Department of Game and Fish, 2006):

- altered hydrology: impoundment, reservoir management and water extraction have led to flow reductions and changes in flow regime (loss of natural flood and drought cycle) and concomitant changes in the sediment regime and channel narrowing;
- altered land use: land and water use for agriculture, mining operations, and cities is leading to declines in water quality due to pollution and sedimentations;
- invasive species: non-native fish and vegetation are altering ecosystems, displacing native species and reducing biodiversity, giant reed and saltcedar are particularly problematic in the Texas WSR segment; and
- recreational users: visitors and associated infrastructure impact the riparian vegetation and protected species; subdivision and building on private lands along the Texas and Mexico segments threaten scenic values and may increase recreational users' impacts.

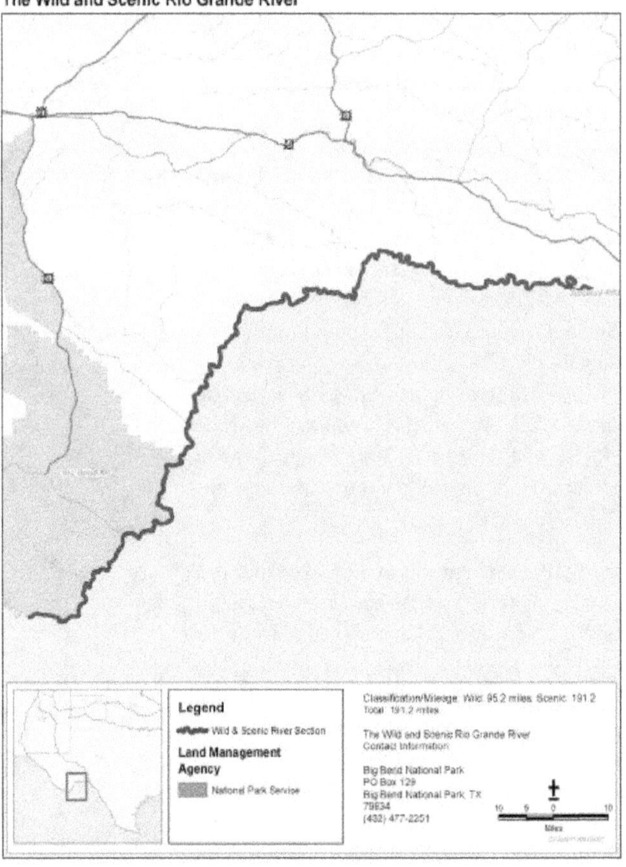

Figure A4.3. The Wild and Scenic portions of the Rio Grande WSR in Texas. Data from USGS, National Atlas of the United States.[20]

All segments of the Rio Grande that are designated as WSR face complex management challenges and multiple stressors on river health, most notably from dams, diversions and other water projects that dot the river and its tributaries, reducing and altering natural flows for much of the river's length. (Fig. A4.4) Although there are no dams on the main stem of the river upstream of the New Mexico WSR corridor, dams and other water projects on major tributaries affect flows downstream. For example, two Bureau of Reclamation projects in Colorado—the Closed Basin (groundwater) Project and the Platoro Dam and Reservoir on the Conejos River—influence downstream flows into New Mexico. The flow regime of the WSR in New Mexico is largely managed by the Bureau of Reclamation, which manages upstream dam and diversion projects based on a century of water rights claims and seasonal fluctuations in available water. The water rights and dams are considered integral to the baseline condition for the WSR, as they were in place prior to the river's designation.

Downstream from El Paso, Texas, the channel of the Rio Grande is effectively dry from diversion for about 80 miles. Because of this "lost reach," the river is more like two separate rivers than one, with management of the Colorado and New Mexico portion having little effect on flows downstream of El Paso. In the past, the river in Colorado and New Mexico normally received annual spring floods from the melting snowpack while the river below Presidio, Texas received additional flood events in the summer through fall from rains in the Rio Conchos Basin, Mexico. However, throughout the Rio Grande these natural cycles of annual floods have been severely disrupted by dams and water extraction.

Management of the Texas Rio Grande WSR still depends on flows entering from Mexico—including the Rio Conchos, which provides 85% of the water to this WSR segment—and which is managed by the International Boundary and Water Commission according to the Rio Grande Compact. Instream

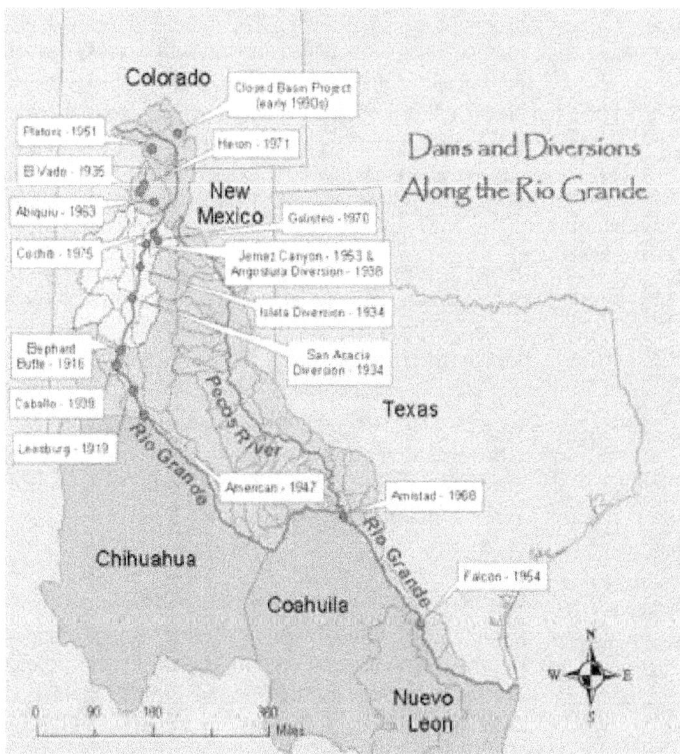

Dams and diversions along the Rio Grande.[27]

Changes in the flow regime of the river are affecting the channel, the floodplain, and the associated aquatic and riparian ecosystems. In the past 90 years, overall stream flow has been reduced more than 50%, and periodic flooding below Presidio has been reduced by 49% (Schmidt, Everitt, and Richard, 2003). Dams in the lower Rio Grande prevent fish migrations so that Atlantic Sturgeon and American Eel no longer reach the WSR.[30] Where native species were dependent on or tolerant of the periodic floods, the new flow regime is apparently giving an edge to invasive, non-native species (National Park Service, 1996). Garrett and Edwards[28] suggest that changes in flow and sedimentation, pollution, simplification of channel morphology and substrates, and increased dominance of non-native plant species can explain recent changes in fish diversity and critical reductions and local extinctions of fish species. Giant reed (*Arundo donax*) and salt cedar (*Tamarix* sp.) are particularly problematic as these exotic species invade the channelized river and further disrupt normal sedimentation, thereby reducing habitats critical to fish diversity.[28] The problems of dams and irregular flows are complicated by local and international water rights issues, and the ecological health of WSR is only one of the many competing needs for limited water resources.

flows in Texas segments of the WSR have decreased 50% in the past 20 years (National Park Service, 2004). During drought years of the late 1990s and into 2004, Mexico did not meet its obligations to the United States under the compact and water levels reached critical lows (Woodhouse, 2005). In 2003, the combination of dams, water extraction and drought were particularly hard on the river, flow essentially ceased, the river became a series of pools in Texas WSR segments and the river failed to reach the ocean.[28]

Inefficient regulation of groundwater contributes to these impacts on the river's flow. The primary source of household water in central New Mexico is groundwater, for which the rate of extraction currently exceeds recharge.[29] Aquifers in the region may not be able to meet demand in twenty years, which will further stress an overburdened surface water resource.

To address pollution issues, BLM, USFS, and NPS managers have reduced pollution to the river from their operations by reducing or eliminating grazing and mining near the river, improving management of recreation sites, and increasing education and outreach. However, as with flow regime, most of the water quality problems are tied to decreases in water quantity and discharge from large-scale agricultural, industrial and urban upstream users.

Federal land managers are making a difference where they can with site-level management. For example, riparian zones are being withdrawn from grazing and mineral leases and are being protected via limited access to sensitive sites and education of backcountry visitors about the values of protected streamside vegetation. Programs are also underway to control erosion in recreation areas and river access points and to improve habitat for protected species (Bureau of Land Management, 2000).

27 **Middle Rio Grande Bosque Initiative**, 2007: Dams and diversions of the Middle Rio Grande. Middle Rio Grande Bosque Initiative Website, http://www.fws.gov/southwest/mrgbi/Resources/Dams/index.html, accessed on 5-17-2007.

28 **Garrett**, G.P. and R.J. Edwards, forthcoming: Changes in fish populations in the Lower Canyons of the Rio Grande. *Proceedings of the Sixth Symposium on Natural Resources of the Chihuahuan Desert Region, Chihuahuan Desert Research Institute.*

29 **New Mexico Office of State Engineer** and Interstate Stream Commission, 2006: *The Impact of Climate Change on New Mexico's Water Supply and Ability to Manage Water Resources.* New Mexico Office of State Engineer/Interstate Stream Commission.

30 **National Park Service**, 2007: Floating the lower canyons. National Park Service, http://www.nps.gov/rigr/planyourvisit/lower_cyns.htm, accessed on 4-14-2007.

A4.2.2 Potential Effects of Climate Change on Ecosystems and Current Management Practices

According to Schmidt *et al.* (2003) the primary drivers of ecosystem change of the Rio Grande are:

- climatic changes that change runoff and influx of sedimentation;
- dam management and water extraction that lead to changes in flow regime (loss of natural flood and drought cycle) and sedimentation;
- changes to the physical structure of the channel and floodplain;
- introduction of exotic species; and
- ecosystem dynamics that cause species to replace other species over time.

The American Southwest in general, including the Rio Grande watershed, seems likely to experience climate extremes in the form of higher temperature, reduced precipitation (including reduced snowpacks), earlier spring melts, and recurring droughts on top of population growth and other existing stressors.[29] While global climate models are inconclusive regarding changes in precipitation for this region, and for the Upper Rio Grande Basin in particular, it seems likely that the projected increase in temperature will result in evaporation rates that more than offset any possible increase in precipitation.[29] In this scenario, the New Mexico WSR segment of the Rio Grande might experience earlier spring floods, with reduced volume and more erratic summer rains.[29] Projections of perhaps 5% decrease in annual precipitation for the middle and lower Rio Grande (see Fig. 6.13 in the Wild and Scenic Rivers chapter) combined with higher temperatures (see Fig. 6.12 in the Wild and Scenic Rivers chapter) suggest that annual flows in the Texas WSR segment may be further reduced, and during severe droughts the water levels may decline to critical levels as has been the case in recent years (National Park Service, 2004). Water quality may be further reduced as the shallower water is susceptible to increased warming due to higher temperatures driven by climate change (Poff, Brinson, and Day, Jr., 2002). These conditions would negatively affect many native species and may favor invasive non-native species, further complicating existing programs to manage for native riparian vegetation and riverine ecosystems (National Park Service, 2004).[29]

A4.2.3 Potential for Altering/Supplementing Current Management to Enable Adaptation to Climate Change

The incorporation of climate change impacts into the planning and management of the WSR corridors of the Rio Grande is complicated by the river's international character, the numerous dams, diversions, and groundwater schemes that already affect its flow regime, and the multiple agencies involved in the river's management within the WSR corridors as well as upstream and downstream. Sustaining the Rio Grande's wild and scenic values under these circumstances will require planning, coordination, monitoring of hydrological trends, and scenario-based forecasting to help river managers anticipate trends and their ramifications. For example, given the probability of reduced snowpack in the headwaters of the Rio Grande, sustaining flows through the New Mexico WSR corridor will likely depend on coordination among the USFS and BLM, which administer this WSR stretch, the Bureau of Reclamation, which manages upstream water projects (both groundwater and surface water) that influence downstream flows, and owners of local and international water rights. Long standing water rights complications make it difficult to predict needed water releases to mimic natural flow regime. In this region, required water deliveries might be met by transferring water rights between watersheds or through credits for future water delivery.

Similarly, the NPS, which administers the Rio Grande WSR corridor in Texas, needs to coordinate with the International Boundary and Water Commission to extract ecological services from regulated flows. This may prove more difficult than securing water for the river in New Mexico. During recent years of drought, Mexico did not meet its obligations to the United States under the compact. With droughts of greater duration expected as temperatures warm, more years of difficulty meeting treaty obligations may arise.

Economic incentives are another approach to securing sufficient clean water needed to meet management objectives of the WSR. Recognizing the value of ecological services, one potential measure, for instance, is to purchase or lease water rights for the river. Additionally, technical assistance and incentives could also be provided to users who improve water efficiency, reduce pollution, and release surplus clean water to the river. Water deliveries could mimic natural flows, including scouring floods to build the channel.

Improving efficiency of agricultural and urban water use and increasing re-use to conserve water and reduce pollution are probably the most cost-effective strategies to make more clean water available in the Rio Grande. If improved water efficiency results in "new" water, the challenge for WSR managers will be to negotiate, purchase or lease water for the river when it is most needed for ecological flows.

A4.3 The Delaware River

The Delaware River runs 330 miles from the confluence of its East and West branches at Hancock, New York to the mouth of the Delaware Bay. Established by Congress in 1978, the Upper Delaware Scenic and Recreational River consists of 73.4 miles (32.1 miles designated as scenic and 50.3 miles as recreational) of the Delaware River between Hancock and Sparrow Bush, New York, along the Pennsylvania-New York border. Although this case study focuses on the Upper

Delaware, there are also 35 miles designated as scenic in the Middle Delaware River in the Delaware Water Gap National Recreational Area and 67.3 miles of Delaware River and tributaries (25.4 scenic and 41.9 recreational) in the Lower Delaware Scenic and Recreational River (Fig. A4.5).

The Upper Delaware Scenic and Recreational River boasts hardwood forests covering over 50% of the river corridor (Conference of the Upper Delaware Townships, 1986). These forests provide lush habitat for diverse fauna including at least 40 species of mammals, such as many of Pennsylvania's remaining river otters and one of the largest populations of black bear in the state. It is one of the most important inland bald eagle wintering habitats in the northeastern United States. Water quality in the Upper Delaware is exceptional and supports abundant cold- and warm-water fish. As the last major river on the Atlantic coast undammed throughout the entire length of its mainstem, the Delaware provides important habitat for migratory fish such as American eel and America shad. In the upper reaches of the Delaware system, rainbow and brown trout are highly sought by anglers. The river and its surrounding ecosystems provide a beautiful setting for recreation including fishing, boating, kayaking, sightseeing and hiking.

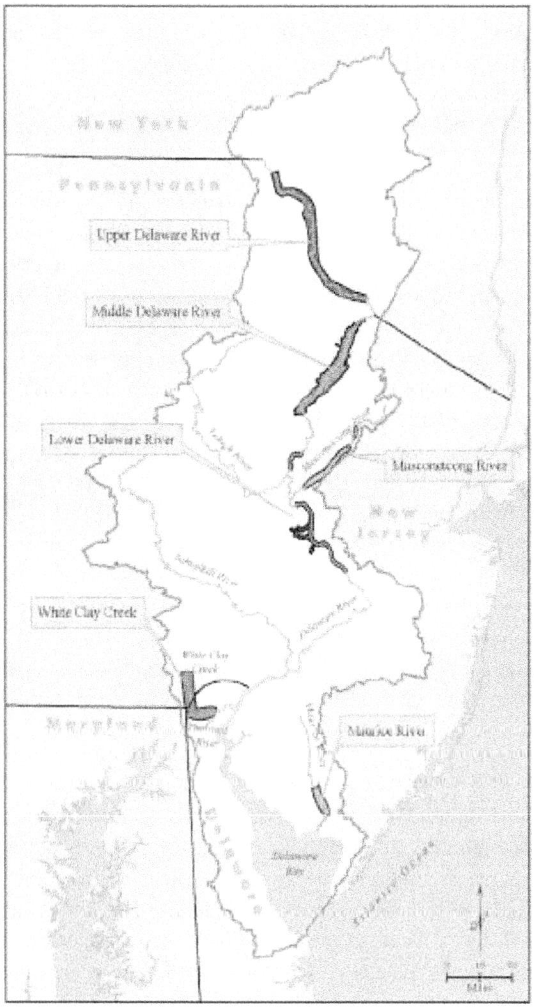

Figure A4.5 Map of Wild and Scenic stretches in the Delaware River basin. Courtesy of Delaware River Basin Commission.[31]

The Upper Delaware Scenic and Recreational River includes a 55,575 acre ridge-top-to- ridge-top (approx. ½ mile wide) corridor, nearly all privately held. The NPS has jurisdiction over 73.4 miles of the river, including a "strand" area along its banks (up to the mean high water mark), but owns only 31 acres within the corridor (Conference of the Upper Delaware Townships, 1986). While the Delaware's main stem remains free flowing, New York City has constructed three reservoirs on major tributaries (the East and West Branches of the Delaware River and the Neversink River) to provide drinking water for more than 17 million people. New York City gets the majority of its water—in fact, its best quality water—from these Catskill reservoirs.

The negligible public ownership, complex private ownership, and significant extraction of water for New York City require that the Upper Delaware be managed as a "Partnership River." The NPS, the Upper Delaware Council (*e.g.*, local jurisdictions), the Delaware River Basin Commission (DRBC, which manages the water releases), the Commonwealth of Pennsylvania, and the State of New York collaborated in preparing the River Management Plan (Conference of the Upper Delaware Townships, 1986) and collaborate in managing the river.

The goals described in the River Management Plan include maintaining or improving water quality and aquatic ecosystems, providing opportunities for recreation, and maintaining scenic values of river corridor and selected historic sites. The rights of private land owners are described in great detail and heavily emphasized throughout the plan, while management actions essential to maintain ecosystem services are more generalized.

The Upper Delaware was chosen as a case study because it exemplifies river ecology for the northeast and management challenges typical of the region, including a significant human population, intense water extraction for enormous urban centers, and its status as a "Partnership River."

A4.3.1 Current Stressors and Management Methods Used to Address Them

The primary stressors in the Upper Delaware include water extraction and unnatural flow regimes associated with reservoir management. Water quality, water temperature, fish and other river biota are negatively affected by these

[31] **Delaware River Basin Commission**, 2007: Wild and Scenic Rivers map. Delaware River Basin Commission Website, http://www.state.nj.us/drbc/wild_scenic_map.htm, accessed on 7-20-2007.

stressors (Mid-Atlantic Regional Assessment Team, 2000). In 2004 to 2006 unusually frequent and severe flooding—three separate hundred-year flood events in a 22-month period—further stressed the river system and added to the management challenges.[32]

Water managers in the Delaware Basin are addressing at least four priority issues: (1) provision of drinking water for major metropolitan areas, (2) flood control, (3) biotic integrity and natural processes of the WSR, and (4) recreation activities, including coldwater fisheries. New York City takes about half of the water available in the Upper Delaware River Basin above the designated WSR. Hence, the primary mechanism remaining to manage the flow regime, water quality, and river ecology and processes in the WSR is dam management, and the secondary mechanism is improved surface water management throughout the Upper Basin. Considering the volume of water extracted, water released from the reservoirs is, overall, significantly below historic flows. Furthermore, while goals for *annual* average releases are met, they do not always conform to the periodicity that stream biologists and anglers say are required for native species and ecological processes. When too little water is released, particularly in the spring and summer, water temperature increases beyond optimal conditions for many species, and pollutants are more concentrated. Aquatic invertebrates decline, trout and other species up the food chain are negatively affected and tourism based on river boating and anglers suffers (Parasiewicz, undated).

Water is also released from the Upper Delaware reservoirs to help maintain river levels adequate to prevent saltwater intrusion from Delaware Bay up river. During droughts in the past 50 years, the "salt front" has moved up river considerably. This intrusion may play a role in the conversion of upland forest areas to marshes, which could affect adjacent river ecosystems.[33] The saltwater is problematic for industries using water along the river front and increases sodium in the aquifer that supplies water to Southern New Jersey. Water conservation in the Delaware Basin and New York City has significantly helped address drought-related water shortages.

Flood control and water quality in the Upper Basin are managed through restoration of stream banks, riparian buffers and floodplain ecosystems and through improved land and water management. The DRBC sets specific objectives for ecosystem management in the basin (Delaware River Basin Commission, 2004). Land use along the river is regulated by Township (PA) and Town (NY) zoning regulations, which are influenced by state regulations and requirements to qualify for FEMA flood insurance. The NPS and other partners work with the towns and townships to promote, through planning and zoning, maintenance of native vegetation in the floodplain and river corridor and to improve stormwater management throughout the watershed.

The NPS and state agencies also manage river recreation, providing access to boaters and hikers and regulating their impacts. Following recent floods, agencies assisted with evacuation of residents in low-lying flood-prone areas; evacuated their own boats, vehicles, and equipment to higher ground; and mobilized post-flood boat patrols to identify hazardous materials (*e.g.,* propane tanks, etc.) left in the floodway and hazards to navigation in the river channel.

NPS and others are beginning to work more closely with the National Weather Service to provide them with data on local precipitation amounts, snowpack, and river ice cover, and to coordinate with their Advanced Hydrologic Prediction Service to enable better forecasting and advanced warning to valley residents of flood crests and times.

A4.3.2 Potential Effects of Climate Change on Ecosystems

Climate in the Delaware Basin can be highly variable, sometimes bringing severe winter ice storms and summer heat-waves. However, there has been a steady increase in mean temperature over the last 50 years as well as an increase in precipitation (Lins and Slack, 1999; Rogers and McCarty, 2000; Najjar *et al.*, 2000). The expectations are for this pattern to continue and, in particular, for there to be the potential for less snowpack that melts earlier in the spring, and rain in the form of more intense rain events that may create greater fluctuations in river levels and greater floods. Severe flood events will likely continue to disrupt the river channel and impact floodplain ecosystems. Furthermore, during periodic droughts there will be increased potential for combinations of shallower water and warmer temperatures, leading to significantly warmer water that could be especially damaging to coldwater invertebrates and fish. It is possible that dam management could offset this warming if water can be drawn from sufficient depths in the reservoir (*e.g.*, with a temperature control device on the dam).

As with any river system, such climate-induced changes in environmental conditions may have serious ecological consequences, including erosion of streambanks and bottom sediments that may decrease the availability of suitable habitat, shifts in the growth rate of species due to thermal and flood-related stresses, and unpredictable changes in ecological processes such as carbon and nitrogen processing (see section 6.4.3 in the Wild and Scenic Rivers chapter).

[32] **Delaware River Basin Commission**, 2006: *Water Resource Program FY2006 – FY 2012*. Available at http://www.state.nj.us/drbc/WRP2006-12.pdf. Delaware River Basin Commission, pp.1-9.

[33] **Partnership for the Delaware Estuary**, 2007: Partnership for the Delaware Estuary, a National Estuary Program homepage. Partnership for the Delaware Estuary Website, http://www.delawareestuary.org/, accessed on 7-12-2007.

A4.3.3 Potential for Altering/Supplementing Current Management to Enable Adaptation to Climate Change

Management of the reservoir levels and dam releases are the most direct methods to maintain riverine ecosystems under increased burdens of climate change. The DRBC Water Resource Program report for 2006–2012[32] identifies the current water management issues for the Basin and their program to address the challenges, including a river flow management program to ensure human and ecosystem needs.[32] A major thrust of the Commission's program is research and modeling to help find a balanced approach to managing the limited water resources. This approach of establishing flow regimes based on sound scientific data, with models and projects extended over decades will serve well in a future impacted by climate change.

Improved watershed management to reduce aberrant flood events and minimize water pollution is one of the most useful long-term tools for managing river resources in a changing climate (Mid-Atlantic Regional Assessment Team, 2000). Federal, state and local authorities can create incentives and pass ordinances to encourage better water and land use that protect the river and its resources. For example, improved efficiency of water use and stormwater management (*e.g.*, household rain barrels and rain gardens, holding ponds), improved use of agrochemicals and soil management, and restoration of wetlands and riparian buffers would combine to reduce severity of floods, erosion damage and water pollution.

Finally, continual improvements in municipal and household water conservation are among the most promising approaches to manage water in the Delaware River Basin. Populations in and around the Delaware Basin will grow, increasing demand on water supplies and river access for recreational uses. Per capita water use in New York City has declined from more than 200 gallons per capita per day around 1990 to 138 gallons per capita per day in 2006.[34] Water pricing can be use to promote further conservation (Mid-Atlantic Regional Assessment Team, 2000). An important component of this approach is educating the public so that consumers better understand the important role that water conservation plays in protecting river ecosystems and future water supplies.

A5 [illegible]

A5.1 The Albemarle-Pamlico Estuarine System

A5.1.1 Introduction

We chose the Albemarle-Pamlico Estuarine System (APES) for our case study. APES provides a range of ecosystem services, extending over a diversity of ecosystem types, which provide the basis for the management goals of the Albemarle-Pamlico National Estuary Program (APNEP). Like other estuaries, the ecosystem services of APES are climate sensitive, and this sensitivity affects the ability to meet management goals. A range of adaptation options exist for climate-sensitive management goals. Many of these adaptation options are applicable across estuarine ecosystems generally. Furthermore, because APNEP represents one of the first national estuaries, documentation of management successes and failures (Korfmacher, 1998; Korfmacher, 2002) exists for its 20-year history. Extensive data and decision support information are available for the system and are likely to continue to be gathered into the future. We highlight a few key climate-related issues in this case study, including warming and altered precipitation patterns, but especially accelerated sea level rise and increased frequency of intense storms.

The rationale for selecting the APES for the in-depth case study is based upon several unique characteristics of this system in addition to the scope of its management challenges related to climate change. First, the shores of the Albemarle and Pamlico Sounds are so gradually sloped that this system possesses more low-lying land within 1.5 m of sea level than any other national estuary. Within the United States, wetlands and coastal lands inundated by sea level rise will be exceeded only on the Louisiana coast of the Mississippi River delta and the Everglades region of South Florida (Titus, 2000; U.S. Climate Change Science Program, 2007). Thus, the incentives here for management adaptation are high. Second, the State of North Carolina passed a Fisheries Reform Act in 1997, which mandated development of a Coastal Habitat Protection Plan (CHPP) for fisheries enhancement. This plan at the state level represents a working example of ecosystem-based management because it engages all the diverse and usually independent state agencies whose mandates involve aspects of the environment that affect fish and their habitat. Consequently, there exists a model opportunity for integrating climate change into an ecosystem-based plan for management adaptation. Third, the Albemarle-Pamlico Sound system faces the daunting management challenges associated with

[34] **New York City Department of Environmental Protection**, 2006: Water Conservation Program. pp.1-54.

projected disintegration of the protective coastal barrier of the Outer Banks of North Carolina (Riggs and Ames, 2003). As a result, the general problem of responding to erosion risk on coastal barriers is of higher urgency here because what is estuary now could become converted to an oceanic bay if the integrity of the banks is breached.

A5.1.2 Historical Context

Like many important estuaries, the Albemarle-Pamlico ecosystem has experienced a long history of human-induced changes including species depletion, habitat loss, water quality degradation, and species invasion (Lotze *et al.*, 2006). About 800 years ago, indigenous Native Americans initiated agriculture in the basin, and approximately 400 years ago Europeans began to colonize and transform the land. Since then, the human population around the estuary has increased by two orders of magnitude from that in 1700 (Lotze *et al.*, 2006). Before European colonization, North Carolina had about 11 million acres of wetlands, of which only 5.7 million remain today. About one-third of the wetland conversion, mostly to managed forests and agriculture, has occurred since the 1950s.[35] Since 1850, the amount of cropland has increased 3.5-fold. More recent land use patterns show that 20% of the basin area consists of agricultural lands, 60% is forested, and relatively little is urbanized (Stanley, 1992). Over the last three decades, the production of swine has tripled and the area of fertilized cropland has almost doubled (Cooper *et al.*, 2004). These changes in land-use patterns and increases in point and non-point nutrient loading have induced multiple changes in water quality, with the greatest changes appearing during the last 50–60 years (Cooper *et al.*, 2004).

Over the last two to three centuries in the Albemarle and Pamlico Sounds, overexploitation, habitat loss, and pollution have resulted in the depletion and loss of many marine species that historically have been of economic or ecological importance (Lotze *et al.*, 2006). Of the 44 marine mammals, birds, reptiles, fish, invertebrates, and plants for which sufficient time series information exists, 24 became depleted (<50% of former abundance), 19 became rare (<90%), and 1 became regionally extinct by 2000 (Lotze *et al.*, 2006). Great losses also occurred among the subtidal bottom habitats. Historical accounts from the late 1800s indicate that bays and waterways near the mainland once had extensive beds of seagrass, while today seagrass is limited to the landward side of the barrier islands (Mallin *et al.*, 2000). Oyster reef acreage has been diminished over the last 100 years as a consequence of overharvesting, habitat disturbance, pollution, and most recently Dermo (*Perkinsus marinus*) infections.[36]

A5.1.3 Geomorphological and Land Use Contexts and Climate Change

Climate change impacts on APES may take numerous forms. Warming in and of itself can alter community and trophic structure through differential species-dependent metabolic, phenological, and behavioral responses. Changes in precipitation patterns also may have species-specific consequences. In combination, warming and precipitation patterns affect evapotranspiration, soil moisture, groundwater use and recharge, and river flow patterns. The current rate of relative rise in mean sea level in this geographic region is among the highest for the Atlantic coast, with estimates commonly over 3 mm per year and in at least one study as high as 4.27 mm per year (Zervas, 2001). The anticipated scenario of increasing frequency of intense storms in combination with rising sea levels creates a likelihood of dramatic physical and biological changes in ecosystem state for APES because the very integrity of the Outer Banks that create the protected estuaries behind them is at risk (Riggs and Ames, 2003; Paerl *et al.*, 2006).

APES is a large and important complex of rivers, tributary estuaries, extensive wetlands, coastal lagoons and barrier islands. Its 73,445 km^2 watershed (Stanley, 1992) is mostly in North Carolina but extends into southern Virginia (Fig. A5.1). The largest water body is Pamlico Sound to the southeast, with two major tributaries, the Neuse and the Tar-Pamlico Rivers. Both rivers empty into drowned river estuaries, the Neuse River Estuary (NRE) and the Pamlico River Estuary (PRE), which connect to Pamlico Sound. Albemarle Sound is farther north with two major tributaries, the Chowan and the Roanoke Rivers, and a number of local tributary estuaries. Other smaller sounds connect the Albemarle and the Pamlico (Roanoke and Croatan Sounds), and the Currituck Sound extends along the northeastern portion of the complex.

The geological framework for coastal North Carolina, including APES has recently been summarized by Riggs and Ames (2003). The system represents several drowned river valley estuaries that coalesce into its large coastal lagoon (Fig. A5.1). The coastal plane, estuaries and sounds have a very gentle slope in which Quarternary sediments are underlain largely by Pliocene sediment. Much of this sediment is organic

[35] **U.S. Geological Survey**, 1999: National water summary on wetland resources: state summary highlights. USGS, http://water.usgs.gov/nwsum/WSP2425/state_highlights_summary.html, accessed on 3-23-2007.

[36] **North Carolina Department of Environmental and Natural Resources**, 2006: Stock status of important coastal fisheries in North Carolina. North Carolina Department of Environmental and Natural Resources, Division of Marine fisheries, http://www.ncfisheries.net/stocks/index.html, accessed on 3-23-2007.

[37] **Albemarle-Pamlico National Estuary Program**, 2007: Albemarle-Pamlico Sounds region. Albemarle-Pamlico National Estuary Program Website, http://www.apnep.org/pages/regions.html, accessed on 7-25-2007.

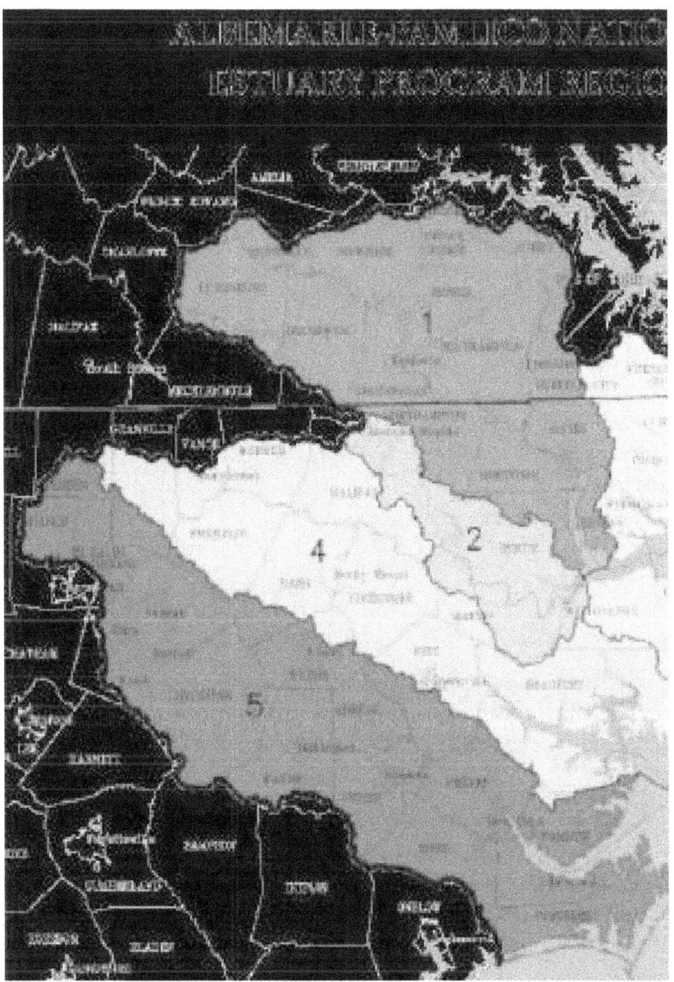

Figure A.1. The Albemarle-Pamlico National Estuary Program region.[38]

rich mud arising from eroding peat of swamps and marshes (Riggs, 1996). The gentle slope has allowed major shifts in position of the shoreline and barrier islands as sea level has risen and fallen. Furthermore, the position and number of inlets has changed along the barrier islands, promoting or limiting the exchange of fresh and seawater.

Much of the watershed is within the coastal plain with low elevations that affect land use. Moorhead and Brinson (1995) estimate that 56% of the peninsula between the Albemarle Sound and PRE is less than 1.5 m in elevation. Fifty-three percent of the peninsula's area is composed of wetlands, and 90% contains hydric soils. Thus, this region of the watershed is sparsely populated and largely rural. In contrast, other regions are more highly developed. The barrier islands, the famous "Outer Banks" of North Carolina, are a mosaic of highly developed lands for tourism and protected natural areas. The southeastern portion of Virginia in the APES basin is highly urbanized, and the piedmont origins of the Neuse and Tar Rivers in North Carolina are highly populated. Agriculture and silvaculture are important land uses and economic drivers in the region. Urban economies dominate much of southeastern Virginia. And a relatively new trend is the development of high-end and retirement subdivisions along the "Inner Banks," the mainland shore zone of the complex. The watershed's population exceeds 3,000,000 people including Virginia. However, only about 25% are found in coastal counties of North Carolina, based on estimates for 2000.[38] A significant portion of this population is considered "vulnerable" to strong storms and thus faces risks from climate change (*i.e.*, people who live in evacuation zones for storm surge or who are subject to risks from high winds by living in mobile homes). With rises in sea level and storm intensities, the low-lying lands and basic nature of services and infrastructure of the rural environment face growing risks of flood damage.

Another characteristic of the system's geomorphology makes it uniquely susceptible to climate change drivers. The exchange of water between the ocean and the sounds is restricted by the few and small inlets that separate the long, thin barrier islands (Giese, Wilder, and Parker, 1985; Riggs and Ames, 2003). This restricted connectivity greatly dampens amplitude of astronomical tides and limits the degree to which seawater is mixed with freshwater. Temperature increases may have significant impacts on the APES because its shallow bays have limited exchange with ocean waters, which serve as a cooling influence in summer.

Water quality has been a recurring management concern for APES and APNEP. The tributary rivers generally have high concentrations of dissolved nutrients. This fosters high primary productivity in tributary estuaries, but under most circumstances nutrient concentrations in the sounds remain relatively low (Peierls, Christian, and Paerl, 2003; Pichler *et al.*, 2004). Most nutrient loading derives from non-point sources, although nitrogen loading from point sources may account for up to 60–70% in summer months (Steel and Carolina, 1991). Nitrogen deposition from the atmosphere may account for an additional 15–32% (Paerl, H.W., Dennis,

[38] **Federal Emergency Management Agency**, 2007: Chapter 01-description of study area. Comprehensive Hurricane Data Preparedness, FEMA Study Web Site, http://chps.sam.usace.army.mil/USHESDATA/NC/Data/chapter1/chapter01_description.html, accessed on 3-23-2007.

and Whitall, 2002). Phosphorus loading to the Pamlico River Estuary was greatly enhanced by phosphate mining, which accounts for about half of the total point source phosphorus loadings to this estuary and officially began in 1964 (Copeland and Hobbie, 1972; Stanley, 1992). Loading has decreased dramatically in recent years as treatment of mine wastes has improved. High surface sediment concentrations of the toxic heavy metals arsenic, chromium, copper, nickel, and lead are found in the Neuse River Estuary, possibly associated with industrial and military operations, while high cadmium and silver levels in the PRE most likely result from phosphate mining discharges (Cooper *et al.*, 2004). In 1960, hypoxia was first reported in the Pamlico River Estuary (Hobbie, Copeland, and Harrison, 1975). Since then, hypoxic and anoxic waters in the PRE and NRE were mostly of short duration (days to weeks) but have resulted in death of benthic invertebrates on the bottom and fish kills (Stanley and Nixon, 1992; Buzzelli *et al.*, 2002; Cooper *et al.*, 2004). Nuisance and toxic algal blooms are reported periodically (Burkholder *et al.*, 1992; Bricker *et al.*, 1999), and about 22 aquatic plants and 116 aquatic animals, of which 22 occur in marine or marine-freshwater habitats, have been identified as non-indigenous species in North Carolina.[39] Increases in temperatures are expected to enhance hypoxia and its negative consequences, through the combined effects of increased metabolism and, to a lesser degree, decreased oxygen solubility.

The interactions between relative sea level rise, shoreline morphology, and bay ravinement could have significant impacts on estuarine water quality and ecosystem function in the APES. Losses of wetlands to inundation could lead to a large shift in function from being a nitrogen sink to being a nitrogen source. Both planktonic and benthic primary producers may be affected by, and mediate, changes in water quality, nutrient and material fluxes across the sediment-water interface that may result from sea level rise (Fig. A5.2). Changes in the water column productivity affect particle composition and concentration, which in turn increases turbidity and feedback to modify further the balance between water column and benthic productivity. Inundated sediments will then be subject to typical estuarine stressors (*e.g.*, salinity, changes in water table, isolation from atmosphere) that can lead to dissolution of particulates, desorption of nutrients or organic matter, and altered redox states. These changes result in fluxes of nutrients and DOC that could radically transform the proportion of productivity and heterotrophic activity in the water above the sediment and in the rest of the estuary. Nutrient management plans generally assume that the frequency and magnitude of bottom water hypoxia will decrease by reducing watershed inputs of dissolved inorganic nitrogen and organic matter that either indirectly or directly fuel water column and benthic respiration (Kemp *et al.*, 1992; Conley *et al.*, 2002). However, factors such as the nutrient and sediment filtration capacity of wetlands under flooded conditions of higher sea levels, and the potential for a large organic matter input from erosion and disintegration of now inundated wetlands, create uncertainty about progress in containing eutrophication across different scales and render the determination of management targets and forecasting of hypoxia extremely difficult.

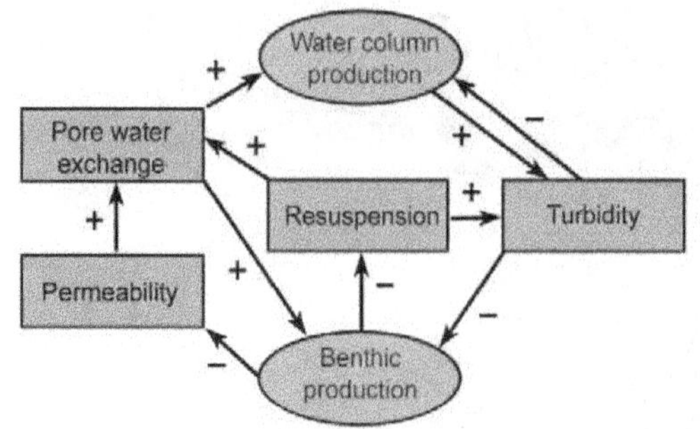

Figure A5.2. Feedbacks between nutrient and sediment exchange and primary production in the benthos and water column. A plus symbol indicates enhancement and a minus symbol indicates suppression.

Because of the large fetch of the major sounds and tributary estuaries, wind tides control water levels and wave energy can be quite high. Wind tides can lead to extended flooding and high erosion rates, especially within the eastern and southern parts of the complex (Brinson, 1991; Riggs and Ames, 2003). Furthermore, the barrier islands are prone to breaching during storms, and geological history demonstrates the fragility of this thin strip of sand and reveals the locations of highest risk of breaching. Formation of persistent inlets within the barrier islands would increase oceanic exchange and thereby the amplitude of astronomical tides. This, in turn, could profoundly alter the ecology of both aquatic and wetland ecosystems in the APES.

The size, geomorphology, and location of the APES complex make it an important source of ecosystem services for the region and the nation. The largest economic contribution of APES today derives from tourism and recreation. The Outer Banks attract people from around the world. Populations during the prime summer season considerably exceed winter populations. The Outer Banks include the most economically important acreage of the complex along with ecologically important natural areas. These coastal barriers are also the most sensitive to the combination of sea level rise and increased frequency of intense storms. Barrier island geomorphology is constantly changing on short and long time scales, increasing and decreasing in width with sand movement and both forming and closing inlets during storms. Inlets have broken through the Outer

[39] **U.S. Geological Survey**, 2005: Nonindigenous aquatic species search page. U.S. Geological Survey, http://nas.er.usgs.gov/queries/default.asp, accessed on 4-9-2007.

Banks repeatedly over the past century and paleo records from the past few thousand years demonstrate dramatic movements in location and character of the barriers as sea level has changed (Riggs and Ames, 2003). But human structures on the islands and human uses of the barrier islands' natural resources have now changed the degree to which natural geological processes occur. Construction and maintenance of Route 12 along the Outer Banks has restricted washover and the movement of sand from the seaward side of the islands to the sound side. Furthermore, the presence of houses, condominiums, hotels, etc. produces conflicts between maintaining the natural geomorphic processes that allow island migration landwards as sea level rises and protecting human infrastructure. Rising sea level and increased frequency of intense storms enhances the potential beach erosion, thereby increasing costs of beach nourishment, and increases risk of island disintegration, leading to increased political pressure to legalize hard structures on the ocean shoreline.

Beaches are a major natural resource and drive many coastal economies. Because the presence of houses, condominiums, and roads and other infrastructure leads to defense of the shoreline position and prevents natural recession, beach erosion now reduces beach widths as sea level is rising. North Carolina prohibits hard structures (*e.g.*, bulkheads, jetties, and permanent sand bags) on the ocean shoreline. Instead, erosion is countered by beach nourishment, in which sand is dredged from offshore. This is a temporary and expensive solution. It also has potentially significant impacts on the living resources of the beach, such as shorebirds and resident invertebrates (Peterson and Bishop, 2005; Peterson *et al.*, 2006). Erosion of beaches tends to occur with the major axis parallel to the islands (*i.e.*, meters or tens of meters of erosion of beach along hundreds to thousands of meters along the beach face). Breaching of new inlets and overwash events penetrate more into the islands. A recent breach occurred on Hatteras Island during Hurricane Isabel, but it was quickly closed by the U.S. Army Corps of Engineers to permit road reconstruction and automobile travel along the Outer Banks. Riggs and Ames (2003) have projected that under higher stands of sea level, future hurricanes may create numerous large, new inlets and break the chain of coastal barriers that forms the eastern edge of the entire APES system. They mapped locations of the paleochannels along the islands and identified these as the most likely locations for such breaches. Such events represent the most dramatic consequences of climate change to APES. Extensive new inlets would lead to an entirely new tidal, salinity, wave, and hydrodynamic regime within APES, and in turn drastically change the ecology of the complex. Wise management for the future must include preparation for the possibility of events such as these and their consequences.

Natural areas in APES have been recognized for their significance as wildlife habitat, nurseries for aquatic species, stop-over sites (flyways) for migratory birds, and important spawning areas for anadromous fish. Recreational fishing and boating add to the attraction of the beaches, barrier islands, and natural areas within the watershed. The nursery services of the complex are also important to fisheries, both locally and along the entire eastern coast of the United States. Cape Hatteras sits at the biogeographic convergence of populations of northern and southern species, and many of these species use the sounds during their life cycles. Thus, the location of APES makes it particularly sensitive to any climate-related changes that alter migratory patterns of both birds and marine organisms.

The wetlands of the Albemarle Pamlico Sound complex are largely non-tidal and subject to irregular wind tides, as described above. In freshwater regions along the rivers and flood plains, swamp forests dominate. Pocosins—peat-forming ombrotrophic wetlands—are found in interstream divides. As sea level rises in oligohaline regions, swamp forests may continue to dominate or be replaced by brackish marshes. Irregularly flooded marshes, dominated by *Juncus roemerianus*, extend over much of the higher-salinity areas. Back barrier island marshes are dominated by *Spartina alterniflora*. The ability of these wetlands to respond to sea level rise is becoming compromised by increased human infrastructure. Roads, residential and urban developments, hard structures for shoreline stabilization, and agricultural ditching are preventing horizontal transgression of wetlands and promoting erosion of edges throughout the complex. Furthermore, development of the barrier islands has prevented natural overwash and inlet forming processes that promote salt marsh development (Christian *et al.*, 2000; Riggs and Ames, 2003).

A5.1.4 Current Management Issues and Climate Change

The APES became part of the NEP (APNEP) in 1987. Initial programmatic efforts focused on assessments of the condition of the system through the Albemarle-Pamlico Estuarine Study. The results of these efforts were used in the stakeholder-based development of a Comprehensive Conservation and Management Plan (CCMP) in 1994. The CCMP presented objectives for plans in five areas: water quality, vital habitats, fisheries, stewardship, and implementation (Box A5.1).[40] For each objective, issues of concern were identified and management actions proposed. None of the issues or proposed actions explicitly included climate change. In 2005, NEP Headquarters conducted its most recent triennial implementation review of APNEP.

[40] **Albemarle-Pamlico National Estuary Program**, 1994: Albemarle-Pamlico NEP Comprehensive Conservation and Management Plan.

□□□A□1. CCMP Objectives for the Albemarle-Pamlico National Estuary Program[40]

□a□e□□□□al□□□□a□□GOAL: Restore, Maintain or Enhance Water Quality in the Albemarle-Pamlico Sounds Region so that it is Fit for Fish, Wildlife and Recreation.

- Objective A: Implement a comprehensive basinwide approach to water quality management.
- Objective B: Reduce sediments, nutrients and toxicants from nonpoint sources.
- Objective C: Reduce pollution from point sources, such as wastewater treatment facilities and industry.
- Objective D: Reduce the risk of toxic contamination to aquatic life and human health.
- Objective E: Evaluate indicators of environmental stress in the estuary and develop new techniques to better assess water quality degradation.

□□al □ab□a□□□□a□□GOAL: Conserve and Protect Vital Fish and Wildlife Habitats and Maintain the Natural Heritage of the Albemarle-Pamlico Sounds Region.

- Objective A: Promote regional planning to protect and restore the natural heritage of the A/P Sounds region.
- Objective B: Promote the responsible stewardship, protection and conservation of valuable natural areas in the A/P Sounds region.
- Objective C: Maintain, restore and enhance vital habitat functions to ensure the survival of wildlife and fisheries.

□□□e□e□□□a□□GOAL: Restore or Maintain Fisheries and Provide for Their Long-Term, Sustainable Use, Both Commercial and Recreational.

- Objective A: Control overfishing by developing and implementing fishery management plans for all important estuarine species.
- Objective B: Promote the use of best fishing practices that reduce bycatch and impacts on fisheries habitats.

□e□a□□□□□□a□□GOAL: Promote Responsible Stewardship of the Natural Resources of the Albemarle-Pamlico Sounds Region.

- Objective A: Promote local and regional planning that protects the environment and allows for economic growth.
- Objective B: Increase public understanding of environmental issues and citizen involvement in environmental policy making.
- Objective C: Ensure that students, particularly in grades K-5, are exposed to science and environmental education.

□□□e□e□a□□□□a□□GOAL: Implement the Comprehensive Conservation and Management Plan in a way that protects environmental quality while using the most cost-effective and equitable strategies.

- Objective A: Coordinate public agencies involved in resource management and environmental protection to implement the recommendations of the CCMP.
- Objective B: Assess the progress and success of implementing CCMP recommendations and the status of environmental quality in the Albemarle-Pamlico Sounds region.

APNEP passed the implementation review and was found eligible for funding through FY 2008.

Although no management objective explicitly identifies climate change or its consequences, water quality, vital habitats, and fisheries are likely to be substantially affected by changes in climate. Recent efforts by APNEP and the State of North Carolina led to more direct consideration of the impacts of climate change. APNEP has identified indicators of condition of the system and begun the process for implementing their use. Multiple indicators assess condition of atmosphere, land, wetland, aquatic, and human components of the system. While some indicators focus on short-term changes in these components, many have meaning only in their long-term trends. Given a changing climate and associated impacts, these indicators place APNEP in position to assess these impacts for wise management. On a broader front, the legislature of North Carolina in 2006 established a commission on climate change to assess how climate change will affect the state and to propose actions to either minimize impacts or take advantage of them.

In 1987 North Carolina passed the Fisheries Reform Act, requiring both development of formal species management plans for each commercially and/or recreationally harvested fishery stock and the development of a CHPP. The CHPP development and implementation process resembles an EBM at the state level because it requires consideration and integrated management of all factors that affect the quality of fish habitats in a synthetic, integrative fashion. To achieve this goal, staff from all appropriate state resource and environmental commissions came together to map coordinated approaches to achieve sustainability of habitat quantity and quality for fishery resources. This partnership among agencies, while only at the state level, addresses one of the biggest goals of EBM (Peterson and Estes, 2001). Commissions and agencies responsible for fisheries management (Marine Fisheries Commission), water quality and wetlands (Environmental Management Commission), and coastal development (Coastal Resources Commission) are the major entities, but the Sedimentation Control Commission and Wildlife Resources Commission also contribute. The CHPP does contemplate several aspects of climate change and human responses to threats such as beach and shoreline erosion, although long-term solutions are elusive. Now that a plan exists, the implementation of its short-term goals has yet to begin and may become contentious.

Other innovative programs and initiatives within North Carolina are the Ecosystem Enhancement Program (EEP), Clean Water Management Trust Fund (CWMTF), and the designation of estuaries as nutrient sensitive. EEP is an agency that coordinates wetland mitigation efforts to maximize their effectiveness. The North Carolina Department of Transportation's mitigation needs are largely met through EEP. The program uses a watershed approach in planning mitigation projects. This allows a broad and comprehensive perspective that should be reconciled with climate change expectations. The CWMTF provides financial support for activities that improve or protect water quality. It offers an opportunity to link consideration of climate change to such activities, although no such link has been an explicit consideration. The designation of nutrient sensitivity allows enhanced controls on nutrient additions and total maximum daily loadings to the Neuse and Tar-Pamlico systems. In fact, regulations have been designed to not only curb expansion of nutrient enrichment but to roll it back with restrictions to both point- and non-point sources.

A5.1.5 Recommendations for Environmental Management in the Face of Climate Change

We make three overarching recommendations for management of estuaries in the face of climate change: (1) maintain an appropriate environmental observing system; (2) educate a variety of audiences on long-term consequences; and (3) pursue adaptation and adaptive management. Each of these is described specifically for APES but has application to other estuaries in whole or part. Furthermore, each involves coordination of multiple initiatives and programs. It is this coordination that should be a major focus of APNEP in particular and the NEP in general.

An appropriate observing system involves a network of programs that detects, attributes and predicts change at multiple scales. It includes sustained monitoring, data and information management, predictive model production, and communication of these products to users. The users include environmental managers, policy makers, and members of the public over a range of economic positions and status. Regulatory and policy needs require a variety of measurements to be made in a sustained way. These measurements extend to variables of physical, chemical, biological, and socioeconomic attributes of APES. Many have been identified by APNEP with its indicator program. These measurements must be made to respond to drivers at different time scales; while these time scales include short-term variation, the most important to this report are long-term trends and infrequent but intense disturbances.

There are other observing system initiatives within coastal North Carolina. These include the North Carolina Coastal Ocean Observing System and Coastal Ocean Research and Monitoring Program. Both have their emphases on the coastal ocean and near real-time products of physical conditions. However, their efforts need to be more directed toward the APES and other estuarine ecosystems to be more valuable to the people of North Carolina. More effort is needed to assess and understand the physical dynamics of the estuarine systems. Observations and analyses should be extended to characterize the physical and geochemical processes of catchment and riverine inflows, which are likely to change dramatically under changing climatic conditions. The systems also need to broaden their observations to include ecological and socioeconomic measurements. These measurements are less likely to be near real-time, but user needs do not require such quick reporting. We recommend that the coastal observing systems be linked explicitly to APNEP indicator activities.

Education is needed across the spectrum of society to produce informed stakeholders and thus facilitate enlightened management adaptations. The need for K–12 education on climate change is obvious, but there is also a lack of general understanding among adults. Education efforts are needed for the general public, policy makers, and even environmental managers. North Carolina has several significant programs that can promote this general understanding. APNEP and the Commission on Climate Change have been mentioned above. Public television and radio have a general mission to educate and have contributed time to the topic. Two other programs are (1) the Partnership for the Sounds, including the Estuarium in Washington, North Carolina, and (2) the North Carolina Aquariums. The latter includes three aquaria along the coast. These programs are in a unique position to teach the general public about climate change. We recommend that coordination among these different programs be fostered to promote education within the state.

Finally, adaptive management and adaptation strategies are essential to respond to the complex implications of climate change. Adaptive management recognizes the need for both sustained monitoring associated with observing systems and adaptive justification of intervention plans that reflect advances in our understanding of impacts of climate change and new insights on what experimental interventions are needed. Adaptive management also recognizes the important role of education that promotes better appreciation of a changing and uncertain world. Adaptive management is explicit within APNEP, CHPP, and EEP. It also is incorporated into controls on nutrient additions to alleviate the impacts of cultural eutrophication. It acknowledges the importance of the ecosystem perspective and breaks the regulatory mold of being specific to an issue, species, single source of pollution, etc. This enhances the ability to meet the challenges of climate change. One aspect of this change is the expectation that landscape units that are controlled by sea level will migrate. Beaches and wetlands will move shoreward. Regulations and policies that foster the ability to retreat from these landscape migrations are part of this adaptive approach. Adaptive management is an established approach in North Carolina, which can serve as a successful example nationally.

A5.1.6 Barriers and Opportunities

APNEP possesses environmental and social barriers to effective implementation of management adaptation to climate change, yet at the same time various social and environmental characteristics represent favorable opportunities for adaptation. Indeed, APNEP was chosen for a case study because it could illustrate both significant barriers and opportunities. Perhaps its greatest single barrier to successful adaptation to climate change is the intractable nature of the challenge of preserving the integrity of the coastal barrier complex of the Outer Banks over the long time scales of a century and longer. These coastal barriers are responsible for creating the APNEP estuarine system, and a major breach in the integrity would ultimately convert the estuary into a coastal ocean embayment (Riggs and Ames, 2003). Current management employs beach nourishment to fortify the barrier, but this method will become increasingly expensive as sea level rises substantially, and thus would be politically infeasible. Construction of a seawall along the entire extent of the barrier complex also does not appear to be a viable option because of financial costs and loss of the beach that defines and enriches the Outer Banks.

Special opportunities for implementation of adaptive management in APNEP include the existence of the CHPP process, a legislatively mandated ecosystem-based management plan for preserving and enhancing coastal fisheries. This plan involves collaborative attentions by all necessary state agencies and thereby can overcome the historic constraints of compartmentalization of management authorities. This plan sets an admirable example for other states. Similarly, the novel state commission on effects of climate change that was legislated in 2005 also provides opportunity for education and participation of legislators in a process of looking forward, well beyond the usual time frames of politics, to serve as an example of proactivity for other states to emulate. Sparse human populations and low levels of development along much of the interior mainland shoreline of the APNEP complex provide opportunities for implementation of policies that protect the ability of the salt marsh and other shallow-water estuarine habitats to be allowed to retreat as sea level rises. Implementing the policies required to achieve this management adaptation would not be possible in places where development and infrastructure are so dense that the economic and social costs of shoreline retreat are high. Special funding to support purchase of rolling easements or other implementation methods can come from the Clean Water Management Trust Fund and the Ecosystem Enhancement Program of North Carolina, two facilitators of large coordinated projects. The State of North Carolina was among the first to establish basin-scale water quality management and has established novel methods of basin-wide capping of nutrient delivery to estuaries, such as the NRE, involving ecosystem-based management through participation of all stakeholders. This too facilitates actions required to manage consequences of climate change to preserve management goals of a national estuary.

A[illegible]

This section includes three U.S. case studies along with an Australian case study for comparison. This report focuses on U.S. federally managed lands and waters to frame the question of adaptation; the goal is to review all types of adaptation options including those developed by non-governmental organizations and internationally that may be implemented to benefit U.S. resources. With regard to climate change impacts and adaptation, coral reefs are the best studied marine system. Because the Great Barrier Reef Marine Park (GBRMP) in Australia is an international leader in addressing climate change impacts to coral reefs, a case study of how this issue is being addressed there is of great value for examining adaptation options that may be transferable to U.S. coral reefs and other U.S. marine systems. Each case study discusses existing management approaches, threats of climate change, and adaptation options. The case studies are located in Florida (Florida Keys National Marine Sanctuary (FKNMS)), Australia (GBRMP), Hawaii (Papahānaumokuākea Marine National Monument (PMNM)), and California (Channel Islands National Marine Sanctuary (CINMS)). These MPAs range in size, species composition, and levels of protection; no-take designations, for example, are 6% (FKNMS), 10% (CINMS), 33% (GBRMP), and 100% (PMNM).

A6.1 The Florida Keys National Marine Sanctuary

A6.1.1 Introduction

The Florida Keys form a limestone archipelago extending southwest over 320 km from the southern tip of the Florida mainland (see Fig. 8.3 in the MPA chapter). The FKNMS surrounds the Florida Reef Tract, one of the world's largest systems of coral reefs and the only bank-barrier reef in the coterminous United States. The FKNMS is bounded by and connected to Florida Bay, the Southwest Florida Continental Shelf, and the Straits of Florida and Atlantic Ocean. It is influenced by the powerful Loop Current/ Florida Current/Gulf Stream system to the west and south, as well as a weaker southerly flow along the West Florida Shelf (Lee *et al.*, 2002). The combined Gulf of Mexico and tropical Atlantic biotic influences make the area one of the most diverse in North America.

The uniqueness of the marine environment and ready access from the mainland by a series of bridges and causeways draws millions of visitors to the Keys, including many from the heavily populated city of Miami and other metropolitan areas of South Florida. Also, in recent years Key West has become a major destination for cruise liners, attracting more than 500 stop-overs annually. The major industry in the Florida Keys has become tourism, including dive shops, charter fishing, and dive boats and marinas as well as hotels and restaurants. There also is an important commercial fishing industry.

National Marine Sanctuaries established at Key Largo in 1975 and Looe Key in 1981 demonstrated that measures to protect coral reefs from direct impacts could be successful using management actions such as mooring buoys, education programs, research and monitoring, restoration efforts, and proactive, interpretive law enforcement. In 1989, mounting threats to the health and ecological future of the coral reef ecosystem in the Florida Keys prompted Congress to take further protective steps. The threat of oil drilling in the mid- to late-1980s off the Florida Keys, combined with reports of deteriorating water quality throughout the region, occurred at the same time as adverse effects of coral bleaching,[41] the Caribbean-wide die-off of the long-spined urchin (Lessios, Robertson, and Cubit, 1984), loss of living coral cover on reefs (Porter and Meier, 1992), a major seagrass die-off (Robblee *et al.*, 1991), declines in reef fish populations (Bohnsack, Harper, and McClellan, 1994; Ault, Bohnsack, and Meester, 1998), and the spread of coral diseases (Kuta and Richardson, 1996). These were already topics of major scientific concern and the focus of several scientific workshops when, in the fall of 1989, three large ships ran aground on the Florida Reef Tract within a brief 18-day period. On November 16, 1990, President Bush signed into law the Florida Keys National Marine Sanctuary and Protection Act. Specific regulations to manage the sanctuary did not go into effect until July 1997, after the final management plan (U.S. Department of Commerce, 1996) had been approved by the Secretary of Commerce and the Governor and Cabinet of the State of Florida. The FKNMS encompasses approximately 9,800 km^2 of coastal and oceanic waters surrounding the Florida Keys (Keller and Causey, 2005) (see Fig. 8.3 in the MPA chapter), including the Florida Reef Tract, all of the mangrove islands of the Florida Keys, extensive seagrass beds and hard-bottom areas, and hundreds of shipwrecks.

Visitors spent $1.2 billion[42] over 12.1 million person-days[43] in the Florida Keys between June 2000 and May 2001. Over that period, visitors and residents spent 5.5 million person-days on natural and artificial reefs. Significantly, visitors (and residents) perceive significant declines in the quality of the marine environment of the Keys.[44]

A6.1.2 Specific Management Goals and Current Ecosystem Stressors Being Addressed

Goal and Objectives of the Florida Keys National Marine Sanctuary

The goal of the FKNMS is "To preserve and protect the physical and biological components of the South Florida estuarine and marine ecosystem to ensure its viability for the use and enjoyment of present and future generations" (U.S. Department of Commerce, 1996). The Florida Keys National Marine Sanctuary and Protection Act as well as the Sanctuary Advisory Council identified a number

[41] **Causey**, B.D., 2001: Lessons learned from the intensification of coral bleaching from 1980-2000 in the Florida Keys, USA. In: *Coral Bleaching and Marine Protected Areas: Proceedings of the Workshop on Mitigating Coral Bleaching Impact Through MPA Design, Bishop Museum, Honolulu, Hawaii, 29-31 May 2001.* [Salm, R.V. and S.L. Coles (eds.)]. Asia Pacific Coastal Marine Program Report #0102, The Nature Conservancy, Honolulu, Hawaii, pp. 60-66.

[42] **Leeworthy**, V.R. and P.C. Wiley, 2003: *Profiles and Economic Contribution: General Visitors to Monroe County, Florida 2000-2001.* National Oceanic and Atmospheric Administration, National Ocean Service, Office of Management and Budget, Special Projects Division, Silver Spring, MD, pp.1-24.

[43] **Johns**, G.M., V.R. Leeworthy, F.W. Bell, and M.A. Bonn, 2003: *Socioeconomic Study of Reefs in Southeast Florida.* Final Report October 19, 2001 as Revised April 18, 2003 for Broward County, Palm Beach County, Miami-Dade County, Monroe County, Florida Fish and Wildlife Conservation Commission, National Oceanic and Atmospheric Administration, Hollywood, FL.

[44] **Leeworthy**, V.R., P.C. Wiley, and J.D. Hospital, 2004: *Importance-Satisfaction Ratings Five-Year Comparison, SPA and ER Use, and Socioeconomic and Ecological Monitoring Comparison of Results 1995-96 to 2000-01.* National Oceanic and Atmospheric Administration, National Ocean Service, Office of Management and Budget, Special Projects Division, Silver Spring, MD, pp.1-59.

of objectives to achieve this goal (Box A6.1). FKNMS management was designed during the 1990s to address local stressors; the subsequent recognition of the significance of regional and global stressors requires that future planning efforts incorporate these larger-scale factors.

Coral Reef and Seagrass Protection

The management plan (U.S. Department of Commerce, 1996) established a channel and reef marking program that coordinated federal, state, and local efforts to mark channels and shallow reef areas. These markers help prevent damage from boat groundings and propeller-scarring.

A mooring buoy program is one of the most simple and effective management actions to protect sanctuary resources from direct impact by boat anchors. By installing mooring buoys in high-use areas, the sanctuary has prevented damage to coral and other sessile invertebrates from the thousands of anchors deployed every week in the Keys.

Marine Zoning

The management plan implemented marine zoning with five categories of zones. The relatively large "no-take" Ecological Reserve at Western Sambo (see Fig. 8.3 in the MPA chapter) was designed to help restore ecosystem structure and function. A second Ecological Reserve was implemented in the Tortugas region in 2001 as the largest no-take areas in U.S. waters at the tiime (U.S. Department of Commerce, 2000; Cowie-Haskell and Delaney, 2003; Delaney, 2003). In addition to the larger Ecological Reserves, there are 18 small, no-take Sanctuary Preservation Areas (SPAs) that protect over 65% of shallow, spur and groove reef habitat. These areas displaced few commercial and recreational fishermen and resolved a user conflict with snorkeling and diving activities in the same shallow reef areas. Four small Research-Only Areas are also no-take; only scientists with permits are allowed access.

In addition, 27 Wildlife Management Areas (WMAs) were established to address human impacts to nearshore habitats such as seagrass flats and mangrove-fringed shorelines. Most of these WMAs only allow non-motorized access. Finally, because the FKNMS Act called for the two existing sanctuaries to be subsumed by the FKNMS, a final type of marine zone, called Existing Management Areas, was used to codify both Key Largo and Looe Key NMS regulations into FKNMS regulations. This was a way to maintain the additional protective resource measures that had been in effect for the Key Largo and Looe Key NMSs since 1975 and 1981,

[illegible] Goal and Objectives of the Florida Keys National Marine Sanctuary (U.S. Department of Commerce, 1996)

[illegible]

To preserve and protect the physical and biological components of the South Florida estuarine and marine ecosystem to ensure its viability for the use and enjoyment of present and future generations.

[illegible]

- Objective 1: Facilitate all public and private uses of the Sanctuary consistent with the primary objective of resource protection.
- Objective 2: Consider temporal and geographic zoning to ensure protection of Sanctuary resources.
- Objective 3: Incorporate regulations necessary to enforce the Water Quality Protection Program.
- Objective 4: Identify needs for research and establish a long-term ecological monitoring program.
- Objective 5: Identify alternative sources of funding needed to fully implement the management plan's provisions and supplement appropriations authorized under the FKNMS and National Marine Sanctuaries Acts.
- Objective 6: Ensure coordination and cooperation between Sanctuary managers and other federal, state, and local authorities with jurisdiction within or adjacent to the Sanctuary.
- Objective 7: Promote education among users of the Sanctuary about coral reef conservation and navigational safety.
- Objective 8: Incorporate the existing Looe Key and Key Largo National Marine Sanctuaries into the FKNMS.

[illegible]

- Objective 1: Encourage all agencies and institutions to adopt an ecosystem and cooperative approach to accomplish the following objectives, including the provision of mechanisms to address impacts affecting Sanctuary resources, but originating outside the boundaries of the Sanctuary.
- Objective 2: Provide a management system that is in harmony with an environment whose long-term ecological, economic, and sociological principles are understood, and which will allow appropriate sustainable uses.
- Objective 3: Manage the FKNMS for the natural diversity of healthy species, populations, and communities.
- Objective 4: Reach every single user of and visitor to the FKNMS with information appropriate to his or her activities.
- Objective 5: Recognize the importance of cultural and historical resources, and managing these resources for reasonable, appropriate use and enjoyment.

respectively. Those areas prohibited spearfishing, marine life collecting, fish trapping, trawling, and a number of other specific activities that posed threats to coral reef resources.

Improvement of Water Quality

The FKNMS Act directed the U.S. Environmental Protection Agency to work with the State of Florida and NOAA to develop a Water Quality Protection Program (WQPP) to address water quality problems and establish corrective actions. The WQPP consists of four interrelated components: 1) corrective actions that reduce water pollution directly by using engineering methods, prohibiting or restricting certain activities, tightening existing regulations, and increasing enforcement; 2) monitoring of water quality, seagrasses, and coral reefs to provide information about status and trends in the sanctuary; 3) research to identify and understand cause-and-effect relationships involving pollutants, transport pathways, and biological communities; and 4) public education and outreach programs to increase public awareness of the sanctuary, the WQPP, and pollution sources and impacts on sanctuary resources.

Research and Monitoring

The FKNMS management plan established a research and monitoring program that focused research on specific management needs. In 2000, staff convened a panel of external peers to review the sanctuary's science program and provide recommendations for improvements.[45] Based on the panel's recommendation that sanctuary managers identify priority research needs, staff prepared a Comprehensive Science Plan to identify priority research and monitoring needs explicitly linked to management objectives (Florida Keys National Marine Sanctuary, 2002).

The three monitoring projects of the WQPP[46] are developing baselines for water quality, seagrass distribution and abundance, and coral cover, diversity, and condition. Such a baseline of information is particularly important to have as the Comprehensive Everglades Restoration Plan (CERP)[47] is implemented just north of the FKNMS. The CERP is designed so that managers can be adaptive to ecological or hydrological changes that are taking place within or emanating from the Everglades, with possible positive or negative influences on communities in the FKNMS (Keller and Causey, 2005).

Additional monitoring comprises the Marine Zone Monitoring Program, which is designed to detect changes in populations, communities, and human dimensions resulting from no-take zoning (Keller and Donahue, 2006). Coupled with environmental monitoring using instrument arrays that provide near-real-time data,[48] routine cruises,[49] remote sensing,[50] and paleoclimatic analyses of coral skeletons, the FKNMS is a relatively data-rich environment for detecting presumptive climate change effects.

Education and Outreach

The management plan for the FKNMS includes an education and outreach program that lays out ways that education efforts can directly enhance the various programs to protect sanctuary resources. Public awareness and understanding are essential to achieve resource protection through cooperation and compliance with regulations.

Regulations and Enforcement

The FKNMS management plan includes regulations that have helped managers protect sanctuary resources while having the least amount of impact on those who enjoy and utilize sanctuary resources in a conscientious way. In order to maximize existing enforcement programs, the management plan contains an enforcement plan that has served to help focus on priority problems within the sanctuary. The program also coordinates all the enforcement agencies in the Keys. Enforcement complements education and outreach in efforts to achieve compliance with regulations.

A6.1.3 Potential Effects of Climate Change on Management

Coral Bleaching

The potential effects of climate change on coral reefs are generally well known (*e.g.*, Smith and Buddemeier, 1992; Hoegh-Guldberg, 1999; Buddemeier, Kleypas, and Aronson, 2004; Hoegh-Guldberg, 2004; Sheppard, 2006), but the fate of individual reef systems such as the Florida Reef Tract will vary based on a combination of factors related to history, geography, and an understanding of processes that explain the patchiness of coral bleaching and subsequent mortality that occurs on reefs. Coral bleaching was first reported in the Florida Keys in 1973 (Jaap, 1979), with at least seven other episodes documented prior to 2000[41] and a major

[45] **Florida Keys National Marine Sanctuary**, 2007: Year 2000 Florida Keys National Marine Sanctuary advisory panel meeting. NOAA Website, http://floridakeys.noaa.gov/research_monitoring/sap2000.html, accessed on 7-27-2007.

[46] **Fish and Wildlife Research Institute**, 2007: Florida Keys National Marine Sanctuary water quality protection program. Fish and Wildlife Research Institute Website, http://ocean.floridamarine.org/fknms_wqpp/, accessed on 7-27-0007.

[47] **U.S. Army Corps of Engineers**, 2007: Official website of the comprehensive Everglades restoration plan. Comprehensive Everglades Restoration Plan Website, http://www.evergladesplan.org/index.aspx, accessed on 5-23-2007.

[48] **National Oceanic and Atmospheric Administration**, 2006: NOAA's coral health and monitoring homepage. NOAA Website, http://www.coral.noaa.gov/seakeys/index.shtml, accessed on 7-27-2007.

[49] **National Oceanic and Atmospheric Administration**, 2007: NOAA's south Florida ecosystem research and monitoring program. NOAA Website, http://www.aoml.noaa.gov/sfp/data.shtml, accessed on 7-27-2007.

[50] **NOAA Coast Watch Program**, 2007: Harmful algae bloom bulletin home page. NOAA Website, Harmful Algae Bloom Bulletin, http://coastwatch.noaa.gov/hab/bulletins_ms.htm, accessed on 7-27-2007.

bleaching event in 2005 that also affected the Caribbean (Miller *et al.*, 2006; Donner, Knutson, and Oppenheimer, 2007). Unfortunately, before-during-and-after sampling has not been conducted during major bleaching events in the Florida Keys (but see Lang *et al.*, 1992 for during- and after-surveys at four sites), which makes assumptions about coral mortality caused by bleaching at best correlative. Hurricanes are an especially confounding factor when they occur during bleaching years, as they did in 1997–98 and 2005. Still, anecdotal evidence suggests that large numbers of corals were killed in 1997–98 when corals remained bleached for two consecutive years.[41] Long-term temperature records do not exist that reveal trends of increasing surface seawater temperature for the Florida Keys, but Williams, Jackson, and Kutzbach (2007), using climate models and IPCC greenhouse gas estimates to forecast how climate zones may change in the next 100 years, identified the southeastern United States as a region with the greatest likelihood of developing novel regional climate conditions that would be associated with temperature increases of several degrees. The consequences of such changes on coral reefs in Florida will be dramatic unless significant adaptation or acclimatization occurs.

Governments and agencies have responded to the crisis of coral bleaching with detailed management plans (Westmacott *et al.*, 2000; Marshall and Schuttenberg, 2006), workshops to develop strategies that support response efforts,[51] and research plans (Marshall and Schuttenberg, 2006; Puglise and Kelty, 2007). Two themes have emerged from these responses. First, effort is needed at local and regional levels to identify and protect bleaching-resistant sites—if they exist. Second, management plans should be developed or modified in the case of the FKNMS to restore or enhance the natural resilience (Hughes *et al.*, 2003; West and Salm, 2003) of coral reefs.

Response plans to coral bleaching events depend upon increasingly accurate predictions to help guide resource assessment and monitoring programs. The NOAA Coral Reef Watch program has increasingly accurate capability to predict the severity, timing, and geographic variability of mass bleaching events, largely using remote sensing technologies.[52] Scientists and managers in Florida have implemented an assessment and monitoring program that specifically addresses bleaching events, including the critical before-during-after sampling that is necessary to quantify the distribution, severity, and consequences of mass bleaching. The Florida Reef Resilience Program (see A6.1.4) is managed by The Nature Conservancy, which trains and coordinates teams of divers in southeastern Florida and the Florida Keys. While such monitoring programs do nothing to prevent coral bleaching, they do provide data that may identify bleaching-resistant sites that, if not already protected, can be considered high priority for management action and protection against local stressors.

Currently in Florida, status and trends monitoring has identified habitat types with higher than average coral cover and abundance, but it is unknown whether these areas are more or less prone to bleaching because only baseline assessments have been conducted.[53] Deeper reefs (to 35 meters) may also exhibit less evidence of mortality caused by coral bleaching (Miller *et al.*, 2001), but even less is known about these habitats—especially related to the distribution and abundance of coral diseases, which can confound assessments of factors causing mortality because the temporal scale of monitoring is sufficient to only assess disease prevalence and not incidence or mortality rates.

No-Take Protection and Zoning for Resistance or Resilience

The use of marine reserves (Sanctuary Preservation Areas, Research-Only Areas, and Ecological Reserves) in the FKNMS has already been adopted as a tool to manage multiple user groups throughout the Sanctuary (U.S. Department of Commerce, 1996), and in the Dry Tortugas to enhance fisheries where positive results have been obtained after only a few years (Ault *et al.*, 2006). Potential exists to use a range of options to identify bleaching-resistant reefs in the Keys, from simply identifying the best remaining sites left and using a decision matrix based on factors that may confer resilience to establish priority sites for protection, to the Bayesian approach of Wooldridge and Done (2005). Only recently have coral community data been obtained at the relevant spatial scales and across multiple habitat types (Smith *et al.*, forthcoming). Whatever approach is used, the results are likely to include sites with high coral cover and abundance, high diversity, connectivity related to current regimes with the potential to transport larvae, and protection from local stressors including overfishing and pollution (Done, 1999; Hughes *et al.*, 2003).[54]

51 **Salm**, R.V. and S.L. Coles (eds.), 2001: Coral Bleaching and Marine Protected Areas. Proceedings of the Workshop on Mitigating Coral Bleaching Impact Through MPA Design, Bishop Museum, Honolulu, Hawaii, 29-31 May 2001. Asia Pacific Coastal Marine Program Report #0102, The Nature Conservancy, Honolulu, Hawaii, pp. 1-118.

52 **NOAA Satellite and Information Service**, 2007: NOAA coral reef watch satellite bleaching monitoring datasets. NOAA Website, National Oceanic and Atmospheric Administration, http://coralreefwatch.noaa.gov/satellite/ge/, accessed on 7-27-2007.

53 **Miller**, S.L., M. Chiappone, L.M. Rutten, D.W. Swanson, and B. Shank, 2005: *Rapid Assessment and Monitoring of Coral Reef Habitats in the Florida Keys National Marine Sanctuary: Quick Look Report: Summer 2005 Keys-Wide Sampling*. National Undersea Research Center, University of North Carolina at Wilmington, Wilimington, NC.

54 See also **Salm**, R.V., S.E. Smith, and G. Llewellyn, 2001: Mitigating the impact of coral bleaching through Marine Protected Area design. In: *Coral Bleaching: Causes, Consequences and Response* [Schuttenberg, H.Z. (ed.)]. Proceedings of the Ninth International Coral Reef Symposium on Coral Bleaching: Assessing and linking ecological and socioeconomic impacts, future trends and mitigation planning, Coastal Management Report 2230, Coastal Resources Center, University of Rhode Island, Narragansett, pp. 81-88.

In the Florida Keys, marine protected areas date to 1960 for the John Pennekamp Coral Reef State Park, 1975 for the Key Largo National Marine Sanctuary, 1981 for Looe Key National Marine Sanctuary, and 1990 for expansion of these sites to include 2,800 square nautical miles of coastal waters that were designated as the Florida Keys National Marine Sanctuary. The Tortugas Ecological Reserve was added in 2001, and six years later a 46-square-mile Research Natural Area was established within Dry Tortugas National Park.[55] While spatial resolution among habitat types from Miami to the Dry Tortugas is not as extensive as in the Great Barrier Reef, work similar to Wooldridge and Done (2005) should be evaluated for application to the Florida Keys. For example, a combination of retrospective sea-surface temperature studies using NOAA Coral Reef Watch products, combined with *in situ* temperature data, water quality monitoring data,[56] and detailed site characterizations might help identify bleaching-resistant sites (if temporally- and spatially-relevant sampling is conducted before, during, and after a bleaching event), identify candidate sites for protection based on resilience criteria, and in general validate the concept of marine reserve networks in the region as a management response to coral bleaching threats.

Geographic Range Extensions of Coral Reefs in Florida

Coral reefs in south Florida represent the northern geographic limit of reef development in the United States. It is reasonable to assume that some northward expansion of either the whole reef community or individual species may occur as a result of warming climate. Indeed, such a northward expansion may already be in progress, but caution is necessary before assigning too much significance to what might be an anomalous event. Specifically, *Acropora cervicornis* was discovered growing in large thickets off Fort Lauderdale in 1998 (Vargas-Ángel, Thomas, and Hoke, 2003) and *A. palmata* was discovered off Pompano Beach in northern Broward county (Precht and Aronson, 2004). It is possible that these populations—over 50 km northward of their previously known northern limit—are a result of recent climate warming known to have occurred in the western Atlantic (Hoegh-Guldberg, 1999; Levitus *et al.*, 2000; Barnett, Pierce, and Schnur, 2001). It is also possible that these reefs represent a remnant population or a chance recruitment event based on a short-term but favorable set of circumstances that will disappear with the next hurricane, cold front, disease epizootic, or bleaching event. Still, the presence of these acroporid reefs is suggestive of what might happen as climate warms. Interestingly, the presence of these northern acroporid populations matches the previous northern extension of reef development in the region during the middle Holocene (Lighty, Macintyre, and Stuckenrath, 1978), when sea surface temperatures were warmer. Reefs up to 10 m thick grew off Palm Beach County in the middle Holocene (Lighty, Macintyre, and Stuckenrath, 1978) and when temperatures started to cool 5,000 years before present reef development moved south to its current location (Precht and Aronson, 2004).

Despite these northern extensions in the geographic distributions of corals seen in the fossil record, predicting future geographic expansions in Florida is complicated by factors other than temperature that influence coral reefs, including light, carbonate saturation state, pollution, disease (Buddemeier, Kleypas, and Aronson, 2004), and a shift from a carbonate to siliciclastic sedimentary regime along with increasing nutrient concentrations up the east coast of Florida (Precht and Aronson, 2004). One thing, however, is certain: geographic shifts of reefs in Florida that result from global warming will not mitigate existing factors that today cause widespread local and regional coral reef decline (Precht and Aronson, 2004). Further, if we assume that the reefs of the mid-Holocene were in better condition than today's reefs, they may not prove to be a good analogue for predicting the future geographic trajectory of today's reefs. Because corals in Florida are already severely impacted by disease, bleaching, pollution, and overfishing, expansion will at best be severely limited compared to what might occur if the ecosystem were intact.

At the global scale and across deep geological time, range extensions to higher latitudes occurred for hard corals that survived the Cretaceous warming period (Kiessling, 2001; Kleypas, 2006), and some coral species today that are found in the Red Sea and Persian Gulf can survive under much greater temperature ranges than they experience throughout the Indo-Pacific (Coles and Fadlallah, 1991). Both of these examples, however, probably reflect long-term adaptation by natural selection and not short-term acclimatization (Kleypas, 2006). At shorter times scales (decades), corals that survive rapid climate warming may be those that are able to quickly colonize and survive at higher latitudes where maximum summer temperatures may be reduced compared to their previous geographic range. An alternative to migration is the situation where corals adapt to increasing temperatures at ecological time scales (decades), and there is some evidence to suggest that this might occur (Guzmán

And **West**, J.M., 2001: Environmental determinants of resistance to coral bleaching: implications for management of marine protected areas. In: *Coral Bleaching and Marine Protected Areas: Proceedings of the Workshop on Mitigating Coral Bleaching Impact Through MPA Design, Bishop Museum, Honolulu, Hawaii, 29-31 May 2001.* [Salm, R.V. and S.L. Coles (eds.)]. Asia Pacific Coastal Marine Program Report #0102, The Nature Conservancy, Honolulu, Hawaii, pp. 40-52.

[55] **National Park Service**, 1-18-2007: Dry Tortugas National Park - research natural area will be effective January 19, 2007. National Park Service Website, http://www.nps.gov/drto/parknews/research-naturalarea.htm, accessed on 7-26-2007.

[56] **Boyer**, J.N. and H.O. Briceño, 2006: *FY2005 Annual Report of the Water Quality Monitoring Project for the Water Quality Protection Program of the Florida Keys National Marine Sanctuary.* Southeast Environmental Research Center, Florida International University, Miami, FL, pp.1-83.

and Cortés, 2001; Podestá and Glynn, 2001). However, the ability to predict if corals will acclimate is complicated because absolute values and adaptive potential are likely to vary across species (Hughes *et al.*, 2003; Kleypas, 2006). Acclimation without range expansion is a topic of great significance related to coral bleaching.

Another question related to the potential for coral reef migration to higher latitudes in Florida is related to understanding factors that currently limit expansion northward. Cold-water temperature tolerances for individual corals are not well known; however, their present-day global distribution generally follows the 18°C monthly minimum seawater isotherm (Kleypas, McManus, and Mendez, 1999; Kleypas, Buddemeier, and Gattuso, 2001; Buddemeier, Kleypas, and Aronson, 2004). South Florida is located between the 18 and 20°C isotherm and is thus significantly affected by severe winter cold fronts, especially for corals in shallow water (Burns, 1985; Walker, Rouse, and Huh, 1987). Well documented coral die-offs due to cold water fronts have occurred repeatedly throughout the Florida Keys (Davis, 1982; Porter, Battey, and Smith, 1982; Walker *et al.*, 1982; Roberts, Rouse, and Walker, 1983; Shinn, 1989); and as far south as the Dry Tortugas (Porter, Battey, and Smith, 1982; Jaap and Hallock, 1990).[57] Porter and Tougas (2001) documented a decreasing trend in generic coral diversity along the east coast of Florida, but a number of coral species extend well beyond the 18°C isotherm with at least two species surviving as far north as North Carolina, likely due to the influence of the Gulf Stream. Thus, climate warming that has the potential to influence the impact of winter cold fronts may influence the range expansion of corals in Florida.

Finally, the above examples have focused mostly on the acroporid corals, which represent only two species out of more than forty that are found regionally (Jaap, 1984). Obviously, when considering range expansion of the total reef system, and not just two coral species, models designed to optimize or anticipate management actions that conserve existing habitat or predict future locations for habitat protection are likely to be exceedingly complicated. In Florida, if reefs are in sufficiently good condition in the future to act as seed populations for range expansion, one management action to anticipate the effects of climate change would be to protect habitats similar to those that thrived during the middle Holocene when coral reefs flourished north of their current distribution (Lighty, Macintyre, and Stuckenrath, 1978). However, existing declines in the acroporids throughout Florida and the Caribbean (Gardner *et al.*, 2003; Precht and Miller, 2006) suggest that at least for these two species, the major framework building species in the region, expansion may not occur unless factors such as disease and coral bleaching are mitigated.

A6.1.4 Adapting Management to Climate Change

The Sanctuary Advisory Council (SAC) is a committee of stakeholder representatives that provides advise to sanctuary managers across a broad range of topics and issues (Keller and Causey, 2005), particularly regarding new issues as they arise. The SAC has a climate change working group, which can work with sanctuary managers to help develop adaptation approaches best suited for the Florida Keys (see also section 8.4.4.2 of the Marine Protected Areas chapter).

Little has been done to restore mangrove habitat in the Florida Keys, where many shorelines were cleared for development. In addition to supporting critical nurseries, mangroves produce tannins and other dissolved organic compounds that absorb ultraviolet radiation. Dependable sources of these compounds from intact mangrove coastlines can provide reefs with some protection from photo-oxidative stress that contributes to bleaching. Mangrove restoration should be considered as a management strategy that may become increasingly important in the context of climate change – for shoreline protection as well as the benefits noted above.

The Great Barrier Reef Marine Park Authority (next section) has a Climate Change Response Program and an action plan (section 8.4.4 of the Marine Protected Areas chapter) that is a model for the FKNMS, which is completing a bleaching response plan, but has not yet developed a broader plan about responding to climate change. Such a plan is a logical next step. At the same time, The Nature Conservancy is leading the Florida Reef Resilience Program[58] to investigate possible patterns of resilience along the Florida Reef Tract and recommend actions.

A6.2 The Great Barrier Reef Marine Park

A6.2.1 Introduction

The Great Barrier Reef (GBR) is a maze of reefs and islands spanning an area of 348,000 km^2 off the Queensland coast in northeast Australia (Fig. A6.1). It spans 14 degrees of latitude, making it the largest coral reef ecosystem in the world and one of the richest in biological diversity. The GBR supports 1,500 species of fish, 350 species of hard corals, more than 4,000 species of mollusks, 500 species of algae, six of the world's seven species of marine turtles, 24 species

[57] See also **Jaap**, W.C. and F.J. Sargent, 1994: The status of the remnant population of *Acropora cervicornis* (Lamarck, 1816) at Dry Tortugas National Park, Florida, with a discussion of possible causes of changes since 1881. In: *Proceedings of the Colloquium on Global Aspects of Coral Reefs: Health, Hazards and History* [Ginsburg, R.N. (ed.)] Rosenstiel School of Marine and Atmospheric Science, University of Miami, Miami, Florida pp. 101-105.

[58] http://www.nature.org/wherewework/northamerica/states/florida/preserves/art17499.html

of seabirds, more than 30 species of whales and dolphins, and the dugong. The GBR was chosen as a case study because it is a large marine protected area that has moderate representation of no-take areas (33%) and has been under a management regime since 1975.

The GBR already appears to have been affected by climate change. The first reports of coral bleaching in the GBR appeared in the literature in the 1980s and have continued to increase in frequency since then (Hoegh-Guldberg, 1999; Done *et al.*, 2003). Coral-coring work done at the Australian Institute of Marine Science detected the earliest growth hiatus associated with mass coral bleaching in 1998 (Lough, 2007). There have been nine bleaching events on the GBR, with three major events in the last decade correlating with elevated sea temperatures and causing damage to parts of the reef. These early signs of climate change, and the extensive research and monitoring data that are available for the GBR, make it a suitable case study for this report.

The conservation values of the GBR are recognized in its status as a World Heritage Area (listed in 1981), and its resources are protected within the Great Barrier Reef Marine Park. The enactment of the Great Barrier Reef Marine Park Act in 1975 established the legal framework for protecting these values. The goal of the legislation is "*...to provide for the protection, wise use, understanding and enjoyment of the Great Barrier Reef in perpetuity through the care and development of the Great Barrier Reef Marine Park.*"

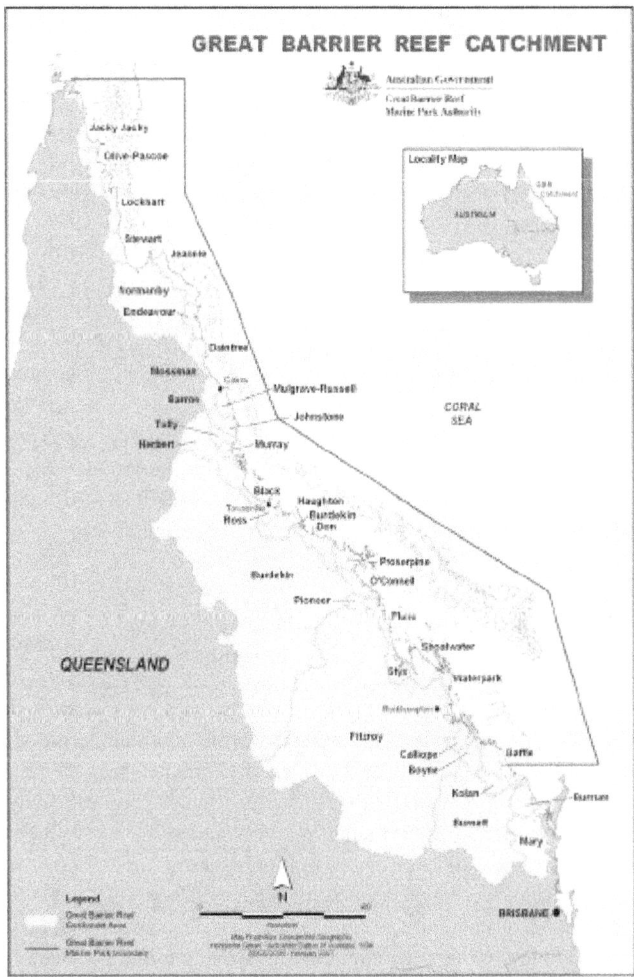

Figure A6.1 Map of the Great Barrier Reef Marine Park showing the adjacent catchment in Queensland. Modified from Haynes (2001) and courtesy of the Great Barrier Reef Marine Park Authority.

A6.2.2 Managing the Great Barrier Reef Marine Park

The Great Barrier Reef Marine Park Authority has management strategies in place to address current stresses on the GBR. Stressors include terrestrial inputs of sediment, nutrients, and pesticides from coastal catchments; fisheries extraction; tourism and recreational activities; and changes to coastal hydrology as a result of coastal development and climate change. Sustainability of the environmental and social values of the Great Barrier Reef depend largely (and in most cases, entirely) on a healthy, self-perpetuating ecosystem. Reducing pressures on this system has been a focus of management activities over the last decade.

The Great Barrier Reef Marine Park was rezoned in 2003 to increase the area of highly protected no-take zones to 33%, with at least 20% protected in each habitat bioregion. These no-take areas aim to conserve biodiversity, increasing the potential of maintaining an intact ecosystem, with larger no-take areas including more representative habitats.

Current Approaches to Management

There are 26 major catchments that drain into the GBR (Fig. A6.1) covering an area of 425,964 km^2. Cropping (primarily of sugar cane), grazing, heavy industry and urban settlement are the main land uses. The *Reef Water Quality Protection Plan* (The State of Queensland and Commonwealth of Australia, 2003) is a joint state and federal initiative that aims to *halt and reverse the decline in the quality of water entering the Reef by 2013*. Under this initiative, diffuse sources of pollution are targeted through a range of voluntary and incentive-driven strategies to address water quality entering the GBR from activities in the catchments.

Important commercial fisheries in the GBR include trawling that mainly targets prawns and reef-based hook-and-line that targets coral trout and sweetlip emperor, inshore fin fish, and three crab fisheries (spanner, blue, and mud). None of these fisheries is considered overexploited; however, there is considerable unused (latent) effort in both the commercial and recreational sectors. Commercial fisheries contribute A$251 million to the Australian economy (Great Barrier Reef Marine Park Authority, 2007). Fisheries management is undertaken by the Queensland Government and includes a range of measures such as limited entry, management plans, catch and effort limits, permits, and industry accreditation. Recreational activities (including fishing) contribute A$623 million per annum to the region (Great Barrier Reef Marine Park Authority, 2007), and recreational fishing is subject to size and bag limits for many species.

Over 1 million tourists visit the GBR annually, contributing A$6.1 billion to the Australian economy (Great Barrier Reef Marine Park Authority, 2007). The Great Barrier Reef Marine Park Authority manages tourism using permits, zoning, and other planning tools such as management plans and site plans (Smith *et al.*, 2004). Visitation is concentrated in the Cairns and Whitsunday Island areas, and an eco-certification program encourages best practices and sustainable tourism (Skeat, 2003).

As one of the fastest growing regions in Australia, the GBR coast is being extensively developed through the addition of tourist resorts, urban subdivisions, marinas, and major infrastructure such as roads and sewage treatment plants. All levels of government regulate coastal development depending on the scale and potential impacts of the development. Local government uses local planning schemes and permits, state government uses the Integrated Planning Act,[59] and in the case of significant developments, the federal government uses the Environment Protection and Biodiversity Conservation Act[60] to assess the environmental impacts of proposals. These efforts have resulted in an increase in biodiversity protection, a multi-stakeholder agreement to address water quality, and a well-managed, multiple-use marine protected area.

Vulnerability of the Great Barrier Reef to Climate Change

Despite these landmark initiatives, the ability of the ecosystem to sustain provision of goods and services is under renewed threat from climate change (Wilkinson, 2004). Climate change is rapidly emerging as one of the most significant challenges facing the GBR and its management. While MPA managers cannot directly control climate, and climate change cannot be fully averted, there is an urgent need to identify possibilities for reducing climate-induced stresses on the GBR (Marshall and Schuttenberg, 2006). The GBR Climate Change Response Program has undertaken an assessment of the vulnerability of the GBR to climate change and is developing strategies to enhance ecosystem resilience, sustain regional communities and industries that rely on the GBR, and provide supportive policy and collaborations.

The Climate Change Response Program used regional GBR climate projections to assess the vulnerability of species, habitats, and key processes to climate change. Some relevant projections emerged. Regional GBR sea temperatures have increased by 0.4°C since 1850 and are projected to increase by a further 1–3°C above present temperatures by 2100 (Fig. A6.2). Sea level rise is projected to be 30–60 cm by 2100, and ocean chemistry is projected to decrease in pH by 0.4–0.5 units by 2100 (Lough, 2007). There is less certainty about: changes to tropical cyclones, with a 5–12% increase in wind speed projected; rainfall and river flow, with projected increases in intensity of droughts and rainfall events; and ENSOs, which will continue to be a source of high interannual variability (Lough, 2007).

Coral Bleaching

The key threats to the GBR ecosystem from climate change manifest in impacts to all components of the ecosystem, from species to populations to habitats and key processes. Although coral reefs represent only 6% of the Great Barrier Reef, they are an iconic component of the system and support a diversity of life. Unusually warm summers caused significant coral bleaching events in the GBR in 1998, 2002, and 2006. More than 50% of reefs were affected by bleaching in the summers of 1998 and 2002, following persistent high sea temperatures throughout the GBR. Fortunately, temperatures cooled soon enough to avoid catastrophic impacts, yet approximately 5% of reefs suffered long-term damage in each year. Stressful temperatures were confined to the southern parts of the GBR in the summer of 2006 and persisted long enough to cause over 40% of the corals to die. Future warming of the world's oceans is projected to increase the frequency and severity of coral bleaching events, making further damage to the GBR inevitable (Hoegh-Guldberg *et al.*, 2007). Continued monitoring efforts—such as those proposed in the GBR Coral Bleaching Response Plan—will be essential for understanding this ecosystem change.

Impacts to Species

Mass mortalities of seabirds and failures of nesting (death of all chicks) have been observed at several key seabird rookeries during anomalously warm summers on the GBR (coinciding with mass coral bleaching). New research is showing that provisioning failure, resulting when adults have to travel too far to find food for their chicks, causes these deaths (Congdon *et al.*, 2007). This is thought to be due to decreased availability of food fish caused by changes in circulation patterns (location and depth of cool water bodies preferred by these fish). Marine turtles are also at risk from climate change, with increasing air temperatures projected to alter the gender ratio of turtle hatchlings; during periods of extremely high temperatures in the past, complete nesting

[59] Number 69 of 1997
[60] Number 91 of 1999

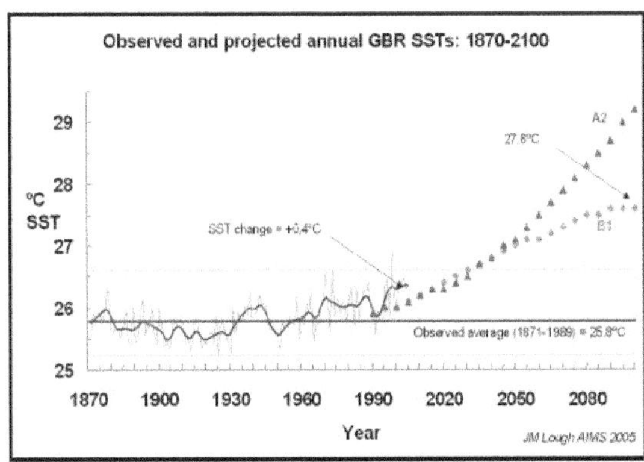

Figure A[illegible] Sea surface temperature (SST) projections for the Great Barrier Reef (GBR) (Lough, 2007).

failures have been observed. Sea level rise also poses a threat to seabirds and turtles, as nesting islands and beaches become inundated and suitability of alternative beaches is reduced by coastal development.

Fish, shark, and ray populations will be most affected by reductions in reef habitat, with resultant decreases in diversity and abundance and changes in community composition (Munday *et al.*, 2007; Chin *et al.*, 2007). Conversely, small increases in sea temperature may benefit larval fish by accelerating embryonic and larval growth and enhancing larval swimming ability. This shows that climate change will not affect all organisms equally, and some populations or groups (such as macroalgae) may in fact benefit by increasing their range or growth rate. However, this will change the distributions of species as they migrate southward or offshore. This in turn would likely result in population explosions of fast growing, 'weed-like' species to the detriment of other species, thereby reducing species diversity. As species and habitats decline, so too does the productivity of the system and its ability to respond to future change.

Impacts to Key Processes

The reef matrix itself is at risk from climate change through loss of coral—not only from coral bleaching but also physical damage from more intense storms and cyclones and reduced coral calcification rates as ocean pH decreases. This is critical from the perspective of the structural integrity of the GBR as well as the services reefs provide to other organisms, such as habitat and food.

Primary productivity, through changes to microbial, plankton, and seagrass communities, is likely to be affected as changes in the carbon cycle occur. Changes in rainfall patterns, runoff, and sea temperature also are likely to change plankton, seagrass, and microbial communities. These changes reduce trophic efficiency, which decreases food quality and quantity for higher trophic levels with a resultant decline in abundance of animals at higher trophic levels. Productivity is also likely to be sensitive to changes in ocean circulation as nutrient transport patterns change, thereby reducing nutrient availability and primary production.

Connectivity is at risk from changes to ocean circulation patterns and ENSO; as ocean currents and upwelling are affected, so too will be the hydrological cycles that transport material latitudinally and across the shelf. Connectivity will also be affected by coastal changes such as sea level rise and altered rainfall regimes, which are likely to have the most influence on coastal connectivity between estuaries and the inshore lagoon of the GBR. As temperature-induced stratification reduces wind-driven upwelling, offshore hydrological cycles are affected, potentially reducing connectivity between offshore reefs. All these changes could interact to affect the survival and dispersal patterns of larvae between reefs.

As biodiversity and connectivity are lost, the system becomes less complex, which initiates a cascade of events that results in long-term change. Simplified systems are generally less resilient and therefore less able to absorb shocks and disturbances while continuing to maintain their original levels of function. Reducing biodiversity and connectivity reduces the number of components and networks that can buffer against poor water quality, overfishing, and climate change. Maintaining a healthy ecosystem requires that ecological processes be preserved and that there is sufficient biodiversity to respond to changes. Larger marine protected areas that include representative habitats and protect biodiversity and connectivity may be more resilient to climate change (Roberts *et al.*, 2006).

A6.2.3 Adapting Management to Climate Change

In the face of these potential climate change impacts, the GBR Climate Change Response Program developed a Climate Change Action Plan in 2006. The action plan has five main objectives:

1. Address climate change knowledge gaps
2. Communicate with and educate communities about climate change implications for the GBR
3. Support greenhouse gas emissions mitigation strategies in the GBR region
4. Enhance resilience of the GBR ecosystem to climate change
5. Support GBR communities and industries to adapt to climate change

Key strategies within the action plan include assessing the vulnerability of the GBR ecological and social systems to climate change; developing an agency-wide communication strategy for climate change; facilitating greenhouse gas emissions reductions using the Reef Guardian incentive project; undertaking resilience mapping for the entire GBR and reviewing management arrangements in light of the relative resilience of areas of the GBR; and working with industries to promote industry-led initiatives to address climate change.

Addressing Information Gaps

The Great Barrier Reef Marine Park Authority (GBRMPA) has been working with scientists to assess the vulnerability of the different components of the GBR ecosystem, industries, and communities to climate change. A resultant publication identifies the key vulnerabilities for all components of the ecosystem, from plankton to corals to marine mammals, and makes management recommendations that aim to maximize the ability of the system to resist or adapt to climate changes (Johnson and Marshall, 2007). Examples of management recommendations include addressing water quality in inshore areas where primary productivity is high (*e.g.*, areas with extensive seagrass meadows or with critical plankton aggregations). Another example is conserving landward areas for migration of mangroves and wetlands as sea level rises, including possible land acquisitions and removal of barrier structures. Finally, protecting sites of specific importance from coral bleaching through shading or water mixing in summer months is an option. Reducing other impacts on critical habitats or species is also recommended (*e.g.*, improving shark fisheries management, reducing disturbance of seabird nesting sites during breeding season, reducing boat traffic and entanglement of marine mammals, protecting key turtle nesting beaches, enhancing resilience of coral reefs by improving water quality, protecting herbivores, and managing other destructive activities such as anchoring and snorkeling). These recommendations will be used to review existing management strategies and incorporate climate change considerations where needed.

Raising Awareness and Changing Behavior

The Climate Change Response Program developed a communication strategy in 2004 that aims to increase public awareness of the implications of climate change for the GBR. This strategy is being amended to include all GBRMPA activities and ensure that all groups consistently present key climate change messages. This is particularly important for groups that are addressing those factors that confer resilience to the ecosystem, such as water quality and fisheries. The key messages of the agency-wide communication strategy are that climate change is real, climate change is happening now, climate change is affecting the GBR, the GBRMPA is working to address climate change, and individuals' actions can make a difference.

The Reef Guardian program is a partnership with schools and local governments in GBR catchments. The program is voluntary and provides resources for schools and councils to incorporate sustainability initiatives into their everyday business. A sustainability and climate change syllabus has been developed for primary schools and will teach students about climate change and the implications for the GBR, as well as provide greenhouse gas emission reductions projects for the schools. The local council participants have been provided with similar information, and in order to be a recognized Reef Guardian, a council must implement a minimum number of sustainability modules. This partnership currently has 180 schools and is incrementally working toward having 20 local councils participating by 2010.

Toward Resilience-Based Management

One of the most significant strategies that coral reef managers can employ in the face of climate change is to enhance the resilience of the ecosystem (West *et al.*, 2006). Working with researchers, the Climate Change Response Program has identified resilience factors that include water quality, coral cover, community composition, larval supply, recruitment success, herbivory, disease, and effective management. These will be used to identify areas of the GBR that have high resilience to climate change and should be protected from other stresses, as well as areas that have low resilience and may require active management to enhance their resilience. Recognized research institutes have provided essential science that has formed the basis of this project and will continue collaborations between GBRMPA and researchers. Ultimately, it is hoped that this information can be used to review existing management regimes (such as planning and permit tools) to protect areas with high resilience as source sites and actively work in areas with low resilience to improve their condition.

Partnering with Stakeholders

The GBRMPA has been working with the GBR tourism industry to facilitate development of the GBR Tourism and Climate Change Action Strategy. This initiative was the result of a workshop with representative tourism operators that generated the GBR Tourism and Climate Change Action Group. This industry-led group has developed the action strategy to identify how climate change will affect the industry, how the industry can respond, and what options are available for the industry to become climate sustainable. The marine tourism industry considers reef-based activities particularly susceptible to the effects of climate change. Loss of coral from bleaching and changes to the abundance and location of fish, marine mammals, and other iconic species are likely to have the greatest impact on the industry. Increasing intensity of cyclones and storms will affect trip scheduling, industry seasonality, tourism infrastructure (particularly on islands), and future tourism industry development. Potential strategies for adapting to climate change include product diversification, new marketing initiatives, and targeting eco-accredited programs.

Managing Uncertainty

A critical component of all these strategies is the ability to manage flexibly and respond to change rapidly. This is important to enable managers to shift focus as new information becomes available or climate impact events occur. In reviewing existing management regimes, there will be a focus on ways of making management more flexible and drawing on management tools as they are needed. This type of adaptive management is essential for addressing the uncertain and shifting climate change impacts on the GBR. Given the scale of the issue and the fact that the cause and many of the solutions lie outside the jurisdiction of GBRMPA managers, effective partnerships with other levels of government and stakeholders to work cooperatively on climate change have been developed and will continue to be integral to adapting management to the climate change challenge.

A6.3 Papahānaumokuākea Northwestern Hawaiian Islands Marine National Monument

A6.3.1 Introduction

The Hawaiian Islands are one of the most isolated archipelagos in the world and stretch for over 2,500 km, from the island of Hawaii in the southeast to Kure Atoll (the world's highest-latitude atoll) in the northwest (Grigg, 1982; 1988; Friedlander *et al.*, 2005). Beginning at Nihoa and Mokumanamana Islands (~7 and 10 million years old, respectively) and extending to Midway and Kure Atolls (~28 million years old), the Northwestern Hawaiian Islands (NWHI) represent the older portion of the emergent archipelago (Grigg, 1988). The majority of the islets, shoals, and atolls are low-lying and remain uninhabited, although Midway, Kure, Laysan Island, and French Frigate Shoals have all been occupied for extended periods over the last century by various government agencies (Shallenberger, 2006). Because of their location in the central Pacific, the NWHI are influenced by large-wave events resulting from extratropical storms passing across the North Pacific each winter that have a profound influence on the geology and biology of the region (Grigg, 1998; Dollar and Grigg, 2004; Jokiel *et al.*, 2004; Friedlander *et al.*, 2005).

Ecosystem Structure

With coral reefs around the world in decline (Jackson *et al.*, 2001; Bellwood *et al.*, 2004; Pandolfi *et al.*, 2005), it is extremely rare to be able to examine a coral reef ecosystem that is relatively free of human influence and consisting of a wide range of healthy coral reef habitats. The remoteness and limited reef fishing and other human activities that have occurred in the NWHI have resulted in minimal anthropogenic impacts (Friedlander and DeMartini, 2002; Friedlander *et al.*, 2005). The NWHI therefore provide a unique opportunity to assess how a "natural" coral reef ecosystem functions in the absence of major localized human intervention.

One of the most striking and unique components of the NWHI ecosystem is the abundance and dominance of large apex predators such as sharks and jacks (Friedlander and DeMartini, 2002; DeMartini, Friedlander, and Holzwarth, 2005). These predators exert a strong top-down control on the ecosystem (DeMartini, Friedlander, and Holzwarth, 2005; DeMartini and Friedlander, 2006) and have been depleted in most other locations around the world (Myers and Worm, 2003; 2005). Differences in fish biomass between the main Hawaiian Islands (MHI) and NWHI represent both near-extirpation of apex predators and heavy exploitation of lower-trophic-level fishes on shallow reefs of the MHI (Friedlander and DeMartini, 2002; DeMartini and Friedlander, 2006).

The geographic isolation of the Hawaiian Islands has resulted in some of the highest endemism of any tropical marine ecosystem on earth (Jokiel, 1987; Kay and Palumbi, 1987; Randall, 1998) (Fig. A6.3). Some of these endemics are a dominant component of the community, resulting in a unique ecosystem that has extremely high conservation value (DeMartini and Friedlander, 2004; Maragos *et al.*, 2004). With species loss in the sea accelerating, the irreplaceability of these species makes Hawaii an important biodiversity hotspot (Roberts *et al.*, 2002; DeMartini and Friedlander, 2006). The coral assemblage in the NWHI contains a large number of endemics (~30%), including at least seven species of acroporid corals (Maragos *et al.*, 2004). Acroporids are the dominant reef-building corals in the Indo-Pacific, but are absent from the MHI (Grigg, 1981; Grigg, Wells, and Wallace, 1981). Kure Atoll is the world's most northern atoll and is referred to as the Darwin Point, where coral growth, subsidence, and erosion balance one another (Grigg, 1982).

The NWHI represent important habitat for a number of threatened and endangered species. The Hawaiian monk seal is one of the most critically endangered marine mammals in the United States (1,300 individuals) and depends almost entirely on the islands of the NWHI for breeding and the surrounding reefs for sustenance (Antonelis *et al.*, 2006). Over 90% of all sub-adult and adult Hawaiian green sea turtles found throughout Hawaii inhabit the NWHI (Balazs and Chaloupka, 2006). Additionally, seabird colonies in the NWHI constitute one of the largest and most important assemblages of seabirds in the world (Friedlander *et al.*, 2005).

In contrast to the MHI, the reefs of the NWHI are relatively free of major human influences. The few alien species known from the NWHI are restricted to the anthropogenic habitats of Midway Atoll and French Frigate Shoals (Friedlander *et al.*, 2005). Disease levels in corals in the NWHI were much lower than those reported from other locations in the Indo-Pacific (Aeby, 2006).

Existing Stressors

Although limited in scale, a number of past and present human activities have negatively affected the NWHI. Marine debris is currently one of the largest threats to the reefs of the NWHI (Boland *et al.*, 2006; Dameron *et al.*, 2007). Marine debris has caused entanglement of a number of protected species and damage to benthic habitats and is a potential vector for invasive species in the NWHI (Dameron *et al.*, 2007). An extensive debris removal effort between 1999 and 2003 has now surpassed the accumulation rate, resulting in a reduction in overall accumulation levels (Boland *et al.*, 2006). However, much of this debris originates thousands of kilometers away in the north Pacific, making the solution to the problem both a national and international issue. Other direct human stresses such as pollution, coastal development, and ship groundings, have had negative consequences in localized areas but have been limited to a small number of locations.

The NWHI are influenced by a dynamic environment that includes large annual water temperature fluctuations, seasonally high wave energy, and strong inter-annual and inter-decadal variations in ocean productivity (Polovina *et al.*, 1994; Grigg, 1998; Polovina *et al.*, 2001; Friedlander *et al.*, 2005). As a result of these influences, natural stressors play an important role in the structure of the NWHI ecosystem. Large swell events generated every winter commonly produce waves up to 10–12 m in vertical height and between 15–20 m about once every decade (Grigg *et al.*, 2007). This limits the growth and abundance of coral communities, particularly on the north and western sides of all the islands. The best-developed reefs on all the islands exist either in the lagoons or off southwestern exposures (Grigg, 1982).

Summer sea surface temperatures (SSTs) along the island chain are generally similar, peaking at about 28°C; however, winter SSTs are much cooler at the northern end of the chain, dipping down to 17°C in some years (Grigg, 1982; Grigg *et al.*, 2007). This represents a 10°C intra-annual difference at the northern end of the chain, while that at the southern end of the NWHI is only half as great: 5°C (22–27°C).

Figure A[illegible] Endemic species from the Hawaiian Islands. A. Masked angelfish, *Genicanthus personatus* (Photo courtesy of J. Watt), B. Rice coral, *Montipora capitata*, and finger coral, *Porites compressa* (photo courtesy of C. Hunter), C. Hawaiian hermit crab, *Calcinus laurentae* (photo courtesy of S. Godwin), D. Red alga, *Acrosymphtyon brainardii* (photo courtesy of P. Vroom).

Compared with most reef ecosystems around the globe, the annual fluctuations of SST of about 10°C at these northerly atolls is extremely high. Cooler water temperatures to the north restrict the growth and distribution of a number of coral species (Grigg, 1982). In addition, the biogeographic distribution of many fish species in the NWHI is influenced by differences in water temperatures along the archipelago (DeMartini and Friedlander, 2004; Mundy, 2005).

Climate Sensitivity

The NWHI ecosystem is sensitive to natural climate variability at a number of spatial and temporal scales. The Pacific Decadal Oscillation (PDO) results in changes in ocean productivity at large spatial and long temporal scales and has been attributed to changes in monk seal pup survival, sea bird fledging success, and spiny lobster recruitment in the NWHI (Polovina *et al.*, 1994; Polovina, Mitchem, and Evans, 1995). Inter-annual variation in the Transition Zone Chlorophyll Front is also known to affect the distribution and survival of a number of species in the NWHI (Polovina *et al.*, 1994; Polovina *et al.*, 2001).

Because of their high latitude location in the central Pacific, the NWHI were thought to be one of the last places in the world to experience coral bleaching (Hoegh-Guldberg, 1999). Hawaiian reefs were unaffected by the 1998 mass bleaching event that affected much of the Indo-Pacific region (Hoegh-Guldberg, 1999; Reaser, Pomerance, and Thomas, 2000; Jokiel and Brown, 2004). The first documented bleaching event in the MHI was reported in 1996 (Jokiel and Brown, 2004). The NWHI were affected by mass coral bleaching in 2002 and again in 2004 (Aeby *et al.*, 2003; Kenyon *et al.*, 2006). Bleaching was most acute at the three northern-most atolls (Pearl and Hermes, Midway, and Kure) and was most severe on backreef habitats (Kenyon and Brainard, 2006). Of the three coral genera that predominate at these atolls, *Montipora* and *Pocillopora* spp. were most affected by bleaching, with lesser incidences observed in *Porites* (Kenyon and Brainard, 2006). The occurrence of two mass bleaching episodes in three years lends credence to the projection of increased frequency of bleaching with climate change.

SST data derived from both remotely sensed satellite observations (Fig. A6.4a) as well as in situ Coral Reef Early Warning System (CREWS) buoys suggest that prolonged, elevated SSTs combined with a prolonged period of anomalously light wind speed led to decreased wind and wave mixing of the upper ocean (Hoeke *et al.*, 2006) (Fig. A6.4b). The reefs to the southeast of the archipelago show smaller positive temperature anomalies compared with the reefs toward the northwest. Research and monitoring efforts should target this pattern to better understand dispersal, bleaching, and other events that might be affected by it.

Potential Impacts of Climate Change

Climate change may increase the intensity of storm events as well as result in changes in ocean temperature, circulation patterns, and water chemistry (Cabanes, Cazenave, and Le Provost, 2001; IPCC, 2001; Caldeira and Wickett, 2003). Warmer temperatures in Hawaii have been shown to cause bleaching mortality (Jokiel and Coles, 1990) and negatively affect fertilization and development of corals. Annual spawning of some species in Hawaii occurs at temperatures near the upper limit for reproduction, so increases in ocean temperature related to climate change may have a profound effect on coral populations by causing reproductive failure. The rate and scale at which bleaching has been increasing in recent decades (Glynn, 1993) points to the likelihood of future bleaching events in Hawaii (Jokiel and Coles, 1990).

Coral disease is currently low in the NWHI (Aeby, 2006), but increases in the frequency and intensity of bleaching events will stress corals and make them more susceptible to disease (Harvell *et al.*, 1999; Harvell *et al.*, 2002). Acroporid corals are prone to bleaching and disease (Willis, Page, and Dinsdale, 2004) and are restricted in range and habitat within the Hawaiian Archipelago to a few core reefs in the NWHI (Grigg, 1981; Grigg, Wells, and Wallace, 1981; Maragos *et al.*, 2004). This combination could lead to the extinction of this genus from Hawaii if mortality associated with climate change becomes severe.

Most of the emergent land in the NWHI is low-lying, highly vulnerable to inundation from storm waves, and therefore vulnerable to sea level rise (Baker, Littnan, and Johnston, 2006). The limited amount of emergent land in the NWHI is critical habitat for the endangered Hawaiian monk seal (Antonelis *et al.*, 2006), the threatened green sea turtle (Balazs and Chaloupka, 2006), and numerous terrestrial organisms and land birds that are found nowhere else on Earth (Rauzon, 2001). The emergent land in the NWHI may shrink by as much as 65% with a 48 cm rise in sea level (Baker, Littnan, and Johnston, 2006). Efforts such as translocation or habitat alteration might be necessary if these species are to be saved from extinction.

At the northern end of the chain, lower coral diversity is linked to lower winter temperatures and lower annual solar radiation (Grigg, 1982). Increases in ocean temperature could therefore change the distribution of corals and other organisms that might currently be limited by lower temperatures. Many shallow-water fish species that are adapted to warmer water are restricted from occurring in the NWHI by winter temperatures that can be as much as 7°C cooler than the MHI (Mundy, 2005). Conversely, some shallow-water species are adapted to cooler water and can be found in deeper waters at the southern end of the archipelago. This phenomenon—known as tropical submergence—is exemplified by species such as the yellowfin soldierfish (*Myripristis chrysonemus*), the endemic Hawaiian grouper (*Epinephelus quernus*), and the masked angelfish (*Genicanthus personatus*), which are found in shallower water at Midway and/or Kure atolls, but are restricted to deeper depths in the MHI (Randall *et al.*, 1993; DeMartini and Friedlander, 2004; Mundy, 2005).

Level/Degree of Management

Administrative jurisdiction over the islands and marine waters is shared by NOAA/NMSP, U.S. Fish and Wildlife Service, and the State of Hawaii. Eight of the 10 NWHI (except Kure and Midway Atolls) have been protected by what is now the Hawaiian Islands National Wildlife Refuge (HINWR) established by President Theodore Roosevelt in 1909. The Northwestern Hawaiian Islands Coral Reef Ecosystem Reserve was created by Executive Orders 13178 and 13196 in December 2000 and amended by Executive Order 13196 in January of 2001 to include the marine waters and submerged lands extending 1,200 nautical miles long and 100 nautical miles wide from Nihoa Island to Kure Atoll.

In June 2006, nearly 140,000 square miles of the marine environment in the NWHI was designated as the Papahānaumokuākea (Northwestern Hawaiian Islands) Marine National Monument (PMNM). This action provided immediate and permanent protection for the resources of the NWHI and established a management structure that requires extensive collaboration and coordination among the three primary co-trustee agencies: the State of Hawaii, the U.S. Fish and Wildlife Service, and NOAA.

Proclamation 8031 states that the monument will:

- Preserve access for Native Hawaiian cultural activities;
- Provide for carefully regulated educational and scientific activities;
- Enhance visitation in a special area around Midway Island;
- Prohibit unauthorized access to the monument;
- Phase out commercial fishing over a five-year period; and
- Ban other types of resource extraction and dumping of waste.

Preservation areas have been established in the PMNM in sensitive areas around all the emergent reefs, islands, and atolls. Vessels issued permits to operate in the PMNM are required to carry approved Vessel Monitoring Systems (VMS).

Program of Monitoring and Research

Long-term monitoring relevant to climate change has been conducted in the NWHI dating back to the 1970s by a variety of agencies (Grigg, 2006). Since 2000, a collaborative interagency monitoring program led by the Coral Reef Ecosystem Division (CRED) of the NOAA Pacific Islands Science Center has conducted integrated assessment and monitoring of coral reef ecosystems in the NWHI and throughout the U.S. Pacific (Wadell, 2005; Friedlander *et al.*, 2005). In conjunction with various state, federal, and academic partners, this program has integrated ecological studies with environmental data to develop a comprehensive ecosystem-based program of assessment and monitoring of U.S. Pacific coral reef ecosystems.

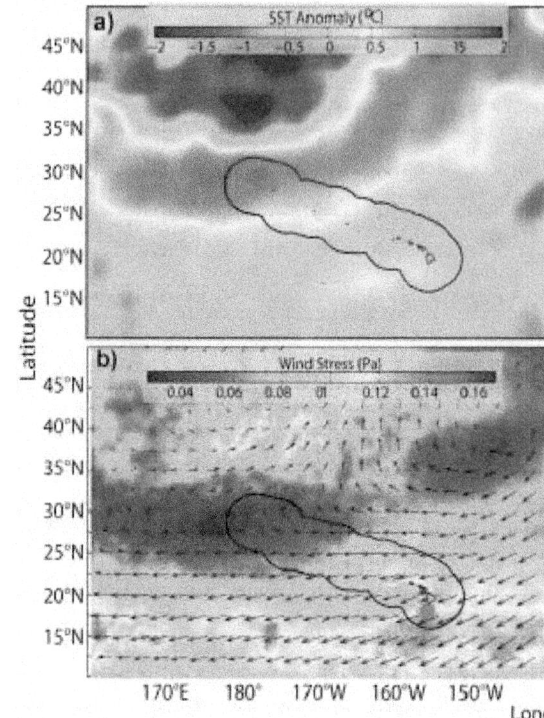

Figure A[illegible] a) NOAA Pathfinder SST anomaly composite during summer 2002 period of NWHI elevated temperatures, July 28–August 29. b) NASA/JPL Quikscat winds (wind stress overlayed by wind vector arrows) composite during summer 2002 period of increasing SSTs, July 16–August 13. The Hawaii Exclusive Economic Zone (EEZ) is indicated with a heavy black line; all island shorelines in the archipelago are also plotted (adapted from Hoeke *et al.*, 2006).

Ocean currents are measured and monitored in the NWHI using shipboard acoustic Doppler current profilers (ADCP), Surface Velocity Program (SVP) current drifters, and APEX profiling drifters (Friedlander *et al.*, 2005; Firing and Brainard, 2006). Spatial maps of ocean currents in the vicinity of the NWHI are also computed from satellite observations of sea surface height from the TOPEX-Poseidon and JASON altimetric satellites (Polovina, Kleiber, and Kobayashi, 1999). Moored ADCPs have been deployed by CRED at several locations to examine temporal variability of ocean currents over submerged banks and reef habitats in the NWHI.

Because of the significant influence of temperature on coral reef ecosystem health, observations of temperature in the NWHI are collected by a wide array of instruments and platforms, including satellite remote sensing (AVHRR) of SST (Smith and Reynolds, 2004), moored surface buoys and subsurface temperature recorders, closely spaced shallow water conductivity-temperature-depth profiles (CTD casts) in nearshore reef habitats, broadly spaced shipboard deep water CTD casts to depths of 500 m, and satellite-tracked SVP drifters. These data are integrated in the Coral Reef Ecosystem Integrated Observing System (CREIOS) as described below.

A6.3.2 Managing the Papahānaumokuākea Marine National Monument

Current Approaches to Research and Monitoring in Support of Management and How Climate Change is Being Examined

Over the past several years, the NOAA Coral Reef Conservation Program has established the Coral Reef Ecosystem Integrated Observing System (CREIOS), which is a cross-cutting collaboration between four NOAA Line Offices (NMFS, OAR, NESDIS, and NOS) focused on mapping, monitoring, and observing ecological and environmental conditions of U.S. coral reefs. At present, the ocean observing system in the NWHI consists of surface buoys measuring SST, salinity, wind, atmospheric pressure, and air temperature (enhanced systems also measure ultraviolet-B (UV-B) and photosynthetically available radiation); surface SST buoys; subsurface Ocean Data Platforms measuring ocean current profiles, wave energy and direction, temperature and salinity; subsurface current meters measuring bottom currents and temperature; and subsurface temperature recorders. Many of the surface platforms provide near real-time data telemetry to the Pacific Islands Fisheries Science Center and subsequent distribution via the World Wide Web. Time series data from subsurface instruments (without telemetry) are typically available every 12 to 24 months, after the instrument has been recovered and the dataset uploaded. Information about available datasets such as geo-location, depth, data format, and other metadata are available for both surface and subsurface instruments at the NOAA Coral Reef Information System (CoRIS) website.[61]

Another component of CREIOS is Coral Reef Watch (NESDIS, Office of Research and Applications) which uses remote sensing, computational algorithms, and artificial intelligence tools in the near real-time monitoring, modeling, and reporting of physical environmental conditions that adversely influence coral reef ecosystems. Satellite remotely sensed data products include near real-time identification of bleaching "hotspots" and identification of low-wind (doldrums) areas over the world's oceans. The CRED long-term moored observing stations are part of the Coral Reef Early Warning System (CREWS) network initiated by the NOAA Coral Health and Monitoring Program, which provides access to near real-time meteorological and oceanographic data from major U.S. coral reef areas. The CREWS buoys deployed by CRED in the NWHI record and telemeter data pertaining to sea-surface temperature, salinity, wind speed and direction, air temperature, barometric pressure, UV-B, and photosynthetically available radiation (Kenyon *et al.*, 2006).[62]

Information from CREIOS serves to alert resource managers and researchers to environmental events considered significant to the health of the surrounding coral reef ecosystem, allowing managers to implement response measures in a timely manner, and allowing researchers to increase spatial or temporal sampling resolution, if warranted. Response measures might include focused monitoring to determine the extent and duration of the event and management actions could include limiting access to these areas until recovery is observed. Information from the Coral Reef Watch Program in summer 2002 indicated conditions favorable for bleaching and resulted in assessments focused on potential bleaching areas during the subsequent research cruise.

Potential for Altering or Supplementing Current Management Practices to Enable Adaptation to Climate Change

To more fully address concerns about the ecological impacts of climate change on coral reef ecosystems and the effect of reef ecosystems on climate change, a number of agencies have proposed a collaborative effort to establish a state-of-the-art ocean observing system to monitor the key parameters of climate change impacting reef ecosystems of the Pacific and Western Atlantic/Caribbean. This proposed system includes:

- Expanding the existing array of oceanographic platforms across the remainder of the U.S. Pacific Islands
- Installing pCO_2 and UV-B sensors to examine long-term changes in carbon cycling and UV radiation
- Establish long-term records of coral reef environmental variability to examine past climate changes using paleoclimatic records of SST and other parameters from coral skeletons. This will allow us to determine if current and future SST stresses are unusual, or part of natural climatic variability.
- Develop/expand integrated *in situ* and satellite based bleaching mapping system
- Continue the development of the Coral Reef Early Warning System, which can be used to develop timely research

[61] **National Oceanic and Atmospheric Administration**, 2007: NOAA's coral reef information system. NOAA Website, http://www.coris.noaa.gov/, accessed on 7-27-2007.

[62] **NOAA National Marine Fisheries Service**, 2007: Coral reef ecosystems - ecological assessment, marine debris removal, oceanography, habitat mapping. NOAA Website, http://www.pifsc.noaa.gov/cred/, accessed on 5-24-2007.

activities to determine the extent and duration of any climate event and management actions that can potentially be implemented to mitigate these events.

In order to better understanding the impact of sea level rise on low-lying emergent areas in the NWHI, data are needed on hydrodynamic and geological characteristics of the region. Detailed information on elevation, bathymetry, waves, wind, tide, etc. is needed to develop predictive models of shoreline change relative to climate change. One possible management measure to counter loss of habitat for monk seals and turtles in the NWHI due to sea level rise might be beach nourishment (Baker, Littnan, and Johnston, 2006). Given the small size of the islets in the NWHI, local sand resources might be sufficient to mitigate sea level rise, but a great deal of research and planning would be required given the remoteness and sensitive nature of the ecosystem (Baker, Littnan, and Johnston, 2006).

A6.3.3 Adapting Management to Climate Change

The draft Monument Management Plan does not address climate and ocean change management actions specifically, but by integrating strategies that focus on climate through research and monitoring, education and outreach, and review and syntheses, management will be better informed and prepared to deal with issues related to climate and ocean change. A comprehensive understanding of the effects of climate change on the NWHI is needed in order to provide managers with the information and tools needed to address these effects. Specific attention should be given to the effects on habitats critical to endemic and protected species.

The continued development and expansion of the Coral Reef Early Warning System and the Ocean Observing System are critical to improve understanding of climate change in the PMNM and the scale and capabilities of these systems should be enhanced. Investigations directed at examining the physiological, ecological, and genetic responses of the entire ecosystem to climate change should be conducted. Continuation and expansion of monitoring programs are important to better understand the ecosystem in time and space and higher-intensity spatial and temporal monitoring and assessment should be initiated in conjunction with disturbance events (*e.g.*, coral bleaching, disease outbreaks, elevated water temperatures).

The draft PMNM science plan calls for a number of specific research activities to examine the effects of climate change on the NWHI ecosystem.

- Determine the effect of climate change on nesting sites of protected species, *e.g.* the effect of sea level rise on nesting site of the green sea turtle and Hawaiian monk seal.
- Determine specific habitats, communities, and populations that will be affected by global climate change (ocean acidification, sea level, temperature, chlorophyll fronts, etc.).
- Understand habitat changes that will result from sea level rise.
- Map areas that will be most affected by extreme wave events.
- Discern anthropogenic impacts from natural variability of the physical environment.

PMNM constituency building and outreach plans should emphasize climate change in its various venues of information dissemination (*e.g.*, websites, brochures, fact sheets, school presentations, meetings, workshops, etc.). Building upon existing NWHI-based curricula developed under the Navigating Change Partnership and the new Hawaii Marine Curriculum, specific study units on climate change should be developed and impacts of climate change incorporated into other study units, where appropriate. By increasing the public's awareness of climate change impacts, the PMNM can provide a societal benefit that extends beyond the boundaries of the monument.

A6.3.4 Conclusions

The nearly pristine condition of the NWHI results in one of the last large-scale, intact, predator-dominated reef ecosystems remaining in the world (Friedlander and DeMartini, 2002; Pandolfi *et al.*, 2005). Top predators can regulate the structure of the entire community and have the potential to buffer some of the ecological effects of climate change (Sala, 2006). Intact ecosystems such as the NWHI are hypothesized to be more resistant and resilient to stressors, including climate change (West and Salm, 2003). Owing to its irreplaceable assemblage of organisms, it possesses extremely high conservation value. The Papahānaumokuākea Marine National Monument is the largest marine protected area (MPA) in the world and provides a unique opportunity to examine the effects of climate change on a nearly intact large-scale marine ecosystem.

A6.4 The Channel Islands National Marine Sanctuary

A6.4.1 Introduction

Ecosystem Structure

Designated in 1980, the Channel Islands National Marine Sanctuary (CINMS) consists of an area of approximately 1,243 nm² of coastal and ocean waters and submerged lands off the southern coast of California (Fig. A6.5). CINMS extends 6 nm offshore from the five northern Channel Islands, including San Miguel, Santa Cruz, Santa Rosa, Anacapa, and Santa Barbara islands. The primary objective of the sanctuary is to conserve, protect, and enhance the biodiversity, ecological integrity, and cultural legacy of marine resources surrounding the Channel Islands for current and future generations. State and federal agencies with overlapping jurisdiction in the CINMS, including the California Department of Fish and Game, the Channel Islands National Park, and the National Marine Fisheries Service, are working together to manage impacts of human activities on marine ecosystems.

The Channel Islands are distributed across a biogeographic boundary between cool temperate waters of the Californian Current and warm temperate waters of the Davidson Current (or California Countercurrent). The California Current is characterized by coastal upwelling of cool, nutrient-rich waters that contribute to high biological productivity. Intertidal communities around San Miguel, Santa Rosa, and part of Santa Cruz islands are characteristic of the cool temperate region, whereas those around Catalina, San Clemente, Anacapa, and Santa Barbara islands are associated with the warm temperate region (Murray and Littler, 1981). Fish communities around the Channel Islands also show a distinctive grouping based on association with western islands (influenced strongly by the California Current) and eastern islands (influenced by the Davidson Current). Rockfish (*Sebastes* spp.), embiotocid species, and pile perch occur more in western islands while Island kelpfish (*Alloclinus holderi*), opaleye (*Girella nigricans*), garibaldi (*Hypsypops rubicundus*), blacksmith (*Chromis punctipinnis*), and kelp bass (*Paralabrax clathratus*) occur more often in the eastern islands (Halpern and Cottenie, 2007).

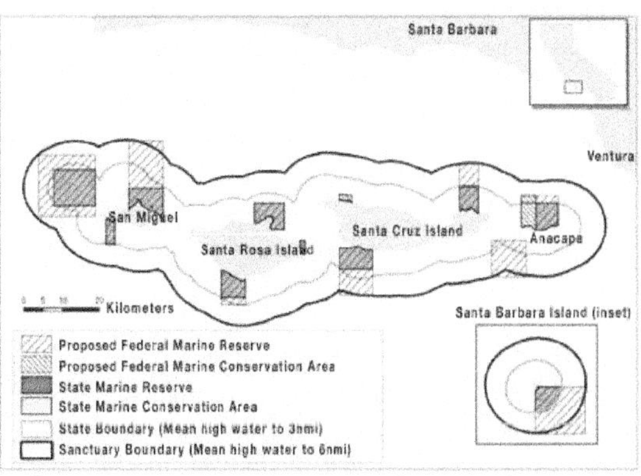

NOAA's preferred alternative for marine zones in the Sanctuary.

Figure A.5 Map of the Channel Islands National Marine Sanctuary showing the location of existing state and proposed federal marine reserves and marine conservation areas.[71]

From Monterey Bay to Baja California, including the Channel Islands, giant kelp (*Macrocystis pyrifera*) is the dominant habitat-forming alga. Giant kelp grows in dense stands on hard rocky substrate at depths of 2–30 m (Foster and Schiel, 1985). Kelp is among the fastest growing of all algae, adding an average of 27 cm/day (in spring) and a maximum of 61 cm/day and reaching lengths of 60 m (200 ft). Giant kelp forests support a diverse community of associated species including marine invertebrates, fishes, marine mammals and seabirds (Graham, 2004). Kelp stocks and fronds may support thousands of invertebrates including amphipods, decapods, polychaetes, and ophiuroids. Some invertebrates such as sea urchins (*Strongylocentrotus* spp.) and abalone (*Haliotis* spp.) rely on bits of drifting kelp as their primary source of food. Fish in the kelp forest community specialize in life at different depths: kelp, black and yellow, and gopher rockfish are found at the base of kelp stocks, while olive, yellowtail, and black rockfish swim in mid-water. Drifting kelp mats at the sea surface provide cover for young fishes that are vulnerable to predation. Marine mammals and seabirds are attracted to abundant fish and invertebrate populations (which serve as their primary prey) associated with kelp forests. Because of their high diversity, California kelp forests are thought to be more resistant and resilient to disturbance than kelp forests elsewhere (Steneck *et al.*, 2002).

Stressors on Marine Ecosystems in the Channel Islands

Kelp forest communities are vulnerable to an array of stressors caused by human activities and natural environmental variation. Using data gathered by the Channel Islands National Park over a period of 20 years, Halpern and Cottenie (2007) documented overall declines in abundance of giant kelp communities over time. These declines were linked with commercial and recreational fishing in the Channel Islands. Overfishing reduces density and average individual size of targeted populations and, consequently, targeted species are more vulnerable to the effects of natural environmental variation. Overfishing also has cascading effects through the marine food web. In areas of the Channel Islands where lobster (*Panulirus interruptus*) and other top predators were fished, purple sea urchin (*Strongylocentrotus purpuratus*) populations were more abundant, overgrazing stands of giant kelp and other algae and resulting in barren reefs devoid of kelp and its associated species (Behrens and Lafferty, 2004).

Kelp forest communities also respond to natural environmental variations, such as increased storm intensity, ocean warming, and shifts in winds associated with ENSO events (Dayton *et al.*, 1992; Ladah, Zertuche-Gonzalez, and Hernandez-Carmona, 1999; Edwards, 2004). Storm intensity, which is known to increase during periods of ocean warming, damages kelp stocks and rips kelp holdfasts from their rocky substrate (Dayton *et al.*, 1992; 1999). In addition to the physical damage from storms, kelp growth may be suppressed by lower levels of nutrients due to relaxation of coastal wind activity and reduction of upwelling during

[63] **Channel Islands National Marine Sanctuary**, 2007: Marine reserves environmental review process. NOAA Website, NOAA, http://channelislands.noaa.gov/marineres/main.html, accessed on 7-1-2007.

ENSO events. Giant kelp forests were decimated during the intense ENSO event of 1982–83 and did not recover to their previous extent for almost two decades. Several other ENSO events, in 1992–93 and 1997–98 also diminished kelp growth. The effects of these ENSO events may have been compounded by a shift (Pacific Decadal Oscillation) in 1977 to a period of slightly warmer waters in the northeastern Pacific Ocean.

Dramatic declines of giant kelp communities are likely the consequence of cumulative impacts of human activities and natural environmental variation. Giant kelp forests in one marine reserve (where fishing has been prohibited since 1978) were more resilient to ocean warming, shifts in winds, and increased storm activity associated with ENSO (Behrens and Lafferty, 2004). Giant kelp forests in the reserve persisted over a period of 20 years, including several intense ENSO events. Kelp forests at all study sites outside of the reserve were overgrazed by dense populations of sea urchins, and their growth was further inhibited by warmer water, increased storm intensity, and lower levels of nutrients, leading to periodic die-backs to a barren reef state. These observations suggest that marine reserves can be used as a management tool to increase resilience of kelp forest communities.

Current Management of the Channel Islands

In 1999, the CINMS and the California Department of Fish and Game (CDFG) developed a partnership and public process (modeled after the Florida Keys National Marine Sanctuary) to consider the use of fully protected marine reserves to protect natural biological communities (Box A6.2). The cooperating agencies engaged a working group of stakeholders through the Sanctuary Advisory Council to evaluate the problem and develop potential solutions. The "Marine Reserves Working Group" developed a problem statement acknowledging that human activities and natural ecological changes contributed to the decline of marine communities in southern California. The working group determined that marine reserves should be established to protect marine habitats and species, to achieve sustainable fisheries and maintain long-term socioeconomic viability, and to protect cultural heritage. The stakeholders, working with marine scientists and economists, created a range of options for marine reserves to meet these goals. Subsequently, the CINMS and CDFG used the two most widely supported options to craft compromise solution that addressed the interests of a broad array of stakeholders.

In 2003, the CDFG established a network of 10 fully protected marine reserves and two conservation areas that allow limited commercial and recreational fishing (Fig. A6.5). The total area protected was 102 nm², approximately 10% of sanctuary waters. The marine reserves and conservation areas included a variety of representative marine habitats characteristic of the region, such as rocky intertidal habitats, sandy beaches, kelp forests, seagrass beds, soft bottom habitats, submerged rocky substrate, and submarine canyons. In 2006, the Pacific Fisheries Management Council designated Essential Fish Habitat to protect benthic communities from bottom contact fishing gear within and adjacent to the state marine protected areas, up to 6 nm offshore. In the same year, the CINMS released a Draft Environmental Impact Statement proposing complementary marine reserves and a marine conservation area extending into federal waters (Fig. A6.5). The Essential Fish Habitat designated by the Council and the marine protected areas proposed by the sanctuary increase the total area of protected marine zones to 19% of the CINMS.

In 2008, data from relevant monitoring programs will be prepared for a review by the California Fish and Game Commission of the first five years of monitoring the Channel Islands state marine reserves. Expectations are that species that were targeted by commercial or recreational fisheries will increase in density and size within marine reserves (Halpern, 2003). Some species are expected to decline if their predators or competitors increase in abundance.

□b□A□□□□Timeline for Establishment of Marine Reserves in the Channel Islands National Marine Sanctuary (CINMS)

- 1998: Sportfishing group initiates discussions about marine reserves in the Channel Islands National Marine Sanctuary
- 1999: California Department of Fish and Game and NOAA develop partnership and initiate community-based Marine Reserves Working Group process
- 2001: Working Group recommendations delivered to California Department of Fish and Game and NOAA
- 2003: California Fish and Game Commission established 10 state marine reserves and 2 state marine conservation areas in state waters of the CINMS
- 2006: Pacific Fisheries Management Council designated Essential Fish Habitat and Habitat of Areas of Particular Concern in adjacent federal waters of the CINMS prohibiting bottom fishing
- 2006: Sanctuary released Draft Environmental Impact Statement to propose marine reserves in federal waters of the CINMS.
- 2007: Pending - NOAA will release Final Environmental Impact Statement and final rule to complete the marine reserves in federal waters
- 2007: Pending - California Fish and Game Commission will take regulatory action to close gaps between state and federal marine protected areas

Potential Effects of Climate Change on Ecosystems in the Channel Islands Region

Coastal SST has increased steadily (by approximately 2°C) since 1950 and is expected to increase further in the coming centuries (IPCC, 2007a). Water temperature affects metabolism and growth (Bayne, Thompson, and Widdows, 1973; Phillips, 2005), feeding behavior (Petraitis, 1992; Sanford, 1999; 2002), reproduction (Hutchins, 1947; Philippart *et al.*, 2003), and rates of larval development (Hoegh-Guldberg and Pearse, 1995; Anil, Desai, and Khandeparker, 2001; Luppi, Spivak, and Bas, 2003; O'Connor *et al.*, 2007) of intertidal and subtidal animals. Shifts in species ranges already have occurred in California with the steady increase of coastal sea surface temperature. The range boundary of *Kelletia kelletii* has shifted north from the late 1970s to the 2000s (Herrlinger, 1981; Zacherl, Gaines, and Lonhart, 2003). Southern species of anthozoans, barnacles, and gastropods increased in Monterey Bay, while northern species of anthozoans and limpets decreased between the 1930s (Hewatt, 1937) and the 1990s (Barry *et al.*, 1995; Sagarin *et al.*, 1999). Holbrook, Schmitt, and Stephens, Jr. (1997) documented an increase of 150% in southern species of kelp forest fish in southern California, and a decrease of 50% in northern species since the 1970s.

Increased ocean temperatures have been linked with outbreaks of marine disease (Hofmann *et al.*, 1999). Populations of black abalone (*Haliotis cracherodii*) in the Channel Islands and north along the California coast to Cambria suffered mass mortalities from "withering syndrome" caused by the intracellular prokaryote *Xenohaliotis californiensis*, between 1986 and 2001. Healthy populations of black abalone persist north of Cambria, where cool waters suppress the disease. Samples of red abalone (*Haliotis rufescens*) from populations around San Miguel Island in 2006 indicated that approximately 58% of the population carries *X. californiensis*, but the red abalone population persists in a thermal refuge within which temperatures are low enough to suppress the expression of the disease. The disease may be expressed during prolonged periods of warming (*e.g.*, over 18°C for several days) associated with ENSO or other warm-water events. In 1992, an ENSO year, an urchin-specific bacterial disease entered the Channel Islands region and spread through dense populations of purple sea urchin (*Strongylocentrotus pupuratus*). Sites located in a marine reserve where fishing was prohibited had more lobster (which prey on urchins), smaller populations of urchins, persistent forests of giant kelp, and a near absence of the disease.[64] During several warm-water events, including the ENSO of 1997–98, scientists observed and documented declines of sea star populations at the Channel Islands due to epidemics of "wasting disease," which disintegrates the animals.

Increased temperature is expected to lead to numerous changes in currents and upwelling activity. As the sea surface warms, thermal stratification will intensify and become more stable, leading to reduced upwelling of cool, nutrient-rich water (Soto, 2001; Field *et al.*, 2001). Reduced upwelling will lead to a decline in primary productivity (McGowan *et al.*, 1998), suppression of kelp growth, and cascading effects through the marine food web.

Introductions of non-native species (such as the European green crab *Carcinus maenas* on the U.S. West Coast) are associated with rising temperatures and altered currents associated with ENSO events (Yamada *et al.*, 2005). The Sanctuary Advisory Council identified non-indigenous species as an emerging issue in the revised Sanctuary Management Plan (U.S. Department of Commerce, 2006). The sanctuary participated in the removal of a non-indigenous alga (*Undaria pinnatifida*) from the Santa Barbara Harbor, but the sanctuary does not support systematic monitoring or removal of non-indigenous species. Introduction of non-indigenous species can disrupt native communities, potentially leading to shifts in community structure.

Sea level may rise up to three feet in the next 100 years, depending on the concentrations of greenhouse gases during this period (Cayan *et al.*, 2006a; IPCC, 2007a). Projections of sea level rise around the Channel Islands indicate little encroachment of seawater onto land due to steep rocky cliffs that form the margins of the islands; however, projections of sea level rise indicate potential saltwater intrusion into low-lying coastal areas such as the Santa Barbara Harbor (where the CINMS Headquarters is located) and the Channel Islands Harbor (where the sanctuary's southern office is located). Changes in sea level may affect the type of coastal ecosystem (Hoffman, 2003). Graham, Dayton, and Erlandson (2003) suggested that sea level rise transformed the Southern California Bight from a productive rocky coast to a less productive sandy coast more than 18,000 years ago.

The severity of storm events is likely to increase with climate change (IPCC, 2001). As described above, storm activity damages kelp stocks and pulls kelp holdfasts from the substrate (Dayton *et al.*, 1992; 1999). Frequent and intense storm activity during the 1982–83 ENSO event decimated populations of giant kelp that once formed extensive beds attached to massive old kelp holdfasts in sandy areas along the mainland coast. Since the old kelp holdfasts were displaced from the mainland coast, young kelp plants have been unable to attach to the sandy substrate and the coastal kelp forests have not returned. At the Channel Islands, kelp forests that were destroyed during the same ENSO event have slowly returned to the rocky reefs around the Channel Islands,

[64] **Lafferty**, K.D. and D. Kushner, 2000: Population regulation of the purple sea urchin, *Strongylocentrotus purpuratus*, at the California Channel Islands. In: *Fifth California Islands Symposium*, Minerals Management Service, Santa Barbara, California, pp. 379-381.

particularly following a Pacific Decadal Oscillation to cooler waters in 1998.

A Shared Vision for the Channel Islands

The CINMS manager and staff work closely with the Sanctuary Advisory Council to identify and resolve resource management issues. As noted above, the Sanctuary Advisory Council consists of representatives from local, state, and federal agencies, which share jurisdiction of resources within the Channel Islands region, and stakeholders with interests in those resources. The Sanctuary Advisory Council offers a unique opportunity to focus attention of regional agencies and stakeholders on the potential threats associated with climate change and to develop a shared vision for how to respond.

The Sanctuary Management Plan (U.S. Department of Commerce, 2006) describes a strategy to work in a coordinated, complementary, and comprehensive manner with other authorities that share similar or overlapping mandates, jurisdiction, objectives, and/or interests. The sanctuary is poised to take a leading role to bring together the relevant agencies and stakeholders to discuss the issue of climate change. The sanctuary can initiate an effort to develop regional plans to adapt to a modified landscape and seascape predicted from climate change models, and mitigate the negative impacts of climate change.

A6.4.2 Management of the Channel Islands National Marine Sanctuary

The Sanctuary Management Plan (U.S. Department of Commerce, 2006) for the CINMS mentions but does not fully address the issue of climate change, with one exception in the strategy for offshore water quality monitoring. The strategy is to better evaluate and understand impacts on water quality from oceanographic and climatic changes and human activities. The proposed actions include continued vessel and staff support for monitoring projects related to water quality. To evaluate the potential impacts of climate change, the sanctuary staff could expand monitoring of—or collaborate with researchers who are monitoring—ocean water temperature, currents, dissolved oxygen, and pH at different depths.

The Sanctuary Management Plan (U.S. Department of Commerce, 2006) describes a strategy to identify, assess, and respond to emerging issues. The plan explicitly identifies noise pollution, non-indigenous species, and marine mammal strikes as emerging issues. Other emerging issues that are not addressed by the management plan, but should be, include ocean warming, sea level rise, shifts in ocean circulation, ocean acidification, spread of disease, and shifts in species ranges.

The Sanctuary Management Plan (U.S. Department of Commerce, 2006) outlined a potential response to emerging issues through consultation with the Sanctuary Advisory Council and local, state, or federal agencies with a leading or shared authority for addressing the issue. With the elevated level of certainty associated with climate change projections (IPCC, 2007a), it is appropriate to bring the topic of climate change to the Sanctuary Advisory Council and begin working with local, state, and federal agencies that share authority in the region to plan for potential impacts of climate change. Regional agency managers may consider and develop strategies to respond to the potential impacts of:

- Ocean warming (contributing to potential shifts in species ranges, changes in metabolic and physiological processes, and accelerated spread of disease);
- Ocean acidification (leading to breakdown of calcareous accretions in corals and shells);
- Shifts in ocean circulation (leading to changes in upwelling activity and possible formation of low oxygen zones); and
- Sea level rise (shifting jurisdictional boundaries, displacing terrestrial and intertidal organisms, leading to salt-water inundation of coastal marshes, lagoons and estuaries, and increasing coastal flood events).

Monitoring and Research in the Channel Islands Region

Monitoring and research are critical for detecting and understanding the effects of climate and ocean change. The Sanctuary Management Plan (U.S. Department of Commerce, 2006) outlines strategies for monitoring and research in the coming years, but the plan does not address climate and ocean change specifically. The current strategies for monitoring and research can be refocused slightly to capture important information about climate and ocean change.

Monitoring of algae, invertebrates, and fishes is needed within and around marine reserves to detect differences between protected and targeted populations in their responses to climate change. One hypothesis is that populations within marine reserves will be more resilient to the effects of climate change than those that are altered by overfishing and other extractive uses. In addition, scientists have determined that local environmental variation causes different populations to respond in different ways to ocean warming (*e.g.*, Helmuth *et al.*, 2006). For example, a population of red abalone at San Miguel Island lives in a "thermal refuge" where waters are cooled by upwelling, preventing spread of disease that is carried in the population. Sustained ocean warming is likely to increase thermal stress of individuals in this population and accelerate the spread of disease through affected populations. Monitoring can be used to detect such changes at individual, population, and regional levels. The CINMS has the capacity to support subtidal monitoring activities from the *RV Shearwater*, aerial surveys of kelp canopy from the sanctuary aircraft, and collaborative research projects with scientists and fishers.

In addition to the ecological monitoring in marine reserves, it will be critical to monitor environmental variables, including ocean water temperature, sea level, currents, dissolved oxygen, and pH at different depths. Any change in these variables should trigger more intensive monitoring to evaluate the ecological impacts of ocean warming, sea level rise, shifts in current patterns, low oxygen, and increased acidification. The sanctuary could benefit from partnerships with scientists who are monitoring ocean changes and who have the capability of ramping up research activities in response to observed changes. For example, before 2002, scientists at Oregon State University, Corvallis, routinely monitored temperature and salinity at stationary moorings off the coast of Oregon. When they detected low oxygen during routine monitoring in 2002, the scientists intensified their monitoring efforts by increasing the number of temperature and salinity sensors and adding oxygen sensors (which transmit data on a daily basis) near the seafloor at a number of locations along the coast. In this way, the scientists can quantify the scope and duration of hypoxic events, which have recurred off the coast of Oregon during the past five years (Barth *et al.*, 2007).

The Sanctuary Management Plan (U.S. Department of Commerce, 2006) describes the need for analysis and evaluation of information from sanctuary monitoring and research. Working with local educational institutions and the National Center for Ecological Analysis and Synthesis, the sanctuary could develop the capacity to catalog and analyze spatial data (maps) that characterize the coastline of the sanctuary and the extent of kelp canopy within the sanctuary, among other types of information. To detect the ecological impacts of climate change, the information from sanctuary monitoring and research should be reviewed at regular intervals (at least annually) by collaborating scientists (such as the Sanctuary Advisory Council's Research Activities Panel), sanctuary staff, and the sanctuary manager. The annual review should compare data from the current year with previous years, from areas inside marine reserves and in surrounding, fished areas. Ecological changes should be placed within the context of El Niño-Southern Oscillation and La Niña cycles and shifts associated with the Pacific Decadal Oscillation. Changes in fisheries or other management regulations also should be considered as part of the evaluation. Any significant shifts away from predictable trends should trigger further evaluation of the data in an effort to understand local and regional ecosystem dynamics and any possible links to climate change.

Communication in the Channel Islands Region

Public awareness and understanding are paramount in the discussion about how to adapt to climate change. The education and outreach strategies described in the Sanctuary Management Plan (U.S. Department of Commerce, 2006) do not focus on the issue of climate change but, with a slight shift in focus, the existing strategies can be used to increase public awareness and understanding of the causes and impacts of climate change on ocean ecosystems. Key strategies are to educate teachers, students, volunteers, and the public using an array of tools, including workshops, public lectures, the sanctuary website and weather kiosks, and a sanctuary publication and brochure, among others. Opportunities to focus the sanctuary education program's activities and products on the issue of climate change include the following:

- Integrate information about climate change into volunteer Sanctuary Naturalist Corps and adult education programs;
- Update the sanctuary website and weather kiosks with information about causes and impacts of climate change;
- Produce a special issue of the sanctuary publication, *Alolkoy*, about the current scientific understanding of climate change and potential impacts on sanctuary resources;
- Develop a brochure about climate change to help members of the community identify opportunities to reduce their contributions to greenhouse gases and other stressors that exacerbate the problem of climate change;
- Expand the sanctuary's Ocean Etiquette program[65] to include consideration and mitigation of individual activities that contribute to climate change;
- Host a teacher workshop on the subject of climate change;
- Prepare web-based curriculum with classroom exercises and opportunities for experiential learning about climate change; and
- Partner with local scientists who study climate change to give public lectures and engage students in monitoring climate change.

A.5 Conclusions about Marine Protected Area Case Studies

The Great Barrier Reef Marine Park has been examined along with the National Marine Sanctuary case studies because it is an example of an MPA that has a relatively highly developed climate change program in place. A Coral Bleaching Response Plan is part of its Climate Change Response Program, which is linked to a Representative Areas Program and a Water Quality Protection Plan in a comprehensive approach to support the resilience of the coral reef ecosystem. In contrast, the Florida Keys National Marine Sanctuary is developing a bleaching response plan. The Florida Reef Resilience Program, under the leadership of The Nature Conservancy, is implementing a quantitative assessment of coral reefs before and after bleaching events. The recently established Papahānaumokuākea (Northwestern Hawaiian Islands) Marine National Monument is the largest MPA in the world and provides a unique opportunity to examine the effects of climate change on a nearly intact

65 http://sanctuaries.noaa.gov/protect/oceanetiquette.html

large-scale marine ecosystem. These three MPAs consist of coral reef ecosystems, which have experienced coral bleaching events over the past two decades.

The Sanctuary Management Plan for the Channel Islands National Marine Sanctuary mentions, but does not fully address, the issue of climate change. The Plan describes a strategy to identify, assess, and respond to emerging issues through consultation with the Sanctuary Advisory Council and local, state, or federal agencies. Emerging issues that are not yet addressed by the management plan include ocean warming, sea level rise, shifts in ocean circulation, ocean acidification, spread of disease, and shifts in species ranges.

Barriers to implementation of adaptation options in MPAs include lack of resources, varying degrees of interest in and concern about climate change impacts, and a need for basic research on marine ecosystems and climate change impacts. National Marine Sanctuary Program staff are hard-pressed to maintain existing management programs, which do not yet include explicit focus on effects of climate change. While the Program's strategic plan does not address climate change, the Program has recently formed a Climate Change Working Group that will be developing recommendations. Although there is considerable research on physical impacts of climate change in marine systems, research on biological effects and ecological consequences is not as well developed.

Opportunities with regard to implementation of adaptation options in MPAs include a growing public concern about the marine environment, recommendations of two ocean commissions, and an increasing dedication of marine scientists to conduct research that is relevant to MPA management. References to climate change as well as MPAs permeate both the Pew Oceans Commission and U.S. Commission on Ocean Policy reports on the state of the oceans. Both commissions held extensive public meetings, and their findings reflect changing public perceptions and attitudes about protecting marine resources from threats of climate change. The interests of the marine science community have also evolved, with a shift from "basic" to "applied" research over recent decades. Attitudes of MPA managers have changed as well, with a growing recognition of the need to better understand ecological processes in order to implement science-based adaptive management.

Con⊠dence Estimates for SAP 4.4 Adaptation Approaches

Authors:
Susan Herrod Julius, U.S. Environmental Protection Agency
Jordan M. West, U.S. Environmental Protection Agency
Jill S. Baron, U.S. Geological Survey and Colorado State University
Brad Grif⊠th, U.S. Geological Survey
Linda A. Joyce, U.S.D.A. Forest Service
Brian D. Keller, National Oceanic and Atmospheric Administration
Margaret Palmer, University of Maryland
Charles Peterson, University of North Carolina
J. Michael Scott, U.S. Geological Survey and University of Idaho

B1. INTRODUCTION

For each adaptation approach, authors were asked to consider two separate but related elements of confidence. The first element is the amount of evidence that is available to assess the effectiveness of a given adaptation approach (indicating that the topic is well-studied and understood). The second is the level of agreement or consensus across the different lines of evidence regarding the effectiveness of the adaptation approach. Authors were asked to rate their confidence according to the following criteria:

High/low amount of evidence

Is this adaptation approach well-studied and understood, or instead is it mostly experimental or theoretical and not well-studied? Does your experience in the field, your analyses of data, and your understanding of the literature and performance of specific adaptation options under this type of adaptation approach indicate that there is a high/low amount of information on the effectiveness of this approach?

High/low amount of agreement

Do the studies, reports, and your experience in the field, analyzing data, or implementing the types of adaptation strategies that comprise this approach reflect a high degree of agreement on the effectiveness of this approach, or does it lead to competing interpretations?

The authors' responses are provided in the following sections, organized by adaptation approach.

B2. ADAPTATION APPROACH: PROTECTING KEY ECOSYSTEM FEATURES

Description: Focusing management protections on structural characteristics, organisms, or areas that represent important "underpinnings" or "keystones" of the overall system.

Confidence: Is strategic protection of key ecosystem features an effective way to preserve or enhance resilience to climate change?

National Forests

Amount of evidence: High

There is ample theoretical and empirical evidence to support the positive relationship between biodiversity and ecosystem resilience. Based on a study in Australian rangeland, Walker, Kinzig, and Langridge (1999) concluded that functional group diversity maintains the resilience of ecosystem structure and function. Resilience is increased when ecosystems have multiple species that fulfill similar "functions" but that respond differently to human actions (Walker, 1995; Fischer, Lindenmayer, and Manning, 2006). Elmqvist *et al.* (2003) concluded that the diversity of responses to management and disturbance enabled by diverse ecosystems "insures the system against the failure of management actions and policies based on incomplete understanding." Brussaard, de Ruiter, and Brown (2007) concluded that soil biodiversity confers resilience against stress and disturbance and protecting it is necessary to sustain agricultural and forestry production. Keystone species and structural elements of ecosystems are particularly important because many species and ecological processes rely on them (Fischer, Lindenmayer, and Manning, 2006). Because keystone species largely "control the future" (*i.e.*, guide the successional trajectories and characteristics) of ecosystems (Walker, 1995; Gunderson, 2000), protecting them (and biodiversity in general) is a fundamental feature of conservation and restoration schemes.

Restoration research currently discussing climate change concludes that key processes may be the only way to address restoration under climate change.

The United States Forest Service (USFS) emphasizes biodiversity conservation and protection of critical habitat and other key ecosystem features in its management of national forests. Some national forest managers currently seek to enhance landscape and species diversity as the most sensible way to adapt to climate change in the absence of contradictory information (see Olympic National Forest case study). Major USFS programs and plans—such as the early detection program for invasive species, the forest health program (which tries to prevent or reduce the impact of insect and disease outbreaks) and the National Fire Plan—also aim to protect key ecosystem features and values. Similarly, efforts to reduce the impacts of fragmentation and create larger, connected landscapes with continuous habitat help conserve keystone species. Maintenance of old-growth habitat and particular characteristics of old-growth is also emphasized in many national forests.

Amount of agreement: Low

Ecologists have engaged in heated debates for the past century about the extent to which diversity begets stability (*i.e.*, resilience). The current state of the debate appears to be somewhat nuanced. Although it appears that "a large number of species is required to sustain the assembly and functioning ecosystems in landscapes subject to increasingly intensive land use," there is still uncertainty about the specific mechanism and details of this dependence on diversity (Loreau *et al.*, 2001). Recent reviews (Loreau *et al.*, 2001; Hooper *et al.*, 2005) note that the debate has become more nuanced because of theoretical and experimental advances (e.g., Tilman, Reich, and Knops, 2006).

Functional groups have been used to explore ecosystem function and the role of suites of species. However, the makeup and composition of these functional groups and their roles in the ecosystem is not always agreed upon by the research community

The inability to accurately define either species or functional groups that ensure the viability of the ecosystem result in an uncertainty and likelihood that as many species as possible must be maintained, a distinct challenge for resource management.

National Parks

Amount of evidence: High

While the large body of literature related to protection of key ecosystem features does not address resiliency in light of climate change, it provides evidence that in the absence of protection of natural flow regimes, natural fire regimes, and physical structures natural processes are compromised.

Protection of soils from erosion using natural materials reduced soil loss, promoted vegetation regrowth, and reduced siltation of streams in northern New Mexico and Colorado (Allen *et al.*, 2002).[1]

Use of wildland fire, mechanical thinning, or prescribed burns where it is documented to reduce risk of anomalously severe fires has been shown to work, but only to work where forest stands are unnaturally dense due to fire suppression such that removal of fuels reduces the risk of anomalous fires.

River systems with minimal disturbance maintain higher levels of native biodiversity than disturbed systems, suggesting the converse is also true, that disturbance of natural flow regimes reduces native biodiversity (Poff *et al.*, 2007).

Studies of certain species, such as whitebark pine in the western United States, show that they are important food sources for many species, including bears and Clark's nutcrackers. In their absence animals find alternative food sources or become locally extirpated (Tomback and Kendall, 2002).

Studies of the effects of reintroducing wolves to Yellowstone ecosystem show a strong cascading positive effect on ecosystem performance, ranging from improved riparian habitat (less trampling by elk), increased beaver activity, and restored habitat leading to increased numbers of migratory birds.

Studies of habitat requirements for bighorn sheep survival and reproduction demonstrated the need for specific vegetation mosaics and densities. In the absence of such vegetation structure (vegetation too dense or too sparse), sheep are exposed to predators and populations decline (Singer, Bleich, and Gudorf, 2000).

---

[1] See also **Sydoriak**, C.A., C.D. Allen, and B.F. Jacobs, 2000: Would ecological landscape restoration make the Bandelier Wilderness more or less of a wilderness? *Proceedings: Wilderness Science in a Time of Change Conference-Volume 5: Wilderness Ecosystems, Threats, and Management*, Proceedings RMRS-P-15-VOL-5, 209-215.

Several papers describe the benefits of maintaining corridors for species migrations (Novacek and Cleland, 2001; Levey *et al.*, 2005).

Amount of agreement: High

There seems to be high agreement, as well as a fair bit of common sense, that maintaining ecosystem structure, including physical structure and natural processes will be at least somewhat protective of ecosystems and their species under climate change, and allow some ability to respond to climate change.

Many papers in the literature that recommend ways to ameliorate the effects of climate change strongly promote protecting features and processes that structure ecosystems as one of their first recommendations (Welch, 2005).

National Wildlife Refuges

Amount of evidence: High

The refuge system has a long history of habitat enhancement to maintain high quality habitat and sustain ecological processes for waterfowl and other aquatic species. There are large number of studies documenting response of species to prescribed burns and altered water regimes. Magnitude of the response varies among species and seasons. Prescribed fire is frequently used for managing grasslands and fire and prescribed cuts for forest lands. The changes projected from climate change are an additional variable. There are many references in the literature to the consequences of altered ecological processes on the integrity, diversity, and health of natural communities. Protection of nesting islands for colonial nesting birds from predators has been shown to positively affect reproductive success of many species. Reintroduction of keystone species such as beavers on refuges significantly alters habitat conditions and population size of other species.

Amount of agreement: High

There is wide agreement that protecting key ecosystem features will preserve or enhance resilience to climate change. Logically, protection will allow more of the resilience capacity to be "dedicated" to climate change because protection will minimize the challenges of non-climate stressors.

Wild and Scenic Rivers

Amount of evidence: Low

It is generally believed that there are no "keystone species" in running water ecosystems. Beaver can affect streams, but they convert them to wetlands and certainly there have been no attempts to protect them.

Headwater streams are the closest thing for WSRs that are "critical" because the rest of the river system is influenced by them and there is growing research evidence showing they have a disproportionate impact on the health of rivers. They should be the focus of protection, but have not been to date.

Amount of agreement: High

This is a difficult question because there is high agreement that headwater streams are disproportionately important, based on studies measuring rates of processes and the impacts of excluding some headwater inputs/processes to downstream reaches. But this research has not been done it a management/protection context. It is all basic research experiments.

National Estuaries

Amount of evidence: Low

There has been much oyster reef restoration, but none testing success in protecting shoreline from erosion.

Managed realignment is good in concept, but no tests exist of its success.

Many tests have been done of how biodiversity affects resilience and observational studies exist relating structural complexity to biodiversity.

No real test exists to assess success of protecting estuarine zones of high biogeochemical functioning.

There is little empirical testing of bulkheads impacts on long enough time scales.

No development or tests of effectiveness of rolling easement concept exist.

Amount of agreement: Low

There are many more failed than successful oyster reef restorations.

Some disagreement exists over need for realignment, due to uncertainty over rate of natural soil accretion in marshes.

Mixed, conflicting results exist in tests of how biodiversity influences resilience.

No data test the success of protecting biogeochemical zones of importance.

There is high conceptual agreement that bulkheads inhibit transgression.

There is high conceptual agreement that many species need corridors but this is of debatable applicability to estuaries, where larval or seed dispersal is almost universal.

The debate over need for rolling easements is only just beginning.

Marine Protected Areas

Amount of evidence: Low

This approach is fundamental to place-based management and MPAs that are designed to protect ecosystems. Palumbi (2002) summarized the situation at the time of his review: "...there are very few data that examine the relative resilience of marine habitats inside and outside reserves, nor are there comprehensive studies available that address whether ecosystems inside reserves can better weather climate shifts." There are some studies that have documented changes in ecosystem features in MPAs (Babcock *et al.* in New Zealand; McClanahan, Mwaguni, and Muthiga in Kenya; Mumby *et al.* in the Bahamas), and Hughes *et al.* (2007) concluded that managing herbivorous fishes is a key component of managing reef resilience. Mumby *et al.* (2007) documented higher coral recruitment rates in a 20-year-old marine reserve, which likely would enhance rates of coral population recovery after disturbances and thus increase resilience compared with areas outside the reserve. One might argue that the evidence is moderate, but "low" was selected to reflect the limited amount of research on this topic directly relevant to resilience to climate change.

Amount of agreement: High

The existing studies, though limited in number, appear consistent. Studies that have not found changes in ecosystem features in MPAs, such as unpublished research in the Florida Keys National Marine Sanctuary, probably reflect the relatively short duration (10 years) of no-take regulations.

ADAPTATION APPROACH: REDUCING ANTHROPOGENIC STRESSES

Definition: Minimizing localized human stressors (*e.g.*, pollution) that hinder the ability of species or ecosystems to withstand climatic events

Key Question: Is reduction of anthropogenic stresses effective at increasing resilience to climate change?

National Forests

Amount of evidence: High

There is considerable literature that current stressors (air quality, invasives, altered fire regimes) increase the stress on plants and animals within ecosystems, and that management to reduce these stressors has a positive impact on ecosystem health.

With respect to air quality impacts, there is extensive literature on the impacts associated with ozone, nitrogen oxides, and mercury; the interactions of these pollutants; and the value of protecting ecosystems from air quality impacts (*e.g.*, National Research Council, 2004). Current levels of ozone exposure are estimated to reduce eastern and southern forest productivity by 5–10% (Joyce *et al.*, 2001; Felzer *et al.*, 2004). In the western United States, increased nitrogen deposition has altered plant communities and reduced lichen and soil mychorriza (Baron *et al.*, 2000; Fenn *et al.*, 2003). Interaction of ozone and nitrogen deposition has been shown to cause major physiological disruption in ponderosa pine trees (Fenn *et al.*, 2003). Mercury deposition negatively affects aquatic food webs, as well as terrestrial wildlife, as a result of bioaccumulation (Chen *et al.*, 2005; Ottawa National Forest, 2006; Driscoll *et al.*, 2007; Peterson *et al.*, 2007). Given that climate change is likely to increase drought, exposure to ozone may further exacerbate the effects of drought on both forest growth and stream health (McLaughlin *et al.*, 2007a; 2007b).

There is considerable literature on the impact of invasives on ecosystems, biodiversity (Stein *et al.*, 1996; Mooney and Hobbs, 2000; Pimentel *et al.*, 2000; Rahel, 2000; Von Holle and Simberloff, 2005). Disturbances such as fire, insects, hurricanes, ice storms, and floods (all of which are likely to increase under climate change), create opportunities for invasive species to become established on areas ranging from multiple stands to landscapes. In turn, invasive plants alter the nature of fire regimes (Williams and Baruch, 2000; Lippincott, 2000; Pimentel *et al.*, 2000; Ziska, Reeves, and Blank, 2005)[2] as well as hydrological patterns (Pimentel *et al.*, 2000), in some cases increasing runoff, erosion, and sediment loads (*e.g.*, Lacey, Marlow, and Lane, 1989). Potential increase in these disturbances under climate change will heighten the challenges of managing invasive species. Climate change is expected to compound the invasive species problem because of its direct influence on native species distributions and because of the effects of its interactions with other stressors (Chornesky *et al.*,

[2] See also **Tausch**, R.J., 1999: Transitions and thresholds: influences and implications for management in pinyon and juniper woodlands. In: *Proceedings: Ecology and Management of Pinyon-Juniper Communities Within the Interior West* US Department of Agriculture, Forest Service, Rocky Mountain Research Station, pp. 361-365.

2005). The need to protect, sustain, and restore ecosystems that are either threatened or impacted by invasives has been recognized by management agencies (USDA Forest Service, 2004).

Adaptation literature describes the value of minimizing these current stressors to reduce ecosystem vulnerability to climate change and to enhance ecosystem resilience to climate change (*e.g.*, Spittlehouse and Stewart, 2003; Schneider *et al.*, 2007; Adger *et al.*, 2007).

Amount of agreement: High

The literature is in agreement that reducing these stressors is an important management strategy.

The literature also agrees that the effectiveness of these restoration approaches is influenced by the current environmental conditions, current condition of the ecosystem, and current status and degree of other human alterations of the ecosystem (*i.e.*, presence of invasives, departure from historical fire regimes, condition of watersheds).

National Parks

Amount of evidence: High

There is a vast amount of literature, plus a lot of common sense, demonstrating that ecosystems and their biota are more resilient to both natural and human-caused disturbances (although not necessarily climate change) when they are not stressed by pollution, habitat alteration, erosion of physical features such as beaches or soil, or prevention of natural disturbance cycles. Some methods may be more effective than others.

The IPCC Working Group II report on coasts offers literature about restoration of natural coastal processes as a way to promote shore, wetland and marsh protection from climate change (IPCC, 2007).

Restoration can protect salmon fisheries from some effects of climate change (Battin *et al.*, 2007).

While there is ample evidence that man-made barriers prevent natural migration of aquatic species, there is also growing evidence that it may not increase ecosystem resilience. Upstream migration of non-native species or diseases may compromise gains made by removal of barriers. Other management activities or land use may similarly compromise gains (U.S. Geological Survey, 2005).

Literature demonstrating that managing visitor use patterns in national parks works to minimize the effects of climate change is not readily available, although there are many examples of where restrictions of use has either been effective in restoring vegetation or enabled birds to nest successfully.

Amount of agreement: High

Reduction of human-caused stressors is the root of restoration ecology, a respected field of applied ecology. Many papers demonstrate recovery of at least some ecosystem attributes when pollutants are removed, including examples of recovery of zooplankton in Ontario lakes recovering from acid rain, increase in lake and stream acid-neutralizing capacity in the Adirondacks and Europe after reductions of SO_2 emissions, and restoration of native fishes after recovery from acid mine drainage or phosphorus reduction.

Removal of non-native fishes in Alberta lakes allowed for natural (and assisted) recovery of natural food webs (Parker and Schindler, 2006).

National Wildlife Refuges

Amount of evidence: High

Management of anthropogenic stresses such as introduced predators, ungulates, etc. has been shown to increase numbers and reproductive success of waterfowl and ground nesting game birds. Reduction in pollutants (*e.g.*, DDT, selenium) has also been shown to increase survival and reproductive success of many species. Control of nest parasites, such as cowbirds, has been widely and successfully used as a management tool for endangered songbirds. The magnitude of the demographic response varies among species and ecological conditions. Provision of contaminant-free food has been used to reduce exposure of carrion feeding birds to lead with mixed success.

Amount of agreement: High

There is wide agreement that reducing anthropogenic stresses will increase resilience to climate change. Reducing anthropogenic stressors will increase the survival, reproductive success, and population size of most organisms (particularly those not dependent on disturbed anthropogenic habitats), and these increases will enhance the resilience capacity of trust species.

Wild and Scenic Rivers

Amount of evidence: High

There have been extensive studies demonstrating that the amount of degradation of a watershed increases directly in relation to human stresses such as deforestation, dam building, urbanization, and agriculture.

There is very strong scientific data to show that when human stresses are reduced, the systems recover. There is also strong scientific evidence that a "healthy" river corridor that has minimal human stress imposed on it is very resilient to new stresses of the magnitude expected in the near term for climate change.

Amount of agreement: High

There are an incredible number of studies showing that reducing impervious cover and agriculture (and other human stressors) impart a healthy, more resilient river. This is probably one of the few areas where there is almost total agreement.

There are many existing and newly forming management actions for rivers that are directly related to the amount of human stress. The management is doing this by capping the total amount of development and land clearing that can occur in a watershed, followed up by data collection.

National Estuaries

Amount of evidence: High

A prodigious amount of research has been conducted to show the role of nutrient loading and organic loading in eutrophication, and to assess BMPs for successful control. It is also clear from many models that climate change will enhance eutrophication in many estuaries.

There is limited but some research on salt water intrusion and groundwater recharge rates with rising sea level.

Amount of agreement: High

There is excellent agreement that reducing one driver of eutrophication will benefit the system and reduce the level of overall eutrophication.

The disagreement applies to models of precipitation change, which provide results that are generally too coarse in scale to project which estuaries will experience increased precipitation and which will receive less.

Marine Protected Areas

Amount of evidence: Low

This theme crops up in reviews dating back to at least Boesch, Field, and Scavia (2000) and Scavia *et al.* (2002), as well as recent works such as Marshall and Schuttenberg (2006) and Marshall and Johnson (2007). The principle is well established, though not well tested. Our understanding of synergistic stressors at a physiological level has substantial evidence for individual species, but the extension to ecosystems is largely through conceptual modeling. This is a logical, common-sense approach, but the hard evidence is limited.

Amount of agreement: High

Although the evidence is low, there appears to be agreement among a number of authors over a long period. On the other hand, the analysis of decline of Indo-Pacific reefs by Bruno and Selig (2007) concluded that high vs. low levels of management did not appear to influence the trajectory of decline.

ADAPTATION APPROACH: REPRESENTATION

Definition: Protecting a portfolio of variant forms of a species or ecosystem so that, regardless of what climatic changes occur, there will be areas that survive and provide a source for recovery.

[illegible] Is representation effective in supporting resilience through preservation of overall biodiversity?

National Forests

Amount of evidence: Low

Reserves and national networks are often established on the premise that additional sites will ensure the persistence of a particular vegetation type. Under a constant climate, this premise for duplication within networks is well accepted.

However, while it is common to duplicate vegetation types, the recent literature on paleoecology demonstrates that plant and animal species respond individualistically and uniquely in time and space, incorporating competition and ecological disturbance as well as climatic factors in their response. Thus, vegetation types are not likely to retain the same composition and structure under change.

If this adaptation were focused on species, the literature would suggest that the evidence is high with respect to this adaptation strategy and its effectiveness.

On the species level, the distributions of species display distinct "leading" edges that are well incised and indistinct "trailing" edges showing the microsites where species can survive locally, but not under the regional climate. This pattern merely displays that there are a myriad of microhabitats outside of the primary range of a species' distribution that will support that species. There is a scale issue regarding the importance of the survival of that species with respect to the overall ecosystem in the region. Survival of the individual species does not necessarily guarantee the survival of the entire ecosystem.

Amount of agreement: Low

While the literature would support agreement on the effectiveness of this approach for species, there is little agreement that this approach is effective for vegetation types or ecosystems. Therefore agreement is low that this approach would increase resilience in the system.

National Parks

Amount of evidence: Low

Multiple representatives of valued populations or systems is a form of bet-hedging and has been shown to protect species of populations when one or more patches or communities are destroyed.

Individual species respond to climate according to specific climate needs. There is at least one paper suggesting multiple representatives of a species within their specific climate niche will have little value in a changing climate (Williams, Jackson, and Kutzbach, 2007). If the different populations all have narrow tolerances to climate, having more of them when all will change beyond their range if viability will not be beneficial.

Amount of agreement: Low

There is insufficient evidence that representation will be effective in promoting resilience of species of ecosystems, although there is ample evidence that having only few populations or representatives of species increases their vulnerability to extinction.

National Wildlife Refuges

Amount of evidence: High

There is a large body of evidence in the literature showing that species that are found on National Wildlife Refuges are more abundant on refuges than on adjacent habitats. Several studies have shown that capturing the full geographical, ecological ,and genetic variation of a species in the wild or in captivity is a hedge against extinction and other losses. Thus, greater numbers of refuges that support higher densities of trust species will reduce the chances that climate change will completely eliminate any trust habitats, populations, or species. Evidence is lacking for most species regarding what degree of representation is sufficient. Each population of a species or ecosystem example on a refuge will experience different effects of climate change. As a result each one is a different entry in the evolutionary sweepstakes under climate change.

Amount of agreement: High

There is wide agreement that increasing representation will be effective in supporting resilience through preservation of overall biodiversity. Logically, and statistically, the broader the range of trust species and/or trust habitats that are included in the refuge system, the lower the likelihood that biodiversity will be lost due to climate change. However, individual refuges or refuge complexes need to be large enough to maintain viable populations to maximize the advantages of increased representation.

Wild and Scenic Rivers

Amount of evidence: High

This is a difficult question because most of the evidence available is from fisheries. If they are becoming threatened, then some areas have been set aside as special conservation areas to ensure some populations remain alive. Then if they do recover, they are released in rivers elsewhere. In the event of climate change, we may need to release fish and other species in to new regions where the climate is now appropriate for them (assuming their old regions are now too warm or otherwise inappropriate). This is a major management strategy that has been around a long time, and in fact Habitat Conservation Plans are required once a riverine species becomes endangered.

Protecting representative running-water ecosystems themselves (*i.e.*, distinguished from species) has not been a management or scientific focus to date in the United States, but it is being tried in Australia. Because of their dire drought situation, many riparian zones along rivers in Australia are losing all of their vegetation. So managers are setting aside some areas where they ensure minimum water needs (through regulating withdrawals and dam releases) to keep the vegetation alive. The idea is then that these plants can be used for "seed" at other sites once the drought is over.

Amount of agreement: High

There are many things coupled together in this management strategy. There is good agreement that maintaining local fish populations when other populations around them (*i.e.*, in different rivers) are dying makes a great deal of sense, and we have the science to support that.

There is not as much agreement on the ecosystem "set-aside" idea, only because it has not been extensively tried. However, most scientists would agree it is a low risk venture—*i.e.*, likely to work.

National Estuaries

Amount of evidence: Low

There is limited study of effects of genetic diversity on resilience of estuarine species (but see Hughes and Stachowicz, 2004).

There has been growing scientific attention to landscape effects of multiple habitats in salt marshes (Minello; Able; Zedler; Grabowski) and some for seagrass beds, but the scope of these studies is limited.

Amount of agreement: High

There is no ambiguity in the theory of natural selection that genetic diversity is the substrate on which adaptation through evolution acts.

The effects of landscape proximity among marsh and other shoreline habitats are reasonably well established, and the importance of habitat edge effects is also becoming clearer.

Marine Protected Areas

Amount of evidence: Low

This is a cornerstone of the zoning approach for the Great Barrier Reef Marine Park (Fernandes *et al.*, 2005)[3]. It is very logical (Salm, Done, and McLeod, 2006) and has been effectively applied to the marine park. Similar approaches for other marine systems are not readily available, although the representative areas approach has broad applicability.

Amount of agreement: High

Although the evidence is low there appears to be agreement among a number of authors (Palumbi, 2002; Sobel and Dahlgren, 2004; Fernandes *et al.*, 2005; Salm, Done, and McLeod, 2006; Roberts *et al.*, 2006; McCook *et al.*, 2007).[1] A contrary line of evidence is not known.

B.5 ADAPTATION APPROACH: REPLICATION

Definition: Maintaining more than one example of each ecosystem or population within a reserve system such that if one area is affected by a disturbance, replicates in another area provide insurance against extinction and a source for recovery of affected areas.

Guiding Question: Is replication effective in supporting resilience by spreading the risks posed by climate change?

National Forests

Amount of evidence: Low

The literature is extensive in terms of the value of maintaining numerous animal and plant populations of species to maintain species viability. The concept is certainly well-supported in both theoretical and experimental (lab) approaches and for some situations in the field. The rationale for maintaining more than one population or ecosystem is often associated with the probability of extreme events, such as drought or fire, that may be associated with future climate change.

A strategy that combines practices to restore vigor and redundancy (Markham, 1996; Noss, 2001) and ecological processes (Rice and Emery, 2003), so that after a disturbance these ecosystems have the necessary keystone species and functional processes to recover to a healthy state even if species composition changes, would be the goal of managing for ecosystem change.

Agreement for this approach is rated as low, however, because few examples have been documented in the field at the ecosystem level.

Amount of agreement: Low

For populations of plants and animals, the literature is in agreement with the effectiveness of this concept.

For ecosystems, less information is available.

Therefore, agreement is low that this approach would increase resilience in the system.

National Parks

Amount of evidence: Low

Multiple representatives of valued populations or systems is a form of bet-hedging and has been shown to protect species of populations when one or more patches or communities are destroyed. This has been a foundation of endangered species protection.

While one paper was found that promotes replication of desired species (Bengtsson *et al.*, 2003), the National Parks chapter does not promote this as a means of building resilience. Human intervention to move species adds a decidedly anthropomorphic slant to natural resources. Only species of interest are considered, while the majority of insects, plants, soil microbes and biota will be ignored.

Species move independently according to their biophysical needs (Williams, Jackson, and Kutzbach, 2007), so that replication of populations with narrow climatic niches may not provide protection against novel climates, or similar climates too far away for effective natural establishment of new colonies.

Amount of agreement: Low

This approach is sanctioned by conservationists, but papers like those of Kutzbach *et al.* (2007) suggest it is insufficient for promoting resilience of ecosystems in novel climates.

---

[3] See also Day, J., L. Fernandes, A. Lewis, G. De'ath, S. Slegers, B. Barnett, B. Kerrigan, D. Breen, J. Innes, J. Oliver, T. Ward, and D. Lowe, 2002: The representative areas program for protecting biodiversity in the Great Barrier Reef World Heritage Area. In: *Proceedings of the Ninth International Coral Reef Symposium* 23, October 2000, pp. 687-696.

National Wildlife Refuges

Amount of evidence: High

A basic principle of conservation by design is redundancy, and this concept is repeatedly addressed in the scientific literature. Having multiple refuges for a trust species or trust habitat in each of the ecological and climate domains in which it occurs provides logical and statistical insurance against loss of a species or habitat from the refuge system due to a catastrophic event at a single refuge. There are several examples of species becoming extinct after storms affected the last known population.

Amount of agreement: High

There is wide agreement in the science community that redundancy in refuges and species populations increases the logical and statistical likelihood that biodiversity will be preserved. There is some discussion regarding how much redundancy is required.

Wild and Scenic Rivers

Amount of evidence: High

The same evidence is available for the last question (fisheries): maintaining multiple populations spreads the risk of total extinction. There is good evidence available for this risk reduction in fisheries. Less evidence is available for river insects and even less for ecosystem processes.

The critical piece of data needed (for fauna other than fish) is how far they disperse and what their dispersal requirements are. This is an important current research area because of the obvious conservation implications—if we know this then we can design the spatial arrangement of the protected "representative ecosystems/populations" in a way that allows organisms to disperse naturally (*i.e.*, no transplants necessary).

Amount of agreement: High

The emerging interest and efforts by nongovernmental organizations to establish freshwater protected areas is a sign of the confidence that this approach is worthwhile.

There has been extensive research in river networks to determine if there are particular configurations of river reaches that minimize extinction risk.

National Estuaries

Amount of evidence: Low

Oyster reef restoration done in replication along a depth gradient was shown to allow fish and crustaceans to survive when environmental degradation occurred that was depth-dependent: the fishes moved to reefs that were not affected and found enough prey to survive (Lenihan *et al.*, 2001).

Migrating shorebirds require replicated estuaries along the flyway so that they can move to more rewarding feeding sites to fuel up for the migration and breeding.

Otherwise, there is little research on replication at the spatial and temporal scales appropriate to project its value in a climate change context.

Amount of agreement: High

There is a high level of agreement, although in part perhaps because so few studies of relevance have been done.

There is agreement in concept that populations of mobile vertebrates such as fishes, birds, and mammals benefit from replication. However, many such species, such as salmon, exhibit high faithfulness to natal sites; replication would not provide much if any benefit for them.

Marine Protected Areas

Amount of evidence: Low

There are numerous modeling studies of reserve networks (*e.g.*, Allison, Lubchenco, and Carr, 1998), but empirical data are lacking. Areas such as the Great Barrier Reef Marine Park and the Channel Islands National Marine Sanctuary should produce relevant results over time. This approach also might be ranked as moderate (per question 1).

Amount of agreement: High

Replication and representation in the marine literature generally go hand-in-hand; please refer to question 3 for literature citations. Again, a contrary line of evidence is not known.

ADAPTATION APPROACH: RESTORATION

Definition: Rebuilding ecosystems that have been lost or compromised.

Question: Is restoration of desired ecological states or ecological processes effective in supporting resilience to climate change?

National Forests

Amount of evidence: High

There is a large body of literature describing and documenting restoration theory and practices across a wide variety of ecosystems and ecological processes.

Amount of agreement: Low

While there is high agreement that the current theories and practices can be used to restore a number of different ecosystems, climate change has the potential to significantly influence the practice and outcomes of ecological restoration under a changing climate (Harris *et al.*, 2006), where the focus is on tying assemblages to one place. The restoration literature is now in discussion about the impact that a changing climate may have on the theories and practices that have been developed. For example, natural resource management, planning, conservation, restoration, and policy are deeply founded on strategies based on the historic range of variability ecological concept (Landres, Morgan, and Swanson, 1999). However, use of such strategies will become increasingly problematic as the potential for a "no analog" futures are realized (Millar, Westfall, and Delany, in press; Williams, Jackson, and Kutzbach, 2007).

The climate sensitivity of best management practices, genetic diversity guidelines, restoration treatments, and regeneration guidelines may need to be revisited. Space for evolutionary development under climate change may be important to incorporate into conservation and restoration programs under a changing climate (Rice and Emery, 2003).

National Parks

Amount of evidence: High

Restoration of some species, such as wolves, into habitats where they have been extirpated has been highly successful by nearly all ecological standards.

There are some examples showing that restoration of natural flow regimes in rivers by dam removal has been successful in restoring reproducing fish populations

There are at least several instances in the literature that decry the lack of restoration standards that allow managers to evaluate the effectiveness of restoration efforts (Bernhardt *et al.*, 2005).

Restoration of wetlands or riparian areas has been shown to bring back some ecosystem services, such as nutrient or pollutant retention, but there is uncertainty among wetland scientists whether restoration activities truly reproduce natural conditions.

Restoration of damaged systems will allow climate change to occur with fewer ecological disruptions than if soils have eroded, invasive species dominate, river banks are trampled, or pollutants contaminate native populations (discussed above in reducing anthropogenic stresses).

Amount of agreement: High

There is an entire professional society devoted to ecological restoration, the Society for Ecological Restoration, with journals that describe the theory behind restoration and practical applications of restoration science.[4]

National Wildlife Refuges

Amount of evidence: Low

Habitat restoration is a widely used tool in relatively small-scale conservation biology activities. There is a large body of literature on the topic, with several journals devoted solely to habitat restoration (*e.g.*, Ecological Management and Restoration, Restoration Ecology) as well as a professional society dedicated to restoration ecology. In Hawaii, restoration of pasture lands to *ohia koa* forests resulted in recolonization by endangered birds. Re-creation of wetlands has been used widely and successfully to restore/attract migratory water birds. However, the magnitude of the site response to restoration can vary due to (1) temporal shifts in habitat use by species, (2) scale of restoration in relation to the desired population goals, (3) introduced species, (4) long-term and large-scale ecological processes, or (5) barriers to recolonization. Further, few restoration studies have been conducted in a controlled experimental design, and reoccupancy of restored habitats by native plants and invertebrates is not well documented. Although there is small-scale evidence for effectiveness of restoration, there is little evaluation or evidence regarding the effectiveness at the larger scales of ecological processes that would be necessary to provide resilience to climate change.

Amount of agreement: Low

There is little general agreement that restoring a desired ecological state or process will be effective in supporting resilience to climate change. There is little logical support for the idea that restoring a state or a process to a historical condition will provide resilience to climate change, because it is expected that the historical restored condition will no longer be appropriate in a changed climate.

Wild and Scenic Rivers

Amount of evidence: Low

Very little rigorous monitoring has been done on stream restoration. This is a very current area of research and data are just starting to come in. The evidence suggests that if the restoration not only repairs the degraded portion of the stream but removes the stress, then the restoration is usually successful. But if the restoration is a local fix, such as regrading streambanks and stabilizing them without taking

[4] Society for Ecological Restoration, http://www.ser.org/about.asp

care of the underlying problem (*e.g.*, inadequate stormwater infrastructure above the reach), then the restoration project will most likely fail or else huge resources will be needed to maintain it.

Amount of agreement: Low

The effectiveness of restoration is a contentious issue. Many scientists are skeptical that most projects work, because many are done poorly or the underlying problem is not addressed. Other scientists point toward data from projects that were adequately monitored and were well done projects—success has clearly been shown. So to a certain extent the low agreement is that some scientists believe we must focus on what is done *in reality* while others focus on what is possible.

National Estuaries

Amount of evidence: High

There are many studies of salt marsh restoration (beginning 40 years ago with *Spartina* methods developed by Seneca, Woodhouse, and Broome).

Similarly, a lot of effort has gone into oyster reef restoration and SAV restoration.

There is not much research on exterminating invasive estuarine species: *Meloluca* is everywhere along Florida waterways; *Phragmites* dominates many areas of East Coast marshes; San Francisco Bay suffers from persistent *Spartina* invasion, etc.

The value of positioning salt marsh restorations where transgressive retreat is possible is strongly supported in concept, although no empirical tests of the effectiveness with sea level rise exist, except for paleontological evidence (*e.g.*, Bertness work) of substantial transgressions of marsh historically.

Amount of agreement: High

There is uniform agreement that salt marsh can be successfully restored.

Some challenges exist in assuring the durability of SAV and oyster reef restorations.

Nevertheless, there is also good agreement that exterminating invasives is generally infeasible for estuaries (although easier for large plants than for mobile animals or microbes).

There is high agreement in concept that building the capacity for transgression will provide a viable means for marshes and other shoreline habitats to become resilient to sea level rise.

Marine Protected Areas

Amount of evidence: Low

Reef restoration following vessel groundings has a long history of application in the Florida Keys (and elsewhere) and more general discussions of restoration are in Marshall and Schuttenberg (2006), Salm, Done, and McLeod (2006), and Precht and Aronson (2006). The discussion has been extended to include restoring herbivory, coral recruitment, and other topics with regard to ecological processes. There is an appreciation by managers that it may be necessary to employ more restoration because of the widespread degradation of marine ecosystems. Nevertheless, it appears that evidence about effectiveness in supporting resilience to climate change is low.

Amount of agreement: Low

There appears to be agreement among several authors (Halpin, 1997; Burke and Maidens, 2004; Salm, Done, and McLeod, 2006; references in Precht and Miller, 2006; Jaap *et al.*, 2006; Gunderson, 2007) but some question the value or potential for success of restoration efforts (Jameson, Tupper, and Ridley, 2002; Hughes *et al.*, 2007). Jameson, Tupper, and Ridley (2002) note that expensive restoration efforts are questionable unless environmental conditions are healthy enough to warrant them.

ADAPTATION APPROACH: REFUGIA

Definition: Using areas relatively less affected by climate change as sources of "seed" for recovery or as destinations for climate-sensitive migrants.

Key Question: Are refugia an effective way to preserve or enhance resilience to climate change at the scale of species, communities or regional networks?

National Forests

Amount of evidence: High

The paleo literature has documented the presence of refugia under past climate changes. Local climate trajectories, local topography, and microclimatology interact in ways that may yield very different climate conditions than those given by broad-scale models. In mountainous terrain especially, the climate landscape is patchy and highly variable, with local inversions, wind patterns, aspect differences, soil relations, storm tracks, and hydrology influencing the weather that a site experiences. Sometimes lower elevations may be refugial during warming conditions, as in inversion-prone basins, deep and narrow canyons, riparian zones, and north slopes. Such patterns, and occupation of them by plants during transitional climate periods, are corroborated in the paleoecological record (Millar and Woolfenden, 1999; Millar *et al.*, 2006). Further, unusual and nutritionally extreme soil types (*e.g.*, acid podsols, limestones etc.) have been noted for their long persistence of species and genetic diversity, resistance to invasive species, and long-lasting community physiognomy compared with adjacent fertile soils (Millar, 1989). During historic periods of rapid climate change and widespread population extirpation, refugial populations persisted on sites that avoided the regional climate impacts and the effects of large disturbance. For example, Camp *et al.* (1995) reported that topographic and site characteristics of old-growth refugia in the Swauk Pass area of the Wenatchee National Forest were uniquely identifiable. These populations provided both adapted germplasm and local seed sources for advance colonization as climates naturally changed toward favoring the species.

Amount of agreement: Low

While the literature has documented these refugia either in the paleo record or on current landscapes, the use of this technique as an adaptation option has been little tested.

National Parks

Amount of evidence: Low

A refugium implies a place where climate conditions will remain similar to present conditions so that species can persist. According to Williams, Jackson, and Kutzbach (2007) many parts of the world will acquire novel climates unseen before on Earth. Selecting, and then protecting, specific habitats for species may in the long run be a matter of chance.

Some very high elevation habitats may provide refugia for cold-loving species such as tundra and pika. High elevation streams where non-native fish can be excluded with natural barriers might provide refugia for cold-water fishes.

Phenological changes that accompany climate change may disrupt mutualistic species associations, regardless of the availability of refugia.

Amount of agreement: Low

Species are currently migrating north and to high elevations as climate changes. Preselecting areas to serve as refuges for individual species or assemblages might or might not work to protect them, with the exception of the high elevations or latitudes where cold-loving species may persist. Therefore, there is low agreement.

National Wildlife Refuges

Amount of evidence: High

Climate refugia, areas where effects of past climate change were minimized, are documented in the paleontological record, and refugia are projected to occur in a changed climate of the future. Historically these refugia were the only areas in which some species survived, and they provided colonization sources when conditions became suitable elsewhere as environmental conditions changed. An analogous situation can be expected to occur with the current episode of climate change. However, large areas of projected climate refugia have no wildlife refuges. There is some evidence that refugia will often be found at the ecological or geographical extremes of species ranges.

Amount of agreement: Low

There is generally low agreement that refugia will be effective at preserving resilience to climate change at all scales, from species to regions. Creating refugia from climate change is not possible; refugia will emerge in response to heterogeneity in landscape characteristics and realized climate change. Further, it is difficult to project the explicit location of future climate change refugia at scales that are ecologically relevant or useful for identifying new sites for strategic growth of the refuge system, particularly at the scale of individual refuges. There may be opportunities to take advantage of emerging refugia, particularly for threatened/endangered species or small scale habitats, but refugia will be difficult to impossible to manage in the adaptive management framework. Predicting species specific responses to potential refugia will be a challenge.

Wild and Scenic Rivers

Amount of evidence: High

There is good evidence that small-scale, local refugia (within-channel such as diverse habitat types) are important to the survival of stream plants and animals, if those areas are protected from significant disturbance events such as unusual floods or droughts. This is directly tied to resilience, because these local refugia act as a protective place from which surviving organisms can disperse. These dispersing individuals then reproduce and re-populate areas denuded of biota.

There is some evidence for plants and fish, but little evidence to date for smaller organisms, that some habitat types, even if widely dispersed, can act as refugia for moderate to large scale (landscape scale) disturbances. Examples include distant floodplains, tributaries that remain intact or undisturbed, or any region that for some reason is protected from the full brunt of a disturbance. Thus, resilience at broad scales (*e.g.*, entire watersheds or perhaps even ecoregions) may depend on setting aside such refuge areas. Since most climate-induced disturbances are expected to be exacerbated by development in a watershed (this makes entire rivers downstream of the development more vulnerable), one form of protection that could be part of a management strategy to provide refugia could include limits to development or protection of floodplains or surrounding forests.

Amount of agreement: High

The only reason there might be some disagreement is if we are considering an organism for which we know nothing or little about its dispersal abilities. If we protect or establish in-stream or regional refugia, but organisms can not move to areas formerly affected by disturbances such as those related to climate change, then the value of the refugia is somewhat reduced. However, because we should be able in most or all cases to transport the biota ourselves (seed, larvae, nymphs, juveniles, etc) using some management programs, this concern is minor. Thus, most river ecologists would strongly agree that provision of refugia is a great way to enhance long term resilience in the face of climate change. In fact, use of such approaches (setting aside "preserves," which are a form of refugia) is already in place in some cases, on the advice of scientific boards in advance of any research or data showing that there is high agreement.

National Estuaries

Amount of evidence: Low

There has been little work done on this topic in estuaries. However, if features such as oyster reefs are restored in replication along a depth gradient or along some other environmental gradient, then when perturbations occur that are depth-dependent or vary in intensity along the gradient, one end of the gradient is more likely to serve as a refugium into which mobile species can escape the threat or impact of the perturbation. This is illustrated by the Lenihan *et al.* (2001) example, in which fish and crabs escape hypoxia/anoxia (which can be climate change-induced) that develops in deep water by retreating to shallow water refugia.

Relative sea level rise does vary geographically, so some salt marsh systems may be able to build soils at rates fast enough to keep up with sea level rise for a relatively long time. However, patterns of geographic distribution in relative rates of sea level rise are too coarse geographically to enable "surviving" estuaries to be successful refugia and sources of migrants. Most estuarine fishes and most marine invertebrates possess highly dispersive planktonic larvae, so there may be some value to refugia at these large distances, but little information is available.

Amount of agreement: Low

There is simply insufficient scientific evidence to determine which marshes may be able to keep up in soil elevation with sea level rise, so a debate will go on.

As regards both oyster reefs and networks of estuaries, virtually no research has been done to assess the effectiveness of refugia, except for the value of alternative estuaries as stop-over sites for migrating shorebirds. Thus, the literature of relevance that exists is relatively speculative and reflects several disagreements.

Marine Protected Areas

Amount of evidence: Low

A number of authors note the potential value of refugia (e.g., McClanahan, Polunin, and Done, 2002; West and Salm, 2003; Coles and Brown, 2003; Salm, Done, and McLeod, 2006; Marshall and Schuttenberg, 2006).[5] Nevertheless, experimental or empirical evidence is limited (*e.g.*, Riegl and Piller, 2003).

Amount of agreement: High

Both the more-speculative as well as at least one empirical study are consistent, so agreement is considered to be high.

ADAPTATION APPROACH: RELOCATION

Definition: Human-facilitated transplanting of organisms from one location to another in order to bypass a barrier (*e.g.*, urban area).

Key Question: Is relocation an effective way to promote system-wide (regional) resilience by moving species that would not otherwise be able to emigrate in response to climate change?

National Forests

Amount of evidence: High

For plants, relocation has been a common technique for commercial plant species. Provenance studies demonstrate the appropriateness of different germplasm, and management is based on the likelihood of planting different provenances across widely scattered landscapes and within landscapes.

5 See also **Salm**, R.V. and S.L. Coles, 2001: Coral bleaching and marine protected areas. In: *Proceedings of the Workshop on Mitigating Coral Bleaching Impact Through MPA Design* [Salm, R.V. and S.L. Coles (eds.)]. Proceedings of the Coral Bleaching and Marine Protected Areas, pp. 1-118.
See chapters in **Johnson**, J. and P. Marshall, 2007: *Climate Change and the Great Barrier Reef: a Vulnerability Assessment.* Great Barrier Reef Marine Park Authority.

For other plant species and for animals, a nascent literature is developing on the advantages and disadvantages of "assisted migration," that is, intentional movement of propagules or juvenile and adult individuals into areas assumed to become their future habitats (Halpin, 1997; Collingham and Huntley, 2000; McLachlan, Hellmann, and Schwartz, 2007). At this point, insufficient data exists to judge the success of such techniques.

Amount of agreement: Low

Protocols for "assisted migration" of species need to be tested and established before approaches are implemented more broadly.

□ational □ar□s

Amount of evidence: Low

Some studies have shown successful colonization of native after removal of invasive species; aggressive control of invasives followed by restoration of native species might be successful in preventing, or slowing, the establishment of unwanted species.

This approach is not well understood, particularly with respect to system-wide resilience.

Amount of agreement: Low

Relocation of desired species may allow that species to persist, but ecosystems are made up of complex webs of living organisms, including insects, soil flora and fauna, and many other types of organisms that would not be relocated.

There is little agreement about whether relocation would increases system resilience.

□ational □ ildlife Refu□es

Amount of evidence: Low

Translocation of species is a very common species-specific management tool. However few of these efforts are conducted with appropriate experimental design. Translocation has been successfully used to introduce game species around the globe. Efforts to use translocation for establishing or re-establishing populations of threatened or endangered species have been highly variable in their success. Synthesis studies indicate that success is very dependent on quality of available habitat and the mitigation of stressors at translocation site prior to relocation. Movement of a species across a dispersal barrier (*e.g.*, fish over dams) assumes that suitable habitat is available beyond the barrier and the uncertainty of climate change challenges that assumption. Climate change projections engender a fear that changes in habitat will result in the loss of species on refuges as conditions become unsuitable and the ability of refuges to mitigate changes is exceeded. The extreme risks would be extinction or extirpation from refuge lands. This presents a very different situation than movement across a barrier (*e.g.*, salamanders, toads and frogs across a highway during dispersal from wintering habitat). Because most evidence has been focused on individual species, the success of species relocation has been variable and there is little to no evidence of the effect of relocated species on recipient communities, there is little evidence that relocation is an effective way to promote system-wide (regional) resilience.

Amount of agreement: Low

There is generally low agreement that relocation will be an effective way to promote system-wide (regional) resilience to climate change. Ethical concerns regarding the unpredictable effects on other species and communities that result from introducing a species into a previously unoccupied habitat are notable; it is not clear that the net effect of translocation will be positive at the system-wide scale. Relocation may be effective at smaller scales; for example, in the case of a threatened or endangered non-disperser that was unlikely to negatively affect a suitable target area.

□ ild and Scenic Rivers

Amount of evidence: Low

While fish have been translocated and are able to survive if put into an appropriate reach, there is no evidence that this will end up promoting system-wide recovery. Most scientists would say the more critical thing for system wide recovery is removing the "insult" to the system. With climate change, that will be pretty hard to do. If you can move the species to a totally new watershed where the climate is appropriate then it is hard to say.

Amount of agreement: Low

Some scientists speculate that we may be able to, for example, shift fish species from lower latitude/altitude places that have become too warm to higher latitude/altitude places that are appropriate under future climates. However, others will argue that even if the temperature is comparable, getting the flow conditions and ecosystem processes that are needed to support the species in the long-run is unlikely.

□ational Estuaries

Amount of evidence: □A

Little, if any, work has been done transplanting estuarine species to overcome dispersal barriers to latitudinal shifts, largely because so many estuarine species are actually highly dispersive at some life stage. Therefore, it is not applicable to rate confidence levels for relocation with regard to estuaries.

Amount of agreement: NA

There is very little agreement that this approach is suitable for most estuarine species. It may, however, play a future role for some reptiles and mammals of salt marshes or mangroves that have limited dispersal capacity, but this requires investigation.

Marine Protected Areas

Amount of evidence: NA

An assessment of "relocation" as a management approach is not made for MPAs because advanced web searches on all the major literature databases result in very little information on the concept of relocation as defined in this report.

Amount of agreement: NA

Since there is virtually no scientific evidence and little discussion of relocation as it would apply to MPAs, it is not applicable to discuss level of agreement in this approach at this time. However, such an approach should not necessarily be written off as a future option; despite the cost, relocation may become an attractive option to managers of small, secluded, higher-impacted reef environments.

Contact Information

Global Change Research Information Office
c/o Climate Change Science Program Office
1717 Pennsylvania Avenue, NW Suite 250
Washington, DC 20006
202-223-6262 (voice)
202-223-3065 (fax)

The Climate Change Science Program incorporates the U.S. Global Change Research Program and the Climate Change Research Initiative.

To obtain a copy of this document, place an order at the Global Change Research Information Office (GCRIO) web site: http://www.gcrio.org/orders

---

Climate Change Science Program and the Subcommittee on Global Change Research

William J. Brennan, Chair
Department of Commerce
National Oceanic and Atmospheric Administration
Acting Director, Climate Change Science Program

Jack Kaye, Vice Chair
National Aeronautics and Space Administration

Allen Dearry
Department of Health and Human Services

Jerry Elwood
Department of Energy

Mary Glackin
National Oceanic and Atmospheric Administration

Patricia Gruber
Department of Defense

William Hohenstein
Department of Agriculture

Linda Lawson
Department of Transportation

Mark Myers
U.S. Geological Survey

Jarvis Moyers
National Science Foundation

Patrick Neale
Smithsonian Institution

Jacqueline Schafer
U.S. Agency for International Development

Joel Scheraga
Environmental Protection Agency

Harlan Watson
Department of State

Executive Office and Other Liaisons

Stuart Levenbach
Office of Management and Budget

Stephen Eule
Department of Energy
Director, Climate Change Technology Program

Katharine Gebbie
National Institute of Standards and Technology

Margaret McCalla
Office of the Federal Coordinator for Meteorology

Bob Rainey
Council on Environmental Quality

Gene Whitney
Office of Science and Technology Policy

www.ingramcontent.com/pod-product-compliance
Lightning Source LLC
Chambersburg PA
CBHW080633180526
45168CB00008B/3147

* 9 7 8 1 5 0 7 8 7 4 0 8 0 *